# Springer Monographs in Mathematics

This series publishes advanced monographs giving well-written presentations of the "state-of-the-art" in fields of mathematical research that have acquired the maturity needed for such a treatment. They are sufficiently self-contained to be accessible to more than just the intimate specialists of the subject, and sufficiently comprehensive to remain valuable references for many years. Besides the current state of knowledge in its field, an SMM volume should ideally describe its relevance to and interaction with neighbouring fields of mathematics, and give pointers to future directions of research.

More information about this series at http://www.springer.com/series/3733

Jørgen Bang-Jensen · Gregory Gutin
Editors

# Classes of Directed Graphs

Springer

*Editors*
Jørgen Bang-Jensen
University of Southern Denmark
Odense
Denmark

Gregory Gutin
Royal Holloway, University of London
Egham
UK

ISSN 1439-7382 ISSN 2196-9922 (electronic)
Springer Monographs in Mathematics
ISBN 978-3-030-10122-0 ISBN 978-3-319-71840-8 (eBook)
https://doi.org/10.1007/978-3-319-71840-8

Mathematics Subject Classification (2010): 05C05, 05C10, 05C15, 05C20, 05C25, 05C38, 05C40, 05C45, 05C70, 05C75, 05C83, 05C85, 05C90, 05C99, 05D40, 68Q25, 68R10, 68W05, 68W25, 68W40, 90C27, 94A60, 94C15



Printed on acid-free paper

This Springer imprint is published by the registered company Springer International Publishing AG part of Springer Nature
The registered company address is: Gewerbestrasse 11, 6330 Cham, Switzerland

# Preface

The two editions of our book *Digraphs: Theory, Algorithms and Applications*, which were published by Springer in 2000 and 2009, respectively, remain the only modern books on graph theory covering more than a small fraction of the theory of directed graphs. We are happy to see that the book has been useful, both for students of advanced courses and to researchers from a wide range of areas, some of which are far from mathematics, such as sociology and medicine.

Since we completed the second edition in 2008, the theory of directed graphs has continued to evolve at a high speed; many important results, including solutions some of the conjectures from *Digraphs*, have appeared and new methods have been developed. So we were faced with the choice of either writing a 3rd edition of our book or developing a new book from scratch. We decided to do the latter for the following main reason: By taking a new, somewhat orthogonal, approach of writing chapters on different and important classes of digraphs, we could give a different viewpoint of digraph theory and include a number of authors whose combined expertise would greatly simplify the process and at the same time increase the quality of the book. We are very happy that the following authors agreed to (co)author chapters for the book: César Hernández-Cruz, Hortensia Galeana-Sánchez, Yubao Guo, Richard Hammack, Frédéric Havet, Jing Huang, Stephan Kreutzer, O-joun Kwon, Marcin Pilipczuk, Michał Pilipczuk, Michel Surmacs, Magnus Wahlström and Anders Yeo.

The book contains more than 120 open problems and conjectures, a feature which should help to stimulate lots of new research. Even though this book should not be seen as an encyclopedia on directed graphs, we have included as many important results as possible. The book contains a considerable number of proofs, illustrating various approaches, techniques and algorithms used in digraph theory.

As was the case with 'Digraphs', one of the main features of this book is its strong emphasis on algorithms. Algorithms on (directed) graphs often play an important role in problems arising in several areas, including computer science and operations research. Secondly, many problems on (directed) graphs are inherently algorithmic. Hence, the book contains many constructive proofs from which one can often extract an efficient algorithm for the problem studied.

To facilitate the use of this book as a reference book and as a graduate textbook, we have added comprehensive symbol and subject indexes. The latter includes separate entries for open problems, conjectures and proof-techniques as well as $\mathcal{NP}$-complete problems. It is our hope that the organization of the book, as well as the detailed subject index, will help many readers to find what they are looking for without having to read through whole chapters.

## Highlights

The book covers the majority of important topics on some of the most important classes of directed graphs, ranging from quite elementary to very advanced results. By organizing the book so as to single out important classes of digraphs, we hope to make it easy for the readers to find results and problems of their interest.

Below we give a brief outline of some of the main highlights of this book. Readers who are looking for more detailed information are advised to consult the list of contents or the subject index at the end of the book.

Chapter 1, by Bang-Jensen and Gutin, contains most of the terminology and notation used in this book as well as several basic results. These are not only used frequently in other chapters, but also serve as illustrations of digraph concepts. Since the terminology and notation used in this book is similar to that in 'Digraphs' some readers may skip parts of this chapter.

Chapter 2, by Bang-Jensen and Havet, deals with tournaments (orientations of complete graphs) and semicomplete digraphs (digraphs whose underlying graphs are complete). Tournaments form undoubtedly the most well-understood class digraphs and they continue to fascinate researchers due to their beautiful theory and a surprisingly large number of difficult open problems. The literature on tournaments is so extensive that one could write a whole book on these, so the chapter attempts to give a comprehensive overview of the theory, various proof techniques and many challenging open problems on tournaments. The chapter contains a number of classical results, including Rédei's theorem that every tournament has an odd number of Hamiltonian paths and Camion's theorem that every strongly connected tournament has a Hamiltonian cycle. It covers important topics such as arc-disjoint in- and out-branchings, decompositions into arc-disjoint Hamiltonian cycles or strong spanning subdigraphs, feedback sets, Seymour's second neighbourhood conjecture, vertex-partitions with prescribed properties, oriented Hamiltonian paths and cycles, (Hamiltonian)-connectivity, Hamiltonian cycles avoiding prescribed arcs, disjoint cycles and finally linkages (disjoint paths with prescribed end vertices). The chapter also contains a beautiful proof, due to Pokrovskiy, of the result that a linear bound on the vertex connectivity suffices to ensure that a semicomplete digraph is $k$-linked.

Chapter 3, by Gutin, deals with acyclic digraphs, that is, digraphs with no directed cycles. This class of digraphs is often used in digraph applications. Thus,

Sections 3.9–3.12 are devoted to four different applications: embedded computing, cryptographic enforcement schemes, project scheduling, and website text analysis. All other sections of the chapter, apart from the last section, consider results on various problems restricted to acyclic digraphs. Most results are on decision problems on subgraphs of acyclic digraphs such as out- and in-branchings, $k$-linkages, and dicuts, but Section 3.5 describes some enumeration results and algebraic techniques to prove them. The last section is devoted to generalizations of acyclic digraphs to the class of edge-coloured graphs. Somewhat surprisingly there are five such generalizations to edge-coloured graphs. This is partially due to the fact that some notions on directed walks, which are equivalent in digraphs, are no longer equivalent for properly coloured walks in edge-coloured graphs.

Chapter 4, by Wahlström, deals with Euler digraphs, that is, digraphs which are connected and in which every vertex has the same number of in-neighbours and out-neighbours. It is well known that the 1736 result of Euler, probably the first result in graph theory, which says that every connected graph in which all degrees are positive even numbers has a closed walk (called an Euler tour) that uses each edge exactly once, extends directly to Euler digraphs. Euler digraphs are interesting, not only because they have a closed Euler tour but also since they can often be viewed as a "half-way" between undirected and directed graphs. Several problems are tractable for undirected graphs, but intractable for directed graphs. Such problems may be either tractable or intractable for Euler digraphs. Good examples of such problems are linkage problems. However, there are exceptions. One of them is the well-known problem on enumerating Euler tours. While this problem is #P-hard on undirected graphs, it is polynomial-time solvable on Euler digraphs by the so-called BEST theorem proved in the chapter.

Chapter 5, by Pilipczuk and Pilipczuk, deals with planar digraphs, that is, digraphs which can be embedded in the plane with no arc crossings. The main goal of the chapter is to show, from multiple angles, how planarity imposes structure on digraphs and how such structure can be used algorithmically. The main focus of the chapter is to show various techniques used in algorithms on planar digraphs. The chapter is not a survey on planar digraphs, instead the authors concentrate on three topics: the $O(n \log n)$-time algorithm for finding a maximum flow between two vertices by Borradaile and Klein, the polynomial-time algorithm, based on advanced algebraic techniques by Schrijver for the $k$-Linkage Problem, and the Directed Grid Theorem.

Chapter 6, by Bang-Jensen, deals with locally semicomplete digraphs and some generalizations of these. A digraph is locally semicomplete if both the out-neighbourhood and the in-neighbourhood of each vertex induces a semicomplete digraph. Locally semicomplete digraphs were discovered by Bang-Jensen in 1988 and have since then been the focus of much attention, including several Ph.D. theses. The reason for this is that many results on tournaments and semicomplete digraphs extend to this much larger class of digraphs whose structure is well understood: they consist of three subclasses, namely semicomplete digraphs, round-decomposable digraphs and finally, so-called evil locally semicomplete digraphs. The last class is by far the most complicated of the two non-semicomplete subclasses of locally semicomplete digraphs. The chapter contains a full proof of the above classification of locally semicomplete digraphs as well as several examples on how to use this classification to extend many results on semicomplete digraphs to locally semicomplete digraphs. These include results on pancyclicity, arc-disjoint in- and out-branchings, decompositions into arc-disjoint strong spanning subdigraphs, feedback sets, (Hamiltonian)-connectivity, disjoint cycles, linkages and finally orientations of locally semicomplete digraphs, that is, digraphs that can be obtained by deleting one arcs from every 2-cycle. The chapter also discusses results on superclasses of the class of locally semicomplete digraphs, such as locally in-semicomplete and path-mergeable digraphs.

Chapter 7, by Yeo, deals with semicomplete multipartite digraphs, that is, digraphs whose underlying undirected graphs are complete multipartite. Clearly semicomplete digraphs form a subclass of this class so it is natural to ask how much of the structure of semicomplete digraphs carries over to semicomplete multipartite digraphs. Moon's book on tournaments from 1968 already contains some results along these lines and in 1976 Bondy initiated the study of cycles intersecting each partite set at least once. In 1988 Gutin solved the Hamiltonian path problem by giving a simple necessary and sufficient condition, and he, Häggkvist and Manoussakis characterized Hamiltonian semicomplete bipartite digraphs. To this date no necessary and sufficient condition for a semicomplete multipartite digraph to be Hamiltonian is known. One of the main results on semicomplete multipartite digraphs is Yeo's irreducible cycle factor theorem from 1997. Using this theorem, many deep results on semicomplete multipartite digraphs have been obtained, e.g. in 1997 Yeo proved a long standing conjecture that every regular semicomplete multipartite digraph is Hamiltonian and Bang-Jensen, Gutin and Yeo used Yeo's theorem to prove the existence of a polynomial algorithm to decide the existence of a Hamiltonian cycle in semicomplete multipartite digraphs. In this chapter Yeo, one of the main contributors to the area, gives a detailed account of the state of the art of results on this important class of digraphs. Besides results on the full class of semicomplete multipartite digraphs and on (almost) regular semicomplete multipartite digraphs, the chapter also contains a number of results on two subclasses, extended semicomplete digraphs and semicomplete bipartite digraphs. For these two classes there is a simple characterization of the length of a longest cycle, leading to a polynomial algorithm to find such a cycle. For the full class of semicomplete multipartite digraphs it is still open whether a polynomial algorithm exists.

Chapter 8, by Galeana-Sánchez and Hernández-Cruz, deals with transitive and quasi-transitive digraphs as well as generalizations of these. A digraph is transitive, respectively, quasi-transitive if it satisfies that whenever $x, y, z$ are distinct vertices so that $xy$ and $yz$ are arcs, then there is also an arc from $x$ to $z$, respectively, between $x$ and $z$. In 1962 Ghouila-Houri proved that a graph $G$ has a quasi-transitive orientation if and only if it has a transitive orientation and hence $G$ is a comparability graph. It was only after 1993, when Bang-Jensen and Huang gave a very useful structural characterization of quasi-transitive digraphs, that research into structural and algorithmic aspects of this class of digraphs flourished. They showed that quasi-transitive digraphs have a recursive structure which allows one to decompose them into smaller pieces, each of which is either a transitive oriented graph or a strong semicomplete digraph. The first non-trivial algorithmic application of the characterization was due to Gutin. The characterization and his approach have led to the study of totally $\Phi$-decomposable digraphs for different choices of digraph classes $\Phi$. These are digraphs which can be decomposed into smaller pieces, each of which belong to the class $\Phi$. This research has revealed that many problems, including the Hamiltonian path and cycle problems and linkage problems, can be solved efficiently for quasi-transitive digraphs and much more general classes of totally $\Phi$-decomposable digraphs. The chapter gives a detailed account of these results as well as results on kings, kernels and the path-partition conjecture by Laborde et al. from 1983. The chapter also contains a number of results on $k$-transitive and $k$-quasi-transitive digraphs. These are classes where the definition of transitive and quasi-transitive digraphs is relaxed.

Chapter 9, by Kreutzer and Kwon, considers structural parameters for digraphs. For undirected graphs, tree-width played a key role in developing an undirected graph structure theory and in designing efficient algorithms for intractable problems restricted to graphs of bounded tree-width. In the chapter, Kreutzer and Kwon classify digraph structural parameter approaches into three categories: tree-width inspired, rank-width inspired and density-based. Each of the approaches has its advantages and disadvantages described in numerous results obtained by various authors. While the great success of tree-width on undirected graphs has not been replicated on directed graphs (in fact, some negative results explain this situation), a number of important positive results and approaches have been obtained recently, such as the Directed Grid Theorem of Kawarabayashi and Kreutzer, Kanté's rank-width concepts, and the directed bounded expansion and nowhere density approaches which generalize their undirected counterparts introduced by Nešetřil and Ossona de Mendez.

Chapter 10, by Hammack, is devoted to products of digraphs. Hammack considers results on four standard digraph products: Cartesian product, direct product, strong product, and lexicographic product. The products have many common properties and several differences. For example, all four products are associative, but only the first three are commutative, and unlike for the three other products, $K_1$ is not a unit for the direct product. While many results on undirected graph products are well known, those on digraph products are less known outside the community of researchers who study the area. We hope this chapter will change the situation.

Chapter 11, by Guo and Surmacs, covers a number of classes of digraphs for which we could not devote a separate chapter, because there are not so many results on the class and also due to space limitations. Several of these classes have applications in interconnection networks and other areas. The classes considered include line digraphs, iterated line digraphs, de Bruijn digraphs, Kautz digraphs, directed cographs, perfect digraphs, arc-locally semicomplete digraphs and finally some generalizations of the latter class. Line digraphs as defined by Harary and Norman in 1960 naturally generalize line graphs of undirected graphs and they have applications in interconnection networks. The chapter contains a number of structural results on line digraphs and two other classes of digraphs, closely related to line digraphs, namely the de Bruijn and Kautz digraphs. These play an important role in network design as they combine the properties of having low out-degree, high connectivity and low diameter, something which is very important in communication networks. The chapter also contains results on directed cographs, in particular on linkages, and on perfect digraphs. The latter is a recent generalization of perfect graphs to digraphs due to Andres and Hochstättler. In contrast to perfect graphs, which can be recognized in polynomial time, recognizing perfect digraphs is $\mathcal{NP}$-complete. In the last sections of the chapter the authors discuss results on arc-locally semicomplete digraphs and some related classes. A digraph is arc-locally semicomplete if, for any choice of 4 distinct vertices $x, y, u, v$ the presence of the arcs $xu, yv$ implies that either none of the pairs $x, y$ and $u, v$ are adjacent or both pairs are adjacent. This class contains all semicomplete and all semicomplete bipartite digraphs and several characterizations, such as those for having a Hamiltonian path or cycle, carry over from semicomplete bipartite digraphs to arc-locally semicomplete digraphs.

Chapter 12, by Huang, deals with orientations of undirected graphs and mixed graphs (which may have both arcs and edges), that is, assigning for each edge $xy$ one of the two possible orientations $x \rightarrow y, y \rightarrow x$. The central topic is deciding whether the given graph has an orientation as an oriented graph which has a certain prescribed property $\Pi$. This could be the property of belonging to a certain class of digraphs, e.g. quasi-transitive digraphs or locally semicomplete digraphs, being strongly connected, or being acyclic and not containing a prescribed set of digraphs as an induced subdigraph. The chapter illustrates a general technique, the lexicographic orientation method, due to Hell and Huang, for achieving such orientations for graphs that are comparability graphs or proper circular arc graphs. When instead the input is a mixed graph $M = (V, E, A)$ with edge set $E$ and arc set $A$ we must orient the edges of $E$ but leave the arcs of $A$ untouched and again the goal is that the final oriented graph has a prescribed property $\Pi$. This is called the $\Pi$-orientation completion problem. It is shown in this chapter that even for semicomplete digraphs there are natural $\mathcal{NP}$-hard $\Pi$-orientation completion problems.

## Technical Remarks

We have tried to unify the book by using common terminology and notation for all chapters. We also used special environments for algorithms and problems, and used a special script for problem names as customary in the modern literature on algorithms. All the above should facilitate the reading of this book.

## Acknowledgements

We wish to thank the following colleagues for helpful reviews of the chapters: Derek Corneil, Peter Dankelmann, Hortensia Galeana-Sánchez, Marc Hellmuth, César Hernández-Cruz, Petr Hliněný, Mamadou Kanté, Stephan Kreutzer, Alessandro Maddaloni, Tilen Marc, Dániel Marx, Marcin Pilipczuk, Michał Pilipczuk, Alexey Pokrovskiy, and Lutz Volkmann.

Last, but most importantly, we wish to thank our families, in particular our wives Lene and Irina, without whose constant support we would never have succeeded in completing this project.

Odense, Denmark — Jørgen Bang-Jensen
London, UK — Gregory Gutin
May 2018

# Contents

# 1. Basic Terminology, Notation and Results

**Jørgen Bang-Jensen and Gregory Gutin**

In this chapter we will provide most of the terminology and notation used in this book. Various examples, figures and results should help the reader to better understand the notions introduced in the chapter. We also prove some basic results on digraphs and provide some fundamental digraph results without proofs. Most of our terminology and notation is standard and agrees with [4]. Thus, some readers may proceed to other chapters after a quick look through this chapter (unfamiliar terminology and notation can be clarified later by consulting the indices supplied at the end of this book).

In Section 1.1 we provide basic terminology and notation on sets and matrices. Digraphs, directed pseudographs, subgraphs, weighted directed pseudographs, neighbourhoods, semi-degrees and other basic concepts of directed graph theory are introduced in Section 1.2. In Section 1.3, we introduce oriented and directed walks, trails, paths and cycles, and related subgraphs. Isomorphism and basic operations on digraphs are considered in Section 1.4. Basic notions and results on strong connectivity are considered in Section 1.5. Section 1.6 provides basic definitions on linkages in digraphs. Undirected graphs and orientations of undirected and directed graphs are considered in Section 1.7. In Section 1.8, we briefly discuss out-branchings or in-branchings. Section 1.9 is devoted to a brief discussion of some results on flows in networks. In the last three sections, we discuss algorithmic approaches and lower bounds for solving $\mathcal{NP}$-hard problems: exponential time algorithms and the Exponential Time Hypothesis, fixed-parameter tractable algorithms and $W$-complexity classes, and approximation algorithms.

## 1.1 Sets, Matrices, Vectors and Hypergraphs

For the sets of real numbers, rational numbers and integers we will use $\mathbb{R}$, $\mathbb{Q}$ and $\mathbb{Z}$, respectively. Also, let $\mathbb{Z}_+ = \{z \in \mathbb{Z} : z > 0\}$ and $\mathbb{Z}_0 = \{z \in \mathbb{Z} : z \geq 0\}$.

---

J. Bang-Jensen
Department of Mathematics and Computer Science,
University of Southern Denmark, Campusvej 55, 5230 Odense M, Denmark
e-mail: jbj@imada.sdu.dk

G. Gutin (✉)
Department of Computer Science,
Royal Holloway University, Egham Hill, TW20 0EX, Egham, England
e-mail: gutin@rhul.ac.uk

 
J. Bang-Jensen and G. Gutin (eds.), *Classes of Directed Graphs*, Springer Monographs in Mathematics, https://doi.org/10.1007/978-3-319-71840-8_1

The sets $\mathbb{R}_+$, $\mathbb{R}_0$, $\mathbb{Q}_+$ and $\mathbb{Q}_0$ are defined similarly. For a positive integer $n$, $[n]$ will denote the set $\{1, 2, \ldots, n\}$.

The main aim of this section is to establish some notation and terminology on finite sets used in this book. We assume that the reader is familiar with the following basic operations for a pair $A, B$ of sets: the **intersection** $A \cap B$, the **union** $A \cup B$ (if $A \cap B = \emptyset$, then we will often write $A + B$ instead of $A \cup B$) and the **difference** $A \backslash B$ (often denoted by $A - B$). Sets $A$ and $B$ are **disjoint** if $A \cap B = \emptyset$.

Often we will not distinguish between a single element set (singleton) $\{x\}$ and the element $x$ itself. For example, we may write $A \cup b$ or $A + b$ instead of $A \cup \{b\}$. The **Cartesian product** of sets $X_1, X_2, \ldots, X_p$ is defined as $X_1 \times X_2 \times \ldots \times X_p = \{(x_1, x_2, \ldots, x_p) : \ x_i \in X_i, \ 1 \leq i \leq p\}$.

For sets $A, B$, $A \subseteq B$ means that $A$ is a subset of $B$; $A \subset B$ stands for $A \subseteq B$ and $A \neq B$. A set $B$ is a **proper subset** of a set $A$ if $B \subset A$ and $B \neq \emptyset$. A collection $S_1, S_2, \ldots, S_t$ of (not necessarily non-empty) subsets of a set $S$ is a **subpartition** of $S$ if $S_i \cap S_j = \emptyset$ for all $1 \leq i \neq j \leq t$. A subpartition $S_1, S_2, \ldots, S_t$ is a **partition** of $S$ if $\cup_{i=1}^{t} S_i = S$. We will often use the name **family** for a collection of sets. A family $\mathcal{F} = \{X_1, X_2, \ldots, X_n\}$ of sets is **covered** by a set $S$ if $S \cap X_i \neq \emptyset$ for every $i \in [n]$. We say that $S$ is a **cover** of $\mathcal{F}$. For a finite set $X$, the number of elements in $X$ (i.e., its **cardinality**) is denoted by $|X|$. We also say that $X$ is an $|\boldsymbol{X}|$**-element set** (or just an $|\boldsymbol{X}|$**-set**). A set $S$ satisfying a property $\mathcal{P}$ is a **maximum** (**maximal**, respectively) set with property $\mathcal{P}$ if there is no set $S'$ satisfying $\mathcal{P}$ and $|S'| > |S|$ ($S \subset S'$, respectively). Similarly, one can define **minimum** **(minimal)** sets satisfying a property $\mathcal{P}$.

In this book, we will also use **multisets** which, unlike sets, are allowed to have repeated (multiple) elements. The **cardinality** $|S|$ of a multiset $M$ is the total number of elements in $S$ (including repetitions). Often, we will use the words 'family' and 'collection' instead of 'multiset'.

For an $m \times n$ matrix $S = [s_{ij}]$ the **transposed matrix** (of $S$) is the $n \times m$ matrix $S^T = [t_{kl}]$ such that $t_{ji} = s_{ij}$ for every $i \in [m]$ and $j \in [n]$. Unless otherwise specified, the vectors that we use are column-vectors. The operation of transposition is used to obtain row-vectors.

A **hypergraph** is an ordered pair $H = (V, \mathcal{E})$ such that $V$ is a set (of **vertices** of $H$) and $\mathcal{E}$ is a family of subsets of $V$ (called **edges** of $H$). The **rank** of $H$ is the cardinality of the largest edge of $H$. For example, $H_0 = (\{1,2,3,4\}, \{\{1,2,3\}, \{2,3\}, \{1,2,4\}\})$ is a hypergraph of rank three. The number of vertices in $H$ is its **order**. We say that $H$ is **2-colourable** if there is a function $f : \ V \rightarrow \{0, 1\}$ such that, for every edge $E \in \mathcal{E}$, there exist a pair of vertices $x, y \in E$ such that $f(x) \neq f(y)$. The function $f$ is called a **2-colouring** of $H$. It is easy to verify that $H_0$ is 2-colourable. In particular, $f(1) = f(2) = 0, \ f(3) = f(4) = 1$ is a 2-colouring of $H_0$. A hypergraph is **uniform** if all its edges have the same cardinality. Thus an undirected graph is just a 2-uniform hypergraph.

## 1.2 Digraphs, Subgraphs, Neighbours, Degrees

A **directed graph** (or just **digraph**[1]) $D$ consists of a non-empty finite set $V(D)$ of elements called **vertices** and a finite set $A(D)$ of ordered pairs of distinct vertices called **arcs**. We call $V(D)$ the **vertex set** and $A(D)$ the **arc set** of $D$. We will often write $D = (V, A)$, which means that $V$ and $A$ are the vertex set and arc set of $D$, respectively. The **order** (**size**) of $D$ is the number of vertices (arcs) in $D$; the order of $D$ will sometimes be denoted by $|D|$. For example, the digraph $D$ in Figure 1.1 is of order and size 6; $V(D) = \{u, v, w, x, y, z\}$, $A(D) = \{(u, v), (u, w), (w, u), (z, u), (x, z), (y, z)\}$. Often the order (size, respectively) of the digraph under consideration is denoted by **n** (**m**, respectively).

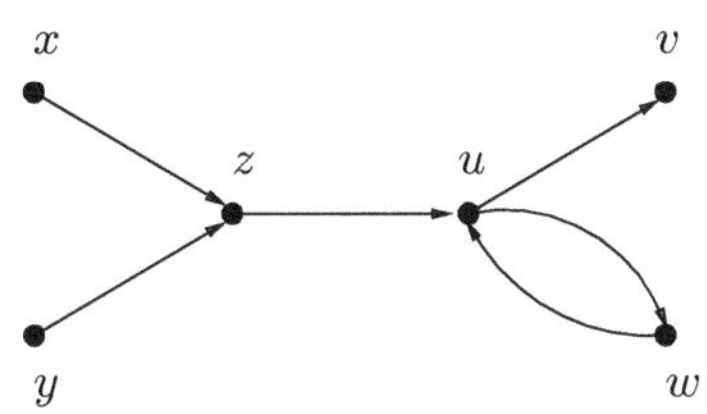


**Figure 1.1** A digraph $D$.

For an arc $(u, v)$ the first vertex $u$ is its **tail** and the second vertex $v$ is its **head**. We also say that the arc $(u, v)$ **leaves** $u$ and **enters** $v$. The head and tail of an arc are its **end-vertices**; we say that the end-vertices, are **adjacent**. If $(u, v)$ is an arc, we also say that $u$ **dominates** $v$ ($v$ **is dominated by** $u$) and denote it by $u \rightarrow v$. We say that a vertex $u$ is **incident** to an arc $a$ if $u$ is the head or tail of $a$. We will often denote an arc $(x, y)$ by $xy$.

For a pair $X, Y$ of vertex sets of a digraph $D$, we define

$$(X, Y)_D = \{xy \in A(D) : \; x \in X, y \in Y\},$$

i.e., $(X, Y)_D$ is the set of arcs with tail in $X$ and head in $Y$. For example, for the digraph $H$ in Figure 1.2, $(\{u, v\}, \{w, z\})_H = \{uw\}$, $(\{w, z\}, \{u, v\})_H = \{wv\}$ and $(\{u, v\}, \{u, v\})_H = \{uv, vu\}$. For disjoint subsets $X$ and $Y$ of $V(D)$, $X \rightarrow Y$ means that every vertex of $X$ dominates every vertex of $Y$. Also, $X \mapsto Y$ stands for $X \rightarrow Y$ and no vertex of $Y$ dominates a vertex in $X$. For example, in the digraph $D$ of Figure 1.1, $u \rightarrow \{v, w\}$ and $\{x, y\} \mapsto z$.

The above definition of a digraph implies that we allow a digraph to have arcs with the same end-vertices (for example, $uv$ and $vu$ in the digraph $H$ in Figure 1.2), but we do not allow it to contain **parallel** (also called **multiple**) arcs, that is, pairs of arcs with the same tail and the same head, or

[1] If we know from the context that $D$ is directed, $D$ may be called a **graph**.

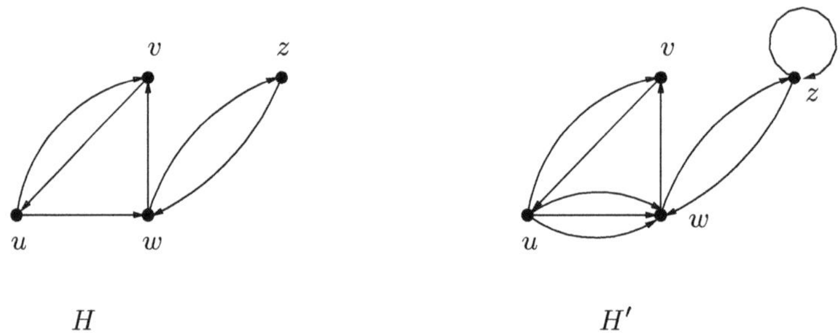


**Figure 1.2** A digraph $H$ and a directed pseudograph $H'$.

**loops** (i.e., arcs whose head and tail coincide). When parallel arcs and loops are admissible we speak of **directed pseudographs**; directed pseudographs without loops are **directed multigraphs**. In Figure 1.2 the directed pseudograph $H'$ is obtained from $H$ by appending a loop $zz$ and two parallel arcs from $u$ to $w$. Clearly, for a directed pseudograph $D$, $A(D)$ and $(X,Y)_D$ (for every pair $X,Y$ of vertex sets of $D$) are multisets (parallel arcs provide repeated elements). We use the symbol $\mu_D(x,y)$ to denote the number of arcs from a vertex $x$ to a vertex $y$ in a directed pseudograph $D$. In particular, $\mu_D(x,y)=0$ means that there is no arc from $x$ to $y$.

We will sometimes give terminology and notation for digraphs only, but we will provide necessary remarks on their extension to directed pseudographs, unless this is trivial.

Below, unless otherwise specified, $D=(V,A)$ is a directed pseudograph. For a vertex $v$ in $D$, we use the following notation:

$$N_D^+(v)=\{u\in V-v:\ vu\in A\},\ N_D^-(v)=\{w\in V-v:\ wv\in A)\}.$$

The sets $N_D^+(v)$, $N_D^-(v)$ and $N_D(v) = N_D^+(v)\cup N_D^-(v)$ are called the **out-neighbourhood, in-neighbourhood** and **neighbourhood** of $v$. We call the vertices in $N_D^+(v)$, $N_D^-(v)$ and $N_D(v)$ the **out-neighbours**, **in-neighbours** and **neighbours** of $v$.

In Figure 1.2, $N_H^+(u)=\{v,w\}$, $N_H^-(u)=\{v\}$, $N_H(u)=\{v,w\}$, $N_{H'}^+(w)=\{v,z\}$, $N_{H'}^-(w)=\{u,z\}$, $N_{H'}^+(z)=\{w\}$. For a set $W\subseteq V$, we let

$$N_D^+(W)=\bigcup_{w\in W}N_D^+(w)-W,\ N_D^-(W)=\bigcup_{w\in W}N_D^-(w)-W.$$

That is, $N_D^+(W)$ consists of those vertices from $V-W$ which are out-neighbours of at least one vertex from $W$. In Figure 1.2, $N_H^+(\{w,z\})=\{v\}$ and $N_H^-(\{w,z\})=\{u\}$.

Recursively, we can define the $i$**th out-neighbourhood** of a set $W$ as follows: $N^{+i}(W)=N^+(N^{+(i-1)}(W))$ for $i\geq 2$. We will denote $N^{+2}(W)$ as $N^{++}(W)$. Similarly, we can define the $i$**th in-neighbourhood** of a set $W$.

The neighbourhoods above are sometimes called **open neighbourhoods**. **Closed neighbourhoods** are defined as follows: For a set $W \subseteq V$ and positive integer $p$, let $N^{+p}[W] = N^{+p}(W) \cup W$ and $N^{-p}[W] = N^{-p}(W) \cup W$.

A digraph is called an **oriented graph** if it has no pair of arcs of the form $xy, yx$. Seymour's Second Neighbourhood Conjecture is one of the most interesting open questions in digraph theory. It has the following simple formulation.

**Conjecture 1.2.1 (Seymour's Second Neighbourhood Conjecture)** *In every oriented graph $D$, there exists a vertex $x$ such that $|N_D^+(x)| \leq |N_D^{++}(x)|$.*

The conjecture is discussed in detail in Chapter 2 including two proofs of the conjecture for tournaments. In addition, recently Gutin and Li [23] proved the conjecture for **quasi-transitive oriented graphs**; a digraph $D$ is called **quasi-transitive** if whenever $x \to y$ and $y \to z$ $(x \neq z)$ we have that $x \to z$ or $z \to x$ (or both). Quasi-transitive digraphs and their generalizations are considered in Chapter 8.

For a set $W \subseteq V$, the **out-degree** of $W$ (denoted by $d_D^+(W)$) is the number of arcs in $D$ whose tails are in $W$ and heads are in $V - W$, i.e., $d_D^+(W) = |(W, V - W)_D|$. The **in-degree** of $W$, $d_D^-(W) = |(V - W, W)_D|$. In particular, for a vertex $v$, the out-degree is the number of arcs, except for loops, with tail $v$. If $D$ is a digraph (that is, it has no loops or multiple arcs), then the out-degree of a vertex equals the number of out-neighbours of this vertex. We call the out-degree and in-degree of a set of vertices $W$ its **semi-degrees**. The **degree** of $W$ is the sum of its semi-degrees, i.e., the number $d_D(W) = d_D^+(W) + d_D^-(W)$. For example, in Figure 1.2, $d_H^+(u) = 2, d_H^-(u) = 1, d_H(u) = 3$, $d_{H'}^+(w) = 2, d_{H'}^-(w) = 4, d_{H'}^+(z) = d_{H'}^-(z) = 1, d_H^+(\{u, v, w\}) = d_H^-(\{u, v, w\}) = 1$. Sometimes, it is useful to count loops in the semi-degrees: the **out-pseudodegree** of a vertex $v$ of a directed pseudograph $D$ is the number of arcs with tail $v$. Similarly, one can define the **in-pseudodegree** of a vertex. In Figure 1.2, both the in-pseudodegree and out-pseudodegree of $z$ in $H'$ are equal to 2.

The **minimum out-degree** (**minimum in-degree**) of $D$ is

$$\delta^+(D) = \min\{d_D^+(x) : \ x \in V(D)\} \quad (\delta^-(D) = \min\{d_D^-(x) : \ x \in V(D)\}).$$

The **minimum semi-degree** of $D$ is

$$\delta^0(D) = \min\{\delta^+(D), \delta^-(D)\}.$$

Finally, the **minimum degree of $D$ is**

$$\delta(D) = \min\{d^+(v) + d^-(v) : \ v \in V(D)\}.$$

Similarly, one can define the **maximum out-degree** of $D$, $\Delta^+(D)$, and the **maximum in-degree**, $\Delta^-(D)$. The **maximum semi-degree** of $D$ is

$$\Delta^0(D) = \max\{\Delta^+(D), \Delta^-(D)\}.$$

We say that $D$ is **regular** if $\delta^0(D) = \Delta^0(D)$. In this case, $D$ is also called **$\delta^0(D)$-regular**.

For degrees, semi-degrees and for other parameters and sets of digraphs, we will usually omit the subscript for the digraph when it is clear which digraph is meant.

Since the number of arcs in a directed multigraph equals the number of their tails (or their heads), we obtain the following very basic result. Recall that $m$ denotes the number of arcs in the digraph under consideration.

**Proposition 1.2.2** *For every directed multigraph $D$ we have*

$$\sum_{x\in V(D)} d^-(x) = \sum_{x\in V(D)} d^+(x) = m.$$

□

Clearly, this proposition is valid for directed pseudographs if in-degrees and out-degrees are replaced by in-pseudodegrees and out-pseudodegrees.

A digraph $H$ is a **subdigraph** (or just **subgraph**) of a digraph $D$ if $V(H) \subseteq V(D)$, $A(H) \subseteq A(D)$ and every arc in $A(H)$ has both end-vertices in $V(H)$. If $V(H) = V(D)$, we say that $H$ is a **spanning subgraph** (or a **factor**) of $D$. The digraph $L$ with vertex set $\{u, v, w, z\}$ and arc set $\{uv, uw, wz\}$ is a spanning subgraph of $H$ in Figure 1.2. If every arc of $A(D)$ with both end-vertices in $V(H)$ is in $A(H)$, we say that $H$ is **induced** by $X = V(H)$ (we write $H = D[X]$ or $H = D\langle X\rangle$ ) and call $H$ an **induced** subgraph of $D$. The digraph $G$ with vertex set $\{u, v, w\}$ and arc set $\{uw, wv, vu\}$ is a subgraph of the digraph $H$ in Figure 1.2; $G$ is neither a spanning subgraph nor an induced subgraph of $H$. The digraph $G$ along with the arc $uv$ is an induced subgraph of $H$. For a subset $A' \subseteq A(D)$ the subgraph **arc-induced** by $A'$ is the digraph $D[A'] = (V', A')$, where $V'$ is the set of vertices in $V$ which are incident with at least one arc from $A'$. For example, in Figure 1.2, $H[\{zw, uw\}]$ has vertex set $\{u, w, z\}$ and arc set $\{zw, uw\}$. If $H$ is a subgraph of $D$, then we say that $D$ is a **supergraph** of $H$.

It is trivial to extend the above definitions of subgraphs to directed pseudographs. To avoid lengthy terminology, we call the 'parts' of directed pseudographs just **subgraphs** (instead of, say, subpseudographs).

For vertex-disjoint subgraphs $H$, $L$ of a digraph $D$, we will often use the shorthand notation $(H, L)_D$ and $H \to L$ instead of $(V(H), V(L))_D$ and $V(H) \to V(L)$, respectively. We may also drop the index $D$ when the digraph is clear from the context.

A **weighted directed pseudograph** is a directed pseudograph $D$ along with a mapping $c : A(D) \to \mathbb{R}$. Thus, a weighted directed pseudograph is a triple $D = (V(D), A(D), c)$. We will also consider **vertex-weighted directed pseudographs**, i.e., directed pseudographs $D$ along with a mapping $c : V(D) \to \mathbb{R}$. (See Figure 1.3.) If $a$ is an **element** (i.e., a vertex or an arc) of a weighted directed pseudograph $D = (V(D), A(D), c)$, then $c(a)$ is called the **weight** or the **cost** of $a$. An (unweighted) directed pseudograph can be viewed as a (vertex-)weighted directed pseudograph whose elements are all of weight 1. For a set $B$ of arcs of a weighted directed pseudograph $D = (V, A, c)$, we define the weight of $B$ as follows: $c(B) = \sum_{a \in B} c(a)$. Similarly, one can define the weight of a set of vertices in a vertex-weighted directed pseudograph. The **weight of a subgraph** $H$ of a weighted (vertex-weighted) directed pseudograph $D$ is the sum of the weights of the arcs in $H$ (vertices in $H$). For example, in the weighted directed pseudograph $D$ in Figure 1.3 the set of arcs $\{xy, yz, zx\}$ has weight 9.5 (here we have assumed that we used the arc $zx$ of weight 1). In the directed pseudograph $H$ in Figure 1.3 the subgraph $U = (\{u, x, z\}, \{xz, zu\})$ has weight 5.

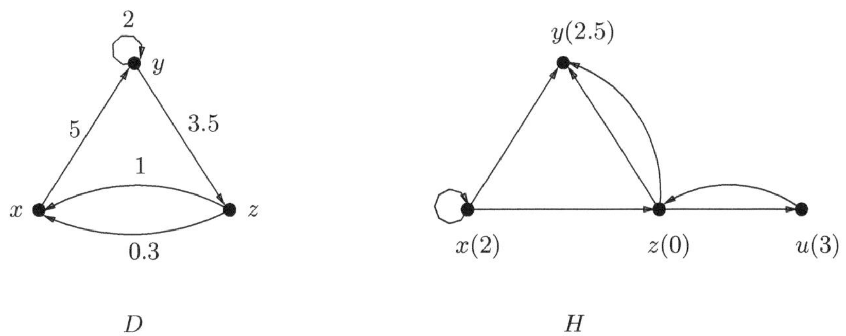


**Figure 1.3** Weighted and vertex-weighted directed pseudographs (the vertex weights are given in brackets).

## 1.3 Walks, Trails, Paths, Cycles and Path-Cycle Subgraphs

In the following, $D$ is always a directed pseudograph, unless otherwise specified. An **oriented walk** (or, just a **walk**) in $D$ is an alternating sequence $W = x_1 a_1 x_2 a_2 x_3 \ldots x_{k-1} a_{k-1} x_k$ of vertices $x_i$ and arcs $a_j$ from $D$ such that $x_i$ and $x_{i+1}$ are end-vertices of $a_i$ for every $i \in [k-1]$. In particular, if $x_i$ and $x_{i+1}$ are the tail and head of $a_i$, respectively, for every $i \in [k-1]$, then $W$ is a **directed walk (diwalk)**. When the fact that $W$ is directed is known from the context, we will often say that $W$ is a walk (this convention extends to every type of walk, i.e. trials, paths and cycles defined below). A walk $W$ is **closed** if $x_1 = x_k$, and **open** otherwise. The set of vertices $\{x_i : i \in [k]\}$ is denoted by $V(W)$; the set of arcs $\{a_j : j \in [k-1]\}$ is denoted by $A(W)$. We say that $W$ is a diwalk **from** $x_1$ **to** $x_k$ or an $(\boldsymbol{x_1, x_k})$**-diwalk**. If a diwalk $W$ is open, then we say that the vertex $x_1$ is the **initial** vertex of $W$, the vertex $x_k$ is the **terminal** vertex of $W$, and $x_1$ and $x_k$ are **end-vertices** of $W$ (the last term can be used for any oriented walk). The **length** of a walk is its number of arcs. Hence the walk $W$ above has length $k-1$; we will say that $W$ is a **$(k-1)$-walk**. A walk is **even** (**odd**) if its length is even (odd). When the arcs of $W$ are defined from the context or simply unimportant, we will denote $W$ by $x_1x_2\dots x_k$.

A **trail** is a walk in which all arcs are distinct. Sometimes, we identify a trail $W$ with the *directed pseudograph* $(V(W), A(W))$, which is a *subgraph* of $D$. If the vertices of the diwalk $W$ are distinct, $W$ is a **directed path (dipath)**. If the vertices $x_1, x_2, \dots, x_{k-1}$ are distinct, $k \geq 3$ and $x_1 = x_k$, $W$ is a **directed cycle (dicycle)**. Note that a loop is a directed cycle of length 1 and a pair of opposite arcs forms a directed cycle of length 2. A digraph is **acyclic** if it has no dicycle. An ordering $v_1, v_2, \dots, v_n$ of the vertices of a digraph $D$ is called an **acyclic ordering** if for every arc $v_iv_j \in A(D)$, we have $i < j$. The following proposition is well-known and not hard to prove (see Chapter 3).

**Proposition 1.3.1** *Every acyclic digraph has an acyclic ordering of its vertices.*

Since paths and cycles are special cases of walks, the **length** of a path and a cycle is already defined. The same remark is valid for other parameters and notions, e.g., an **$(x, y)$-path**. A directed path $P$ is an **$[x, y]$-path** if $P$ is a path between $x$ and $y$, e.g., $P$ is either an $(x, y)$-path or a $(y, x)$-path. A **longest** (**shortest**) $(x, y)$-dipath in a digraph $D$ is a $(x, y)$-dipath of maximum (minimum) length in $D$. The **distance** $\mathrm{dist}(x, y)$ from a vertex $x$ to a vertex $y$ is the length of a shortest $(x, y)$-dipath. If in a digraph $D$ there is a dipath from every vertex to every other vertex (i.e., $D$ is strongly connected, see Section 1.5), then the **diameter** of $D$ is the maximum of the distances $\mathrm{dist}(x, y)$ over all vertices $x$ and $y$ in $D$. If $D$ is not strongly connected, the diameter of $D$ is $\infty$. An $(x, y)$-dipath $P$ is a **minimal** **$(x, y)$-dipath** if it is the only $(x, y)$-dipath in $D[V(P)]$.

When $W$ is a cycle and $x$ is a vertex of $W$, we say that $W$ is a cycle **through** $x$. The **girth** $g(D)$ of $D$ is the length of a shortest dicycle in $D$. If $D$ does not have a cycle, we define $g(D) = \infty$. A digraph $D$ is **vertex-$k$-cyclic** (**arc-$k$-cyclic**, respectively) if every vertex (arc, respectively) of $D$ is contained in a directed $k$-cycle. A digraph $D$ is **pancyclic** if it has a $k$-cycle for every $k \in \{3, 4, \dots, n\}$; $D$ is **vertex-pancyclic** (**arc-pancyclic**, respectively) if $D$ is vertex-$k$-cyclic (arc-$k$-cyclic, respectively) for every $k \in \{3, 4, \dots, n\}$.

For subsets $X, Y$ of $V(D)$, a directed $(x, y)$-path $P$ is an **$(X, Y)$-path** if $x \in X$, $y \in Y$ and $V(P) \cap (X \cup Y) = \{x, y\}$. Note that if $X \cap Y \neq \emptyset$, then a vertex $x \in X \cap Y$ forms an $(X, Y)$-path by itself. Sometimes we will talk about an $(H, H')$-path when $H$ and $H'$ are subgraphs of $D$. By this we mean a $(V(H), V(H'))$-path in $D$.

For a cycle $C = x_1x_2\dots x_px_1$, the subscripts are considered modulo $p$, i.e., $x_s = x_i$ for every $s$ and $i$ such that $i \equiv s \bmod p$. As pointed out above (for trails), we will often view paths and cycles as subgraphs. We can also consider paths and cycles as digraphs themselves. Let $\vec{P}_n$ ($\vec{C}_n$) denote a dipath (a dicycle) with $n$ vertices, i.e., $\vec{P}_n = ([n], \{(1,2),(2,3),\dots,(n-1,n)\})$ and $\vec{C}_n = \vec{P}_n + (n,1)$.

A directed walk (path, cycle) $W$ is a **Hamilton** (or **Hamiltonian**) walk (path, cycle) if $V(W) = V(D)$. A digraph $D$ is **Hamiltonian (traceable)** if $D$ contains a Hamilton dicycle (Hamilton dipath). A directed trail $W$ is an **Euler** (or **Eulerian**) trail if $W$ is closed, $V(W) = V(D)$ and $A(W) = A(D)$; a directed multigraph $D$ is **Eulerian** if it has an Euler trail.

To illustrate these definitions, consider Figure 1.4.

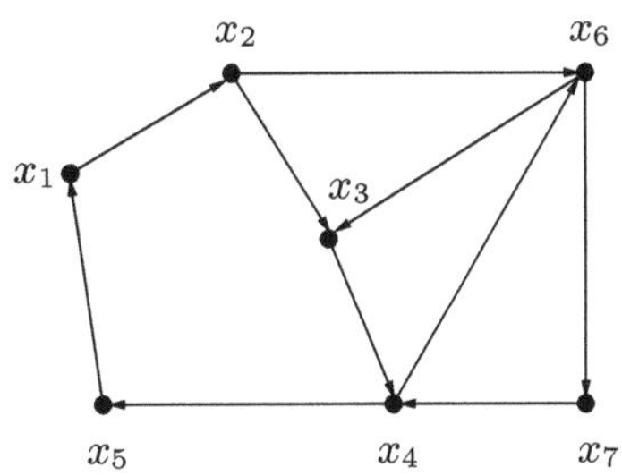


**Figure 1.4** A directed graph $H$.

The walk $x_1x_2x_6x_3x_4x_6x_7x_4x_5x_1$ is a Hamiltonian diwalk in $D$. The path $x_5x_1x_2x_3x_4x_6x_7$ is a Hamiltonian dipath in $D$. The path $x_1x_2x_3x_4x_6$ is an $(x_1,x_6)$-path and $x_2x_3x_4x_6x_3$ is an $(x_2,x_3)$-trail. The cycle $x_1x_2x_3x_4x_5x_1$ is a 5-cycle in $D$. The girth of $D$ is 3 and the longest dicycle in $D$ has length 6.

Let $W = x_1x_2\dots x_k$, $Q = y_1y_2\dots y_t$ be a pair of walks in a digraph $D$. The walks $W$ and $Q$ are **disjoint** if $V(W) \cap V(Q) = \emptyset$ and **arc-disjoint** if $A(W) \cap A(Q) = \emptyset$. If $W$ and $Q$ are open walks, they are called **internally disjoint** if $\{x_2, x_3, \dots, x_{k-1}\} \cap V(Q) = \emptyset$ and $V(W) \cap \{y_2, y_3, \dots, y_{t-1}\} = \emptyset$.

We will use the following notation for a path or a cycle $W = x_1x_2\dots x_k$ (recall that $x_1 = x_k$ if $W$ is a cycle):

$$W[x_i, x_j] = x_ix_{i+1}\dots x_j.$$

It is easy to see that $W[x_i, x_j]$ is a path for $x_i \neq x_j$; we call it the **subpath** of $W$ from $x_i$ to $x_j$. If $1 < i \leq k$, then the **predecessor** of $x_i$ on $W$ is the vertex $x_{i-1}$. If $1 \leq i < k$, then the **successor** of $x_i$ on $W$ is the vertex $x_{i+1}$.

**Proposition 1.3.2** *Let $D$ be a digraph and let $x, y$ be a pair of distinct vertices in $D$. If $D$ has an $(x,y)$-diwalk $W$, then $D$ contains an $(x,y)$-dipath $P$*

*such that $A(P) \subseteq A(W)$. If $D$ has a closed $(x,x)$-diwalk $W$, then $D$ contains a dicycle $C$ through $x$ such that $A(C) \subseteq A(W)$.*

**Proof:** Consider a diwalk $P$ from $x$ to $y$ of minimum length among all $(x,y)$-diwalks whose arcs belong to $A(W)$. We show that $P$ is a path. Let $P = x_1x_2\dots x_k$, where $x = x_1$ and $y = x_k$. If $x_i = x_j$ for some $1 \le i < j \le k$, then the walk $P[x_1, x_i]P[x_{j+1}, x_k]$ is shorter than $P$; a contradiction. Thus, all vertices of $P$ are distinct, so $P$ is a dipath with $A(P) \subseteq A(W)$.

Let $W = z_1z_2\dots z_k$ be a diwalk from $x = z_1$ to itself ($x = z_k$). Since $D$ has no loop, $z_{k-1} \neq z_k$. Let $y_1y_2\dots y_t$ be a shortest diwalk from $y_1 = z_1$ to $y_t = z_{k-1}$. We have proved above that $y_1y_2\dots y_t$ is a dipath. Thus, $y_1y_2\dots y_ty_1$ is a dicycle through $y_1 = x$. □

An **oriented** graph is a digraph with no cycle of length two. A **tournament** is an oriented graph where every pair of distinct vertices are adjacent. In other words, a digraph $T$ with vertex set $\{v_i : i \in [n]\}$ is a tournament if exactly one of the arcs $v_iv_j$ and $v_jv_i$ is in $T$ for every $i \neq j \in [n]$. In Figure 1.5, one can see a pair of tournaments. It is easy to see that each of them contains a Hamilton dipath. Actually, this is not a coincidence due to the following theorem of Rédei [32].

**Theorem 1.3.3 (Redei's theorem)** *Every tournament $T$ is traceable.*

**Proof:** Let $x_1, \dots, x_n$ be an ordering of the vertices of $T$ such that the number of **forward** arcs, i.e. arcs of the form $x_ix_j$ $(i < j)$, is maximal. Observe that $x_i \to x_{i+1}$ for each $i \in [n-1]$. Indeed, if we had $x_{i+1} \to x_i$ for some $i$, we could swap vertices $x_i$ and $x_{i+1}$ in the ordering and obtain one more forward arc, a contradiction. Thus, $x_1 \dots x_n$ is a Hamiltonian dipath. □

In fact, Rédei proved a stronger result: every tournament contains an odd number of Hamiltonian dipaths (see Theorem 2.6.1).

**Figure 1.5** Tournaments.

A **directed $q$-path-cycle subgraph** $\mathcal{F}$ of a digraph $D$ is a collection of $q$ dipaths $P_1, \dots, P_q$ and $t$ dicycles $C_1, \dots, C_t$ such that all of $P_1, \dots, P_q, C_1, \dots, C_t$ are pairwise disjoint (possibly, $q = 0$ or $t = 0$). We will denote $\mathcal{F}$ by $\mathcal{F} = P_1 \cup \dots \cup P_q \cup C_1 \cup \dots \cup C_t$ (the paths always being listed first). Quite often, we will consider **directed $q$-path-cycle factors**, i.e., spanning directed $q$-path-cycle subgraphs. If $t = 0$, $\mathcal{F}$ is a **directed $q$-path subgraph** and it is a **directed $q$-path factor** (or just a **directed path factor**) if it is spanning. If $q = 0$, we say that $\mathcal{F}$ is a **directed $t$-cycle subgraph** (or just a **directed cycle subgraph**) and it is a **directed $t$-cycle factor** (or just a **directed cycle factor**) if it is spanning. In Figure 1.6, $abc \cup defd$ is a directed 1-path-cycle factor, and $abcea \cup dfd$ is a directed cycle factor (or, more precisely, a directed 2-cycle factor).

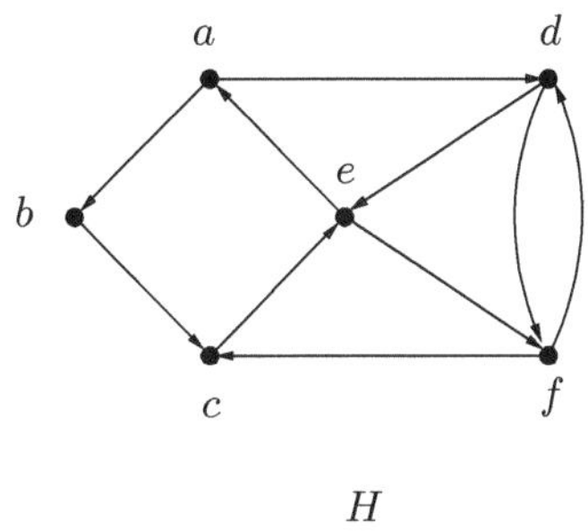


**Figure 1.6** A digraph H.

A **multipartite tournament** is a digraph obtained from a complete multipartite undirected graph by replacing every edge by an arc with the same end-vertices. The following extension of Redei's theorem (Theorem 1.3.3) to multipartite tournaments was proved by Gutin [22].

**Theorem 1.3.4** *A multipartite tournament has a Hamilton dipath if and only if it contains a 1-path-cycle factor.*

Chapter 7 is devoted to multipartite tournaments and their generalization, semicomplete multipartite digraphs.

The **path covering number** $\text{pc}(D)$ of $D$ is the minimum positive integer $k$ such that $D$ contains a $k$-path factor. In particular, $\text{pc}(D) = 1$ if and only if $D$ is traceable. The **path-cycle covering number** $\text{pcc}(D)$ of $D$ is the minimum positive integer $k$ such that $D$ contains a $k$-path-cycle factor. Clearly, $\text{pcc}(D) \leq \text{pc}(D)$. The following simple yet helpful assertion on the path covering number is not hard to show and so it is left without a proof.

**Proposition 1.3.5** *Let $D$ be a digraph, and let $k$ be a positive integer. Then the following statements are equivalent:*

1. $\text{pc}(D) = k$.
2. *There are $k - 1$ (new) arcs $e_1, \ldots, e_{k-1}$ such that $D + \{e_1, \ldots, e_{k-1}\}$ is traceable, but there is no set of $k - 2$ arcs with this property.*
3. *$k - 1$ is the minimum integer $s$ such that addition of $s$ new vertices to $D$ together with all possible arcs between $V(D)$ and these new vertices results in a traceable digraph.* □

## 1.4 Isomorphism and Basic Operations on Digraphs

Suppose $D = (V, A)$ is a directed multigraph. A directed multigraph obtained from $D$ by **deleting multiple arcs** is a digraph $H = (V, A')$ where $xy \in A'$ if and only if $\mu_D(x, y) \geq 1$. Let $xy$ be an arc of $D$. By **reversing the arc** $xy$, we mean that we replace the arc $xy$ by the arc $yx$. That is, in the resulting directed multigraph $D'$ we have $\mu_{D'}(x, y) = \mu_D(x, y) - 1$ and $\mu_{D'}(y, x) = \mu_D(y, x) + 1$.

A pair of (unweighted) directed pseudographs $D$ and $H$ are **isomorphic** (denoted by $D \cong H$) if there exists a bijection $\phi : V(D) \to V(H)$ such that $\mu_D(x, y) = \mu_H(\phi(x), \phi(y))$ for every ordered pair $x, y$ of vertices in $D$. The mapping $\phi$ is an **isomorphism**. Quite often, we will not distinguish between isomorphic digraphs or directed pseudographs. For example, we may say that there is only one digraph on a single vertex and there are exactly three digraphs with two vertices. Also, there is only one digraph of order 2 and size 2, but there are two directed multigraphs and six directed pseudographs of order and size 2. For a set $\Psi$ of directed pseudographs, we say that a directed pseudograph $D$ **belongs** to $\Psi$ or is a **member** of $\Psi$ (denoted $D \in \Psi$) if $D$ is isomorphic to a directed pseudograph in $\Psi$. Since we usually do not distinguish between isomorphic directed pseudographs, we will often write $D = H$ instead of $D \cong H$ for isomorphic $D$ and $H$.

In case we do want to distinguish between isomorphic digraphs, we speak of **labelled digraphs**. In this case, a pair of digraphs $D$ and $H$ is indistinguishable if and only if they completely coincide (i.e., $V(D) = V(H)$ and $A(D) = A(H)$). In particular, there are four labeled digraphs with vertex set $\{1, 2\}$. Indeed, the labeled digraphs $(\{1, 2\}, \{(1, 2)\})$ and $(\{1, 2\}, \{(2, 1)\})$ are distinct, even though they are isomorphic.

The **converse** of a directed multigraph $D$ is the directed multigraph $H$ which one obtains from $D$ by reversing all arcs. It is easy to verify, using only the definitions of isomorphism and converse, that a pair of directed multigraphs are isomorphic if and only if their converses are isomorphic. To obtain subdigraphs, we use the following operations of **deletion**. For a directed multigraph $D$ and a set $B \subseteq A(D)$, the directed multigraph $D - B$ (sometimes denoted by $D \setminus B$) is the spanning subgraph of $D$ with arc set $A(D) \setminus B$. If $X \subseteq V(D)$, the directed multigraph $D - X$ (sometimes denoted by $D \setminus X$) is the subgraph induced by $V(D) \setminus X$, i.e., $D - X = D\langle V(D) - X\rangle$. For a subgraph $H$ of $D$, we define $D - H = D - V(H)$. Since we do not distinguish between a single element set $\{x\}$ and the element $x$ itself, we will often write $D - x$ rather than $D - \{x\}$. If $H$ is a non-induced subgraph of a digraph $D$ and $xy \in A(D) - A(H)$ with $x, y \in V(H)$, we can construct another subgraph $H'$ of $D$ by adding the arc $xy$ of $H$; $H' = H + xy$.

Let $G$ be a subgraph of a directed multigraph $D$. The **contraction** of $G$ in $D$ is a directed multigraph $D/G$ with $V(D/G) = \{g\} \cup (V(D) - V(G))$, where $g$ is a 'new' vertex not in $D$, and $\mu_{D/G}(x, y) = \mu_D(x, y)$, and for all distinct vertices $x, y \in V(D) - V(G)$ we have

$$\mu_{D/G}(x,g) = \sum_{v\in V(G)} \mu_D(x,v),\ \mu_{D/G}(g,y) = \sum_{v\in V(G)} \mu_D(v,y).$$

(Note that there is no loop in $D/G$.) Let $G_1, G_2, \ldots, G_t$ be vertex-disjoint subgraphs of $D$. Then

$$D/\{G_1, G_2, \ldots, G_t\} = (\ldots((D/G_1)/G_2)\ldots)/G_t.$$

Clearly, the resulting directed multigraph $D/\{G_1, G_2, \ldots, G_t\}$ does not depend on the order of $G_1, G_2, \ldots, G_t$. Contraction can be defined for sets of vertices, rather than subgraphs. It suffices to view a set of vertices $X$ as a subgraph with vertex set $X$ and no arcs. Figure 1.7 depicts a digraph $H$ and the contraction $H/L$, where $L$ is the subgraph of $H$ induced by the vertices $y$ and $z$.

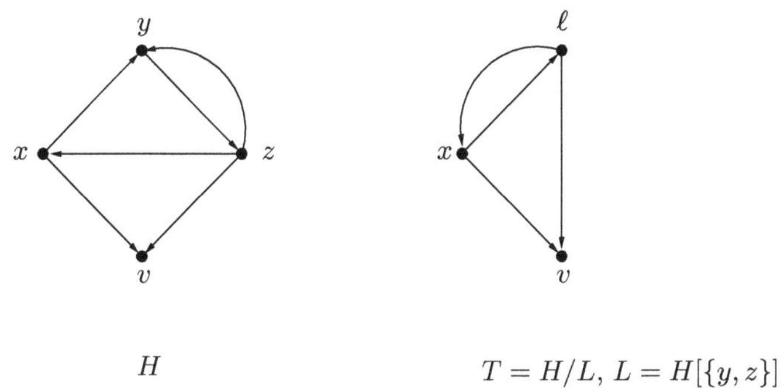


**Figure 1.7** Contraction.

We will often use the following variation of the operation of contraction. This operation is called **path-contraction** and is defined as follows. Let $P$ be a directed $(x,y)$-path in a directed multigraph $D = (V, A)$. Then $D//P$ stands for the directed multigraph with vertex set $V(D//P) = V \cup \{z\} - V(P)$, where $z \notin V$, and $\mu_{D//P}(u,v) = \mu_D(u,v)$, $\mu_{D//P}(u,z) = \mu_D(u,x)$, $\mu_{D//P}(z,v) = \mu_D(y,v)$ for all distinct $u, v \in V - V(P)$. In other words, $D//P$ is obtained from $D$ by deleting all vertices of $P$ and adding a new vertex $z$ such that every arc with head $x$ (tail $y$) and tail (head) in $V - V(P)$ becomes an arc with head (tail) $z$ and the same tail (head). Observe that a path-contraction in a digraph results in a digraph (no parallel arcs arise). We will often consider path-contractions of paths of length one, i.e., arcs $e$. Clearly, a directed multigraph $D$ has a directed $k$-cycle ($k \geq 3$) through an arc $e$ if and only if $D//e$ has a cycle through $z$. Observe that the obvious analogue of path-contraction for undirected multigraphs does not have this nice property, which is of use in this section. The difference between (ordinary) contraction (which is also called **set-contraction**) and path-contraction is reflected in Figure 1.8.

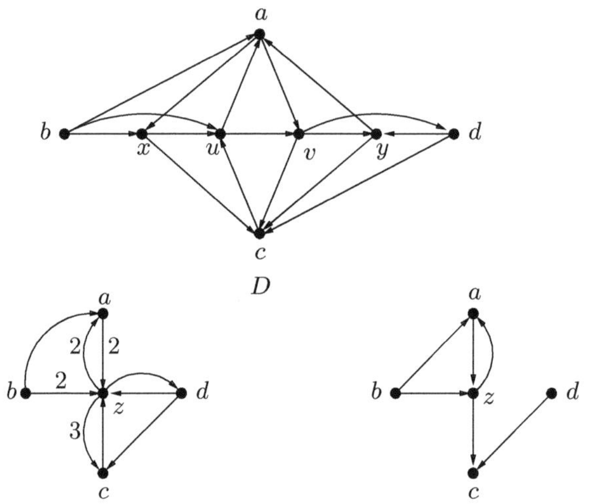


$D/\{x, u, v, y\}$ $D//P, P = xuvy$

**Figure 1.8** The two different kinds of contraction, set-contraction and path-contraction. The integers 2 and 3 indicate the number of corresponding parallel arcs.

As for set-contraction, for vertex-disjoint paths $P_1, P_2, \ldots, P_t$ in $D$, the path-contraction $D//\{P_1, \ldots, P_t\}$ is defined as the directed multigraph $(\ldots((D//P_1)//P_2)\ldots)//P_t$; clearly, the result does not depend on the order of $P_1, P_2, \ldots, P_t$.

To construct 'bigger' digraphs from 'smaller' ones, we will often use the following operation called **composition**. Let $D$ be a digraph with vertex set $\{v_i : i \in [n]\}$, and let $G_1, G_2, \ldots, G_n$ be digraphs which are pairwise vertex-disjoint. The composition $D[G_1, G_2, \ldots, G_n]$ is the digraph $L$ with vertex set $V(G_1) \cup V(G_2) \cup \ldots \cup V(G_n)$ and arc set $(\cup_{i=1}^{n} A(G_i)) \cup \{g_i g_j : g_i \in V(G_i), g_j \in V(G_j), v_i v_j \in A(D)\}$. Figure 1.9 shows the composition $T[G_x, G_l, G_v]$, where $G_x$ consists of a pair of vertices and an arc between them, $G_l$ has a single vertex, $G_v$ consists of a pair of vertices and the pair of mutually opposite arcs between them, and the digraph $T$ is from Figure 1.7.

If $D = H[S_1, \ldots, S_h]$ and none of the digraphs $S_1, \ldots S_h$ has an arc, then $D$ is an **extension** of $H$. This notion is also used for classes of digraphs. Hence an **extended tournament** is any digraph $D = T[S_1, \ldots, S_t]$ that can be obtained from a tournament $T$ by substituting each vertex $i$ of $T$ by an independent set $S_i$. Distinct vertices $x, y$ are **similar** if $x, y$ have the same in- and out-neighbours in $D$. For every $i \in [h]$, the vertices of $S_i$ are similar in $D$.

Chapter 10 is devoted to digraph products. Here we will consider just one such product. The **Cartesian product** of a family of digraphs $D_1, D_2, \ldots, D_n$,

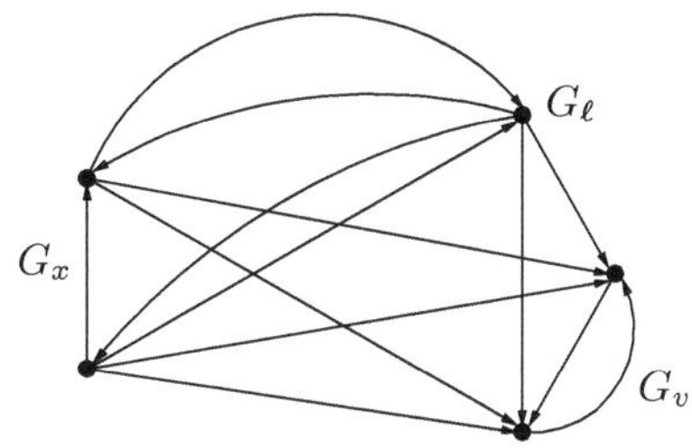


**Figure 1.9** $T[G_x, G_\ell, G_v]$.

denoted by $D_1 \square D_2 \square \ldots \square D_n$ or $\square_{i=1}^n D_i$, where $n \geq 2$, is the digraph $D$ having

$$\begin{aligned} V(D) &= V(D_1) \times V(D_2) \times \ldots \times V(D_n) \\ &= \{(w_1, w_2, \ldots, w_n) : \; w_i \in V(D_i), i \in [n]\} \end{aligned}$$

and a vertex $(u_1, u_2, \ldots, u_n)$ dominates a vertex $(v_1, v_2, \ldots, v_n)$ of $D$ if and only if there exists an $r \in [n]$ such that $u_r v_r \in A(D_r)$ and $u_i = v_i$ for all $i \in [n] \setminus \{r\}$. (See Figure 1.10.)

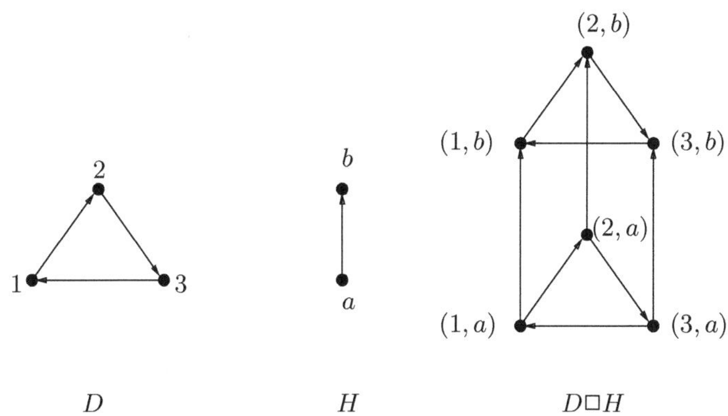


**Figure 1.10** The Cartesian product of two digraphs.

The operation of **splitting** a vertex $v$ of a directed multigraph $D$ consists of replacing $v$ by two new vertices $v', v''$, replacing all arcs of the form $xv$ by an arc $xv'$, replacing all arcs of the form $vy$ by an arc $v''y$ and finally adding the arc $v'v''$. The **subdivision** of an arc $uv$ of $D$ consists of replacing $uv$ by two arcs $uw, wv$, where $w$ is a new vertex. If $H$ can be obtained from $D$ by subdividing one or more arcs (here we allow subdividing arcs that are already subdivided), then $H$ is a **subdivision** of $D$. For a positive integer $p$ and a digraph $D$, the ***p*th power** $D^p$ of $D$ is defined as follows: $V(D^p) = V(D)$,

$x \rightarrow y$ in $D^p$ if $x \neq y$ and there are $k \leq p-1$ vertices $z_1, z_2, \ldots, z_k$ such that $x \rightarrow z_1 \rightarrow z_2 \rightarrow \ldots \rightarrow z_k \rightarrow y$ in $D$. According to this definition $D^1 = D$. For example, for the digraph $H_n = ([n], \{(i, i+1) : i \in [n-1]\})$, we have $H_n^2 = ([n], \{(i,j) : 1 \leq i < j \leq i+2 \leq n\} \cup \{(n-1, n)\})$. See Figure 1.11 for the second power of a digraph.

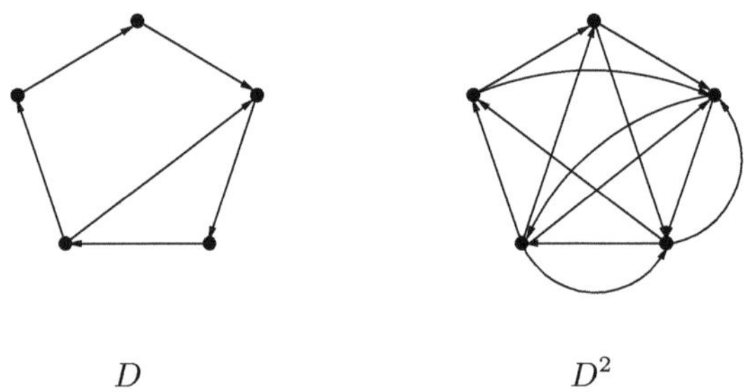


**Figure 1.11** A digraph $D$ and its second power $D^2$.

Let $H$ and $L$ be a pair of directed pseudographs. The **union** $H \cup L$ of $H$ and $L$ is the directed pseudograph $D$ such that $V(D) = V(H) \cup V(L)$ and $\mu_D(x,y) = \mu_H(x,y) + \mu_L(x,y)$ for every pair $x, y$ of vertices in $V(D)$. Here we assume that the function $\mu_H$ is naturally extended, i.e., $\mu_H(x,y) = 0$ if at least one of $x, y$ is not in $V(H)$ (and similarly for $\mu_L$). Figure 1.12 illustrates this definition.

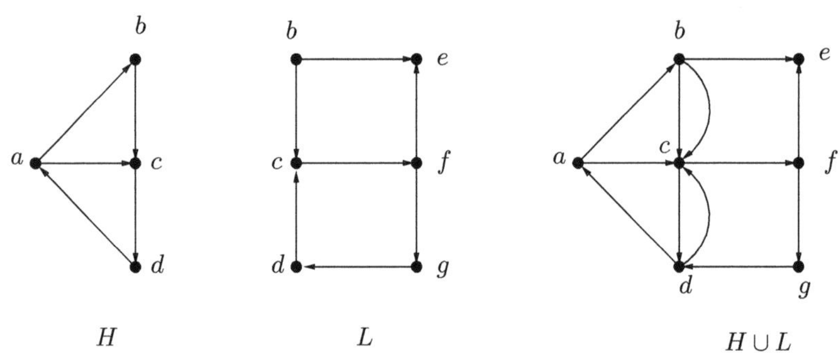


**Figure 1.12** The union $D = H \cup L$ of the directed pseudographs $H$ and $L$.

## 1.5 Strong Connectivity

In a digraph $D$ a vertex $y$ is **reachable** from a vertex $x$ if $D$ has an $(x, y)$-diwalk. In particular, a vertex is reachable from itself. By Proposition 1.3.2,

$y$ is reachable from $x$ if and only if $D$ contains an $(x, y)$-dipath. A digraph $D$ is **strongly connected** (or, just, **strong**) if, for every pair $x, y$ of distinct vertices in $D$, there exists an $(x, y)$-diwalk and a $(y, x)$-diwalk. In other words, $D$ is strong if every vertex of $D$ is reachable from every other vertex of $D$. We define a digraph with one vertex to be strongly connected. It is easy to see that $D$ is strong if and only if it has a closed Hamiltonian diwalk. As $\vec{C}_n$ is strong, every Hamiltonian digraph is strong.

Recall that a digraph $D$ is vertex-pancyclic if for every $x \in V(D)$ and every integer $k \in \{3, 4, \ldots, n\}$, there exists a $k$-cycle through $x$ in $D$. The following basic result on tournaments is due to Moon [30] and is proved in Chapter 2.

**Theorem 1.5.1** *Every strong tournament is vertex-pancyclic.*

A digraph $D$ is **semicomplete** if there is an arc between every pair of vertices in $D$. The class of semicomplete digraphs is a generalization of tournaments and many results for tournaments can be extended to semicomplete digraphs. In particular, it follows from Theorem 1.7.3 and Moon's theorem that every strong semicomplete digraph is vertex-pancyclic. A digraph $D$ is **complete** if, for every pair $x, y$ of distinct vertices of $D$, both $xy$ and $yx$ are in $D$. The complete digraph on $n$ vertices will be denoted by $\overleftrightarrow{K}_n$.

A digraph $D$ is **locally in-semicomplete** (**locally out-semicomplete**, respectively) if, for every vertex $x$ of $D$, all in-neighbours (out-neighbours, respectively) of $D$ induce a semicomplete digraph. It follows from Moon's theorem that every strong tournament is Hamiltonian. The following is an extension of this result by Bang-Jensen, Huang and Prisner [6].

**Theorem 1.5.2** *Every strong locally in-semicomplete digraph is Hamiltonian.*

As the converse of every locally out-semicomplete digraph is locally in-semicomplete and the converse of a Hamiltonian dicycle is a Hamiltonian dicycle, Theorem 1.5.2 holds for locally out-semicomplete digraphs as well. Chapter 6 is devoted to results on locally in- and out-semicomplete digraphs.

For a strong digraph $D = (V, A)$, a set $S \subset V$ is a **separator** (or a **separating set**) if $D - S$ is not strong. A digraph $D$ is **$k$-strongly connected** (or **$k$-strong**) if $|V| \geq k+1$ and $D$ has no separator with less than $k$ vertices. It follows from the definition of strong connectivity that a complete digraph with $n$ vertices is $(n-1)$-strong, but is not $n$-strong. The largest integer $k$ such that $D$ is $k$-strongly connected is the **vertex-strong connectivity** of $D$ (denoted by $\kappa(D)$). If a digraph $D$ is not strong, we set $\kappa(D) = 0$. For a pair $s, t$ of distinct vertices of a digraph $D$, a set $S \subseteq V(D) - \{s, t\}$ is an **$(s, t)$-separator** if $D - S$ has no $(s, t)$-dipaths.

For a strong digraph $D = (V, A)$, a set of arcs $W \subseteq A$ is a **cut** (or a **cutset**) if $D - W$ is not strong. Clearly, every minimal cut is of the form $(X, \bar{X})$, where $X \subset V$ and $\bar{X} = V - X$. A cut $(X, \bar{X})$ is called a **$(u, v)$-**

**cut** if $u \in X$ and $v \in \bar{X}$. A digraph $D$ is ***k*-arc-strong** (or ***k*-arc-strongly connected**) if $D$ has no cut with less than $k$ arcs. The largest integer $k$ such that $D$ is $k$-arc-strongly connected is the **arc-strong connectivity** of $D$ (denoted by $\lambda(D)$). If $D$ is not strong, we set $\lambda(D) = 0$. Note that $\lambda(D) \geq k$ if and only if $d^+(X), d^-(X) \geq k$ for all proper subsets $X$ of $V$. A collection $\mathcal{P}$ of paths is called **arc-disjoint** if no pair of paths in $\mathcal{P}$ has common arcs.

The following theorem is one of the most fundamental results in graph theory.

**Theorem 1.5.3 (Menger's theorem)***[29] Let $D$ be a directed multigraph and let $u, v \in V(D)$ be a pair of distinct vertices. Then the following holds:*

*(a) The maximum number of arc-disjoint $(u, v)$-dipaths equals the minimum number of arcs covering all $(u, v)$-dipaths and this minimum is attained for some $(u, v)$-cut $(X, \bar{X})$.*
*(b) If the arc $uv$ is not in $A(D)$, then the maximum number of internally disjoint $(u, v)$-dipaths equals the minimum number of vertices in a $(u, v)$-separator.*

A **strong component** of a digraph $D$ is a maximal induced subgraph of $D$ which is strong. If $D_1,\dots,D_t$ are the strong components of $D$, then clearly $V(D_1) \cup \dots \cup V(D_t) = V(D)$ (recall that a digraph with only one vertex is strong). Moreover, we must have $V(D_i) \cap V(D_j) = \emptyset$ for every $i \neq j$ as otherwise all the vertices $V(D_i) \cup V(D_j)$ are reachable from each other, implying that the vertices of $V(D_i) \cup V(D_j)$ belong to the same strong component of $D$. We call $V(D_1) \cup \dots \cup V(D_t)$ the **strong decomposition** of $D$. The **strong component digraph** $SC(D)$ of $D$ is obtained by contracting the strong components of $D$ and deleting any parallel arcs obtained in this process. In other words, if $D_1,\dots,D_t$ are the strong components of $D$, then $V(SC(D)) = \{v_i : i \in [t]\}$ and $A(SC(D)) = \{v_i v_j : (V(D_i), V(D_j))_D \neq \emptyset\}$. The subgraph of $D$ induced by the vertices of a dicycle in $D$ is strong, and hence is contained in a strong component of $D$. Thus, $SC(D)$ is acyclic. By Proposition 3.1.2 in Chapter 3, the vertices of $SC(D)$ have an acyclic ordering. This implies that the strong components of $D$ can be labelled $D_1,\dots,D_t$ such that there is no arc from $D_j$ to $D_i$ unless $j < i$. We call such an ordering an **acyclic ordering** of the strong components of $D$. The strong components of $D$ corresponding to the vertices of $SC(D)$ of in-degree (out-degree) zero are the **initial (terminal) strong components** of $D$. The remaining strong components of $D$ are called the **intermediate strong components** of $D$. Figure 1.13 shows a digraph $D$ and its strong component digraph $SC(D)$.

It is easy to see that the strong component digraph of a tournament $T$ is an acyclic tournament. Thus, there is a unique acyclic ordering of the strong components of $T$, namely, $T_1,\dots,T_t$ such that $T_i \to T_j$ for every $i < j$. Clearly, every tournament has only one initial (terminal) strong component.

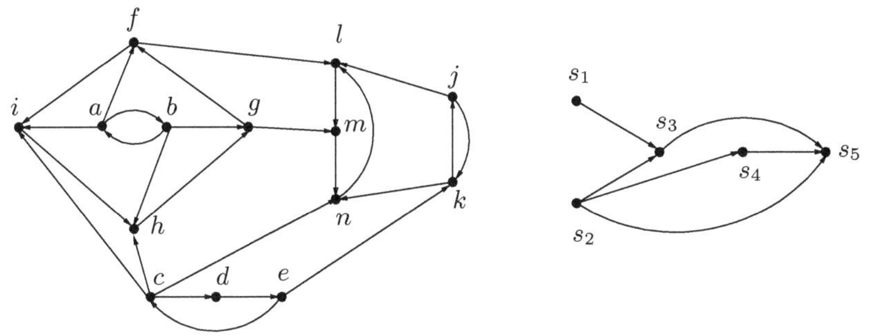


**Figure 1.13** A digraph $D$ and its strong component digraph $SC(D)$. The vertices $s_1, s_2, s_3, s_4, s_5$ are obtained by contracting the sets $\{a,b\}, \{c,d,e\}, \{f,g,h,i\}, \{j,k\}$ and $\{l,m,n\}$ which correspond to the strong components of $D$. The digraph $D$ has two initial components, $D_1, D_2$ with $V(D_1) = \{a,b\}$ and $V(D_2) = \{c,d,e\}$. It has one terminal component $D_5$ with vertices $V(D_5) = \{l,m,n\}$ and two intermediate components $D_3, D_4$ with vertices $V(D_3) = \{f,g,h,i\}$ and $V(D_4) = \{j,k\}$.

## 1.6 Linkages

Let $D = (V,A)$ be a digraph and let $s_1, \dots, s_k, t_1, \dots, t_k$ be a collection of (not necessarily distinct) vertices of $D$. A **k-linkage** from $(s_1, \dots, s_k)$ to $(t_1, \dots, t_k)$ is a collection of $k$ internally disjoint dipaths $P_1, \dots, P_k$ such that, for each $i \in [k]$, $P_i$ is an $(s_i, t_i)$-dipath if $s_i \neq t_i$ and a dicycle containing $s_i$ if $s_i = t_i$ and $s_i, t_i$ are not internal vertices of $P_j$ for any $j \neq i$. In the case of a cycle $C$ containing $s_i$, the term internally disjoint means the same as for paths, i.e., no other path or cycle contains vertices $V(C) - \{s_i, t_i\}$. Note that a dicycle with just one vertex must be a loop, not just a vertex itself. A **weak k-linkage** from $(s_1, \dots, s_k)$ to $(t_1, \dots, t_k)$ is a collection of $k$ arc-disjoint dipaths $P_1, \dots, P_k$ such that, for each $i \in [k]$, $P_i$ is an $(s_i, t_i)$-dipath if $s_i \neq t_i$ and a dicycle containing $s_i$ if $s_i = t_i$. The next two problems on linkages are fundamental and of central importance in digraph theory.

$k$-LINKAGE
**Input:** A digraph $D = (V,A)$ and not necessarily distinct vertices $s_1, \dots, s_k, t_1, \dots, t_k$.
**Question:** Does $D$ contain a $k$-linkage from $(s_1, \dots, s_k)$ to $(t_1, \dots, t_k)$?

WEAK $k$-LINKAGE
**Input:** A digraph $D = (V, A)$ and not necessarily distinct vertices $s_1, \ldots, s_k, t_1, \ldots, t_k$.
**Question:** Does $D$ contain a weak $k$-linkage from $(s_1, \ldots, s_k)$ to $(t_1, \ldots, t_k)$?

A digraph is **k-linked** (**weakly k-linked**, respectively) if it has a $k$-linkage (a weak $k$-linkage, respectively) for every choice of vertices as above.

Kühn and Osthus [27] proved the following:

**Theorem 1.6.1** *Let $k \geq 2$ be an integer. Every digraph $D$ of order $n \geq 400k^3$ which satisfies $\delta^0(D) \geq n/2 + k - 1$ is $k$-linked.*

The $k$-LINKAGE and the WEAK $k$-LINKAGE problems are both $\mathcal{NP}$-hard even for $k = 2$ [20], Still, somewhat surprisingly, weakly $k$-linked digraphs are easy to classify due to the following result of Shiloach [35]. Its proof, due to Shiloach, is a beautiful application of Edmonds' branching theorem (Theorem 1.8.2), see [35].

**Theorem 1.6.2** *A digraph is weakly $k$-linked if and only if it is $k$-arc-strong. Furthermore, there is a polynomial algorithm for finding a weak $k$-linkage from $(s_1, \ldots, s_k)$ to $(t_1, \ldots, t_k)$, for any choice of these vertices, in a $k$-arc-strong digraph.*

So for weak $k$-linkages, the interesting case is when the arc-strong connectivity is less than $k$.

## 1.7 Undirected Graphs and Orientations of Undirected and Directed Graphs

An **undirected graph** $G = (V, E)$ consists of a non-empty finite set $V = V(G)$ of elements called **vertices** and a finite set $E = E(G)$ of unordered pairs of distinct vertices called **edges**. We call $V(G)$ the **vertex set** and $E(G)$ the **edge set** of $G$. In other words, an edge $\{x, y\}$ is a 2-element subset of $V(G)$. We will often denote $\{x, y\}$ just by $xy$. If $xy \in E(G)$, we say that the vertices $x$ and $y$ are **adjacent**. Notice that, in the above definition of an undirected graph, we do not allow loops (i.e., pairs consisting of the same vertex) or parallel edges (i.e., multiple pairs with the same end-vertices). The **complement** $\overline{G}$ of an undirected graph $G$ is the undirected graph with vertex set $V(G)$ in which two vertices are adjacent if and only if they are not adjacent in $G$.

When parallel edges and loops are admissible we speak of **undirected pseudographs**; pseudographs with no loops are **multigraphs**. For a pair $u, v$ of vertices in a pseudograph $G$, $\mu_G(u, v)$ denotes the number of edges between $u$ and $v$. In particular, $\mu_G(u, u)$ is the number of loops at $u$.

A multigraph $G$ is **complete** if every pair of distinct vertices in $G$ are adjacent (that is, $\mu_G(u,v) > 0$ for all $u, v \in V$, $u \neq v$). We will denote the complete undirected graph on $n$ vertices (which is unique up to isomorphism) by $K_n$. Its complement $\overline{K}_n$ has no edge.

A multigraph $H$ is ***p*-partite** if there exists a partition $V_1, V_2, \ldots, V_p$ of $V(H)$ into $p$ **partite sets** (i.e., $V(H) = V_1 \cup \ldots \cup V_p$, $V_i \cap V_j = \emptyset$ for every $i \neq j$) such that every edge of $H$ has its end-vertices in different partite sets. The special case of a $p$-partite graph when $p = 2$ is called a **bipartite graph.** We often denote a bipartite graph $B$ by $B = (V_1, V_2; E)$. A $p$-partite multigraph $H$ is **complete *p*-partite** if, for every pair $x \in V_i$, $y \in V_j$ $(i \neq j)$, an edge $xy$ is in $H$. A complete graph on $n$ vertices is clearly a complete $n$-partite graph for which every partite set is a singleton. We denote the complete $p$-partite *graph* with partite sets of cardinalities $n_1, n_2, \ldots, n_p$ by $K_{n_1,n_2,\ldots,n_p}$. Complete $p$-partite graphs for $p \geq 2$ are also called **complete multipartite graphs.**

To obtain short proofs of various results on subgraphs of a directed multigraph $D = (V, A)$ the following transformation to the class of bipartite (undirected) multigraphs is extremely useful. Let $BG(D) = (V', V''; E)$ denote the bipartite multigraph with partite sets $V' = \{v' : v \in V\}$, $V'' = \{v'' : v \in V\}$ such that $\mu_{BG(D)}(u', w'') = \mu_D(u, w)$ for every pair $u, w$ of vertices in $D$. We call $BG(D)$ the **bipartite representation** of $D$; see Figure 1.14.

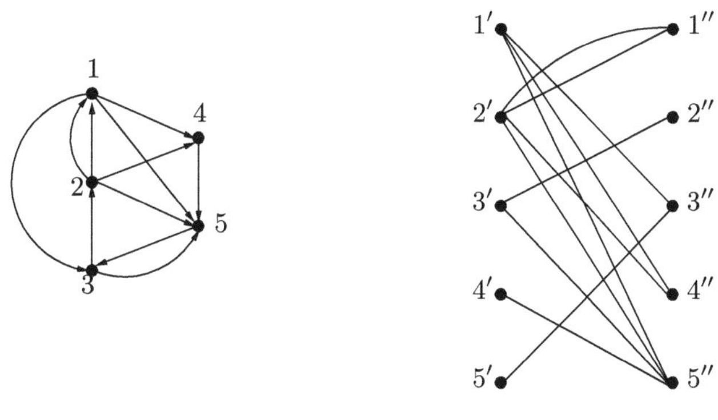


$D$ $BG(D)$

**Figure 1.14** A directed multigraph and its bipartite representation.

An **orientation** of an undirected graph $G$ is an oriented graph $H$ obtained from $G$ by replacing every edge $xy$ by either arc $(x, y)$ or arc $(y, x)$. Let $D$ be a directed multigraph. The **underlying multigraph** $UMG(D)$ of $D$ is an undirected multigraph obtained from $D$ by replacing every arc $(x, y)$ with the edge $xy$. The **underlying graph** $UG(D)$ of $D$ is obtained from $UMG(D)$ by deleting all multiple edges between every pair of vertices apart from one.

For example, for a digraph $H$ with vertices $u, v$ and arcs $uv, vu$, $UG(H)$ has one edge and $UMG(H)$ has two parallel edges. Chapter 12 is devoted to underlying graphs of digraphs.

A digraph $D = (V, A)$ is **symmetric** if $xy \in A$ implies $yx \in A$. For an undirected graph $G$, the **complete biorientation** of $G$ is a symmetric digraph $\overleftrightarrow{G}$ obtained from $G$ by replacing each edge $\{x, y\}$ with the pair $xy, yx$ of arcs. Clearly, $D$ is symmetric if and only if $D$ is the complete biorientation of some graph.

An undirected pseudograph $G$ is **connected** if its complete biorientation $\overleftrightarrow{G}$ is strongly connected. Similarly, $G$ is ***k*-connected** if $\overleftrightarrow{G}$ is $k$-strong. Strong components in $\overleftrightarrow{G}$ are **connected components**, or just **components** in $G$. A **bridge** in an undirected pseudograph $G$ is an edge whose deletion from $G$ increases the number of connected components. An undirected pseudograph $G$ is ***k*-edge-connected** if the graph obtained from $G$ after deletion of at most $k-1$ edges is connected. Clearly, a connected undirected pseudograph is bridgeless if and only if it is 2-edge-connected. The **neighbourhood** $N_G(x)$ of a vertex $x$ in $G$ is the set of vertices adjacent to $x$. The **degree** $d(x)$ of a vertex $x$ is the number of edges except loops having $x$ as an end-vertex The **minimum (maximum) degree** of $G$ is

$$\delta(G) = \min\{d(x) : x \in V(G)\} \ (\Delta(G) = \max\{d(x) : x \in V(G)\}).$$

We say that $G$ is **regular** (or **$\delta(G)$-regular**) if $\delta(G) = \Delta(G)$. A pair of undirected graphs $G$ and $H$ is **isomorphic** if $\overleftrightarrow{G}$ and $\overleftrightarrow{H}$ are isomorphic.

A digraph is **connected** if its underlying graph is connected. The following well-known theorem is due to Robbins [33]. This theorem is a special case of Theorem 1.7.3.

**Theorem 1.7.1** *A connected graph $G$ has a strongly connected orientation if and only if $G$ has no bridge.*

Here is a well-known characterization of Eulerian directed multigraphs (clearly, the deletion of loops in a directed pseudograph $D$ does not change the property of $D$ of being Eulerian or otherwise): A directed multigraph $D$ is Eulerian if and only if $D$ is connected and $d^+(x) = d^-(x)$ for every vertex $x$ in $D$ [4]. Eulerian directed multigraphs are considered in Chapter 4.

The notions of walks, trails, paths and cycles in undirected pseudographs are analogous to those for directed pseudographs (we merely disregard orientations). An ***xy*-path** in an undirected pseudograph is a path whose end-vertices are $x$ and $y$. An undirected graph is a **forest** if it has no cycle. A connected forest is a **tree**. It is easy to see that every connected undirected graph has a **spanning tree**, i.e., a spanning subgraph, which is a tree.

A **matching** $M$ in a directed (an undirected) pseudograph $G$ is a set of arcs (edges) with no common end-vertices. We also require that no element of $M$ is a loop. If $M$ is a matching, then we say that the edges (arcs) of

$M$ are **independent**. A matching $M$ in $G$ is **maximum** if $M$ contains the maximum possible number of edges. A maximum matching is **perfect** if it has $n/2$ edges, where $n$ is the order of $G$. A set $Q$ of vertices in a directed or undirected pseudograph $H$ is **independent** if the graph $H\langle Q\rangle$ has no edges (arcs). The **independence number**, $\alpha(H)$, of $H$ is the maximum integer $k$ such that $H$ has an independent set of cardinality $k$. A **(proper) colouring** of a directed or undirected graph $H$ is a partition of $V(H)$ into (disjoint) independent sets. The minimum number, $\chi(H)$, of independent sets in a proper colouring of $H$ is the **chromatic number** of $H$.

In Section 1.4, the operation of composition of digraphs was introduced. Similarly, we can define the operation of **composition** of undirected graphs. Let $H$ be a graph with vertex set $\{v_i : i \in [n]\}$, and let $G_1, G_2, \ldots, G_n$ be graphs which are pairwise vertex-disjoint. The composition $H[G_1, G_2, \ldots, G_n]$ is the graph $L$ with vertex set $V(G_1) \cup V(G_2) \cup \ldots \cup V(G_n)$ and edge set

$$\cup_{i=1}^{n} E(G_i) \cup \{g_i g_j : \; g_i \in V(G_i), g_j \in V(G_j), v_i v_j \in E(H)\}.$$

If none of the graphs $G_1, \ldots, G_n$ in this definition of $H[G_1, \ldots, G_n]$ have edges, then $H[G_1, \ldots, G_n]$ is an **extension** of $H$.

We conclude this section with the notion of an orientation of a digraph, which extends the notion of an orientation of an undirected graph. An **orientation** of a digraph $D$ is a subgraph of $D$ obtained from $D$ by deleting exactly one arc between $x$ and $y$ for every pair $x \neq y$ of vertices such that both $xy$ and $yx$ are in $D$. See Figure 1.15 for an illustration of this definition.

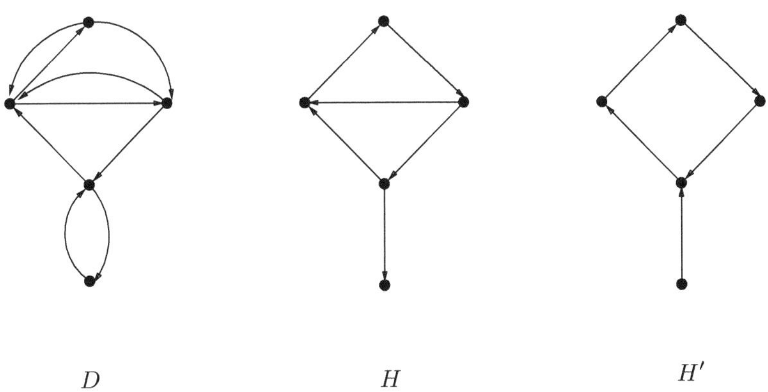


**Figure 1.15** A digraph $D$ and subgraphs $H$ and $H'$ of $D$. The digraph $H$ is an orientation of $D$ but $H'$ is not.

**Lemma 1.7.2** *Let $D$ be a strong digraph and $x, y$ vertices of $D$ such that both $xy$ and $yx$ are arcs. Then either $D - xy$ or $D - yx$ is strong if and only if $e$ is not a bridge in $UG(D)$.*

**Proof:** If $e$ is a bridge in $UG(D)$, then clearly neither $D - xy$ nor $D - yx$ is strong. Assume that $e$ is not a bridge in $UG(D)$ and consider $D' = D - \{xy, yx\}$. If $D'$ is strong, then clearly both $D - xy$ and $D - yx$ are strong. Thus, assume that $D'$ is not strong. Since $e$ is not a bridge, $D'$ is connected. Let $L_1, L_2, \ldots, L_k$ be strong components of $D'$. Since $D$ is strong, there is only one initial strong component, say $L_1$, and only one terminal strong component, say $L_k$. Since $D$ is strong, one of the vertices $x$ and $y$ is in $L_1$ and the other in $L_k$. Without loss of generality, $x$ is in $L_1$ and $y$ is in $L_k$. Then $D - xy$ is strong. □

This lemma immediately implies the following theorem of Boesch and Tindell [12], which generalizes Theorem 1.7.1.

**Theorem 1.7.3** *A strong digraph $D$ has a strong orientation if and only if $UG(D)$ has no bridge.*

## 1.8 Trees in Digraphs

A digraph $D$ is an **oriented forest** (**tree**) if $D$ is an orientation of a forest (tree). A digraph $T$ is an **out-tree** (an **in-tree**) if $T$ is an oriented tree with just one vertex $s$ of in-degree zero (out-degree zero). The vertex $s$ is the **root** of $T$. A digraph $F$ is an **out-forest** (an **in-forest**) if $F$ is the vertex disjoint union of out-trees (in-trees).

If an out-tree (in-tree) $T$ is a spanning subgraph of $D$, $T$ is called an **out-branching** (an **in-branching**). (See Figure 1.16.)

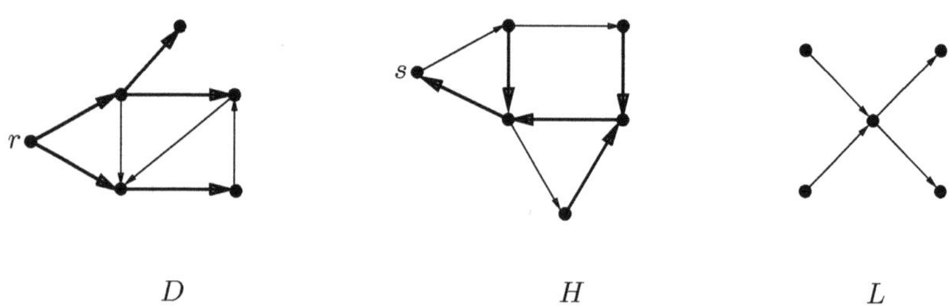


**Figure 1.16** The digraph $D$ has an out-branching with root $r$ (shown in bold); $H$ contains an in-branching with root $s$ (shown in bold); $L$ possesses neither an out-branching nor an in-branching.

Since each spanning oriented tree $R$ of a connected digraph is acyclic, $R$ has at least one vertex of out-degree zero and at least one vertex of in-degree zero (see Proposition 3.1.1 of Chapter 3). Hence, the out-branchings and in-branchings capture the important cases of uniqueness of the corresponding vertices. The following is a characterization of digraphs with in-branchings (out-branchings).

**Proposition 1.8.1** *A connected digraph $D$ contains an out-branching (in-branching) if and only if $D$ has only one initial (terminal) strong component.*

**Proof:** We prove this characterization only for out-branchings since the second claim follows from the first one by considering the converse of $D$.

Assume that $D$ contains at least two initial strong components and suppose that $D$ has an out-branching $T$. Observe that the root $r$ of $T$ is an initial strong component of $D$. Let $x$ be a vertex in another initial strong component of $D$. Since $r$ is the root of $T$, there is a path from $r$ to $x$ in $T$ and, thus, in $D$, which is a contradiction to the assumption that $r$ and $x$ are in different initial strong components of $D$.

Now we assume that $D$ contains only one initial strong component $D_1$, and $r$ is an arbitrary vertex of $D_1$. We prove that $D$ has an out-branching rooted at $r$. In $SC(D)$, the vertex $x$ corresponding to $D_1$ is the only vertex of in-degree zero and, hence every vertex $v$ of $SC(D)$ is reachable from $x$ (the longest path to $v$ must start at $x$). Thus, every vertex of $D$ is reachable from $r$. We construct an oriented tree $T$ as follows. In the first step $T$ consists of $r$. In Step $i \geq 2$, for every vertex $y$ appended to $T$ in the previous step, we add to $T$ a vertex $z$, such that $y \to z$ and $z \notin V(T)$, together with the arc $yz$. We stop when no vertex can be included in $T$. Since every vertex of $D$ is reachable from $r$, $T$ is spanning. Clearly, $r$ is the only vertex of in-degree zero in $T$. Hence, $T$ is an out-branching. □

The following theorem is a very important result, which can be viewed as just a fairly simple generalization of Menger's theorem. However, it has many important consequences, see the book [4] by Bang-Jensen and Gutin for many such applications of the theorem.

**Theorem 1.8.2 (Edmonds' branching theorem)** *[citeedmonds1973] A directed multigraph $D = (V, A)$ with a special vertex $z$ has $k$ arc-disjoint out-branchings rooted at $z$ if and only[2] if*

$$d^-(X) \geq k \ \ \forall \ \emptyset \neq X \subseteq V - z. \tag{1.1}$$

*There exists a polynomial algorithm for finding $k$ arc-disjoint out-branchings from a given root $s$ in a directed multigraph which satisfies (1.1).*

A **leaf** in an out-tree (in-tree) is a vertex of out-degree (in-degree) zero. The minimum (maximum, respectively) number of leaves in an out-branching

[2] By Menger's theorem (Theorem 1.5.3), (1.1) is equivalent to the existence of $k$ arc-disjoint dipaths from $z$ to every other vertex of $D$.

of a digraph $D$ will be denoted by $\ell_{\min}(D)$ ($\ell_{\max}(D)$, respectively). Clearly, the problem of finding $\ell_{\min}(D)$ is $\mathcal{NP}$-hard as even the problem of deciding whether $\ell_{\min}(D) = 1$ is $\mathcal{NP}$-complete as it is equivalent to the Hamilton dipath problem. The following theorem of Las Vergnas gives a bound to the minimum number of leaves in an out-branching. Recall that for a digraph $D$, $\alpha(D)$ denotes the maximum number of vertices without an arc between them.

**Theorem 1.8.3** ([28]) *Let $D$ be a digraph and let $\ell_{\min}(D)$ be the minimum number of leaves in an out-branching of $D$. Then $\ell_{\min}(D) \leq \alpha(D)$.*

This theorem implies the Gallai–Milgram theorem (Theorem 1.8.4), for a proof of this fact see the paper [5] by Bang-Jensen and Gutin.

The problem of finding $\ell_{\max}(D)$ is $\mathcal{NP}$-hard; Alon, Fomin, Gutin, Krivelevich and Saurabh showed that it in fact remains $\mathcal{NP}$-hard when restricted to acyclic digraphs [1]. Daligault and Thomassé [17] designed a 92-approximation algorithm for the (general) problem and Daligault, Gutin, Kim and Yeo [16] obtained an $O^*(3.72^k)$-time algorithm for deciding whether a digraph $D$ contains an out-branching with at least $k$ leaves.

Rédei's theorem (Theorem 2.2.4) can be rephrased as saying that every digraph with independence number one has a Hamiltonian dipath and hence has path covering number one. Gallai and Milgram generalized this as follows.

**Theorem 1.8.4 (Gallai–Milgram theorem)** *[21] For every digraph $D$ the path covering number is at most its independence number, that is $pc(D) \leq \alpha(D)$.*

In fact, the following stronger result holds. It can be useful in certain applications, see, e.g., Section 3.10.3.

**Theorem 1.8.5 (Gallai–Milgram theorem)** *[21] Let $D$ be a digraph, let $P = P_1 \cup \ldots P_\ell$ be a dipath factor of $D$, and let $I(P)$ and $T(P)$ denote the sets of initial and terminal vertices, respectively, of dipaths of $P$. If $\ell > \alpha(D)$, then $D$ contains a dipath factor $P'$ with $\ell - 1$ paths and such that $I(P') \subset I(P)$ and $T(P') \subset T(P)$.*

## 1.9 Flows in Networks

A **network** $\mathcal{N}$ is a digraph $D = (V, A)$ in which each arc $a$ is associated with a **capacity** $u(a)$. A **flow** in a network $\mathcal{N}$ associates each arc $a$ of $\mathcal{N}$ with a non-negative number which must not exceed the capacity $u(a)$ of the arc. Flows in networks are widely used to model systems in which some quantity passes through channels (arcs in the network) that meet at junctions (vertices); examples include traffic in a road system, fluids in pipes, or electrical current in circuits. Here is a formal definition of networks and flows in these.

A **network** is a tuple $\mathcal{N} = (V, A, l, u, c)$, where $D = (V, A)$ is a digraph with vertex set $V$ and arc set $A$, and $l : A \to \mathbb{Z}_0$, $u : A \to \mathbb{Z}_0$ and $c : A \to \mathbb{R}$ are functions. Intuitively, $l$ and $u$ represent **lower bounds** and **capacities** (also called **upper bounds**), respectively, on how much flow can pass through each arc, and $c$ represents the **cost** associated with each unit of flow in each arc. If there are no costs specified and $l(a) = 0$ for each $a \in A$, then we omit the relevant letters from the notation. For example, if $\mathcal{N} = (V, A, u, c)$, then $l(a) = 0$ for each $a \in A$. Sometimes we also specify a function $b : V \to \mathbb{Z}$ such that $\sum_{v \in V} b(v) = 0$. This is called a **balance vector** and if this is also specified, we denote the network by $\mathcal{N} = (V, A, l, u, c, b)$.

Given a network $\mathcal{N} = (V, A, l, u, c)$ (or $\mathcal{N} = (V, A, l, u, c, b)$), a function $x : A \to \mathbb{R}_0$ is called a **flow** in $\mathcal{N}$; it is an **integer flow** if $x(a) \in \mathbb{Z}_0$ for each $a \in A$. For a flow $x$, define the **balance vector** $b_x$ as follows: $b_x(v) = \sum_{v' \in N^+(v)} x(vv') - \sum_{v' \in N^-(v)} x(v'v)$ for every $v \in V$. For two distinct vertices $s, t \in V$, a flow $x$ is an **$(s, t)$-flow** if $b_x(s) = -b_x(t) \geq 0$ and $b_x(v) = 0$ for each $v \in V \setminus \{s, t\}$. The **value** of an $(s, t)$-flow $x$ is $|x| = b_x(s)$. A flow $x$ is a **circulation** if $b_x(v) = 0$ for every $v \in V$. The **cost** of a flow $x$ is given by $c(x) = \sum_{vv' \in A} c(vv')x(vv')$. A flow $x$ is **feasible** in $\mathcal{N} = (V, A, l, u, c, b)$ if the following conditions are satisfied:

(a) $l(a) \leq x(a) \leq u(a)$ for every $vv' \in A$;
(b) $b_x(v) = b(v)$ for every $v \in V$.

If no balance constraint is specified, that is, $\mathcal{N} = (V, A, l, u, c)$, then a feasible flow in $\mathcal{N}$ just has to satisfy (a) above.

See Figure 1.17 for an example of a feasible flow.

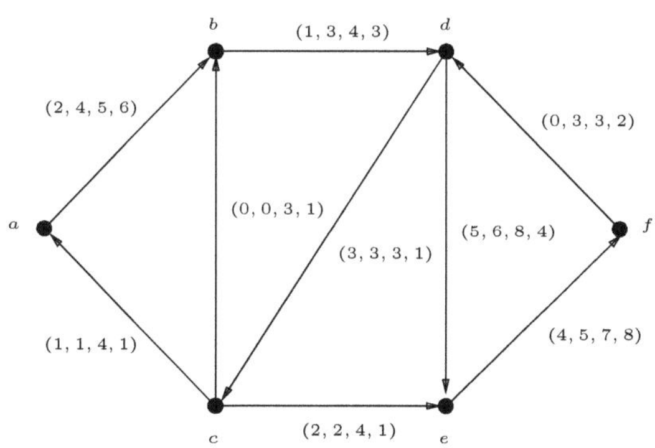


**Figure 1.17** A network $\mathcal{N} = (V, A, l, u, c)$ with a feasible flow $x$ specified. The specification on each arc $uv$ is $(l(vw), x(vw), u(vw), c(vw))$. The cost of the flow is 109.

The following two simple propositions allow us to reduce problems about general feasible flows to problems about feasible $(s,t)$-flows. See [4, Section 4.2].

**Proposition 1.9.1** *Let $\mathcal{N} = (V, A, l, u, b, c)$ be a network.*

*(a) Suppose that the arc $ij \in A$ has $l(ij) > 0$. Let $\mathcal{N}'$ be obtained from $\mathcal{N}$ by making the following changes: $b(j) := b(j) + l(ij)$, $b(i) := b(i) - l(ij)$, $u(ij) := u(ij) - l(ij)$, $l(ij) := 0$. Then every feasible flow $x$ in $\mathcal{N}$ corresponds to a feasible flow $x'$ in $\mathcal{N}'$ and vice versa. Furthermore, the costs of these two flows are related by $c(x) = c(x') + l(ij)c(ij)$.*
*(b) There exists a network $\mathcal{N}_{l\equiv 0}$ in which all lower bounds are zero such that every feasible flow $x$ in $\mathcal{N}$ corresponds to a feasible flow $x'$ in $\mathcal{N}_{l\equiv 0}$ and vice versa. Furthermore, the costs of these two flows are related by $c(x) = c(x') + \sum_{ij\in A} l(ij)c(ij)$.*

**Proposition 1.9.2** *Let $\mathcal{N} = (V, A, l \equiv 0, u, b, c)$ be a network. Let $M = \sum_{\{v:b(v)>0\}} b(v)$ and let $\mathcal{N}_{st}$ be the network defined as follows: $\mathcal{N}_{st} = (V \cup \{s,t\}, A', l' \equiv 0, u', b', c')$, where*

*(a) $A' = A \cup \{sr : b(r) > 0\} \cup \{rt : b(r) < 0\}$,*
*(b) $u'(ij) = u(ij)$ for all $ij \in A$, $u_{sr} = b(r)$ for all $r$ such that $b(r) > 0$ and $u(qt) = -b(q)$ for all $q$ such that $b(q) < 0$,*
*(c) $c'(ij) = c(ij)$ for all $ij \in A$ and $c' = 0$ for all arcs leaving $s$ or entering $t$,*
*(d) $b'(v) = 0$ for all $v \in V$, $b'(s) = M$, $b'(t) = -M$.*

*Then every feasible flow $x$ in $\mathcal{N}$ corresponds to a feasible flow $x'$ in $\mathcal{N}_{st}$ and vice versa. Furthermore, the costs of $x$ and $x'$ are the same.*

For a function $f : A \to \mathbb{Z}$ and a proper subset $X$ of $V$, let $\overline{X} = V \setminus X$ and $f(X, \overline{X}) = \sum_{yz\in(X,\overline{X})} f(yz)$. It is not hard to see that given a network $\mathcal{N} = (V, A, l, u)$ if $l(\overline{S}, S) > u(S, \overline{S})$ then $\mathcal{N}$ has no feasible circulation. Hoffman [26] proved that the converse holds as well.

**Theorem 1.9.3 (Hoffman's circulation theorem)** *Let $\mathcal{N} = (V, A, l, u)$ be a network with lower bounds on the arcs, then $\mathcal{N}$ has a feasible circulation if and only if the following holds for every proper subset $S$ of $V$:*

$$l(\overline{S}, S) \leq u(S, \overline{S}). \tag{1.2}$$

## 1.10 Polynomial and Exponential Time Algorithms, SAT and ETH

Unless explicitly stated otherwise, when we say that an algorithm is polynomial, respectively that a problem is polynomial, we mean that the running time of the algorithm is polynomial in the size of the input, respectively that there exists a polynomial algorithm for solving the problem.

Recall that a **CNF formula** is a conjunction of clauses. Each **clause** is a disjunction of literals, each of which is either a variable or its negation. A CNF formula $F$ is **satisfiable** if there is a truth assignment to the variables of $F$ such that every clause contains at least one literal equal true. In **k-CNF** formula every clause has exactly $k$ literals. For $k \geq 2$, the problem **k-SAT** is stated as follows: Given a $k$-CNF formula $F$, decide whether $F$ is satisfiable. It is well-known that while 2-SAT is polynomial-time solvable (see e.g. Section 17.5 in [4]), $k$-SAT is $\mathcal{NP}$-complete for every $k \geq 3$. The following variations of 3-SAT are also $\mathcal{NP}$-hard. In **NAE-3-SAT**, we are to decide whether there is a truth assignment for which each clause of a 3-CNF formula $F$ has a literal equal true and a literal equal false. The problem **monotone-NAE-3-SAT** is a special case of NAE-3-SAT in which a 3-CNF formula contains no negations of variables. Finally, in **1-in-3-SAT**, given a 3-CNF formula $F$, decide whether there is a truth assignment making exactly one literal true in each clause of $F$.

It is widely believed that $\mathcal{P} \neq \mathcal{NP}$ and thus there are no polynomial time algorithms for $\mathcal{NP}$-complete problems. Unfortunately, many problems in graph theory are $\mathcal{NP}$-complete and just declaring them intractable seems too simplistic. In this and the next two sections we will briefly consider modern approaches for dealing with $\mathcal{NP}$-hard problems. We will consider only theory-based methods largely ignoring many heuristic approaches, which are of great interest in graph theory applications, but unfortunately are outside the scope of this book.

It seems that the oldest practical way to deal with $\mathcal{NP}$-hard problems is to use exponential time algorithms such as branch-and-bound. The theoretical foundations of such algorithms have been largely ignored for a while, but in the last two decades the situation has changed and many approaches and results on exponential-time algorithms have been obtained, see, e.g., [19] which is the only monograph on the topic. One such example is Schöning's randomized $k$-SAT algorithm [34] and its derandomization by Moser and Scheder [31]. The runtimes of Schöning's algorithm and of its derandomization are $O^*((\frac{2(k-1)}{k})^n)$ and $O^*((\frac{2(k-1)}{k} + \varepsilon)^n)$, where $n$ is the number of variables and $\varepsilon$ is an arbitrary positive number. As customary in the area of exponential algorithms, we used above $O^*$ which hides not only constant factors, but also polynomial ones. Note that the obvious brute-force algorithm for $k$-SAT is of runtime $O^*(2^n)$.

Recently many lower bound results for the complexity of exponential time algorithms have been proved under the assumption that the Exponential Time Hypothesis (ETH) (see [15]) holds. ETH claims that there exists a real number $\delta > 0$ such that 3-SAT cannot be solved in time $O(2^{\delta n})$, where $n$ is the number of variables in the CNF formula of 3-SAT. For example, Cygan, Fomin, Golovnev, Kulikov, Mihajlin, Pachocki and Socala [14] proved that, subject to ETH, there is no $2^{o(n \log n)}$-time algorithm deciding whether an $n$-vertex graph $H$ is a subgraph of another $n$-vertex graph $G$ (the obvious brute-force algorithm solves this problem in time $2^{O(n \log n)}$).

## 1.11 Parameterized Algorithms and Complexity

Parameterized algorithms and complexity is one of the approaches for dealing with $\mathcal{NP}$-hard problems. The main idea of this approach is that using only the size of the problem in the complexity bound for the problem is often too simplistic as the instances of the problem under consideration which are of our interest, often have some small parameter $k$ (such as the maximum semi-degree of a digraph or the treewidth of an undirected graph). Problems with parameters are called **parameterized** problems; an instance of a parameterized problem is a pair $(I, k)$, where $I$ is an instance of the problem (no parameter) and $k$ is the value of the parameter.For a parameterized problem with parameter $k$, an algorithm of runtime $O^*(f(k)) := O(f(k)n^c)$, where $f(k)$ is an arbitrary computable function, $n$ is the size of the problem and $c$ is a constant (independent of $k$ and $n$), can be viewed as a generalization of a polynomial algorithm and, thus, an efficient algorithm (especially when $f(k)$ grows relatively slowly and $c$ is of moderate value). Such algorithms are called **fixed-parameter tractable (FPT)** and parameterized problems admitting such algorithms are also called FPT. The class of FPT problems is denoted by FPT.

From the practical point of view, the chosen parameters should be relatively small on practically-interesting instances of the problem under consideration. The Directed rural postman problem (DRPP) is formulated as follows: Given a strongly connected directed multigraph $D = (V, A)$ with nonnegative integral weights on the arcs, a subset $R$ of required arcs and a nonnegative integer $\ell$, we are to decide whether $D$ has a closed directed walk of weight at most $\ell$ containing every arc of $R$. DRPP is $\mathcal{NP}$-hard. Let $k$ be the number of connected components in the subgraph of $UG(D)$ induced by $R$. In [37] Sorge, van Bevern, Niedermeier and Weller commented that "$k$ is presumably small in a number of applications" and Sorge [36] noted that in planning for snow plowing routes for Berliner Stadtreinigung, $k$ is only between 3 and 5. Gutin, Wahlström and Yeo [25] developed an $O^*(2^k)$-time randomized algorithm for DRPP. Unfortunately, the existence of a deterministic FPT algorithm for DRPP parameterized by $k$ still remains "a more than thirty years open ... question with significant practical relevance" (see [37]).

When the runtime $O(f(k)n^c)$ is replaced by the much more powerful $n^{O(f(k))}$, we obtain the class **XP** where each problem is polynomial-time solvable for any fixed value of $k$. There are a number of parameterized complexity classes between FPT and XP (for each integer $t \geq 1$, there is a class W[$t$]) and they form the following tower:

$$FPT \subseteq W[1] \subseteq W[2] \subseteq \cdots \subseteq W[P] \subseteq XP.$$

Here W[P] is the class of all parameterized problems (with parameter $k$) that can be solved in $f(k)n^{O(1)}$ time by a non-deterministic Turing machine that makes at most $f(k)\log n$ non-deterministic steps for some function $f$. For the definition of classes W[$t$], see, e.g., the monographs [15] by Cygan, Fomin, Kowalik, Lokshtanov, Marx, Pilipczuk, Pilipczuk and Saurabh, and [18] by Downey and Fellows. It is widely believed that FPT $\neq$ W[1]. One reason for this is that if FPT = W[1], then ETH fails, see, e.g., [18]. The problem of deciding whether a graph has a clique with $k$ vertices is W[1]-complete [15, 18], so it is highly unlikely that the problem is FPT.

For parameterized problems $\Pi$ and $\Pi'$, a **bikernelization** is a polynomial algorithm that maps an instance $(I, k)$ of $\Pi$ to an instance $(I', k')$ of $\Pi'$ (the **bikernel**) such that (i) $(I, k) \in \Pi$ if and only if $(I', k') \in \Pi'$, (ii) $k' \leq g(k)$, and (iii) $|I'| \leq g(k)$ for some function $g$. The function $g(k)$ is called the **size** of the bikernel. When $\Pi' = \Pi$, a bikernel is called a **problem kernel** or just a **kernel**. It is well-known that a parameterized problem $\Pi$ is fixed-parameter tractable if and only if it is decidable and admits a kernelization [15, 18]. The same holds if "kernel" is replaced by a "bikernel" (see [2] by Alon, Gutin, Kim, Szeider and Yeo).

Due to applications, low degree polynomial size kernels are of main interest. Unfortunately, many FPT problems do not have kernels of polynomial size unless $\mathcal{NP} \subseteq \text{co}\mathcal{NP}/\text{poly}$, which is highly unlikely as $\mathcal{NP} = \text{co}\mathcal{NP}/\text{poly}$ would imply that the polynomial hierarchy collapses to its third level; for definitions and more information, see, e.g., [15, 18]. In particular, the problem of whether a digraph contains a $k$-dipath is FPT but has no polynomial kernel unless $\text{co}\mathcal{NP} \subseteq \mathcal{NP}/\text{poly}$ [11]. Binkele-Raible, Fernau, Fomin, Lokshtanov, Saurabh and Villanger [10] proved that the problem of deciding whether a digraph $D$ and a vertex $v \in V(D)$ has an out-tree rooted at $v$ with least $k$ leaves admits a problem kernel with at most $O(k^3)$ vertices (and, hence, at most $O(k^6)$ arcs). Interestingly, Binkele-Raible *et al.* [10] also proved that if we allow the out-tree to be rooted at any vertex of $D$, then the "unrooted" problem does not admit a polynomial kernel unless $\text{co}\mathcal{NP} \subseteq \mathcal{NP}/\text{poly}$. For further background and terminology on parameterized complexity we refer the reader to the monographs [15, 18].

Let us consider a couple of recent results on parameterized complexity of problems on digraphs.

Bang-Jensen and Yeo [7] asked whether the following problem is FPT.

| | |
|---|---|
| CONNECTIVITY PRESERVING PATH CONTRACTIONS | **Parameter:** $k$ |
| **Input:** A strongly connected digraph $D$. | |
| **Question:** Can we path-contract $k$ arcs from $D$ such that $D$ remains strongly connected? | |

Gutin, Ramanujan, Reidl and Wahlström [24] proved that the problem is, in fact, W[1]-hard. However, the problem is FPT if the operation of path-contraction is replaced by deletion, which was proved by Basavaraju, Misra, Ramanujan and Saurabh [8].

We complete this section with an open questions on the parameterized complexity of the following digraph problem introduced by Bezáková, Curticapean, Dell and Fomin [9].

**Problem 1.11.1** *For given vertices $s$ and $t$ of a digraph $D$, and an integer (parameter) $k$, decide whether $D$ has an $(s,t)$-path in $D$ that is at least $k$ longer than a shortest $(s,t)$-path.*

If "at least" is replaced by "exactly", then the problem is FPT [9]. However, it is unknown whether the original problem is even in XP.

## 1.12 Approximation Algorithms

There are several situations when the use of exact optimization algorithms does not seem to be a good idea. One is when the time is greatly limited or the problem should be solved online. Another is when the data is not exact or the objective function is not well-defined and, thus, we cannot get an optimal solution even by exhaustive search. In such situations, we can use approximation algorithms for finding a solution that is often not optimal, but we have some performance guarantee in each case.

Let $P$ be a combinatorial optimization problem, and let $\mathcal{A}$ be an approximation algorithm for $P$. Let $X(I)$ denote the set of all feasible solutions for some instance $I \in P$ and let $|I|$ be the size of $I$. We denote the solution obtained by $\mathcal{A}$ for an instance $I$ of $P$ by $x(I)$. Furthermore let $opt(I)$ denote the optimal solution of $I$. The weight of a solution $y$ of $P$ will be denoted by $w(y)$.

The theoretical performance of an approximation algorithm is normally measured by the (worst case) **performance ratio**. Usually, upper or lower bounds for the worst case performance ratio are obtained, where the performance ratio is defined as

$$\max_{I \in P: |I| = n} \left\{ \frac{w(x(I))}{w(opt(I))}, \frac{w(opt(I))}{w(x(I))} \right\}$$

The performance ratio defined in this way has its advantage in the fact that it is always at least 1 (for both minimization and maximization problems).

We normally require that an approximation algorithm has a polynomial running time. Some approximation algorithms provide a good performance guarantee. For example, the well-known Christofides algorithm [13] for the symmetric TSP[3] with triangle inequality (i.e., $w_{ij} + w_{jk} \geq w_{ik}$ for every

[3] The symmetric TSP is the problem of finding a minimum weight Hamilton cycle in a weighted complete undirected graph.

triple $i, j, k$ of vertices, where $w_{ij}$ is the weight of an edge $ij$) has performance ratio 1.5. Unless $\mathcal{P}$=$\mathcal{NP}$, there are no approximation algorithms of constant performance ratio for the (general) symmetric TSP [3].

A **polynomial-time approximation scheme (PTAS)** is an algorithm which takes an instance of a minimization problem $\mathcal{Q}$ and a parameter $\varepsilon > 0$ and, in polynomial time, returns a solution that is within a factor $1 + \varepsilon$ of being optimal. The definition remains the same for maximization problems, but the solution must be within a factor $1 - \varepsilon$ of being optimal. It is well-known that MaxSNP-hard problems do not admit PTAS unless $\mathcal{P} = \mathcal{NP}$.

For many results on approximation algorithms and in-approximability, see, e.g., the monograph [38] by Williamson and Shmoys.

## References


1. N. Alon, F.V. Fomin, G. Gutin, M. Krivelevich, and S. Saurabh. Spanning directed trees with many leaves. *SIAM J. Discrete Math.*, 23(1):466–476, 2009.
2. N. Alon, G. Gutin, E.J. Kim, S. Szeider, and A. Yeo. Solving MAX-$r$-SAT above a tight lower bound. *Algorithmica*, 61(3):638–655, 2011.
3. G. Ausiello, P. Crescenzi, G. Gambosi, V. Kann, A. Marchetti-Spaccamela, and M. Protasi. *Complexity and Approximation.* Springer-Verlag, Berlin, 1999.
4. J. Bang-Jensen and G. Gutin. *Digraphs: Theory, Algorithms and Applications.* Springer-Verlag, London, 2nd edition, 2009.
5. J. Bang-Jensen and G. Gutin. Out-branchings with extremal number of leaves. *Ramanujan Math. Soc. Lect. Notes*, 13:91–99, 2010.
6. J. Bang-Jensen, J. Huang, and E. Prisner. In-tournament digraphs. *J. Combin. Theory Ser. B*, 59(2):267–287, 1993.
7. J. Bang-Jensen and A. Yeo. The minimum spanning strong subdigraph problem is fixed parameter tractable. *Discrete Appl. Math.*, 156:2924–2929, 2008.
8. M. Basavaraju, P. Misra, M.S. Ramanujan, and S. Saurabh. On finding highly connected spanning subgraphs. *CoRR*, arXiv:1701.02853, 2017.
9. I. Bezáková, R. Curticapean, H. Dell, and F.V. Fomin. Finding detours is fixed-parameter tractable. In *ICALP 2017*, volume 80 of *LIPIcs*, pages 54:1–54:14, 2017.
10. D. Binkele-Raible, H. Fernau, F.V. Fomin, D. Lokshtanov, S. Saurabh, and Y. Villanger. Kernel(s) for problems with no kernel: On out-trees with many leaves. *ACM Trans. Algorithms*, 8(4):38:1–38:19, 2012.
11. H.L. Bodlaender, R.G. Downey, M.R. Fellows, and D. Hermelin. On problems without polynomial kernels. *J. Comput. Syst. Sci.*, 75(8):423–434, 2009.
12. F. Boesch and R. Tindell. Robbins's theorem for mixed multigraphs. *Amer. Math. Mon.*, 87(9):716–719, 1980.
13. N. Christofides. Worst-case analysis of a new heuristic for the traveling salesman problem. Technical Report CS-93-13, Carnegie Mellon University, 1976.
14. M. Cygan, F.V. Fomin, A. Golovnev, A.S. Kulikov, I. Mihajlin, J. Pachocki, and A. Socala. Tight bounds for graph homomorphism and subgraph isomorphism. In *SODA 2016: 27th ACM-SIAM Symposium on Discrete Algorithms*, pages 1643–1649, 2016.
15. M. Cygan, F.V. Fomin, L. Kowalik, D. Lokshtanov, D. Marx, M. Pilipczuk, M. Pilipczuk, and S. Saurabh. *Parameterized Algorithms.* Springer, 2015.

16. J. Daligault, G. Gutin, E.J. Kim, and A. Yeo. FPT algorithms and kernels for the directed $k$-leaf problem. *J. Comput. Syst. Sci.*, 76:144–152, 2010.
17. J. Daligault and S. Thomassé. On finding directed trees with many leaves. In *IWPEC 2009*, volume 5917 of *Lect. Notes Comput. Sci.*, pages 86–97. Springer, 2009.
18. R.G. Downey and M.R. Fellows. *Fundamentals of Parameterized Complexity.* Springer, 2013.
19. F.V. Fomin and D. Kratsch. *Exact Exponential Algorithms.* Springer, 2010.
20. S. Fortune, J.E. Hopcroft, and J. Wyllie. The directed subgraph homeomorphism problem. *Theor. Comput. Sci.*, 10:111–121, 1980.
21. T. Gallai and A.N. Milgram. Verallgemeinerung eines graphentheoretischen Satzes von Rédei. *Acta Sci. Math. Szeged*, 21:181–186, 1960.
22. G. Gutin. Characterization of complete $n$-partite digraphs that have a Hamiltonian path. *Kibernetika (Kiev)*, no. 1:107–108, 1988.
23. G. Gutin and R. Li. Seymour's second neighbourhood conjecture for quasi-transitive oriented graphs. *CoRR*, abs/1704.01389, 2017.
24. G. Gutin, M.S. Ramanujan, F. Reidl, and M. Wahlström. Path-contractions, edge deletions and connectivity preservation. In *ESA 2017: 25th Annual European Symposium on Algorithms*, volume 87 of *LIPIcs*, pages 47:1–47:13, 2017.
25. G. Gutin, M. Wahlström, and A. Yeo. Rural postman parameterized by the number of components of required edges. *J. Comput. Syst. Sci.*, 83(1):121–131, 2017.
26. A.J. Hoffman. Some recent applications of the theory of linear inequalities to extremal combinatorial analysis. In R. Bellman and M. Hall, editors, *Combinatorial Analysis*, pages 113–128. American Mathematical Society, Providence, RI, 1960.
27. D. Kühn and D. Osthus. Linkedness and ordered cycles in digraphs. *Combin. Prob. Comput.*, 17:411–422, 2008.
28. M. Las Vergnas. Sur les arborescences dans un graphe orienté. *Discrete Math.*, 15(1):27–39, 1976.
29. K. Menger. Zur allgemeinen Kurventheorie. *Fund. Math.*, 10:96–115, 1927.
30. J.W. Moon. On subtournaments of a tournament. *Can. Math. Bull.*, 9:297–301, 1966.
31. R.A. Moser and D. Scheder. A full derandomization of Schöning's $k$-SAT algorithm. In *STOC 2011: 43rd ACM Symposium on Theory of Computing*, pages 245–252, 2011.
32. L. Rédei. Ein kombinatorischer Satz. *Acta Litt. Szeged*, 7:39–43, 1934.
33. H.E. Robbins. A theorem on graphs with an application to a problem on traffic control. *Amer. Math. Mon.*, 46:281–283, 1939.
34. U. Schöning. A probabilistic algorithm for $k$-SAT and constraint satisfaction problems. In *FOCS 1999: 40th IEEE Symposium on Foundations of Computer Science*, pages 410–414, 1999.
35. Y. Shiloach. Edge-disjoint branching in directed multigraphs. *Inform. Process. Lett.*, 8(1):24–27, 1979.
36. M. Sorge. Some algorithmic challenges in arc routing, May 2013. Talk at NII Shonan Seminar no. 18.
37. M. Sorge, R. van Bevern, R. Niedermeier, and M. Weller. A new view on rural postman based on eulerian extension and matching. *J. Discrete Algor.*, 16:12–33, 2012.
38. D.P. Williamson and D.B. Shmoys. *The Design of Approximation Algorithms.* Cambridge University Press, 2011.

# 2. Tournaments and Semicomplete Digraphs

**Jørgen Bang-Jensen and Frédéric Havet**

The class of tournaments is by far the most well-studied class of digraphs with many deep and important results. Since Moon's pioneering book in 1968 [146], the study of tournaments and their properties has flourished. A search in May 2017 on MathSciNet for 'tournament' and 05C20 gives more than 900 hits. Clearly we can only cover a small fraction of the research on tournaments, but we believe that our coverage will stimulate new research on this beautiful class of digraphs.

Being a super-class of tournaments, the class of semicomplete digraphs inherits many of the properties of tournaments, but there are important differences and we shall try to point out such when relevant. Due to space limitations we will not mention all places where a result for tournaments extends to semicomplete digraphs. Note that the results of Section 2.3 imply that results for $k$-strong tournaments often imply similar results for $(3k-2)$-strong semicomplete digraphs.

In Section 2.1 we introduce some special tournaments that occur in several proofs and results in the chapter. Section 2.2 gives some basic properties of tournaments and semicomplete digraphs such as the fact that they are always traceable. The short Section 2.3 is about spanning tournaments of high connectivity in highly connected semicomplete digraphs. In Section 2.4 we give two very different proofs for the tournament case of the conjecture of Seymour (and Dean in the case of tournaments) that every oriented graph has a vertex with distance 2 to at least as many vertices as it has out-neighbours. Section 2.5 deals with linkages and disjoint cycles in tournaments and semicomplete digraphs. In Section 2.6 we discuss further topics related to Hamiltonian paths and cycles and give a proof of Redéi's theorem that every tournament has an odd number of Hamiltonian paths. Section 2.7 is devoted to oriented subgraphs in tournaments, in particular to oriented Hamiltonian paths and cycles in tournaments. In Section 2.8 we study vertex-partitions

J. Bang-Jensen (✉)
Department of Mathematics and Computer Science,
University of Southern Denmark, Campusvej 55, DK5230, Odense M, Denmark
e-mail: jbj@imada.sdu.dk

F. Havet
Projet COATI, INRIA, Sophia-Antipolis,
2004 Route de Lucioles BP93, 06902, Sophia-Antipolis Cedex, France
e-mail: frederic.havet@inria.fr

 
J. Bang-Jensen and G. Gutin (eds.), *Classes of Directed Graphs*, Springer Monographs in Mathematics, https://doi.org/10.1007/978-3-319-71840-8_2

of semicomplete digraphs where each part has to have certain properties, e.g. being strongly connected or being acyclic. Section 2.9 deals with results of feedback sets, that is, sets of vertices or arcs whose deletion makes the resulting digraph acyclic. Even for tournaments, finding such a set of minimum cardinality is $\mathcal{NP}$-complete. In Section 2.10 we study the problem of how many arcs one may delete from a $k$-(arc)-strong tournament without reducing the connectivity of the resulting digraph. The answer is that we may delete surprisingly many. Section 2.11 is also on connectivity, but this time the operation we consider is that of either reversing arcs or of deorienting arcs, that is, adding an arc oppositely oriented to an existing arc. In Section 2.12 we consider arc-disjoint spanning subdigraphs of semicomplete digraphs. This includes the famous Kelly conjecture that the arc set of every regular tournament decomposes into Hamiltonian cycles. Section 2.13 is on minors of semicomplete digraphs. It turns out that for this class of digraphs the notion of a minor, defined as being any digraph that can be obtained by contracting strong subdigraphs, leads to results in the same vein as the graph minor theory of Roberson and Seymour. Finally, in Section 2.14 we briefly survey a few further topics on tournaments.

We will use the shorthand names **n-tournament** and **n-semicomplete** digraph for a tournament, resp. semicomplete digraph on $n$ vertices. Throughout this chapter, except for Section 2.7, paths and cycles are always assumed to be directed.

## 2.1 Special Tournaments

We first define a number of special tournaments that will be referred to later. Let $n \geq 1$ be an integer. The unique acyclic $n$-tournament is the **transitive tournament**, denoted $TT_n$. This has an ordering $(v_1, v_2, \ldots, v_n)$ of its vertices so that $v_iv_j$ is an arc whenever $1 \leq i < j \leq n$.

A tournament is **almost transitive** if it is obtained from a transitive tournament with acyclic ordering $(v_1, v_2, \ldots, v_n)$ (i.e., $v_i \rightarrow v_j$ for all $1 \leq i < j \leq n$) by reversing the arc $v_1v_n$.

The **random $n$-tournament** $RT_n$ is the (random) digraph one obtains from the complete graph $K_n$ by choosing one from each of the two possible orientations of each edge $uv$ of $K_n$ with probability $\frac{1}{2}$ for each of the two possible orientations.

Recall that an $n$-tournament is **regular** if $n = 2k + 1$ for some $k \geq 1$ and every vertex has in- and out-degree $k$. Below we describe two important examples of classes of regular tournaments.

Let $\mathbb{Z}_{2k+1}$ be the set of integers modulo $2k + 1$ and let $J$ be a subset of $\mathbb{Z}_{2k+1} \setminus \{0\}$ such that for every $i \in \mathbb{Z}_{2k+1} \setminus \{0\}$, we have $i \in J$ if and only if $-i \notin J$. Then the **circulant tournament** $CT_{2n+1}(J)$ is the tournament whose vertex set is $\mathbb{Z}_{2k+1}$ and $ij$ is an arc if and only if $j - i \in J$. For some examples of papers on circulant tournaments, see [14, 47, 136, 149].

For each prime power $q$ of the form $q = 4k+3$, the **Paley tournament** $\mathbb{P}_q$ is the $q$-tournament whose vertices are the elements of the finite field $GF(q)$ with $q$ elements. There is an arc from $x$ to $y$ if and only if $y-x$ is a non-zero square in the field. E.g. when $q = 7$ the vertex set of $\mathbb{P}_7$ is $\{0,1,2,3,4,5,6\}$ and $ij$ is an arc of $\mathbb{P}_7$ if and only if $((j-i) \mod 7) \in \{1,2,4\}$. For examples of papers dealing with Paley tournaments, see e.g. [44, 46, 47, 51].

## 2.2 Basic Properties of Tournaments and Semicomplete Digraphs

We start with a very simple but important observation which is proved by a simple counting argument.

**Proposition 2.2.1** *Every semicomplete digraph on $n$ vertices contains a vertex with out-degree at least $\lfloor \frac{n}{2} \rfloor$ and a vertex with in-degree at least $\lfloor \frac{n}{2} \rfloor$.*

**Proof:** Let $T$ be a semicomplete digraph on $n$ vertices. We have

$$\sum_{v \in V(T)} d^+(v) = \sum_{v \in V(T)} d^-(v) = |A(T)| \geq \binom{n}{2} = n \cdot \frac{n-1}{2}.$$

Thus there is a vertex with out-degree (resp. in-degree) at least $\lceil \frac{n-1}{2} \rceil = \lfloor \frac{n}{2} \rfloor$.
$\square$

**Proposition 2.2.2** *Let $k$ be a positive integer. Every semicomplete digraph has at most $2k-1$ vertices of out-degree less than $k$.*

**Proof:** Let $D$ be a semicomplete digraph and let $X$ be the set of vertices of out-degree less than $k$ in $T$. The number of arcs in the subdigraph $D[X]$ is at most $|X|(k-1)$. On the other hand, $D[X]$ has at least $\binom{|X|}{2}$ arcs. Hence,

$$\frac{|X|(|X|-1)}{2} \leq |A(D[X])| \leq |X|(k-1),$$

implying that $|X| \leq 2k-1$.

Using Proposition 2.2.1 we can now give a lower bound on the largest transitive subtournament in any tournament.

**Proposition 2.2.3** *Every $n$-tournament contains a transitive subtournament $TT_k$ with $k \geq \lceil \log n \rceil$.*

**Proof:** The following algorithm produces such a transitive subtournament: Let $T' := T$ and $R = \emptyset$. While $T'$ has at least one vertex: let $v$ be a vertex of maximum out-degree in $T'$ and let $R := R \cup \{v\}$. By Proposition 2.2.1,

$|N^+_{T'}(v)| \geq \lfloor \frac{n}{2} \rfloor$. Hence, letting $T' := T'[N^+(v)]$, the new $T'$ has size at least $\lfloor \frac{n}{2} \rfloor$. Repeat the step above for $T'$.

Clearly the set $R$ returned by this algorithm induces a transitive subtournament of $T$. To see that $R$ has size at least $\lceil \log n \rceil$, consider the integer $r$ satisfying $2^r \leq n < 2^{r+1}$. After step number $i$ in the algorithm above we have $2^{r-i} \leq |V(T')|$, from which it follows that $|R| \geq r + 1 \geq \lceil \log n \rceil$ holds at the end. □

One of the first results on tournaments is the following, due to Rédei. See Section 2.6 for a beautiful generalization of this, also due to Rédei.

**Theorem 2.2.4 (Rédei's Theorem** [158]) *Every tournament contains a Hamiltonian dipath.*

**Proof:** By induction on the number of vertices. The statement is trivial for the 1-tournament. Let $n \geq 2$, let $T$ be an $n$-tournament and let $v$ be a vertex of $T$. By the induction hypothesis, $T\langle N^-(v)\rangle$ and $T\langle N^+(v)\rangle$ have Hamiltonian directed paths $P^-$ and $P^+$. Thus $P^- v P^+$ is a Hamiltonian dipath of $T$[1]. □

Since we can obtain a tournament from a semicomplete digraph by removing an arbitrary arc from each 2-cycle, we obtain that Theorem 2.2.4 also holds for semicomplete digraphs (and this can also be proved directly with the same proof as above).

**Corollary 2.2.5** *Every semicomplete digraph has a Hamiltonian path.*

There is no analogue to Theorem 2.2.4 for Hamiltonian dicycles since the transitive tournaments are acyclic and in particular have no Hamiltonian dicycle. More generally, no non-strong tournament has a Hamiltonian dicycle because it has a vertex-partition $(L, R)$ such that $L \mapsto R$ (e.g. if we take $L$ to be the vertices of the initial strong component and $L$ to be the remaining vertices). In contrast, all strong tournaments have a Hamiltonian directed cycle as shown by Camion [56].

**Theorem 2.2.6 (Camion's Theorem** [56]) *Every strong tournament has a Hamiltonian dicycle.*

A simple proof of Camion's Theorem, due to Moon [144], actually proves a stronger result.

**Theorem 2.2.7 (Moon's Theorem** [144]) *Every strong tournament is vertex-pancyclic.*

---

1 Note that here we allowed one of the two tournaments to be empty, in which case the corresponding path is also empty

**Proof:** Let $x$ be a vertex in a strong tournament $T$ on $n \geq 3$ vertices. The proof is by induction on $k$. We first prove that $T$ has a 3-cycle through $x$. Since $T$ is strong, each of the sets $O = N^+(x)$ and $I = N^-(x)$ are non-empty and the set $(O, I)$ is also non-empty. Let $yz \in (O, I)$. Then $xyzx$ is a 3-cycle through $x$. Let $C = x_0x_1 \dots x_t$ be a dicycle in $T$ with $x = x_0 = x_t$ and $t \in \{3, 4, \dots, n-1\}$. We prove that $T$ has a $(t+1)$-cycle through $x$.

If there is a vertex $y \in V(T) \setminus V(C)$ which dominates a vertex in $C$ and is dominated by a vertex in $C$, then it is easy to see that there exists an index $i$ such that $x_i \rightarrow y$ and $y \rightarrow x_{i+1}$. Therefore, $C[x_0, x_i]yC[x_{i+1}, x_t]$ is a $(t+1)$-cycle through $x$. Thus, we may assume that every vertex outside of $C$ either dominates every vertex in $C$ or is dominated by every vertex in $C$. The vertices of $V(T) \setminus V(C)$ that dominate all vertices of $V(C)$ form a set $R$; the rest of the vertices in $V(T) \setminus V(C)$ form a set $S$. Since $T$ is strong, both $S$ and $R$ are non-empty and the set $(S, R)$ is non-empty. Hence, taking $sr \in (S, R)$, we see that $x_0srC[x_2, x_t]$ is a $(t+1)$-cycle through $x = x_0$. □

The following is an easy consequence of Theorem 1.7.3.

**Proposition 2.2.8** *Every strong semicomplete digraph on $n \geq 3$ vertices contains a strong spanning tournament.*

Together with Moon's theorem, Proposition 2.2.8 implies the following.

**Theorem 2.2.9** *Every strong semicomplete digraph is vertex-pancyclic.* □

This easily implies the following.

**Corollary 2.2.10** *Every strong semicomplete digraph $D$ on at least four vertices has two distinct vertices $v_1, v_2$ such that $D - v_i$ is strong for $i \in [2]$.*

This is the best possible as shown by the tournament that one obtains from a transitive tournament $TT_k$ on at $k \geq 3$ vertices by reversing the arcs of the unique Hamiltonian path.

### 2.2.1 Median Orders, a Powerful Tool

Now we introduce a very useful tool for proving results about tournaments and other classes of digraphs.

A **median order** of a digraph $D$ is a linear order $(v_1, v_2, \dots, v_n)$ of its vertex set such that $|\{(v_i, v_j) : i < j\}|$ (the number of arcs directed from left to right) is as large as possible. In the case of a tournament, such an order can be viewed as a ranking of the players which minimizes the number of upsets (matches won by the lower-ranked player). As we shall see, median orders of tournaments reveal a number of interesting structural properties.

Let us first note two basic properties of median orders of tournaments whose easy proofs are left to the reader.

**Lemma 2.2.11** *Let $T$ be a tournament and $(v_1, v_2, \dots, v_n)$ a median order of $T$. Then, for any two indices $i, j$ with $1 \leq i < j \leq n$:*

*(M1) The suborder $(v_i, v_{i+1}, \dots, v_j)$ is a median order of the induced subtournament $T\langle\{v_i, v_{i+1}, \dots, v_j\}\rangle$ ;*
*(M2) The vertex $v_i$ dominates at least half of the vertices $v_{i+1}, v_{i+2}, \dots, v_j$, and vertex $v_j$ is dominated by at least half of the vertices $v_i, v_{i+1}, \dots, v_{j-1}$.*

In particular, each vertex $v_i$, $1 \leq i < n$, dominates its successor $v_{i+1}$. The sequence $v_1v_2 \dots v_n$ is thus a Hamiltonian directed path, providing an alternative proof of Rédei's Theorem (Theorem 2.2.4).

### 2.2.2 Kings

The **second out-neighbourhood** of a vertex $v$ in a digraph $D$, denoted by $N_D^{++}(v)$ or simply $N^{++}(v)$, is the set of vertices at distance 2 from $v$. In other words, it is the set of vertices that are dominated by an out-neighbour of $v$ and are not in $v \cup N^+(v)$. The dual notion of **second in-neighbourhood** of a vertex $v$ in a $D$ is defined similarly and is denoted by $N_D^{--}(v)$ or simply $N^{--}(v)$.

A **king** in a tournament $T$ is a vertex $v$ such that $\{v\} \cup N^+(v) \cup N^{++}(v) = V(T)$. Landau [129] proved that every tournament has a king.

**Theorem 2.2.12** ([129]) *Every tournament has a king. More precisely, every vertex with maximum out-degree is a king.*

**Proof:** Let $v$ be a vertex of maximum out-degree in a tournament $T$. Suppose by way of contradiction that $v$ is not a king. Then there exists a vertex $w$ in $T$ that is dominated by no vertex of $N^+(v) \cup \{v\}$. Hence $w$ dominates $N^+(v) \cup \{v\}$ and $d^+(w) \geq d^+(v) + 1$, a contradiction. □

Havet and Thomassé demonstrated that the existence of a king in a tournament can also be proved using median order.

**Lemma 2.2.13** ([109]) *Let $T$ be a tournament. If $(v_1, v_2, \dots, v_n)$ is a median order of $T$, then $v_1$ is a king of $T$.*

**Proof:** Consider $v_i$ for $2 \leq i \leq n$. We shall prove that $v_i \in N^+(v_1) \cup N^{++}(v_1)$. Assume that $v_i$ is not in $N^+(v_1)$. Then it dominates $v_1$. By the property (M2) of Lemma 2.2.11, $v_1$ dominates at least half of the vertices $\{v_2, \dots, v_i\}$, and so, since $v_1$ is dominated by $v_i$, it dominates more than half the vertices of $\{v_2, \dots, v_{i-1}\}$. Similarly, $v_i$ is dominated by more than half the the vertices of $\{v_2, \dots, v_{i-1}\}$. Therefore, there is a vertex in $\{v_2, \dots, v_{i-1}\}$, which dominates $v_i$ and is dominated by $v_1$. Hence $v_i \in N^{++}(v_1)$. □

Since every tournament admits a median order, Lemma 2.2.13 directly implies Theorem 2.2.12. Moon [145] proved that a tournament has at least three kings, provided that it has no **source** (that is, a vertex with in-degree 0 and thus dominating all other vertices). Observe this condition is necessary: if a tournament contains a source, then this vertex is its unique king.

**Corollary 2.2.14** ([145]) *Every tournament $T$ with $\delta^-(T) \geq 1$ has at least three kings.*

**Proof:**
We give a proof due to Havet and Thomassé [109]. Assume that $\delta^-(T) \geq 1$. Let $(v_1, v_2, \ldots, v_n)$ be a median order of $T$. By Lemma 2.2.13, vertex $v_1$ is a king. Let $i$ be the smallest index such that $v_i$ is an in-neighbour of $v_1$, and let $j$ be the smallest index such that $v_j$ is an in-neighbour of $v_i$. Those vertices exist since $T$ has no source. We claim that both $v_i$ and $v_j$ are kings of $T$. First, observe that $1 < j < i$ by (M2). Now, by (M1), $v_i, \ldots, v_n$ is a median order of $T' = T\langle\{v_i, \ldots, v_n\}\rangle$, and so, by Lemma 2.2.13, $v_i$ is a king of $T'$. Moreover, via $v_1$, which dominates all vertices in $v_2, \ldots, v_{i-1}$ (by the choice of $i$), $v_i$ is also a king of $T\langle\{v_1, \ldots, v_i\}\rangle$. Hence $v_i$ is a king of $T$. Similarly, $v_j$ is a king of $T\langle\{v_j, \ldots, v_n\}\rangle$, and, via $v_i$, which dominates all vertices in $v_1, \ldots, v_{j-1}$ (by the choice of $j$), is a king of $T$. □

The above results have been generalized to arc-coloured tournaments. A **monochromatic king** in an arc-coloured tournament is a vertex $v$ such that for every vertex $w$, one can find a monochromatic $(v, w)$-dipath. There are many examples of arc-coloured tournaments with no monochromatic king. Firstly, a tournament with no source and with all its arcs coloured differently has no monochromatic king. Secondly, if there is a partition $(V_1, V_2, V_3)$ of the vertex set of a tournament $T$ such that $V_1 \to V_2 \to V_3 \to V_1$, then $T$ has no monochromatic king. Shen gave a simple necessary condition for the existence of a monochromatic king in arc-coloured tournaments.

**Theorem 2.2.15** ([169]) *If we colour the arcs of a tournament $T$ in such a way that no subtournament of order 3 gets three different colours on its arcs, then there exists a monochromatic king.*

**Proof:** The proof is by induction on the number of vertices. Remove a vertex $x_1$ from $T$. By the induction hypothesis, one can find a monochromatic king $x_2$ in $T - x_1$. If $x_2 \to x_1$, then $x_2$ is a monochromatic king in $T$. Therefore, we may assume $x_1 \to x_2$. Repeating the process for $x_2$, and so on, either we find a monochromatic king in $T$, or we find a directed cycle $C = x_k \ldots x_\ell x_k$ such that $x_i$ is a monochromatic king in $T - x_{i-1}$ (with $x_{k-1} = x_\ell$). If $C$ does not span $T$, then by the induction hypothesis, there is a monochromatic king in $T\langle C\rangle$, say $x_i$. Thus there is a monochromatic $(x_i, x_{i-1})$-dipath in $T\langle C\rangle$. Because, $x_i$ is a monochromatic king in $T - x_{i-1}$, it follows that $x_i$ is also a monochromatic king in $T$. Henceforth, we assume that $C = x_1 \ldots x_n x_1$

is Hamiltonian in $T$. If the arcs of $C$ are monochromatic, the conclusion holds, so there is one particular $x_i$ such that $x_{i-1}x_i$ and $x_ix_{i+1}$ have different colours, say $c_1$ and $c_2$. By the induction hypothesis, there is a monochromatic dipath $P$ from $x_{i+1}$ to $x_{i-1}$. If $P$ is coloured by $c_1$ or $c_2$, then either $x_{i+1}$ or $x_i$ respectively is a monochromatic king in $T$. Henceforth, we may assume that $P$ is coloured by $c_3$. Set $P = y_1 \dots y_q$ with $y_1 = x_{i+1}$ and $y_q = x_{i-1}$. Let $j$ be the smallest index such that the arc $a_j$ between $x_i$ and $y_j$ is not coloured $c_2$. Such a $j$ exists because $y_qx_i$ is coloured $c_1$. Since $T\langle\{y_{j-1}, y_j, x_i\}\rangle$ does not have three different colours on its arcs, necessarily $a_j$ is coloured $c_3$. If $a_j = y_jx_i$ (resp. $a_j = x_iy_j$), then there is a $c_3$-monochromatic $(x_{i+1}, x_i)$-dipath (resp. $(x_i, x_{i-1})$-dipath) and $x_{i+1}$ (resp. $x_i$) is a monochromatic king in $T$. □

In Shen's paper the following question was asked: is it true that no matter how we colour the arcs of a tournament, there is either a trichromatic 3-cycle or a monochromatic king. This was disproved by Galeana-Sánchez and Rojas-Monroy in [93].

### 2.2.3 Scores and Landau's Theorem

Let $T$ be a tournament. Its **score sequence** is the sequence of the out-degrees of its vertices in non-decreasing order. Hence, if $V(T) = \{v_1, v_2, \dots, v_n\}$ with $d^+(v_1) \leq d^+(v_2) \leq \dots \leq d^+(v_n)$, then the score sequence of $T$ is $(d^+(v_1), d^+(v_2), \dots, d^+(v_n))$.

Consider a score sequence $\mathbf{s}$ of some $n$-tournament $T$. Any $k$ vertices of $T$ induce a subtournament $S$ and, hence, the sum of the scores in $T$ of these $k$ vertices must be at least the sum of their scores in $S$, which is just the total number of arcs in $S$, that is, $\binom{k}{2}$. Hence $\sum_{i \in I} s_i \geq \binom{|I|}{2}$ for all $I \subseteq \{1, 2, \dots, n\}$, with equality for $I = \{1, 2, \dots, n\}$. In particular, $\sum_{i=1}^{k} s_i \geq \binom{k}{2}$, for all $1 \leq k \leq n$ with equality for $k = n$. Landau proved that this obvious necessary condition is actually also sufficient.

**Theorem 2.2.16 (Landau [129])** *The sequence* $\mathbf{s} = (s_1 \leq s_2 \leq \dots \leq s_n)$ *of integers is the score sequence of an n-tournament if and only if*

$$\sum_{i=1}^{k} s_i \geq \binom{k}{2}, \quad \textit{for all } 1 \leq k \leq n, \quad \textit{with equality for } k = n. \tag{2.1}$$

There are many known proofs of Landau's theorem (see [52, 97, 140, 160, 185]). Many of these proofs are discussed in the survey [160] by Reid. The proof we present here is due to Griggs and Reid [97].

**Proof:** The specific sequence $\mathbf{t} = (0, 1, 2, \dots, n-1)$ satisfies conditions (2.1) as it is the score sequence of the transitive $n$-tournament. If a sequence $\mathbf{s} \neq \mathbf{t}$ satisfies (2.1), then since $s_1 \geq 0$ and $s_n \leq n-1$, $s$ contains a repeated term.

The object of this proof is to produce a new sequence $\mathbf{s}'$ from $\mathbf{s}$ which also satisfies (2.1), is 'closer' to $\mathbf{t}$ than is $\mathbf{s}$, and is a score sequence if and only if $\mathbf{s}$ is a score sequence. Toward this end, define $j$ to be the smallest index for which $s_j = s_{j+1}$, and define $m$ to be the number of occurrences of the term $s_j$ in $\mathbf{s}$. Note that $j \geq 1$ and $m \geq 2$, and that either $j + m - 1 = n$ or $s_j = s_{j+1} = \cdots = s_{j+m-1} < s_{j+m}$. Define $\mathbf{s}'$ as follows:

$$\text{for } 1 \leq i \leq n, \quad s'_i = \begin{cases} s_i - 1, \text{ if } i = j, \\ s_i + 1, \text{ if } i = j + m - 1, \\ s_i, \qquad \text{otherwise.} \end{cases}$$

Clearly, $s'_1 \leq s'_2 \leq \cdots \leq s'_n$.

Let us show that $\mathbf{s}'$ a score sequence if and only if $\mathbf{s}$ is a score sequence. If $\mathbf{s}'$ is the score sequence of some $n$-tournament $T'$ in which vertex $v_i$ has out-degree $s'_i$, $1 \leq i \leq n$, then, since $s'_{j+m-1} > s'_j$, there is a vertex in $T'$, say $v_p$, for which $v_{j+m-1} \to v_p$ and $v_p \to v_j$. The reversal of those two arcs in $T'$ yields an $n$-tournament with score sequence $\mathbf{s}$. Conversely, if $\mathbf{s}$ is the score sequence of some $n$-tournament $T$ in which vertex $v_i$ has score $s_i$, $1 \leq i \leq n$, then we may suppose that $v_j$ dominates $v_{j+m-1}$ in $T$, for otherwise, interchanging the labels on these two vertices does not change $\mathbf{s}$. The reversal of the arc $v_jv_{j+m-1}$ in $T$ yields an $n$-tournament with score sequence $\mathbf{s}'$.

To conclude the inductive proof, since $\mathbf{s}'$ is closer to $\mathbf{t}$ than $\mathbf{s}$, it remains to show that $\mathbf{s}'$ satisfies (2.1). By definition of $\mathbf{s}'$, one needs to show that $\sum_{i=1}^{k} s_i \geq \binom{k}{2} + 1$ for all $j \leq k \leq j + m - 2$. The proof is by induction on $k \geq j$. The case $k = j$ is very similar to the induction step and is omitted. Suppose that for some $k$, $j \leq k < j + m - 2$, $\sum_{i=1}^{k} s_i \geq \binom{k}{2} + 1$. We shall prove that $\sum_{i=1}^{k+1} s_i \geq \binom{k+1}{2} + 1$. Suppose by way of contradiction that this is not the case. Then by (2.1),

$$\sum_{i=1}^{k+1} s_i = \binom{k+1}{2}. \tag{2.2}$$

Now since $j < k + 2 \leq j + m - 1$, by definition of $j$ and $m$ and the above equation, we have

$$s_{k+1} = s_{k+2} = \sum_{i=1}^{k+2} s_i - \sum_{i=1}^{k+1} s_i \geq \binom{k+2}{2} - \binom{k+1}{2} = k + 1.$$

Consequently, by the induction hypothesis,

$$\sum_{i=1}^{k+1} s_i = s_{k+1} + \sum_{i=1}^{k} s_i \geq s_{k+1} + \binom{k}{2} + 1 \geq k + 1 + \binom{k}{2} + 1 \geq \binom{k+1}{2} + 1.$$

This contradicts (2.2). □

## 2.3 Spanning $k$-Strong Subtournaments of Semicomplete Digraphs

Theorem 1.7.3 asserts that every strong digraph $D$ without a bridge contains a spanning strong oriented graph (obtained by deleting one arc from every 2-cycle in $D$). It is then natural to ask whether there exists, for each non-negative integer $k$, a minimum integer $f(k)$ such that every $f(k)$-strong digraph contains a spanning $k$-strong oriented graph. Because every $k$-strong oriented graph has at least $2k+1$ vertices and the complete digraph on $r+1$ vertices is $r$-strong, we have $f(k) \geq 2k$ for all $k \geq 2$. Jackson and Thomassen (see [178]) conjectured that this lower bound is indeed tight.

**Conjecture 2.3.1 (Jackson and Thomassen** [178]**)** *Every $2k$-strong digraph contains a spanning $k$-strong oriented graph.*

This conjecture is still widely open for general digraphs, even in the case when $k = 2$. It was verified by Thomassen [186] for the special case when $k = 2$ and $D$ is a symmetric digraph (all arcs are in 2-cycles), thus improving on a result of Jordán [115] establishing the existence of a spanning 2-strong oriented graph in every 18-strong symmetric digraph. For all $k \geq 3$ it is still open whether there is a function $g(k)$ such that every $g(k)$-strong symmetric digraph has a spanning $k$-strong oriented subdigraph.

Even for the class of semicomplete digraphs the conjecture is open when $k \geq 3$. The case $k = 2$ and $D$ semicomplete follows from the next result.

Improving an earlier bound of $5k$, due to Bang-Jensen and Thomassen, Guo proved the following, which implies that the case $k = 2$ of Conjecture 2.3.1 holds for semicomplete digraphs.

**Theorem 2.3.2** ([99]) *Let $k$ be a positive integer. Every $(3k-2)$-strong tournament contains a spanning $k$-strong tournament.*

Bang-Jensen and Jordán proved that the function $3k - 2$ is not the best possible when $k = 2$.

**Theorem 2.3.3** ([30]) *Every 3-strong semicomplete digraph on at least 5 vertices contains a spanning 2-strong tournament. There is a polynomial algorithm for constructing a spanning 2-strong tournament of a given 3-strong semicomplete digraph.*

Bang-Jensen and Jordán conjectured that the bound $(3k - 2)$ can be improved as follows.

**Conjecture 2.3.4** ([30]) *For each $k \geq 1$, every $(2k-1)$-strong semicomplete digraph on at least $2k+1$ vertices contains a spanning $k$-strong tournament.*

The number $(2k-1)$ would be the best possible as seen from the following construction from [30]: Let $k \geq 2$ be an integer, let $U$ and $W$ be disjoint copies of the complete digraph $\overleftrightarrow{K}_{2k-2}$ with vertex sets $\{u_1, \ldots, u_{2k-2}\}$ and $\{w_1, \ldots, w_{2k-2}\}$, respectively, and let $H'$ be the semicomplete digraph obtained from these by adding the arcs of a matching $\{u_iw_i | i \in [2k-2]\}$ oriented from $U$ to $W$ and the arcs $\{w_iu_j | i,j \in [n] \text{ and } i \neq j\}$ from $W$ to $U$. It is easy to check that $H'$ is $(2k-2)$-strong and that $H'$ cannot contain a spanning $k$-strong tournament, because when we delete one arc from every 2-cycle there is some vertex of $U$ which will have out-degree at most $k-1$. By taking an arbitrary tournament $C$ and adding all arcs from $W$ to $C$ and from $C$ to $U$, we obtain an infinite family of $(2k-2)$-strong semicomplete digraphs containing no spanning $k$-strong tournament.

## 2.4 The Second Neighbourhood Conjecture

One of the (apparently) simplest open questions concerning digraphs is Seymour's Second Neighbourhood Conjecture, asserting that one can always find, in an oriented graph $D$, a vertex whose second out-neighbourhood is at least as large as its out-neighbourhood (see [69]).

**Conjecture 2.4.1 (Seymour's Second Neighbourhood Conjecture)** *In every oriented graph $D$, there exists a vertex $x$ such that $|N_D^+(x)| \leq |N_D^{++}(x)|$.*

Observe that this conjecture is false for digraphs in general. Consider for example $\overleftrightarrow{K}_n$, the complete digraph on $n$ vertices: for every vertex $v$, $N^+(v) = V(\overleftrightarrow{K}_n) \setminus \{v\}$ while $N^{++}(v) = \emptyset$.

Kaneko and Locke [116] proved Conjecture 2.4.1 for oriented graphs with minimum out-degree at most 6. Fidler and Yuster [79] proved that it holds for oriented graphs $D$ with minimum degree $|V(D)|-2$, tournaments minus a star, and tournaments minus the arc set of a subtournament. Cohn, Godbole, Wright Harkness, and Zhang [66] proved that the conjecture holds for random oriented graphs. Gutin and Li proved Conjecture 2.4.1 for quasi-transitive oriented graphs [102].

One approach to Conjecture 2.4.1 is to determine the maximum value $\lambda$ such that in every oriented graph $D$, there exists a vertex $x$ such that $|N_D^+(x)| \leq \lambda |N_D^{++}(x)|$. The conjecture is that $\lambda = 1$. Chen, Shen, and Yuster [60] proved that $\lambda \geq \gamma$ where $\gamma = 0.657298...$ is the unique real root of $2x^3 + x^2 - 1 = 0$. They also claim a slight improvement to $\lambda \geq 0.67815...$

For tournaments, Seymour's Second Neighbourhood Conjecture was also known as Dean's conjecture [69] and was first solved by Fisher [80].

**Theorem 2.4.2** ([80]) *In any tournament, there is a vertex $v$ such that $|N^+(v)| \leq |N^{++}(v)|$.*

The original proof of Fisher used a sort of weighted version of the problem via probability distributions. It is presented in the next subsection. A more elementary proof using median orders was then given by Havet and Thomassé [109]. Their proof also yields the existence of two vertices $v$ such that $|N^+(v)| \leq |N^{++}(v)|$ under the condition that no vertex is a **sink** (that is, a vertex of out-degree 0). This is detailed in Subsection 2.4.2.

### 2.4.1 Fisher's Original Proof

A **(probability) distribution** on a digraph $D$ is a function $p$ that assigns to each vertex a non-negative real number such that $p(V(D)) = \sum_{v\in V(D)} p(v) = 1$. For every subset $S$ of $V(D)$, we set $p(S) = \sum_{v\in S)} p(v)$. A distribution is **losing** if $p(N^-(v)) \leq p(N^+(v))$ for all $v \in V(D)$.

Let $D$ be an oriented graph with $n$ vertices $v_1, \dots, v_n$. The **adjacency matrix** of $D$, denoted by $\mathbf{A}_D$, is the $n \times n$ matrix defined by $(\mathbf{A}_D)_{i,j} = 1$ if $v_i \to v_j$, $(\mathbf{A}_D)_{i,j} = -1$ if $v_j \to v_i$ and $(\mathbf{A}_D)_{i,j} = 0$ otherwise. Observe that $\mathbf{A}_D^T = -A_D$.

We shall use the following well-known lemma, due to Farkas, see e.g. [92, Lemma 1].

**Lemma 2.4.3 (Farkas's Lemma)** *Let* $\mathbf{A}$ *be an* $m \times n$ *matrix and* $\mathbf{b}$ *an* $m$*-dimensional real vector. Then exactly one of the following two statements is true:*

1. *There exists a* $\mathbf{x} \in \mathbb{N}^n$ *such that* $\mathbf{A}\mathbf{x} = \mathbf{b}$ *and* $\mathbf{x} \geq \mathbf{0}$*;*
2. *There exists a* $\mathbf{y} \in \mathbb{N}^m$ *such that* $\mathbf{A}^T\mathbf{y} \geq \mathbf{0}$ *and* $\mathbf{b}^T\mathbf{y} < \mathbf{0}$.

**Theorem 2.4.4** ([80]) *Every digraph has a losing distribution.*

**Proof:** Let $D$ be a digraph with $n$ vertices $v_1, \dots, v_n$. To each distribution $p$ of $D$, we can associate the vector $\mathbf{w}_p = (p(v_1), \dots, p(v_n))^T$. Observe that $\mathbf{w}_p \geq \mathbf{0}$ and $\mathbf{1}^T\mathbf{w}_p = p(V(D)) = 1$. Furthermore, $(\mathbf{A}_D\mathbf{w}_p)_i, = p(N_D^+(v_i)) - p(N_D^-(v_i))$. Hence $p$ is a losing distribution if $\mathbf{A}_D\mathbf{w}_p \geq \mathbf{0}$.

Suppose $D$ has no losing distribution. Since $\mathbf{A}_D^T = -\mathbf{A}_D$, the following system has no solutions. ($\mathbf{I}$ denotes the identity $n \times n$ matrix.)

$$\begin{bmatrix} \mathbf{A}_D^T & \mathbf{I} \\ \mathbf{1}^T & \mathbf{0}^T \end{bmatrix} \begin{pmatrix} \mathbf{w} \\ \mathbf{z} \end{pmatrix} = \begin{pmatrix} \mathbf{0} \\ 1 \end{pmatrix} \text{ with } \begin{pmatrix} \mathbf{w} \\ \mathbf{z} \end{pmatrix} \geq \begin{pmatrix} \mathbf{0} \\ \mathbf{0} \end{pmatrix}.$$

Farkas's Lemma implies that there exists an $n$-dimensional vector $u$ and a real number $t$ such that

$$\begin{bmatrix} \mathbf{A}_D & \mathbf{1} \\ \mathbf{I} & \mathbf{0} \end{bmatrix} \begin{pmatrix} \mathbf{u} \\ t \end{pmatrix} \geq \begin{pmatrix} \mathbf{0} \\ \mathbf{0} \end{pmatrix} \text{ with } (\mathbf{0}^T\ 1) \begin{pmatrix} \mathbf{u} \\ t \end{pmatrix} < 0.$$

Thus $\mathbf{u} \geq \mathbf{0}$, $\mathbf{A}_D\mathbf{u} + t\mathbf{1} \geq \mathbf{0}$ and $t < 0$. Hence $\mathbf{A}_D\mathbf{u} > \mathbf{0}$, so $\frac{1}{\mathbf{1}^T\mathbf{u}}\mathbf{u}$ is the vector associated to a losing distribution, a contradiction. □

We shall now give some properties of losing distributions.

**Lemma 2.4.5** *Let $D$ be a digraph and $p$ a losing distribution. If $p(v) > 0$, then $p(N^+(v)) = p(N^-(v))$.*

**Proof:** We use the notation of the previous proof.

Since $p$ is a losing distribution, then $\mathbf{A}_D\mathbf{w}_p \geq \mathbf{0}$ and $\mathbf{w}_p \geq \mathbf{0}$. Hence $(\mathbf{w}_p)_i(\mathbf{A}_D\mathbf{w}_p)_i \geq 0$ for all $i$. But, since $\mathbf{A}_D$ is skew-symmetric, $(\mathbf{w}_p)^T\mathbf{A}_D\mathbf{w}_p = 0$, so $(\mathbf{w}_p)_i(\mathbf{A}_D\mathbf{w}_p)_i = 0$ for all $i$. Therefore if $(\mathbf{w}_p)_i = p(v_i) > 0$, necessarily, $0 = (\mathbf{A}_D\mathbf{w}_p)_i = p(N^+(v_i)) - p(N^-(v_i))$. In other words, if $w(v_i) > 0$, then $p(N^+(v_i)) = p(N^-(v_i))$. □

**Lemma 2.4.6** *Let $p$ be a losing distribution on a tournament $T$. Then $p(N^-(v)) \leq p(N^{--}(v))$ for every vertex $v$.*

**Proof:** Let $v$ be a vertex of $T$. Since $p$ is a losing distribution, $p(N^-(v)) \leq \frac{1}{2}$. If $p(N^{--}(v)) \geq \frac{1}{2}$, then we are done, so we may assume that $p(N^{--}(v)) < \frac{1}{2}$. Set $R = N^-(v) \cup N^{--}(v)$ and $Q = V(T) \setminus R$. We have $p(R) < 1$ and so $p(Q) > 0$.

Now

$$\sum_{w\in Q} p(w)p(N^-_{T\langle Q\rangle}(w)) = \sum_{w\in Q}\ \sum_{u\in N^-_{T\langle Q\rangle}(w)} p(w)p(u) = \sum_{u\in Q}\ \sum_{w\in N^+_{T\langle Q\rangle}(u)} p(w)p(u)$$
$$= \sum_{u\in Q} p(u)p(N^+_{T\langle Q\rangle}(u)).$$

Hence, there is a vertex $w \in Q$ with $p(w) > 0$ such that $p(N^-_{T\langle Q\rangle}(w)) \geq p(N^+_{T\langle Q\rangle}(w))$. By Lemma 2.4.5, $p(N^+_T(v)) = p(N^-_T(v))$. Since $w$ is not in $N^{--}(v)$, it is dominated by $N^-_T(v)$. Thus $p(N^-_T(w)) \geq p(N^-_{T\langle Q\rangle}(w)) + p(N^-_T(v))$ and $p(N^+_T(w)) \leq p(N^+_{T\langle Q\rangle}(w)) + p(N^{--}_T(v))$. Hence

$$p(N^-_{T\langle Q\rangle}(w)) + p(N^-_T(v)) \leq p(N^+_{T\langle Q\rangle}(w)) + p(N^{--}_T(v)).$$

Since $p(N^-_{T\langle Q\rangle}(w)) \geq p(N^+_{T\langle Q\rangle}(w)$, we obtain $p(N^-(v)) \leq p(N^{--}(v))$.

□

We are now ready to prove Theorem 2.4.2.

**Proof of Theorem** 2.4.2: Let $T$ be a tournament. By Theorem 2.4.4, it admits a losing distribution $p$.

Set $E^+ = \sum_{v\in V(T)} p(v)|N^+(v)|$ and let $E^{++} = \sum_{v\in V(T)} p(v)|N^{++}(v)|$. Since $w \in N^+(v)$ if and only if $v \in N^-(w)$, we have

$$E^+ = \sum_{v\in V(T)} p(v)|N^+(v)| = \sum_{v\in V(T)}\ \sum_{w\in N^+(v)} p(v) = \sum_{w\in V(T)}\ \sum_{v\in N^-(w)} p(v)$$
$$= \sum_{w\in V(T)} p(N^-(w)).$$

Similarly, since $w \in N^{++}(v)$ if and only if $v \in N^{--}(w)$, we have

$$E^{++} = \sum_{w \in V(T)} p(N^{--}(w)).$$

Now, as $p$ is a losing distribution, it follows from Lemma 2.4.6 that we have $p(N^-(w)) \leq p(N^{--}(w))$ for every vertex $w$. Hence $E^+ \leq E^{++}$. Consequently, there must be a vertex $v$ such that $|N^+(v)| \leq |N^{++}(v)|$. □

### 2.4.2 Proof Using Median Orders

**Theorem 2.4.7** ([109]) *Let $T$ be a tournament and $\sigma = (v_1, v_2, \dots, v_n)$ be a median order of $T$. Then $|N_T^+(v_n)| \leq |N_T^{++}(v_n)|$.*

**Proof:** We distinguish two types of vertices of $N^-(v_n)$: a vertex $v_j \in N^-(v_n)$ is **$\sigma$-good** if there exists a vertex $v_i \in N^+(v_n)$, with $i < j$, such that $v_i \to v_j$; otherwise $v_j$ is **$\sigma$-bad**. We denote by $G_\sigma$ the set of $\sigma$-good vertices. Observe that $G_\sigma \subseteq N_T^{++}(v_n)$.

We shall prove by induction on $n$ that $|N_T^+(v_n)| \leq |G_\sigma|$ which directly implies the result. The case $n = 1$ holds vacuously. Assume now $n > 1$. If there is no $\sigma$-bad vertex, then $G_\sigma = N^-(v_n)$. Moreover, by the property (M2) of Lemma 2.2.11, $|N^+(v_n)| \leq |N^-(v_n)|$, so the conclusion holds. Assume now that there exists a $\sigma$-bad vertex. Let $i$ be the smallest integer $i$ such that $v_i$ is $\sigma$-bad. Set $T_r = T\langle\{v_{i+1}, \dots, v_n\}\rangle$. By the property (M1) of Lemma 2.2.11, $\sigma_r = (v_{i+1}, \dots, v_n)$ is a median order of $T_r$. By the induction hypothesis, $|N_{T_r}^+(v_n)| \leq |G_{\sigma_r}|$. Since every $\sigma_r$-good vertex is also $\sigma$-good, we get

$$|N_T^+(v_n) \cap \{v_{i+1}, \dots, v_n\}| \leq |G_\sigma \cap \{v_{i+1}, \dots, v_n\}|. \tag{2.3}$$

By the minimality of the index of $i$, every vertex of $\{v_1, \dots, v_{i-1}\}$ is either in $G_\sigma$ or in $N^+(v_n)$. Moreover, since $v_i$ is $\sigma$-bad, we have $N^+(v_n) \cap \{v_1, \dots, v_i\} \subseteq N^+(v_i) \cap \{v_1, \dots, v_i\}$, so $G_\sigma \cap \{v_1, \dots, v_i\} \supseteq N^-(v_i) \cap \{v_1, \dots, v_i\}$. Now by property (M2) of Lemma 2.2.11, $|N^-(v_i) \cap \{v_1, \dots, v_i\}| \geq |N^+(v_i) \cap \{v_1, \dots, v_i\}|$. Hence

$$|N_T^+(v_n) \cap \{v_1, \dots, v_i\}| \leq |N^-(x_i) \cap \{v_1, \dots, v_i\}| \leq |G_\sigma \cap \{v_1, \dots, v_i\}| \tag{2.4}$$

Equations (2.3) and (2.4) yield $|N_T^+(v_n)| \leq |G_\sigma|$. □

A natural question is to seek another vertex $v$ **with large second out-neighbourhood**, i.e. such that $|N^+(v)| \leq |N^{++}(v)|$. Obviously, this is not always possible: consider, for instance, a regular tournament dominating a single vertex, or simply a transitive tournament. In both cases, the sole vertex $v$ with $|N^{++}(v)| \geq |N^+(v)|$ is the sink. Still using median orders, Havet and Thomassé [109] proved that a tournament always has two vertices with large second out-neighbourhood, provided that every vertex has out-degree at least 1.

**Theorem 2.4.8** ([109]) *A tournament with no sink has at least two vertices $v$ such that $|N^+(v)| \leq |N^{++}(v)|$.*

To prove this result, we need the notion of the **sedimentation** of a median order $\sigma = (v_1, \dots, v_n)$ of a tournament $T$, denoted by $Sed(\sigma)$. If $|N^+(v_n)| < |G_\sigma|$, then $Sed(\sigma) = \sigma$. If $|N^+(v_n)| = |G_\sigma|$, we denote by $b_1, ..., b_k$ the $\sigma$-bad vertices and by $w_1, ..., w_{n-1-k}$ the vertices of $N^+(v_n) \cup G_\sigma$, both enumerated in increasing order with respect to $\sigma$. In this case, $Sed(\sigma)$ is the order $(b_1, \dots, b_k, v_n, w_1, \dots, w_{n-1-k})$.

**Lemma 2.4.9** *If $\sigma$ is a median order of a tournament $T$, then $Sed(\sigma)$ is also a median order of $T$.*

**Proof:** Let $\sigma = (v_1, \dots, v_n)$ be a median order of $T$. If $Sed(\sigma) = \sigma$, there is nothing to prove. So we assume it is not the case, that is, $|N^+(v_n)| = |G_\sigma|$.

The proof is by induction on the number $k$ of $\sigma$-bad vertices. If $k = 0$, all the vertices are $\sigma$-good or in $N^+(v_n)$, in particular $N^-(v_n) = G_\sigma$. Thus, $|N^+(v_n)| = |N^-(v_n)|$ and the order $Sed(\sigma) = (v_n, v_1, \dots, v_{n-1})$ is a median order of $T$. Assume now that $k$ is a positive integer. Let $i$ be the smallest index (wrt. $\sigma$) of a $\sigma$-bad vertex.

For convenience, for any set $S$, we denote by $S[i,j]$ the set $S \cap \{v_i, \dots, v_j\}$. By Equation (2.3), $|G_\sigma[i+1,n]| \geq |N_T^+(v_n)[i+1,n]|$, and by Equation (2.4), $|G_\sigma[1,i]| \geq |N_T^+(v_n)[1,i]|$. Now by assumption, $|G_\sigma| = |N^+(x_n)|$, that is, $|G_\sigma[1,i]| + |G_\sigma[i+1,n]| = |N_T^+(v_n)[1,i]| + |N_T^+(v_n)[i+1,n]|$. Hence $|G_\sigma[1,i]| = |N_T^+(v_n)[1,i]|$ and $|G_\sigma[i+1,n]| = |N_T^+(v_n)[i+1,n]|$. But since $x_i$ is $\sigma$-bad, $N^+(v_n)[1,i] \subseteq N^+(v_i)[1,i]$ and so $N^-(v_i)[1,i-1] \subseteq N^-(v_n)[1,i-1]$. Moreover, by property (M2) of Lemma 2.2.11, $|N^+(v_i)[1,i]| \leq |N^-(v_i)[1,i]|$ and by defintion of $i$, $N^-(v_n)[1,i-1] = G_\sigma[1,i-1] = G_\sigma[1,i]$. Hence,

$$|G_\sigma[1,i]| \leq |N^+(v_i)[1,i]| \leq |N^-(v_i)[1,i]| = G_\sigma[1,i].$$

Thus $|N^+(v_i)[1,i]| \leq |N^-(v_i)[1,i]|$, and so $(v_i, v_1, \dots, v_{i-1}, v_{i+1}, \dots, v_n)$ is a median order of $T$. Applying the induction hypothesis to the median order $(v_1, \dots, v_{i-1}, v_{i+1}, \dots, v_n)$, which has one bad vertex less than $\sigma$, we obtain the result. □

**Proof of Theorem** 2.4.8**:** Let $\sigma = (v_1, \dots, v_n)$ be a median order of $T$. By Theorem 2.4.7, $v_n$ has a large second neighbourhood, so we need to find another vertex with this property.

Observe that if $(u_1, \dots, u_{n-1})$ is a median order of $T - v_n$, then the order $(u_1, \dots, u_{n-1}, v_n)$ is a median order of $T$, and consequently $u_{n-1} \rightarrow v_n$.

Set $T^* = T - v_n$ Assume first that $T^*$ has a median order $\sigma^* = (u_1, \dots, u_{n-1})$ such that $\sigma^* = Sed(\sigma^*)$. Then

$$|N_T^+(u_{n-1})| = |N_{T^*}^+(u_{n-1})| + 1 \leq |G_{\sigma^*}| \leq |N_{T^*}^{++}(u_{n-1})| \leq |N_T^{++}(u_{n-1})|.$$

Assume now that for every median order $\sigma^*$ of $T^*$, $\sigma^* \neq Sed(\sigma^*)$. Define now inductively $\sigma_0 = (v_1, \dots, v_{n-1})$ and $\sigma_{q+1} = Sed(\sigma_q)$. By property (M1) of Lemma 2.2.11, $\sigma_0$ is a median order of $T^*$; Lemma 2.4.9 and an easy induction imply that $\sigma_q$ is a median order of $T^*$ for every positive integer $q$. Since $T$ has no dominated vertex, $v_n$ has an out-neighbour $v_j$. As observed above, for every integer $q$, the last vertex of $\sigma_q$ dominates $v_n$. So $v_j$ is not the last vertex of any $\sigma_q$. Observe also that there is a $q$ such that $v_j$ is $\sigma_q$-bad, for otherwise the index of $x_j$ would always increase. Let $\sigma_q = (u_1, \dots, u_{n-1})$. We have

$$|N_T^+(u_{n-1})| = |N_{T^*}^+(u_{n-1})| + 1 = |G_{\sigma_q}| + 1.$$

Moreover $u_{n-1} \to v_n \to v_j$, so the second neighbourhood of $u_{n-1}$ has at least $|G_{\sigma_q}| + 1$ elements. Hence $|N_T^+(u_{n-1})| \leq |N_T^{++}(u_{n-1})|$. □

### 2.4.3 Relation with Other Conjectures

One of the most celebrated problems concerning digraphs is the Caccetta–Häggkvist conjecture.

**Conjecture 2.4.10 (Caccetta and Häggkvist [54])** *Every digraph $D$ on $n$ vertices and with minimum out-degree at least $n/k$ has a directed cycle of length at most $k$.*

Since every non-transitive tournament contains a directed 3-cycle, this conjecture easily holds for tournaments. However, little is known about this problem, and, more generally, questions concerning digraphs and involving the minimum out-degree tend to be intractable. As a consequence, many open problems flourished in this area, see [175] for a survey. The Hoàng–Reed conjecture [112] is one of these.

A **directed-cycle-tree** is either a singleton or consists of a set of directed cycles $C_1, \dots, C_k$ such that $|V(C_i) \cap (V(C_1) \cup \dots \cup V(C_{i-1}))| = 1$ for all $i = 2, \dots, k$, where $V(C_j)$ is the set of vertices of $C_j$. A less explicit, yet concise, definition is simply that a directed-cycle-tree is a digraph in which there exists a unique directed $(x, y)$-path for every choice of distinct vertices $x$ and $y$. A vertex-disjoint union of directed-cycle-trees is a **directed-cycle-forest**. When all directed cycles have length 3, we speak of a **triangle-tree**. For short, a $k$-directed-cycle-forest is a directed-cycle-forest consisting of $k$ directed cycles.

**Conjecture 2.4.11 (Hoàng and Reed [112])** *Every digraph $D$ has a $\delta^+(D)$-directed-cycle-forest.*

In the case $\delta^+(D) = 2$, Thomassen proved in [187] that every digraph with minimum out-degree 2 has two directed cycles intersecting on a vertex (i.e. contains a directed-cycle-tree with two directed cycles). Welhan [192] proved Conjecture 2.4.11 for $\delta^+(D) = 3$. The motivation of the Hoàng-Reed conjecture is that it would imply the Caccetta-Häggkvist conjecture, as the reader can easily check.

Havet, Thomassé and Yeo [111] proved Conjecture 2.4.11 for tournaments. This result does not yield a better understanding of Hoàng–Reed conjecture. However, it gives a little bit of insight into the triangle-structure of a tournament $T$, that is, the 3-uniform hypergraph on the vertex set $V(T)$ whose hyperedges are the directed 3-cycles of $T$.

Indeed, by the fact that every directed cycle in a tournament induces a strong subtournament that contains a directed 3-cycle through any given vertex, if a tournament $T$ has a $\delta^+(T)$-directed-cycle-forest, then $T$ also has a $\delta^+(T)$-triangle-forest. Observe that a $\delta^+(T)$-triangle-forest spans exactly $2\delta^+(T) + c$ vertices, where $c$ is the number of components of the triangle-forest. When $T$ is a regular tournament with out-degree $\delta^+(T)$, hence with $2\delta^+(T) + 1$ vertices, a $\delta^+(T)$-triangle-forest of $T$ is necessarily a spanning $\delta^+(T)$-triangle-tree. Havet, Thomassé and Yeo [111] established the existence of such a tree for every tournament.

**Theorem 2.4.12** ([111]) *Every tournament $T$ has a $\delta^+(T)$-triangle-tree.*

## 2.5 Disjoint Paths and Cycles

We now turn to results on linkages and weak linkages in semicomplete digraphs. The reader may wish to recall the definitions of these from Section 1.6.

### 2.5.1 Polynomial Algorithms for Linkage and Weak Linkage

> WEAK $k$-LINKAGE
> **Input:** A digraph $D = (V, A)$ and not necessarily distinct vertices $s_1, \ldots, s_k, t_1, \ldots, t_k$.
> **Question:** Does $D$ contain a weak $k$-linkage from $(s_1, \ldots, s_k)$ to $(t_1, \ldots, t_k)$?

Recall that for general digraphs the WEAK $k$-LINKAGE problem is $\mathcal{NP}$-complete already when $k = 2$ [84]. Bang-Jensen [16] solved the WEAK $k$-LINKAGE problem for semicomplete digraphs by giving a polynomial algorithm and a complete characterization of those semicomplete digraphs that do not have a weak linkage from $(s_1, s_2)$ to $(t_1, t_2)$ for given vertices $s_1, s_2, t_1, t_2$ where we may have $s_{3-i} = t_i$ for $i = 1$ or $i = 2$ but all other vertices are distinct (all the remaining cases are easy for semicomplete digraphs).

Fradkin and Seymour [85] generalized the algorithmic part of these results in two ways: from weak 2-linkage to weak $k$-linkage for any fixed integer $k$ and from semicomplete digraphs to digraphs of bounded independence number.

**Theorem 2.5.1** ([85]) *For every fixed positive integer $\alpha$ the* WEAK $k$-LINKAGE *problem is polynomially solvable for every fixed $k$, when we consider digraphs with independence number at most $\alpha$.*

A key ingredient in the proof of this theorem is the notion of the cutwidth of a digraph. Let $D = (V, A)$ be a digraph and let $\mathcal{O} = (v_1, v_2, \ldots, v_n)$ be an ordering of the vertices of $D$. We say that $\mathcal{O}$ has **cutwidth** at most $\theta$ if for all $j \in \{2, 3, \ldots, n\}$ there are at most $\theta$ arcs $uv$ with $u \in \{v_1, \ldots, v_{j-1}\}$ and $v \in \{v_j, \ldots, v_n\}$ and we say that $D$ has cutwidth at most $\theta$ if there exists an ordering $\mathcal{O}$ of $V(D)$ which has cutwidth at most $\theta$. The minimum $\theta$ such that $D$ has cutwidth at most $\theta$ is called the **cutwidth** of $D$ and is denoted by $cw(D)$.

Barbero, Paul and Pilipczuk proved that, even for semicomplete digraphs, cutwidth is not an easy parameter to determine.

**Theorem 2.5.2** ([37]) *Determining the cutwidth of a semicomplete digraph is $\mathcal{NP}$-hard.*

Single exponential FPT algorithms were obtained in [82, 152]. Pilipczuk found an approximation algorithm for the cutwidth of semicomplete digraphs.

**Theorem 2.5.3** ([152]) *There exists an $O(n^2)$ algorithm for computing an ordering $\mathcal{O}$ of an $n$-semicomplete digraph $D$ whose cutwidth is at most $O(cw(D)^2)$.*

In fact, it is shown in [152] (see also [153]) that just sorting the vertices according to their out-degrees achieves the bound above. See [153] for a discussion of which properties of a semicomplete digraph forces high cutwidth. One such example is the result that if a semicomplete digraph $D$ contains a set $S$ of $4k + 2$ vertices such that the maximum difference between the out-degrees of any pair of vertices in $S$ is at most $k$, then $cw(D) \geq k/2$ holds. Many other results on cutwidth of semicomplete digraphs can be found in the paper [81] by Fomin and Pilipczuk and in Pilipczuk's thesis [154].

For tournaments the situation is much better. Barbero, Paul and Pilipczuk proved the following.

**Theorem 2.5.4** ([37]) *One can determine the cutwidth of a tournament in polynomial time. Furthermore, if $cw(T) = p$, then $T$ contains a subtournament $T'$ whose number of vertices is linear in $p$ and such that $cw(T) = cw(T')$.*

Fradkin and Seymour also solved the WEAK $k$-LINKAGE problem for the class of directed pseudographs that one obtains from semicomplete digraphs by adding arcs and loops.

**Theorem 2.5.5 (Fradkin and Seymour** [85]) *The* WEAK $k$-LINKAGE *problem is solvable in polynomial time for every fixed $k$, when we consider directed pseudographs that are obtained from a semicomplete digraph by replacing some arcs with multiple copies of those arcs and adding any number of loops.*

We now turn to vertex-disjoint linkages.

> $k$-LINKAGE
> **Input:** A digraph $D = (V, A)$ and not necessarily distinct vertices $s_1, \ldots, s_k, t_1, \ldots, t_k$.
> **Question:** Does $D$ contain a $k$-linkage from $(s_1, \ldots, s_k)$ to $(t_1, \ldots, t_k)$?

Below we shall always assume that all the terminals to be linked (that is, $s_1, \ldots, s_k, t_1, \ldots, t_k$) are distinct. Bang-Jensen and Thomassen solved the 2-linkage problem for semicomplete digraphs.

**Theorem 2.5.6** ([34]) *The 2-linkage problem is solvable in time $O(n^5)$ for semicomplete digraphs.*

Bang-Jensen and Thomassen also proved that if $k$ is part of the input, then the $k$-linkage problem is $\mathcal{NP}$-complete already for tournaments.

Besides the trivial case $k = 1$, the value 2 remained the only $k$ for which the $k$-linkage problem was solved for semicomplete digraphs until Chudnovsky, Seymour and Scott [62] found a polynomial algorithm for the $k$-linkage problem for any fixed $k$ in semicomplete digraphs. In fact, their algorithm works for a more general class of digraphs which they call $d$-path dominant. A digraph $D = (V, A)$ is **d-path-dominant** if, for every minimal path $P$ on $d$ vertices, every vertex $v \in V - V(P)$ is adjacent to at least one vertex of $V(P)$. Thus $D$ is 1-path dominant if and only if it is semicomplete and 2-path dominant if and only if it is semicomplete multipartite. Hence this is a very general class of digraphs.

**Theorem 2.5.7** ([62]) *For all fixed $d, k$ there is a polynomial algorithm for the $k$-linkage problem in $d$-path-dominant digraphs.*

Following [62], for a given sequence $\mathbf{x} = (x_1, \ldots, x_k)$ of positive integers, we say that the digraph $D$ has an **x-linkage** from $(s_1, s_2, \ldots, s_k)$ to $(t_1, t_2, \ldots, t_k)$ if it has a collection of disjoint paths $P_1, P_2, \ldots, P_k$ such that $P_i$ is an $(s_i, t_i)$-path and has $x_i$ vertices. A sequence $\mathbf{x} = (x_1, \ldots, x_k)$ of positive integers is then a **quality** of $(D, s_1, s_2, \ldots, s_k, t_1, t_2, \ldots, t_k)$ if $D$ has an $\mathbf{x}$-linkage from $(s_1, s_2, \ldots, s_k)$ to $(t_1, t_2, \ldots, t_k)$. A quality $\mathbf{x}$ of $(D, s_1, s_2, \ldots, s_k, t_1, t_2, \ldots, t_k)$ is a **key quality** if there is no other quality $\mathbf{y} \neq \mathbf{x}$ with $y_i \leq x_i$ for all $i \in [k]$. The main result of [62] is the following.

**Theorem 2.5.8** ([62]) *For all integers $d, k \geq 1$ there exists a polynomial algorithm for the following problem: Given a $d$-path-dominant digraph $D = (V, A)$*

*and vertices $s_1, s_2, \ldots, s_k, t_1, t_2, \ldots, t_k$, compute the set of key qualities of $(D, s_1, s_2, \ldots, s_k, t_1, t_2, \ldots, t_k)$. The algorithm runs in time $O(n^{6k^2 d(k+d)+13k})$.*

**Corollary 2.5.9** ([62]) *For all integers $d, k \geq 1$ there exists a polynomial algorithm for the following problem: Given a d-path-dominant digraph $D = (V, A)$, vertices $s_1, s_2, \ldots, s_k, t_1, t_2, \ldots, t_k$ and integers $x_1, x_2, \ldots, x_k \geq 1$, decide whether $D$ contains disjoint paths $P_1, P_2, \ldots, P_k$ such that $P_i$ is an $(s_i, t_i)$-path and has at most $x_i$ vertices.*

The proof of Theorem 2.5.7 is long but the main idea is simple: as in the algorithm for $k$-linkage in acyclic digraphs (see Section 3.4) one can define an auxiliary digraph $H$ with two special vertices $s_0, t_0$ such that $H$ has an $(s_0, t_0)$-path if and only if $D$ has the desired $k$-linkage.

The following problem is open even for $k = 2$ and independence number 2.

**Problem 2.5.10** *Determine the complexity of the $k$-linkage problem for digraphs with bounded independence number.*

A special case of digraphs with independence number at most $p$ is the class of digraphs that have $p$-partition $(V_1, V_2, \ldots, V_p)$ such that $D[V_i]$ is a semicomplete digraph. For this class Chudnovsky, Scott and Seymour recently found a solution.

**Theorem 2.5.11** ([63]) *For every pair of fixed positive integers $k, p$, the $k$-linkage problem is polynomially solvable for digraphs which have a $p$-partition each part of which is semicomplete and provided we are given such a partition as part of the input.*

For an application of that result, see the discussion around Theorem 6.11.3.

### 2.5.2 Sufficient Conditions for a Tournament to be $k$-Linked

We now turn to sufficient conditions in terms of connectivity for a semicomplete digraph to be $k$-linked. Bang-Jensen determined the minimum connectivity implying 2-linkedness.

**Theorem 2.5.12** ([17]) *Every 5-strong semicomplete digraph is 2-linked. Furthermore, there exists an infinite class of 4-strong tournaments which are not 2-linked (see Figure 2.1).*

We leave it to the reader to check that one can generalize the example in Figure 2.1 to an infinite family of 4-strong semicomplete digraphs none of which is 2-linked (see also [17]).

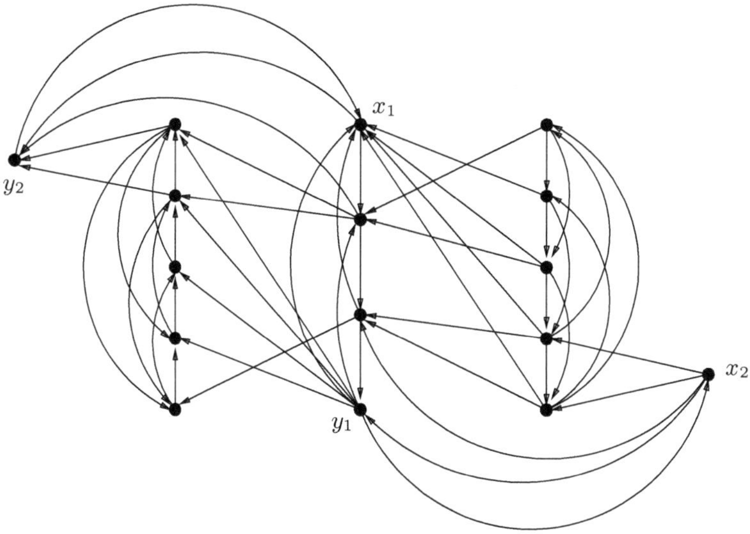


**Figure 2.1** A 4-strong non-2-linked semicomplete digraph $T$. All arcs not shown go from left to right and $x_1y_2x_1, x_2y_1x_2$ are the only 2-cycles in $T$. There is no pair of disjoint $(x_1, y_1)$-,$(x_2, y_2)$-paths in $T$. The tournament which results from $T$ by deleting the arcs $y_2x_1$ and $y_1x_2$ is also 4-strong

Thomassen [179] proved the existence of a function $f(k)$ such that every $f(k)$-strong tournament is $k$-linked. Clearly $f(1) = 1$ and by Theorem 2.5.12 we have $f(2) = 5$. Thomassen's function $f(k)$ grows exponentially in $k$. This was first improved to a polynomial in $k$ by Kühn, Lapinskas, Osthus and Patel [124] and recently Pokrovskiy showed that a linear function suffices. We will give the main details in the proof of that result below.

A key ingredient in Pokrovskiy's proof of Theorem 2.5.15 is the following interesting result which illustrates the richness of tournament structure.

**Theorem 2.5.13** ([156]) *Let $n, p$ be positive integers satisfying $p \leq n/11$. Every $n$-tournament contains two disjoint sets of vertices $\{x_1, \ldots, x_p\}$ and $\{y_1, \ldots, y_p\}$ such that for every permutation $\sigma$ of $[p]$, $T$ contains vertex-disjoint paths $P_1, \ldots, P_p$ such that $P_i$ is an $(x_i, y_{\sigma(i)})$-path.*

For later reference, we call the sets $\{x_1, \ldots, x_p\}$, $\{y_1, \ldots, y_p\}$ in the above theorem an **all-linkable pair**.

We need some more concepts which were introduced by Kühn, Lapinskas, Osthus and Patel in [124]. A set of vertices $X$ **in-dominates** (**out-dominates**) another set $Y$ in a digraph $D$ if every $y \in Y \setminus X$ has an out-neighbour (in-neighbour) in $X$. The definition implies that any set in-dominates (out-dominates) itself. An **in-dominating** (**out-dominating**) set in $D$ is then a set which in-dominates (out-dominates) $V(D)$.

Below we focus on semicomplete digraphs. Every $n$-semicomplete digraph contains an in-dominating (out-dominating) set of size $\lceil \log n \rceil$. Such a set $X$ can be constructed from the empty set by repeatedly adding a vertex $v$ of maximum in-degree (out-degree) in the current semicomplete digraph $D$ to $X$ and then deleting $v$ together with its in-neighbourhood (out-neighbourhood) from $D$.

In a semicomplete digraph a vertex $x$ may be both an in- and an out-neighbour of a vertex $v$, so we needed to adjust the definition below a bit compared to [124]. For a vertex $v$ of a semicomplete digraph we define the sets $N^{+!}(v), N^{-!}(v)$ as follows: $N^{+!}(v) = V \setminus (N^-(v) \cup \{v\})$ and $N^{-!}(v) = V \setminus (N^+(v) \cup \{v\})$.

A sequence of vertices $(v_1, v_2, \ldots, v_k)$ of a semicomplete digraph $D$ is a **partial greedy in-dominating sequence** if $v_1$ has maximum in-degree in $D$ and for each $i$, the vertex $v_i$ has maximum in-degree in $D[N^{+!}(v_1) \cap \ldots \cap N^{+!}(v_{i-1})]$. Similarly, $(v_1, v_2, \ldots, v_k)$ is a **partial greedy out-dominating sequence** if $v_1$ has maximum out-degree in $D$ and for each $i$, $v_i$ has maximum out-degree in $D[N^{-!}(v_1) \cap \ldots \cap N^{-!}(v_{i-1})]$.

Note that if at some point the set $N^{+!}(v_1) \cap \ldots \cap N^{+!}(v_{i-1})$ $(N^{-!}(v_1) \cap \ldots \cap N^{-!}(v_{i-1}))$ becomes empty, then the sequences above may have less than $k$ vertices. This will not affect the validity of the proof below.

As we saw above, if $k = \lceil \log n \rceil$ then every partial greedy in-dominating (resp. out-dominating) sequence on $k$ vertices is an in-dominating (resp. out-dominating) sequence. The following very nice property of partial greedy in- and out-dominating sequences, which was first observed by Kühn *et al.* [124] and later reformulated by Pokrovskiy [156], shows that already for much smaller values of $k$, partial greedy dominating sequences are useful (as illustrated in the proof below).

**Lemma 2.5.14** ([124, 156]) *Let $X = (v_1, v_2, \ldots, v_k)$ be a partial greedy in-dominating (resp. out-dominating) sequence in a semicomplete digraph $D$. Let $Y$ be the set of vertices which are not in-dominated (resp. out-dominated) by $X$. Then every $y \in Y$ satisfies $d^+(y) \geq 2^{k-1}|Y|$ $(d^-(y) \geq 2^{k-1}|Y|)$.*

We are now ready to state and prove the main result of [156].

**Theorem 2.5.15** ([156]) *Every $452k$-strong semicomplete digraph is $k$-linked.*

**Proof:** Pokrovskiy did not express his result for semicomplete digraphs, but his proof, which we give below, is also valid for semicomplete digraphs. Let $D$ be a $452k$-strong semicomplete digraph. In particular this means that $\delta^0(D) \geq 452k$. Let $x_1, \ldots, x_k, y_1, \ldots, y_k$ be an arbitrary collection of $2k$ distinct vertices of $D$. We shall construct disjoint paths $R_1, \ldots, R_k$ so that $R_i$ is an $(x_i, y_i)$-path for $i \in [k]$. Let $D' = D \setminus \{x_1, \ldots, x_k, y_1, \ldots, y_k\}$.

Let $I_1^-$ be a partial greedy in-dominating set on two vertices of $D'$ and for each $i = 2, \ldots, 55k$, let $I_i^-$ be a partial greedy in-dominating set of $D' \setminus (I_1^- \cup \ldots \cup I_{i-1}^-)$. Finally, let $D'' = D' \setminus (I_1^- \cup \ldots \cup I_{55k}^-)$. Denote the vertices of $I_i^-$ by $u_i^-, v_i^-$, $i \in [55k]$, where $u_i^-$ is the first vertex chosen. Note that if at some point the first vertex we choose is already an in-dominating set, then $I_i^- = \{u_i^-\}$ and we let $v_i^- = u_i^-$. Otherwise $I_i^- = \{u_i^-, v_i^-\}$ and $u_i^-$ dominates $v_i^-$. Now let $O_1^+$ be a partial greedy out-dominating set on two vertices of $D''$ and for each $i = 2, \dots, 55k$ let $O_{i-1}^+$ be a partial greedy out-dominating set on two vertices of $D'' \setminus (O_1^+ \cup \dots \cup O_{i-1}^+))$. As above we denote $O_i^+$ by $\{u_i^+, v_i^+\}$, where possibly $v_i^+ = u_i^+$ and otherwise $v_i^+$ dominates $u_i^+$.

Let $X = I_1^- \cup \dots \cup I_{55k}^- \cup O_1^+ \cup \dots \cup O_{55k}^+ \cup \{x_1, \dots, x_k, y_1, \dots, y_k\}$. By construction, $|X| \leq 222k$. Note that we may not have equality since, by the remark above, some of the sets constructed may have size one instead of two. For each $i \in [55k]$ denote by $E_i^-$ (resp. $E_i^+$) the sets of those vertices of $D-X$ that are not in-dominated by $I_i^-$ (resp. out-dominated by $O_i^+$). By Lemma 2.5.14, each vertex in $v \in E_i^-$ (resp. $w \in E_i^+$) satisfies $d^+(v) \geq 2|E_i^-|$ (resp. $d^-(w) \geq 2|E_i^+|$).

Let $V^- = \{v_1^-, \dots, v_{55k}^-\}$ and $V^+ = \{v_1^+, \dots, v_{55k}^+\}$. By Theorem 2.5.13, applied to $D[V^-]$ (resp. $D[V^+]$), we can find two sets $X^-, Y^-$ (resp. $X^+, Y^+$) both of order $5k$ in $V^-$ (resp. $V^+$) which form an all-linkable pair in $D[V^-]$ (resp. $D[V^+]$). Now relabel $I_1^-, \dots, I_{55k}$ and $O_1^+, \dots, O_{55k}^+$ so that $X^- = \{v_1^-, \dots, v_{5k}^-\}$ and $Y^+ = \{v_1^+, \dots, v_{5k}^+\}$.

By assumption, $D$ is $452k$-strong so Menger's theorem (Theorem 1.5.3) implies that $D[(V - X) \cup Y^- \cup X^+)]$ has $5k$ disjoint paths $Q_1, \dots, Q_{5k}$ which all start in $Y^-$ and end in $X^+$. As $|X| \leq 222k$, for each $i \in [k]$ there exist distinct vertices $x'_1, \dots, x'_k, y'_1, \dots, y'_k \in V \setminus X$ such that $x'_i$ is dominated by $x_i$ and $y'_i$ dominates $y_i$ for $i \in [k]$. Let $X' = X \cup \{x'_1, \dots, x'_k, y'_1, \dots, y'_k\}$.

Now we consider the vertices of $E_i^-$ and $E_i^+$, $i \in [55k]$. We saw above that each vertex in $v \in E_i^-$ (resp. $w \in E_i^+$) satisfies $d^+(v) \geq 2|E_i^-|$ (resp. $d^-(w) \geq 2|E_i^+|$). We also have $d^+(v) \geq 452k \geq 2|X'| + 4k$ (resp. $d^-(w) \geq 452k \geq 2|X'| + 4k$) so by averaging these two lower bounds we get that $d^+(v) \geq |E_i^-| + |X'| + 2k$ for every $v \in E_i^-$ and similarly $d^-(w) \geq |E_i^+| + |X'| + 2k$ for every $w \in E_i^+$. This implies that every $v \in E_i^-$ (resp. $w \in E_i^+$) has at least $2k$ out-neighbours (resp. in-neighbours) outside of $E_i^- \cup X'$ ($E_i^+ \cup X'$). For each $i \in [k]$ define $x''_i$ (resp. $y''_i$) as follows: If $x'_i \notin E_i^-$ (resp. $y'_i \notin E_i^+$), then $x'_i$ dominates (resp. $y'_i$ is dominated by) at least one vertex of $I_i^-$ ($O_i^+$) and we let $x''_i = x'_i$ (resp. $y''_i = y'_i$). Otherwise $x'_i \in E_i^-$ (resp. $y'_i \in E_i^+$) and now we let $x''_i$ (resp. $y''_i$) be an out-neighbour of $x'_i$ (resp. $y'_i$) in $D-(E_i^- \cup X')$ ($D - (E_i^+ \cup X')$). By the remark above, we can choose the $2k$ vertices (some of which may not be new) $x''_1, \dots, x''_k, y''_1, \dots, y''_k$ so that these are all distinct.

Note that $x''_i$ dominates (resp. $y''_i$ is dominated by) at least one of the vertices in $I_i^-$ (resp. $O_i^+$) for $i \in [k]$. Thus, for each $i \in [k]$ we can take the $(x''_i, v_i^-)$-path $Q_i^-$ to be either the arc $x''_i v_i^-$ or the path $x''_i u_i^- v_i^-$. Similarly, we can take the $(v_i^+, y''_i)$-path $Q_i^+$ to be either the arc $v_i^+ y''_i$ or the path $v_i^+ u_i^+ y''_i$. By construction, all the paths $Q_1^-, \dots, Q_k^-, Q_1^+, \dots, Q_k^+$ are disjoint.

At least $k$ of the paths $Q_1, \dots, Q_{5k}$ do not intersect any of the paths $Q_1^-, \dots, Q_k^-, Q_1^+, \dots, Q_k^+$ so fix such a set $Q'_1, \dots, Q'_k$ to be such paths. Since

$Q_i^-$ ends in $X^-$ and $Q'_i$ starts in $Y^-$, Theorem 2.5.13 implies that we can find disjoint paths $P_1^-, \ldots, P_k^-$ in $D[V^-]$ such that $P_i^-$ starts in $v_i^-$ and ends in the initial vertex of $Q'_i$. Similarly, we can find disjoint paths $P_1^+, \ldots, P_k^+$ in $D[V^+]$ such that $P_i^+$ starts in the terminal vertex of $Q'_i$ and ends in $v_i^+$.

Let $R_i = x_i x'_i Q_i^- P_i^- Q'_i P_i^+ y'_i y_i$ for $i \in [k]$. By the above arguments, $R_1, \ldots, R_k$ form the desired linkage. □

The value $452k$ is probably far from being best possible and the real answer could be close to $2k$. By Theorem 2.5.12, $f(k) > 2k$, at least when $k = 2$.

**Proposition 2.5.16** ([156]) *For all $n \geq 6k$, there exists a $(2k-2)$-strong $n$-tournament $T$ which is not $k$-linked.*

Note also that Theorem 2.6.15 gives a better bound when $k < 449$ and even guarantees that there is a linkage that spans all vertices of $T$.

Pokrovskiy conjectures that when the minimum semi-degree is sufficiently high, already $2k$-strong should be sufficient to guarantee a $k$-linkage for every choice of terminals.

**Conjecture 2.5.17** ([156]) *For every $k$ there exists an integer $d = d(k)$ such that every $2k$-strong tournament $T$ with $\delta^0(T) \geq d$ is $k$-linked.*

### 2.5.3 The Bermond–Thomassen Conjecture for Tournaments

We now turn to disjoint directed cycles. We only discuss the celebrated Bermond–Thomassen conjecture. For more results on disjoint cycles, see Section 2.8.

Thomassen [42, 180] proved that every digraph $D$ with $\delta^+(D) \geq 3$ has two disjoint cycles. Inspired by this, Bermond and Thomassen posed the following difficult conjecture.

**Conjecture 2.5.18 (Bermond–Thomassen [42])** *For every positive integer $k$, every digraph $D$ with $\delta^+(D) \geq 2k+1$ has $k$ disjoint cycles.*

This difficult conjecture is wide open. Lichiardopol, Pór and Sereni [134] have verified the conjecture for $k = 3$. Alon [4] was the first to prove that a linear bound suffices. He obtained the following result.

**Theorem 2.5.19** *There exists an absolute constant $C$ such that $f(k) \leq Ck$ for all $k$. In particular, $C = 64$ will do.* ◇

We now consider tournaments and semicomplete digraphs. By Moon's Theorem (2.2.7), a tournament has $k$ disjoint cycles if and only if it has $k$ disjoint 3-cycles so the following result, due to Bang-Jensen, Bessy and Thomassé, shows that Conjecture 2.5.18 holds for tournaments.

**Theorem 2.5.20** ([19]) *Every tournament $T$ with $\delta^+(T) \geq 2k - 1$ has $k$ disjoint 3-cycles.*

Bang-Jensen, Bessy and Thomassé showed how to improve this bound on the minimum out-degree for tournaments with large minimum out-degree. Roughly speaking, a tournament $T$ with $\delta^+(T) > 1.5k$ and $k$ large enough contains $k$ disjoint 3-cycles. More precisely, they proved the following.

**Theorem 2.5.21** ([19]) *For every real number $\alpha > 1.5$, there exists a constant $k_\alpha$ such that, for every $k \geq k_\alpha$, every tournament $T$ with $\delta^+(T) \geq \alpha k$ has $k$ disjoint 3-cycles.*

The constant 1.5 is the best possible as shown by the circulant tournaments $CT_{2p+1}(\{1, 2, \ldots, p\})$: when $2p + 1 \equiv 0 \mod 3$, every vertex has out-degree $p = \lfloor \frac{3}{2}k \rfloor$, where $k = \frac{2p+1}{3}$, and $CT_{2p+1}(\{1, 2, \ldots, p\})$ has a cycle factor consisting of $k$ disjoint 3-cycles covering all its vertices [19].

It is important to note that the following obvious idea does not lead to a proof of Conjecture 2.5.18 for tournaments: find a 3-cycle $C$ which is not dominated by any vertex of $V(T) \setminus V(C)$, remove $C$ and apply induction. This approach does not work because of the following.[2]

**Proposition 2.5.22** ([19]) *For infinitely many $k \geq 3$ there exists a tournament $T$ with $\delta(T) = 2k - 1$ such that every 3-cycle $C$ is dominated by at least one vertex of minimum out-degree.*

**Proof:** Consider the Paley tournament $\mathbb{P}_{11}$. It has vertex set $V(\mathbb{P}_{11}) = \{1, 2, \ldots, 11\}$ and arc set $A(\mathbb{P}_{11}) = \{(i, i + p \pmod{11}) \mid i \in [11], p \in \{1, 3, 4, 5, 9\}\}$. The possible types of 3-cycles in $T$ are $i \to i + 1 \to i + 10 \to i, i \to i + 1 \to i + 6 \to i, i \to i + 3 \to i + 6 \to i, i \to i + 3 \to i + 7 \to i$, where the indices are taken modulo 11. These are dominated by the vertices $i - 3, i - 3, i + 2, i + 2$, respectively. By substituting an arbitrary tournament for each vertex of $\mathbb{P}_{11}$, we can obtain a tournament with arbitrarily many vertices which has the property that every 3-cycle is dominated by some vertex of minimum out-degree. □

On the other hand, removing a 2-cycle from a digraph $D$ with $\delta^+(D) \geq 2k - 1$ clearly results in a new digraph $D'$ with $\delta^+(D') \geq 2(k - 1) - 1$ and hence, when trying to prove Conjecture 2.5.18, we may always assume that the digraph in question has no 2-cycles. In particular, the following is a direct consequence of Theorem 2.5.20.

**Corollary 2.5.23** *Every semicomplete digraph $D$ with $\delta^+(D) \geq 2k - 1$ contains $k$ disjoint cycles.* □

---

[2] See also [9, Section 9.1].

For regular tournaments Lichiardopol proved the following, which strengthens Theorem 2.5.20 when $r$ is larger than 20.

**Theorem 2.5.24** ([132]) *Every* $(2r-1)$*-regular tournament contains at least* $\frac{7}{6}r - \frac{7}{3}$ *disjoint cycles.*

Lichiardopol posed the following conjecture, which he proved for $k = 2$. The complete digraph on $g(k) + 1$ vertices shows that $g(k) \geq \frac{k^2+3k-2}{2}$.

**Conjecture 2.5.25** ([131]) *For every* $k \geq 2$*, there exists an integer* $g(k)$ *such that every digraph* $D$ *with* $\delta^+(D) \geq g(k)$ *has* $k$ *disjoint cycles of different lengths.*

Bensmail, Harutyunyan, Le, Li and Lichiardopol [40] confirmed the conjecture for tournaments.

**Theorem 2.5.26** ([40]) *Every tournament* $T$ *with* $\delta^+(T) \geq \frac{k^2+4k-3}{2}$ *contains* $k$ *disjoint cycles of different lengths.*

It is natural to ask for the minimum function $g_T(k)$ such that every tournament $T$ with $\delta^+(T) \geq \frac{k^2+4k-3}{2}$ contains $k$ disjoint cycles of different lengths. The regular tournaments on $n = 2g_T(k) + 1$ vertices show that $g_T(k) \geq \frac{k^2+5k-2}{4}$.

Finally, we point out that already for tournaments it is difficult to find the maximum number of disjoint cycles. The following recent result is due to Bessy, Marin and Thiebaut. The authors also showed that there is no polynomial time approximation scheme for the problem unless $\mathcal{P} = \mathcal{NP}$.

**Theorem 2.5.27** ([43]) *Finding the maximum number of disjoint 3-cycles in a tournament is* $\mathcal{NP}$*-hard.*

## 2.6 Hamiltonian Paths and Cycles

In this section we discuss results on the number of Hamiltonian paths in tournaments, Hamiltonian paths with prescribed end vertices and Hamiltonian cycles containing or avoiding a set of prescribed arcs.

### 2.6.1 Redei's Theorem

Rédei proved an interesting generalization of Theorem 2.2.4 concerning the parity of the number of Hamiltonian directed paths;

**Theorem 2.6.1 (Rédei** [158]) *Every tournament contains an odd number of Hamiltonian directed paths.*

The proof of Theorem 2.6.1 is established by means of a proof technique known as the **Inclusion-Exclusion Principle**, or the **Möbius Inversion Formula**, an inversion formula with applications throughout mathematics. We present here a simple version which suffices for our purpose. We refer the interested reader to Chapter 21 of Handbook of Combinatorics by Gessel and Stanley [96].

**Lemma 2.6.2 (Inclusion-Exclusion Principle)** *Let $Z$ be a finite set and $f : 2^Z \to \mathbb{N}$ a real-valued function defined on the subsets of $Z$. Define the function $g : 2^Z \to \mathbb{N}$ by $g(X) = \sum_{\{Y|X\subseteq Y\subseteq Z\}} f(Y)$. Then*

$$f(X) = \sum_{\{Y|X\subseteq Y\subseteq Z\}} (-1)^{|Y|-|X|} g(Y).$$

**Proof:** By the Binomial Theorem,

$$\sum_{\{Y|X\subseteq Y\subseteq W\}} (-1)^{|Y|-|X|} = \sum_{k=|X|}^{|W|} \binom{|W|-|X|}{k-|X|} (-1)^{k-|X|} = (1-1)^{|W|-|X|}$$

which is equal to 0 if $X \subset W$, and to 1 if $X = W$. Therefore,

$$\begin{aligned} f(X) &= \sum_{\{W|X\subseteq W\subseteq Z\}} f(W) \sum_{\{Y|X\subseteq Y\subseteq W\}} (-1)^{|Y|-|X|} \\ &= \sum_{\{Y|X\subseteq Y\subseteq Z\}} (-1)^{|Y|-|X|} \sum_{\{W|Y\subseteq W\subseteq Z\}} f(W) \\ &= \sum_{\{Y|X\subseteq Y\subseteq Z\}} (-1)^{|Y|-|X|} g(Y). \end{aligned}$$

□

**Proof of Theorem** 2.6.1 Let $T = (V, A)$ be a tournament with vertex set $V = \{1, 2, \ldots, n\}$ and denote by $h(T)$ the number of Hamiltonian paths in $T$. For any permutation $\sigma$ of $V$, let $A_\sigma = A \cap \{\sigma(i)\sigma(i+1) \mid 1 \le i \le n-1\}$. Then $A_\sigma$ induces a subdigraph of $T$ each of whose components is a directed path.

For any subset $X$ of $A$, let us define $f(X) = |\{\sigma \in \mathcal{S}_n \mid X = A_\sigma\}|$ and $g(X) = |\{\sigma \in \mathcal{S}_n \mid X \subseteq A_\sigma\}|$. Then $g(X) = \sum_{X\subseteq Y\subseteq A} f(Y)$, so by the Inclusion-Exclusion Principle

$$f(X) = \sum_{X\subseteq Y\subseteq A} (-1)^{|Y|-|X|} g(Y).$$

Observe that $g(Y) = r!$ if and only if the spanning subdigraph of $T$ with arc set $Y$ is the disjoint union of $r$ directed paths. Thus $g(Y)$ is odd if and only if $Y$ induces a Hamiltonian directed path of $T$. Hence, defining $h(X) =$

$|\{H \in \mathcal{H} \mid X \subseteq A(H)\}|$ with $\mathcal{H}$ the set of Hamiltonian directed paths of $T$, we obtain

$$f(X) \equiv \sum_{\{H \in \mathcal{H} | X \subseteq A(H)\}} (-1)^{n-1-|X|} \equiv h(X) \pmod{2}.$$

The theorem is true for transitive tournaments as there is a unique Hamiltonian directed path. Since any $n$-tournament may be obtained from the transitive $n$-tournament by reversing the orientation of appropriate arcs, it suffices to prove that the parity of the number of Hamiltonian directed paths $h(T)$ is unaltered by the reversal of any one arc $e$.

Let $T'$ be the tournament obtained from $T$ by reversing $e$. Then $h(T') = h(T) + f(\{e\}) - h(\{e\})$. Since $f(\{e\}) \equiv h(\{e\}) \pmod{2}$, we have $h(T') \equiv h(T) \pmod{2}$. □

### 2.6.2 Hamiltonian Connectivity

Recall that an $[x, y]$-path in a digraph $D = (V, A)$ is a directed path which either starts at $x$ and ends at $y$ or oppositely. We say that $D$ is **weakly Hamiltonian-connected** if it has a Hamiltonian $[x, y]$-path (also called an **$[x, y]$-Hamiltonian path**) for every choice of distinct vertices $x, y \in V$. Thomassen found the following characterization of weakly Hamiltonian-connected tournaments.

**Theorem 2.6.3** ([184]) *Let $D = (V, A)$ be a tournament and let $x_1, x_2$ be distinct vertices of $D$. Then $D$ has an $[x_1, x_2]$-Hamiltonian path if and only if none of the following holds.*

*(a) $D$ is not strong and either none of $x_1, x_2$ belongs to the initial strong component of $D$ or none of $x_1, x_2$ belongs to the terminal strong component of $D$.*
*(b) $D$ is strong and for $i = 1$ or $2$, $D - x_i$ is not strong and $x_{3-i}$ belongs to neither the initial nor the terminal strong component of $D - x_i$.*
*(c) $D$ is isomorphic to one of the two tournaments in Figure 2.2 (possibly after interchanging the names of $x_1$ and $x_2$).*

For semicomplete digraphs there is also a characterization which can be read out of Theorem 6.7.3 (as every semicomplete digraph is also locally semicomplete).

**Corollary 2.6.4** ([184]) *Let $D$ be a strong tournament and let $x, y, z$ be distinct vertices of $D$. Then $D$ has a Hamiltonian path connecting two of the vertices in the set $\{x, y, z\}$.* □

**Corollary 2.6.5** ([184]) *A tournament $T$ with at least three vertices is weakly Hamiltonian-connected if and only if it satisfies (1)–(3) below.*

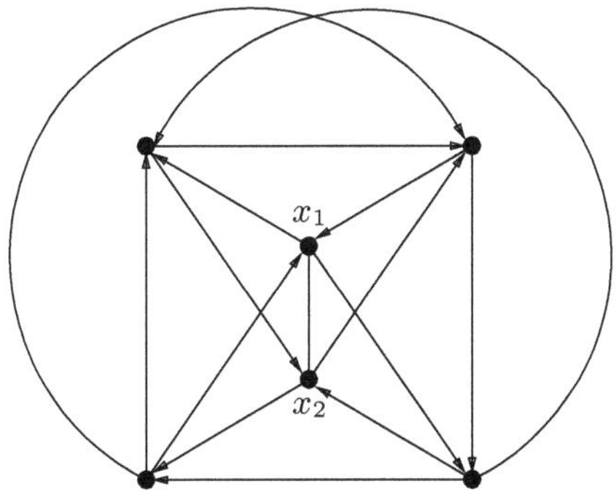


**Figure 2.2** The exceptional tournaments in Theorem 2.6.3. The edge between $x_1$ and $x_2$ can be oriented arbitrarily

*(1) $T$ is strong.*
*(2) For every vertex $v \in V(T)$, $T - v$ has at most two strong components.*
*(3) $T$ is not isomorphic to any of the two tournaments in Figure 2.2.*

We now turn to Hamiltonian paths with specified initial and terminal vertices. An **$(x, y)$-Hamiltonian path** is a Hamiltonian path from $x$ to $y$. A digraph $D = (V, A)$ is **Hamiltonian-connected** if $D$ has an $(x, y)$-Hamiltonian path for every choice of distinct vertices $x, y \in V$. The following result of Thomassen gives a sufficient condition for a semicomplete digraph to have an $(x, y)$-Hamiltonian path.

**Theorem 2.6.6 (Thomassen** [184]) *Let $D = (V, A)$ be a 2-strong semicomplete digraph with distinct vertices $x, y$. Then $D$ contains an $(x, y)$-Hamiltonian path if either (a) or (b) below is satisfied.*

*(a) $D$ contains three internally disjoint $(x, y)$-paths each of length at least 2.*
*(b) $D$ contains a vertex $z$ which is dominated by every vertex of $V \setminus \{x\}$ and $D$ contains two internally disjoint $(x, y)$-paths each of length at least 2.* □

Theorem 2.6.6 and Menger's theorem (Theorem 1.5.3) immediately imply the following result.

**Theorem 2.6.7** ([184]) *If a semicomplete digraph $D$ is 4-strong, then $D$ is Hamiltonian-connected.* □

Thomassen constructed an infinite family of 3-strongly connected tournaments with two vertices $x, y$ for which there is no $(x, y)$-Hamiltonian path [184]. Hence, from a connectivity point of view, Theorem 2.6.7 is the best possible.

Theorem 2.6.7 has several important consequences. Thomassen has shown in several papers how to use Theorem 2.6.7 to obtain results on spanning collections of paths and cycles in semicomplete digraphs. See, e.g., the papers [179, 181] and also Section 2.6.3.

The next theorem of Bang-Jensen, Manoussakis and Thomassen generalizes Theorem 2.6.6 (when $n \geq 10$). Recall that for specified distinct vertices $s, t$, an **$(s, t)$-separator** is a subset $S \subseteq V \setminus \{s, t\}$ such that $D - S$ has no $(s, t)$-path. An $(s, t)$-separator is **trivial** if either $s$ has out-degree 0 or $t$ has in-degree 0 in $D - S$.

**Theorem 2.6.8** ([32]) *Let $D$ be a 2-strong semicomplete digraph on at least ten vertices and let $x, y$ be vertices of $D$ such that $xy \notin A(D)$. Suppose that both of $D - x$ and $D - y$ are 2-strong. If all $(x, y)$-separators consisting of two vertices (if any exist) are trivial, then $D$ has an $(x, y)$-Hamiltonian path.* □

Based on Theorem 2.6.8 and several other structural results on 2-strong semicomplete digraphs Bang-Jensen, Manoussakis and Thomassen proved the following.

**Theorem 2.6.9** ([32]) *The $(x, y)$-Hamiltonian path problem is solvable in polynomial time for semicomplete digraphs.*

The algorithm uses a divide-and-conquer approach and cannot be easily modified to find a longest $(x, y)$-path in a semicomplete digraph. There also does not seem to be any simple reduction of this problem to the problem of deciding the existence of a Hamiltonian path from $x$ to $y$. Bang-Jensen and Gutin conjectured that there exists a polynomial algorithm for the problem.

**Conjecture 2.6.10** ([23]) *There exists a polynomial algorithm that, given a semicomplete digraph $D$ and two distinct vertices $x$ and $y$ of $D$, finds a longest $(x, y)$-path.*

Note that if we ask for the longest $[x, y]$-path in a tournament, then this can be answered using Theorem 2.6.3. We leave the details to the interested reader.

The following result, due to Bang-Jensen, Maddaloni and Simonsen, shows that if we generalize the $(x, y)$-Hamiltonian path problem in a natural way, we obtain an $\mathcal{NP}$-complete problem.

**Theorem 2.6.11** ([31]) *The following problem is $\mathcal{NP}$-complete: given a strong tournament $T$, a $p$-partition $(V_1, \ldots, V_p)$ of $V(T)$ and distinct vertices $x, y$ of $T$; determine whether $T$ has an $(x, y)$-path which intersects each of the sets $V_i$, $i \in [p]$.*

### 2.6.3 Hamiltonian Cycles Containing or Avoiding Prescribed Arcs

We now turn our attention to Hamiltonian cycles in digraphs with the extra condition that these cycles must either contain or avoid all arcs from a prescribed subset $A'$ of the arcs. As we shall see, problems of this type are quite difficult even for semicomplete digraphs. If we have no restriction on the size of $A'$, then we may easily formulate the Hamiltonian cycle problem for arbitrary digraphs as an avoiding problem for semicomplete digraphs. Hence the avoiding problem without any restrictions is certainly $\mathcal{NP}$-complete. Below, we study both types of problems from a connectivity as well as from a complexity point of view. Bang-Jensen and Gutin [24] showed that when the number of arcs to be avoided, respectively, contained in a Hamiltonian cycle, is some constant, then, from a complexity point of view, the avoiding version is no harder than the containing version.

Consider the following problem.

| HAMILTONIAN CYCLE THROUGH $k$-PRESCRIBED ARCS ($k$-HCA) |
|---|
| **Input:** A digraph $D$ and prescribed arcs $e_1, e_2, \ldots, e_k$ |
| **Question:** Does $D$ have a Hamiltonian cycle containing all of these arcs? |

Clearly this is $\mathcal{NP}$-complete for general digraphs, but even for semicomplete digraphs this is a difficult problem. For $k = 1$ the $k$-HCA problem is a special case of the $(x, y)$-Hamiltonian path problem and hence it is polynomial for semicomplete digraphs by Theorem 2.6.9. The problem is open for semicomplete digraphs for all other values of $k$.

Based on the evidence from Theorem 2.6.9, Bang-Jensen, Manoussakis and Thomassen posed the following conjecture.

**Conjecture 2.6.12** ([32]) *For each fixed $k$, the $k$-HCA problem is polynomial time solvable for semicomplete digraphs.*

Bang-Jensen and Thomassen proved that when $k$ is not fixed the $k$-HCA problem becomes $\mathcal{NP}$-complete even for tournaments [34]. The proof of this result in [34] contains an interesting idea which was generalized by Bang-Jensen and Gutin in [24]. Consider a digraph $D$ containing a set $W$ of $k$ vertices such that $D - W$ is semicomplete. Construct a new semicomplete digraph $D_W$ as follows. First, split every vertex $w \in W$ into two vertices $w_1, w_2$ such that all arcs entering $w$ now enter $w_1$ and all arcs leaving $w$ now leave $w_2$. Let $W_i = \{w_i | w \in W\}$, $i = 1, 2$. For each $w_1 \in W_1, w'_2 \in W_2$ add the arc $w_1 w'_2$ except if the arc $w'_2 w_1$ is already present. Add all edges between distinct vertices of $W_i$ for $i = 1, 2$ and orient these arbitrarily. Finally, add all arcs of the kind $w_1 z$ and $z w_2$, where $w \in W$ and $z \in V(D) - W$. See Figure 2.3. It is easy to show that the following proposition holds:

**Proposition 2.6.13** ([24]) *Let $W$ be a set of $k$ vertices of a digraph $D$ such that $D - W$ is a semicomplete digraph. Then $D$ has a cycle of length $c \geq k$ containing all vertices of $W$ if and only if the semicomplete digraph $D_W$ has a cycle of length $c + k$ through the arcs $\{w_1 w_2 : w \in W\}$.*

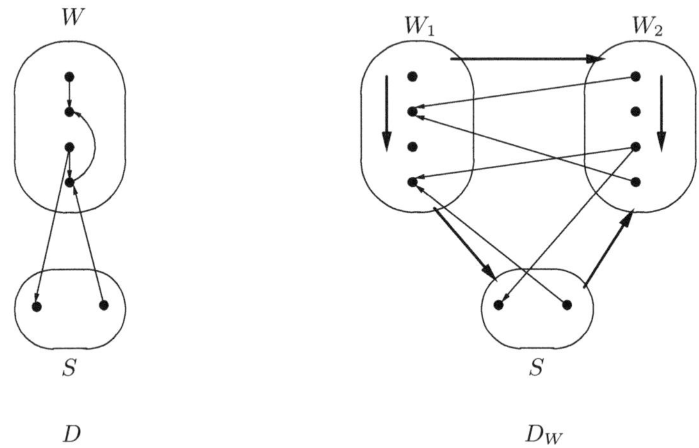


**Figure 2.3** The construction of $D_W$ from $D$ and $W$. The bold arc from $W_1$ to $W_2$ indicates that all arcs not already going from $W_2$ to $W_1$ (as copies of arcs in $D$) go in the direction shown. The four other bold arcs indicate that all possible arcs are present in the shown direction

Bang-Jensen and Gutin observed that the following is equivalent to Conjecture 2.6.12.

**Conjecture 2.6.14** ([24]) *Let $k$ be a fixed positive integer. There exists a polynomial algorithm to decide if there is a Hamiltonian cycle in a given digraph $D$ which is obtained from a semicomplete digraph by adding at most $k$ new vertices and some arcs.*

The truth of this conjecture when $k = 1$ follows from Proposition 2.6.13 and Theorem 2.6.9. Surprisingly, when $|W| = 2$ the problem already seems to be very difficult.

Using Theorem 2.6.7 Thomassen [179] proved the existence of a function $h(k)$ such that for every $h(k)$-strong semicomplete digraph $D$ and every choice of distinct vertices $x_1, y_1, ..., x_k, y_k$ $D$ has $k$-path factor $P_1 \cup P_2 \cup ... \cup P_k$ such that $P_i$ is an $(x_i, y_i)$-path for $i = 1, ..., k$. The function $h(k)$ is super-exponential. Recently Kim, Kühn and Osthus improved this to a polynomial.

**Theorem 2.6.15** ([121]) *Let $k$ be a positive integer, and let $T$ be a $(k^2+3k)$-strong tournament. For any set $\{x_1, y_1, ..., x_k, y_k\}$ of distinct vertices, $T$ has a $k$-path factor $P_1 \cup P_2 \cup ... \cup P_k$ such that $P_i$ is an $(x_i, y_i)$-path for $i = 1, ..., k$.*

Note that Theorem 2.6.15 gives a better bound than Theorem 2.5.15 when $k < 449$ and even guarantees a $k$-linkage that spans all vertices of the tournament.

**Corollary 2.6.16** ([121]) *If $a_1, ..., a_k$ are arcs with no common head or tail in a $(k^2+3k)$-strong tournament $T$, then $T$ has a Hamiltonian cycle containing $a_1, ..., a_k$ in that cyclic order.*

Pokrovskiy [155] showed that the bound in Theorem 2.6.15 can be replaced by a linear function, thus answering a question of Thomassen from [179]. The constant $C$ below is very large, which is why we also stated Theorem 2.6.15, which gives a better bound as long as $k$ is not very large.

**Theorem 2.6.17** ([155]) *There exists a constant $C$ such that for every $Ck$-strong tournament $T$ and every set $\{x_1, y_1, ..., x_k, y_k\}$ of distinct vertices, $T$ has a $k$-path-factor $P_1, P_2, \ldots, P_k$ such that $P_i$ is an $(x_i, y_i)$-path for $i = 1, ..., k$.*

By Theorem 2.3.2, similar results hold for semicomplete digraphs.

Recall that a set of arcs is **independent** if no two of the arcs share a vertex. Combining the ideas of avoiding and containing, Thomassen proved the following (below we have replaced his exponential function by the one from Theorem 2.6.15).

**Theorem 2.6.18** ([179]) *Let $T$ be a $(k^2+3k)$-strong tournament. For any set $A_1$ of at most $k$ arcs in $T$ and for any set $A_2$ of at most $k$ independent arcs of $T \setminus A_1$, the digraph $T \setminus A_1$ has a Hamiltonian cycle containing all arcs of $A_2$.*

Even though tournaments have a lot of structure and the Hamiltonian cycle problem is almost trivial, the situation changes dramatically if we delete just a few arcs from a tournament. For some tournaments, such as the almost transitive tournaments, the answer is that even one missing arc may destroy all Hamiltonian cycles. If there is exactly one arc entering (resp. leaving) a vertex, then deleting that arc clearly suffices to destroy all Hamiltonian cycles. However, it is not just a simple degree question since, for every $p$, there exists an infinite set $\mathcal{S}$ of strong tournaments in which $\delta^0(T) \geq p$ for every $T \in \mathcal{S}$ and yet there is some arc of $T$ which is on every Hamiltonian cycle of $T$ ([22, Exercise 7.19]). It follows from Theorem 2.6.19 below that all such tournaments are strong but not 2-strong.

Obviously, if a $k$-strong tournament $T$ has $\delta^0(T) = k$ (this is the smallest possible by the connectivity assumption), we may again kill all Hamiltonian cycles by removing just $k$ arcs. Thomassen [181] conjectured that in a $k$-strong tournament, $k$ is the minimum number of arcs one can delete in order to destroy all Hamiltonian cycles. The next theorem due to Fraisse and Thomassen answers this in the affirmative.

**Theorem 2.6.19 (Fraisse and Thomassen** [87]**)** *For every $k$-strong tournament $T$ and every set $A' \subset A(T)$ such that $|A'| \leq k-1$, there is a Hamiltonian cycle $C$ in $T \setminus A'$.*

The proof is long and non-trivial; in particular it uses Theorem 2.6.7. Below we describe a stronger result due to Bang-Jensen, Gutin and Yeo [25].

**Theorem 2.6.20** ([25]) *Let $T = (V, A)$ be a $k$-strong $n$-tournament, and let $X_1, X_2, \ldots, X_p$ ($p \geq 1$) be a partition of $V$ such that $1 \leq |X_1| \leq |X_2| \leq \ldots \leq |X_p|$. Let $D$ be the digraph obtained from $T$ by deleting all arcs which have both head and tail in the same $X_i$ (i.e., $D = T \setminus \bigcup_{i=1}^{p} A(T[X_i])$). If $|X_p| \leq n/2$ and $k \geq |X_p| + \sum_{i=1}^{p-1} \lfloor |X_i|/2 \rfloor$, then $D$ is Hamiltonian. In other words, $T$ has a Hamiltonian cycle which avoids all arcs with both head and tail in some $X_i$. Furthermore, the bound on $k$ is sharp.*

The proof of Theorem 2.6.20 in [25] uses results on irreducible cycle factors in multipartite tournaments, in particular Yeo's irreducible cycle factor theorem (Theorem 7.3.2).

The main idea of the proof is as follows: By construction (deleting all arcs inside several disjoint sets) $D$ is a multipartite tournament. The goal is to apply Theorem 7.3.2 to $D$. Hence we need to establish that $D$ is strong and has a cycle factor. Both of these are true and the latter can be proved using Hoffman's circulation theorem. Now we can apply Theorem 7.3.2 to prove that every irreducible cycle factor in $D$ is a Hamiltonian cycle. This last step is non-trivial.

**Problem 2.6.21** ([25]) *Which sets $B$ of edges of the complete graph $K_n$ have the property that every $k$-strong orientation of $K_n$ induces a Hamiltonian digraph on $K_n - B$?*

The Fraisse–Thomassen theorem says that this is the case whenever $B$ contains at most $k - 1$ edges. Theorem 2.6.20 says that a union of disjoint cliques of sizes $r_1, \ldots, r_p$ has the property whenever $\sum_{i=1}^{l} \lfloor r_i/2 \rfloor + \max_{1 \leq i \leq l} \{\lceil r_i/2 \rceil\} \leq k$. As shown in [25] this is the best possible result for unions of cliques.

See [22, pages 293–294] for a proof that Theorem 2.6.20 implies Theorem 2.6.19. Note that if $A'$ induces a tree and possibly some disjoint edges in $UG(T)$, then Theorem 2.6.20 is no stronger than Theorem 2.6.19. In all other cases Theorem 2.6.20 provides a stronger bound.

How easy is it to decide, for a given semicomplete digraph $D = (V, A)$ and a subset $A' \subseteq A$, whether $D$ has a Hamiltonian cycle $C$ which avoids all arcs of $A'$? As we mentioned earlier, this problem is $\mathcal{NP}$-complete if we pose no restriction on the arcs in $A'$. In the case when $A'$ is precisely the set of those arcs that lie inside the sets of some partition $X_1, X_2, \ldots, X_r$ of $V$, then the existence of $C$ can be decided in polynomial time. This follows from the fact that $D \setminus A'$ is a semicomplete multipartite digraph and, by Theorem 7.6.1, the Hamiltonian cycle problem is polynomial for semicomplete multipartite digraphs. The same argument also covers the case when $k = 1$ in the conjecture below.

**Conjecture 2.6.22** ([22]) *For every fixed positive integer $k$, there exists a polynomial algorithm which, for a given semicomplete digraph $D$ and a subset $A' \subseteq A(D)$ such that $|A'| = k$, decides whether $D$ has a Hamiltonian cycle that avoids all arcs in $A'$.*

At first glance, cycles that avoid certain arcs seem to have very little to do with cycles that contain certain specified arcs. Hence, somewhat surprisingly, if Conjecture 2.6.12 is true, then so is Conjecture 2.6.22 as observed by Thomassen[3]: Suppose that Conjecture 2.6.12 is true. Then it follows from the discussion above on Hamiltonian cycles containing prescribed arcs that Conjecture 2.6.14 also holds. Hence, for fixed $k$, there is a polynomial algorithm $\mathcal{A}_k$ which, given a digraph $D$ and a subset $W \subseteq V(D)$ for which $D-W$ is semicomplete and $|W| \leq k$, decides whether or not $D$ has a Hamiltonian cycle. Let $k$ be fixed and $D$ be a semicomplete digraph and let $A'$, $|A'| \leq k$, be a prescribed set of arcs in $D$. Let $W$ be the set of all vertices such that at least one arc of $A'$ has head or tail in $W$. Then $|W| \leq 2|A'|$ and $D$ has a Hamiltonian cycle avoiding all arcs in $A'$ if and only if the digraph $D \setminus A'$ has a Hamiltonian cycle. By the above remark, we can test this using the polynomial algorithm $\mathcal{A}_r$, where $r = |W|$.

## 2.7 Oriented Subgraphs of Tournaments

A digraph is **$n$-unavoidable** if it is contained in every $n$-tournament and simply **unavoidable** if there exists some $n$ such that it is $n$-unavoidable. Redei's Theorem states that the directed $n$-path is $n$-unavoidable. A natural question is which digraphs are unavoidable? Because the transitive tournaments are acyclic, every digraph containing a directed cycle is not unavoidable. On the other hand, we now prove that every acyclic digraph is unavoidable.

**Theorem 2.7.1 (Folklore)** *A digraph is unavoidable if and only if it is acyclic. Moreover, every acyclic $n$-digraph is $2^{n-1}$-unavoidable.*

**Proof:** We already mentioned that every non-acyclic digraph is not unavoidable. Reciprocally, we need to prove that every acyclic digraph is unavoidable, and more precisely that every acyclic $n$-digraph $2^{n-1}$-unavoidable. As every acyclic $n$-digraph is a subdigraph of the transitive $n$-tournament $TT_n$, it suffices to prove the result for $TT_n$. This follows directly from Proposition 2.2.3. □

Now, for each acyclic (and hence unavoidable) digraph $D$, it is natural to ask for the minimum integer unvd$(D)$ such that $D$ is unvd$(D)$-unavoidable. Since an acyclic $n$-digraph is contained in $TT_n$ and so unvd$(D) \leq$ unvd$(TT_n)$,

[3] private communication, August 1999.

the first interesting case is that of transitive tournaments, which also yields a good estimate of unvd($D$) for digraphs $D$ with many arcs. The unavoidability of transitive tournaments is detailed in Subsection 2.7.1. We then study the unavoidability of acyclic digraphs with few arcs, namely oriented paths (Subsection 2.7.2), oriented cycles (Subsection 2.7.3), and oriented trees (Subsection 2.7.4).

### 2.7.1 Transitive Subtournaments

Erdős and Moser [76] ask for the value of $\mathrm{unvd}(TT_n)$.

**Problem 2.7.2** ([76]) *What is* $\mathrm{unvd}(TT_n)$*?*

Theorem 2.7.1 yields $\mathrm{unvd}(TT_n) \leq 2^{n-1}$. This upper bound is almost tight, as shown by the following result due to Erdős and Moser [76].

**Theorem 2.7.3** ([76]) *There exists a tournament on* $2^{(n-1)/2}$ *vertices which contains no* $TT_n$.

**Proof:** The proof is probabilistic and uses the First Moment Method. (For more on the Probabilistic Method and in particular the First Moment Method, we refer the reader to the book of Alon and Spencer [8].) Set $N = 2^{(n-1)/2}$ and consider $T = RT_N$, a random tournament on $N$ vertices.

For an ordered $n$-tuple $(v_1, v_2, \dots, v_n)$ the probability that $T\langle\{v_1, \dots, v_n\}\rangle$ is a transitive tournament with Hamiltonian directed path $v_1v_2\dots v_n$ is $\left(\frac{1}{2}\right)^{\binom{n}{2}}$. Hence the expected number of transitive $n$-subtournaments is

$$\frac{N!}{(N-n)!}\left(\frac{1}{2}\right)^{\binom{n}{2}} < N^n\left(\frac{1}{2}\right)^{\binom{n}{2}} \leq 1$$

because $N \leq 2^{(n-1)/2}$. Hence by the First Moment Principle, there exists an $N$-tournament with less than 1 (i.e. no) $n$-subtournament. □

In the same way, for every acyclic $n$-digraph $D$ with $m$ arcs one can show that $\mathrm{unvd}(D) > 2^{\frac{m}{n}}$. This gives a meaningful lower bound for digraphs with sufficiently many arcs, namely at least $n \log n$ arcs.

Clearly, $\mathrm{unvd}(TT_1) = 1$, $\mathrm{unvd}(TT_2) = 2$ and $\mathrm{unvd}(TT_3) = 4$. Also $\mathrm{unvd}(TT_4) = 8$ because the Paley tournament $\mathbb{P}_7$ contains no $TT_4$. Moreover, Reid and Parker [162] showed that $\mathrm{unvd}(TT_5) = 14$ and $\mathrm{unvd}(TT_6) = 28$ and Sanchez-Flores [167] showed $\mathrm{unvd}(TT_7) \leq 54$. A similar induction as in the proof of Theorem 2.7.1 yields that $\mathrm{unvd}(TT_n) \leq 54 \times 2^{n-7}$ if $n \geq 7$.

In addition, for $1 \leq n \leq 6$ it has been shown [162, 167] that there is a unique tournament of order $\mathrm{unvd}(TT_n) - 1$ that contains no $TT_n$. This leads us to the following conjecture:

**Conjecture 2.7.4 (Havet, 2008)** *For every $n$, there is a unique tournament on* $\mathrm{unvd}(TT_n) - 1$ *vertices that contains no* $TT_n$.

### 2.7.2 Oriented Paths in Tournaments

An **oriented path** is an orientation $P$ of an undirected path $x_1 \cdots x_n$. We say that $x_1$ is the **origin** of $P$ and $x_n$ is the **terminus** of $P$. If $x_1 \to x_2$, $P$ is an **out-path**, otherwise $P$ is an **in-path**. The **directed out-path** of order $n$ is the orientation of $x_1 \cdots x_n$ in which $x_i \to x_{i+1}$ for all i, $1 \leq i < n$; the dual notion is **directed in-path**. The **length** of a path is its number of arcs. We denote by $^*P$ (resp. $P^*$) the oriented path obtained from $P$ by removing its origin (resp. terminus). The **blocks** of $P$ are the maximal directed subpaths of $P$. We enumerate the blocks of $P$ from the origin to the terminus. The first block of $P$ is denoted by $B_1(P)$ and its length by $b_1(P)$. Likewise, the $i$th block of $P$ is denoted by $B_i(P)$ and its length by $b_i(P)$. The path $P$ is totally described by the signed sequence $sgn(P)(b_1(P), b_2(P), \cdots, b_k(P))$ where $k$ is the number of blocks of $P$ and $sgn(P) = +$ if $P$ is an out-path and $sgn(P) = -$ if $P$ is an in-path. An **antidirected** path is an oriented path in which all blocks have length 1.

Let $X$ be a set of vertices of $T$. The **out-section** generated by $X$ in $T$ is the set of vertices $y$ to which there exists a directed out-path from some $x \in X$; we denote this set by $S^+(X)$ (note that $X \subseteq S^+(X)$ since we allow paths of length zero). We abbreviate $S^+(\{x\})$ to $S^+(x)$ and $S^+(\{x, y\})$ to $S^+(x, y)$. The dual notion, the **in-section**, is denoted by $S^-(X)$. We also write $s^+(X)$ (resp. $s^-(X)$) for the number of vertices of $S^+(X)$ (resp. $S^-(X)$). If $X \subseteq Y \subseteq V$, we write $S^+_Y(X)$ instead of $S^+_{T[Y]}(X)$. An **out-generator** of $T$ is a vertex $x \in T$ such that $S^+(x) = V(T)$, the dual notion is an **in-generator**.

Redei's Theorem states that the directed $n$-out-path is $n$-unavoidable. It is then a natural question to ask whether the other oriented $n$-paths are also $n$-unavoidable. Grünbaum [98] proved that this is the case for antidirected paths except for three exceptions, the paths $\pm(1, 1)$ which is not contained in the directed 3-cycle $\vec{C}_3$, $\pm(1, 1, 1, 1)$ which is not contained in the regular 5-tournament $R_5$, and $\pm(1, 1, 1, 1)$ which is not contained in the Paley 7-tournament $\mathbb{P}_7$. A year later, in 1972, Rosenfeld [165] gave an easier proof of a stronger result: in a tournament on at least 9 vertices, each vertex is the origin of an antidirected Hamiltonian path. He also made the following conjecture: there is an integer $N > 7$ such that every tournament on $n$ vertices, $n \geq N$, contains any orientation of the Hamiltonian path. The condition $N > 7$ results from Grünbaum's counterexamples. Several papers gave partial answers to this conjecture: for paths with two blocks (Alspach and Rosenfeld [13], Straight [174]), and for paths having the $i$th block of length at least $i+1$ (Alspach and Rosenfeld [13]); interestingly Forcade [83] proved in a way similar to the proof of Theorem 2.6.1 that there is always an odd number of Hamiltonian paths of any type in tournaments with $2^n$ vertices. Rosenfeld's conjecture was verified by Thomason, who proved in [176] that $N$ exists and is less than $2^{128}$. While he did not make any attempt to sharpen this bound, he wrote that $N = 8$ should be the right value. The problem was finally closed by Havet and Thomassé [110] who proved the following theorem.

**Theorem 2.7.5 (Havet and Thomassé** [110]) *Apart from Grünbaum's exceptions, every $n$-tournament contains every oriented $n$-path.*

The proof of Havet and Thomassé relies on sufficient conditions for vertices to be an origin of a given oriented path in a tournament. An easy condition for a vertex $x$ to be an origin of an oriented out-path $P$ is that $s^+(x) \geq b_1(P)+1$. It is sometimes sufficient: for example, this condition says that an origin of a Hamiltonian directed out-path in a tournament must be a out-generator, and one can easily show that it is also sufficient.

**Proposition 2.7.6** *In a tournament $T$, a vertex $v$ is an origin of a Hamiltonian directed out-path in $T$ if and only if $v$ is an out-generator of $T$.*

In contrast, for other Hamiltonian oriented paths, the condition $s^+(x) \geq b_1(P) + 1$ is not sufficient to guarantee $x$ being an origin of $P$. However, Havet and Thomassé [110] proved that among two distinct vertices $x, y$ such that $s^+(x, y) \geq b_1(P)+1$, there must be an origin of $P$ except in some exceptional cases that they completely characterized. The proof of this result is by induction and is tedious because of a long case analysis due to the exceptional cases (51 small ones plus 17 infinite families). However, the general idea of the proof is the same as that of the following weaker theorem about oriented $n$-paths in $(n+1)$-tournaments.

**Theorem 2.7.7** ([110]) *Let $T$ be a tournament of order $n+1$, $P$ an out-path of order $n$ and $x, y$ two distinct vertices of $T$. If $s^+(x, y) \geq b_1(P)+1$, then $x$ or $y$ is an origin of $P$ in $T$.*

**Proof:** We prove the statement and its directional dual (where $P$ is an in-path and $s^-(x,y) \geq b_1(P)+1$) by induction on $n$, the result holding trivially for $n = 1$. Let $x$ and $y$ be two vertices of a tournament $T = (V, A)$ such that $x \rightarrow y$ and $s^+(x, y) \geq b_1(P) + 1$. We distinguish two cases:

Case 1 : $b_1(P) \geq 2$. If $d^+(x) \geq 2$, let $z \in N^+(x)$ be an out-generator of $T\langle S^+(x)\backslash\{x\}\rangle$ and let $t \in N^+(x)$, $t \neq z$. By definition of $z$, $s^+_{V\backslash\{x\}}(t, z) = s^+(x) - 1 > b_1(^*P)$. Since $^*P$ is an out-path, by the induction hypothesis, either $t$ or $z$ is an origin of $^*P$ in $T - x$. Thus $x$ is an origin of $P$ in $T$.

So we may assume that $y$ is the unique out-neighbour of $x$. Let $z$ be an out-generator of $T\langle N^+(y)\rangle$ ($z$ exists since $s^+(x, y) \geq 3$). Then $z \rightarrow x$ and $z$ is an out-generator of $T\langle S^+(x, y)\backslash\{y\}\rangle$. It follows that $s^+_{V\backslash\{y\}}(x, z) = s^+(x, y) - 1$, so by the induction hypothesis, either $x$ or $z$ is an origin of $^*P$ in $T - y$. Since $d^+_{V\backslash\{y\}}(x) = 0$, this origin is certainly $z$. We conclude that $y$ is an origin of $P$ in $T$.

Case 2 : $b_1(P) = 1$. Assume first that $d^+(x) \geq 2$. We denote by $X$ the set $S^-_{V\backslash\{x\}}(N^+(x))$. Consider the partition $(X, Y, \{x\})$ of $V$ where $Y = V\backslash(X \cup \{x\})$. We have $Y \rightarrow x$, $X \rightarrow Y$ and $y \in X$. If $|X| \geq b_2(P) + 1$, then $x$ is an origin of $P$ in $T$; indeed, let $z \in N^+(x)$ be an in-generator of $T\langle X\rangle$ and let $u \in N^+(x)$ $u \neq z$. By the induction hypothesis, $z$ or $u$ is an origin of $^*P$ in $T - x$. Hence $x$ is an origin of $P$ in $T$. If $|X| \leq b_2(P)$, we have $|Y| > 1$ since $b_2(P) \leq n - 2$ and $|X| + |Y| = n$. Let $w \in Y$ be an in-generator of $T\langle Y\rangle$. Notice that since $d^+(x) > 1$, $S^-_{V\backslash\{y\}}(w) = V\backslash\{y\}$. Let $u \in Y - w$. By the induction hypothesis, $u$ or $w$ is an origin of $^*P$ in $T - y$, consequently $y$ is an origin of $P$ in $T$.

Now assume that $d^+(x) = 1$, thus $N^+(x) = \{y\}$. If $d^+(y) < 2$, then $N^-_{V\backslash\{x\}}(y)$ has at least $n-2$ vertices. By the induction hypothesis, one can find $^{**}P$ in $T\langle N^-_{V\backslash\{x\}}(y)\rangle$, thus $x$ is an origin of $P$ in $T$. If $d^+(y) \geq 2$, denote $S^-_{V\backslash\{y\}}(N^+(y))$ by $Y$ and consider the partition $(X, Y, \{x\}, \{y\})$ of $V$ with $X = V\backslash(Y \cup \{x, y\})$. By definition, $X \rightarrow \{x, y\}$, $Y \rightarrow X \cup \{x\}$. If $|Y| \geq b_2(P) + 1$, then $y$ is an origin of $P$ by the previous argument. If $|Y| \leq b_2(P)$, then $b_2(P) \geq d^+(y) \geq 2$. If $|X| \geq 2$, let $z \in X$ be an in-generator of $T - \{x, y\}$ and let $u \in X$ $u \neq z$. Since $b_2(P) \geq 2$, $^{**}P$ is an in-path and by the induction hypothesis, $z$ or $u$ is an origin of $^{**}P$ in $T - \{x, y\}$. Thus $x$, (via $y$) is an origin of $P$ in $T$. Finally, if $|X| = 1$ then $|Y| = n - 2$ and since $n - 2 \geq b_2(P) \geq |Y|$ we have $b_2(P) = n - 2$. This means that $^*P$ is a directed in-path. Since $y$ is an in-generator of $T - x$, $x$ is an origin of $P$ in $T$. □

The following result, due to Thomason, is an easy consequence of Theorem 2.7.7.

**Corollary 2.7.8** ([176]) *Every tournament $T$ of order $n + 1$ contains each oriented path $P$ of order $n$. Moreover, any subset of $b_1(P)+1$ vertices contains an origin of $P$. In particular, at least two vertices of $T$ are origins of $P$ .*

### 2.7.3 Oriented Cycles in Tournaments

As we did for paths, we can seek arbitrary orientations of cycles, i.e. **oriented cycles**. Observe that by Camion's Theorem (2.2.6) a tournament has a directed Hamiltonian cycle if and only if it is strong. A natural equation is then whether every tournament contains all Hamiltonian non-directed cycles. The existence of Grünbaum's exceptions implies the existence of tournaments that do not contain certain Hamiltonian oriented cycles. Indeed $\overrightarrow{C}_3$, $R_5$ and $\mathbb{P}_7$ do not contain the cycle obtained from a Hamiltonian antidirected path by adding an arc between its terminus and its origin. Moreover, the tournaments of order $n$ that have a subtournament on $n - 1$ vertices isomorphic to one of $\overrightarrow{C}_3$, $R_5$ and $\mathbb{P}_7$ do not contain a Hamiltonian antidirected cycle. (Similarly to paths, an **antidirected cycle** is a cycle in which every block has length 1.)

However, as for oriented paths, Rosenfeld [164] conjectured that there is an integer $N > 8$ such that every tournament of order $n \geq N$ contains every non-directed cycle of order $n$. This was settled by Thomason [176] for tournaments of order $n \geq 2^{128}$. While Thomason made no attempt to sharpen this bound, he indicated that it should be true for tournaments of order at least 9.

**Conjecture 2.7.9 (Rosenfeld–Thomason)** Every tournament of order $n \geq 9$ contains every non-directed cycle of order $n$.

Havet [107] improved Thomason's result by showing that this conjecture is true for $n \geq 68$.

**Theorem 2.7.10** ([107]) *Every tournament of order $n \geq 68$ contains every non-directed cycle of order $n$.*

The proof is based on complementary lemmas: Some establish the existence of an oriented cycle in every tournament whose strong connectivity is small compared to the length of its longer block; others show the existence of an oriented cycle in every tournament whose strong connectivity is large compared to the lengths of all blocks. In particular, Conjecture 2.7.9 is true if the tournament is either not 2-strong or 8-strong [107]. The conjecture is also true if the tournament is either 5-strong and of order at least 43 or 4-strong and of order at least 65.

Better results are also known for particular types of directed cycles. Conjecture 2.7.9 has been proved for cycles with a block of length $n-1$ by Grünbaum [98], for antidirected cycles by Thomassen [177] ($n \geq 50$), Rosenfeld [164] ($n \geq 28$) and Petrović [151] ($n \geq 16$), and for cycles with just two blocks by Benhocine and Wojda [39].

### 2.7.4 Trees in Tournaments

As we did for paths and cycles, we can seek an arbitrary orientation of trees, i.e. **oriented trees**. Observe that an oriented tree of order $k$ is an acyclic digraph and thus it is $2^{k-1}$-unavoidable by Theorem 2.7.1. However this bound $2^{k-1}$ is far from tight as an oriented tree has very few arcs compared to the transitive tournament of the same order.

**Conjecture 2.7.11 (Sumner, 1972)** Every oriented tree with $k > 1$ vertices is $(2k-2)$-unavoidable.

If true, this conjecture would be tight since the **out-star** $S_k^+$, which is the out-tree of order $k$ with a root dominating $k-1$ leaves, is not contained in any regular tournament of order $2k-3$, because all vertices of such a tournament have out-degree $k-2$.

The first linear bound was given by Häggkvist and Thomason [104]. Havet and Thomassé [109] proved that the conjecture holds for out-trees (and thus also for in-trees).

**Theorem 2.7.12** ([109]) *Every tournament of order $2k-2$ contains every out-tree of order $k>1$.*

**Proof:** Let $(v_1, v_2, \ldots, v_{2k-2})$ be a median order of a tournament $T$ on $2k-2$ vertices, and let $A$ be an out-tree on $k$ vertices. Consider the intervals $(v_1, v_2, \ldots, v_i)$, $1 \leq i \leq 2k-2$. We show, by induction on $k$, that there is a copy of $A$ in $T$ whose vertex set includes at least half the vertices of any such interval.

This is clearly true for $k=2$. Suppose, then, that $k \geq 3$. Delete a leaf $y$ of $A$ to obtain an out-tree $A'$ on $k-1$ vertices, and set $T' := T - \{v_{2k-3}, v_{2k-2}\}$. By (M1), $(v_1, v_2, \ldots, v_{2k-4})$ is a median order of the tournament $T'$, so there is a copy of $A'$ in $T'$ whose vertex set includes at least half the vertices of any interval $v_1, v_2, \ldots, v_i$, $1 \leq i \leq 2k-4$. Let $x$ be the predecessor of $y$ in $A$. Suppose that $x$ is located at vertex $v_i$ of $T'$. In $T$, by (M2), $v_i$ dominates at least half of the vertices $v_{i+1}, v_{i+2}, \ldots, v_{2k-2}$, thus at least $k-1-i/2$ of these vertices. On the other hand, $A'$ includes at least $(i-1)/2$ of the vertices $v_1, v_2, \ldots, v_{i-1}$, thus at most $k-1-(i+1)/2$ of the vertices $v_{i+1}, v_{i+2}, \ldots, v_{2k-2}$. It follows that, in $T$, there is an out-neighbour $v_j$ of $v_i$, where $i+1 \leq j \leq 2k-2$, which is not in $A'$. Locating $y$ at $v_j$, and adding the vertex $y$ and the arc $xy$ to $A'$, we now have a copy of $A$ in $T$. It is readily checked that this copy of $A$ satisfies the required additional property. □

The same method can be easily adapted to prove that every oriented tree of order $k$ is $(4k-4)$-unavoidable. At each step of the induction, we add two vertices to the right and two vertices to the left of the ordering and we insist that at each step for each vertex $v$ at least half of the vertices to the right of $v$ are unused and half of the vertices to the left are unused. El Sahili [73] used it in a clever way to show that every oriented tree of order $k$ is $(3k-3)$-unavoidable. Recently, Kühn, Mycroft and Osthus [125] proved that Sumner's conjecture is true for all sufficiently large $k$.

**Theorem 2.7.13 (Kühn, Mycroft and Osthus** [125]) *There exists a $k_0$ such that every oriented tree with $k \geq k_0$ vertices is $(2k-2)$-unavoidable.*

Their complicated proof makes use of the directed version of Szemeredi's Regularity Lemma.

As we mentioned above, Sumner's conjecture is tight for out-stars. On the other hand, it is not tight for paths which are trees with two leaves. Consequently, Havet and Thomassé made the following conjecture, which directly implies Sumner's conjecture because a tree of order $n$ has at most $n-1$ leaves.

**Conjecture 2.7.14 (Havet and Thomassé, 1996)** *If $A$ is an oriented tree with $n$ vertices and $k$ leaves, then it is $(n+k-1)$-unavoidable.*

If true this conjecture would be tight because of out-stars, but also because of Grünbaum's exceptions. Conjecture 2.7.14 holds for $k=2$, as trees with two leaves are paths, Ceroi and Havet [57] proved it for $k=3$, and it easily holds for $k=n-1$, that is, when the tree is an oriented star. Havet [106] proved that it holds for a large class of oriented trees. Häggkvist and Thomason [104] proved that there is a function $g$ such that every tree with $n$ vertices and $k$ leaves is $(n+g(k))$-unavoidable.

Instead of looking for a fixed oriented tree in tournaments, one may also seek an oriented tree having certain properties. In this vein, Lu [137] proved that there exists an out-branching of height 2, in which all nodes except the root have small out-degree.

**Theorem 2.7.15** ([137]) *Every tournament $T$ is has an out-branching of height 2 and whose vertices on level 1 have out-degree at most 2.*

**Proof:** The proof we give here is due to Bondy [49]. Let $x$ be a vertex of maximal out-degree. By Theorem 2.2.12, $x$ is a king, so $(\{x\}, N^+(x), N^{++}(x))$ is a partition of $V(T)$. Note that, by the choice of $x$ and since in every $k$-tournament there is a vertex with out-degree at least $\lfloor k/2 \rfloor$, for every $A \subseteq N^{++}(x)$ we have $2|A^- \cap N^+(x)| \geq |A|$. By Hall's theorem, one can cover $N^{++}(x)$ by two directed matchings from $N^+(x)$ to $N^{++}(x)$. This gives the desired out-branching. □

### 2.7.5 Largest $n$-Unavoidable Digraphs

Let $\mathrm{lu}(n)$ be the largest $m$ such that there is an $n$-unavoidable digraph with $m$ arcs. Linial, Saks and Sós [135] showed the following.

**Theorem 2.7.16** ([135]) *There exist positive constants $c_1$ and $c_2$ such that for all positive integers $n$, $n\log n - c_1 n \log\log n \leq \mathrm{lu}(n) \leq n \log n - c_2 n$.*

The upper bound comes from a simple counting argument working over all labelled $n$-tournaments. The lower bound follows from several propositions that allow an inductive construction of an $n$-unavoidable, weakly connected spanning digraph with $n\log n - c_1 n \log\log n$ arcs.

### 2.7.6 Generalization to $k$-Chromatic Digraphs

A tournament is an orientation of a complete graph, and the complete graph $K_k$ is the easiest example of a graph with chromatic number $k$. Recall that the **chromatic number** of a digraph $D$, denoted by $\chi(D)$, is the chromatic number of its underlying undirected graph. A digraph is **$k$-chromatic** if its chromatic number is $k$. One can then wonder whether some results on tournaments can be extended to digraphs with large chromatic number. This is in particular the case with Rédei's Theorem (2.2.4), which has been generalized to the following theorem, often referred to as the Gallai–Roy Theorem, even if it was independently proved by four researchers: Gallai [94], Hasse [105], Roy [166] and Vitaver [191].

**Theorem 2.7.17 (Gallai–Hasse–Roy–Vitaver** [94, 105, 166, 191]) *Every $k$-chromatic digraph contains a directed path of order $k$.*

Theorem 2.7.17 has many proofs. One of them is based on median orders (see [50] Chapter 14). We present here a proof due to El-Sahili and Kouider [74]. It is based on the concept of **out-forests**, which are disjoint unions of out-trees. An out forest of $D$ is spanning if it covers all vertices of $D$.

Let $F$ be a spanning out-forest of $D$. The **level** of $x$ is the number of vertices of a longest directed path of $F$ ending at $x$. For instance, the level 1 vertices are the roots of the out-trees of $F$. We denote by $F_i$ the set of vertices with level $i$ in $F$. A vertex $y$ is a **descendant** of $x$ in $F$ if there is a directed path from $x$ to $y$ in $F$.

If there is an arc $xy$ in $D$ from $F_i$ to $F_j$, with $i \geq j$, and $x$ is not a descendant of $y$, then the out-forest $F'$ obtained by adding $xy$ and removing the arc of $F$ with head $y$ (if such exists, that is, if $j > 1$) is called an **elementary improvement** of $F$. An out-forest $F'$ is an **improvement** of $F$ if it can be obtained from an out-forest $F$ by a sequence of elementary improvements. The key-observation is that if $F'$ is an improvement of $F$ then the level of every vertex in $F'$ is at least its level in $F$. Moreover, at least one vertex of $F$ has its level in $F'$ strictly greater than its level in $F$. Thus, one cannot perform infinitely many improvements. A spanning out-forest $F$ is **final** if there is no elementary improvement of $F$.

The following proposition follows immediately from the definition of a final spanning out-forest:

**Proposition 2.7.18 (El Sahili and Kouider** [74]) *Let $D$ be a digraph and $F$ a final spanning out-forest of $D$. If a vertex $x \in F_i$ dominates in $D$ a vertex $y \in F_j$ for $j \leq i$, then $x$ is a descendant of $y$ in $F$. In particular, every level of $F$ is an independent set in $D$.*

**Proof of Theorem** 2.7.17: Consider a final spanning out-forest of a $k$-chromatic digraph $D$. Since every level is an independent set by Proposition 2.7.18, there are at least $k$ levels. Hence $D$ contains a directed path of order at least $k$. □

More generally, one can ask which digraphs are **$k$-universal**, i.e. contained in every $k$-chromatic digraph. A result of Erdős [75] states that for every choice of positive integers $k$ and $g$, there exist $k$-chromatic graphs with no cycle of length less than $g$. Consequently, $k$-universal digraphs must be oriented trees.

Bondy conjectured the following generalization of Theorem 2.7.5.

**Conjecture 2.7.19 (Bondy, 1995)** *For sufficiently large $k$, every oriented path on $k$ vertices is $k$-universal.*

As support for this conjecture, El Sahili proved [72] that every oriented path of order 4 is 4-universal and that the antidirected path of order 5 is 5-universal. Addario-Berry, Havet, and Thomassé [2] proved that every oriented path of order $k \geq 4$ with two blocks is $k$-universal. Their proof use the notion of a final spanning out-forest.

Burr [53] generalized Sumner's Conjecture as follows.

**Conjecture 2.7.20 (Burr [53], 1980)** *Every oriented tree on $k$ vertices is $(2k-2)$-universal.*

Burr [53] showed that every oriented tree of order $k$ is $(k-1)^2$-universal. This was slightly improved by Addario-Berry, Havet, Linhares Sales, Reed, and Thomassé [1].

**Theorem 2.7.21** ([1]) *Every oriented tree on $k$ vertices is $(k^2/2 - k/2 + 1)$-universal.*

Addario-Berry *et al.* [1] proved that every oriented tree on $k$ vertices is contained in every acyclic digraph of order $n$. They also established that every antidirected tree of order $k \geq 3$ is $(5k-9)$-universal. An **antidirected tree** is an oriented tree in which every vertex has either in-degree 0 or out-degree 0.

Finally, Havet and Thomassé generalized Conjecture 2.7.14 about unavoidabiity to universality.

**Conjecture 2.7.22 (Havet and Thomassé, 2000)** *If $A$ is an oriented tree with $n$ vertices and $k$ leaves, then it is $(n+k-1)$-universal.*

Let us now consider cycles. As we already saw, they cannot be universal because there are digraphs with no cycles of small length having arbitrarily large chromatic number, as stated by a result of Erdős [75]. However, Bondy generalized Camion's Theorem (2.2.6) to digraphs with large chromatic number.

**Theorem 2.7.23 (Bondy [48])** *Every strong digraph of chromatic number at least $k$ contains a directed cycle of length at least $k$.*

A directed cycle of length at least $k$ may be seen as a subdivision of the directed $k$-cycle $\vec{C}_k$. Recall that a **subdivision** of a digraph $D$ is a digraph obtained from $D$ by replacing each arc $ab$ of $D$ by a directed $(a,b)$-path. Hence a natural question is to ask whether Theorem 2.7.10 can be generalized, or if at least every non-directed cycle $C$ is $k$-universal for some large enough $k$. This was answered in the negative by Cohen, Havet, Lochet, and Nisse.

**Theorem 2.7.24** ([65]) *Let $C$ be an oriented cycle. There exist digraphs with arbitrarily large chromatic number that contains no subdivision of $C$.*

However, they conjectured that, as for the directed cycle, if we require the digraph to be strongly connected, the picture is different.

**Conjecture 2.7.25** ([65]) *Let $C$ be an oriented cycle $C$. There exists a constant $h(C)$ such that every strong digraph with chromatic number at least $h(C)$ contains a subdivision of $C$.*

As partial evidence, Cohen, Havet, Lochet, and Nisse [65] proved this conjecture for cycles with two blocks and the antidirected cycle of order 4. In particular, they proved that for $C(k, \ell)$ the cycle on two blocks, one of length $k$ and the other of length $\ell$, $h(C(k, \ell)) = O((k + \ell)^4)$. This bound on the value was recently improved by Kim, Kim, Ma and Park [122] who proved $h(C(k, \ell)) = O((k + \ell)^2)$.

## 2.8 Vertex-Partitions of Semicomplete Digraphs

In this section, we consider properties of vertex-partitions of semicomplete digraphs. A **k-partition** of a digraph $D = (V, A)$ is a partition $(V_1, V_2, \ldots, V_k)$ of $V$ into $k$ non-empty disjoint sets.

### 2.8.1 2-Partitions into Strong Semicomplete Digraphs

Being strongly connected is one of the basic properties of a digraph. Hence, it is natural to determine which (semicomplete) digraphs $D$ have a $k$-partition into strong subdigraphs, that is, a partition $(V_1, \ldots, V_k)$ such that $D[V_i]$ is strong for $i = 1, \ldots, k$. Bang-Jensen, Cohen and Havet proved [21] that this problem is $\mathcal{NP}$-complete for general digraphs already when $k = 2$. The papers [21, 26] provide a complete characterization of the complexity of a number of related problems where we wish to partition $V(D)$ into two sets such that each of these have prescribed properties (e.g. both are strongly connected).

We now turn to semicomplete digraphs. Recall that a cycle factor is a spanning collection of disjoint cycles. Since every strongly connected semicomplete digraph is Hamiltonian, a semicomplete digraph has a 2-partition into two strong subdigraphs if and only if it has a cycle factor with two cycles. A pair of cycles forming a cycle factor with two cycles is also called a pair of **complementary cycles**.

Reid proved that every 2-strong $n$-tournament with $n \geq 8$ has a 2-partition into strong subtournaments, one of which has order 3. Song extended this result by showing that there exists such a partition with one subtournament of any fixed order $k$ for any $3 \leq k \leq n-3$.

**Theorem 2.8.1** ([161, 171]) *Every 2-strong tournament $D$ on at least 8 vertices has a 2-partition $(V_1, V_2)$ such that $D[V_i]$ is strong for $i = 1, 2$ and $|V_1| = k$ for every $3 \leq k \leq n-3$.*

Theorem 2.8.1 also holds for 2-strong tournaments on 6 vertices and the only exception on 7 vertices is the Paley tournament $\mathbb{P}_7$ (see [161]). Furthermore, there are infinite families of tournaments $T$ with $\kappa(T) = 1$ which do not have complementary cycles. One such example was given in [130] by Li and Shu. Those families show that Theorem 2.8.1 cannot be extended to strong tournaments. However, Li and Shu proved that strong tournaments with sufficiently large minimum in- or out-degree have a partition into strong subtournaments.

**Theorem 2.8.2** ([130]) *Let $T$ be a strong tournament on at least 6 vertices. If $\max\{\delta^-(T), \delta^+(T)\} \geq 3$ and $T$ is not isomorphic to the Paley tournament $\mathbb{P}_7$, then $T$ has a 2-partition into strong subtournaments.* □

It follows from Theorem 6.9.2 that Theorem 2.8.1 also holds for semicomplete digraphs. For semicomplete digraphs Bang-Jensen and Nielsen solved the problem from a complexity point of view.

**Theorem 2.8.3** ([33]) *There exists a polynomial algorithm that, given semicomplete digraph $D$, finds a 2-partition $(V_1, V_2)$ such that $D[V_i]$ is strong for $i = 1, 2$, or correctly reports that no such pair exists.*

If we require more structure on the digraphs $D[V_i]$, such as requiring each of these to induce a tournament, then the problem becomes very difficult, even when the input is a semicomplete digraph. The following result is due to Bang-Jensen and Christiansen.

**Theorem 2.8.4** ([20]) *It is $\mathcal{NP}$-complete to decide whether a given semicomplete digraph $D$ has a 2-partition $(V_1, V_2)$ such that $D[V_i]$ is a strong tournament for $i = 1, 2$.*

In an attempt to generalize Theorem 2.8.1, Bollobás asked whether every sufficiently large $k$-strong tournament has a cycle factor with $k$-cycles or equivalently a $k$-partition into strong subtournaments (see [161]). This was answered in the positive by Chen, Gould and Li.

**Theorem 2.8.5** ([59]) *Every $k$-strong tournament on $n \geq 8k$ vertices has a $k$-partition into strong subtournaments.*

Furthermore, Kühn, Osthus and Townsend proved that if the tournament is $r$-strong for $r$ sufficiently high, then one can prescribe the sizes of the strong subtournaments of the $k$-partition. This answers a question by Song [171].

**Theorem 2.8.6** ([127]) *Let $T$ be a tournament on $n$ vertices, let $k \geq 2$ and let $n_1, n_2, \ldots, n_k \geq 3$ satisfy $n = n_1 + n_2 + \ldots + n_k$. If $T$ is $10^{10}k^4 \log k$-strong, then it has a partition $(V_1, \ldots, V_k)$ into strong subtournaments such that $|V_i| = n_i$ for $i \in [k]$.*

### 2.8.2 Partition into Highly Strong Subtournaments

As a generalization of Theorem 2.8.1, Thomassen (see [161]) conjectured that for all positive integers $k_1, k_2$ there exists an integer $f(k_1, k_2)$ such that every $f(k_1, k_2)$-strong tournament $T$ has a 2-partition $(V_1, V_2)$ so that $T[V_i]$ is $k_i$-strong, $i = 1, 2$. This is clearly equivalent to the existence, for all integers $k, t$, of an integer $g(k, t)$ such that every $g(k, t)$-strong tournament $T$ has a $t$-partition $(V_1, \ldots, V_t)$ so that $T[V_i]$ is $k$-strong, $i \in [t]$. The existence of such a $g(k, t)$ was established by Kühn, Osthus and Townsend [127].

**Theorem 2.8.7** ([127]) *Let $k, t \geq 1$ be integers. Every tournament $T$ which is $(10^7 k^6 t^3 \log(kt^2))$-strong has a $t$-partition $(V_1, \ldots, V_t)$ such that $T[V_i]$ is $k$-strong for $i \in [t]$.*

Kim, Kühn and Osthus proved that when the connectivity is sufficiently high we can get an even stronger type of 2-partition. For a digraph $D$ and a 2-partition $(V_1, V_2)$, we denote by $D[V_1, V_2]$ the bipartite subdigraph induced by the arcs with one end in $V_1$ and the other in $V_2$.

**Theorem 2.8.8** ([121]) *Every $10^9 k^6 \log(2k)$-strong tournament has a 2-partition $(V_1, V_2)$ such that each of $T[V_1], T[V_2], T[V_1, V_2]$ is a $k$-strong digraph.*

See Theorem 2.8.18 for a related partition result for out-degrees.

### 2.8.3 2-Partitions With Prescribed Minimum Degrees

We now turn to 2-partitions where we want a certain minimum out-, in- or semi-degree in each of the parts. E.g. a $(\delta^+ \geq 1, \delta^+ \geq 1)$-partition is a 2-partition $(V_1, V_2)$ where the digraph induced by each set has minimum out-degree at least 1. Bang-Jensen, Cohen and Havet proved in [21] that when we want the chosen parameter among $\{\delta^+, \delta^-, \delta^0\}$ to be at least 1 in each side of the partition, then we obtain an $\mathcal{NP}$-complete problem for general digraphs, except in the case of $(\delta^+ \geq 1, \delta^+ \geq 1)$- and $(\delta^- \geq 1, \delta^- \geq 1)$-partitions for which easy polynomial algorithms exist. Furthermore, Bang-Jensen and Christiansen proved that the $(\delta^+ \geq 1, \delta^+ \geq 2)$-partition problem (that is, deciding whether there is a 2-partition $(V_1, V_2)$ of $D$ such that $\delta^+(D[V_i]) \geq i$ for $i = 1, 2$) is already $\mathcal{NP}$-complete [20].

A suprisingly difficult problem is the following conjecture due independently to Alon and Stiebitz.

**Conjecture 2.8.9** ([4, 173]) *There exists a function $f(k, \ell)$, where $k, \ell$ are positive integers, such that every digraph $D$ with $\delta^+(D) \geq f(k, \ell)$ has a $(\delta^+ \geq k, \delta^+ \geq \ell)$-partition.*

It is easy to see that a digraph with minimum out-degree $k + \ell$ has a $(\delta^+ \geq k, \delta^+ \geq \ell)$-partition if and only if it has two disjoint subdigraphs with minimum out-degree at least $k$ and $\ell$. Thomassen [180] proved that every digraph $D$ with $\delta^+(D) \geq 3$ has two disjoint cycles, hence Conjecture 2.8.9 holds for $k = \ell = 1$ and $f(1, 1) = 3$ because the second power $C^2_{2r+1}$ of an odd cycle has no $(\delta^+ \geq 1, \delta^+ \geq 1)$-partition. But even the existence of $f(1, 2)$ is still open.

In the remaining part of this section, we shall see that the situation is a lot simpler for semicomplete digraphs: Conjecture 2.8.9 holds for semicomplete digraphs and the problem of deciding whether a semicomplete digraph has a $(\delta^+ \geq k, \delta^+ \geq \ell)$-partition can be solved in polynomial time. A crucial notion here is that of an out-critical set. A set $X$ of vertices of a digraph $D$ is ***$k$-out-critical*** if $\delta^+(D\langle X\rangle) = k$ and for every proper subset $S \subset X$, $\delta^+(D\langle S\rangle) < k$. Let $X$ be a set of vertices in a digraph $D$. A set $X' \subseteq V(D)$ is called ***$(X, k)$-out-critical*** if $X \subseteq X'$, $\delta^+(D[X']) \geq k$ and $\delta^+(D[Y]) < k$ for every $X \subseteq Y \subset X'$. Note that if $\delta^+(D[X]) \geq k$, then $X$ is the only $(X, k)$-out-critical set in $D$. By definition, a digraph of minimum out-degree at least $k$ contains at least one $(X, k)$-out-critical set for every subset $X$ of vertices (including the empty set). The key fact is that the number of $(X, k)$-out-critical sets is bounded since their size is bounded. This was first observed by Lichiardopol for tournaments. In fact, his result holds for semicomplete digraphs as well.

**Lemma 2.8.10** ([133]) *Let $k$ be a positive integer, let $D$ be a semicomplete digraph with minimum degree at least $k$, and let $X \subseteq V(D)$. If $X'$ is an $(X, k)$-out-critical set in $D$, then $|X'| \leq \frac{k^2+3k+2}{2} + |X|$. In particular, every $k$-out-critical digraph in $D$ has order at most $\frac{k^2+3k+2}{2}$.*

**Proof:** By induction on $|V(D)|$. If $|V(D)| \leq \frac{k^2+3k+2}{2} + |X|$ we are done, so assume $|V(D)| > \frac{k^2+3k+2}{2} + |X|$. Let $M$ be the set of vertices that have out-degree $k$ in $T$ and let $m = |M|$.

Since $T[M]$ is semicomplete, we have

$$|N^+[M]| \leq m + mk - \frac{m(m-1)}{2} = -\frac{m^2}{2} + \left(\frac{3}{2} + k\right) m =: P(m).$$

Now $P(m)$ has global maximum at $(3/2+k)$ and maximum for $m$ integer at $k+1$ and $k+2$ with $P(k+1) = P(k+2) = \frac{k^2+3k+2}{2}$. Hence $|N^+[M]| \leq \frac{k^2+3k+2}{2}$ and since $|V(D)| > \frac{k^2+3k+2}{2} + |X|$ there exists a vertex $u \in V(D) \setminus (N^+[M] \cup X)$. Then $\delta^+(T-u) \geq k$ and the result follows by induction. □

**Corollary 2.8.11** ([133]) *For any pair of integers $k, \ell \geq 1$, every semicomplete digraph $D$ with $\delta^+(D) \geq (k^2+3k+2)/2 + \ell$ has a $(\delta^+ \geq k, \delta^+ \geq \ell)$-partition. Furthermore, such a partition can be constructed in polynomial time.*

**Proof:** This follows easily from Lemma 2.8.10 by taking a $k$-out-critical set $V_1$ (which has size at most $(k^2+3k+2)/2$) and taking $V_2 = V \setminus V_1$.

Let us describe a polynomial algorithm, due to Bang-Jensen and Christiansen, for deciding whether a given semicomplete digraph has a $(\delta^+ \geq k, \delta^+ \geq \ell)$–partition.

**Theorem 2.8.12** ([20]) *For every fixed pair of integers $k$ and $\ell$, there exists a polynomial algorithm that either constructs a $(\delta^+ \geq k, \delta^+ \geq \ell)$-partition of a given semicomplete digraph $D$ or correctly outputs that none exists.*

**Proof:** Let $O$ be the set of vertices with out-degree less than $k+\ell-1$. For a given partition $(O_1, O_2)$ of $O$ we let $X$ be an $(O_1, k)$-out-critical set such that $X \subseteq V \setminus O_2$ (if no such set exists, we stop considering the pair $(O_1, O_2)$). The following subalgorithm $\mathcal{B}$ will decide whether there exists a $(\delta^+ \geq k, \delta^+ \geq \ell)$-partition $(V_1, V_2)$ with $X \subseteq V_1, O_2 \subseteq V_2$: Starting from the partition $(V_1, V_2) = (X, V \setminus X)$, and moving one vertex at a time, the algorithm will move vertices $v$ of $V_2 \setminus O_2$ such that $d^+_{T[V_2]}(v) < \ell$ to $V_1$. If, at any time, this results in a vertex $v \in O_2$ having $d^+_{T[V_2]}(v) < \ell$, or $V_2 = \emptyset$, then there is no $(\delta^+ \geq k, \delta^+ \geq \ell)$-partition with $O_i \subseteq V_i, = 1,2$ and $\mathcal{B}$ terminates. Otherwise $\mathcal{B}$ will terminate with $O_2 \subseteq V_2 \neq \emptyset$ and hence it has found a $(\delta^+ \geq k, \delta^+ \geq \ell)$-partition $(V_1, V_2)$ with $O_i \subseteq V_i$, $i = 1, 2$.

The correctness of $\mathcal{B}$ follows from the fact that we only move vertices that are not in $O$ and each such vertex has at least $k+\ell-1$ out-neighbours in $D$. Hence, when moved, a vertex has less than $\ell$ out-neighbours in $V_2$, so it has at least $k$ out-neighbours in $V_1$. Thus $\delta^+(D[V_1]) \geq k$ holds throughout the execution of $\mathcal{B}$.

By Proposition 2.2.2, $|O| \leq 2k+2\ell-3$ and hence the number of $(O_1, O_2)$-partitions is at most $2^{2k+2\ell-3}$ which is a constant when $k$ and $\ell$ are fixed. Furthermore, by Lemma 2.8.10, the size of every $O_1$-critical set is also bounded by a function of $k$ and hence each $(O_1, O_2)$-partition induces only a polynomial number of $O_1$-critical sets. Thus we obtain the desired polynomial algorithm by running the subalgorithm $\mathcal{B}$ for (at most) all possible partitions $(O_1, O_2)$ of $O$ and all possible $(O_1, k)$-out-critical sets. □

Lichiardopol proved an analogue of Corollary 2.8.11 for partitions with prescribed lower bounds on semi-degrees in tournaments. His result can easily be extended to semicomplete digraphs.

**Theorem 2.8.13** ([133]) *For any choice of integers $k, \ell \geq 1$, every semicomplete digraph $D$ with $\delta^0(D) \geq (k^2+3k+2)+\ell$ has a $(\delta^0 \geq k, \delta^0 \geq \ell)$-partition. Furthermore, such a partition can be constructed in polynomial time.*

The complexity of finding 2-partitions with prescribed minimum semi-degrees has been studied by Bang-Jensen and Christiansen. Recall that for general digraphs it is $\mathcal{NP}$-complete to decide the existence of a $(\delta^0 \geq k, \delta^0 \geq \ell)$-partition when $k + \ell \geq 2 \rightarrow k, \ell \geq 1$ [21]. Bang-Jensen and Christiansen showed that for semicomplete digraphs the situation is better, at least when $k = \ell = 1$.

**Theorem 2.8.14** ([20]) *There exists a polynomial algorithm that given a semicomplete digraph $D$ either finds a $(\delta^0 \geq 1, \delta^0 \geq 1)$-partition of $D$ or correctly returns that none exists.*

**Problem 2.8.15** *For any fixed positive integers $k, \ell$, what is the complexity of deciding whether a semicomplete digraph has a $(\delta^0 \geq k, \delta^0 \geq \ell)$-partition ?*

One may also study all other possible variants, for example $(\delta^+ \geq k, \delta^- \geq \ell)$-partitions. The associated complexity problem is the following.

**Problem 2.8.16** *For any fixed positive integers $k, \ell$, what is the complexity of deciding whether a semicomplete digraph has a $(\delta^+ \geq k, \delta^- \geq \ell)$-partition?*

Bang-Jensen, Cohen and Havet proved that Problem 2.8.16 is $\mathcal{NP}$-complete for general digraphs already when $k = \ell = 1$. Bang-Jensen and Christiansen [20] proved that a semicomplete digraph $D$ has a $(\delta^+ \geq 1, \delta^- \geq 1)$-partition if and only if it has two disjoint cycles. Since one can find such a pair of disjoint cycles if one exists in polynomial time, one can decide in polynomial time whether a semicomplete digraph $D$ has a $(\delta^+ \geq 1, \delta^- \geq 1)$-partition. The following partial result on Problem 2.8.16 was obtained by Bang-Jensen and Christiansen.

**Theorem 2.8.17** ([20]) *For every fixed integer $k \geq 1$ there exists a polynomial algorithm that either constructs a $(\delta^+ \geq 1, \delta^- \geq k)$-partition of a given semicomplete digraph $D$ or correctly outputs that none exists.*

### 2.8.4 2-Partitions with Restrictions Both Inside and Between Sets

For a 2-partition $(V_1, V_2)$ we denote by $D[V_1, V_2]$ the digraph induced by the arcs between $V_1$ and $V_2$. We now consider the degree analogue of Theorem 2.8.8, that is, we seek a 2-partition $(V_1, V_2)$ so that each of

$D[V_1], D[V_2], D[V_1, V_2]$ has minimum out-degree at least some prescribed number. The following results are due to Alon, Bang-Jensen and Bessy.

**Theorem 2.8.18** ([6]) *Except for the Paley tournament $\mathbb{P}_7$ every semicomplete digraph $D$ with minimum out-degree at least 3 has a 2-partition $(V_1, V_2)$ such that $D[V_1], D[V_2], D[V_1, V_2]$ has minimum out-degree at least one. Furthermore, when $D \neq \mathbb{P}_7$ one can always find such a 2-partition which is balanced, that is, $||V_1| - |V_2|| \leq 1$.*

For higher values of the degree bounds the authors obtained the following.

**Theorem 2.8.19** ([6]) *There exist two absolute positive constants $c_1, c_2$ such that the following holds.*

1. *Let $T = (V, E)$ be a semicomplete digraph with minimum out-degree at least $2k + c_1\sqrt{k}$. Then there is a balanced a 2-partition $(V_1, V_2)$ of $V$ such that $\delta^+(D[V_1]), \delta^+(D[V_2])$ and $\delta^+(D[V_1, V_2])$ are all at least $k$.*
2. *For infinitely many values of $k$ there is a tournament with minimum out-degree at least $2k+c_2\sqrt{k}$ such that for any 2-partition $(V_1, V_2)$ of $V$ at least one of the quantities $\delta^+(D[V_1]), \delta^+(D[V_2])$ and $\delta^+(D[V_1, V_2])$ is smaller than $k$.*

We only give the proof of the second part of Theorem 2.8.19. The proof illustrates one of the remarkable properties of the Paley tournaments: They behave almost like random tournaments.

Recall that for a prime $q$ which is congruent to 3 modulo 4, the Paley tournament $\mathbb{P}_q$ is the tournament whose vertices are the integers modulo $p$ where $(i, j)$ is a directed edge if and only if $i - j$ is a quadratic residue modulo $q$.

**Lemma 2.8.20** *Let $\mathbb{P}_q = (V, A)$ be the Paley tournament on $q$ vertices. Then for any function $f : V \to \{-1, 1\}$ there is a vertex $v \in V$ such that $|\sum_{u \in N^+(v)} f(u)| > \frac{1}{2}\sqrt{q}$.*

**Proof:** It is easy and well known (c.f., e.g., [10], Chapter 9) that every vertex of $\mathbb{P}_q$ has out-degree and in-degree $(q-1)/2$ and any two vertices of it have exactly $(q-3)/4$ common in-neighbours (and out-neighbours). Let $A = A_q$ be the adjacency matrix of $\mathbb{P}_q$, that is, the 0/1 matrix whose rows and columns are indexed by the vertices of $\mathbb{P}_q$, where $A_{ij} = 1$ if and only if $(i, j)$ is an arc. By the above comment, each diagonal entry of $A^tA$ is $(q-1)/2$ and each other entry is $(q-3)/4$. Thus the eigenvalues of $A^tA$ are $(q-1)/2 + (q-1)(q-3)/4 = (q-1)^2/4$ (with multiplicity 1) and $(q-1)/2 - (q-3)/4 = (q+1)/4$ (with multiplicity $(q-1)$). This implies that

$$||Af||_2^2 = f^tA^tAf \geq (q+1)/4||f||_2^2 = q(q+1)/4.$$

It follows that there is an entry of $Af$ whose square is at least $(q+1)/4$, completing the proof. □

Note that, by Lemma 2.8.20, for any partition of the vertices of $\mathbb{P}_q$ into two disjoint (not necessarily nearly equal) sets $V_1$ and $V_2$ there is a vertex $v$ of $\mathbb{P}_q$ such that the number of its out-neighbours in $V_1$ differs from that in $V_2$ by more than $\sqrt{q}/4$ (if there are $x$ more neighbours in one set than in the other, then the sum in the lemma is $|\sum_{u \in N^+(v)} f(u)| = 2x$). This implies the assertion of part (ii) of Theorem 2.8.19 for infinitely many values of $k$.

### 2.8.5 Partitioning into Transitive Tournaments

A **$k$-dicolouring** of a digraph $D$ is a $k$-partition $(V_1, \ldots, V_k)$ of its vertex set such that $D\langle V_i \rangle$ is acyclic. The **dichromatic number** of $D$, denoted by $\vec{\chi}(D)$, is the smallest positive integer such that $D$ admits a $k$-dicolouring. This notion was first treated by Neumann-Lara [148] and was independently introduced by Mohar [142] two decades later. Note that if $G$ is an undirected graph, and $D$ is the symmetric digraph obtained from $G$ by replacing each edge by the pair of oppositely directed arcs joining its end vertices, then $\chi(G) = \vec{\chi}(D)$ since any two adjacent vertices in $D$ induce a directed 2-cycle. Observe, moreover, that the dichromatic number of a tournament $T$ is the minimum integer $k$ such that $T$ can be partitioned into $k$ transitive subtournaments.

Finding the dichromatic number of a tournament is $\mathcal{NP}$-hard. Chen, Hu, and Zhang [61] proved that it is in fact already $\mathcal{NP}$-complete to decide whether a tournament has dichromatic number 2.

**Theorem 2.8.21** ([61]) *Deciding whether a tournament has a 2-partition into two transitive subtournaments is $\mathcal{NP}$-complete.*

**Proof:** The original proof by Chen *et al.* was a reduction from NAE-3-SAT. We present here a simpler reduction from MONOTONE NAE-3-SAT (recall that monotone means that there are no negated variables).

Let $\mathcal{F} = C_1 \wedge \cdots \wedge C_m$ be an instance of MONOTONE NAE-3-SAT on $n$ variables $x_1, \ldots, x_n$. We construct a tournament $T_{\mathcal{F}}$ as follows. Its vertex set is the union of $X = \{x_1, \ldots, x_n\}$, a set $Y = \{y_1, y_2, y_3\}$ and $Z = \bigcup_{j=1}^m Z_j$, where $Z_j = \{z_j^1, z_j^2, z_j^3\}$. For $z \in Z$, we define $x_z$ as follows: let $j$ and $\ell$ be the indices such that $z = z_j^\ell$, and let $i$ be the index such that $x_i$ is the $\ell$th literal of $C_j$; then $x_z = x_i$.

Let $\sigma$ be the following ordering of $V(T_{\mathcal{F}})$

$$(x_1, \ldots, x_n, y_1, y_2, y_3, z_1^1, z_1^2, z_1^3, z_2^1, z_2^2, z_2^3, \ldots, z_m^1, z_m^2, z_m^3).$$

All the arcs of $T_{\mathcal{F}}$ agree with $\sigma$ (i.e. if $u$ precedes $v$ in $\sigma$, then $u \to v$) except for a set $B = B_Y \cap B_Z \cap B'$ of backward arcs, where $B_Y = \{y_3 y_1\}$, $B_Z = \{z_j^3 z_j^1 \mid 1 \le j \le m\}$ and $B' = \{z x_z \mid z \in Z\}$.

Observe that every directed cycle in $T_{\mathcal{F}}$ is either the 3-cycle $y_1y_2y_3y_1$, or the 3-cycle $z_j^1z_j^2z_j^3z_j^1$ for some $1 \leq j \leq m$, or contains an arc in $B$.

We shall now prove that $\mathcal{F}$ has an NAE-assignment if and only if $T_{\mathcal{F}}$ has a 2-partition $(V_1, V_2)$ such that $T[V_i]$ is transitive.

Let us assume that $\mathcal{F}$ has an NAE-assignment $\phi$. Let $X_1 = \{x_i \mid \phi(x_i) = true\}$, $X_2 = \{x_i \mid \phi(x_2) = false\}$, $Z_1 = \{z \mid x_z \in X_2\}$ and $Z_2 = \{z \mid x_z \in X_1\}$. Setting $V_1 = X_1 \cup \{y_1\} \cup Z_1$ and $V_1 = X_2 \cup \{y_2\} \cup Z_2$, one can easily check that $(V_1, V_2)$ is a partition of $T_{\mathcal{F}}$ into two transitive tournaments. Indeed, the arcs of $B$ have their end vertices in different part, each $\{z_j^1, z_j^2, z_j^3\}$ contains at least one vertex in $V_1$ and one in $V_2$ because $\phi$ is an NAE-assignment.

Assume now that $T_{\mathcal{F}}$ admits a partition $(V_1, V_2)$ into two transitive subtournaments. Since $Y$ induces a 3-cycle, at least one vertex of $Y$ is in $V_1$ and another one is in $V_2$. Without loss of generality, we may assume $y_1 \in V_1$ and $y_2 \in V_2$. Similarly, each $Z_j$, $1 \leq j \leq m$ has a vertex in $V_1$ and a vertex in $V_2$. Now consider an arc $zx_z$ in $B'$. The two vertices $z$ and $x_z$ are not in the same $V_k$ ($k \in \{1, 2\}$) for otherwise $zx_zy_kz$ would be a directed 3-cycle. Now one checks easily that the truth assignment $\phi$ defined by $\phi(x_i) = true$ if $x_i \in V_1$ and $\phi(x_i) = false$ if $x_i \in V_2$ is an NAE-assignment. □

Theorem 2.8.21 implies that it is unlikely to find a characterization of tournaments with dichromatic number $k$. However, it is interesting to find properties of such tournaments. A natural question, in the same flavour as unavoidablity (see Section 2.7), is to ask which subtournaments must appear in every tournament with sufficiently large dichromatic number. Such a tournament is called a **hero**. Clearly, transitive tournaments are heroes, since every tournament of order $n$ contains a transitive subtournament of order at least $\log_2 n$ by Proposition 2.2.3. Moreover, Theorem 2.2.7 implies that the directed 3-cycle is contained in every tournament of dichromatic number at least 2. Observe moreover that if $H$ is a hero, then every subtournament of $H$ is also a hero.

When $P$, $Q$ are tournaments, we denote by $C(P, Q)$ the tournament that one obtains from disjoint copies of $P$ and $Q$, by adding a new vertex $x$ dominating $P$ and dominated by $Q$, and adding all the arcs from $P$ to $Q$ (thus $C(P, Q) = C_3[P, Q, \{x\}]$).

Let us define the sequence $(A_i)_{i \in \mathbb{N}}$ of tournaments inductively as follows:

- $A_1$ is the tournament with one vertex and no arcs;
- $A_{i+1} := C(A_i, A_i)$.

The proof of the next proposition is left to the reader.

**Proposition 2.8.22** *If $i \geq 1$, then $\vec{\chi}(A_i) = i$.*

The tournaments $(A_i)_{i \in \mathbb{N}}$ imply that strongly connected heroes must have a special form.

**Lemma 2.8.23** *Every strongly connected hero is of the form $C(P,Q)$, where $P$ and $Q$ are heroes.*

**Proof:** Let $H$ be a hero. Then, by Proposition 2.8.22, for $i$ sufficiently large, $A_i$ contains $H$. Let $k$ be the minimum integer $i$ such that $A_i$ contains $H$. Let us denote by $L$ and $R$ the copies of $A_{k-1}$ in $A_k$ such that all arcs are from $L$ to $R$ and let $x$ be the vertex of $A_k - (L \cup R)$. By definition of $k$, neither $L$ nor $R$ contains $H$, so the copy of $H$ in $A_k$ must contain $x$. Now, since $A_k$ is strong, $H$ must contain at least one vertex of $L$ and one vertex of $R$. Therefore $H$ is of the desired form.

In fact, the tournaments that are heroes have been completely characterized by Berger, Choromanski, Chudnovsky, Fox, Loebl, Scott, Seymour and Thomassé.

**Theorem 2.8.24** ([41])

- *A tournament $T$ is a hero if and only if all its strong components are heroes;*
- *a strong tournament $T$ is a hero if and only if $T = C(P, TT_r)$ or $T = C(TT_r, P)$ for some hero $P$ and some $r \geq 1$.*

## 2.9 Feedback Sets

Feedback sets in a digraph are sets of vertices or arcs whose removal leaves the digraph acyclic. Formally, a **feedback vertex set** in a digraph $D$ is a set $S$ of vertices such that $D - S$ is acyclic, and a **feedback arc set** in a digraph $D$ is a set $F$ of arcs such that $D \setminus F$ is acyclic.

| Feedback vertex set | **Parameter:** $k$ |
|---|---|
| **Input:** A digraph $D = (V, A)$ | |
| **Question:** Does $D$ have a vertex set $X$ of size at most $k$ such that $D - X$ is acyclic? | |

| Feedback Arc Set | **Parameter:** $k$ |
|---|---|
| **Input:** A digraph $D = (V, A)$ | |
| **Question:** Does $D$ have a set of arcs $A'$ of size at most $k$ such that $D \setminus A'$ is acyclic? | |

A feedback vertex (resp. arc) set is **minimal** if none of its proper subsets is also a feedback vertex (resp. arc) set. A feedback vertex (resp. arc) set is **minimum** if it is of minimum size. The minimum size of a feedback vertex set (resp. feedback arc set) in $D$ is denoted by $\mathrm{fvs}(D)$ (resp. $\mathrm{fas}(D)$).

We are then interested in the optimization versions of Feedback vertex set and Feedback arc set where one wishes to determine $\mathrm{fvs}(D)$

and fas($D$), respectively, for a given digraph $D$, as well as their restriction to tournaments FEEDBACK VERTEX SET IN TOURNAMENT (FVST for short) and FEEDBACK ARC SET IN TOURNAMENT (FAST for short).[4] These problems are very fundamental and have many practical applications. For example, FEEDBACK ARC SET IN TOURNAMENTS models the problem of ranking the teams of a round-robin sport tournament and the problem of clustering webpages (see e.g. the paper [190] by van Zuylen and Williamson).

An **ordering associated to** a feedback vertex set $S$ (resp. feedback arc set $F$) is an acyclic ordering of $D - S$ (resp. $D \setminus F$). Observe that if $(v_1, \ldots, v_n)$ is an ordering associated to a feedback arc set $F$ of $D$, then $\{v_iv_j \in A(D) \mid i > j\}$ is a feedback arc set contained in $F$. Therefore, every minimum feedback arc set induces an acyclic digraph. In contrast, a feedback vertex set is usually not acyclic: a digraph has an acyclic feedback vertex set if and only if its dichromatic number is at most 2 (see Subsection 2.8.5). Some papers studied feedback vertex sets with a certain property $\mathbb{P}$, this is the same as studying a 2-partition $(V_1, V_2)$ of a digraph $D$ such that $D[V_1]$ has property $\mathbb{P}$ and $D[V_2]$ is acyclic. See e.g. the papers of Bang-Jensen, Cohen and Havet [21, 26].

**Proposition 2.9.1** *Let $F$ be a minimum feedback arc set in a digraph $D$. The digraph obtained from $D$ by reversing all arcs of $F$ is acyclic.*

**Proof:** Let $(v_1, \ldots, v_n)$ be an acyclic ordering associated to $F$. Observe that every arc $a$ of $F$ is of the form $v_iv_j$ with $i > j$ for otherwise $F \setminus \{a\}$ would also be a feedback arc set with $(v_1, \ldots, v_n)$ associated to it, contradicting the minimality of $F$. Therefore reversing the arcs of $F$ results in an acyclic digraph with acyclic ordering $(v_1, \ldots, v_n)$. □

Proposition 2.9.1 implies that fas($D$) is the minimum size of a set $F$ of arcs whose reversal yields an acyclic digraph.

### 2.9.1 Feedback Vertex Sets

FEEDBACK VERTEX SET is one of the the first problems shown to be $\mathcal{NP}$-complete listed by Karp in [118]. Its easy reduction from VERTEX COVER is the following. Let $(G, k)$ be an instance of VERTEX COVER. Let $D$ be the symmetric digraph associated to $G$, that is, the digraph obtained from $G$ by replacing each edge by a directed 2-cycle. One can easily check that $G$ has a vertex cover of size $k$ if and only if $D$ has a feedback vertex set of size $k$.

It is also not very hard to show that FVST is $\mathcal{NP}$-complete. this was shown independently by Speckenmeyer [172] and by Bang-Jensen and Thomassen [34]. The proof below is from [34].

[4] For simplicity and because they are polynomially equivalent, we do not distinguish between the decision and the optimization versions of these problems.

**Theorem 2.9.2** ([34]) FEEDBACK VERTEX SET IN TOURNAMENT *is* $\mathcal{NP}$-*complete.*

**Proof:** Reduction from INDEPENDENT SET which is well-known to be $\mathcal{NP}$-complete [118]. Let $G$ be an undirected graph with vertices $v_1^0, \ldots, v_n^0$. Let $T$ be the tournament defined as follows. $V(T) = V(G) \cup \{v_i^j \mid 1 \leq i \leq n \text{ and } 1 \leq j \leq n+1\}$ and there is an arc $(v_{i_1}^{j_1}, v_{i_2}^{j_2})$ whenever $i_1 > i_2$ or $i_1 = i_2$ and $j_1 > j_2$, unless $j_1 = j_2 = 0$ and $v_{i_1}^0 v_{i_2}^0$ is an edge of $G$, in which case $T$ contains the arc $(v_{i_2}^0, v_{i_1}^0)$. One can easily check that a vertex set $S$ is a maximum independent set in $G$ if and only if $V(G) \setminus S$ is a minimum feedback vertex set in $T$. □

FVST has a trivial 3-approximation algorithm, which proceeds as follows. As long as the tournament $T$ is not transitive, find a directed 3-cycle $C$, delete its vertices from $T$ and add them to the feedback vertex set $S$. Cai, Deng, and Zang [55] gave a 5/2-approximation. Recently, a 7/3-approximation was found by Mnich, Vassilevska Williams, and Végh [141].

For general digraphs, no non-trivial upper bound on the number of minimal feedback vertex sets is known. In contrast, we have some bounds for tournaments. Let $\#\text{fvs}(n)$ denote the maximum over all $n$-tournaments of the number of minimal feedback vertex sets. Note that $\#\text{fvs}(n)$ is also the maximum of the number of maximal transitive subtournaments since in a tournament $T$, a set $S$ is a feedback vertex set if and only if $T - S$ is transitive. Moon [143] was the first to give bounds on $\#\text{fvs}(n)$. He proved $1.4757^n \leq \#\text{fvs}(n) \leq 1.7170^n$. This was later improved by Gaspers and Mnich [95]

$$1.5548^n < 21^{n/7} \leq \#\text{fvs}(n) \leq 1.6740^n.$$

To get the lower bound, consider the tournament $T$ on $n = 7k$ vertices obtained from a transitive $k$-tournament by blowing up each vertex into a copy of the Paley tournament $\mathbb{P}_7$ on 7 vertices. The minimal feedback vertex sets of $\mathbb{P}_7$ are also minimum feedback vertex sets and have size 4. Furthermore, there are 21 of them. Hence $T$ has $21^{n/7}$ minimal feedback vertex sets.

The upper bound relies on an enumeration algorithm, based on iterative compression (see the proof of Theorem 2.9.6 for an example of iterative compression), that enumerates in $1.6740^n$-time all minimal feedback vertex sets in tournaments. Since a minimum feedback vertex set is also minimal, this algorithm allows us to solve FVST in $1.6740^n$-time.

### 2.9.2 Feedback Arc Sets

FEEDBACK ARC SET is also one of the the first problems known to be $\mathcal{NP}$-Complete listed by Karp in [118]. The easy reduction due to Karp and Lawler is from VERTEX COVER. Given a graph $G$, let $D$ be the digraph defined by

$V(D) = V(G) \times \{0, 1\}$.
$A(D) = \{((v, 0), (v, 1)) \mid v \in V(G)\} \cup \{((u, 1), (v, 0)) \mid (u, v) \in E(G)\}$.

We easily check that $G$ has a vertex cover of size $k$ if and only if $D$ has a feedback arc set of size $k$.

In contrast, FEEDBACK ARC SET IN TOURNAMENTS was conjectured to be $\mathcal{NP}$-complete in 1992 by Bang-Jensen and Thomassen [34]. Ailon, Charikar, and Newman [3] proved it is $\mathcal{NP}$-hard under randomized reductions. Shortly after, it was proved under deterministic reductions independently by Alon in [5] and by Charbit, Thomassé and Yeo in [58].

**Theorem 2.9.3** FEEDBACK ARC SET IN TOURNAMENTS *is $\mathcal{NP}$-complete.*

The proofs of Alon [5] and Charbit, Thomassé, and Yeo [58] both use the same reduction based on the existence of bipartite tournaments with large minimum feedback arc sets. Here, by large, we mean close to the trivial upper bound. Indeed, consider a bipartite tournament $B$ with both partite sets $R, S$ of size $k$. Considering the set of arcs from $R$ to $S$ and the one from $S$ to $R$, we obtain trivially that $\text{fas}(B) \leq \frac{k^2}{2}$. Hence by a large minimum feedback arc set, we mean a minimum feedback arc set of size close to $\frac{k^2}{2}$.

**Lemma 2.9.4** ([58]) *Let $\ell$ be a positive integer $\ell$ and set $k = 2^{3\ell}$. There exists a bipartite tournament $B_k$ with both partite sets of size $k$ and* $\text{fas}(B_k) \geq \frac{k^2}{2} - 2k^{5/3}$.

**Proof of Theorem** 2.9.3 **using Lemma** 2.9.4**:** The reduction is from FEEDBACK ARC SET in general digraphs.

Let $D$ be a digraph. We may assume that $D$ has no directed cycle of length at most 2, as deleting such a cycle decreases fas by exactly 1. Let $V(D) = \{v_1, \ldots, v_n\}$ and set $k = 2^{6\lceil 1+\log_2 n \rceil}$. Observe that $k = O(n^6)$ and $k \geq 64n^6$.

Let $B_k$ be the bipartite tournament defined in Lemma 2.9.4, and let $\{r_1, \ldots, r_k\}$ and $\{s_1, \ldots, s_k\}$ be the partite set of $B_k$.

Let $T$ be the tournament obtained by blowing up every vertex of $D$ by a transitive tournament, and adding copies of $B_k$ between blow-ups of non-adjacent vertices. To be precise, the vertex set of $T$ is $\{w_a^i \mid 1 \leq a \leq n$ and $1 \leq j \leq k\}$ and its arc set is $A_a \cup A_b \cup A_c$, where

$$
\begin{aligned}
A_a &= \{w_a^i w_a^j \mid 1 \leq a \leq n \text{ and } 1 \leq i < j \leq k\},\\
A_b &= \{w_a^i w_b^j \mid v_a v_b \in A(D) \text{ and } 1 \leq i, j \leq k\}, \text{and}\\
A_c &= \{w_a^i w_b^j \mid ab, ba \notin A(D) \text{ and } 1 \leq a < b \leq n \text{ and } r_i s_j \in A(B_k)\}\\
&\quad \cup \{w_b^j w_a^i \mid ab, ba \notin A(D) \text{ and } 1 \leq a < b \leq n \text{ and } s_j r_i \in A(B_k)\}.
\end{aligned}
$$

Let us now bound $\text{fas}(T)$. Without loss of generality, we may assume that $(v_1, \ldots, v_n)$ is an acyclic ordering associated to a minimum feedback arc set of $D$. Observe that since a minimum feedback arc set induces an acyclic digraph, Lemma 2.9.4 implies that the arcs of $A_c$ contribute at least $(\binom{n}{2} - |A(D)|)(\frac{k^2}{2} - 2k^{5/3})$ and at most $(\binom{n}{2} - |A(D)|)(\frac{k^2}{2} + 2k^{5/3})$ to $\text{fas}(T)$.

Considering the ordering $(w_1^1 \dots, w_1^k, w_2^1, \dots, w_2^k, w_3^1 \dots, w_n^k)$, we get

$$\text{fas}(T) \leq k^2 \,\text{fas}(D) + \left(\binom{n}{2} - |A(D)|\right)\left(\frac{k^2}{2} + 2k^{5/3}\right). \qquad (2.5)$$

Consider now a minimum feedback arc set of $T$. For any integers $i_1, \dots, i_n$ in $\{1, \dots, k\}$, at least $\text{fas}(D)$ arcs of $T\langle\{w_1^{i_1}, w_2^{i_2}, \dots, w_n^{i_n}\}\rangle$ are in $F$ because this digraph is isomorphic to $D$. Summing over all possible values of $i_1, \dots, i_n$ we get at least $k^n \,\text{fas}(D)$ arcs, where each arc can be counted at most $k^{n2}$ times. Hence

$$\text{fas}(T) \geq \frac{k^n \,\text{fas}(D)}{k^{n-2}} + \left(\binom{n}{2} - |A(D)|\right)\left(\frac{k^2}{2} - 2k^{5/3}\right). \qquad (2.6)$$

Now as $k^{1/3} \geq 64^{1/3}n^2$, we get that $(\binom{n}{2} - |A(D)|) \cdot 2k^{5/3} < \frac{k^2}{2}$. Hence Equations (2.5) and (2.6) imply the following.

$$\text{fas}(D) - \frac{1}{2} < \frac{\text{fas}(T)}{k^2} - \frac{1}{2}\left(\binom{n}{2} - |A(D)|\right) < \text{fas}(D) + \frac{1}{2}.$$

Hence if we could compute $\text{fas}(T)$ in polynomial time, we could also compute $\text{fas}(D)$. □

For general digraphs, the best known approximation algorithm for FEEDBACK ARC SET has performance guarantee $O(\log n \log \log n)$. The existence of such a feedback arc set is due to Seymour [168] and the algorithmic part is due to Even, Naor, Schieber and Sudan [77]. In contrast, for tournaments van Zuylen and Williamson [190] proposed a 2-approximation. Their algorithm is based on a linear programming relaxation of the problem and a nice rounding procedure. This procedure is a derandomization of the algorithm by Ailon, Charikar and Newman given in [3] based on the so-called 'pivot'.

The dual maximization problem consisting in finding an acyclic spanning subdigraph of a digraph $D$ with the maximum number of arcs is easy to approximate. A trivial 2-approximation consists in considering any ordering $(v_1, \dots, v_n)$ of the vertices of $D$ and the subdigraphs $D^+$ and $D^-$ with arc set $A^+ = \{v_iv_j \in A(D) \mid i < j\}$ and $A^- = \{v_jv_i \in A(D) \mid i > j\}$. These two digraphs are acyclic, and each of them has at least $|A(D)|/2$ arcs. There exist polynomial time approximation schemes (PTAS) for this problem in tournaments, see the papers [15] by Arora, Frieze and Kaplan and [91] by Frieze and Kannan.

By the above upper bound on $\text{fas}(D)$, for every $n$-tournament $T$, we have $\text{fas}(T) \leq \frac{n(n-1)}{4}$. This upper bound is almost tight as shown below.

**Theorem 2.9.5** *For every $n \geq 3$, there exists a tournament $T$ of order $n$ such that* $\mathrm{fas}(T) \geq \frac{n(n-1)}{4} - \frac{1}{2}\sqrt{n^3 \log_e n}$.

**Proof:** Consider a random tournament $RT_n$ on vertices $1, 2, \ldots, n$. Observe that for every pair $i \neq j \in \{1, 2, \ldots, n\}$, $ij \in A(RT_n)$ with probability 1/2.

For every pair $i < j \in \{1, 2, \ldots, n\}$, define the random variable $x_{i,j}$ by

$$x_{i,j} := \begin{cases} +1 \text{ if } ij \in A(RT_n) \\ -1 \text{ otherwise.} \end{cases}$$

Let $N = \binom{n}{2}$. With respect to the ordering $\pi = 1, 2, \ldots, n$, the number of forward arcs minus the number of backward arcs equals

$$\sum_{1 \leq i < j \leq n} x_{i,j} = S_N.$$

Then, $E_\pi := \{|S_N| > a\}$ denotes the event that, in one of the two orderings $\pi = \pi(1), \pi(2), \ldots, \pi(n) (= 1, 2, \ldots, n)$ and $\pi^* = \pi(n), \pi(n-1), \ldots, \pi(1) (= n, n-1, \ldots, 1)$, the number of forward arcs exceeds $n(n-1)/4 + a/2$. On the other hand, $S_N$ is the sum of $\binom{n}{2}$ random independent variables taking values $+1$ and $-1$, each with probability 1/2. By the Chernoff bound (Corollary A.2 in the book of Alon and Spencer [8]),

$$\mathrm{Prob}(|S_N| > a) \leq 2 \exp\left(-\frac{a^2}{2N}\right), \tag{2.7}$$

for every positive number $a$.

Observe that the event $E$ that for at least one permutation of $1, 2, \ldots, n$, the number of forward arcs exceeds $n(n-1)/4 + a/2$ equals the union of the events $E_\pi$ for all permutations $\pi$ of $1, 2, \ldots, n$, whose total number is $n!$. Put $a = \sqrt{n^3 \log_e n}$. Applying (2.7) we obtain

$$\begin{aligned} \mathrm{Prob}(E) &\leq 2n! \exp(-n \log_e n) \\ &\leq 2n! n^{-n} \\ &< 1 \end{aligned}$$

for every $n \geq 3$. This means that with positive probability the event $E$ does not hold, i.e. for every permutation of $1, 2, \ldots, n$, the number of forward arcs does not exceed $\frac{n(n-1)}{4} + \frac{1}{2}\sqrt{n^3 \log_e n}$. By the definition of $RT_n$, it follows that there exists a tournament of order $n$ with the above-mentioned property. □

A slightly better result was obtained by de la Vega in [70] who proved that $\sqrt{\log_e n}$ in the inequality of Theorem 2.9.5 can be replaced by a constant.

### 2.9.3 FPT Algorithms for FEEDBACK VERTEX SET IN TOURNAMENTS

Downey, Langston, Niedermeier, Raman, and Saurabh [157] proved that FVST is FPT by giving a $O(2.42^k \cdot n^{O(1)})$-time algorithm that solves it. This running time was improved by Fernau [78] who gave a $O(2.18^k \cdot n^{O(1)})$-time algorithm to solve FVST. We present below a faster FPT algorithm in $O(2^k \cdot n^{O(1)})$ time due to Dom, Guo, Hüffner, Niedermeier and Truß [71]. Very recently, an even faster FPT algorithm in $O(1.618^k + n^{O(1)})$ time was shown by Kumar and Lokshtanov [128].

**Theorem 2.9.6** ([71]) FEEDBACK VERTEX SET IN TOURNAMENTS *can be solved in time* $O(2^k \cdot n^3)$.

**Proof:** We present an algorithm solving FVST in $O(2^k \cdot n^3)$ time. This algorithm uses the method, called **iterative compression**, which was introduced by Reed, Smith, and Vetta [159]. The key part of this algorithm is a **compression routine** which, given a tournament and a feedback vertex set of size $k+1$, computes a feedback vertex set of size $k$ or proves that none exists.

Using such a compression routine FVST can be solved as follows. Let $\{v_1, \dots, v_n\} = V(T)$, and let $T_i = T\langle\{v_1, \dots, v_i\}\rangle$. We start with $S_2 = \emptyset$, which is a minimum feedback vertex set of $T_2$. Now for $i = 3$ to $n$, we compute a minimum feedback vertex set of $T_i$ using $S_{i-1}$. Observe that $S_{i-1} \cup \{v_i\}$ is a feedback vertex set of $T_i$, so a minimum feedback vertex set of $T_i$ has size $|S_{i-1}|$ or $|S_{i-1}|+1$. Therefore, using the compression routine, we either find a feedback vertex set $S_i$ of $T$ of size $|S_{i-1}|$, or we prove that none exists, in which case we set $S_i = S_{i-1} \cup \{v_i\}$. At the end, after $n-2$ calls to the compression routine, we obtain $S_n$, a minimum feedback vertex set of $T$.

Let us now describe the compression routine running in $O(2^k \cdot n^2))$ time Let $T$ be a tournament and $S$ a feedback vertex set of size $k+1$. By brute-force, we enumerate all $O(2^k)$ partitions $(X, S \setminus X)$ of $S$, and for each of them we only look for feedback vertex sets that contain all vertices of $S \setminus X$ and none of $X$.

We delete the vertices of $S \setminus X$, i.e. $T' := T - (S \setminus X)$. Observe that $T$ has a feedback vertex set of size $k$ that contains all vertices of $S \setminus X$ and none of $X$ if and only if $T'$ has a feedback vertex set of size $|X| - 1$ disjoint from $X$. If $T'\langle X\rangle$ is not acyclic, we stop as there cannot be any feedback vertex set of $T'$ disjoint from $X$. Hence we may assume that $T'\langle X\rangle$ is acyclic. Note also that $T' - X = T - S$ is acyclic.

We shall now determine the minimum size $s$ of a feedback vertex set of $T'$ disjoint from $X$. In fact, we compute $|T'| - s$, which is the maximum size of an acyclic subtournament of $T'$ containing all of $X$. Such a tournament has an acyclic ordering which can be thought of as resulting from the insertion of a subset of $V(T') \setminus X$ into the acyclic ordering $(x_1, \dots, x_{|X|})$ of $X$.

We first determine the set $P$ of vertices $v$ that we can insert into $X$, that are the vertices such that $T'\langle X \cup \{v\}\rangle$ is acyclic. Note that such a vertex has a unique possible position in $L$: there is an index $i(v) = i$ such that $N^-(v) \cap X = \{x_1, \dots, x_i\}$ and $N^+(v) \cap X = \{x_{i+1}, \dots, x_n\}$. Note that for each vertex $v \in V(T') \setminus X$, deciding if $v \in P$ and, if so, computing $i(v)$ can be done in $O(n^2)$ time. Let $L$ be an acyclic ordering of $T' - X$ (it exists because $T' - X$ is acyclic), and let $R = (r_1, \dots, r_{|P|})$ be the ordering of $P$ in which the vertices are ordered in increasing order of $i(v)$ and according to $L$ as tie-breaker: for any $j < \ell$, either $i(r_j) < i(r_\ell)$, or $i(r_j) = i(r_\ell)$ and $r_j$ is before $r_\ell$ in $L$. Now a largest acyclic subtournament of $T'$ containing all of $X$ is obtained from $X$ by adding a longest common subsequence of $L$ and $P$. Since $L$ and $P$ are permutations of each other, finding a longest common subsequence reduces to finding a longest increasing subsequence of the intersection. This can be done in $O(n \log n)$ time [90]. □

Dom, Guo, Hüffner, Niedermeier and Truß [71] also proved that FVST admits a cubic kernel. The idea of the proof is to transform an instance of FVST $(T, k)$ into an equivalent instance $(H, k)$ of HITTING SET, where $H$ is the 3-uniform hypergraph with vertex set $V(T)$ and hyperedge set the sets of 3-cycles in $T$. Then applying the kernelization algorithm given by Niedermeier and Rossmanith [150] for HITTING SET, one can show that the resulting instance has cubic size.

### 2.9.4 FPT Algorithms for FEEDBACK ARC SET IN TOURNAMENTS

Downey, Langston, Niedermeier, Raman, and Saurabh [157] proved that FAST is FPT providing a $O(2.42^k \cdot n^{O(1)})$-time algorithm for this problem. Alon, Lokshtanov and Saurabh [7] gave a faster algorithm that runs in $2^{O(\sqrt{k} \log^2 k)} + n^{O(1)}$ time. Their algorithm combines the colour coding technique (initiated in [11]) with a divide-and-conquer algorithm and a quadratic kernel for FAST. The existence of such a kernel was established by Dom, Guo, Hüffner, Niedermeier and Truß [71].

**Theorem 2.9.7** ([71]) FEEDBACK ARC SET IN TOURNAMENTS *admits a quadratic kernel. In particular, it is FPT.*

**Proof:** Here we use the fact that fas$(D)$ is the minimum size of a set of arcs whose reversal makes the digraph acyclic (see Proposition 2.9.1). The kernelization procedure `FastKer`$(T, k)$ proceeds as follows.

1. If a vertex $v$ is in no directed 3-cycle, then return `FastKer`$(T - v, k)$;
2. If $|T| = 0$, then return a 'Yes' instance;
3. If $k = 0$, then return a 'No' instance;
4. If there is an arc $a$ in more than $k$ directed 3-cycles, then let $T'$ be the tournament obtained from $T$ by reversing $a$ and return `FastKer`$(T', k-1)$;
5. If $|T| \le k(k+1)$, return $(T, k)$, otherwise return a 'No' instance.

Clearly, FastKer$(T, k)$ runs in $O(kn^3)$ time. Clearly, Steps 1 to 4 of FastKer are valid, since a feedback arc set of size $k$ must contain each arc which is in more than $k$ directed 3-cycles. At Step 5, all arcs are in less than $k$ directed 3-cycles. Hence if $T$ has a feedback arc set of size $k$ it has at most $k(k-1)$ directed 3-cycles, spanning at most $k(k+1)$ vertices. Since every vertex is in a directed 3-cycle after Step 1, $|T| \leq k(k+1)$. Hence Step 5 is valid. □

Finally, the existence of a linear-size kernel for FAST has been proved by Cuzzocrea, Taniar, Bessy, Fomin, Gaspers, Paul, Perez, Saurabh, and Thomassé [67].

## 2.10 Small Certicates for $k$-(Arc)-Strong Connectivity

By a **certificate** for the $k$-(arc)-strong connectivity of a digraph $D$, we mean a spanning $k$-(arc)-strong subdigraph $D'$ of $D$. Already for $k = 1$ it is $\mathcal{NP}$-hard to find a certificate with the minimum number of arcs, as this number is $|V(D)|$ if and only if $D$ is Hamiltonian. Since every vertex in a $k$-(arc)-strong digraph has out-degree at least $k$, an optimal certificate for $k$-(arc)-strong connectivity has at least $kn$ arcs.

Together with Edmonds' branching theorem (Theorem 1.8.2) the next result implies that, in polynomial time, one can find a certificate for $k$-arc-strong connectivity with at most twice the size of an optimal certificate.

**Proposition 2.10.1** *Every $k$-arc-strong digraph contains a spanning $k$-arc-strong subdigraph with at most $2k(n-1)$ arcs. Furthermore, such a certificate can be constructed in polynomial time.*

**Proof:** Let $D = (V, A)$ be a $k$-arc-strong and let $s \in V$ be arbitrary. By Edmonds' branching theorem, $D$ has $k$ arc-disjoint out-branchings $B^+_{s,1}, \ldots, B^+_{s,k}$ and $k$ arc-disjoint in-branchings $B^-_{s,1}, \ldots, B^-_{s,k}$. The union of the arcs of these $2k$ branchings is clearly $k$-arc-strong and it has exactly $2k(n-1)$ arcs. The complexity claim follows from Theorem 1.8.2. □

For all $k \geq 1$ and $n \geq 5k + 2$, we define $\mathcal{T}_{n,k}$ as the class of tournaments that can be obtained from a transitive tournament $A = TT_k$ on $k$ vertices and two $k$-arc-strong tournaments $B, C$ as shown in Figure 2.4. It is not difficult to show that each tournament in $\mathcal{T}_{n,k}$ is $k$-arc-strong.

Let $T$ be any member of $\mathcal{T}_{n,k}$. Observe that every $k$-arc-strong subdigraph $D$ of $T$ must contain at least $k(k+1)/2$ arcs from $B$ to $A$ and exactly $k$ arcs from $C$ to $B$ (there are no more). Hence we have $[\sum_{x \in B} d^+_D(x)] - [\sum_{x \in B} d^-_D(x)] \geq k(k+1)/2 - k$, implying that $\sum_{x \in B} d^+_D(x) \geq k|B| + k(k-1)/2$. This implies that $D$ has at least $nk + k(k-1)/2$ arcs. Thus the tournaments in $\mathcal{T}_{n,k}$ show the existence of $k$-arc-strong tournaments for which every certificate has at least $nk + ck^2$ arcs for some constant $c > 0$ and hence the

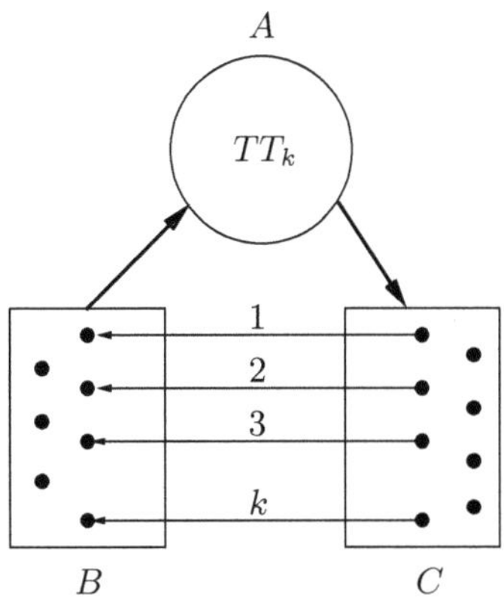


**Figure 2.4** The structure of the tournaments in $\mathcal{T}_{n,k}$. The tournament $A$ is the transitive tournament on $k$ vertices, $B$ and $C$ are arbitrary $k$-arc-strong tournaments. The bold arcs $B \to A, A \to C$ indicate that all possible arcs are present in that direction. There are exactly $k$ arcs from $C$ to $B$ and all other arcs go from $B$ to $C$

following result of Bang-Jensen, Huang and Yeo is the best possible in terms of the exponent on $k$.

**Theorem 2.10.2** ([28]) *For any $n \geq 3$ and $k \geq 1$, every $k$-arc-strong tournament on $n$ vertices $T$ contains a spanning $k$-arc-strong subdigraph with at most $nk + 136k^2$ arcs.*

The following result can be shown using network flows.

**Proposition 2.10.3** ([28]) *Every $k$-arc-strong tournament contains a spanning subdigraph $D$ on at most $nk + k(k-1)/2$ arcs such that $\delta^0(D) \geq k$.*

By the remark in Theorem 2.10.2, the truth of the following conjecture, due to Bang-Jensen, Huang and Yeo, would imply that the right bound in Theorem 2.10.2 would be $nk + k(k-1)/2$.

**Conjecture 2.10.4** ([28]) *For every $k$-arc-strong tournament $T$, the minimum number of arcs in a $k$-arc-strong spanning subdigraph of $T$ is equal to the minimum number of arcs in a spanning subdigraph of $T$ with the property that every vertex has in- and out-degree at least $k$.*

Bang-Jensen asked [18] whether a result similar to Theorem 2.10.2 would also hold for vertex connectivity. This was confirmed recently by Kang, Kim, Kim and Suh.

**Theorem 2.10.5** ([117]) *For $k \geq 1$, every $k$-strong $n$-tournament $T$ has a $k$-strong spanning subdigraph with at most $nk + 750k^2 \log(k+1)$ arcs.*

The proof of this result is long and uses several results on linkages in tournaments. Some of the methods are very similar to those used in [156].

Below we prove a weaker, yet interesting, result from [117] which is used in the proof of Theorem 2.10.5 in [117].

Let $t, k$ be positive integers with $t \geq k$. For a given ordering $\mathcal{O} = (v_1, v_2, \ldots, v_n)$ of the vertices of a digraph $D = (V, A)$ we denote by $F_{\mathcal{O}}$ the set of arcs $v_i v_j$ with $i < j$ and call such arcs **forward** arcs wrt. $\mathcal{O}$. An ordering $\mathcal{O} = (v_1, v_2, \ldots, v_n)$ of the vertices of a digraph $D$ is $(k, t)$-**good** if $D_{\mathcal{O}} = (V, F_{\mathcal{O}})$ satisfies

(a) $d^+_{D_F}(v_i) \geq k$ for all $i \in [n-t]$,
(b) $d^-_{D_F}(v_j) \geq k$ for all $t+1 \leq j \leq n$.

The following lemma is a special case of a lemma proved by Kang, Kim, Kim and Suh [117].

**Lemma 2.10.6** ([117]) *Let $k \geq 1$ be an integer and let $T$ be an $n$-tournament. Then there exists an ordering $\mathcal{O}$ of $V(T)$ and a spanning subdigraph $D'$ of $T_{\mathcal{O}}$ such that $D'$ is $(k, 2k-1)$-good and $|A(D')| \leq kn - k$.*

The following lemma is similar to Theorem 2.5.13.

**Lemma 2.10.7** ([117]) *Let $k \geq 1$ and $n \geq 5k$ be integers. Every $n$-tournament $T$ contains disjoint sets of vertices $X, Y$, each of size $k$ such that, for any set $S$ of $k-1$ vertices, the tournament $T - S$ has an $(x, y)$-path for every choice of $x \in X \setminus S, y \in Y \setminus S$.*

Let $v$ be a vertex of a $k$-strong digraph $D$ and let $Z = \{z_1, z_2, \ldots, z_k\}$ be a set of $k$ vertices in $V(D) \setminus v$. A $(\mathbf{v}, \mathbf{Z})$-**fan** (resp. $(\mathbf{Z}, \mathbf{v})$-**fan**) is a collection of internally disjoint paths $P_1, \ldots, P_k$ (resp. $Q_1, Q_2, \ldots, Q_k$) such that $P_i$ (resp. $Q_i$) is a $(v, z_i)$-path (resp. $(z_i, v)$-path). It is an easy consequence of Menger's theorem that every $k$-strong digraph has such a fan for arbitrary $v$ and $Z$ as above. We denote it by $F^+_{v,Z}$ (resp. $F^-_{Z,v}$). Note that it has at most $n-1$ arcs it is an out-tree (resp. in-tree) in $D$.

**Theorem 2.10.8** ([117]) *Let $k$ be a positive integer. Every $k$-strong $n$-tournament $T$ contains a $k$-strong spanning subdigraph $D$ with $|A(D)| \leq (5k-2)n + \binom{5k}{2}$.*

**Proof:** Set $V := V(T)$. If $n \leq 5k$, we let $D$ be $T$ itself. So assume $n > 5k$ and let $V' \subset V$ be an arbitrary set of $5k$ vertices. By Lemma 2.10.7, we can find two disjoint $k$-sets $X, Y$ such that for every $S \subset V$ with $|S| = k-1$ and every choice of $x \in X \setminus S, y \in Y \setminus S$ the tournament $T[V' \setminus S]$ has an $(x, y)$-path. Applying Lemma 2.10.6, we obtain an ordering $\mathcal{O}$ of $V(T)$ and a spanning $(k, 2k-1)$-good subdigraph $D'$ of $D_{\mathcal{O}}$ such that $|A(D')| \leq kn - k$. For each $n - 2k + 2 \leq i \leq n$, let $F_{v_i,X}$ be a $(v_i, X)$-fan in $T$, and, for each $1 \leq i \leq 2k-1$, let $F_{Y,v_i}$ be a $(Y, v_i)$-fan. Now define the spanning digraph $D^* = (V, A^*)$ to be the union of all the arcs in $T[V']$, $D'$, $F_{v_{n-2k+2},X}, \ldots, F_{v_n,X}$, $F_{Y,v_1}, \ldots, F_{Y,v_{2k-1}}$. By the remark on the size of fans above, it is easy to check that $|A(D^*)| \leq (5k-2)n + \binom{5k}{2}$. We now prove that $D^*$ is $k$-strong. To show this, let $S$ be any subset of $k-1$ vertices and let $u, v \in V \setminus S$ be arbitrary. We need to show that $D^* - S$ has a $(u,v)$-path. Because $D'$ is $(k, 2k-1)$-good, in $D' - S$ there is a $(u, v_i)$-path $P$ for some $n-2k+2 \leq i \leq n$ and a $(v_j, v)$-path $P'$ for some $j \in [2k-1]$ (recall that $D'$ is acyclic so every directed path moves forward in the ordering). After deleting the vertices of $S$ from the fans $F_{v_i,X}$ and $F_{Y,v_j}$ there still remains at least one intact path in each of these (as there are $k$ internally disjoint such paths). Let $x_s \in X, y_s \in Y$ be such that $F_{v_i,X} - S$ contains a $(v_i, x_s)$-path $P_{v_i,x_s}$ and $F_{Y,v_j} - S$ contains a $(y_s, v_j)$-path $P_{y_s,v_j}$. By Lemma 2.10.7, $T[V' \setminus S]$ has an $(x_s, y_s)$-path $P''$. Now the subdigraph of $D^* - S$ formed by the arcs of $P, P', P'', P_{v_i,x_s}$ and $P_{y_s,v_j}$ contains a $(u,v)$-path and we are done. □

## 2.11 Increasing Connectivity by Adding or Reversing Arcs

In this section we consider the following problems for semicomplete digraphs

(1) Given a digraph $D = (V, A)$ on at least $k+1$ vertices for some positive integer $k$, find a minimum set $F$ of new arcs such that the digraph $D' = (V, A \cup F)$ is $k$-strong. Let $a_k(D) = |F|$.
(2) Given a digraph $D = (V, A)$ on at least $2k+1$ vertices for some positive integer $k$, find a minimum set $F \subset A$ of arcs in $D$ such that the digraph $D'$ obtained from $D$ by reversing every arc in $F$ is $k$-strong. Let $r_k(D) = |F|$.

Clearly,

$$a_k(D) \leq r_k(D), \tag{2.8}$$

since, instead of reversing arcs in $D$, we may add exactly those new arcs we would obtain by reversing and keep the original ones.

Frank and Jordán showed that $a_k(D)$ can be computed in polynomial time [88, 89]. The number $r_1(D)$ can be calculated via submodular flows (see e.g. [22, Section 13.1]). For $k \geq 2$, it is not clear how we can decide whether $r_k(D)$ even exists for a given arbitrary digraph $D$, let alone find an optimal reversal (unless we try all possibilities, which clearly requires exponential time). Indeed, this seems to be a very difficult problem.

We will now show that for semicomplete digraphs $D$, the function $r_k(D)$ behaves nicely. Note that, since we are dealing with vertex-connectivity, we gain nothing by reversing arcs that are contained in 2-cycles. Hence below we only consider arcs that are not contained in 2-cycles for possible reversal.

**Proposition 2.11.1** ([29]) *If a semicomplete digraph $D$ has at least $2k+1$ vertices, then $r_k(D)$ exists and is bounded by a function depending only on $k$.*

**Proof:** To see this it suffices to use the following two simple observations; the proof of the first one is left to the reader, and the second one follows directly from Proposition 2.2.2 and its directional dual.

(a) If $D$ is a $k$-strong digraph and $D'$ is obtained from $D$ by adding a new vertex $x$ and arcs from $x$ to every vertex in a set $X$ of $k$ distinct vertices of $D$ and arcs from every vertex of a set $Y$ of $k$ distinct vertices of $D$ to $x$, then $D'$ is also $k$-strong.
(b) If $D$ is a semicomplete digraph on at least $4k-1$ vertices, then $D$ contains a vertex with in-degree and out-degree at least $k$.

By observations (a) and (b), for every semicomplete digraph $D$, $r_k(D) \leq r_k(D')$ for some induced subdigraph $D'$ of $D$ with $|V(D')| \leq 4k-2$. We can find such a subdigraph $D'$ as follows: Continue removing vertices as long as the current semicomplete digraph graph has at least $2k+2$ vertices and a vertex of in-degree and out-degree at least $k$. When this process stops, we have $2k+1 \leq |V(D')| \leq 4k-2$ in the current semicomplete digraph $D'$. Then we can make $D'$ $k$-strong by reversing some arcs and add back each of the removed vertices in the reverse order of the deletion. This provides a simple upper bound for $r_k(D)$ (and hence for $a_k(D)$) as a function of $k$: we need to reverse at most half of the arcs in $D'$, that is, at most $\frac{(4k-2)(4k-3)}{4}$ arcs. □

The process above may not lead to an optimal reversal for the original semicomplete digraph (in terms of the number of arcs to reverse), not even if we reverse optimally in $D'$.

It is easy to see that $r_k(TT_n) = k(k+1)/2$ when $n \geq 2k+1$. Bang-Jensen conjectured that no other tournament needs more reversals.

**Conjecture 2.11.2 (Bang-Jensen** [22]) *For every tournament $T$ with $|V(T)| = n \geq 2k+1$, we have $r_k(T) \leq k(k+1)/2$.*

Since every semicomplete digraph contains a spanning tournament, if Conjecture 2.11.2 is true, this implies that the same conclusion holds for semicomplete digraphs on at least $2k+1$ vertices.

Bang-Jensen and Jordán showed that as soon as the number of vertices in the given semicomplete digraph $D$ is sufficiently high (depending only on $k$), the minimum number of arcs in $D$ we need to reverse in order to achieve a $k$-strong semicomplete digraph equals the minimum number of new arcs we need to add to $D$ to obtain a $k$-strong semicomplete digraph.

**Theorem 2.11.3** ([29]) *Let $k \geq 2$ be an integer. If $D$ is a semicomplete digraph on at least $3k-1$ vertices, then $a_k(D) = r_k(D)$.*

The idea, which also leads to a polynomial algorithm for finding the desired reversal (see [29]), is to show that $r_k(D) \leq a_k(D)$, by demonstrating that a certain optimal augmenting set $F$ of $D$ has the property that, if we reverse the existing (opposite) arcs of $F$ in $D$, then we obtain a $k$-strong semicomplete digraph. It was shown in [29] that $3k-1$ is the best possible bound for semicomplete digraphs. However, in the case when $D$ is a tournament, the question as to whether or not the bound is the best possible was left open and the following conjecture was implicitly formulated.

**Conjecture 2.11.4** ([29]) *For every tournament $T$ on at least $2k+1$ vertices, we have $a_k(T) = r_k(T)$.*

Now we turn to arc-strong connectivity, where we shall see that the analogous problem to the one above has been solved.

Let $r_k^{deg}(D)$ be the minimum number of arcs one needs to reverse in a directed multigraph $D$ in order to obtain a directed multigraph $D'$ with $\delta^0(D') \geq k$. If no such reversal exists, we let $r_k^{deg}(D) = \infty$. Determining $r_k^{deg}$ for a given digraph can be formulated as a feasibility flow problem and is thus polynomial (see e.g. [22, Section 14.5.1]). Analogously define $r_k^{arc}(D)$ to be the minimum number of arcs one needs to reverse in $D$ in order to obtain a $k$-arc-strong directed multigraph.

By the Nash-Williams orientation Theorem [147], $r_k^{arc}(D) < \infty$ precisely when $UMG(D)$ is $2k$-edge-connected and one can calculate $r_k^{arc}(D)$ (including detecting whether $r_k^{arc}(D) = \infty$) in polynomial time using submodular flows (see e.g. [22, Section 11.8]). It follows from the results below that for tournaments the function $r_k^{arc}$ can be calculated just using standard maximum-flow calculations.

The following result by Bang-Jensen and Yeo shows that for tournaments $r_k^{deg}(T)$ and $r_k^{arc}(T)$ are always bounded by a function that depends only on $k$.

**Theorem 2.11.5** ([36]) *Let $T$ be an $n$-tournament, with $n \geq 2k+1$. The following hold:*

*(i)* $r_k^{deg}(T) \leq k(k+1)/2$.
*(ii)* $r_k^{arc}(T) = \max\{k-\lambda(T), r_k^{deg}(T)\}$.

Observe that combining (i) and (ii) of Theorem 2.11.5, we obtain $r_k^{arc}(T) \leq k(k+1)/2$ which provides support to Conjecture 2.11.2. Recall again that the transitive tournaments show that this is the best possible.

The proof in [36] of Theorem 2.11.5 can be turned into a polynomial algorithm for finding a set of $q$ arcs whose reversal makes $T$ $k$-arc-strong using just maximum-flow calculations.

We now consider the operation of deorienting an arc. Let $xy$ be an arc of a digraph $D$ which is not in a 2-cycle. By **deorienting** $xy$ we mean the operation which adds the arc $yx$ to $D$. Clearly, deorienting arcs is equivalent to adding new arcs with the restriction that we can only add an arc which is opposite to an existing arc and we cannot create parallel arcs. Hence we may view deorienting arcs as a restricted version of the arc addition operation.

Let $deor_k^{deg}(D)$ denote the minimum number of arcs we need to deorient in $D$ in order to obtain a digraph $D'$ with $\delta^0(D') \geq k$. Using flows one can determine $deor_k^{deg}(D)$ for an arbitrary digraph $D$ ([22, Exercise 14.18]). Clearly $deor_k^{deg}(D) \leq r_k^{deg}(D)$ for every oriented graph, in particular for every tournament. The example in Figure 2.5 shows that this inequality does not always hold for semicomplete digraphs.

Bang-Jensen and Yeo proved that for tournaments deorienting arcs is generally no better than reversing in terms of obtaining a desired minimum degree.

**Theorem 2.11.6** ([36]) *Let $T$ be a tournament on at least $2k+1$ vertices. Then $deor_k^{deg}(T) = r_k^{deg}(T)$. In particular, $deor_k^{deg}(T) \leq k(k+1)/2$.*

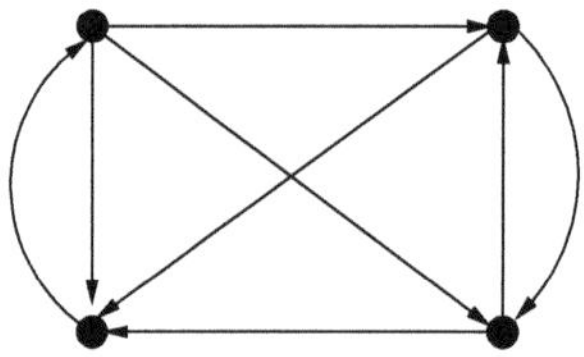

**Figure 2.5** A semicomplete digraph $D$ for which $1 = r_2^{arc}(D) < deor_2^{arc}(D) = 2$

Analogously define $deor_k^{arc}(D)$ to be the minimum number of arcs one needs to deorient in $D$ in order to obtain a $k$-arc-strong digraph. It is easy to see that $deor_k^{arc}(D) < \infty$ if and only if $UG(D)$ is $k$-edge-connected. Furthermore, if $D$ is an oriented graph (in particular, if $D$ is a tournament), then we have $deor_k^{arc}(D) \leq r_k^{arc}(D)$ since instead of reversing an optimal set $A'$ of arcs we may deorient these arcs and obtain a digraph with minimum semi-degree at least $k$. Figure 2.5 shows that the inequality above may not hold when $D$ contains 2-cycles.

The following is a corollary of the Lucchesi–Younger theorem [138] about covering of directed cuts in a digraph.

**Theorem 2.11.7** *Let $D$ be a non-strong digraph for which $UG(D)$ is 2-edge-connected. Then $deor_1^{arc}(D) = r_1^{arc}(D)$.* □

When $k \geq 2$ and $D$ is an arbitrary digraph, we do not know how to determine $deor_k^{arc}(D)$ efficiently, but as we show below, this is possible when $D$ is a tournament.

One might expect that $deor_k^{arc}(D) < r_k^{arc}(D)$ for most oriented graphs. The next result, due to Bang-Jensen and Yeo, shows that for tournaments the two numbers are equal and hence, with respect to increasing the arc-strong connectivity of a tournament, there is no gain from deorienting arcs rather than reversing arcs.

**Theorem 2.11.8** ([36]) *For every tournament $T$ on at least $2k+1$ vertices we have $deor_k^{arc}(T) = r_k^{arc}(T)$.*

**Proof:** We saw in Theorem 2.11.5 that $r_k^{arc}(T) = \max\{k - \lambda(T), r_k^{deg}(T)\}$. If $r_k^{arc}(T) = r_k^{deg}(T)$, then, by Theorem 2.11.6, we have

$$\begin{aligned} deor_k^{arc}(T) &\leq r_k^{arc}(T) \\ &= r_k^{deg}(T) \\ &= deor_k^{deg}(T) \\ &\leq deor_k^{arc}(T), \end{aligned}$$

implying that $deor_k^{arc}(T) = r_k^{arc}(T)$. So we may assume that $r_k^{arc}(T) = k - \lambda(T)$. Now the claim follows from the easy fact that $deor_k^{arc}(T) \geq k - \lambda(T)$. □

We argued above that, in polynomial time, for a given tournament $T$, we can find a set of arcs $A' \subset A(T)$ of size $r_k^{arc}(T)$ such that reversing the arcs of $A'$ results in a $k$-arc-strong tournament. Thus it follows from Theorem 2.11.8 that, in polynomial time, we can determine $deor_k^{arc}(T)$ and find a set of $deor_k^{arc}(T)$ arcs to deorient such that the resulting semicomplete digraph is $k$-arc-strong. One optimal set of arcs to deorient is simply a set that would form an optimal reversal.

**Problem 2.11.9** ([36]) *Let $k \geq 1$ be a fixed integer. Is there a polynomial algorithm for determining the number $deor_k^{arc}(D)$ for a given input $D$?*

As we saw above the answer is yes if either $k = 1$ or if $D$ is a tournament, but even the case of semicomplete digraphs and $k = 2$ is open. We also do not know whether there exists a polynomial algorithm for general oriented graphs when $k = 2$. Recall that for any digraph $D$ and positive integer $k$ the number $deor_k^{deg}(D)$ can be calculated in polynomial time via flows.

Analogously to the definition of $deor_k^{arc}(D)$ we may define $deor_k(D)$ to denote the minimum number of arcs we need to deorient in $D$ in order to obtain a $k$-strong digraph. Clearly $deor_k(D) < \infty$ precisely when $UG(D)$ is $k$-connected. We have $deor_1(D) = deor_1^{arc}(D)$ for every digraph but for higher values of $k$ nothing is known about $deor_k(D)$ for general digraphs. Notice that when $D$ is a semicomplete digraph on at least $3k-1$ vertices we have $deor_k(D) = a_k(D)$ by Theorem 2.11.3.

## 2.12 Arc-Disjoint Spanning Subdigraphs of Semicomplete Digraphs

Below we discuss results on arc-disjoint Hamiltonian cycles, strong spanning subdigraphs and in- and out-branchings.

### 2.12.1 Arc-Disjoint Hamiltonian Paths and Cycles

Let $T$ be a non-strong tournament and let $T_1, T_2, \ldots, T_k$ be the acyclic ordering of its strong components. Two components $T_i, T_{i+1}$ are called **consecutive** for $i = 1, 2, \ldots, k-1$.

Thomassen [181] completely characterized tournaments having a pair of arc-disjoint Hamiltonian paths.

**Theorem 2.12.1** ([181]) *A tournament $T$ fails to have two arc-disjoint Hamiltonian paths if and only if $T$ has a strong component which is an almost transitive tournament of odd order or $T$ has two consecutive strong components of order 1.* ◇

Thomassen posed the following problem.

**Problem 2.12.2** ([181]) *What is the complexity of deciding whether a tournament has a Hamiltonian path $P$ and a Hamiltonian cycle $C$ which are arc-disjoint?*

Thomassen solved this problem for tournaments that are arc-3-cyclic (that is, every arc is contained in a 3-cycle) [181]. Moon proved that almost all tournaments are arc-3-cyclic [146] so Thomassen's result covers almost all tournaments.

**Theorem 2.12.3** ([181]) *Every arc-3-cyclic $n$-tournament with $n \geq 6$ has a Hamiltonian path and a Hamiltonian cycle which are arc-disjoint.*

Observe that Theorem 2.12.1 implies that every 2-arc-strong tournament has two arc-disjoint Hamiltonian paths. Thomassen [181] conjectured the existence of a function $h(k)$ such that every $h(k)$-strong tournament contains $k$ arc-disjoint Hamiltonian cycles. He proved that $h(2) \geq 3$ and conjectured that equality holds. The existence of $h(k)$ was recently verified by Kühn, Lapinskas, Osthus and Patel [124] who proved that $h(k) \in O(k^2 \log^2(k))$ suffices. They conjectured that $h(k) \in O(k^2)$ would suffice. This was confirmed by Pokrovskiy.

**Theorem 2.12.4** ([155]) *There exists a constant $C$ such that every $Ck^2$-strong tournament contains $k$ arc-disjoint Hamiltonian cycles.*

By Theorem 2.6.19, $h(2) = 3$ would follow from the following conjecture due to Bang-Jensen and Yeo.

**Conjecture 2.12.5** ([35]) *Every tournament $T$ either contains two arc-disjoint Hamiltonian cycles or a set $A'$ of at most two arcs such that $T \setminus A'$ has no Hamiltonian cycle.*

Confirming a conjecture of Erdős, Kühn and Osthus proved the following. Here 'almost all' means that the probability of a random $n$-tournament having the desired property tends to 1 as $n$ tends to infinity.

**Theorem 2.12.6** ([126]) *Almost all tournaments have $\delta^0(T)$ arc-disjoint Hamiltonian cycles.*

Now we turn to decompositions into arc-disjoint Hamiltonian cycles. Clearly any digraph which has an arc-decomposition into Hamiltonian cycles must be regular. Tillson characterized when one can decompose the arc set of the complete digraph into arc-disjoint Hamiltonian cycles.

**Theorem 2.12.7** ([189]) *The complete digraph $\overleftrightarrow{K}_k$ can be decomposed into arc-disjoint Hamiltonian cycles if and only if $k \neq 4, 6$.*

The following conjecture, due to Kelly (see [146]), is the most famous open problem on tournaments.

**Conjecture 2.12.8 (Kelly, 1968)** *Every regular $n$-tournament can be partitioned into $(n-1)/2$ Hamiltonian cycles.*

This conjecture has attracted a lot of attention and a number of partial or closely related results have been obtained, e.g. [42, 103, 113, 119, 181, 183, 196]

The major breakthrough on the Kelly conjecture was made by Kühn and Osthus who proved the conjecture for (very) large $n$.

**Theorem 2.12.9** ([126]) *For $k$ sufficiently large, every $k$-regular tournament decomposes into $k$ arc-disjoint Hamiltonian cycles.*

The proof in [126] is very long, almost 100 pages. It still remains a major challenge to prove Conjecture 2.12.8 in full.

For $k$-regular semicomplete digraphs, we do not necessarily have $k$-arc-disjoint Hamiltonian cycles. For $k = 2$, one such example is obtained from a 4-cycle by adding a 2-cycle between the two pairs of vertices of distance 2 along the cycle.

**Problem 2.12.10** *What is the complexity of deciding whether a given regular semicomplete digraph has a decomposition into arc-disjoint Hamiltonian cycles?*

It follows from Theorem 2.12.9 that for regular tournaments there is a polynomial algorithm to decide whether the given tournament has a decomposition into Hamiltonian cycles. Of course, if Kelly's conjecture is true, then there is a trivial algorithm, because the answer will always be 'yes'.

Let $T$ bethe tournament on $n = 4m+2$ vertices obtained from two regular tournaments $T_1$ and $T_2$, each on $2m+1$ vertices, by adding all arcs from the vertices of $T_1$ to $T_2$. Clearly $T$ is not strong and so has no Hamiltonian cycle. The minimum semi-degree of $T$ is $m = \frac{n-2}{4}$. One can easily prove that every

$n$-tournament with $\delta^0(T) \geq \frac{n}{4}$ is strongly connected. Bollobás and Häggkvist [45] showed that if we increase the minimum semi-degree slightly, then, not only do we obtain many arc-disjoint Hamiltonian cycles, we also obtain a very structured set of such cycles provided that the tournament has enough vertices.

**Theorem 2.12.11** ([45]) *For every $\epsilon > 0$ and every positive integer $k$, there is an integer $n(\epsilon, k)$ with the following property. If $T$ is a tournament of order $n > n(\epsilon, k)$ such that $\delta^0(T) \geq (\frac{1}{4} + \epsilon)n$, then $T$ contains the $k$th power of a Hamiltonian cycle.* ◇

### 2.12.2 Arc-Disjoint Spanning Strong Subdigraphs

In this subsection, we study the decomposition of digraphs into strong subdigraphs. Since adding an arc to a strong digraph results in another strong digraph, a digraph decomposes into $k$ arc-disjoint spanning strong subdigraphs if and only of it contains $k$ arc-disjoint spanning strong subdigraphs.

Bang-Jensen and Yeo posed the following conjecture, which contains the Kelly conjecture (Conjecture 2.12.8) as the special case when $n = 2k + 1$.

**Conjecture 2.12.12 (Bang-Jensen and Yeo** [35]) *A tournament $T$ can be decomposed into $k$ arc-disjoint spanning strong subdigraphs if and only if $T$ is $k$-arc-strong.*

They proved this conjecture for $k = 2$ and also characterized the 2-strong semicomplete digraphs that have an arc decomposition into two spanning strong subdigraphs.

Let $S_{2k}$ be the semicomplete digraph which one obtains from two disjoint copies of the complete digraph $\overleftrightarrow{K}_k$ by adding a perfect matching oriented from one copy to the other and adding all remaining arcs in the opposite direction.

**Lemma 2.12.13** ([35]) *The semicomplete digraph $S_{2k}$ decomposes into $k$-arc-disjoint spanning strong subdigraphs except when $k = 2$.*

The following theorem implies that Conjecture 2.12.12 holds for $k = 2$.

**Theorem 2.12.14** ([35]) *Let $D$ be a 2-arc-strong semicomplete digraph, on $n$ vertices. Then $D$ decomposes in two arc-disjoint spanning strong subdigraphs if and only if it is not isomorphic to $S_4$.*

Below we shall give a proof Conjecture 2.12.12 for the class of $k$-arc-strong tournaments which have a non-trivial $k$ arc-cut (Theorem 2.12.17). The proof, which is due to Bang-Jensen and Yeo, uses Theorem 2.12.7 and Theorem 2.12.16, which can be deduced from the following result of Smetanuik on completion of partial Latin squares.

**Theorem 2.12.15** ([170]) *Let $B$ be a complete bipartite graph (undirected), with $n$ vertices in each partite set, and let $R$ be a set of edges in $B$ such that $|R| \leq n-1$. Then we can decompose $E(B)$ into $n$ edge-disjoint matchings $M_1, M_2, ..., M_n$ such that $|M_i \cap R| \leq 1$ for all $i = 1, 2, ..., n$.*

**Theorem 2.12.16** ([170]) *Let $B = (X, Y, E)$ be an undirected complete bipartite graph with $|X| = t$, $|Y| = s$ and $t > s$. Let $R$ be a set of edges in $B$ such that $|R| \leq s$. Then we can colour the edges of $B$ by $|R|$ colours in such a way that all edges in $R$ receive distinct colours and every vertex in $X \cup Y$ is incident with all $|R|$ colours.*

**Theorem 2.12.17** *Let $k \geq 1$ and let $D$ be a $k$-arc-strong semicomplete digraph such that there exists a set $S \subset V(D)$ with $2 \leq |S| \leq |V(D)| - 2$ and $d^+(S) = k$. There exist $k$ arc-disjoint strong spanning subgraphs of $D$ except if $D = S_4$.*

**Proof:** By Lemma 2.12.13 we may assume that $D$ is not isomorphic to $S_4$.

It is not difficult to show that $k \leq |S| \leq n-k$ (by showing that $|S| \geq k$ and $|V(D) - S| \geq k$, respectively). If $|S| = |V - S| = 2$ then $D$ contains $S_4$ as a proper spanning subdigraph and it is easy to check that adding any arc to $S_4$ will result in a digraph with two arc-disjoint strong spanning subdigraphs. Hence we may assume that $n \geq 5$. Let $e_1, e_2, \dots, e_k$ be the $k$ arcs from $S$ to $V(D) - S$, and let $e_i = x_iy_i$, for $i = 1, 2, \dots, k$. Let $X = \{x_1, x_2, \dots, x_k\}$ and $Y = \{y_1, y_2, \dots, y_k\}$. Note that we may have $|X| < k$ or $|Y| < k$ or both. We may assume, by reversing all arcs if necessary, that $|V - S| \geq |S|$.

By Lemma 2.12.13 and the remark above, we may assume that $|V - S| > |S|$ if $|S| = k$. By Theorem 2.12.16 (with $R = \{e_1, e_2, \dots, e_k\}$) we can colour all arcs between $S$ and $V(D) - S$ with $k$ colours such that the arcs from $S$ to $V(D) - S$ get different colours and every vertex in $V$ is incident with arcs of all $k$ colours. Note that if $|V - S| = |S| > k$ this follows from Theorem 2.12.15.

Assume, without loss of generality, that the arc $x_iy_i$ is coloured with colour $i$, and let $F_i$ contain all arcs between $S$ and $V(D) - S$ of colour $i$.

By Theorem 1.8.2 there exists $k$ arc-disjoint out-branchings $U_1, U_2, \dots, U_k$, in $D[V(D) - S]$ such that $U_i$ is rooted at $y_i$, for $i = 1, 2, \dots, k$ (consider $k$ arc-disjoint out-branchings from any vertex in $S$. Each of these must contain exactly one of the arcs $e_1, e_2, \dots, e_k$. Thus the out-branching that contains the arc $e_i$ must contain an out-branching from $y_i$ in $D[V(D) - S]$). Analogously, there exists $k$ arc-disjoint in-branchings $V_1, V_2, \dots, V_k$, in $D[S]$ such that $V_i$ is rooted at $x_i$, for $i = 1, 2, \dots, k$. Let $T_i = V_i \cup U_i \cup F_i$, for $i = 1, 2, \dots, k$. Clearly $T_1, T_2, \dots, T_k$ are arc-disjoint and spanning. Each $T_i$ is furthermore strong: by the construction of the colouring, every vertex in $V$ is incident to an arc of colour $i$, every vertex in $V(D) - S - y_i$ is the tail of an arc in $T_i$ into $S$, and hence every vertex in $V$ can reach $y_i$ (via $V_i$ and the arc $x_iy_i$) and every vertex in $S - x_i$ is the head of an arc from $V(U_i)$ in $T_i$, implying that in $T_i$ all vertices can be reached by $y_i$ and reach $x_i$. This completes the proof. □

The following theorem, due to Bang-Jensen and Yeo, implies that we can always obtain about $\frac{1}{37}\lambda(T)$ arc-disjoint spanning strong subdigraphs in any tournament $T$. Note that in the case when $\lambda(T) < 37k$ the result below follows from Theorem 2.12.17.

**Theorem 2.12.18** ([35]) *Let $T$ be a $k$-arc-strong tournament, with minimum semi-degree $\delta^0(T) \geq 37k$. Then there exists $k$ arc-disjoint spanning strong subgraphs in $T$.*

### 2.12.3 Arc-Disjoint In- and Out-Branchings

We now turn to branchings and consider the following problem

ARC-DISJOINT IN- AND OUT-BRANCHINGS
**Input:** A digraph $D$ and vertices $u, v$ (not necessarily distinct).
**Question:** Does $D$ have a pair of arc-disjoint branchings $B_u^+, B_v^-$ such that $B_u^+$ is an out-branching rooted at $u$ and $B_v^-$ is an in-branching rooted at $v$?

The following result was proved by Thomassen [16].

**Theorem 2.12.19** ARC-DISJOINT IN- AND OUT-BRANCHINGS *is $\mathcal{NP}$-complete for arbitrary digraphs.*

Bang-Jensen and Huang showed that, if the vertex that is to be the root is adjacent to all other vertices in the digraph and is not in any 2-cycle, then the problem becomes polynomially solvable.

**Theorem 2.12.20** ([27]) *Let $D = (V, A)$ be a strongly connected digraph and $v$ a vertex of $D$ such that $v$ is not on any 2-cycle and $V(D) = \{v\} \cup N^-(v) \cup N^+(v)$. Let $\mathcal{A} = \{A_1, A_2, \ldots, A_k\}$ ($\mathcal{B} = \{B_1, B_2, \ldots, B_r\}$) denote the set of terminal (initial) components in $D\langle N^+(v)\rangle$ ($D\langle N^-(v)\rangle$). Then $D$ contains a pair of arc-disjoint branchings $B_v^+, B_v^-$ such that $B_v^+$ is an out-branching rooted at $v$ and $B_v^-$ is an in-branching rooted at $v$ if and only if there exist two disjoint arc sets $E_{\mathcal{A}}, E_{\mathcal{B}} \subset A$ such that all arcs in $E_{\mathcal{A}} \cup E_{\mathcal{B}}$ go from $N^+(v)$ to $N^-(v)$ and every $A_i \in \mathcal{A}$ ($B_j \in \mathcal{B}$) is incident with an arc from $E_{\mathcal{A}}$ ($E_{\mathcal{B}}$). Furthermore, there exists a polynomial algorithm to find the desired branchings, or demonstrate the non-existence of such branchings.*

This implies the following result due to Bang-Jensen.

**Corollary 2.12.21** ([16]) *A tournament $T = (V, A)$ has arc-disjoint branchings $B_v^+, B_v^-$ rooted at a specified vertex $v \in V$ if and only if $T$ is strong and for every arc $a \in A$ the digraph $T - a$ contains either an out-branching or an in-branching with root $v$.*

When $u \neq v$, ARC-DISJOINT IN- AND OUT-BRANCHINGS becomes much harder even for semicomplete digraphs. Bang-Jensen [16] found a complete characterization for the case of tournaments. This characterization, which is quite complicated, implies the tournament case of the following Theorem by Bang-Jensen and Yeo.

**Theorem 2.12.22** ([35]) *Every 2-arc-strong semicomplete digraph $T = (V, A)$ contains arc-disjoint in- and out-branchings $B_r^-, B_s^+$ for every choice of vertices $r, s \in V$.*

**Proof:** This follows from Lemma 2.12.13 since it is easy to show that the semicomplete digraph $S_4$, which is the unique exception to that theorem, has arc-disjoint in- and out-branchings $B_u^-, B_v^+$ for every choice of $u, v \in V(S_4)$. □

Bang-Jensen found a polynomial algorithm for ARC-DISJOINT IN- AND OUT-BRANCHINGS in the case of tournaments.

**Theorem 2.12.23** ([16]) *There is a polynomial algorithm for checking whether a given tournament with specified distinct vertices $u, v$ has arc-disjoint branchings $B_u^+, B_v^-$ and finding such branchings if they exist.* □

Thomassen conjectured that every digraph which has sufficiently high arc-strong connectivity has arc-disjoint in- and out-branchings for every choice of roots.

**Conjecture 2.12.24** ([178]) *There exists a positive integer $N$ such that every digraph $D$ which is $N$-arc-strong has arc-disjoint branchings $B_v^+, B_v^-$ for every choice of $v \in V(D)$.*

Bang-Jensen and Yeo generalized this as follows.

**Conjecture 2.12.25** *There exists a positive integer $N$ such that every digraph $D$ which is $N$-arc-strong has two arc-disjoint spanning strong subdigraphs.*

Theorem 2.12.14 implies that the conjecture holds with $N = 3$ for semicomplete digraphs and with $N = 2$ for tournaments. The following consequence of Theorem 2.12.18 verifies a conjecture by Bang-Jensen and Gutin [23].

**Theorem 2.12.26** ([35]) *Let $T$ be $74k$-arc-strong tournament. Then $T$ has $2k$ arc-disjoint branchings $B_{v,1}^+, \ldots, B_{v,k}^+, B_{v,1}^-, \ldots, B_{v,k}^-$ such that $B_{v,1}^+, \ldots, B_{v,k}^+$ are out-branchings rooted at $v$ and $B_{v,1}^-, \ldots, B_{v,k}^-$ are in-branchings rooted at $v$, for every vertex $v \in V(T)$.*

Note that if Conjecture 2.12.12 is true then we may replace $74k$ by $2k$.

**Conjecture 2.12.27** ([35]) *Theorem 2.12.26 also holds if we replace* $74k$ *by* $2k$.

## 2.13 Minors of Semicomplete Digraphs

The most important advance in graph theory in the last few decades is certainly the Robertson–Seymour minor theory and by now the minor relation for graphs is well-established. However it is not clear how it should be extended to digraphs. A **minor** of a graph $G$ is usually defined as a graph that can be obtained from a subgraph of $G$ by contracting edges. Unfortunately, in digraphs, contracting an arc may yield a directed cycle, even when we are starting from an acyclic digraph, and this seems undesirable for a theory of excluded minors. One way to avoid this is to permit the contraction only of certain special arcs; for instance, in the paper [114] by Johnson, Robertson, Seymour and Thomas, an arc $uv$ can be contracted if it is either the only arc with tail $u$ or the only arc with head $v$. Another way, called **shallow directed minors**, has been introduced by Kreutzer and Tazari in [123]. A third approach comes from the observation that a minor of a graph $G$ can also be defined as a graph that can be obtained from a subgraph of $G$ by contracting connected subgraphs. Therefore Kim and Seymour [120] defined a **minor** of a digraph $D$ as a digraph that can be obtained from a subdigraph of $D$ by contracting strong subdigraphs.

An important property of minors for graphs is that they define a **well quasi-order** as shown by Robertson and Seymour [163]. (Recall that a quasi-order $\leq$ is a reflexive and transitive relation, and that it is a well quasi-order if for every infinite sequence $q_1, q_2, \dots$ there exist $j > i$ such that $q_i \leq q_j$.) The analogous statement is not true for directed minors. For example, a directed cycle is not a minor of a bigger directed cycle, and so if we take an infinite set of directed cycles, all of different lengths, then this set is an infinite antichain under the minor order. However, Kim and Seymour [120] proved that minor containment defines a well quasi-order for the class of all semicomplete digraphs, and therefore the same is true for the class of all tournaments.

**Theorem 2.13.1** ([120]) *Minor containment is a well quasi-order on the class of all semicomplete digraphs.*

Kim and Seymour [120] also showed that this result cannot be generalized to larger classes of digraphs. In particular, they showed that minor containment is not a well quasi-order on the class of all digraphs with independence number 2. Indeed, consider the digraphs $D_i$, $i \geq 2$, defined as follows:

- $V(D_i)$ is the disjoint union of $C_i$, $C'_i$, $T_i$ and $T'_i$;
- $D_i\langle C_i\rangle$ and $D_i\langle C'_i\rangle$ are directed 3-cycles;

- $D_i\langle T_i\rangle$ and $D_i\langle T_i'\rangle$ are transitive tournaments with Hamiltonian directed paths $t_1 \dots t_i$ and $t_i' \dots t_1'$, respectively;
- $C_i \to T_i$ and $T_i' \to C_i'$;
- there is exactly one arc with tail in $C_i'$ and head in $C_i$;
- for every $1 \le j \le i$, $\{t_j, t_{j+1}\} \to t_j'$ with $t_{i+1} = t_1$.

One can check that there do not exist $j > i \ge 2$ such that $D_i$ is a minor of $D_j$.

In [64], Chudnovsky and Seymour proved that immersion is a well quasi-order on the class of all tournaments, by using the parameter cutwidth (see the definition in Section 2.5.1). This was recently extended to the class of all semicomplete digraphs by Barbero, Paul and Pilipczuk [38].

Kim and Seymour proved Theorem 2.13.1 by using another parameter called path-width. For a digraph $D$, a sequence $(W_1, \dots, W_r)$ of subsets of $V(D)$ is a **path decomposition** of $D$ if it satisfies the following conditions:

- $\bigcup_{i=1}^{r} W_i = V(D)$;
- for $1 \le h < i < j \le r$, $W_h \cap W_j \subseteq W_i$; and
- if $uv \in A(D)$, then $u \in W_i$ and $v \in W_j$ for some $i \ge j$.

The **width** of such a path decomposition is defined to be the number $\max\{|W_i| - 1 \mid 1 \le i \le r\}$. The **path-width** of $D$ is the smallest width of a path-decomposition. For example, if $v_1, \dots, v_n$ is an acyclic ordering of an acyclic digraph, then $(\{v_1\}, \dots, \{v_n\})$ is a path-decomposition of this digraph of width 0. Hence every acyclic digraph has path-width 0.

Having bounded path-width is a minor-closed property.

**Lemma 2.13.2** ([120]) *If a digraph has path-width at most $k$, then so do all its minors.*

**Proof:** Let $(W_1, \dots, W_r)$ be a path-decomposition of a digraph $D$ with width at most $k$. It is also a path-decomposition of $D \setminus a$ for every arc $a \in A(D)$ and $(W_1 \setminus \{v\}, \dots, W_r \setminus \{v\})$ is a path-decomposition of $D - v$ for every vertex $v \in V(D)$. Hence, it remains to show that for every strong subdigraph $H$, the digraph $D/H$ obtained from $D$ by contracting $H$ into a vertex $w$ has path-width at most $k$.

Let $I_H = \{i \mid W_i \cap V(H) \neq \emptyset\}$. One can check that $i_H$ is an interval and that the path-decomposition $(W_1', \dots, W_r')$ defined by $W_i' = (W_i \setminus V(H)) \cup \{w\}$ if $i \in i_H$ and $W_i' = W_i$ otherwise is a path-decomposition of $G/H$ of width at most $k$. □

The $k$-**triple** is the digraph $T_k$ defined by

$$V(T_k) = \{a_1, \dots, a_k, b_1, \dots, b_k, c_1, \dots, c_k\}, \text{and}$$
$$E(T_k) = \{a_i b_j \mid 1 \le i \le k, 1 \le j \le k\} \cup \{b_i c_j \mid 1 \le i \le k, 1 \le j \le k\} \cup \{c_i a_i \mid 1 \le i \le k\}.$$

Observe that every semicomplete digraph with $k$ vertices is a minor of the $k$-triple $T_k$. Indeed, set $B = \{b_1, \dots, b_k\}$; then $D\langle\{a_i, b_i, c_i\}\rangle$ is strong for each $i$. The digraph $D'$ obtained from $D$ by contracting $D\langle\{a_i, b_i, c_i\}\rangle$ to a vertex for each $i$ is the complete symmetric digraph of order $k$.

A theorem of Fradkin and Seymour [86] says that a semicomplete digraph $D$ has large path-width if and only if it contains a large $k$-triple.

**Theorem 2.13.3** ([86]) *Let $\mathcal{S}$ be a set of semicomplete digraphs. The following two statements are equivalent:*

1. *There exists a positive integer $k_1$ such that for each $D \in \mathcal{S}$, there is no $k_1$-triple in $D$.*
2. *There exists a positive integer $k_2$ such that each $D \in \mathcal{S}$ has path-width at most $k_2$.*

Hence in order to prove Theorem 2.13.1, Kim and Seymour [120] proved the following result.

**Theorem 2.13.4** ([120]) *Minor containment is a well quasi-order on the class of all semicomplete digraphs with path-width at most $k$.*

**Proof of** 2.13.1 **assuming Theorem** 2.13.4**:** Let $D_1, D_2, \dots$ be an infinite sequence of semicomplete digraphs. We may assume that $D_1$ is not a minor of $D_i$ for each $i \geq 2$. Set $k_1 = |D_1|$. By the above observation $D_1$ is a minor of $T_{k_1}$, so $T_{k_1}$ is not contained in $D_i$ for each $i \geq 2$. Hence, by Theorem 2.13.3, there exists a $k_2$ such that every $D_i$, $i \geq 2$, has path-width at most $k_2$. Thus, by Theorem 2.13.4, there exists $j > i \geq 2$ such that $G_i$ is a minor of $G_j$. □

## 2.14 Miscellaneous Topics

In the next few subsections we briefly survey results and problems on a few further topics on tournaments.

### 2.14.1 Arc-Pancyclicity

As mentioned earlier, Moon proved that almost all tournaments are arc-3-cyclic [146], so for tournaments this is not a very hard restriction.

Tian, Wu and Zhang characterized all tournaments that are arc-3-cyclic but not arc-pancyclic. See Figure 2.6 for the definition of the classes $\mathcal{D}_6, \mathcal{D}_8$.

**Theorem 2.14.1** ([188]) *An arc-3-cyclic tournament is arc-pancyclic unless it belongs to one of the families $\mathcal{D}_6, \mathcal{D}_8$ (in which case the arc $yx$ belongs to no Hamiltonian cycle).*

**Corollary 2.14.2** ([188]) *Every arc-3-cyclic tournament has at most one arc which is not in cycles of all lengths $3, 4, \dots, n$.*

The following result due to Wu, Zhang and Zou is also a corollary of Theorem 2.14.1.

**Corollary 2.14.3** ([193]) *A tournament is arc-pancyclic if and only if it is arc-3-cyclic and arc-n-cyclic.*

The following result due to Alspach is also an easy corollary:

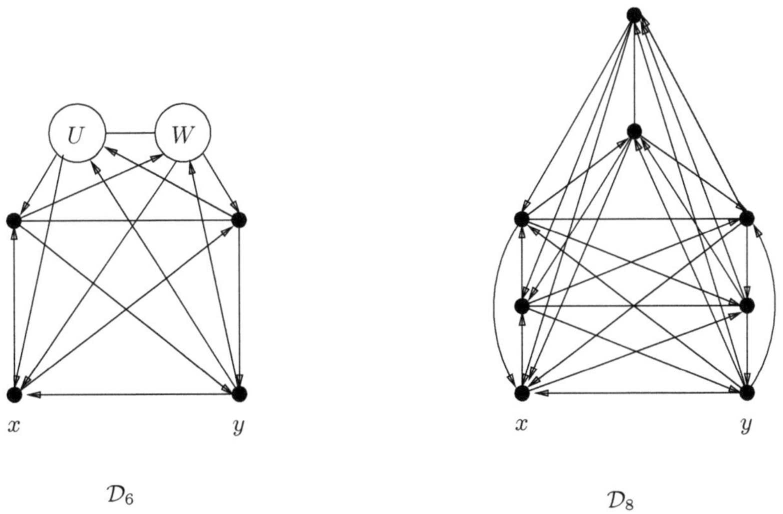


**Figure 2.6** The two families of non-arc-pancyclic arc-3-cyclic tournaments. Each of the sets $U$ and $W$ induces an arc-3-cyclic tournament. All edges that are not already oriented may be oriented arbitrarily, but all arcs between $U$ and $W$ have the same direction

**Corollary 2.14.4** ([12]) *Every regular tournament is arc-pancyclic.*

Finally, observe that since each tournament in the infinite family $\mathcal{D}_6$ is 2-strong and the arc $yx$ is not in any Hamiltonian cycle we obtain the following result due to Thomassen:

**Theorem 2.14.5** ([184]) *There exist infinitely many 2-strong tournaments containing an arc which is not in any Hamiltonian cycle.*

**Problem 2.14.6** *Characterize arc-pancyclic semicomplete digraphs.*

A partial result on this problem was obtained by Darrah, Liu and Zhang [68].

A vertex $u$ in a digraph $D$ is **out-pancyclic** if every arc whose tail is $u$ is contained in cycles of all lengths $3, 4, \ldots, |V(D)|$. Clearly $D$ is arc-pancyclic if and only if every vertex of $D$ is out-pancyclic and hence it is of interest to study out-pancyclic vertices in tournaments and semicomplete digraphs.

When $T$ is a strong tournament with $\delta^+(T) \geq 2$, Yao, Guo and Zhang [194] call a vertex $v \in V(T)$ a **bridgehead** if there is a 2-partition $(V_1, V_2)$ of $V(T)$ such that $|V_1| \geq 2$, $T[V_1]$ is strong and there is no arc from $V_1 \setminus \{v\}$ to $V_2$. It is easy to check that every tournament of minimum out-degree at least 2 contains a vertex which is not a bridgehead.

**Theorem 2.14.7** ([194]) *Let $T$ be a strong tournament on $n$ vertices and let $u_1, u_2, \ldots, u_n$ be a labelling of its vertices so that $d^+(u_1) \leq d^+(u_2) \leq \ldots \leq d^+(u_n)$. Let $u$ be the vertex $u_1$ if $d^+(u_1) = 1$ and otherwise $u$ is a vertex of minimum out-degree among those that are not bridgeheads. Then $u$ is an out-pancyclic vertex.*

**Corollary 2.14.8** ([194]) *Every strong tournament has an out-pancyclic vertex.*

Yao *et al.* [194] constructed an infinite family of strong tournaments with exactly one out-pancyclic vertex.

**Conjecture 2.14.9** ([194]) *Every $k$-strong tournament has at least $k$ out-pancyclic vertices.*

When $r_i \geq k$, $i \in [3]$ the tournament $C_3[TT_{r_1}, TT_{r_2}, TT_{r_3}]$ is $k$-strong and has exactly 3 out-pancyclic vertices, namely the vertices with out-degree 0 in each of the three transitive tournaments [195]. Yeo conjectured that every 2-strong tournament contains three out-pancyclic vertices and this was confirmed by Guo, Guo, Li, Li and Zhao.

**Theorem 2.14.10** ([100, 101]) *Every strong tournament $T$ with $\delta^+(T) \geq 2$ contains at least three out-pancyclic vertices and this is the best possible.*

See [108, 195] for results and conjectures by Havet and Yeo on the number of pancyclic arcs in tournaments as well as the number of pancyclic arcs contained in the same Hamiltonian cycle.

### 2.14.2 Critically $k$-Strong Tournaments

A digraph is **critically k-strong** if $D$ is $k$-strong but $\kappa(D - v) = k - 1$ for all $v \in V$. When $k = 1$ such digraphs are also called **critically strong**. The structure of critically strong digraphs is surprisingly complicated, see the paper [139] by Mader. By Corollary 2.2.10 the only critically strong semicomplete digraph is the 3-cycle. For larger connectivities Thomassen gave a construction which shows that the situation is quite different.

**Theorem 2.14.11 (Thomassen [22] Section 5.7)** *For every $k \geq 3$, there are infinitely many critically $k$-strong tournaments.*

See [22, Figure 5.9] for an infinite family of critically 3-strong tournaments. Let us call a tournament $T$ **minimally k-strong** if $T$ is $k$-strong but no proper subtournament of $T$ is $k$-strong. We saw above that there are arbitrarily large critically-$k$-strong tournaments. Lichiardopol conjectured [133] that this is not the case for minimally $k$-strong tournaments.

**Conjecture 2.14.12** ([133]) *For every integer $k \geq 1$ there exists a function $f(k)$ such that every minimally $k$-strong tournament has at most $f(k)$ vertices.*

### 2.14.3 Subdivisions and Linkages

A famous conjecture due to Lovász (see e.g. [182, page 262]) states that for every positive integer $k$ there exists an integer $r(k)$ such that for every pair of vertices $x, y$ in a $r(k)$-connected graph $G$ we can find an induced $(x, y)$-path $P$ such that $G - V(P)$ is $k$-connected. Thomassen proved the following tournament version of Lovász's conjecture.

**Theorem 2.14.13** ([179]) *Let $k$ be a positive integer and let $T$ be a $(k+4)$-strong tournament. Then for every pair of vertices $x, y$ and every shortest $(x, y)$-path $P$ the tournament $T - V(P)$ is $k$-strong.*

Kim, Kühn and Osthus generalized this as follows. Theorem 2.14.13 is obtained by taking $d = 2$ and $m = 1$.

**Theorem 2.14.14** ([121]) *Let $k, d, m$ be positive integers. Suppose that $T$ is a $(k + m(d + 2))$-strong tournament, that $X$ is a set of $d$ vertices of $T$, that $H$ is a digraph on $d$ vertices and $m$ arcs and that $\phi$ is a bijection from $V(H)$ to $X$. Then $T$ contains a subdivision $H^*$ of $H$ such that*

*(i) for each $h \in V(H)$, the branch vertex of $H^*$ corresponding to $h$ is $\phi(h)$,*
*(ii) $T - V(H^*)$ is $k$-strong,*
*(iii) for every arc $a$ of $H$, the path $P_a$ of $H^*$ is minimal.*

## References

1. L. Addario-Berry, F. Havet, C.L. Sales, B. Reed, and S. Thomassé. Oriented trees in digraphs. *Discrete Math.*, 313(8):967–974, 2013.
2. L. Addario-Berry, F. Havet, and S. Thomassé. Paths with two blocks in $n$-chromatic digraphs. *J. Combin. Theory Ser. B*, 97:620–626, 2007.
3. N. Ailon, M. Charikar, and A. Newman. Aggregating inconsistent information: ranking and clustering. In *STOC 2005: 37th Annual ACM Symp. on Theory of Computing*, pages 684–693. ACM Press, 2005.
4. N. Alon. Disjoint directed cycles. *J. Combin. Theory Ser. B*, 68(2):167–178, 1996.
5. N. Alon. Ranking tournaments. *SIAM J. Discrete Math.*, 20:137–142, 2006.

6. N. Alon, J. Bang-Jensen, and S. Bessy. Out-colourings of digraphs, manuscript 2017.
7. N. Alon, D. Lokshtanov, and S. Saurabh. Fast FAST. In *ICALP 2009: 36th International Colloquium on Automata, Languages and Programming, Part I*, volume 5555 of *Lect. Notes Comput. Sci.*, pages 49–58, 2009.
8. N. Alon and J.H. Spencer. *The Probabilistic Method.* John Wiley & Sons, New York, 1992. With an appendix by Paul Erdős.
9. N. Alon and J.H. Spencer. *The Probabilistic Method.* John Wiley & Sons, New York, 2nd edition, 2000.
10. N. Alon and J.H. Spencer. *The Probabilistic Method.* John Wiley & Sons, New York, 4th edition, 2016.
11. N. Alon, R. Yuster, and U. Zwick. Color-coding. *J. Assoc. Comput. Mach.*, 42:844–856, 1995.
12. B. Alspach. Cycles of each length in regular tournaments. *Can. Math. Bull.*, 10:283–285, 1967.
13. B. Alspach and M. Rosenfeld. Realization of certain generalized paths in tournaments. *Discrete Math.*, 34:199–202, 1981.
14. G. Araujo-Pardo and M. Olsen. A conjecture of Neumann-Lara on infinite families of r-dichromatic circulant tournaments. *Discrete Math.*, 310:489–492, 2010.
15. S. Arora, A. Frieze, and H. Kaplan. A new rounding procedure for the assignment problem with applications to dense graph arrangement problems. *Math. Program.*, 92(1):1–36, 2002.
16. J. Bang-Jensen. Edge-disjoint in- and out-branchings in tournaments and related path problems. *J. Combin. Theory Ser. B*, 51(1):1–23, 1991.
17. J. Bang-Jensen. On the 2-linkage problem for semicomplete digraphs. In *Graph theory in memory of G. A. Dirac (Sandbjerg, 1985)*, volume 41 of *Ann. Discrete Math.*, pages 23–37. North-Holland, 1989.
18. J. Bang-Jensen. Problems and conjectures concerning connectivity, paths, trees and cycles in tournament-like digraphs. *Discrete Math.*, 309:5655–5667, 2009.
19. J. Bang-Jensen, S. Bessy, and S. Thomassé. Disjoint 3-cycles in tournaments: a proof of the Bermond-Thomassen conjecture for tournaments. *J. Graph Theory*, 75:284–302, 2014.
20. J. Bang-Jensen and T.M. Christiansen. Degree constrained 2-partitions of semicomplete digraphs. manuscript.
21. J. Bang-Jensen, N. Cohen, and F. Havet. Finding good 2-partitions of digraphs II. Enumerable properties. *Theor. Comput. Sci.*, 640:1–19, 2016.
22. J. Bang-Jensen and G. Gutin. *Digraphs: Theory, Algorithms and Applications.* Springer-Verlag, London, 2nd edition, 2009.
23. J. Bang-Jensen and G. Gutin. Generalizations of tournaments: A survey. *J. Graph Theory*, 28:171–202, 1998.
24. J. Bang-Jensen and G. Gutin. Paths, trees and cycles in tournaments. *Congr. Numer.*, 115:131–170, 1996. Surveys in graph theory (San Francisco, 1995).
25. J. Bang-Jensen, G. Gutin, and A. Yeo. Hamiltonian cycles avoiding prescribed arcs in tournaments. *Combin. Prob. Comput.*, 6(3):255–261, 1997.
26. J. Bang-Jensen and F. Havet. Finding good 2-partitions of digraphs I. Hereditary properties. *Theor. Comput. Sci.*, 636:85–94, 2016.
27. J. Bang-Jensen and J. Huang. Quasi-transitive digraphs. *J. Graph Theory*, 20(2):141–161, 1995.
28. J. Bang-Jensen, J. Huang, and A. Yeo. Spanning $k$-arc-strong subdigraphs with few arcs in $k$-arc-strong tournaments. *J. Graph Theory*, 46(4):265–284, 2004.

29. J. Bang-Jensen and T. Jordán. Adding and reversing arcs in semicomplete digraphs. *Combin. Prob. Comput.*, 7(1):17–25, 1998.
30. J. Bang-Jensen and T. Jordán. Spanning 2-strong tournaments in 3-strong semicomplete digraphs. *Discrete Math.*, 310:1424–1428, 2010.
31. J. Bang-Jensen, A. Maddaloni, and S. Simonsen. Quasi-hamiltonian paths in semicomplete multipartite digraphs. *Discrete Appl. Math.*, 161:889–898, 2013.
32. J. Bang-Jensen, Y. Manoussakis, and C. Thomassen. A polynomial algorithm for Hamiltonian-connectedness in semicomplete digraphs. *J. Algor.*, 13(1):114–127, 1992.
33. J. Bang-Jensen and M.H. Nielsen. Finding complementary cycles in locally semicomplete digraphs. *Discrete Appl. Math.*, 146(3):245–256, 2005.
34. J. Bang-Jensen and C. Thomassen. A polynomial algorithm for the 2-path problem for semicomplete digraphs. *SIAM J. Discrete Math.*, 5:366–376, 1992.
35. J. Bang-Jensen and A. Yeo. Decomposing $k$-arc-strong tournaments into strong spanning subdigraphs. *Combinatorica*, 24(3):331–349, 2004.
36. J. Bang-Jensen and A. Yeo. Making a tournament $k$-arc-strong by reversing or deorienting arcs. *Discrete Appl. Math.*, 136(2-3):161–171, 2004.
37. F. Barbero, C. Paul, and M. Pilipczuk. Exploring the Complexity of Layout Parameters in Tournaments and Semi-Complete Digraphs. In *ICALP 2017: 44th International Colloquium on Automata, Languages, and Programming*, volume 80 of *LIPIcs*, pages 70:1–70:13, 2017.
38. F. Barbero, C. Paul, and M. Pilipczuk. Strong immersion is a well-quasi-ordering for semicomplete digraphs. arXiv:1707.03563.
39. A. Benhocine and A.P. Wojda. On the existence of specified cycles in a tournament. *J. Graph Theory*, 7:469–473, 1983.
40. J. Bensmail, A. Harutyunyan, N.K. Le, B. Li, and N. Lichiardopol. Disjoint cycles of different lengths in graphs and digraphs. arXiv:1510.06667v2.
41. E. Berger, K. Choromanski, M. Chudnovsky, J. Fox, M. Loebl, A. Scott, P.D. Seymour, and S. Thomassé. Tournaments and colouring. *J. Combin. Theory Ser. B*, 103(1):1 – 20, 2013.
42. J.-C. Bermond and C. Thomassen. Cycles in digraphs—a survey. *J. Graph Theory*, 5(1):1–43, 1981.
43. S. Bessy, M. Bourgeret, and J. Thiebaut. Triangle packing in (sparse) tournaments: approximation and kernelization. arXiv:1707.04220.
44. A. Blass, G. Exoo, and F. Harary. Paley graphs satisfy all first-order adjacency axioms. *J. Graph Theory*, 5:435–439, 1981.
45. B. Bollobás and R. Häggkvist. Powers of Hamilton cycles in tournaments. *J. Combin. Theory Ser. B*, 50(2):309–318, 1990.
46. B. Bollobás and A. Thomason. Graphs which contain all small graphs. *European J. Combin.*, 2:13–15, 1981.
47. A. Bonato, P. Gordinowicz, and P. Pralat. Bounds and constructions for $n$-e.c. tournaments. *Contributions to Discrete Math.*, 5:52–66, 2010.
48. J.A. Bondy. Diconnected orientations and a conjecture of Las Vergnas. *J. London Math. Soc. (2)*, 14(2):277–282, 1976.
49. J.A. Bondy. Short proofs of classical theorems. *J. Graph Theory*, 44(3):159–165, 2003.
50. J.A. Bondy and U.S.R. Murty. *Graph Theory*, volume 244 of *Graduate Texts in Mathematics.* Springer-Verlag, Berlin, 2008.
51. M. Bozóki. Nontransitive dice sets realizing the Paley tournaments for solving Schüte's tournament problem. *Miskolc Math. Notes*, 15:39–50, 2014.
52. R.A. Brualdi and K. Kiernan. Landau's and Rado's Theorems and partial tournaments. *Elec. J. Combin.*, 16:# N2, 2009.

53. S.A. Burr. Subtrees of directed graphs and hypergraphs. *Congress. Numer.*, 28:227–239, 1980.
54. L. Caccetta and R. Häggkvist. On minimal digraphs with given girth. *Congr. Numer.*, 21:181–187, 1978.
55. M.-C. Cai, X. Deng, and W. Zang. An approximation algorithm for feedback vertex sets in tournaments. *SIAM J. Comput.*, 30(6):1993–2007, 2001.
56. P. Camion. Chemins et circuits hamiltoniens des graphes complets. *C. R. Acad. Sci. Paris*, 249:2151–2152, 1959.
57. S. Ceroi and F. Havet. Trees with three leaves are (n+1)-unavoidable. *Discrete Appl. Math.*, 141(13):19–39, 2004. Brazilian Symposium on Graphs, Algorithms and Combinatorics.
58. P. Charbit, S. Thomassé, and A. Yeo. The minimum feedback arc set problem is NP-hard for tournaments. *Combin. Prob. Comput.*, 16:1–4, 2007.
59. G. Chen, R.J. Gould, and H. Li. Partitioning Vertices of a Tournament into Independent Cycles. *J. Combin. Theory Ser. B*, 83:213–220, 2001.
60. G. Chen, J. Shen, and R. Yuster. Second neighborhood via first neighborhood in digraphs. *Ann. Combin.*, 7(1):15–20, 2003.
61. X. Chen, X. Hu, and W. Zang. A min-max theorem on tournaments. *SIAM J. Comput.*, 37(3):923–937, 2007.
62. M. Chudnovsky, A. Scott, and P.D. Seymour. Disjoint paths in tournaments. *Adv. Math.*, 270:582–597, 2015.
63. M. Chudnovsky, A. Scott, and P.D. Seymour. Disjoint paths in unions of tournaments. arXiv:1604.02317, 2016.
64. M. Chudnovsky and P.D. Seymour. A well-quasi-order for tournaments. *J. Combin. Theory Ser. B*, 101(1):47–53, 2011.
65. N. Cohen, F. Havet, W. Lochet, and N. Nisse. Subdivisions of oriented cycles in digraphs with large chromatic number. arXiv:1605.07762, May 2016.
66. Z. Cohn, A. Godbole, E. Wright Harkness, and Y. Zhang. The Number of Seymour Vertices in Random Tournaments and Digraphs. *Graphs Combin.*, pages 1–12, 2016.
67. A. Cuzzocrea, D. Taniar, S. Bessy, F. Fomin, S. Gaspers, C. Paul, A. Perez, S. Saurabh, and S. Thomassé. Kernels for feedback arc set in tournaments. *J. Comput. Syst. Sci.*, 77(6):1071–1078, 2011.
68. M. Darrah, Y.-P. Liu, and C.-Q. Zhang. Cycles of all lengths in arc-3-cyclic semicomplete digraphs. *Discrete Math.*, 173(1-3):23–33, 1997.
69. N. Dean and B.J. Latka. Squaring the tournament—an open problem. *Congr. Numer.*, 109:73–80, 1995.
70. W.F. de la Vega. On the maximum cardinality of a consistent set of arcs in a random tournament. *J. Combin. Theory Ser. B*, 35:328–332, 1983.
71. M. Dom, J. Guo, F. Hüffner, R. Niedermeier, and A. Truß. Fixed-parameter tractability results for feedback set problems in tournaments. In *CIAC 2006: 6th Italian Conference on Algorithms and Complexity*, pages 320–331, Berlin, Heidelberg, 2006. Springer-Verlag.
72. A. El Sahili. Paths with two blocks in $k$-chromatic digraphs. *Discrete Math.*, 287:151–153, 2004.
73. A. El Sahili. Trees in tournaments. *J. Combin. Theory Ser. B*, 92(1):183–187, 2004.
74. A. El Sahili and M. Kouider. About paths with two blocks. *J. Graph Theory*, 55:221–226, 2007.
75. P. Erdős. Graph theory and probability. *Can. J. Math.*, 11:34–38, 1959.
76. P. Erdős and L. Moser. On the representation of directed graphs as unions of orderings. *Magyar Tud. Akad. Mat. Kutató Int. Közl.*, 9:125–132, 1964.

77. G. Even, J. Naor, B. Schieber, and M. Sudan. Approximating minimum feedback sets and multicuts in directed graphs. *Algorithmica*, 20(2):151–174, 1998.
78. H. Fernau. A top-down approach to search-trees: Improved algorithmics for 3-hitting set. *Algorithmica*, 57(1):97–118, 2010.
79. D. Fidler and R. Yuster. Remarks on the second neighborhood problem. *J. Graph Theory*, 55(3):208–220, 2007.
80. D.C. Fisher. Squaring a tournament: a proof of Dean's conjecture. *J. Graph Theory*, 23(1):43–48, 1996.
81. F.V. Fomin and M. Pilipczuk. Jungles, bundles and fixed-parameter tractability. In *SODA 2013: 24th Annual ACM-SIAM Symposium on Discrete Algorithms*, pages 396–413, 2013.
82. F.V. Fomin and M. Pilipczuk. Subexponential parameterized algorithm for computing the cutwidth of a semi-complete digraph. In *ESA 2013: 21st Annual European Symposium on Algorithms*, pages 505–516, 2013.
83. R. Forcade. Parity of paths and circuits in tournaments. *Discrete Math.*, 6:115–118, 1973.
84. S. Fortune, J.E. Hopcroft, and J. Wyllie. The directed subgraph homeomorphism problem. *Theor. Comput. Sci.*, 10:111–121, 1980.
85. A. Fradkin and P.D. Seymour. Edge-disjoint paths in digraphs with bounded independence number. *J. Combin. Theory Ser. B*, 110:19–46, 2015.
86. A. Fradkin and P.D. Seymour. Tournament pathwidth and topological containment. *J. Combin. Theory Ser. B*, 103(3):374–384, 2013.
87. P. Fraisse and C. Thomassen. Hamiltonian dicycles avoiding prescribed arcs in tournaments. *Graphs Combin.*, 3(3):239–250, 1987.
88. A. Frank and T. Jordán. Directed vertex-connectivity augmentation. *Math. Program. Ser. B*, 84:537–553, 1999.
89. A. Frank and T. Jordán. Minimal edge-coverings of pairs of sets. *J. Combin. Theory Ser. B*, 65(1):73–110, 1995.
90. M.L. Fredman. On computing the length of longest increasing subsequences. *Discrete Math.*, 11(1):29–35, 1975.
91. A. Frieze and R. Kannan. Quick approximation to matrices and applications. *Combinatorica*, 19(2):175–220, 1999.
92. D. Gale, H. Kuhn, and A.W. Tucker. Linear programming and the theory of games. In *Activity analysis of production and allocation*, chapter XII. John Wiley and sons, New York, 1951.
93. H. Galeana-Sánchez and R. Rojas-Monroy. A counterexample to a conjecture on edge-coloured tournaments. *Discrete Math.*, 282(1-3):275–276, 2004.
94. T. Gallai. On directed paths and circuits. In *Theory of Graphs (Proc. Colloq., Tihany, 1966)*, pages 115–118. Academic Press, 1968.
95. S. Gaspers and M. Mnich. Feedback vertex sets in tournaments. *J. Graph Theory*, 72(1):72–89, 2013.
96. I.M. Gessel and R.P. Stanley. Algebraic enumeration. In *Handbook of combinatorics, Vol. 1, 2*, pages 1021–1061. Elsevier, Amsterdam, 1995.
97. J.R. Griggs and K.B. Reid. Landau's theorem revisited. *Australas. J. Combin.*, 20:19–24, 1999.
98. B. Grünbaum. Antidirected Hamiltonian paths in tournaments. *J. Combin. Theory Ser. B*, 11:249–257, 1971.
99. Y. Guo. Spanning local tournaments in locally semicomplete digraphs. *Discrete Appl. Math.*, 79(1-3):119–125, 1997.
100. Q. Guo, Y. Li, S. Guo, and H. Li. Out-arc pancyclicity of vertices in tournaments. *Discrete Appl. Math.*, 158:996–1005, 2010.
101. Q. Guo, S. Li, H. Li, and H. Zhao. The number of vertices whose out-arcs are pancyclic in a strong tournament. *Graphs Combin.*, 30:1163–1173, 2014.

102. G. Gutin and R. Li. Seymour's second neighbourhood conjecture for quasi-transitive oriented graphs. *CoRR*, abs/1704.01389, 2017.
103. R. Häggkvist. Hamilton cycles in oriented graphs. *Combin. Prob. Comput.*, 2(1):25–32, 1993.
104. R. Häggkvist and A.G. Thomason. Trees in tournaments. *Combinatorica*, 11(2):123–130, 1991.
105. M. Hasse. Zur algebraischen Begründung der Graphentheorie. I. *Math. Nachr.*, 28:275–290, 1964/1965.
106. F. Havet. On unavoidability of trees with $k$ leaves. *Graphs Combin.*, 19(1):101–110, 2003.
107. F. Havet. Oriented Hamiltonian cycles in tournaments. *J. Combin. Theory*, 80:1–31, 2000.
108. F. Havet. Pancyclic arcs and connectivity in tournaments. *J. Graph Theory*, 47:87–110, 2004.
109. F. Havet and S. Thomassé. Median orders of tournaments: a tool for the second neighborhood problem and Sumner's conjecture. *J. Graph Theory*, 35(4):244–256, 2000.
110. F. Havet and S. Thomassé. Oriented hamiltonian paths in tournaments: a proof of Rosenfeld's conjecture. *J. Combin. Theory Ser. B*, 78:243–273, 2000.
111. F. Havet, S. Thomassé, and A. Yeo. Hoàng–Reed Conjecture holds for tournaments. *Discrete Math.*, 308:3412–3415, 2008.
112. C.T. Hoang and B. Reed. A note on short cycles in digraphs. *Discrete Math.*, 66(1-2):103–107, 1987.
113. B. Jackson. Long paths and cycles in oriented graphs. *J. Graph Theory*, 5(2):145–157, 1981.
114. T. Johnson, N. Robertson, P.D. Seymour, and R. Thomas. Directed Tree-Width. *J. Combin. Theory Ser. B*, 82(1):138–154, 2001.
115. T. Jordán. On the existence of $k$ edge-disjoint 2-connected spanning subgraphs. *J. Combin. Theory Ser. B*, 95(2):257–262, 2005.
116. Y. Kaneko and S.C. Locke. The minimum degree approach for Paul Seymour's distance 2 conjecture. *Congress. Numer.*, 148:201–206, 2001.
117. D.Y. Kang, J. Kim, Y. Kim, and G. Suh. Sparse spanning $k$-connected subgraphs in tournaments. arXiv:1603.02474.
118. R.M. Karp. Reducibility among combinatorial problems. In *Complexity of computer computations (Proc. Symp., IBM Thomas J. Watson Res. Center, Yorktown Heights, N.Y., 1972)*, pages 85–103. Plenum, 1972.
119. P. Keevash, D. Kühn, and D. Osthus. An exact minimum degree condition for Hamilton cycles in oriented graphs. *J. London Math. Soc.*, 79:144–166, 2009.
120. I. Kim and P.D. Seymour. Tournament minors. *J. Combin. Theory Ser. B*, 112:138–153, 2015.
121. J. Kim, D. Kühn, and D. Osthus. Bipartitions of highly connected tournaments. *SIAM J. Discrete Math.*, 30:895–911, 2016.
122. R. Kim, S.-J. Kim, J. Ma, and B. Park. Cycles with two blocks in $k$-chromatic digraphs. *CoRR*, abs/1610.05839, October 2016.
123. S. Kreutzer and S. Tazari. Directed nowhere dense classes of graphs. In *SODA 2012: 23rd Annual ACM-SIAM Symposium on Discrete Algorithms*, pages 1552–1562, 2012.
124. D. Kühn, D. Lapinskas, J. Osthus, and V. Patel. Proof of a conjecture of Thomassen on Hamilton cycles in highly connected tournaments. *Proc. London Math. Soc.*, 109:733–762, 2014.
125. D. Kühn, R. Mycroft, and D. Osthus. A proof of Sumner's universal tournament conjecture for large tournaments. *Proc. London Math. Soc.*, 102:731–766, 2011.

126. D. Kühn and D. Osthus. Hamilton decompositions of regular expanders: a proof of Kelly's conjecture for large tournaments. *Adv. Math.*, 237:62–146, 2013.
127. D. Kühn, D. Osthus, and T. Townsend. Proof of a tournament partition conjecture and an application to 1-factors with prescribed cycle length. *Combinatorica*, 36:451–469, 2016.
128. M. Kumar and D. Lokshtanov. Faster exact and parameterized algorithm for Feedback Vertex Set in tournaments. In *STACS 2016: 33rd Symposium on Theoretical Aspects of Computer Science*, volume 47 of *LIPIcs*, pages 49:1–49:13. Schloss Dagstuhl - Leibniz-Zentrum fuer Informatik, 2016.
129. H.G. Landau. On dominance relations and the structure of animal societies III. The condition for a score structure. *Bull. Math. Biophys.*, 15:143–148, 1953.
130. H. Li and J. Shu. The partition of a strong tournament. *Discrete Math.*, 290:211–220, 2005.
131. N. Lichiardopol. Proof of a conjecture of Henning and Yeo on vertex-disjoint directed cycles. *SIAM J. Discrete Math.*, 28:1618–1627, 2014.
132. N. Lichiardopol. Vertex-disjoint cycles in regular tournaments. *Discrete Math.*, 312:1927–1930, 2012.
133. N. Lichiardopol. Vertex-disjoint subtournaments of prescribed minimum outdegree or minimum semidegree: Proof for tournaments of a conjecture of Stiebitz. *Intern. J. Combin.*, Article ID 273416:1–9, 2012.
134. N. Lichiardopol, A. Pór, and J.-S. Sereni. A step towards the Bermond–Thomassen conjecture about disjoint cycles in digraphs. Manuscript, 2007.
135. N. Linial, M. Saks, and V.T. Sós. Largest digraphs contained in all $n$-tournaments. *Combinatorica*, 3(1):101–104, 1983.
136. B. Llano and V. Neumann-Lara. Circulant tournaments of prime order are tight. *Discrete Math.*, 308:6056–6063, 2008.
137. X. Lu. On avoidable and unavoidable trees. *J. Graph Theory*, 22:335–346, 1996.
138. C.L. Lucchesi and D.H. Younger. A minimax theorem for directed graphs. *J. London Math. Soc. (2)*, 17(3):369–374, 1978.
139. W. Mader. Critically $n$-connected digraphs. In *Graph theory, combinatorics, and applications. Vol. 2 (Kalamazoo, MI, 1988)*, Wiley-Intersci. Publ., pages 811–829. Wiley, New York, 1991.
140. E.S. Mahmoodian. A critical case method of proof in combinatorial mathematics. *Bull. Iranian Math. Soc.*, 8:1L–26L, 1978.
141. M. Mnich, V. Vassilevska Williams, and L.A. Végh. A 7/3-Approximation for Feedback Vertex Sets in Tournaments. In *The 24th Annual European Symposium on Algorithms (ESA 2016)*, volume 57 of *LIPIcs*, pages 67:1–67:14, 2016.
142. B. Mohar. Circular colorings of edge-weighted graphs. *J. Graph Theory*, 43(2):107–116, 2003.
143. J.W. Moon. On maximal transitive subtournaments. *Proc. Edinburgh Math. Soc. (Ser. 2)*, 17:345–349, 12 1971.
144. J.W. Moon. On subtournaments of a tournament. *Can. Math. Bull.*, 9:297–301, 1966.
145. J.W. Moon. Solution to problem 463. *Math. Mag.*, 35:189, 1962.
146. J.W. Moon. *Topics on tournaments.* Holt, Rinehart & Winston, New York, 1968.
147. C.St.J.A. Nash-Williams. On orientations, connectivity and odd-vertex-pairings in finite graphs. *Can. J. Math.*, 12:555–567, 1960.

148. V. Neumann-Lara. The dichromatic number of a digraph. *J. Combin. Theory Ser. B.*, 33:265–270, 1982.
149. V. Neumann-Lara. Vertex critical 4-dichromatic circulant tournaments. *Discrete Math.*, 170:289–291, 1997.
150. R. Niedermeier and P. Rossmanith. An efficient fixed-parameter algorithm for 3-hitting set. *J. Discrete Algor.*, 1(1):89–102, 2003.
151. V. Petrović. Antidirected Hamiltonian circuits in tournaments. In *Graph theory (Novi Sad, 1983)*, pages 259–269. Univ. Novi Sad, Novi Sad, 1984.
152. M. Pilipczuk. Computing cutwidth and pathwidth of semi-complete digraphs via degree orderings. In *STACS 2013: 30th International Symposium on Theoretical Aspects of Computer Science*, pages 197–208, 2013.
153. M. Pilipczuk. Computing cutwidth and pathwidth of semi-complete digraphs. In *Encyclopedia of Algorithms*, pages 412–415. Springer Verlag, 2016.
154. M. Pilipczuk. *Tournaments and optimality.* PhD thesis, University of Bergen, Normay, 2013.
155. A. Pokrovskiy. Edge disjoint Hamilton cycles in highly connected tournaments. *Intern. Math. Research Notes*, 2015:19 pp, 2015.
156. A. Pokrovskiy. Highly linked tournaments. *J. Combin. Theory Ser B*, 115:339–347, 2015.
157. V. Raman and S. Saurabh. Parameterized complexity of feedback set problems and their duals in tournaments. *Theor. Comput. Sci.*, 351:446–458, 2006.
158. L. Rédei. Ein kombinatorischer Satz. *Acta Litt. Szeged*, 7:39–43, 1934.
159. B. Reed, K. Smith, and A. Vetta. Finding odd cycle transversals. *Oper. Res. Lett.*, 32(4):299–301, 2004.
160. K.B. Reid. Tournaments: scores, kings, generalizations and special topics. *Congr. Numer.*, 115:171–211, 1996. Surveys in graph theory (San Francisco, 1995).
161. K.B. Reid. Two complementary circuits in two-connected tournaments. In *Cycles in graphs (Burnaby, B.C., 1982)*, volume 115 of *North-Holland Math. Stud.*, pages 321–334. North-Holland, 1985.
162. K.B. Reid and E.T. Parker. Disproof of a conjecture of Erdős and Moser on tournaments. *J. Combin. Theory*, 9:225–238, 1970.
163. N. Robertson and P.D. Seymour. Graph Minors. XX. Wagner's conjecture. *J. Combin. Theory Ser. B*, 92(2):325–357, 2004.
164. M. Rosenfeld. Antidirected Hamiltonian cycles in tournaments. *J. Combin. Theory Ser. B*, 16:234–242, 1974.
165. M. Rosenfeld. Antidirected Hamiltonian paths in tournaments. *J. Combin. Theory Ser. B*, 12:93–99, 1971.
166. B. Roy. Nombre chromatique et plus longs chemins d'un graphe. *Rev. Fr. Inf. Rech. Opér.*, 1(5):129–132, 1967.
167. A. Sanchez-Flores. On tournaments free of large transitive subtournaments. *Graphs Combin.*, 14(2):181–200, 1998.
168. P.D. Seymour. Packing directed circuits fractionally. *Combinatorica*, 15(2):281–288, 1995.
169. M.G. Shen. On monochromatic paths in $m$-coloured tournaments. *J. Combin. Theory Ser. B*, 45(1):108–111, 1988.
170. B. Smetaniuk. A new construction on Latin squares. *Ars. Combin.*, 11:155–172, 1981.
171. Z.M. Song. Complementary cycles of all lengths in tournaments. *J. Combin. Theory Ser. B*, 57(1):18–25, 1993.
172. E. Speckenmeyer. On feedback problems in digraphs. In *WG 1989*, volume 411 of *Lect. Notes Comput. Sci.*, pages 218–231. Springer, 1989.

173. M. Stiebitz. Decomposition of graphs and digraphs. In *KAM Series in Discrete Mathematics-Combinatorics-Operations Research-Optimization 95-309*. Charles University Prague, 1995.
174. H.J. Straight. The existence of certain type of semi-walks in tournaments. *Congr. Numer.*, 29:901–908, 1980.
175. B.D. Sullivan. A summary of problems and results related to the Caccetta-Haggkvist conjecture. arXiv:math/0605646, May 2006.
176. A. Thomason. Paths and cycles in tournaments. *Trans. Amer. Math. Soc.*, 296(1):167–180, 1986.
177. C. Thomassen. Antidirected Hamilton circuits and paths in tournaments. *Math. Ann.*, 201:231–238, 1973.
178. C. Thomassen. Configurations in graphs of large minimum degree, connectivity, or chromatic number. *Annals of the New York Academy of Sciences*, 555:402–412, 1989.
179. C. Thomassen. Connectivity in tournaments. In *Graph theory and combinatorics (Cambridge, 1983)*, pages 305–313. Academic Press, 1984.
180. C. Thomassen. Disjoint cycles in digraphs. *Combinatorica*, 3(3-4):393–396, 1983.
181. C. Thomassen. Edge-disjoint Hamiltonian paths and cycles in tournaments. *Proc. London Math. Soc. (3)*, 45(1):151–168, 1982.
182. C. Thomassen. Graph decomposition with constraints on the connectivity and minimum degree. *J. Graph Theory*, 7:165–167, 1983.
183. C. Thomassen. Hamilton circuits in regular tournaments. In *Cycles in graphs (Burnaby, B.C., 1982)*, volume 115 of *North-Holland Math. Stud.*, pages 159–162. North-Holland, 1985.
184. C. Thomassen. Hamiltonian-connected tournaments. *J. Combin. Theory Ser. B*, 28(2):142–163, 1980.
185. C. Thomassen. Landau's characterization of tournament score sequences. In *The theory and applications of graphs (Kalamazoo, Mich., 1980)*, pages 589–591. Wiley, New York, 1981.
186. C. Thomassen. Strongly 2-connected orientations of graphs. *J. Combin. Theory Ser. B*, 110:67–78, 2015.
187. C. Thomassen. The 2-linkage problem for acyclic digraphs. *Discrete Math.*, 55(1):73–87, 1985.
188. F. Tian, Z.S. Wu, and C.Q. Zhang. Cycles of each length in tournaments. *J. Combin. Theory Ser. B*, 33(3):245–255, 1982.
189. T.W. Tillson. A Hamiltonian decomposition of $K^*_{2m}$, $2m \geq 8$. *J. Combin. Theory Ser. B*, 29(1):68–74, 1980.
190. A. van Zuylen and D.P. Williamson. Deterministic pivoting algorithms for constrained ranking and clustering problems. *Math. Oper. Res.*, 34(3):594–620, 2009.
191. L.M. Vitaver. Determination of minimal coloring of vertices of a graph by means of Boolean powers of the incidence matrix. *Dokl. Akad. Nauk SSSR*, 147:758–759, 1962.
192. M. Welhan. The Hoàng-Reed Conjecture for $\delta^+ = 3$ . *Discrete Math.*, 310(13-14):1932 –1939, 2010.
193. Z.S. Wu, K.M. Zhang, and Y. Zou. A necessary and sufficient condition for arc-pancyclicity of tournaments. *Sci. Sinica Ser. A*, 25:249–254, 1982.
194. T. Yao, Y. Guo, and K. Zhang. Pancyclic out-arcs of a vertex in a tournament. *Discrete Appl. Math.*, 99:245–249, 2000.
195. A. Yeo. The number of pancyclic arcs in a $k$-strong tournament. *J. Graph Theory*, 50:211–219, 2005.

196. C.Q. Zhang. Every regular tournament has two arc-disjoint hamiltonian cycles. *J. Qufu Normal College*, Special Issue Oper. Res.:70–81, 1980.

# 3. Acyclic Digraphs

**Gregory Gutin**

Recall that a digraph $D$ is acyclic if it has no dicycle. Acyclic digraphs form a well-studied family of digraphs of great interest in graph theory, algorithms and applications. This chapter is organised as follows. We start with the terminology used throughout this chapter. In the next section we consider some basic results on acyclic digraphs including the existence of an acyclic ordering (called a topologic ordering in most of the literature). In Section 3.2, we introduce transitive digraphs, and the transitive closure and transitive reduction of a digraph. In particular, we prove that an acyclic digraph has a unique transitive reduction.

Results on out- and in-branchings of acyclic digraphs are discussed in Section 3.3. The $k$-LINKAGE problem for acyclic digraphs in studied in Section 3.4. Enumeration results are considered in Section 3.5. In Section 3.6 we consider bounds on maximum dicuts in acyclic digraphs and compare them to those in general digraphs. Section 3.7 is devoted to the problems of finding maximum spanning and induced acyclic subdigraphs. We consider some parameterizations of the problems and bounds on oriented planar graphs for the maximum induced acyclic subgraph and dichromatic number. Section 3.8 is devoted to the multicut problem, where given a digraph and a set of pairs $(s_i, t_i)$ of vertices of $D$, our goal is to eliminate all dipaths from $s_i$ to $t_i$ for every pair $(s_i, t_i)$.

The next four sections are devoted to applications of acyclic digraphs.[1] Convex sets of acyclic digraphs, which are of interest in embedded computing, are considered in Section 3.9. Section 3.10 is devoted to recently introduced cryptographic enforcement schemes based on spanning out-forests in transitive acyclic digraphs. In Section 3.11 we consider a classical area of acyclic digraph applications – project scheduling. In Section 3.12, we intro-

[1] In this chapter, we consider only some applications of acyclic digraphs. There are many others and some even appeared while the chapter was being written, see e.g. [6], where Antoniou, Araújo, Bustamante and Gibali used basic properties of acyclic digraphs to design an algorithm for disassembling toy models produced by Engino®.

G. Gutin (✉)
Department of Computer Science,
Royal Holloway University, Egham Hill, TW20 0EX, Egham, England
e-mail: gutin@rhul.ac.uk



J. Bang-Jensen and G. Gutin (eds.), *Classes of Directed Graphs*, Springer Monographs in Mathematics, https://doi.org/10.1007/978-3-319-71840-8_3

duce a problem related to analyzing how short, distinctive phrases (typically, parts or mutations of quotations) spread to various news sites and blogs. In the problem, we are to find a minimum-weight set of arcs whose deletion leaves connected components with unique vertices of out-degree zero. Finally, Section 3.13 is devoted to generalisations of acyclic digraphs by edge-coloured undirected graphs.

A vertex is called a **source** (**sink**, respectively) if it is of in-degree zero (of out-degree zero, respectively). Recall that an **out-tree** is an orientation of a tree which has only one source, called the **root**. An **out-forest** is a collection of vertex-disjoint out-trees. Similarly, one defines in-trees and in-forests, but in an in-tree the root is a sink.

## 3.1 Acyclic Orderings and Longest and Shortest Paths

Recall that an ordering $v_1, v_2, \ldots, v_n$ of vertices of a digraph $D$ is called **acyclic** if for every arc $v_iv_j \in A(D)$, we have $i < j$. We will show that every acyclic digraph has an acyclic ordering of vertices.

**Proposition 3.1.1** *Every acyclic digraph has at least one source and at least one sink.*

**Proof:** Let $D$ be a digraph in which all vertices have positive out-degrees. We show that $D$ has a cycle. Choose a vertex $v_1$ in $D$. Since $d^+(v_1) > 0$, there is a vertex $v_2$ such that $v_1 \to v_2$. As $d^+(v_2) > 0$, $v_2$ dominates some vertex $v_3$. Proceeding in this manner, we obtain a walk of the form $v_1v_2 \ldots v_k$. As $V(D)$ is finite, there exists a least $k > 2$ such that $v_k = v_i$ for some $1 \leq i < k$. Clearly, $v_iv_{i+1} \ldots v_k$ is a closed walk and thus contains a cycle. Therefore, an acyclic digraph $D$ has a source. Since the converse of $D$ is also acyclic, $D$ has a sink as well. □

The procedure in the above proof allows one to decide whether a digraph $D$ is acyclic. However, there is another, $O(n+m)$-time, algorithm for verifying whether a digraph is acyclic based on a depth-first search, see, e.g., [9]. (Recall that $n$ and $m$ denote the order and size of $D$.) The algorithm allows us to find an acyclic ordering of vertices in an acyclic digraph. Such an ordering exists due to the following:

**Proposition 3.1.2** *Every acyclic digraph has an acyclic ordering of vertices.*

**Proof:** We give a constructive proof by describing a procedure that generates an acyclic ordering of the vertices in an acyclic digraph $D$. At the first step, we choose a source $v$ (such a vertex exists by Proposition 3.1.1). Set $x_1 = v$ and delete $x_1$ from $D$. At the $i$th step, we find a source $u$ in the remaining acyclic digraph, set $x_i = u$ and delete $x_i$ from the remaining acyclic digraph. The procedure has $n$ steps.

Suppose that $x_i \to x_j$ in $D$, but $i > j$. As $x_j$ was chosen before $x_i$, it means that $x_j$ was not of in-degree zero at the $j$th step of the procedure; a contradiction. □

Now it is easy to see the following:

**Corollary 3.1.3** *A digraph $D$ is acyclic if and only if every subgraph of $D$ has a source.*

Let $D = (V, A, c)$ be an acyclic digraph with weight function $c: A(D) \rightarrow \mathbb{R}$. We will show that shortest and longest paths from a vertex $s$ to the rest of the vertices can be found quite easily, using dynamic programming. Without loss of generality, we may assume that $s$ is a source. Let $\mathcal{L} = v_1, v_2, \ldots, v_n$ be an acyclic ordering of the vertices of $D$ such that $v_1 = s$. Clearly, $\mathrm{dist}(s, v_1) = 0$. For every $i$, $2 \leq i \leq n$, we have

$$\mathrm{dist}(s, v_i) = \begin{cases} \min\{\mathrm{dist}(s, v_j) + c(v_j, v_i) : \ v_j \in N^-(v_i)\} & \text{if } N^-(v_i) \neq \emptyset \\ \infty & \text{otherwise.} \end{cases} \tag{3.1}$$

The correctness of this formula can be shown by the following argument. We may assume that $v_i$ is reachable from $s$. Since the ordering $\mathcal{L}$ is acyclic, the vertices of a shortest path $P$ from $s$ to $v_i$ belong to $\{v_1, v_2, \ldots, v_i\}$. Let $v_k$ be the vertex dominating $v_i$ in $P$. By induction, $\mathrm{dist}(s, v_k)$ is computed correctly using (3.1). The term $\mathrm{dist}(s, v_k) + c(v_k, v_i)$ is one of the terms in the right-hand side of (3.1). Clearly, it provides the minimum.

The algorithm has two phases: the first finds an acyclic ordering, the second implements Formula (3.1). The complexity of this algorithm is $O(n + m)$ since the first phase runs in time $O(n + m)$ and the second phase requires the same asymptotic time due to the formula $\sum_{x \in V} d^-(x) = m$ in Proposition 1.2.2. Hence we have shown the following:

**Theorem 3.1.4** *The lengths of shortest paths from a fixed vertex $s$ to all other vertices can be found in time $O(n + m)$ for acyclic digraphs.*

We can also find the length of longest $(s, x)$-paths in linear time in any acyclic digraph, by replacing the weight $c(uv)$ of every arc $uv$ with $-c(uv)$. In particular, we can see immediately that the longest path problem for acyclic digraphs is solvable in polynomial time. In fact, a longest path of an acyclic digraph can always be found in linear time:

**Theorem 3.1.5** *For acyclic digraphs a longest path can be found in time $O(n + m)$.*

## 3.2 Transitive Acyclic Digraphs

A digraph $D$ is **transitive** if, for every pair $xy$ and $yz$ of arcs in $D$ with $x \neq z$, the arc $xz$ is also in $D$. Transitive digraphs form the underlying graph-theoretical model in a number of applications. For example, transitive acyclic digraphs are graphic representations of partial orders. The aim of this

Section is to give a brief overview of some properties of transitive digraphs as well as transitive closures and reductions of acyclic digraphs.

The following result is the famous Dilworth's theorem [29] formulated in the language of transitive acyclic digraphs. Recall that for a digraph $D$, $\mathrm{pc}(D)$ and $\alpha(D)$ denote the minimum number of vertex-disjoint dipaths covering $V(D)$ and the maximum number of vertices without an arc between them.

**Theorem 3.2.1 (Dilworth's theorem)** *For every transitive acyclic digraph $D$, we have $\mathrm{pc}(D) = \alpha(D)$.*

Theorem 3.2.1 is equivalent to Theorem 3.3.2 proved below (for details, see [9, Theorem 13.5.8]).

The **transitive closure** $TC(D)$ of a digraph $D$ is a digraph with $V(TC(D)) = V(D)$ and, for distinct vertices $u, v$, the arc $uv \in A(TC(D))$ if and only if $D$ has a $(u, v)$-path. The uniqueness of the transitive closure of an arbitrary digraph is obvious. To compute the transitive closure of a digraph we can use the fact discovered by a number of researchers (see, e.g., the paper [32] by Fisher and Meyer, or [35] by Furman) that the transitive closure problem and the matrix multiplication problem are closely related: there exists an $O(n^a)$-algorithm, with $a \geq 2$, to compute the transitive closure of a digraph of order $n$ if and only if the product of two boolean $n \times n$ matrices can be computed in $O(n^a)$ time. Coppersmith and Winograd [21] showed that there exists an $O(n^\omega)$-algorithm for the matrix multiplication with $\omega < 2.3755$. Le Gall [54] improved the bound on $\omega$ to 2.3729. Goralcikova and Koubek [38] designed an $O(nm_{red})$-algorithm to find the transitive closure of an acyclic digraph $D$ with $n$ vertices and $m_{red}$ arcs in the transitive reduction of $D$ (the notion of transitive reduction is introduced below). This algorithm was also studied and improved by Mehlhorn [64] and Simon [79].

An arc $uv$ in a digraph $D$ is **redundant** if there is a $(u, v)$-path in $D$ which does not contain the arc $uv$. A **transitive reduction** of a digraph $D$ is a spanning subgraph $H$ of $D$ with no redundant arc such that the transitive closures of $D$ and $H$ coincide. Not every digraph $D$ has a unique transitive reduction. Indeed, if $D$ has a pair of Hamiltonian directed cycles, then each of these cycles is a transitive reduction of $D$. Below we show that a transitive reduction of an acyclic digraph is unique, i.e., we may speak of *the* transitive reduction of an acyclic digraph.

The **intersection of digraphs** $D_1, \ldots, D_k$ with the same vertex set $V$ is the digraph $H$ with vertex set $V$ and arc set $A(D_1) \cap \ldots \cap A(D_k)$. Similarly one can define the union of digraphs with the same vertex sets (see Section 1.4). Let $\mathcal{S}(D)$ be the set of all spanning subdigraphs $L$ of $D$ for which $TC(L) = TC(D)$.

**Theorem 3.2.2** ([2]) *For an acyclic digraph $D$, there exists a unique digraph $D'$ with the property that $TC(D') = TC(D)$ and every proper subdigraph $H$ of $D'$ satisfies $TC(H) \neq TC(D')$. The digraph $D'$ is the intersection of digraphs in $\mathcal{S}(D)$.*

The proof of this theorem, which is due to Aho, Garey and Ullman [2], follows from Lemmas 3.2.3 and 3.2.4.

**Lemma 3.2.3** *Let $D$ and $H$ be a pair of acyclic digraphs on the same vertex set such that $TC(D) = TC(H)$ and $A(D) - A(H) \neq \emptyset$. Then, for every $e \in A(D) - A(H)$, we have $TC(D - e) = TC(D)$.*

**Proof:** Let $e = xy \in A(D) - A(H)$. Since $e \notin A(H)$, $H$ must have an $(x, y)$-dipath passing through some other vertex, say $z$. Hence, $D$ has an $(x, z)$-dipath $P_{xz}$ and a $(z, y)$-dipath $P_{zy}$. If $P_{xz}$ contains $e$, then $D$ has a $(y, z)$-dipath. The existence of this path contradicts the existence of $P_{zy}$ and the hypothesis that $D$ is acyclic. Similarly, one can show that $P_{zy}$ does not contain $e$. Therefore, $D-e$ has an $(x, y)$-dipath. Hence, $TC(D-e) = TC(D)$. □

**Lemma 3.2.4** *Let $D$ be an acyclic digraph. Then the set $\mathcal{S}(D)$ is closed under union and intersection.*

**Proof:** Let $G, H$ be a pair of digraphs in $\mathcal{S}(D)$. Since $TC(G) = TC(H) = TC(D)$, $G \cup H$ is a subgraph of $TC(D)$. The transitivity of $TC(D)$ now implies that $TC(G \cup H)$ is a subgraph of $TC(D)$. Since $G$ is a subgraph of $G \cup H$, we have that $TC(D)$ $(= TC(G))$ is a subgraph of $TC(G \cup H)$. Thus, we conclude that $TC(G \cup H) = TC(D)$ and $G \cup H \in \mathcal{S}(D)$.

Now let $e_1, \ldots, e_p$ be the arcs of $G - A(G \cap H)$. By repeated application of Lemma 3.2.3, we obtain $TC(G - e_1 - e_2 - \ldots - e_p) = TC(G)$. This means that $TC(G \cap H) = TC(G) = TC(D)$, hence $G \cap H \in \mathcal{S}(D)$. □

Aho, Garey and Ullman [2] also proved that there exists an $O(n^a)$-algorithm, with $a \geq 2$, to compute the transitive closure of an arbitrary digraph $D$ of order $n$ if and only if a transitive reduction of $D$ can be constructed in time $O(n^a)$. Therefore, we have

**Proposition 3.2.5** *For an arbitrary digraph $D$, the transitive closure and a transitive reduction can be computed in time $O(n^{2.376})$.* ◇

## 3.3 Out-branchings and in-branchings

The next simple but useful result follows immediately from Proposition 1.8.1. However, it also follows from the fact that if we choose an in-arc of every vertex of non-zero in-degree in an acyclic digraph $D$, then we obtain a spanning out-forest. The number of out-trees in this out-forest equals the number of vertices of in-degree zero in $D$.

**Proposition 3.3.1** *An acyclic digraph has an out-branching if and only if it has only one source.*

In the MINIMUM WEIGHT OUT-BRANCHING problem, given a digraph $D$ with non-negative weights on its arcs and a vertex $s \in V(D)$, we are required to find an out-branching of $D$ rooted at $s$ of minimum total weight. Edmonds [30] found a polynomial time algorithm for solving the problem for general digraphs. In the case of acyclic digraphs, we can simply apply the following greedy algorithm used by Crampton, Farley, Gutin, Jones and Poettering [24, 25], see Section 3.10. Let $D$ be an acyclic digraph containing a unique source $s$. For every vertex $v \in V(D) - \{s\}$ choose an arc to $v$ of minimum weight. Clearly, the chosen arcs form a minimum weight out-branching.

In the next two subsections, we consider out-branchings with the minimum and maximum number of leaves, and with bounded out-degrees.

### 3.3.1 Extremal number of leaves

Recall that a **leaf** in an out-branching is a vertex of out-degree zero. We will denote the minimum and maximum number of leaves in an out-branching of a digraph $D$ by $\ell_{\min}(D)$ and $\ell_{\max}(D)$, respectively. The problems of finding out-branchings with $\ell_{\min}(D)$ and $\ell_{\max}(D)$ leaves are both $\mathcal{NP}$-hard for general digraphs. The problem of finding $\ell_{\max}(D)$ restricted to acyclic digraphs remains $\mathcal{NP}$-hard [5] and moreover it was proved by Schwartges, Spoerhase and Wolff [78] that there is no PTAS for the restricted problem. However, Schwartges *et al.* [78] obtained a 2-approximation algorithm for the restricted problem. For the general problem the currently best known approximation ratio is 92, obtained by Daligault and Thomassé [27].

The algorithm of Schwartges *et al.* [78] inputs an acyclic digraph $D$ with a unique source $r$, outputs an out-tree $T$, and consists of two phases. At Phase 1, called **expansion**, the algorithm starts from an empty out-forest $F$ and empty sets $V$ and $A$. All vertices of $D$ apart from $r$ are initially unmarked ($r$ is marked). For every $v \notin V$, the algorithm adds $v$ to $V$ and if $v$ has at least two unmarked out-neighbours, then all arcs from $v$ to such out-neighbours are added to $A$ and such out-neighbours are all marked. At Phase 2, called **connection**, for every unmarked vertex $u$ the algorithm marks $u$ and adds an incoming arc to $u$ to $A$. The output $T$ is the subgraph of $D$ induced by $A$.

The problem of finding $\ell_{\min}(D)$ becomes polynomial-time solvable when restricted to acyclic digraphs. This result is of interest as it has an application to the area of database systems, see, e.g., the US patent [28] by Demers and Downing. Before describing the polynomial-time algorithm, we will prove the following result, which is equivalent to Dilworth's theorem (Theorem 3.2.1) as shown by Bang-Jensen and Gutin [10].

Recall that $\alpha(D)$ denotes the maximum number of vertices in a digraph $D$ with no arc between any pair of them.

**Theorem 3.3.2** *If $D$ is a transitive acyclic digraph with a unique source $s$, then $\ell_{\min}(D) = \alpha(D)$.*

**Proof:** By Proposition 3.3.1 and Las Vergnas' theorem (Theorem 1.8.3), $D$ contains an out-branching $B$ with $k \leq \alpha(D)$ leaves. Observe that $B$ is rooted at $s$ and the vertices of every dipath in $B$ starting at $s$ and terminating at a leaf induce a clique in $UG(D)$. Thus, the vertices of $UG(D)$ can be covered by $k$ cliques and, hence, $\alpha(D) \leq k$. We conclude that $\ell_{\min}(D) = \alpha(D)$. □

Let $D = (V, A)$ be an acyclic digraph. Let us assume that $D$ has a unique source $r$ as, by Proposition 3.3.1, this is a necessary and sufficient condition for $D$ to have an out-branching. Let $V' = \{v' : v \in V\}$ and let us define a bipartite graph $B$ with partite sets $X$ and $X'$ as follows: $X = V, X' = V' \setminus \{r'\}$ and $E(B) = \{xy' : x \in X, y' \in X', xy \in A\}$. Let $M$ be a maximum matching in $B$ and let $M^*$ be obtained from $M$ by adding to it any edge $uv' \in E(B)$ for each $v'$ not covered by $M$. We have the following lemma of Gutin, Razgon and Kim [45].

**Lemma 3.3.3** *The minimum number of leaves in an out-branching of $D$ equals the number of isolated vertices in the subgraph of $B$ induced by $M^*$.*

*Proof* A set $N$ of edges of $B$ is called **nice** if each vertex of $X'$ is incident to exactly one edge in $N$ and $N$ contains an edge incident to $r$. Let $T$ be an out-branching of $D$ and let $N(T) = \{xy' : xy \in A(T)\}$. We will prove that $T \to N(T)$ is a bijection between all out-branchings of $D$ and all nice edge sets of $B$. Indeed, if $P$ is an out-branching, then clearly $N(P)$ is a nice edge set. Let $N$ be a nice edge set and let $Q$ be a spanning subgraph of $D$ constructed as follows: $xy \in A(Q)$ if and only if $xy' \in N$. Since every vertex of $X'$ is incident to exactly one edge of $N$, we have $d_Q^-(z) = 1$ for each $z \in V(Q) \setminus \{r\}$. Since $Q$ is acyclic with a unique source, by Proposition 3.3.1, $Q$ is an out-branching.

Let $T$ be an out-branching of $D$ and let $B[N(T)]$ be the subgraph of $B$ induced by the edge set $N(T)$. Observe that the leaves of $T$ are exactly the isolated vertices of $B[N(T)]$. Thus, an out-branching of $D$ with the minimum number of leaves corresponds to a nice set $N$ such that $B[N]$ has the minimum possible number of isolated vertices.

Let $N$ be a nice edge set in $B$, let $m(N)$ denote the maximum size of a matching in $B[N]$ and let $H$ be a matching in $B[N]$ of size $m(N)$. Let $y' \in X'$ be a vertex of $B$ not incident to an edge of $H$ and let $xy' \in N$. Since $H$ is maximum, $x$ is incident to an edge of $H$. Notice that $r$ is covered by $H$. Indeed, there exists a vertex $u$ such that $r$ is the only in-neighbour of $u$ in $D$. Hence if $r$ was not covered by $H$ then $u'$ would not be covered by $H$ either, which means we could extend $H$ by $ru'$, a contradiction. Thus, the number of isolated vertices in $B[N]$ equals $|X| - m(N)$. Hence, the number of leaves in $T$ is minimum if and only if $m(N(T)) = \max\{m(N) : N \text{ is nice}\}$. Notice that we constructed $M^*$ in such a way as to guarantee that $m(M^*) = \max\{m(N) : N \text{ is nice}\}$ and, thus, $\ell_{\min}(D)$ is the number of isolated vertices in $B[M^*]$. □

This lemma and its proof leads to the following algorithm for finding a minimum leaf out-branching $T$ in an input acyclic digraph $D$.

**Algorithm 1** MINLEAF

*Input:* An acyclic digraph $D$ with vertex set $V$.
*Output:* A minimum leaf out-branching $T$ of $D$ if $\ell_{\min}(D) > 0$ and "NO", otherwise.

1: Find a source $r$ in $D$. If there is another source in $D$, return "no out-branching". Let $V' = \{v' : v \in V\}$.
2: Construct a bipartite graph $B = B(D)$ of $D$ with partite sets $V, V' - r'$ and edge $xy'$ for each arc $xy \in A(D)$.
3: Find a maximum matching $M$ in $B$.
4: $M^* := M$. For all $y' \in V'$ not covered by $M$, set $M^* := M^* \cup \{$an arbitrary edge incident with $y'\}$.
5: $A(T) := \emptyset$. For all $xy' \in M^*$, set $A(T) := A(T) \cup \{xy\}$.
6: Return $T$.

Figure 3.1 illustrates MINLEAF.

It is not hard to implement MINLEAF such that its time complexity is $O(m + \sqrt{mn^3})$, where $n = |V|$ and $m = |A|$ (see, e.g., [45]). Thus, we have the following result proved by Gutin, Razgon and Kim [45].

**Theorem 3.3.4** *Let $D$ be an acyclic digraph with $n$ vertices and $m$ arcs. We can decide whether $D$ has an out-branching and find one with the minimum number of leaves in time $O(m + n^{1.5}\sqrt{m})$.*

### 3.3.2 Bounded out-degrees

For acyclic directed multigraphs, Bang-Jensen, Thomassé and Yeo [12] gave a complete characterization for the existence of an out-branching satisfying given (not necessarily uniform) restrictions on the out-degree of each vertex. For a set $X$ of vertices, let $X^- = \bigcup_{x \in X} N^-(x)$.

**Theorem 3.3.5** ([12]) *Let $D = (V, A)$ be an acyclic directed multigraph and let $f : V \to \mathbb{Z}_0$. Suppose that $D$ has precisely one source $s$. Then $D$ has an out-branching $B$ rooted at $s$ satisfying*

$$d_B^+(v) \le f(v) \text{ for all } v \in V \tag{3.2}$$

*if and only if*

$$\sum_{x \in X^-} f(x) \ge |X| \text{ for all } X \subset V - s. \tag{3.3}$$

*Furthermore, there exists a polynomial algorithm which, given an acyclic directed multigraph and a non-negative integer assignment to its vertices, either finds an out-branching satisfying (3.2) or a set $X$ of vertices violating (3.3).*

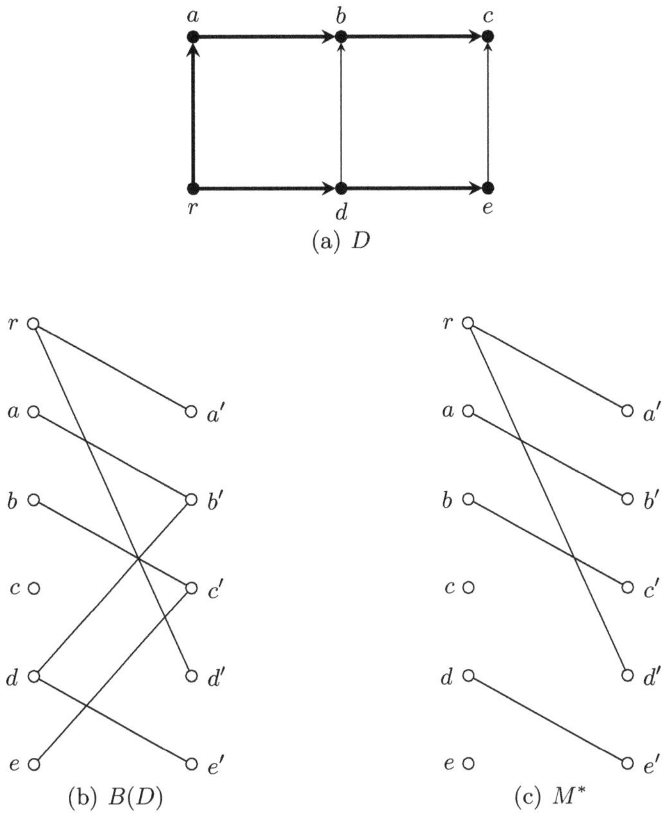


**Figure 3.1** Example for MINLEAF. In $D$, the thick arcs belong to $T$.

**Proof:** We start by constructing a flow network $N$ as follows. The vertex set of $N$ consists of two copies $v', v''$ of each vertex $v \in V - s$, one copy $s''$ of $s$ and finally a new vertex $z$. The arc set of $N$ is $A(N) = \{u''v' : uv \in A\} \cup \{v'z : v \in V - s\} \cup \{zv'' : v \in V\}$. We have the following upper and lower bounds on the arcs:

- all arcs to $z$ have upper and lower bound equal to one.
- all arcs $u''v'$ corresponding to arcs in $D$ have lower bound zero and infinite upper bound.
- all arcs from $zv''$ have lower bound zero and upper bound equal to $f(v)$.

We claim that $D$ has an out-branching $T$ rooted at $s$ satisfying (3.2) if and only if $N$ has a feasible circulation. First assume that $T$ is an out-branching rooted at $s$ in $D$ which satisfies (3.2). Since the in-degree of every vertex except $s$ is precisely one in $T$ it follows that the following $x$ is a feasible circulation in $N$:

- $x(u''v') = 1$ if $uv$ is an arc of $T$.

- $x(v'z) = 1$ for all arcs to $z$.
- $x(zv'') = d_T^+(v)$ for all arcs from $z$.

Conversely, if $x$ is a feasible integer-valued circulation in $N$, then let $A'$ be the set of those arcs $uv$ in $D$ for which $x(u''v') = 1$. It is easy to see that these arcs form a spanning acyclic subgraph $T'$ of $D$ with $n-1$ arcs and in which $s$ is the only source. Thus, by Proposition 3.3.1 $T'$ is an out-branching rooted at $s$. Furthermore, by the capacity constraint on the arcs from $z$, no vertex $v$ is the tail of more than $f(v)$ arcs in $T'$.

Now we are ready to prove the first claim of the theorem. Since every vertex except $s$ has in-degree one in an out-branching from $s$, it is easy to see that (3.3) must hold if $D$ has an out-branching satisfying (3.2). Suppose now that $D$ and $f$ satisfy (3.3). By the arguments above it suffices to prove that $N$ has a feasible circulation.

Assume this is not the case. Then by Hoffman's circulation theorem (Theorem 1.9.3) there exists some partition $S, \bar{S}$ of $V(N)$ such that the sum $l(S,\bar{S})$ of the lower bounds on the arcs from $S$ to $\bar{S}$ is strictly larger than the sum $u(\bar{S},S)$ of the capacities on the arcs from $\bar{S}$ to $S$. Since only arcs into $z$ have a non-zero lower bound we have $z \in \bar{S}$. Let $X'$ be the set of those $v'$ that belong to $S$ and let $X$ be the corresponding set of vertices in $D$. Note that $l(S,\bar{S}) = |X'| = |X|$. By the choice of capacities in $N$ we see that every vertex $w''$ which has an arc to a vertex in $X'$ must belong to $S$. Since $z$ has an arc to all such vertices $w''$ with capacity $f(w)$ and each such arc contributes to $u(\bar{S},S)$ we have

$$l(S,\bar{S}) > u(\bar{S},S) \geq \sum_{w \in X^-} f(w) \geq |X| = l(S,\bar{S}),$$

a contradiction. Hence $N$ has a feasible circulation and the desired out-branching exists in $D$.

The second part of the theorem follows from the fact that our proof can be turned into an algorithm for checking the existence of the desired branching (and finding one if it exists) by using flow techniques to search for a feasible integer-valued circulation in the corresponding network $N$. □

It is easy to see that if an acyclic digraph $D$ has arc-disjoint branchings $F^+, F^-$, where $F^+$ is an out-branching rooted at $s$ and $F^-$ is an in-branching rooted at $t$, then $s$ ($t$) must be the unique vertex of in-degree (out-degree) zero in $D$. The following result characterises when an acyclic digraph contains such a pair of arc-disjoint branchings.

**Corollary 3.3.6** ([12]) *Let $D$ be an acyclic digraph such that there is exactly one source $s$ and exactly one sink $t$ in $D$. Then $D$ contains arc-disjoint branchings $F^+$ and $F^-$ where the first is an out-branching rooted at $s$ and the second is an in-branching rooted at $t$ if and only if we have*

$$\sum_{x \in X^-} (d^+(x) - 1) \geq |X| \text{ for all } X \subseteq V - s. \tag{3.4}$$

*Furthermore, it can be decided in polynomial time whether $D$ has such branchings.*

**Proof:** As remarked above, an acyclic digraph $H$ has an in-branching rooted at a vertex $z$ if and only if $z$ is the unique sink in $H$. Now we see that $D$ has the desired branchings if and only if $D$ has an out-branching rooted at $s$ which satisfies (3.2) with respect to $f(v) = d^+(v) - 1$ for $v \neq t$ and $f(t) = 0$. By Theorem 3.3.5 this is equivalent to requiring that (3.4) must hold.

The complexity claim follows from the last part of Theorem 3.3.5. □

The complexity part of Corollary 3.3.6 was also obtained by Bérczi, Fujishige and Kamiyama [14].

## 3.4 The $k$-Linkage problem

When the digraph considered is acyclic there is enough structure to allow an efficient solution of the $k$-LINKAGE problem for every fixed $k$. Perl and Shiloach [68] proved that the 2-LINKAGE problem is solvable in polynomial time for acyclic digraphs. In their elegant proof they showed how to reduce the 2-linkage problem for a given acyclic digraph to a simple path finding problem in another digraph. Fortune, Hopcroft and Wyllie extended Perl and Shiloach's result to arbitrary $k$. The proof of this result below is an extension of the proof by Perl and Shiloach (see also Thomassen's survey [84]).

**Theorem 3.4.1** ([33]) *For each fixed $k$, the $k$-LINKAGE problem is solvable in polynomial time for acyclic digraphs.*

**Proof:** Let $D = (V, A)$ be an acyclic digraph for which we wish to find a $k$-linkage from $(x_1, x_2, \ldots, x_k)$ to $(y_1, y_2, \ldots, y_k)$. We may assume that $d^-_D(x_i) = d^+_D(y_i) = 0$ for all $i \in [k]$, since arcs from $y_i$ and to $x_i$ play no role in the problem and may thus be deleted. We may also assume that vertices $x_1, x_2, \ldots, x_k, y_1, y_2, \ldots, y_k$ are all distinct. Indeed, if there are $p$ appearances of $x_i$ in the sequence $x_1, x_2, \ldots, x_k, y_1, y_2, \ldots, y_k$ then we will replace $x_i$ by $p$ copies such that they have the same in- and out-neighbours as $x_i$.

Form a new digraph $D' = (V', A')$ whose vertex set is the set of all $k$-tuples of distinct vertices of $V$. For any such $k$-tuple $(v_1, v_2, \ldots, v_k)$ there is at least one vertex, say $v_r$, which cannot be reached from any of the other $v_i$ by a path in $D$. (Here we used the fact that $D$ is acyclic.) For each out-neighbour $w$ of $v_r$ such that $w \notin \{v_1, v_2, \ldots, v_k\}$, we let $A'$ contain an arc from $(v_1, v_2, \ldots, v_{r-1}, v_r, v_{r+1}, \ldots, v_k)$ to $(v_1, v_2, \ldots, v_{r-1}, w, v_{r+1}, \ldots, v_k)$. Only arcs as those described above are in $A'$.

We claim that $D'$ has a directed path from the vertex $(x_1, x_2, \ldots, x_k)$ to the vertex $(y_1, y_2, \ldots, y_k)$ if and only if $D$ contains disjoint paths $P_1, P_2, \ldots, P_k$ such that $P_i$ is an $(x_i, y_i)$-path for each $i \in [k]$.

Suppose first that $D'$ has a path $P$ from $(x_1, x_2, \dots, x_k)$ to $(y_1, y_2, \dots, y_k)$. By definition, every arc of $P$ corresponds to one arc in $D$. Hence we get a collection of paths $P_1, P_2, \dots, P_k$ such that $P_i$ is an $(x_i, y_i)$-path for each $i \in [k]$ by letting $P_i$ contain those arcs that correspond to a shift in the $i$th vertex of a $k$-tuple. Suppose two of these paths, $P_i, P_j$ are not disjoint. Then it follows from the assumption that $d_D^-(x_i) = d_D^+(y_i) = 0$ for all $i \in [k]$ and the definition of $D'$ that there is some vertex $u \in V - \{x_1, x_2, \dots, x_k, y_1, y_2, \dots, y_k\}$ such that $u \in V(P_i) \cap V(P_j)$. Let $w$ $(z)$ be the predecessor of $u$ on $P_i$ $(P_j)$. We may assume without loss of generality that the arc on $P$ corresponding to $wu$ is used before that corresponding to $zu$. This means that at the time we change from $w$ to $u$ in the $i$th coordinate, the $j$th coordinate corresponds to a vertex $z'$ which can reach $u$ in $D$ (through $z$). Now it follows from the definition of the arcs in $A'$ that we could not have changed the $i$th coordinate again before we have used the arc corresponding to $zu$ in $D'$. However, that would lead to a $k$-tuple which contains two copies of the same vertex $u$ from $D$, contradicting the definition of $D'$. Hence $P_i$ and $P_j$ must be disjoint.

Suppose now that $D$ contains disjoint paths $Q_1, Q_2, \dots, Q_k$ such that $Q_i$ is an $(x_i, y_i)$-path for all $i \in [k]$. Then we can construct a path from $(x_1, x_2, \dots, x_k)$ to $(y_1, y_2, \dots, y_k)$ in $D'$ as follows. Start with the tuple $(x_1, x_2, \dots, x_k)$. At any time we choose a coordinate $j$ of the current $k$-tuple $(z_1, z_2, \dots, z_k)$ such that the vertex $z_j$ is not in $\{y_1, y_2, \dots, y_k\}$ and $z_j$ cannot be reached in $D$ by any other vertex from the tuple. Note that such a vertex exists since $D$ is acyclic and $d^+(y_i) = 0$ for all $i \in [k]$. It is easy to show by induction that we will always have $z_j \in V(Q_j)$. Now we use the arc $z_j w$ corresponding to the arc out of $z_j$ on $Q_j$ and change the $j$'th coordinate from $z_j$ to $w$. If follows from the fact that $Q_1, \dots, Q_k$ are disjoint that this will produce a path from $(x_1, x_2, \dots, x_k)$ to $(y_1, y_2, \dots, y_k)$ in $D'$.

Given any instance $(D, x_1, x_2, \dots, x_k, y_1, y_2, \dots, y_k)$ we can produce the digraph $D'$ in time $O(k!n^{k+2})$ by forming all possible $k$-tuples and deciding which arcs to add based on the definition of $D'$. Then we can decide the existence of a path from $(x_1, x_2, \dots, x_k)$ to $(y_1, y_2, \dots, y_k)$ in polynomial time using, say, a breadth-first search on $D'$. This proves that the $k$-LINKAGE problem is polynomial for each fixed $k$. □

Note that we do not actually have to construct $D'$ in advance. It suffices to introduce the vertices and arcs when they become relevant for the search for a path from $(x_1, x_2, \dots, x_k)$ to $(y_1, y_2, \dots, y_k)$ in $D'$.

It is not difficult to see that we can also use the approach above to find the cheapest collection of $k$ disjoint paths where the $i$th path is an $(x_i, y_i)$-path in a given acyclic digraph with non-negative costs on the arcs. Here the goal is to minimize the total cost of the arcs used by the paths.

Producing $D'$ in the proof above cannot lead to an FPT algorithm for the $k$-LINKAGE problem parameterized by $k$. A natural question is whether there is an FPT algorithm for the problem. Unfortunately, it is highly unlikely as was shown by Slivkins:

**Theorem 3.4.2** ([81]) *The $k$-LINKAGE problem parameterized by $k$ is W[1]-hard for acyclic digraphs.*

Bang-Jensen and Kriesell [11] considered a generalisation of the $k$-LINKAGE problem when some of the paths may be oriented, not necessarily directed. They proved that even the generalisation of 2-LINKAGE is $\mathcal{NP}$-hard when one of the paths is required to be directed and the other oriented.

It is not hard to see that Theorem 3.4.1 also holds for the WEAK $k$-LINKAGE problem, where we require the paths to be arc-disjoint rather than internally vertex-disjoint. Indeed, consider an acyclic digraph $D = (V, A)$ and two $k$-tuples $x_1, x_2, \ldots, x_k$ and $y_1, y_2, \ldots, y_k$ of distinct vertices of $D$ for which we wish to find a weak (i.e. arc-disjoint) $k$-linkage from $(x_1, x_2, \ldots, x_k)$ to $(y_1, y_2, \ldots, y_k)$. Add to $D$ extra vertices $\{x'_i, y'_i : i \in [k]\}$ and arcs $A_X \cup A_Y$, where $A_X = \{x'_i x_i : i \in [k]\}$ and $A_Y = \{y_i y'_i\ i \in [k]\}$. Observe that the line digraph of the resulting graph has a $k$-linkage from the vertices corresponding to the arcs of $A_X$ to the vertices corresponding to the arcs $A_Y$ if and only if $D$ has a weak $k$-linkage from $(x_1, x_2, \ldots, x_k)$ to $(y_1, y_2, \ldots, y_k)$. Thus, we have the following:

**Corollary 3.4.3** *For each fixed $k$,* WEAK $k$-LINKAGE *is polynomial-time solvable for acyclic digraphs.*

A collection $\mathcal{F}$ of flows in a network $N = (A, A)$ is called vertex-disjoint if for every vertex $x \in V$ at most one flow from $\mathcal{F}$ has a positive value on any arc incident to $x$. Bang-Jensen and Bessy [8] considered vertex-disjoint flows in acyclic networks, i.e. in networks whose digraph is acyclic. Generalising the proof of Theorem 3.4.1, they proved the following:

**Theorem 3.4.4** *For every fixed collection of integers $k, v_1, \ldots, v_k, U$, there is a polynomial-time algorithm for deciding whether an acyclic network $N = (V, A, u)$ with $u(ij) \in [U]$ for all $ij \in A$ has vertex-disjoint flows $x_1, \ldots, x_k$ such that $x_i$ is an $(s_i, t_i)$-flow of value $v_i$ for $i \in [k]$, where $s_1, \ldots, s_k, t_1, \ldots, t_k$ are distinct vertices of $V$.*

## 3.5 Enumeration

In this section we consider enumeration results for acyclic digraphs. We start from theorems on enumeration of labelled acyclic digraphs and the fact that the number of $n \times n$ (0,1)-matrices whose eigenvalues are positive real numbers equals the number of labelled acyclic digraphs. We then briefly discuss enumeration of unlabelled acyclic digraphs. Finally, we consider dipath enumeration in acyclic digraphs.

Labelled acyclic digraphs were first counted by Robinson [75, 77], and independently by Stanley [82]. The short proof below is by Liskovets [57]. The proof uses Corollary 3.1.3.

**Theorem 3.5.1** *Let $a_n$ be the number of acyclic digraphs on $n \geq 1$ labelled vertices. Then $a_n = \sum_{t=0}^{n-1}(-1)^{n-t-1}\binom{n}{t}2^{t(n-t)}a_t$.*

**Proof:** Let $[n]$ be the set of vertices of $n$-vertex acyclic digraphs, let $X \subseteq [n]$, and let $a_n(X)$ be the number of $n$-vertex acyclic digraphs in which every vertex from $X$ is a source. Observe that $a_n(X) = 2^{t(n-t)}a_t$, where $t = n-|X|$. Using the inclusion-exclusion principle, we obtain

$$\sum_{t=0}^{n}(-1)^{n-t}\binom{n}{t}2^{t(n-t)}a_t = 0,$$

which implies the claimed formula. □

Gessel [36] enumerated acyclic digraphs with specified numbers of sources and sinks. Acyclic digraphs are the basic representation of the structure underlying Bayesian networks. In many practical applications, such as the reverse engineering of gene regulatory networks, the reconstruction of the network is of great interest. Such reconstructions can be obtained if we can generate acyclic digraphs randomly and uniformly. Kuipers and Moffa [53] showed how Theorem 3.5.1 can be used to generate large acyclic digraphs randomly and uniformly.

McKay, Foggier, Royle, Sloane, Wanless and Wilf [63] associated acyclic digraphs with (0,1)-matrices whose eigenvalues are positive real numbers and proved the following:

**Theorem 3.5.2** *The number of acyclic digraphs on $n \geq 1$ labelled vertices equals the number of $n \times n$ (0,1)-matrices whose eigenvalues are positive real numbers.*

**Proof:** Let $D$ be an acyclic digraph on vertex set $[n]$ such that $1, 2, \ldots, n$ is an acyclic ordering. Let $A$ be the adjacency matrix of $D$ and $B = I + A$, where $I$ is the $n \times n$ identity matrix. We claim that $B$ has only positive eigenvalues. Indeed, $B$ is upper triangular with 1's on the diagonal. Hence all of its eigenvalues are equal to 1.

Conversely, let $B$ be a (0, 1)-matrix whose eigenvalues $\lambda_i$ are all positive real numbers. Then we have

$$\begin{aligned} 1 &\geq \frac{1}{n}\text{trace}(B) \\ &= \frac{1}{n}(\lambda_1 + \lambda_2 + \cdots + \lambda_n) \\ &\geq (\lambda_1\lambda_2 \ldots \lambda_n)^{1/n} \\ &= (\det(B))^{1/n} \\ &\geq 1. \end{aligned}$$

The last equality is due to the fact that $\det(B - \lambda I) = \prod(\lambda_i - \lambda)$ and so setting $\lambda$ to zero, we obtain $\det(B) = \lambda_1\lambda_2 \ldots \lambda_n$. The last inequality holds since $\det(B)$ is a positive integer.

Thus, $\lambda_1\lambda_2\ldots\lambda_n = 1$ and $\lambda_1 = \lambda_2 = \cdots = \lambda_n$ due to the equality between arithmetic and geometric means. Therefore, $\lambda_i = 1$ for each $i \in [n]$.

Now view $B$ as the adjacency matrix of a directed pseudograph $H$, which has a loop at every vertex but no parallel arcs. Since

$$\mathrm{trace}(B^k) = \sum_{i=1}^{k} \lambda_i^k = \sum_{i=1}^{k} 1 = n,$$

for all $k$, the number of closed diwalks in $H$, of each length $k$, is $n$.

Since $\mathrm{trace}(B) = n$, all diagonal entries of $B$ are 1's. Thus we account for all $n$ of the closed diwalks of length $k$ that exist in $H$ by the loops only. Hence there are no closed diwalks of any length that use an arc of $H$ other than the loops at the vertices. Set $A = B - I$. Then $A$ is a (0, 1)-matrix that is the adjacency matrix of an acyclic digraph. □

Robinson [77] enumerated unlabelled (i.e., non-isomorphic) acyclic digraphs using standard enumeration techniques for unlabelled graphs [47]. Later Robinson [76] came up with a more efficient method to count unlabelled acyclic digraphs using the inclusion-exclusion principle.

Recall that a **multipartite tournament** is an orientation of a complete multipartite undirected graph. Gutin [39] enumerated "almost" unlabelled acyclic multipartite tournaments using bijections from classes of multipartite tournaments to sets of integral sequences. An **almost unlabelled** $p$-partite tournament is an ordered $(p+1)$-tuple $(T, V_1, \ldots, V_p)$, where $T$ is a $p$-partite tournament and $(V_1, \ldots, V_p)$ an ordered $p$-tuple of its partite sets. If the partite sets of $T$ are of size $n_1, \ldots, n_p$ respectively ($n_i > 0$, $i \in [p]$), then $T$ is called an **$(n_1, \ldots, n_p)$-tournament**. We say that almost unlabelled $(n_1, \ldots, n_p)$-tournaments $(T, V_1, \ldots, V_p)$ and $(M, U_1, \ldots, U_p)$ are **equivalent** if there exists an isomorphism $f$ from $T$ to $M$ such that $f(V_i) = U_i$ for every $i \in [p]$. Intuitively, this means that vertices in the partite set are interchangeable, but the partite sets themselves are not. Gutin [39] proved that the number of non-equivalent almost unlabelled acyclic $(n_1, \ldots, n_p)$-tournaments equals the multinomial coefficient $\binom{n}{n_1, \ldots, n_p}$.

Stanley [83] studied dipath enumeration in acyclic digraphs. A simple technical lemma is followed by the main result. An $n \times n$ matrix $B = [b_{ij}]$ is called **special** if $b_{ij} = 1$ for all $i \geq j$.

**Lemma 3.5.3** *Let $B = [b_{ij}]$ be a special $n \times n$ matrix. Then*

$$\det B = \prod_{i=1}^{n-1} (1 - b_{i,i+1}).$$

**Proof:** We will use induction on $n$, the case $n = 1$ being trivial. Expand $\det B$ by the first row. By induction, the first two terms are

$$\prod_{i=2}^{n-1}(1-b_{i,i+1}) - b_{12}\prod_{i=2}^{n-1}(1-b_{i,i+1}) = \prod_{i=1}^{n-1}(1-b_{i,i+1}).$$

In the remaining terms, the first two columns of the cofactor are equal, so the terms are all zero. □

**Theorem 3.5.4** ([83]) *Let $H$ be an acyclic digraph with an acyclic ordering $v_1, v_2, \ldots, v_n$ of its vertices. Let $A = [a_{ij}]$ be an $n \times n$ (0,1)-matrix with $a_{ij} = 0$ if and only if $v_i \to v_j$ and let $D = \mathrm{diag}(x_1, \ldots, x_n)$, where $X = \{x_1, \ldots, x_n\}$ is a set of variables. Then*

$$\det(I + zDA) = \sum_{j=0}^{n}(\sum_{P} x_{k_1}\ldots x_{k_j})z^j,$$

*where $P$ ranges over all dipaths $v_{k_1}\ldots v_{k_j}$ of $H$ with $j$ vertices.*

**Proof:** The coefficient of $z^j$ in $\det(I+zDA)$ is the sum of the principal $j \times j$ minors of $DA$. The rows and columns of a principal submatrix $DA[W]$ are indexed by a $j$-element subset $W$ of $[n]$. We claim that

$$\det(DA[W]) = \begin{cases} \prod_{j\in W} x_j, & \text{if } W \text{ is a set of vertices of a dipath,} \\ 0, & \text{otherwise.} \end{cases}$$

from which the theorem immediate follows. Observe that

$$\det(DA[W]) = (\prod_{j\in W} x_j)\det A[W].$$

Therefore, we need to prove that

$$\det A[W] = \begin{cases} 1, & \text{if } W \text{ is a set of vertices of a dipath} \\ 0, & \text{otherwise} \end{cases}$$

Since $v_1, v_2, \ldots, v_n$ is an acyclic ordering, $A[W]$ is special. Let $W$ be a set of vertices of a dipath. Then all the entries of $A[W]$ on the diagonal just above the main diagonal are equal to 0, and so by Lemma 3.5.3 the above formula for $\det A[W]$ holds. Suppose now that $W$ is not the set of vertices of a dipath. Since at least one entry of $A[W]$ on the diagonal just above the main diagonal is equal to 1, by Lemma 3.5.3 the above formula for $\det A[W]$ holds as well. □

In the **dipath polynomial** $P_H(z) = \sum c_j z^j$ of a digraph $H$, every $c_j$ is the number of $j$-vertex dipaths in $H$. For instance, if $H$ is the digraph of proper relations of a poset $Q$ (i.e., $(x, y)$ is an arc of $H$ if $x < y$ in $Q$), then $c_j$ is just the number of $j$-element chains of $Q$. By setting $x_j = 1$ for every $j \in [n]$ in the formula of Theorem 3.5.4, we have the following:

**Corollary 3.5.5** *Let $H$ be an acyclic digraph with an acyclic ordering $v_1$, $v_2, \ldots, v_n$ of its vertices and let $A$ be as in Theorem 3.5.4. Then $P_H(z) = \det(I + zA)$.*

## 3.6 Maximum Dicuts

The Max Cut problem has been extensively studied. In this problem, as in other related problems, the term **cut** is understood as a minimal cut, i.e. a cut of the form $(X, Y) = \{xy : x \in X, y \in Y\}$, where $X$ and $Y$ partition the vertices of a directed or undirected graph $G$. The **size** of a cut is the number of arcs or edges in it. Max Cut is the problem of finding a cut of maximum size. The size of a maximum cut of an undirected graph $G$ will be denoted by $f(G)$. Let $f(m)$ be the minimum of $f(G)$ over all undirected graphs $G$ with $m$ edges. Edwards [31] proved that

$$f(m) \geq \frac{m}{2} + \sqrt{\frac{m}{8} + \frac{1}{64}} - \frac{1}{8}. \tag{3.5}$$

This bound is sharp as the equality holds for complete graphs of odd order. The size of a maximum cut of a digraph $D$ will be denoted by $g(D)$, and the minimum $g(D)$ over all digraphs $D$ with $m$ arcs will be denoted by $g(m)$. It follows from (3.5) that

$$g(m) \geq \frac{m}{4} + \sqrt{\frac{m}{32} + \frac{1}{256}} - \frac{1}{16}.$$

This bound is also sharp as the equality holds for tournaments in which the out-degree and in-degree of every vertex coincide.

In this section we consider this problem for acyclic digraphs, where the question of the lower bound seems more complicated. Let $h(m)$ be the minimum $g(D)$ over all acyclic digraphs $D$ with $m$ arcs. Clearly, $h(m) \geq f(m)/2$, and so $h(m) \geq m/4$. A natural question (with an application mentioned in [4]) is whether there is a constant $c > 1/4$ such that $h(m) \geq cm$. Alon, Bollobás, Gyárfás, Lehel and Scott [4] provided a negative answer to this question by proving the following:

**Theorem 3.6.1** *For $m \geq 1$, $h(m) \leq \frac{m}{4} + O(m^{4/5})$.*

**Proof:** Fix $n \geq 1$ and let $r = \lfloor n^{1/3} \rfloor$. We will construct an acyclic digraph $D'$ with $m = m(n) = (1 + o(1))n^{5/3}$ arcs, and no dicut of size more than $\frac{m}{4} + O(m^{4/5})$.

We first define a digraph $D$ as follows. By a well-known theorem of Singer [80], there exists a set $A$ of $r$ natural numbers such that all differences $a - b$, with $a, b \in A$ and $a \neq b$, are distinct and $\max A \leq (1 + o(1))r^2$. Let $V(D) = \mathbb{Z}_n$ and $A(D) = \{(i + a, i + b) : i \in \mathbb{Z}_n, a, b \in A, a < b\}$ (the sums are taken modulo $n$). By the definition of $A$, there are no multiple arcs, and so $|A(D)| = n\binom{r}{2}$.

Let $G$ be the underlying graph of $D$. Since $G$ is a union of $n$ copies of $K_r$ (where the $i$th copy has vertex set $\{i + a : a \in A\}$), we have $f(G) \leq nf(K_r) \leq nr^2/4$. Since $D$ is Eulerian, $g(D) = \frac{1}{2}f(G) \leq nr^2/8$.

To obtain an acyclic digraph $D'$ from $D$ delete all arcs $(i,j)$ with $i > j$ (where we identify the vertices of $Z_n$ with the integers $0,\dots,n-1$). Since $\max A \le (1+o(1))r^2$, we have deleted at most

$$(1+o(1))r^2\binom{r}{2} \le (1+o(1))\frac{r^4}{2}$$

arcs. Thus the number of arcs in $D'$ is

$$m \ge n\binom{r}{2} - (1+o(1))\frac{r^4}{2} = \frac{nr^2}{2} - \frac{nr}{2} - (1+o(1))\frac{r^4}{2}$$

and

$$\begin{aligned} g(D') &\le nr^2/8 \\ &\le \frac{m}{4} + \frac{nr}{8} + (1+o(1))\frac{r^4}{8} \\ &\le \frac{m}{4} + (1+o(1))\frac{n^{4/3}}{4}. \end{aligned}$$

Observe that $m = (1+o(1))n^{5/3}/2$ and thus

$$g(D') \le \frac{m}{4} + (1+o(1))\frac{m^{4/5}}{2^{6/5}}.$$

□

Alon *et al.* [4] also proved that $h(m) \ge \frac{m}{4} + \Omega(m^{2/3})$. They posed the following question:

**Problem 3.6.2** What is the infimum of $d$ such that, for $m \ge 1$, $h(m) = \frac{m}{4} + O(m^d)$?

## 3.7 Acyclic Subdigraphs

We may consider two types of subdigraphs: spanning and induced. Subsection 3.7.1 studies the problem of finding an acyclic subgraph with maximum number of arcs. In Subsection 3.7.1, we also mention an interesting sharp result on partitioning the arc set of a digraph into subsets which induce acyclic subgraphs with bounded out-degree of each vertex. Subsection 3.7.2 mainly addresses the problem of partitioning of vertices of a digraph into the minimum number of subsets such that each subset induces an acyclic subgraph. The main problems on the subsections are dual to the FEEDBACK ARC SET problem and FEEDBACK VERTEX SET problem, respectively. In FEEDBACK ARC SET, given a digraph $D$ and an integer $k$, we are to decide whether $D$ contains $k$ arcs whose deletion makes $D$ acyclic. DIRECTED FEEDBACK VERTEX SET is the same problem, but instead of $k$ arcs, we can delete $k$ vertices.

### 3.7.1 Spanning Acyclic Subdigraphs

The goal of the MAXIMUM ACYCLIC SUBDIGRAPH problem is to find an acyclic subgraph $H$ with maximum number of arcs in a given digraph $D$. Recall that MAXIMUM ACYCLIC SUBDIGRAPH is dual to the FEEDBACK ARC SET problem and hence $\mathcal{NP}$-hard even for tournaments, see Theorem 2.9.3. In this subsection, we will consider results for some parameterizations of MAXIMUM ACYCLIC SUBDIGRAPH and FEEDBACK ARC SET.

Let $D = (V, A)$ be a digraph with $m$ arcs and vertices $v_1, \ldots, v_n$. Consider subdigraphs $(V, A_1)$ and $(V, A_2)$ of $D$ such that $A_1 = \{v_iv_j \in A : \ i < j\}$ and $A_2 = \{v_iv_j \in A : \ i > j\}$. Observe that both subgraphs are acyclic, and as $|A_1| + |A_2| = m$, at least one of them has at least $m/2$ arcs. Replacing every edge $xy$ of an undirected graph by two arcs, $xy$ and $yx$, we obtain a digraph $D$ whose maximum acyclic subgraph has just half of the arcs of $D$. Thus the lower bound $m/2$ is tight. This bound is no longer tight if $D$ has no directed 2-cycle.

Since MAXIMUM ACYCLIC SUBDIGRAPH is $\mathcal{NP}$-hard, it is natural to study parameterized versions of the problem. Let us consider such a parameterization for a weighted version of MAXIMUM ACYCLIC SUBDIGRAPH: every arc $(i, j) \in A$ is assigned an integral positive integer $w_{ij}$ and we are to decide whether $D$ contains an acyclic subgraph of total weight at least $W/2 + k$, where $W$ is the total weight of $D$ and $k$ is the parameter.[2] Mahajan, Raman, and Sikdar [62] asked whether this parameterized problem is FPT for the special case when all arcs are of weight 1. Gutin, Kim, Szeider and Yeo [44] solved the parameterized weighted problem by obtaining a quadratic kernel. Let us consider their solution.

A random variable is **discrete** if its distribution function has a finite or countable number of positive increases. A random variable $X$ is **symmetric** if $-X$ has the same distribution as $X$. If $X$ is discrete, then $X$ is symmetric if and only if $\text{Prob}(X = a) = \text{Prob}(X = -a)$ for each real $a$. Let $X$ be a symmetric variable for which the first moment $\mathbb{E}(X)$ exists. Then $\mathbb{E}(X) = \mathbb{E}(-X) = -\mathbb{E}(X)$ and, thus, $\mathbb{E}(X) = 0$. We will use below the following easy-to-prove result [44].

**Lemma 3.7.1** *If $X$ is a symmetric random variable and $\mathbb{E}(X^2)$ is finite, then*

$$\text{Prob}(\ X \geq \sqrt{\mathbb{E}(X^2)}\ ) > 0.$$

Consider the following simple reduction rule. For every directed 2-cycle $iji$, (i) delete the cycle if $w_{ij} = w_{ji}$, and (ii) delete the arc $ji$ and replace $w_{ij}$ by $w_{ij} - w_{ji}$ if $w_{ij} > w_{ji}$. Clearly, after applying the reduction rule, we obtain an oriented graph (i.e., a digraph with no directed 2-cycle). Clearly,

[2] Note that in this parameterization $k$ is not necessarily an integer, but an integer divided by 2, such that $W/2 + k$ is an integer. A similar remark also holds for the other parameterization of this section.

the original input $D$ is a positive instance of the parameterized weighted problem if and only if so is the oriented graph obtained by the reduction rule.

**Theorem 3.7.2** ([44]) *The weighted version of* MAXIMUM ACYCLIC SUBDIGRAPH *has a kernel with* $O(k^2)$ *arcs.*

**Proof:** Consider a random ordering: $\alpha : V \to [n]$ and a random variable $X(\alpha) = \frac{1}{2}\sum_{ij\in A} x_{ij}(\alpha)$, where $x_{ij}(\alpha) = w_{ij}$ if $\alpha(i) < \alpha(j)$ and $x_{ij}(\alpha) = -w_{ij}$, otherwise. It is easy to see that $X(\alpha) = \sum\{w_{ij} : \ ij \in A, \alpha(i) < \alpha(j)\} - W/2$. Thus, the answer to the parameterized weighted problem is YES if and only if there is an ordering $\alpha : V \to [n]$ such that $X(\alpha) \geq k$.

By the reduction rule, we may assume that the input of the parameterized weighted problem is an oriented graph $D = (V, A)$. Let $\alpha : V \to [n]$ be a random ordering. Since $X(-\alpha) = -X(\alpha)$, where $-\alpha(i) = n + 1 - \alpha(i)$, $X$ is a symmetric random variable and, thus, we can apply Lemma 3.7.1. It was proved in [44] that $\mathbb{E}(X^2) \geq m/12$. By this inequality and Lemma 3.7.1, we have Prob( $X \geq \sqrt{m/12}$ ) $> 0$. Thus, if $\sqrt{m/12} \geq k$, there is an ordering $\beta : V \to [n]$ such that $X(\beta) \geq k$ and so the answer to the parameterized weighted problem is YES. Otherwise, $\sqrt{m/12} \leq k$ implying $m \leq 12k^2$ and we are done. □

Kim and Williams [51] showed that in the case of all weights equal to 1 the parameterized weighted problem admits a kernel with a linear number of vertices.

Let us revert to the unweighted MAXIMUM ACYCLIC SUBDIGRAPH, but consider a lower bound stronger than $m/2$ in many cases. Poljak and Turzík [70] proved that every connected oriented graph $D$ contains an acyclic subgraph with at least $\frac{m}{2} + \frac{n-1}{4}$ arcs. To see that this bound is tight, consider a directed path $x_1x_2 \dots x_{2t+1}$ and add to it arcs $x_3x_1, x_5x_3, \dots, x_{2t+1}x_{2t-1}$. This oriented graph $H_t$ consists of $t$ directed 3-cycles and has $2t + 1$ vertices and $3t$ arcs. Thus, $\frac{m}{2} + \frac{n-1}{4} = 2t$ and $2t$ is the maximum size of an acyclic subgraph of $H_t$: we have to delete an arc from every directed 3-cycle as the cycles are arc-disjoint.

The following natural question was asked by Raman and Saurabh [72]: what is the parameterized complexity of deciding whether a connected oriented graph $D$ has an acyclic subgraph with at least $\frac{m}{2} + \frac{n-1}{4} + k$ arcs, where $k$ is the parameter. (Note that for connected oriented graphs, the parameter $k$ in this parameterization is smaller than that in the $m/2+k$ one, which means that the former parameterization is of greater interest.) Crowston, Gutin and Jones [26] answered this question by proving the following result.

**Theorem 3.7.3** *The parameterization of Raman and Saurabh admits an algorithm of runtime* $O^*((12k)!)$ *and a kernel with* $O(k^2)$ *vertices and* $O(k^2)$ *arcs.*

Mnich, Philip, Saurabh, and Suchý [65] proved independently that the parameterization of Raman and Saurabh is FPT, but they did not obtain a polynomial size kernel.

The following open problem is of interest.

**Problem 3.7.4** Is there a kernel for the parameterization of Raman and Saurabh with a linear number of vertices?

The lower bound $\frac{m}{2}+\frac{n-1}{4}$ can be easily generalized to oriented graphs with $c$ connected components: $\frac{m}{2}+\frac{n-c}{4}$. It is not hard to generalize the results of [26] and [65] and prove that we can decide in FPT time whether an oriented graph with $c$ connected components contains an acyclic subgraph with at least $\frac{m}{2}+\frac{n-c}{4}+k$ arcs: For each connected component $H$ find the maximum integer $p_H := \frac{m_H}{2}+\frac{n_H-1}{4}+k_H$ such that $H$ contains an acyclic subgraph with at least $p_H$ arcs, where $m_H$ and $n_H$ is the number of arcs and vertices in $H$ and $k_H \le k$. This will require $O(\log k)$ applications of an FPT algorithm. It remains to compare $\sum_H p_H$, where the sum runs over all connected components of $D$, and $\frac{m}{2}+\frac{n-c}{4}+k$.

We can apply the Poljak–Turzík bound (and the corresponding parameterization) to general digraphs after the reduction rule is carried out (note that the reduction rule may increase the number of connected components).

Consider now the standard parameterization of Feedback arc set: given a digraph $D$ and an integer $k$, we are to decide whether $D$ contains $k$ arcs whose deletion makes $D$ acyclic, where $k$ is the parameter. It had been a challenging open problem to decide whether the parameterized Feedback arc set problem is FPT untill Chen, Liu, Lu, O'Sullivan and Razgon [17, 18] proved the following:

**Theorem 3.7.5** *The parameterized* Feedback arc set *and* Feedback vertex set *problems are FPT.*

It is not hard to show that Feedback arc set is FPT if and only if so is Feedback vertex set. Thus, it was sufficient for Chen *et al.* [18] to prove the theorem only for Feedback vertex set.

Let us finish this subsection with the following interesting result of Wood [87], which is the best possible.

**Theorem 3.7.6** *For every integer $s \ge 2$, arcs of every digraph $D$ can be partitioned into $s$ subsets such that each subset $A_i$ induces an acyclic digraph $D[A_i]$ and the out-degree of every vertex $v$ in $D[A_i]$ is at most $\lceil d^+(v)/(s-1) \rceil$.*

### 3.7.2 Induced Acyclic Subgraphs

A set $X$ of vertices of a digraph $D$ is called **acyclic** if $D[X]$ is acyclic. The **dichromatic number** of a digraph $D$ is the minimum number $\chi_A(D)$ such that $V(D)$ can be partitioned into $\chi_A(D)$ acyclic sets. A digraph $D$ is **weakly $k$-degenerate** if every subgraph $H$ of $D$ contains a vertex $v$ with either $d^+_H(v) \leq k$ or $d^-_H(v) \leq k$.

Neumann-Lara [67] conjectured the following:

**Conjecture 3.7.7** *Every oriented planar graph is of dichromatic number 1 or 2.*

This conjecture has not been resolved, but the following weaker result was proved by Bokal, Fijavz, Juvan, Kayll, and Mohar [15].

**Theorem 3.7.8** *Every oriented planar graph is of dichromatic number at most 3.*

This theorem is a corollary of the fact that every planar undirected graph has a vertex of degree at most 5 and the following result of [15].

**Theorem 3.7.9** *If a digraph $D$ is weakly $k$-degenerate, then $\chi_A(D) \leq k+1$.*

**Proof:** Let $v_1, \ldots, v_n$ be the vertices of $D$ ordered so that for every $i \in [n]$ the vertex $v_i$ has either in-degree or out-degree at most $k$ in the induced subdigraph $D_i = D[\{v_1, \ldots, v_i\}]$. We can obtain such an ordering by first choosing $v_n$ with $d^+(v_n) \leq k$ or $d^-(v_n) \leq k$, then choosing $v_{n-1}$ in $D - v_n$, etc. Now define $A_0, \ldots, A_k$ as follows. Start with empty sets. For every $i \in [n]$, there is a set $A_j$, with $j = j(i)$, such that $A_j$ contains either no out-neighbours or no in-neighbours of $v_i$ in $D_i$. Then put $v_i$ in $A_j$.

Suppose that one of the resulting sets $A_j$ contains a dicycle $C$. If $v_i$ is the vertex on $C$ with largest index $i$, then $v_i$ has an in- and an out-neighbour among the other vertices on $C$, which is impossible by the construction of the sets $A_0, \ldots, A_k$. Therefore, each $A_i$ is acyclic. □

Harutyunyan and Mohar [48], confirmed Conjecture 3.7.7 for all oriented planar graphs of girth at least 5.

Theorem 3.7.8 implies that every oriented planar graph of order $n$ has an acyclic set of size at least $n/3$. If confirmed Conjecture 3.7.7 would imply that every oriented planar graph has an acyclic set of size at least $n/2$. Harutyunyan and Mohar [48] conjectured an even stronger bound which, if true, would be tight.

**Conjecture 3.7.10** *Every oriented planar graph of order $n$ has an acyclic set of size at least $3n/5$.*

Golowich and Rolnick [37] proved this conjecture for all oriented planar graphs of girth at least 8.

## 3.8 The Multicut Problem

We now consider the following problem, called MULTICUT.[3]

> MULTICUT **Parameter:** $p$
> **Input:** A digraph $D$, a set $\mathcal{T} = \{(s_i, t_i) : i \in [r]\}$ of pairs of **terminal vertices**, and an integer $p$
> **Question:** Does $D$ contain a set $Z$ of at most $p$ non-terminal vertices of $D$ that **separate** the terminal pairs (i.e. in $D - Z$ there is no directed $(s_i, t_i)$-path for any $i \in [r]$)?

Clearly, for $r = 1$, MULTICUT is a well-understood and polynomial-solvable problem. The situation is already different for $r = 2$, see e.g. Corollary 3.8.4 below. Pilipczuk and Wahlström [69] proved that MULTICUT parameterized by $p$ is already W[1]-hard for $r = 4$. The problem is FPT for $r = 2$ as proved by Chitnis, Hajiaghayi and Marx [20], but its paramerized complexity for $r = 3$ is currently unknown.

For acyclic digraphs, Kratsch, Pilipczuk, Pilipczuk, and Wahlström [52] proved the following:

**Theorem 3.8.1** MULTICUT *on acyclic digraphs can be solved in time* $O^*(2^{O(r^2p+r2^{O(p)})})$.

Thus, MULTICUT on acyclic digraphs is FPT when parameterized by two parameters, $r$ and $p$. Unless FPT=W[1], which is highly unlikely, by the next theorem the problem is not FPT when parameterized by $p$ only.

**Theorem 3.8.2** ([52]) MULTICUT *on acyclic digraphs parameterized by $p$ only is W[1]-hard.*

The SKEW MULTICUT problem is a special case of MULTICUT, where we are given $d$ sources $s_i$ and $d$ sinks $t_i$ such that the set of terminal pairs is $\mathcal{T} = \{(s_i, t_j) : 1 \le i \le j \le d\}$.

It is even less likely that MULTICUT on acyclic digraphs is FPT when parameterized by $r$ only, due to Corollary 3.8.4, which immediately follows from the next theorem. Note that SKEW MULTICUT is of interest as it was used in the proof of Theorem 3.7.5.

**Theorem 3.8.3** ([52]) SKEW MULTICUT *on acyclic digraphs is $\mathcal{NP}$-complete even if $d = 2$.*

**Corollary 3.8.4** ([52]) MULTICUT *on acyclic digraphs is $\mathcal{NP}$-complete even if $r = 2$.*

[3] There is an arc version of MULTICUT, where arcs are to separate the terminal pairs of vertices. However, the vertex and arc versions have the same classical and parameterized complexity for digraphs [20].

The MULTICUT problem has several applications, see, e.g., [52]. Let us consider one of them for classes of digraphs close to acyclic. Let FEEDBACK ARC SET be the following problem: given a digraph $D$ and natural number $p$, decide whether $D$ has at most $p$ arcs whose deletion makes $D$ acyclic. Let $\mathcal{D}_r$ be the set of digraphs which have at most $r$ vertices whose deletion makes the digraphs acyclic. Bang-Jensen and Yeo [13] proved the following dichotomy.

**Theorem 3.8.5** FEEDBACK ARC SET *is polynomial-time solvable on* $\mathcal{D}_1$ *and* $\mathcal{NP}$*-complete already on* $\mathcal{D}_2$.

## 3.9 Convex Sets and Embedded Computing

A non-empty set $X$ of vertices in an acyclic digraph $D$ is **convex** if for any $x, y \in X$, all vertices of every $(x, y)$-dipath are in $X$. In the graph depicted in Figure 3.2, sets $A$ and $B$ are convex, but $C$ is not. Let conv$(D)$ denote the number of convex sets in $D$. Let $\overline{X} = V(D) \setminus X$. A vertex $y \in \overline{X}$ is called an **input vertex** of $X$ if $y$ dominates a vertex in $X$. A vertex $x \in X$ is an **output vertex** of $X$ if $x$ dominates a vertex in $\overline{X}$. In Figure 3.2, set $A$ has three input vertices and one output vertex. For $B$ the numbers are four and two.

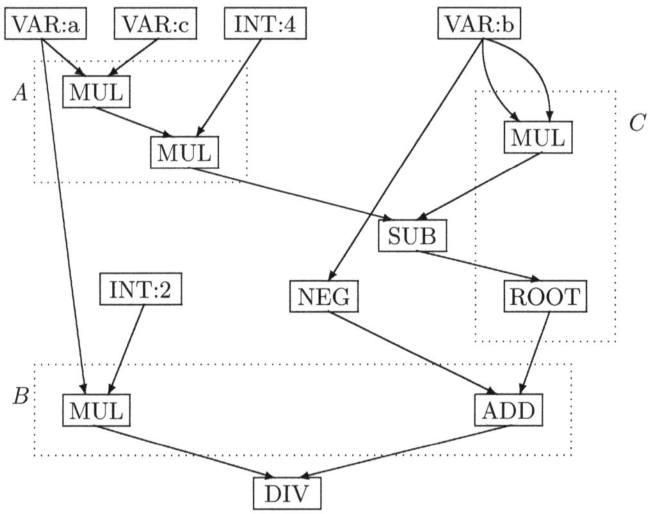


**Figure 3.2** Data dependency graph for $\frac{-b+\sqrt{b^2-4ac}}{2a}$.

An embedded system is a computer system with a dedicated function within a larger mechanical or electrical system. Embedded systems control many devices in common use today. One of the major design choices for any new computer processor is the selection of the machine instruction set. In an embedded system, the processor will only execute a single fixed program during its lifetime, and significant efficiency gains can be made by choosing the machine instruction set, and associated hardware, to support the program that will be executed.

In particular there exist extensible general purpose processors that can be customised for specific applications by the addition of custom-designed machine instructions and supporting hardware. The approach is to choose a set of application specific machine instructions by examination of the target program; candidate instructions are likely to involve a combination of several basic computations. For example, a program solving a system of linear equations may find it useful to have a single instruction to perform matrix inversion on a set of values held in registers.

Candidate instruction identification is carried out on **data dependency graphs** (DDGs), which are obtained from the application program by first splitting it into **basic blocks**, regions of sequential computation with no control transfer into their bodies, and then creating vertices for each instruction. The resulting DDGs are acyclic and any convex subset of vertices is a candidate for a custom instruction which could be implemented in hardware. Figure 3.2 depicts a DDG.

Thus, algorithms generating convex sets in acyclic digraphs are of interest and such an algorithm, presented below, was designed by Balister, Gerke, Gutin, Johnstone, Reddington, Scott, Soleimanfallah and Yeo [7]. However, in practice a given hardware application will have specific, and usually small, input and output constraints. This significantly reduces the size of the solution space and thus presents an opportunity for a more efficient enumeration algorithm. Furthermore, certain instructions, such as writes to main memory, cannot be combined into a custom instruction, and thus certain vertices in the acyclic digraph can be designated as **forbidden** from the point of view of inclusion in a candidate set. Therefore, we are interested in finding all convex sets which have specified upper bounds, $n_{in}$ and $n_{out}$, on the numbers of input and output vertices and which do not contain any vertices from a specified forbidden set $F$. We call such convex sets **valid** convex sets.

Bonzini and Pozzi [16] and Chen, Maskell and Sun [19] proved that with the two constraints above there are at most $O(n^{n_{in}+n_{out}})$ valid convex sets in an $n$-vertex acyclic digraph $D$. Note that in practice, $n_{in}$ and $n_{out}$ are small constants. Both papers above designed algorithms for generating all valid convex sets. However, while the algorithm in [16] has running time $O(n^{n_{in}+n_{out}+1})$, it does not produce all valid convex sets, see [40, 73]. The algorithm of [19] is correct but it is given without an upper bound of its complexity; its performance was tested in computational experiments. Pozzi, Atasu and Ienne [71] obtained an algorithm of running time $O(n^{n_{in}+n_{out}+1})$. Gutin, Johnstone, Reddington, Scott and Yeo [40] designed another correct algorithm, its running time is $O(m \cdot n_{in}^2(\mathrm{vconv}(D)+n^{n_{out}}))$, where vconv($D$) is the number of valid convex sets in $D$. Reddington, Gutin, Johnstone, Scott and Yeo [74] designed a modification of the algorithm in [40], which while having a worse upper bound on running time, performs better than all the above algorithms in computational experiments.

As the valid convex set algorithms above are quite complicated, we will not describe them in this chapter. In the following subsection, we will consider bounds on the number of convex sets which induce connected subgraphs. In the subsection afterwards we will look at the algorithm introduced in [7] for generating all convex sets.

### 3.9.1 Bounds for the Number of Connected Convex Sets

For a set $X$ of a digraph $D$, the subgraph $D[X]$ is called **connected convex** (or just a **cc-subgraph**) if it is connected and $X$ is convex. The number of cc-subgraphs in an acyclic digraph $D$ is denoted by $\mathrm{cc}(D)$. The following results obtained by Gutin and Yeo [46] give tight bounds for $\mathrm{cc}(D)$ of an $n$-vertex acyclic digraph $D$.

**Theorem 3.9.1** *For every connected acyclic digraph $D$ of order $n$, $\mathrm{cc}(D) \geq n(n+1)/2$. If an acyclic digraph $D$ of order $n$ has a Hamilton dipath, then $\mathrm{cc}(D) = n(n+1)/2$.*

Consider a complete bipartite graph $K_{a,b}$ with partite sets $A, B$ ($|A| = a$, $|B| = b$) and orient all its edges from $A$ to $B$. The resulting bipartite tournament will be denoted by $BT_{a,b}$.

**Theorem 3.9.2** *Let $f(n) = 2^n+n+1-d_n$, where $d_n = 2\cdot 2^{n/2}$ for every even $n$ and $d_n = 3 \cdot 2^{(n-1)/2}$ for every odd $n$. For every connected acyclic digraph $D$ of order $n$, $\mathrm{cc}(D) \leq f(n)$. We also have $\mathrm{cc}(BT_{a,n-a}) = f(n)$ provided $|n-2a| \leq 1$.* □

A proof of Theorem 3.9.1 can be found in [46] and [9]. In the remainder of this subsection, we give a proof of Theorem 3.9.2.

**Lemma 3.9.3** *Let $n = a+b$. We have $cc(BT_{a,b}) = 2^{a+b} - 2^a - 2^b + a + b + 1$ and*

$$\max\{cc(BT_{a,b}) : \ a+b=n\} = 2^n + n + 1 - d_n,$$

*where $d_n = 2 \cdot 2^{n/2}$ for every even $n$ and $d_n = 3 \cdot 2^{(n-1)/2}$ for every odd $n$.*

**Proof:** Let $g(a,b) = 2^{a+b} - 2^a - 2^b + a + b + 1$. Since all non-empty sets of vertices of $BT_{a,b}$, excluding those that are subsets of $A$ or $B$ of cardinality at least 2, induce cc-subgraphs, we have $cc(BT_{a,b}) = g(a,b)$. It remains to observe that $\max\{g(a,b) : \ a+b=n\}$ is obtained when $a$ and $b$ differ by at most 1. □

In the following theorem, we will show that the bipartite tournaments $BT_{a,n-a}$ with $|n-2a| \leq 1$ have the maximum possible number of cc-subgraphs.

**Theorem 3.9.4** *Let $H$ be a connected acyclic digraph of order $n$ and let $f(n) = 2^n + n + 1 - d_n$, where $d_n = 2 \cdot 2^{n/2}$ for every even $n$ and $d_n = 3 \cdot 2^{(n-1)/2}$ for every odd $n$. Then $cc(H) \leq f(n)$.*

**Proof:** Clearly, we may assume that $n \geq 3$. Suppose that $H$ has a dipath of length 2. We will prove that $cc(H) \leq f(n)$. If $xyz$ is a dipath of length 2 in $H$, then we have the following:

(C1) There are at most $2^{n-2}$ cc-subgraphs containing $x$ but not $z$.
(C2) There are at most $2^{n-2}$ cc-subgraphs containing $z$ but not $x$.
(C3) There are at most $2^{n-2} - 1$ cc-subgraphs containing neither $x$ nor $z$.
(C4) There are at most $2^{n-3}$ cc-subgraphs containing $x$ and $z$.

(C4) is true as if $x$ and $z$ belong to a cc-subgraph, then $y$ has to belong to it as well. Therefore there are at most $7 \cdot 2^{n-3} - 1$ cc-subgraphs. Observe that $7 \cdot 2^{n-3} - 1 \leq f(n)$ for every $n \geq 3$ apart from $n = 5$. Indeed, it is not difficult to prove that $f(n) - 7 \cdot 2^{n-3} + 1 > 2^{\frac{n}{2}+1}(2^{\frac{n}{2}-4} - 1)$ for every even $n$ and that $f(n) - 7 \cdot 2^{n-3} + 1 > 2^{\frac{n-1}{2}}(2^{\frac{n-5}{2}} - 3)$ for every odd $n$. These two inequalities imply $7 \cdot 2^{n-3} - 1 \leq f(n)$ for each $n \geq 8$. The cases $n = 3, 4, 6, 7$ can be easily checked separately. Thus, it remains to consider the case $n = 5$.

Suppose that $H$ has a dipath $P$ with $n-1$ vertices and let $u$ be the vertex not on $P$. Then by Theorem 3.9.1, $cc(H - u) = n(n-1)/2$. There are at most $2^{n-1}$ induced subgraphs of $H$ containing $u$. Thus, $cc(H) \leq 2^{n-1} + n(n-1)/2$. Observe that $2^{n-1} + n(n-1)/2 \leq f(n)$ for every $n \geq 5$. Thus, we may assume that if $n \geq 5$, then $H$ has no directed path with $n - 1$ vertices.

Let $n = 5$ and let $u \in V(H) \setminus \{x, y, z\}$. By (C4), at most $2^{n-3}$ subgraphs containing $x$ and $z$ are not cc-subgraphs. Observe that $(2^n - 1 - 2^{n-3}) - f(n) = 1$ for $n = 5$. Thus, to show that $cc(H) \leq f(5)$, it suffices to find a non-empty non-cc-subgraph of $H$ that does not contain at least one of the vertices $x$ and $z$. Since $H$ has no dipath of length 3, $u$ is not adjacent with at least one of the vertices $x, y, z$. The subgraph induced by any such pair of non-adjacent vertices is not a cc-subgraph.

So we may now assume that there is no dipath of length 2. This means that the vertices can be partitioned into sets $A$ and $B$ such that $A$ contains all vertices with in-degree zero and $B$ contains all the vertices with out-degree zero. Observe that now every connected induced subgraph of $H$ is a cc-subgraph. This implies that $cc(H)$ is maximum when there is an arc from $a$ to $b$ for each $a \in A$, $b \in B$. Now our result follows from Lemma 3.9.3. □

### 3.9.2 Algorithm for Generating Convex Sets

Let $D$ be a connected acyclic digraph of order $n$. The family of all convex sets of $D$ will be denoted by $\mathcal{CONV}(D)$; thus $\text{conv}(D) = |\mathcal{CONV}(D)|$.

To obtain all convex sets of $D$ (and $\emptyset$, which is not convex by definition), we call the following recursive procedure with the original digraph $D$ and with $F = \emptyset$. This call yields an algorithm whose properties are studied below.

In general, the procedure $\mathcal{CS}$ takes as input an acyclic digraph $D = (V, A)$ and a set $F \subseteq V$ and outputs all convex sets of $D$ which contain $F$. The procedure $\mathcal{CS}$ outputs $V$ and then considers all sources and sinks of the graph that are not in $F$. For each such source or sink $s$, we call $\mathcal{CS}(D - s, F)$ and then add $s$ to $F$. Thus, for each sink or source $s \in V \setminus F$ we consider all sets that contain $s$ and all sets that do not contain $s$.

---

**Algorithm 2** $\mathcal{CS}(D = (V, A), F)$

*Input:* An acyclic digraph $D = (V, A)$ and a set $F \subseteq V$
*Output:* all convex sets of $D$ which contain $F$.

1: **output** $V$; set $X := V \setminus F$
2: **for all** $s \in X$ with $|N^+(s)| = 0$ or $|N^-(s)| = 0$ **do**{
3: **for all** vertices $v$ find $N^+_{D-s}(v)$ and $N^-_{D-s}(v)$
4: call $\mathcal{CS}(D - s, F)$; set $F := F \cup \{s\}$
5: **for all** vertices $v$ find $N^+_D(v)$ and $N^-_D(v)$ }

---

**Correctness of the procedure.** Proposition 3.9.6 and Theorem 3.9.7 imply that the procedure $\mathcal{CS}$ is correct. We first show that all sets generated in line 1 are, in fact, convex sets. To this end, we use the following lemma whose proof is left as an exercise.

**Lemma 3.9.5** *Let $D$ be an acyclic graph, let $X$ be a convex set of $D$ and let $s \in X$ be a source or sink of $D[X]$. Then $X \setminus \{s\}$ is a convex set of $D$.* ◇

Now we can prove the following proposition.

**Proposition 3.9.6** *Let $D = (V, A)$ be an acyclic digraph and let $F \subseteq V$. Then every set output by $\mathcal{CS}(D, F)$ is convex.*

**Proof:** We prove the result by induction on the number of vertices of the output set. The entire vertex set $V$ is convex and is output by the procedure. Now assume all sets of size $n - i \geq 2$ that are output by the procedure are convex. We will show that all sets of size $n - i - 1$ that are output are also convex. When a set $C$ is output the procedure $\mathcal{CS}(D[C], F')$ was called for some set $F' \subseteq V$. The only way $\mathcal{CS}(D[C], F')$ can be invoked is that there exist a set $C' \subset V$ and a source or sink $c$ of $D[C']$ with $C = C' \setminus \{c\}$. Moreover, $C'$ will be output by the procedure and, thus, by our assumption is convex. The result now follows from Lemma 3.9.5. □

**Theorem 3.9.7** *Let $D = (V, A)$ be an acyclic digraph and let $F \subseteq V$. Then every convex set of $D$ containing $F$ is output exactly once by $\mathcal{CS}(D, F)$.*

**Proof:** Let $C$ be a convex set of $D$ containing $F$. We first claim that there exist vertices $c_1, c_2, \ldots, c_t \in V$ with $V = \{c_1, c_2, \ldots, c_t\} \cup C$ and $c_i$ is a source or sink of $D\langle C \cup \{c_i, c_{i+1}, \dots, c_t\}\rangle$ for all $i \in [t]$. To prove the claim we will show that for every convex set $H$ with $C \subset H \subseteq V$, there exists a source or sink $s \in H \setminus C$ of the digraph $D[H]$. This will prove our claim as by Lemma 3.9.5 $H \setminus \{s\}$ is a convex set of $D$ and we can repeatedly apply the claim.

If there exists no arc from a vertex of $C$ to a vertex of $D[H \setminus C]$, then any source of $H \setminus C$ is a source of $D[H]$. Note that $D[H \setminus C]$ is an acyclic digraph and, thus, has at least one source (and sink). Thus we may assume that there is an arc from a vertex $u$ of $C$ to a vertex $v$ of $H \setminus C$. Consider a longest path $v = v_1v_2 \dots v_r$ in $D[H \setminus C]$ leaving $v$. Observe that $v_r$ is a sink of $D[H \setminus C]$ and, moreover, there is no arc from $v_r$ to any vertex of $C$ since otherwise there would be a directed path from $u \in C$ to a vertex in $C$ containing vertices in $H \setminus C$, which is impossible as $C$ is convex. Hence $v_r$ is a sink of $D[H]$ and the claim is shown.

Next note that a sink or source remains a sink or source when vertices are deleted. Thus when $\mathcal{CS}(D, F)$ is executed and a source or sink $s$ is considered, then we distinguish the cases when $s = c_i$ for some $i \in [t]$ or when this is not the case. If $s = c_i$ and we currently consider the digraph $D'$ and the fixed set $F'$, then we follow the execution path calling $\mathcal{CS}(D' - s, F')$. Otherwise we follow the execution path that adds $s$ to the fixed set. When the last $c_i$ is deleted, we call $\mathcal{CS}(D[C], F'')$ for some $F''$ and the set $C$ is output. It remains to show that there is a unique execution path yielding $C$. To see this, note that when we consider a source or sink $s$, then either it is deleted or moved to the fixed set $F$. Thus every vertex is considered at most once and then deleted or fixed. Therefore each time we consider a source or sink there is a unique decision that finally yields $C$. □

**Running time of $\mathcal{CS}$.** We will use the following data structure for a set $Y = \{y_1, y_2, \dots, y_{|Y|}\} \subseteq \{1, 2, \dots, n\}$ that supports unit time element insertion and deletion, unit time checking whether $Y$ is empty, and allows us to iterate over the elements of $Y$ in $O(|Y|)$ time. We maintain arrays of integers $\textsc{succ}$ and $\textsc{pred}$ indexed from 0 to $|Y|$ where $\textsc{succ}_k = k$ and $\textsc{pred}_k = k$ if and only if $k \notin Y$. If $Y = \emptyset$, then $\textsc{pred}_0 = \textsc{succ}_0 = 0$. If $Y \neq \emptyset$, then $\textsc{pred}_i$ ($\textsc{succ}_i$) hold $y_{i-1}$ ($y_{i+1}$), where $i-1$ and $i+1$ are taken modulo $|Y|$, and we can iterate over the elements of $V$ by following the chain of links from $\textsc{succ}_0$. Notice that $\textsc{succ}_0$ holds $y_1$ and $\textsc{pred}_0$ holds $y_{|Y|}$.

By analogy with conventional doubly-linked list insertion and deletion, we have

| *insert*($k$) | *delete*($k$) |
|---|---|
| $\textsc{succ}_k \leftarrow 0$ | $\textsc{succ}_{\textsc{pred}_k} \leftarrow \textsc{succ}_k$ |
| $\textsc{pred}_k \leftarrow \textsc{pred}_0$ | $\textsc{pred}_{\textsc{succ}_k} \leftarrow \textsc{pred}_k$ |
| $\textsc{succ}_{\textsc{pred}_0} \leftarrow k$ | $\textsc{pred}_k \leftarrow k$ |
| $\textsc{pred}_0 \leftarrow k$ | $\textsc{succ}_k \leftarrow k$ |

We can use this data structure for sets $V, X, N_D^+(v), N_D^-(v), v \in V$, and $F$ for the input acyclic digraph $D = (V, A)$ of order $n$. We can initialize the data structures for all these sets in time $O(n^2)$ using, say, the adjacency matrix of $D$. Observe that we output the vertex set of $D$ as one convex set. Thus, it suffices to show that the running time of $\mathcal{CS}(D, F)$ without the recursive calls is $O(|V|)$. This will yield the running time $O(\sum_{C \in \mathcal{CONV}(D)} |C|)$ of $\mathcal{CS}$ by Theorem 3.9.7.

Using our data structure, we can determine *all* sources and sinks in $O(|V|)$ time. For the recursive calls of $\mathcal{CS}$ we delete one vertex and have to update the number of in-, respectively, out-neighbours of all neighbours of the deleted vertex $s$ by iterating over $V$. The vertex $s$ has at most $|V| - 1$ neighbours and we can charge the cost of the updating information to the call of $\mathcal{CS}(D - s, F)$. Moreover we store the neighbours of $s$ so that we can reintroduce them after the call of $\mathcal{CS}(D - s, F)$. Moving the sinks and sources to $F$ needs constant time for each source or sink and thus we obtain $O(|V|)$ time in total.

In summary we initially need $O(n^2)$ time, and then each call of $\mathcal{CS}(D, F)$ is charged with $O(|V|)$ before it is called and then additionally with $O(|V|)$ time during its execution. Since we output a convex set of size $O(|V|)$, the total running time is $O(n^2) + O(\sum_{C \in \mathcal{CONV}(D)} |C|)$. Since $\sum_{C \in \mathcal{CONV}(D)} |C| = \Omega(n^2)$ by Theorem 3.9.1, the running time of $\mathcal{CS}$ is $O(\sum_{C \in \mathcal{CONV}(D)} |C|)$.

## 3.10 Out-forest-based Cryptographic Enforcement Schemes

In this section, we discuss results due to Crampton, Farley, Gutin, Jones and Poettering [23–25] in the area of cryptographic access control, which gives rise to some interesting problems connected with acyclic digraphs. We describe these results from the perspective of digraphs using digraph notation and terminology (rather than the notation and terminology associated with partially ordered sets that was used in the original papers). In particular, we use the following notation for a vertex $v$ of an $n$-vertex digraph $D$: $N_D^+[v] = N_D^+(v) \cup \{v\}$; $N_D^{-p}[v]$ is the set of vertices that can reach $v$ by dipaths of length at most $p$; in particular, $N_D^{-n+1}[v]$ is the set of vertices that can reach $v$, and $N_D^-[v] = N_D^{-1}[v]$.

Cryptographic access control provides a method of regulating access to sensitive resources by users, without the use of a trusted software component. This form of access control is particularly suitable in environments where the data is stored by a third party who cannot be trusted to enforce the desired access control policy. We focus on the use of symmetric cryptographic primitives to enforce information flow policies. In this setting, each user and resource is associated with a security label taken from a partially ordered set of security labels $(X, \leqslant)$. A user $u$ is authorized to access resource $r$ if the label of $u$, denoted $\lambda(u)$, is greater than or equal to the label of $r$, $\lambda(r)$.

Resource $r$ is encrypted with the key associated with $\lambda(r)$, which we denote by $\kappa(\lambda(r))$, before being transmitted to the storage provider. Hence, in order to decrypt the resources for which a user $u$ is authorised, $u$ must be able to derive $\kappa(y)$ for all $y \leqslant \lambda(u)$. Henceforth, we write $U_x$ to denote the set of users assigned to label $x \in X$.

In many schemes for this type of cryptographic access control, each user $u \in U_x$ is supplied with the key $\kappa(x)$ and this key, together with public information, is used to derive $\kappa(y)$ for $y < x$. In many schemes, key derivation is performed by successively deriving keys associated with labels on a directed path from $x$ to $y$ in the Hasse diagram of $(X, \leqslant)$. Crampton, Daud and Martin [22] suggested an alternative approach in which each user is supplied with several keys, the trade-off being that no public information is required for key derivation. Thus, there are three important parameters that characterise schemes for cryptographic access control: the number of keys required by users; the amount of public information required for key derivation; and the number of key derivation operations that are required.

Note that $(X, \leqslant)$ can be represented as a transitive acyclic digraph $D = (X, A)$, where $xy \in A$ if and only if $x \geqslant y$, and Crampton *et al.* [22] partition $X$ into a disjoint collection $P$ of dipaths. If a user $u$ is assigned a label (vertex) $x$ in $P_i \in P$ then $u$ can derive every key reachable from $x$ along the path $P_i$ *using only a one-way function with the keys for successive vertices as input.* The important point here is that in the path factor $P$ of $D$ no vertex has more than one in-neighbour, which enables the use of a one-way function for key derivation. However, Crampton, Daud and Martin [22] did not study the problem of finding an optimal path factor, which, e.g. is a factor that minimises the total number of keys required by users. This optimisation problem was solved by Crampton, Farley, Gutin and Jones [23], who obtained a polynomial-time algorithm for the problem. Moreover, Crampton *et al.* extended eligible spanning subgraphs $F$ of $D$ under consideration while preserving the above property that every vertex of $F$ has at most one in-neighbour [24, 25]. This means that $F$ can be any **spanning out-forest**, i.e. a collection of vertex-disjoint out-trees covering $X$. Crampton *et al.* [24, 25] designed a polynomial-time algorithm to find a spanning out-forest with minimum total number of keys.

The rest of this section is organised as follows. In Subsection 3.10.1 we consider the optimal key allocation for a given spanning out-forest. In Subsections 3.10.2 and 3.10.3, we describe algorithms for finding a spanning out-forest with the minimum total number of keys and a dipath factor with the minimum total number of keys, respectively.

### 3.10.1 Optimal Key Allocation

After an optimal spanning out-forest is found, one also has to allocate keys to users in an optimal way. Crampton, Farley, Gutin, Jones and Poettering [24, 25] showed that this can be done in polynomial time as well. Let us start from this phase, i.e. let us assume that we are given a spanning out-forest $F$ (not necessarily optimal) and we wish to find an optimal key allocation for $F$, i.e. an assignment with the minimum total number of keys (for $F$).

A key allocation can be viewed as a function $\psi_F : X \to 2^X$ such that $\psi_F(x)$ is the set of labels (vertices) which are assigned certain keys such that every user in $U_x$ can access all authorised resources as described above. Informally, $\psi_F$ represents the set of starting points for key derivation; alternatively, one might view the pair $(x, y)$, for each $y \in \psi_F(x) \setminus \{x\}$, as a "shortcut" arc to "compensate" for arcs deleted from $D$ to obtain $F$. A key allocation $\psi$ has to satisfy two properties: (a) if $y \in N_D^+[x]$ then there exists a $z \in \psi_F(x)$ such that $y \in N_F^+[z]$, and (b) if $y \notin N_D^+[x]$ then there is no $z \in \psi_F(x)$ such that $y \in N_F^+[z]$.

In the rest of this section, let $D = (X, A)$ be a transitive acyclic digraph, $F$ a spanning out-forest of $D$, and $n = |X|$. For a vertex $z$ of $D$, note that if $N_F^-(z) \neq \emptyset$ then $N_F^-(z)$ is a singleton consisting of the vertex of $F$ dominating $z$. For simplicity, $N_F^-(z)$ will often denote the vertex itself.

Crampton *et al.* [25] showed that an optimal function $\psi_F$, denoted $\phi_F$, can be computed as follows. Let $x, y \in X$ be such that $y \in N_D^+[x]$. This means that for some $p \in \{0, 1, \ldots, n\}$, $N_D^+[x] \cap N_F^{-p}[y] \neq \emptyset$ and for every such $p$, $N_D^+[x] \cap N_F^{-p}[y]$ is a vertex on a dipath of $F$ terminating at $y$. Now let the vertex for the maximum such $p$ be denoted by $\alpha(xy)$. Then

$$\phi_F(x) = \{\alpha(xy) : \ y \in N^+[x]\}.$$

Figure 3.3 depicts the transitive reduction $H$ of a transitive acyclic digraph $D$ (recall that such a reduction is unique by Theorem 3.2.2) and a spanning out-forest $F$ of $D$ (induced by the thick arcs). Table 3.1 contains the values of $\phi_F(x)$ for all vertices of $D$.

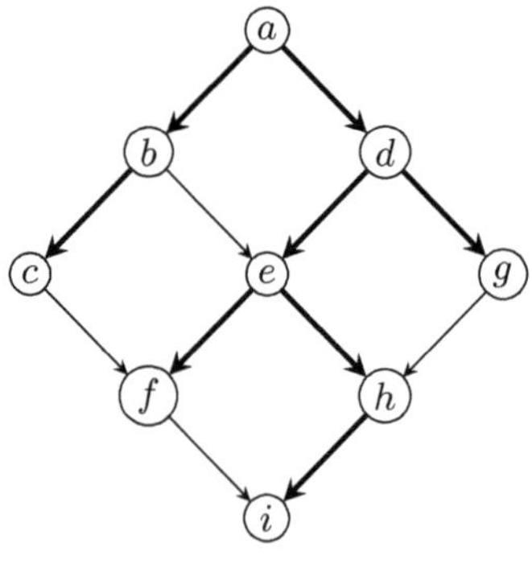


**Figure 3.3** Spanning out-forest.

Since for a user $u \in U_x$, $|\phi_F(x)|$ is the minimum number of keys that needs to be allocated, the minimum total number of keys to be allocated for

**Table 3.1** Values of $\phi_F(x)$

| $x$ | $a$ | $b$ | $c$ | $d$ | $e$ | $f$ | $g$ | $h$ | $i$ |
|---|---|---|---|---|---|---|---|---|---|
| $\phi_F(x)$ | $a$ | $b, e$ | $c, f, i$ | $d$ | $e$ | $f, i$ | $g, h$ | $h$ | $i$ |

$F$ is

$$\mathcal{K}(F) = \sum_{x \in X} |\phi_F(x)||U_x|. \tag{3.6}$$

Below we will use the following characterisation of $\phi_F$.

**Lemma 3.10.1** *For every $x \in X$ and every $z \in X$, we have $z \in \phi_F(x)$ if and only if exactly one of the following conditions hold:*

*(i)* $z = x$;
*(ii)* $xz \in A$, $N_F^-(z) \neq \emptyset$, *and* $(x, N_F^-(z)) \notin A$;
*(iii)* $xz \in A$ *and* $d_F^-(z) = 0$.

**Proof:** Suppose $xz \in A$ and $(x, N_F^-(z)) \notin A$. Since $(x, N_F^-(z)) \notin A$ but $(N_F^-(z), z) \in A(F)$, we have $z = \alpha(xz)$. Similarly, if $d_F^-(z) = 0$ or $z = x$, then $z = \alpha(xz)$. Thus, $z \in \phi_F(x)$.

Conversely, if $z \in \phi_F(x)$, then $x = z$ or $xz \in A$, by definition, and $\alpha(xz) = z$. Thus, $(x, N_F^-(z)) \notin A$ if $d_F^-(z) > 0$. Otherwise, $N_F^-(z) \in N_D^+[x] \cap N_F^{-n}[z]$ and $z \neq \alpha(xz)$. □

**Proposition 3.10.2** *We can compute $\phi_F$ in time $O(n^2)$.*

**Proof:** By Lemma 3.10.1, for all $x \in X$, besides $x$ itself, we add all those elements $z \in X$, $xz \in A$, to $\phi_F(x)$ that are of zero in-degree in $F$ or, if not, satisfy $(x, N_F^-(z)) \notin A$. In both cases, we must determine whether $xz \in A$ for some $z \in X$.

After $O(n^2)$ time preprocessing, we may assume that we have data structures allowing us to check whether $xz \in A$ in $O(1)$ time, and test whether $z$ is of zero in-degree in $F$ (and compute $N_F^-(z)$ otherwise) in $O(1)$ time. Hence, we can compute $\phi_F$ in $O(n^2)$ time. □

### 3.10.2 Optimal Spanning Out-forests

Now we consider how to find an optimal spanning out-forest, i.e. a spanning out-forest with the minimum total number of keys. We will minimise $\mathcal{K}(F)$ given in (3.6) over all spanning out-forests $F$ of $D$. Let us define $\gamma_D(yz) = \{x \in X : x \in N_D^-[z], x \notin N_D^-[y]\}$ for an arc $yz$ of $D = (X, A)$. We will need the following:

**Lemma 3.10.3** *For every dipath $xyz$ of $D$, $\gamma_D(yz) \subset \gamma_D(xz)$.*

**Proof:** Let $t \in \gamma_D(yz)$. Then $tz \in A$ and $ty \notin A$. Now if $tx \in A$, we would have $ty \in A$ by transitivity of $D$. Thus, $tx \notin A$ and hence $t \in \gamma_D(xz)$. Moreover, $y \in \gamma_D(xz)$, since $yz \in A$ and $yx \notin A$, and $y \notin \gamma_D(yz)$, so the inclusion is strict. □

If $y = N_F^-(z)$ then we will write $\gamma_D(yz)$ as $\gamma_F(z)$. Now we will define a weight function $\Omega_F : X \to \mathbb{N}$, where

$$\Omega_F(z) = \begin{cases} \sum\limits_{x \in N_D^-[z]} |U_x| & \text{if } d_F^-(z) = 0, \\ \sum\limits_{x \in \gamma_F(z)} |U_x| & \text{otherwise.} \end{cases}$$

The following result can be derived from formula (3.6) using Lemma 3.10.1; its proof can be found in [25].

**Lemma 3.10.4** *We have*

$$\mathcal{K}(F) = \sum_{z \in X} \Omega_F(z). \tag{3.7}$$

Define a weight function $\omega_D : A \to \mathbb{N}$ as follows: for $yz \in A$, $\omega_D(yz) = \sum_{x \in \gamma_D(yz)} |U_x|$. We will prove the following:

**Theorem 3.10.5** *Let $H = (X, A')$ be the transitive reduction of $D$. We can compute an optimal spanning out-forest $\hat{F}$ of $D$ in time*[4] *$O(|A'| + n^2)$.*

**Proof:** Let $R(\hat{F})$ be the set of roots of out-trees forming an optimal spanning out-forest $\hat{F}$. Observe that if $d_D^-(z) > 0$, then $z \notin R(\hat{F})$. Indeed, suppose $d_{\hat{F}}^-(z) = 0$ and let $y \in N_D^-(z)$. Then $\Omega_{\hat{F}}(z) > \Omega_{F'}(z)$, where $F'$ is obtained from $\hat{F}$ by adding arc $yz$, since $\gamma_{F'}(z) \subset N^-[z]$; the inclusion is strict since $y$ is in the second set but not the first. Thus, $R(\hat{F}) = \{x \in X : d_D^-(x) = 0\}$.

Now to obtain an optimal spanning out-forest $\hat{F}$ it remains to find the in-neighbour in $\hat{F}$ of every vertex $z \notin R(\hat{F})$. Note that $\Omega_{\hat{F}}(z) = \omega_D(N_{\hat{F}}^-(z)z)$. By Lemma 3.10.3, we have $\gamma_D(yz) \subset \gamma_D(xz)$ for a dipath $xyz$ of $D$. It follows that $\omega_D(yz) \leq \omega_D(xz)$, the inequality being strict if we assume that at least one user is assigned to each vertex in $X$. Thus it suffices to consider only in-neighbours of $z$ in $H$ when constructing $\hat{F}$ and choose among them a vertex $y$ with minimum $\omega_D(yz)$.

Finally, we analyze the running time to compute $\hat{F}$. We can compute $\omega_D(yz)$ for each $z \notin R(\hat{F})$ and each in-neighbour $y$ of $z$ in $H$ in time $O(n^2)$ using an algorithm similar to that used in the proof of Proposition 3.10.2. This allows us to find, in time $O(|A'|)$, for each $z \notin R(\hat{F})$ an in-neighbour $y$ in $H$ such that $\omega_D(yz)$ is minimum. □

---

4 We can clearly reduce $O(|A'| + n^2)$ to $O(n^2)$; the reason we keep $|A'|$ is to stress that we can consider only arcs of $H$.

Let us find an optimal spanning out-forest for the transitive acyclic digraph $D$ whose transitive reduction $H = (X, A')$ is depicted in Figure 3.3. We assume that $|U_x| = 1$ for every vertex $x$ of $D$. By the proof of Theorem 3.10.5, we first find the weight $\omega_D(yz)$ of every arc of $H$. Since $|U_x| = 1$ for every vertex $x$ of $D$, we have $\omega_D(yz) = |\gamma_D(yz)|$ for every $yz \in A'$. Every arc of the dipaths $abc$ and $adg$ is of weight 1, every arc of the dipaths $cfi$ and $ghi$ is of weight 3, and every other arc is of weight 2. Thus, the spanning out-forest $F$ of $D$ depicted in Figure 3.3 with thick arcs is optimal. Recall that the optimal key allocation for $F$ is given in Table 3.1, which implies that the minimum total number of keys for a spanning out-forest of $D$ is 14. The sum of the weights of $F$ is 13, but it is not a contradiction since one key has to be allocated to the root $a$ of $F$.

### 3.10.3 Optimal Dipath Factors

We call a dipath factor **optimal** if it requires the minimum total number of keys among all dipath factors of $D$. We first show the following somewhat unexpected result: The number of keys required when a dipath factor $P$ is used depends only on the terminal vertices of paths in $P$. This in turn implies that there exists an optimal dipath factor which has the minimum possible number of paths among all dipath factors.

In this subsection, as above, $D = (X, A)$ is a transitive acyclic digraph. In what follows, let $P = P_1 \cup \cdots \cup P_\ell$ be a dipath factor of $D$ and let $P_i = z_1^i \dots z_{c_i}^i$, $i \in [\ell]$.

**Lemma 3.10.6** *We have*

$$\mathcal{K}(P) = \sum_{i=1}^{\ell} \sum_{x \in N^-[z_{c_i}^i]} |U_x|. \tag{3.8}$$

**Proof:** Observe that $N_D^-[z_{c_i}^i]$ is the disjoint union of sets $X_i$, $i \in [c_i]$, where $X_1 = \{x \in X : x \in N_D^-[z_1^i]\}$ and $X_j = \{x \in X : x \notin N_D^-[z_{j-1}^i], x \in N_D^-[z_j^i]\}$, $2 \le j \le c_i$. Observe that $X_j = \{x \in X : x \in \gamma_P(z_j^i)\}$, $2 \le j \le c_i$. This decomposition of $N_D^-[z_{c_i}^i]$ into sets $X_i$, $i \in [c_i]$, will be used in the following derivation. By (3.7) and the definition of $\Omega_P(z)$,

$$\begin{aligned}
\mathcal{K}(P) &= \sum_{z \in X} \Omega_P(z) \\
&= \sum_{i=1}^{\ell} \sum_{x \in X_1} |U_x| + \sum_{i=1}^{\ell} \sum_{j=2}^{c_i} \sum_{x \in X_j} |U_x| \\
&= \sum_{i=1}^{\ell} \sum_{x \in N^-[z_{c_i}^i]} |U_x|.
\end{aligned}$$

□

By Dilworth's theorem (Theorem 3.2.1), the minimum number of paths in a dipath factor of $D$ is $\alpha(D)$. Let us show that $D$ contains an optimal directed $\alpha(D)$-path factor. Let $P = P_1 \cup \cdots \cup P_\ell$ be an optimal dipath factor with the minimum number of paths and suppose that $\ell > \alpha(D)$. By the Gallai-Milgram theorem (Theorem 1.8.5) $D$ contains a directed $(\ell - 1)$-path factor $P'$ such that the terminal vertices of the paths of $P'$ are a subset of the terminal vertices of the paths of $P$. By (3.8) and optimality of $P$, $\mathcal{K}(P') = \mathcal{K}(P)$, a contradiction.

Figure 3.4 depicts an optimal directed $\alpha(D)$-path factor (induced by the thick arcs) of a transitive acyclic digraph $D$ (all arcs form the transitive reduction of $D$).

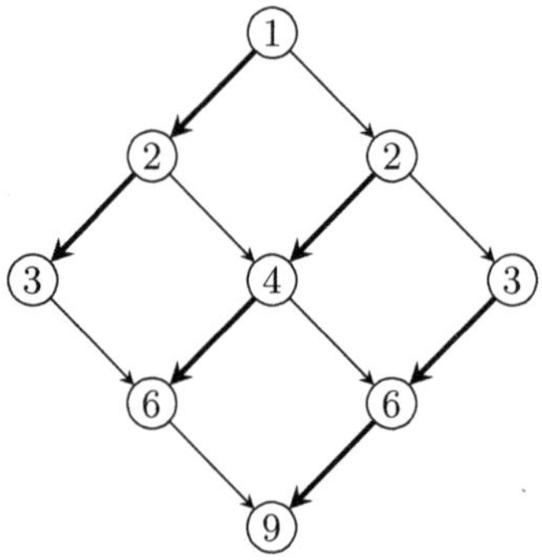


**Figure 3.4** Optimal directed $\alpha(D)$-path factor. The number given for every vertex $x$ is $|N^-[x]|$.

One can find an optimal directed $\alpha(D)$-path in polynomial time as follows. Transform $D$ into a flow network[5] by adding adding new vertices $s$ and $t$ to $D$ such that $s$ dominates every vertex in $D$ and every vertex of $D$ dominates $t$. Let us set lower and upper bounds equal to one for every vertex[6] in $X$ and assign a cost of $\sum_{x \in N^-[v]} |U_x|$ to arc $vt$ for each $v \in X$. The cost of all other arcs is zero. It is not hard to see that an optimal directed $\alpha(D)$-path factor corresponds to a minimum cost $(s,t)$-flow of value $\alpha(D)$. For more details, see [23, 25]. Thus, we have the following:

**Theorem 3.10.7** *Every $D$ contains an optimal directed $\alpha(D)$-path factor, which can be found in polynomial time.*

---

[5] For the basics on network flows, see Section 1.9 of Chapter 1.

[6] We can assign bounds to vertices rather than arcs as every vertex can be split, i.e., replaced by an arc, see Section 1.4 of Chapter 1.

## 3.11 PERT/CPM in Project Scheduling

Often a large project consists of many activities, some of which can be done in parallel, others can start only after certain activities have been accomplished. In such cases, the **critical path method (CPM)** and **Program Evaluation and Review Technique (PERT)** are of interest. They allow us to predict when the project will be finished and monitor the progress of the project. They allow one to identify certain activities which should be finished on time if the predicted completion time is to be achieved.

CPM and PERT were developed independently in the late 1950s. They have many features in common and several others that distinguish them. However, over the years the two methods have practically merged into one combined approach often called PERT/CPM. Notice that PERT/CPM has been used in a large number of projects including a new plant construction, NASA space exploration, movie production and ship building (see, e.g., the book [49] by Hillier and Lieberman). PERT/CPM has many tools for project management, but we will restrict ourselves only to a brief introduction and refer the reader to various books on operations research such as [49, 50, 66] for more information on the method.

We will introduce PERT/CPM using an example, see Figure 3.5. Suppose the tasks to complete the construction of a house are as follows (in brackets we give their abbreviation and duration in days): Wiring (Wi, 5), Plumbing (Pl, 8), Walls & Ceilings (WC, 10), Floors (Fl, 4), Exterior Decorating (ED, 3) and Interior Decorating (ID, 12). We cannot start doing Walls & Ceilings or Floors before Wiring and Plumbing are accomplished, we cannot do Exterior Decorating before Walls & Ceilings are completed, and we cannot do Interior Decorating before Walls & Ceilings and Floors are accomplished. How much time do we need to accomplish the construction?

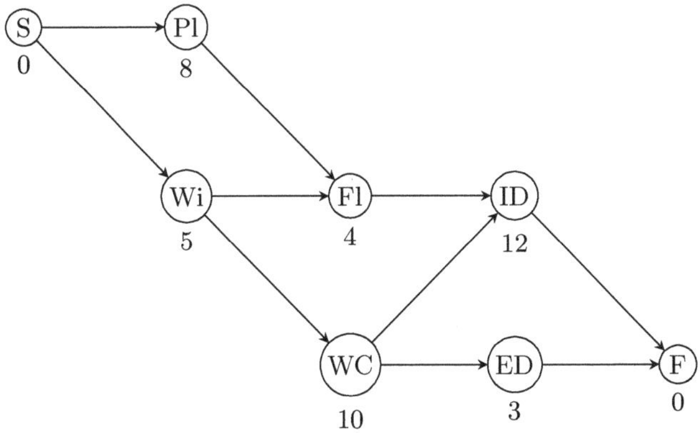


**Figure 3.5** House construction network

To solve the problem we first construct a digraph $N$, which is called an **activity-on-node (AON)** project network.[7] We associate the vertices of $N$ with the starting and finishing points of the projects (vertices $S$ and $F$) and with the activities described above, i.e., Wiring ($Wi$), Plumbing ($Pl$), Floors ($Fl$), Walls & Ceiling ($WC$), Interior Decoration ($ID$) and Exterior Decorating ($ED$). The network $N$ is a vertex-weighted digraph, where the weights of $S$ and $F$ are 0 and the weight of any other vertex is the duration of the corresponding activities. Observe that the duration of the house construction project equals the maximum weight of an $(S, F)$-path.

As in the example above, in the general case, an AON network $D$ is a vertex-weighted digraph with the starting and finishing vertices $S$ and $F$. Our initial aim is to find the maximum weight of an $(S, F)$-path in $D$. Since $D$ is an acyclic digraph, this can be done in linear time using the algorithm of Theorem 3.1.5 after a vertex splitting procedure, which replaces every vertex $x$ by arc $x'x''$ such that $N^+(x'') = \{y' : y \in N^+(x)\}$ and $N^-(x') = \{y'' : y \in N^-(x)\}$. We can also use dynamic programming directly: for a vertex $x$ of $D$ let $t(x)$ be the earlier time when the activity corresponding to $x$ can be accomplished. Then $t(S) = 0$ and for any other vertex $x$, we have $t(x) = \ell(x) + \max\{t(y) : y \in N^-(x)\}$, where $\ell(x)$ is the duration of the activity associated with $x$. To ensure that we know the value of $t(y)$ for each in-neighbour of $y$ of $x$, we consider the vertices of $D$ in an acyclic ordering.

It is easy to see that the maximum weight of an $(S, F)$-path in $N$ (our example) is 27 days and the path is $(S, Wi, WC, ID, F)$. Every maximum weight $(S, F)$-path is called **critical** and every vertex (and the corresponding activity) belonging to a critical path is **critical**. Observe that to ensure that the project takes no longer than required, no critical activity should be delayed. At the same time, delay with non-critical activities may not affect the duration of the project. For example, if we do Plumbing in 13 days instead of 8 days, the project will be finished in 27 days anyway. This means that the project manager mainly has to monitor critical activities and may delay non-critical activities in order to enforce critical ones (e.g., by moving workforce from a non-critical activity to a critical one).

The manager may want to expedite the project (if, for example, earlier completion will result in a considerable bonus) by spending more money on it. This issue can be investigated using linear programming, see Hillier and Lieberman [49].

[7] The original versions of PERT and CPM used another type of network, **activity-on-arc (AOA)** project networks, but AOA networks are significantly harder to construct and change than AON networks and it makes more sense to use AON networks rather than AOA networks.

## 3.12 One-sink Partitioning

For an acyclic digraph $D$, an arc set $P$ is called a **partitioning set** if every connected component in $D - P$ has just one sink. Consider the following problem introduced by Leskovec, Backstrom and Kleinberg [55]:

> DIRECTED ACYCLIC DIGRAPH (DAG) PARTITIONING
> **Input:** An arc-weighted acyclic digraph $D = (V, A)$.
> **Find:** a partitioning set $P$ of minimum weight.

The problem is motivated by our interest in analyzing how short, distinctive phrases (typically, parts or mutations of quotations) spread to various news sites and blogs. To demonstrate their approach, Leskovec *et al.* [55] collected and analyzed phrases from 90 million articles that appeared during the time of the 2008 United States presidential elections; the results were featured in the New York Times [61]. Leskovec *et al.* [55] create an arc-weighted acyclic digraph with phrases as vertices and an arc from phrase $p$ to phrase $q$ if $p$ presumably originates from $q$. (Unfortunately, the description of the weight of an arc in [55] is not precise.)

Leskovec *et al.* [55] proved that DAG PARTITIONING is $\mathcal{NP}$-hard and suggested a heuristic to solve it. Alamdari and Mehrabian [3] proved that the problem is hard to approximate: for fixed $\varepsilon > 0$, it is $\mathcal{NP}$-hard to approximate the minimum weight of a partitioning set within a factor of $O(n^{1-\varepsilon})$. This result holds even for digraphs with unit weights, maximum out-degree 3 and with just two sinks. van Bevern, Bredereck, Chopin, Hartung, Hüffner, Nichterlein and Suchý [86] studied DAG PARTITIONING parameterized by the weight $k$ of a partitioning set, i.e., we are to decide whether an arc-weighted acyclic digraph $D = (V, A)$ has a partitioning set of weight at most $k$. They assumed that no arc weight is smaller than 1. They proved that DAG PARTITIONING admits an $O(2^k|V|^2)$-time algorithm and that there is no algorithm for the problem of running time $2^{o(k)}|V|^{O(1)}$ unless the Exponential Time Hypothesis fails. Also, they proved that the problem does not admit a kernel with size polynomial in $k$, unless $NP \subseteq coNP/poly$.

In a newer version [85] of [86] it is proved that DAG PARTITIONING admits an asymptotically faster $O(2^k(|V| + |A|))$-time algorithm. Also, they demonstrated, in computational experiments, that using their faster algorithm and some polynomial-time preprocessing rules, within five minutes one can solve instances of the problem with more than $10^7$ arcs and $k \leq 190$. Even such an algorithm may be too slow to be practical, but it allowed van Bevern *et al.* [85] to evaluate[8] the heuristic of Leskovec *et al.* [55].

Let us describe the FPT algorithm of [86] as it is simpler than the one in [85]. The algorithm of [86] is based on the following:

[8] The idea of using FPT algorithms to evaluate heuristics was likely coined first by Gutin, Karapetyan and Razgon [43].

**Proposition 3.12.1** ([85]) *For an acyclic digraph $D$ and its minimal partitioning set $P$, a vertex $v$ is a sink in $D - P$ if and only if $v$ is a sink in $D$.*

**Proof:** Observe that arc deletion cannot make a sink become a non-sink. Thus, it suffices to show that no sink of $D - P$ is a non-sink in $D$. Suppose that $u$ is a non-sink in $D$. Then $u$ dominates some vertex $v$ in $D$. Let $C_u$ and $C_v$ be the vertices of connected components of $D - P$ containing $u$ and $v$, respectively, and let $v'$ be a sink in $C_v$. Then $D[C_u \cup C_v]$ has the unique sink $v'$. This means that $P \setminus \{uv\}$ is a partitioning set, contradicting the assumption that $P$ is minimal. □

In the rest of this section we will assume that the input $D = (V, A)$ of DAG PARTITIONING is connected. Otherwise, we can deal with its connected component separately. The algorithm of [86] uses the fact that no new sinks are created after deleting a minimal partitioning set. Assume that there is a vertex $v$ with a unique out-neighbour $u$; then the arc $vu$ is not in any minimal partitioning set. This observation leads to the following reduction rule.

**Reduction Rule.** *Let $v \in V$ be of out-degree one and let $u$ be the unique out-neighbour of $v$. Then for each arc $wv \in A$, add an arc $wu$ of the same weight. If an arc $wu$ was already in $D$, increase the weight of this arc by the weight of $wv$. Finally, delete $v$.*

Clearly, the resulting digraph has a solution for DAG PARTITIONING if and only if the original $D = (V, A)$ does. Let $n = |V|$. It is not hard to see that for each appropriate $v$ we can apply this rule in time $O(n)$. Thus, in time $O(n^2)$, we can obtain a digraph, which is irreducible by the reduction rule and either has at most two vertices or has no vertices of out-degree zero or one. This fact will be used in the following:

**Theorem 3.12.2** ([85]) *The* DAG PARTITIONING *problem can be solved in time $O(2^k n^2)$.*

**Proof:** Recall that $D$ is connected. We may assume that $D$ is irreducible. Let $S$ be the set of sinks of $D$. Assume that $D$ has at least three vertices. Then $|S| \geq 2$ and consider a sink $t$ in $D - S$. Let $d = d_D^+(t)$. Then every minimal partitioning set contains exactly $d-1$ arcs leaving $t$. Thus, we may branch by considering all $d$ possibilities of keeping just one arc leaving $t$. Since no arc weight is smaller than 1, at each branch we reduce the value of $k$ by at least $d - 1$. Thus, the running time $T_k$ as a function of $k$ satisfies the following: $T_k \geq dT_{k-d+1}$. Observe that $2^k$ satisfies this inequality as long as $d \geq 2$, which holds as $D$ is irreducible. It remains to observe that fully executing the reduction rule and finding vertices $t$ can be done in time $O(n^2)$. □

## 3.13 Acyclic edge-coloured graphs

In this section, we consider **edge-coloured graphs** which are undirected graphs with a colour assigned to every edge. An edge-coloured graph is $c$**-edge-coloured** if all colours are from the set $[c]$. For 2-edge-coloured graphs we use colours blue and red instead of 1 and 2. A walk $W = v_1e_1v_2 \ldots v_{p-1}e_{p-1}v_p$ is **properly coloured (PC)** if edges $e_i$ and $e_{i+1}$ are of different colours for every $i \in [p-2]$ and, in addition, if $W$ is closed then edges $e_{p-1}$ and $e_1$ are of different colours. PC walks are of interest in graph theory applications and in graph theory itself as generalisations of walks in digraphs. Indeed, consider the standard transformation from a digraph $D$ into a 2-edge-coloured graph $G$ by replacing every arc $uv$ of $D$ by a path with blue edge $uw_{uv}$ and red edge $w_{uv}v$, where $w_{uv}$ is a new vertex. Clearly, every diwalk in $D$ corresponds to a PC walk in $G$ and vice versa. There is an extensive literature on PC walks: for a detailed survey of pre-2009 publications, see [9, Chapter 16], and more recent papers include [1, 34, 41, 42, 56, 58–60].

It is well-known and easy to prove that every directed graph with no dicycles has no closed diwalks either. This is not the case for PC cycles and PC walks. In fact, the properties of having no PC cycles, having no PC closed trails, and having no PC closed walks, are all distinct, as we will see below.

The following notion of a monochromatic vertex will often be used in this section. A vertex $v$ in an edge-coloured graph $G$ is called $G$**-monochromatic** if all edges incident to $v$ in $G$ are of the same colour. Clearly, a PC closed walk has no $G$-monochromatic vertex.

In order to better understand the structure of acyclic edge-coloured graphs, following Gutin, Jones, Sheng, Wahlström and Yeo [41] we introduce five types of PC acyclicity as follows.

**Definition 3.13.1** *Let $G$ be an edge-coloured undirected graph. An ordering $v_1, v_2, \ldots, v_n$ of vertices of $G$ is* **of type**

1. *if for every $i \in [n]$, all edges from $v_i$ to each connected component of $G[\{v_{i+1}, v_{i+2}, \ldots, v_n\}]$ have the same colour;*
2. *if for every $i \in [n]$, all edges from $v_i$ to $\{v_{i+1}, v_{i+2}, \ldots, v_n\}$ which are not bridges in $G[\{v_i, v_{i+1}, \ldots, v_n\}]$ have the same colour.*
3. *if for every $i \in [n]$, all edges from $v_i$ to $\{v_{i+1}, v_{i+2}, \ldots, v_n\}$ have the same colour;*
4. *if for every $i \in [n]$, all edges from $v_i$ to $\{v_{i+1}, v_{i+2}, \ldots, v_n\}$ have the same colour and all edges from $v_i$ to $\{v_1, v_2, \ldots, v_{i-1}\}$ have the same colour;*
5. *if for every $i \in [n]$, all edges from $v_i$ to $\{v_{i+1}, v_{i+2}, \ldots, v_n\}$ have the same colour and all edges from $v_i$ to $\{v_1, v_2, \ldots, v_{i-1}\}$ have the same colour but different from the colour of edges from $v_i$ to $\{v_{i+1}, v_{i+2}, \ldots, v_n\}$.*

**Definition 3.13.2** *Let $i \in [5]$. $G$ is* **PC acyclic of type** *$i$ if it has an ordering $v_1, v_2, \ldots, v_n$ of vertices of type $i$.*

Clearly, the class of edge-coloured acyclic graphs of type $i$ contains the class of edge-coloured acyclic graphs of type $i+1$, $i \in [4]$. We will see below that the containments are proper. We will also see that edge-coloured acyclic graphs of type $i \in [3]$ are precisely those without PC cycles (for $i = 1$), without PC trails (for $i = 2$), and without PC walks (for $i = 3$). One reason to study the five types instead of just the first three is the fact that the first three types have quite a complicated structure, e.g., Menger's theorem does not hold for them. An example from [41] of an acyclic edge-coloured graph of type 3 for which Menger's theorem does not hold is depicted in Fig. 3.6. Let $x = v_1$ and $y = v_8$; note that any PC path between $x$ and $y$ uses at least two blue edges, thus there is at most one internally vertex-disjoint PC path between $x$ and $y$. However, after deleting any vertex apart from $\{x, y\}$ the remaining graph will still have a PC path between $x$ and $y$. However, Menger's theorem holds for type 4 [41].

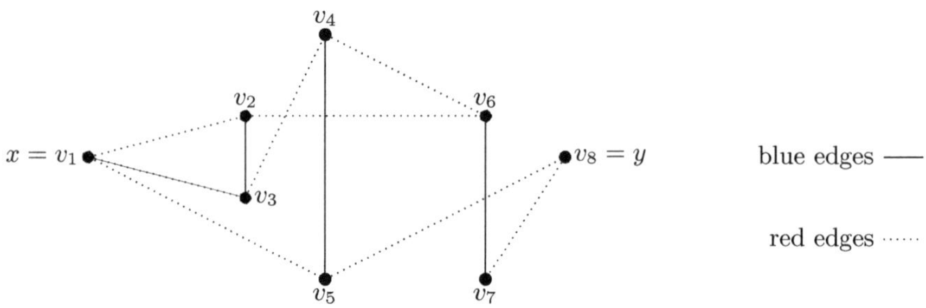


**Figure 3.6** Menger's theorem fails on $G$; the blue edges are $v_2v_3, v_4v_5, v_6v_7$; $v_8v_7v_6v_1v_2v_3v_4v_5$ is an ordering of type 3.

Let us consider a characterisation of types 1, 2 and 3 obtained by Gutin *et al.* [41].

**Theorem 3.13.3** *An edge-coloured graph $G$ is PC acyclic of type $i \in [3]$ if and only if $G$ has no PC cycle (for $i = 1$), no PC trail (for $i = 2$), no PC walk (for $i = 3$).*

Theorem 3.13.3 for $i = 1$ is an easy consequence of the following result of Yeo [88].

**Theorem 3.13.4** (**Yeo's theorem**) *If an edge-coloured graph $G$ has no PC cycle then $G$ has a vertex $z$ such that every connected component of $G - z$ is joined to $z$ by edges of the same colour.*

Using Yeo's theorem it is not hard to design a polynomial time algorithm for deciding whether an edge-coloured graph is acyclic of type 1. Theorem 3.13.3 for $i = 2$ easily follows from a theorem of Abouelaoualim, Das, Faria, Manoussakis, Martinhon and Saad [1] which states that an edge-coloured graph $G$ has either a bridge or a PC closed trail or a $G$-monochromatic vertex. Theorem 3.13.3 for $i = 3$ can easily be derived from the following proposition.

**Proposition 3.13.5** ([41]) *If an edge-coloured graph $G$ has no PC closed walk then $G$ has a $G$-monochromatic vertex.*

**Proof:** We call an edge-coloured graph $H$ an **extension** of an edge-coloured graph $G$ if $H$ is obtained from $G$ by replacing every vertex $u$ by a set $I_u$ of independent vertices with the same adjacencies and edge colours as $u$. Observe that $G$ has no PC closed walk if and only if no extension of $G$, in which $I_u$ is sufficiently large, has a PC cycle. Now apply Yeo's theorem to an extension $H$ of a connected edge-coloured graph $G$ in which $|I_u| > 1$ for each $u \in V(G)$, and note that for every vertex $z \in V(H)$, $H - z$ is connected. □

To see that containment is proper between PC acyclicities of type 1 and type 2, consider a graph with vertex set $\{v_1, v_2, x, u_1, u_2\}$ and edge set $\{v_1v_2, v_1x, v_2x,\ u_1u_2, u_1x, u_2x\}$, where $v_1v_2, u_1x, u_2x$ are coloured red, and $u_1u_2, v_1x, v_2x$ are coloured blue. Clearly, this 2-edge-coloured graph $G$ has no PC cycle, but it has a PC closed trail. Thus, $G$ is PC acyclic of type 1 but not PC acyclic of type 2.

To see that containment is proper between PC acyclicities of type 2 and type 3, consider the following graph $G$ with $V(G) = \{a_1, a_2, a_3, b_1, b_2, b_3\}$, with blue edges $a_1b_1$, $a_2b_2$ and $a_3b_3$ and red edges $a_1a_2$, $b_1a_2$, $a_3b_2$ and $b_3b_2$. In $G$ we have a PC closed walk $a_1a_2b_2b_3a_3b_2a_2b_1a_1$. This walk uses the edge $a_2b_2$ twice. There is no PC closed trail in $G$: as $a_2b_2$ is a bridge it does not belong to a closed trail and removing $a_2b_2$ makes it obvious that there is no PC closed trail in the remainder.

To see that containment is proper between PC acyclicities of type 3 and type 4, consider a complete graph on three vertices with two blue edges and one red edge. It is easy to find an ordering of type 3 and to see that there is no ordering of type 4. Finally, to see that containment is proper between PC acyclicities of type 4 and type 5, consider any non-bipartite 2-edge-coloured graph with all edges being blue.

Using Yeo's theorem, the theorem of Abouelaoualim *et al.* and Proposition 3.13.5 it is easy to see that one can decide in polynomial time whether an edge-coloured graph is acyclic of type 1, 2 and 3, respectively. We are not aware of a nice characterisation of type 4. In fact, it is $\mathcal{NP}$-complete to decide whether a 2-edge-coloured graph is acyclic of type 4 [41]. Interestingly, we can decide in polynomial time whether an edge-coloured graph is acyclic of type 5 [41].

## References


1. A. Abouelaoualim, K.Ch. Das, L. Faria, Y. Manoussakis, C.A. Martinhon, and R. Saad. Paths and trails in edge-colored graphs. *Theor. Comput. Sci.*, 409:497–510, 2008.
2. A.V. Aho, M.R. Garey, and J.D. Ullman. The transitive reduction of a directed graph. *SIAM J. Comput.*, 1(2):131–137, 1972.
3. S. Alamdari and A. Mehrabian. On a DAG partitioning problem. In *WAW 2012*, volume 7323 of *Lect. Notes Comput. Sci.*, pages 17–28. Springer, 2012.
4. N. Alon, B. Bollobàs, A. Gyàrfàs, J. Lehel, and A. Scott. Maximum directed cuts in acyclic digraphs. *J. Graph Theory*, 55:1–13, 2007.
5. N. Alon, F.V. Fomin, G. Gutin, M. Krivelevich, and S. Saurabh. Spanning directed trees with many leaves. *SIAM J. Discrete Math.*, 23(1):466–476, 2009.
6. E.N. Antoniou, A. Araújo, M.D. Bustamante, and A. Gibali. Physically feasible decomposition of Engino® toy models: A graph theoretic approach. *CoRR*, abs/1707.09040, 2017.
7. P. Balister, S. Gerke, G. Gutin, A. Johnstone, J. Reddington, E. Scott, A. Soleimanfallah, and A. Yeo. Algorithms for generating convex sets in acyclic digraphs. *J. Discrete Algor.*, 7:509–518, 2009.
8. J. Bang-Jensen and S. Bessy. (Arc-)disjoint flows in networks. *Theor. Comput. Sci.*, 526:28–40, 2014.
9. J. Bang-Jensen and G. Gutin. *Digraphs: Theory, Algorithms and Applications.* Springer-Verlag, London, 2nd edition, 2009.
10. J. Bang-Jensen and G. Gutin. Out-branchings with extremal number of leaves. *Ramanujan Math. Soc. Lect. Notes*, 13:91–99, 2010.
11. J. Bang-Jensen and M. Kriesell. Disjoint directed and undirected paths and cycles in digraphs. *Theor. Comput. Sci.*, 410:5138–5144, 2009.
12. J. Bang-Jensen, S. Thomassé, and A. Yeo. Small degree out-branchings. *J. Graph Theory*, 42(4):297–307, 2003.
13. J. Bang-Jensen and A. Yeo. The complexity of multicut and mixed multicut problems in (di)graphs. *Theor. Comput. Sci.*, 520, 2014.
14. K. Bérczi, S. Fujishige, and N. Kamiyama. A linear-time algorithm to find a pair of arc-disjoint spanning in-arborescence and out-arborescence in a directed acyclic graph. *Inform. Process. Lett.*, 109(23-24):1227–1231, 2009.
15. D. Bokal, G. Fijavz, M. Juvan, P.M. Kayll, and B. Mohar. The circular chromatic number of a digraph. *J. Graph Theory*, 46(3):227–240, 2004.
16. P. Bonzini and L. Pozzi. Polynomial-time subgraph enumeration for automated instruction set extension. In *DATE 2007: Conference on Design, Automation and Test in Europe*, pages 1331–1336, 2007.
17. J. Chen, Y. Liu, S. Lu, B. O'Sullivan, and I. Razgon. A fixed-parameter algorithm for the directed feedback vertex set problem. *J. Assoc. Comput. Mach.*, 55(5):21:1–21:19, 2008.
18. J. Chen, Y. Liu, S. Lu, B. O'Sullivan, and I. Razgon. Directed Feedback Vertex Set is Fixed-Parameter Tractable. In *STOC 2008: 40th Annual ACM Symposium on Theory of Computing.* ACM Press, 2008.
19. X. Chen, D.L. Maskell, and Y. Sun. Fast identification of custom instructions for extensible processors. *IEEE Trans. Computer-Aided Design Integr. Circuits Syst.*, 26:359–368, 2007.
20. R.H. Chitnis, M. Hajiaghayi, and D. Marx. Fixed-parameter tractability of directed multiway cut parameterized by the size of the cutset. *SIAM J. Comput.*, 42(4):1674–1696, 2013.

21. D. Coppersmith and S. Winograd. Matrix multiplication via arithmetic progressions. In *STOC 1987: 19th Ann. ACM Symp. on Theory of Computation*, pages 1–6, ACM Press, 1987.
22. J. Crampton, R. Daud, and K.M. Martin. Constructing key assignment schemes from chain partitions. In *DBSEC 2010: the 24th Conference on Data and Applications Security and Privacy*, pages 130–145, 2010.
23. J. Crampton, N. Farley, G. Gutin, and M. Jones. Optimal constructions for chain-based cryptographic enforcement of information flow policies. In *DBSec 2015*, volume 9149 of *Lect. Notes Comput. Sci.*, pages 330–345. Springer, 2015.
24. J. Crampton, N. Farley, G. Gutin, M. Jones, and B. Poettering. Cryptographic enforcement of information flow policies without public information. In *ACNS 2015*, volume 9092 of *Lect. Notes Comput. Sci.*, pages 389–408. Springer, 2015.
25. J. Crampton, N. Farley, G. Gutin, M. Jones, and B. Poettering. Cryptographic enforcement of information flow policies without public information via tree partitions. *J. Comput. Security*, 25:511–535, 2017.
26. R. Crowston, G. Gutin, and M. Jones. Directed acyclic subgraph problem parameterized above the Poljak-Turzik bound. In *FSTTCS 2012*, volume 18 of *LIPIcs*, pages 400–411, 2012.
27. J. Daligault and S. Thomassé. On finding directed trees with many leaves. In *IWPEC 2009*, volume 5917 of *Lect. Notes Comput. Sci.*, pages 86–97. Springer, 2009.
28. A. Demers and A. Downing. Minimum leaf spanning tree. US Patent no. 6,105,018, August 2000.
29. R.P. Dilworth. A decomposition theorem for partially ordered sets. *Ann. Math.*, 51:161–166, 1950.
30. J. Edmonds. Optimum branchings. *J. Res. Natl. Bur. Stand. Sect. B*, 71B:233–240, 1967.
31. C.S. Edwards. Some extremal properties of bipartite subgraphs. *Can. J. Math.*, 25:475–485, 1973.
32. M.J. Fisher and A.R. Meyer. Boolean matrix multiplication and transitive closure. In *12th Ann. ACM Symp. on Switching and Automata Theory*, pages 129–131. ACM Press, 1971.
33. S. Fortune, J.E. Hopcroft, and J. Wyllie. The directed subgraph homeomorphism problem. *Theor. Comput. Sci.*, 10:111–121, 1980.
34. S. Fujita and C. Magnant. Properly colored paths and cycles. *Discrete Appl. Math.*, 159:1391–1397, 2011.
35. M.E. Furman. Application of a method of fast multiplication of matrices in the problem of finding the transitive closure of a graph. *Sov. Math. Dokl.*, 11:1252, 1970.
36. I.M. Gessel. Counting acyclic digraphs by sources and sinks. *Discrete Math.*, 160:253–258, 1996.
37. N. Golowich and D. Rolnick. Acyclic subgraphs of planar digraphs. *Electr. J. Combin.*, 22(3):P3.7, 2015.
38. A. Goralcikova and V. Koubek. A reduct-and-closure algorithm for graphs. In *MFCS 1979: 8th Symp. on Math. Foundations of Computer Science*, volume 74 of *Lect. Notes Comput. Sci.*, pages 301–307. Springer-Verlag, 1979.
39. G. Gutin. A note on the cardinality of certain classes of unlabeled multipartite tournaments. *Discrete Math.*, 186(1-3):277–280, 1998.
40. G. Gutin, A. Johnstone, J. Reddington, E. Scott, and A. Yeo. An algorithm for finding input/output constrained convex sets in an acyclic digraph. *J. Discrete Algor.*, 13:47–58, 2012.
41. G. Gutin, M. Jones, B. Sheng, M. Wahlström, and A. Yeo. Acyclicity in edge-colored graphs. *Discrete Math.*, 340:1–8, 2017.

42. G. Gutin, M. Jones, B. Sheng, M. Wahlström, and A. Yeo. Chinese Postman Problem on edge-colored multigraphs. *Discrete Appl. Math.*, 217, part 2:196–202, 2017.
43. G. Gutin, D. Karapetyan, and I. Razgon. Fixed-parameter algorithms in analysis of heuristics for extracting networks in linear programs. In *IWPEC 2009*, volume 5917 of *Lect. Notes Comput. Sci.*, pages 222–233. Springer, 2009.
44. G. Gutin, E. J. Kim, S. Szeider, and A. Yeo. A probabilistic approach to problems parameterized above or below tight bounds. *J. Comput. Syst. Sci.*, 77(2):422–429, 2011.
45. G. Gutin, I. Razgon, and E.J. Kim. Minimum leaf out-branching and related problems. *Theor. Comput. Sci.*, 410:4571–4579, 2009.
46. G. Gutin and A. Yeo. On the number of connected convex subgraphs of a connected acyclic digraph. *Discrete Appl. Math.*, 157(7):1660–1662, 2009.
47. F. Harary and E.M. Palmer. *Graphical Enumeration.* Academic Press, 1973.
48. A. Harutyunyan and B. Mohar. Planar digraphs of digirth five are 2-colorable. *J. Graph Theory*, 84(4):408–427, 2017.
49. F.S. Hillier and G.J. Lieberman. *Introduction to Operations Research.* McGraw Hill, 2001.
50. F.W. Horsley. *Means Scheduling Manual: On-Time, On-Budget Construction Up-To-Date Computerized Scheduling.* Roberts Means Co., 1991.
51. E.J. Kim and R. Williams. Improved parameterized algorithms for above average constraint satisfaction. In *IPEC 2011*, volume 7112 of *Lect. Notes Comput. Sci.*, pages 118–131. Springer, 2012.
52. S. Kratsch, M. Pilipczuk, M. Pilipczuk, and M. Wahlström. Fixed-parameter tractability of multicut in directed acyclic graphs. *SIAM J. Discrete Math.*, 29(1):122–144, 2015.
53. J. Kuipers and G. Moffa. Uniform random generation of large acyclic digraphs. *Stat. Comput.*, 25:227–242, 2015.
54. F. Le Gall. Powers of tensors and fast matrix multiplication. In *ISSAC 2014: International Symposium on Symbolic and Algebraic Computation*, pages 296–303. ACM, 2014.
55. J. Leskovec, L. Backstrom, and J. Kleinberg. Meme-tracking and the dynamics of the news cycle. In *KDD 2009: 15th ACM SIGKDD International Conference on Knowledge Discovery and Data Mining*, pages 497–506, 2009.
56. H. Li and G. Wang. Color degree and alternating cycles in edge-colored graphs. *Discrete Math.*, 309:4349–4354, 2009.
57. V.A. Liskovets. On the number of maximal vertices of a random acyclic digraph. *Theory Prob. Appl.*, 20(2):401–409, 1976.
58. A. Lo. A Dirac type condition for properly colored paths and cycles. *J. Graph Theory*, 76:60–87, 2014.
59. A. Lo. An edge-colored version of Dirac's theorem. *SIAM J. Discrete Math.*, 28:18–36, 2014.
60. A. Lo. Properly coloured Hamiltonian cycles in edge-coloured complete graphs. *Combinatorica*, 36:471–492, 2016.
61. S. Lohr. Study measures the chatter of the news cycle. New York Times, July 13th:B1, 2009. New York edition.
62. M. Mahajan, V. Raman, and S. Sikdar. Parameterizing above or below guaranteed values. *J. Comput. Syst. Sci.*, 75(2):137–153, 2009.
63. B.D. McKay, F.E. Foggier, G.F. Royle, N.J.A. Sloane, I.M. Wanless, and H.S. Wilf. Acyclic digraphs and eigenvalues of (0, 1)-matrices. *J. Integer Sequences*, 7, 2004. article 04.3.3.
64. K. Mehlhorn. *Data Structures and Algorithms 2: Graph Algorithms and NP-completeness.* Springer-Verlag, Berlin, 1984.

65. M. Mnich, G. Philip, S. Saurabh, and O. Suchý. Beyond max-cut: $\lambda$-extendible properties parameterized above the Poljak-Turzík bound. *J. Comput. Syst. Sci.*, 80(7):1384–1403, 2014.
66. K.G. Murty. *Network programming*. Prentice Hall, Englewood Cliffs, NJ, 1992.
67. V. Neumann-Lara. Vertex colorings in digraphs. some problems., 1985. Seminar notes, University of Waterloo.
68. Y. Perl and Y. Shiloach. Finding two disjoint paths between two pairs of vertices in a graph. *J. Assoc. Comput. Mach.*, 25:1–9, 1978.
69. M. Pilipczuk and M. Wahlström. Directed multicut is $W[1]$-hard, even for four terminal pairs. In *SODA 2016*, pages 1167–1178. SIAM, 2016.
70. S. Poljak and D. Turzík. A polynomial time heuristic for certain subgraph optimization problems with guaranteed worst case bound. *Discrete Math.*, 58(1):99–104, 1986.
71. L. Pozzi, K. Atasu, and P. Ienne. Exact and approximate algorithms for the extension of embedded processor instruction sets. *IEEE Trans. Computer-Aided Design Integr. Circuits Syst.*, 25:1209–1229, 2006.
72. V. Raman and S. Saurabh. Parameterized complexity of feedback set problems and their duals in tournaments. *Theor. Comput. Sci.*, 351:446–458, 2006.
73. J. Reddington. *Improvements to instruction identification for custom instruction set design.* PhD thesis, Royal Holloway, University of London, 2008.
74. J. Reddington, G. Gutin, A. Johnstone, E. Scott, and A. Yeo. Better than optimal: Fast identification of custom instruction candidates. In *CSE 2009: 12th IEEE International Conference on Computational Science and Engineering*, pages 17–24, 2009.
75. R.W. Robinson. Counting labeled acyclic digraphs. In F. Harary, editor, *New Directions in the Theory of Graphs*, pages 239–273. Acad. Press, 1973.
76. R.W. Robinson. Counting unlabeled acyclic digraphs. In C.H.C. Little, editor, *Combinatorial Mathematics V: 5th Australian Conference, 1976*, pages 28–43. Springer, 1977.
77. R.W. Robinson. Enumeration of acyclic digraphs. In *The 2nd Chapel Hill Conf. on Combinatorial Mathematics and its Applications*, pages 391–399, 1970.
78. N. Schwartges, J. Spoerhase, and A. Wolff. Approximation algorithms for the maximum leaf spanning tree problem on acyclic digraphs. In *WAOA 2011*, volume 7164 of *Lect. Notes Comput. Sci.*, pages 77–88. Springer, 2012.
79. K. Simon. An improved algorithm for transitive closure on acyclic digraphs. *Theor. Comput. Sci.*, 58:325–346, 1988.
80. J. Singer. A theorem in finite projective geometry and some applications to number theory. *Trans. Am. Math. Soc.*, 43:377–385, 1938.
81. A. Slivkins. Parameterized tractability of edge-disjoint paths on directed acyclic graphs. *SIAM J. Discrete Math.*, 24(1):146–157, 2010.
82. R.P. Stanley. Acyclic orientations of graphs. *Discrete Math.*, 5:171–178, 1973.
83. R.P. Stanley. A matrix for counting paths in acyclic digraphs. *J. Combin. Th. Ser. A*, 74:169–172, 1996.
84. C. Thomassen. Paths, circuits and subdivisions. In *Selected topics in graph theory Vol. 3*, pages 97–131. Academic Press, 1988.
85. R. van Bevern, R. Bredereck, M. Chopin, S. Hartung, F. Hüffner, A. Nichterlein, and O. Suchý. Fixed-parameter algorithms for DAG partitioning. *CoRR*, abs/1611.08809, 2016.
86. R. van Bevern, R. Bredereck, M. Chopin, S. Hartung, F. Hüffner, A. Nichterlein, and O. Suchý. Parameterized complexity of DAG partitioning. In *CIAC 2013*, volume 7878 of *Lect. Notes Comput. Sci.*, pages 49–60. Springer, 2013.

87. D.R. Wood. Bounded degree acyclic decompositions of digraphs. *J. Combin. Theory*, 90:309–313, 2004.
88. A. Yeo. A note on alternating cycles in edge-coloured graphs. *J. Combin. Theory Ser. B*, 69:222–225, 1997.

# 4. Euler Digraphs

**Magnus Wahlström**

An Euler digraph is a connected digraph where every vertex has in-degree equal to its out-degree. The name, of course, comes from the directed version of Euler's theorem. Recall than an Euler tour in a digraph is a directed closed walk that uses each arc exactly once. Then in this terminology, by the famous theorem of Euler, a digraph admits an Euler tour if and only if it is an Euler digraph.

However, beyond this point of historical interest, Euler digraphs are also interesting since they form a class of intermediate complexity between undirected graphs and fully general digraphs for many problems. For example, consider the $k$-LINKAGE and WEAK $k$-LINKAGE problems. Recall that in these problems, the input is a digraph $D = (V, A)$ together with $k$-tuples $(s_1, \ldots, s_k)$ and $(t_1, \ldots, t_k)$ of vertices, and the goal is to find internally vertex-disjoint paths (respectively, arc-disjoint paths) from $s_i$ to $t_i$ for every $i \in [k]$. For undirected graphs, both problems are famously FPT parameterized by $k$, as a central result of the graph minor theory of Robertson and Seymour [36]. For general digraphs, both variants are $\mathcal{NP}$-hard already for $k = 2$ as shown by Fortune, Hopcroft and Wyllie [15]. For Euler digraphs, the $k$-LINKAGE PROBLEM is in general $\mathcal{NP}$-hard, but the WEAK $k$-LINKAGE PROBLEM is in $\mathcal{P}$ at least up to $k = 3$, and it is a long open question whether the WEAK $k$-LINKAGE PROBLEM is in $\mathcal{P}$ for every fixed $k$ or even FPT. (We discuss these problems later in this chapter.)

For another example, consider the concepts regarding classes of graphs and digraphs of restricted structure, e.g., bounded width. For undirected graphs, although many alternatives have been considered, arguably the established standard width notion is **bounded treewidth**, and the related notion of **bounded pathwidth**. These width measures have several desirable properties, not least including algorithmic applications such as linear-time FPT algorithms for a multitude of problems when parameterized by the width $k$. On the other hand, for directed graphs, although directed analogues of these basic width notions exist, not only are the basic definitions significantly more complex, but the algorithmic implications are also typically weaker, e.g., most problems would not be FPT parameterized by directed pathwidth. We will see that in the general case there is no significant difference between the di-

M. Wahlström (✉)
Department of Computer Science,
Royal Holloway University, Egham Hill, Egham TW20 0EX, England
e-mail: Magnus.Wahlstrom@rhul.ac.uk

 
J. Bang-Jensen and G. Gutin (eds.), *Classes of Directed Graphs*, Springer Monographs in Mathematics, https://doi.org/10.1007/978-3-319-71840-8_4

rected width measures on Euler digraphs and on general digraphs, but (unlike for general digraphs) if we additionally impose that an Euler digraph is of bounded degree, then the undirected and directed versions of pathwidth and treewidth coincide up to a constant factor.

However, interestingly, there are also a few problems which are easier to deal with on Euler digraphs than on undirected graphs (even undirected Euler graphs). We will see two main examples of this. The first is the so-called BEST theorem, which states that the number of Euler tours in a digraph can be counted efficiently; the same is not true for undirected graphs, where the corresponding problem is #P-hard. The second, less well known example is Arc Multiway Cut (see later for definitions), which is $\mathcal{NP}$-hard both for undirected Euler graphs and for general digraphs, but which admits a simple polynomial-time algorithm on Euler digraphs. We will cover both of these results in the following.

This chapter is structured as follows. We begin with some basic constructions and observations in Section 4.1; in Section 4.2 we consider the BEST theorem and other questions of Euler tours; in Section 4.3 we consider notions of Euler digraphs of bounded width; in Section 4.4 we review problems related to packing and hitting cycles, and in Section 4.5 we review general problems of path-packing and linkages.

## 4.1 Basic Constructions and Properties

To relax the definition of Euler digraphs slightly, let a digraph $D$ be **balanced** if every vertex has in-degree equal to its out-degree, but with no requirement that $D$ be connected. Most of our algorithmic results will apply to balanced digraphs as well as Euler digraphs, and it will sometimes be convenient to not have to require connectivity. In fact, we will frequently gloss over the difference between the two notions. Also note that a balanced digraph is strongly connected if and only if it is connected.

We review two basic properties of balanced digraphs. First, we note that a balanced digraph can be exhaustively decomposed into simple directed cycles. Second, we note that for every balanced digraph $D = (V, A)$ and every vertex $v \in V$, there are $d_D^+(v)$ pairwise arc-disjoint cycles through $v$. Both of these results follow via simple induction (in the latter case using the existence of an Euler tour for the component).

### 4.1.1 Cuts in Euler Digraphs

The following observation is the underlying source of many of the tractability results in this chapter.

**Proposition 4.1.1** *Let $D = (V, A)$ be a balanced digraph. For any set $S \subseteq V$, it holds that $d^+(S) = d^-(S)$.*

**Proof:** Note that $\sum_{v \in S} d^+(v) = \sum_{v \in S} d^-(v)$ since $D$ is balanced. Thus we have

$$0 = \sum_{v \in S} d^+(v) - \sum_{v \in S} d^-(S) = (d^+(S) + |(S,S)|) - (d^-(S) + |(S,S)|,$$

so $d^+(S) = d^-(S)$. □

The following is an easy but important consequence.

**Proposition 4.1.2** *Let $D = (V, A)$ be a balanced digraph and let $G = UMG(D)$ be the underlying multigraph of $D$. Then for any $S \subseteq V$, we have $d_D^+(S) = d_D^-(S) = d_G(S)/2$.*

This implies that tools developed for edge cuts in undirected graphs will transfer directly to arc-cuts in Euler digraphs. In particular, for an Euler digraph $D = (V, A)$ with two vertices $s, t \in V$, using the **treewidth reduction theorem** of Marx, O'Sullivan and Razgon [32] it follows that all minimal $s$-$t$- and $t$-$s$ arc-cuts in $D$ of size at most $k$ are contained in a subgraph of $D$ of treewidth bounded as a function of $k$. Although we will not need this result in the remainder of the chapter, it is a powerful tool for FPT algorithms in undirected graphs and worth observing for its potential applications. The same can be said for other advanced methods for producing FPT algorithms for undirected graph problems. In particular, there is a method of designing FPT algorithms via **recursive understanding**, which was pioneered by Kawarabayashi and Thorup for $k$-Way Cut [28] and was developed further and made more efficient using the method of **randomized contractions** by Chitnis, Cygan, Hajiaghayi, Pilipczuk and Pilipczuk [6]. Another related work is the special tree decomposition used by Cygan, Lokshtanov, Pilipczuk, Pilipczuk and Saurabh for the Minimum Bisection problem [10]. All of these cases represent advanced and successful methods for designing FPT algorithms for cut problems on undirected graphs, which a priori seem not to transfer in a useful way to general digraphs, but which may be worth considering for the case of Euler digraphs.

### 4.1.2 Hardness Constructions

We review two simple constructions that will be useful in showing problem hardness on Euler digraphs. Recall that for an undirected graph $G = (V, E)$, the **complete biorientation** of $G$ is a digraph $\overleftrightarrow{G}$ with vertex set $V$ and with a pair of arcs $uv, vu$ for every edge $\{u, v\} \in E$. Clearly, $\overleftrightarrow{G}$ is balanced, and Euler if $G$ is connected. This construction can frequently be used to show that problems on Euler digraphs are "at least as hard" as the corresponding problem on undirected graphs.

The second construction is as follows. Let $D = (V, A)$ be a connected digraph, and define $b_D(v) = d_D^+(v) - d_D^-(v)$ as the **balance number** of $v$

in $D$. The **Euler two-vertex extension** of $D$ is the directed multigraph obtained by adding two vertices $x, y \notin V$ to $D$, then for every vertex $v \in V$ adding $|b(v)|$ arcs $vx$ if $b(v) < 0$ and $b(v)$ arcs $yv$ if $b(v) > 0$, then finally adding $d^-(x)$ arcs $xy$. It is clear that the resulting directed multigraph is Euler. If the situation does not allow parallel arcs, we may simply subdivide all arcs into and out of $x$ and $y$. The significance of this construction is that by simply deleting $x$ or $y$ from $D'$, we eliminate all paths not present in the original graph $D$.

Above all, this means that for many problems which come in a "vertex version" and an "arc version", the vertex version is usually equally difficult on an Euler digraph as on general digraphs, while the arc version can be significantly easier. For example, considering the problems mentioned in the introduction, it is trivial to show that VERTEX MULTIWAY CUT (i.e., the vertex-deletion version), the $k$-LINKAGE PROBLEM and the VERTEX-DISJOINT CYCLE PACKING problem are all as hard on Euler digraphs as on general digraphs (in the case of the $k$-LINKAGE PROBLEM increasing $k$ by one), but ARC MULTIWAY CUT, the WEAK $k$-LINKAGE PROBLEM and ARC-DISJOINT CYCLE PACKING are all significantly easier on Euler digraphs, as we see in the results surveyed in this chapter.

### 4.1.3 Splitting Off and Other Operations

One frequently used operation in this chapter is the **splitting-off** operation. Let $D = (V, A)$ be an Euler digraph and $uv$, $vw$ a pair of arcs in $D$. Then **splitting off** $uv$ **and** $vw$ **in** $D$ refers to the operation of deleting the arcs $uv$ and $vw$, and creating a new arc $uw$. It is clear that this operation preserves balance (if not necessarily connectivity). If all vertices $u, v, w$ are distinct, and if the result is a digraph (as opposed to a directed multigraph or pseudograph), then we refer to $uv$ and $vw$ as a **simple splitting pair**. Let us make a simple observation.

**Proposition 4.1.3** *Let $D = (V, A)$ be a balanced digraph with no simple splitting pair. Then every connected component of $D$ is a complete digraph.*

**Proof:** If $D$ has no simple splitting pair, then for every pair of arcs $uv$ and $vw$ such that $u \neq w$, the arc $uw$ already exists, i.e., $D$ is transitive. It is well known (and easy to see) that a strongly connected transitive digraph must be complete. Since every connected component of a balanced digraph is also strongly connected, the result follows. □

Other operations that preserve the balance property of a digraph include arc contractions and the removal of a balanced subgraph.

We will also occasionally need the notion of a **minor** of a digraph. There are two variants of this definition, **butterfly minors** and **topological minors**. Let $D = (V, A)$ be a digraph and $uv \in A$ an arc. The arc $uv$ is **butterfly contractible** in $D$ if either $uv$ is the only arc out of $u$ or the only arc into $v$. A **butterfly minor of** $D$ is obtained from a subgraph of $D$ by contracting butterfly contractible arcs. Alternatively, a **topological minor of** $D$ is produced from arc contractions of a subgraph of $D$ by contracting an arc $uv$ only if either $u$ or $v$ has in-degree and out-degree 1.

## 4.2 Problems Regarding Euler Tours

Let us begin as a warm-up with the so-called **BEST theorem**, showing that Euler tours of an Euler digraph can be counted in polynomial time.

The BEST theorem was implicit in the work of Tutte and Smith [41] and was shown in full by van Aardenne-Ehrenfest and de Bruijn [42]; the theorem takes its name from the authors.

**Theorem 4.2.1 (BEST theorem)** *Let $D = (V, A)$ be an Euler digraph, and $w \in V$ an arbitrary vertex. The number of Euler tours in $D$ is*

$$t_D(w) \prod_{v \in V} (d(v) - 1)!,$$

*where $t_D(w)$ is the number of out-branchings of $D$ rooted in $w$. In particular, there is a polynomial-time algorithm for counting the number of directed Euler tours.*

**Proof:** Let $ww' \in A$, and let $T$ be an Euler tour of $D$. Observe that $T$ induces a permutation $\pi_v$ of the out-arcs of $v$ for every $v \in V$, according to the order in which these arcs are visited in $T$, starting the count from $ww'$. Also note that $T$ can be recovered from this collection of permutations, and conversely, every such collection of permutations $\{\pi_v\}_{v \in V}$ defines a closed trail in $D$ containing the arc $ww'$, although not every collection of permutations induces an Euler tour. For a collection of permutations $P = \{\pi_v\}_{v \in V}$, let $F(P) = \{\pi_v(d^+(v)) \mid v \in V, v \neq w\}$ be the set containing the last outgoing arc from every vertex except $w$. We claim that $P$ defines an Euler tour if and only if $F(P)$ is an in-branching in $D$ rooted in $w$.

On the one hand, let $P$ be defined via an Euler tour $T$. The set $F(P)$ forms a digraph where every vertex except $w$ has out-degree 1; hence $F(P)$ forms an in-branching rooted in $w$ if and only if it is acyclic. We claim that for every arc $uv \in F(P)$, $v \neq w$, the last out-arc of $u$ is visited before the last out-arc of $v$ in $T$, if we begin the counting from $ww'$. Indeed, $uv$ is the last out-arc of $u$ visited in $T$ by definition, and clearly whatever arc follows $uv$ in $T$ is an out-arc of $v$ visited after $uv$. Since the out-degree of $w$ in $F(P)$ is zero, it follows that $F(P)$ is acyclic, and that $F(P)$ is an in-branching rooted in $w$.

On the other hand, let $P$ be a collection of permutations such that $F(P)$ forms an in-branching, and let $T$ be the closed tour defined by $P$ starting from the arc $ww'$. Then $D - T$ is a balanced digraph. Let $H$ be a connected component of $D - T$. Clearly, for every vertex $v$ that is not of degree zero in $D - T$, the out-arc of $v$ in $F(P)$ is contained in $D - T$. But then the set $F_H$ of out-arcs of $F(P)$ of vertices in $H$ forms a subgraph of $H$ where every vertex has out-degree 1, which necessarily contains a cycle. This contradicts our assumption on $P$.

The formula follows from this claim. For every in-branching $B$ of $D$ rooted in $w$, there are exactly $\prod_{v \in V}(d^+(v) - 1)!$ collections $P$ of permutations such that $F(P) = B$: for every vertex $v \neq w$, the in-branching $B$ fixes the last out-arc of $v$ in $P$, whereas for $w$, the out-arc $ww'$ is the first out-arc of $w$ by definition. Any choice of a permutation $\pi_v$ on the remaining arcs does not affect $F(P)$. Finally, it is well known that the number of rooted in-branchings can be counted in polynomial time using Tutte's matrix-tree theorem, see e.g. [2]. □

Interestingly, both the Tutte–Smith paper and the van Aardenne-Ehrenfest and de Bruijn paper have as their main interests something other than counting Euler cycles. Tutte and Smith considered the problem of tracing a 4-regular undirected planar drawing without lifting the pen, in such a way that the line traced never crosses itself, and showed that this can be reduced to a question of counting Euler tours in a digraph (using arguments similar to those used in the polynomial-time algorithm for counting planar matchings). On the other hand, van Aardenne-Ehrenfest and de Bruijn arrived at the question via their study of string problems, specifically De Bruijn sequences, cyclic sequences over an alphabet $\Sigma$ that contain every $n$-tuple over $\Sigma$ exactly once.

Other questions on Euler tours include the following. We say that two Euler tours are **compatible** if they use only distinct transitions at every vertex, i.e., for every vertex $v$ with an in-arc $uv$ and out-arc $vw$, at most one of the tours contains the transition from $uv$ to $vw$. Fleischner and Jackson [14] showed that every Euler digraph $D$ of minimum degree $2k$ contains $\lfloor \frac{k}{2} \rfloor$ pairwise compatible Euler tours, and conjectured that the bound can be improved to $k - 2$.

## 4.3 Euler Digraphs of Bounded Width

The notion of **width measures** has been a highly successful approach for studying graphs of restricted structure, especially for undirected graphs. The rough idea is that a graph of simple structure can be **decomposed** recursively into pieces that interact with each other only in a limited way, where each piece is either very simple (e.g., constant size) or can itself be further decomposed. For undirected graphs, arguably the default notion of bounded structure is having bounded **treewidth**, motivated by a plethora of algorithmic results, e.g., methods such as Courcelle's theorem [9] or other dynamic programming results for efficiently solving $\mathcal{NP}$-hard problems on a graph, given a decomposition of the graph that shows it has bounded treewidth. However, many other width notions also exist, both less expressive ones such as **pathwidth** or **treedepth**, and more expressive ones such as **rankwidth**.

For directed graphs, the story of bounded width measures has arguably been more limited in terms of algorithmic applications. Although we do have a growing understanding of the structure of graphs of small or large width under various natural directed width notions, compared to the undirected case, the algorithmic implications of bounded directed widths are generally weaker (see Chapter 9 for results on bounded width measures on digraphs). In the undirected case, it is a common occurrence that a problem is FPT parameterized by treewidth, in fact often with a running time such as $O(f(k)\cdot n)$ that is linear in the order of the graph. For directed width notions, it is far more common that a parameterized problem is $W[1]$-hard – meaning that, while it may be polynomial-time tractable on graphs of (say) bounded directed treewidth, the running time is of the less appealing form $O(n^{f(k)})$ and FPT algorithms are not expected to exist (see Section 1.11). In fact, one could argue that the width measure that has had the widest success for digraphs in terms of FPT algorithms is simply the undirected treewidth, i.e., the treewidth of the underlying undirected graph.

In view of this, it is natural to ask about the structure of Euler digraphs in this perspective. In particular, to what extent do the directed and undirected notions of treewidth and pathwidth differ for Euler digraphs?

We prove a simple result in this direction. First, we may observe that if a width notion is closed under taking induced subgraphs, then there is no sense in studying it in full generality for Euler digraphs, since (by the Euler vertex-extension) every digraph on $n$ vertices is an induced subgraph of a (not necessarily simple) Euler digraph on $n+2$ vertices. Since most width measures are closed under taking induced subgraphs and closed or approximately closed under subdividing parallel arcs, in general Euler digraphs of bounded width will not have any extra structure that is not present in other digraphs. On the other hand, we may observe that the above reduction creates a vertex of unbounded degree, and ask whether the situation changes under a combined bound of bounded width and bounded maximum degree. Indeed, it is known that an Euler digraph of maximum degree $d$ and directed treewidth $k$ has undirected treewidth at most $O(dk)$ [26]. Hence for bounded-degree Euler digraphs, the difference between directed and undirected width disappears. (This is certainly not true for general digraphs, as, e.g., an acyclic grid has total degree 4, unbounded undirected treewidth, and directed pathwidth 0, as we will see below.)

For more information on digraphs of bounded width, see Chapter 9 of this book. In the following, we omit the technical details of directed treewidth, and prove a simpler result that relates the undirected and directed notions of bounded pathwidth.

### 4.3.1 Cutwidth and Bounded Pathwidth

We begin by studying the directed pathwidth of Euler digraphs. Let us first recall the notions.

We need two variants of pathwidth, undirected and directed. We begin with the undirected version. Let $G = (V, E)$ be an undirected graph. A **path decomposition** of $G$ is a sequence of vertex sets $X_1, \ldots, X_s$ called **bags** such that $\bigcup_{i \in [s]} X_i = V$, where the following hold.

1. For every edge $uv \in E$, there is a bag $X_i$, $i \in [s]$, such that $u, v \in X_i$, and
2. for every triple of indices $i < j < k$, $i, j, k \in [s]$, we have $X_i \cap X_k \subseteq X_j$. Equivalently, $\{i \in [s] \mid v \in X_i\}$ forms an interval for every $v \in V$.

The **width** of the decomposition is $\max_{i \in [s]} |X_i| - 1$. The **pathwidth of** $G$ is the minimum width of a path decomposition of $G$. We let the **undirected pathwidth** of a digraph $D = (V, A)$ refer to the pathwidth of its underlying undirected graph $UG(D)$.

Analogously, let $D = (V, A)$ be a digraph. A **directed path decomposition** is a sequence of vertex sets $X_1, \ldots, X_s$, again called bags, such that $\bigcup_{i \in [s]} X_i = V$, where the following hold.

1. For every arc $uv \in E$, there are indices $i \leq j$, $i, j \in [s]$ such that $u \in X_i$ and $v \in X_j$, and
2. as in the undirected case, for every triple of indices $i < j < k$, $i, j, k \in [s]$ we have $X_i \cap X_k \subseteq X_j$.

The **width** of the decomposition is $\max_{i \in [s]} |X_i| - 1$ and the **directed pathwidth of** $D$ is the minimum width of a directed path decomposition of $D$.

We will also need a width measure that is less commonly used in general, but which will be highly relevant for Euler digraphs. Let $D = (V, A)$ be a digraph, and let $\sigma = v_1 \ldots v_n$ be an ordering of $V$. The **undirected cutwidth of** $\sigma$ equals the maximum over $i$ of the number of arcs between $\{v_1, \ldots, v_i\}$ and $\{v_{i+1}, \ldots, v_n\}$. The **undirected cutwidth of** $D$ is the minimum undirected cutwidth over all orderings of $V$. Note that this is by its nature an undirected width measure, i.e., we do not distinguish arcs by their direction.

We make a few simple observations.

1. For any digraph $D = (V, A)$, the directed pathwidth is at most the undirected pathwidth, which in turn is at most the undirected cutwidth. Indeed, the former is trivial, and given any ordering $\sigma$ of undirected cutwidth $k$ it is easy to produce a path decomposition of width $k$.
2. Both inequalities are strict. Indeed, a star has undirected pathwidth 1 but unbounded undirected cutwidth, and an acyclic grid has directed pathwidth 0 (and constant degree) but unbounded undirected pathwidth.
3. A digraph has directed pathwidth 0 if and only if it is an acyclic digraph. In particular, the only Euler digraph of directed pathwidth 0 is a single vertex.

We now observe formally (as already sketched) that there are Euler digraphs of constant directed pathwidth but unbounded undirected pathwidth.

**Lemma 4.3.1** *For every $k \geq 1$, there is an Euler digraph with undirected treewidth at least $k$ but directed pathwidth 1.*

**Proof:** Let $D$ be a $k \times k$ acyclic grid, with all arcs oriented downwards and to the right. Add one additional vertex as in-neighbour of the top-left grid vertex, and another as out-neighbour of the bottom-right grid vertex, and observe that $|b_D(v)| \leq 1$ for every vertex in the resulting graph $D$. Hence we can complete $D$ into an Euler digraph by adding a single vertex $x$ and for every unbalanced vertex $v$ either an arc $vx$ or $xv$, as required. Then $UG(D)$ contains a $k \times k$ grid and hence has treewidth at least $k$, whereas $D$ has a path decomposition of width 1 formed by simply adding $x$ to every bag in the decomposition of the acyclic digraph $D - x$. □

Finally, we have the following positive result.

**Lemma 4.3.2** *If $D$ is an Euler digraph with directed pathwidth $k$, then $D$ has undirected cutwidth at most $k \cdot \Delta(D)$.*

**Proof:** Let $X_1, \ldots, X_s$ be a directed path decomposition of $D$ of width $k$, and construct a linear ordering $\sigma$ of $V(D)$ by first arbitrarily arranging the vertices of $X_1$, then the vertices of $X_2 \setminus X_1$, and so on until $X_s$. Let $d = \Delta^+(D)$. We claim that the ordering $\sigma$ has undirected cutwidth at most $2dk$.

Let $(L_i, R_i)$ be the vertex cut corresponding to some position $i$ of $\sigma$, i.e., $L_i$ is the set of vertices ordered at or before position $i$ in $\sigma$, and $R_i = V(D) \setminus L_i$. Assume that $(L_i, R_i)$ cuts through the bag $X_j$, $j \in [s]$, and let $A = X_j \cap L_i$ and $B = X_j \cap R_i$; assume $A \neq \emptyset$. We claim that every arc of $(R_i, L_i)_D$ has either its tail in $B$ or its head in $A$. Indeed, by definition every such arc has its tail in $R_i$, and any such arc with tail in $R_i \setminus B$ has its head in a vertex still present in a bag $X_{j'}$ for some $j' > j$; hence the head is contained in $X_j$. Since there are $|A|d$ arcs with head in $A$ and $|B|d$ arcs with tail in $B$, and $|A| + |B| \leq k$, there are at most $dk$ such arcs. On the other hand, since $D$ is Euler we have $d^+(L_i) = d^-(L_i) \leq dk$; hence the undirected cutwidth of $\sigma$ is at most $2dk = k\Delta(D)$. □

Since undirected pathwidth is sandwiched between undirected cutwidth and directed pathwidth, we also get that the undirected pathwidth of Euler digraphs of bounded degree and bounded directed pathwidth is bounded, as promised.

As noted, the above result also holds for the more general notion of **directed treewidth**: If $D$ is an Euler digraph with maximum degree $d$ and the directed treewidth of $D$ is at most $k$, then $D$ has undirected treewidth at most $O(dk)$ [26]. Directed treewidth is the most general of the various width-measures that serve as directed analogues of (undirected) treewidth, and has some interesting structural properties. See, in particular, the **directed grid theorem** [27] proved by Kawarabayashi and Kreutzer, which shows that every digraph of large enough directed treewidth contains a particular canonical obstacle known as a **cylindrical grid** of large order as a butterfly minor (see Theorem 9.3.14). Hence this shows that any Euler digraph that does not contain a large cylindrical grid as a butterfly minor, and which has bounded degree, also has bounded undirected treewidth.

## 4.4 Cycle-Packing and Cycle-Hitting

In this section, we consider problems of cycle-hitting and cycle-packing in Euler digraphs. More properly, we consider the following two problems. A **feedback arc set** of a digraph $D$ is a set $F$ of arcs of $D$ such that $D - F$ is acyclic. The problem FEEDBACK ARC SET takes as input a digraph $D$ and an integer $k$, and asks whether $D$ has a feedback arc set of cardinality $k$. Dually to this, given a digraph $D$ and an integer $k$, ARC-DISJOINT CYCLES asks whether $D$ contains a packing of $k$ pairwise arc-disjoint cycles. For both problems, the vertex versions (FEEDBACK VERTEX SET, respectively VERTEX-DISJOINT CYCLES) are defined in the natural way.

Before we proceed, we observe that these vertex-versions are not easier in Euler digraphs than in general graphs.

**Lemma 4.4.1** *For both* FEEDBACK VERTEX SET *and* VERTEX-DISJOINT CYCLES, *there are polynomial-time reductions from the versions on general digraphs to the versions on Euler digraphs that increase the value of $k$ by only 1.*

**Proof:** The reduction is the same for both problems. Let $D$ be a given digraph $D$, and let $D'$ be its Euler two-vertex extension with added vertices $x$, $y$. Add an additional pair of arcs $xy$, $yx$ (and if needed, subdivide parallel arcs to acquire a simple digraph). For FEEDBACK VERTEX SET, it is now easy to observe that for every $X \subseteq V(D)$, $X$ is a feedback vertex set of $D$ if and only if $X+x$ is a feedback vertex set of $D'$, and that there is a minimum feedback vertex set $X'$ of $D'$ for which $X' \cap (V(D') \setminus V(D)) = \{x\}$. For VERTEX-DISJOINT CYCLES, note that every cycle that intersects $V(D') \setminus V(D)$ intersects both $x$ and $y$. Let $C$ be the cycle on $x$ and $y$, contained in $V(D') \setminus V(D)$. Then there exists an optimal cycle-packing that contains $C$, and having included $C$, the cycle-packing problem on $D' - V(C)$ is equivalent to that on $D$. □

Regarding the hardness of these problems, we recall that FEEDBACK VERTEX SET is $\mathcal{NP}$-complete but FPT on digraphs by the algorithm of Chen, Liu, Lu, O'Sullivan, and Razgon [5], whereas VERTEX-DISJOINT-CYCLES on general digraphs is in XP, i.e., has an algorithm with running time $O(n^{f(k)})$ for some function $f(k)$ due to Reed, Robertson, Seymour and Thomas [35], but is W[1]-hard due to the results of Slivkins [39]. The problem also has an **FPT approximation** due to Grohe and Grüber [18], building on the results of Reed *et al.* – i.e., a parameterized algorithm running in FPT time, which either reports that $D$ does not contain $k$ vertex-disjoint cycles, or returns $g(k)$ vertex-disjoint cycles, for some growing, unbounded function $g(k)$. By the above reduction, all these statements hold for general digraphs as well as for the restriction to Euler digraphs (for the vertex versions).

In the rest of this section, we will consider the arc-versions of these problems, which differ significantly in behavior on Euler digraphs.

### 4.4.1 Feedback Arc Set

First, let us consider Feedback arc set. We begin by showing $\mathcal{NP}$-hardness; the reduction is easy, but its correctness proof is revealing. We need the following observation. Relative to an ordering $(v_1, \ldots, v_n)$ of the vertices of an Euler digraph $D = (V, A)$, the **backward arcs** are arcs $v_iv_j \in A$ with $i > j$. Note that for any such ordering, the backward arcs form a feedback arc set, and conversely, if $F$ is a minimal feedback arc set of $D$, then $F$ is exactly the set of backward arcs for some acyclic ordering of $D - F$.

**Lemma 4.4.2** *The number of backward arcs of an ordering of vertices of an Euler digraph is invariant under cyclic shifts, i.e., for any Euler digraph $D = (V, A)$ and any ordering $(v_1, \ldots, v_n)$ of $V$, the orderings $(v_1, \ldots, v_n)$ and $(v_2, \ldots, v_n, v_1)$ have the same number of backward arcs.*

**Proof:** When $v_1$ is in the first position, every in-arc of $v_1$ is a backward arc but no out-arc is; when $v_1$ is in the last position, the opposite statement holds. Every other arc is unaffected by the change. Since $d^+(v_1) = d^-(v)$, the two orderings have the same number of backward arcs. □

An important corollary is that for every vertex $v$ of an Euler digraph $D$, there exists a minimum feedback arc set of $D$ that contains all out-arcs of $v$. Indeed, if $F$ is a feedback arc set, let $v_1, \ldots, v_n$ be an acyclic ordering of the vertices of $D - F$, and rotate the ordering until it ends with $v$. Then the backward arcs of the new ordering form a feedback arc set $F'$ of $D$, with $|F| = |F'|$ and where every out-arc of $v$ is contained in $F'$. The $\mathcal{NP}$-hardness reduction is now trivial. Recall that $\text{fas}(D)$ denotes the size of a minimum feedback arc set of $D$.

**Lemma 4.4.3** Feedback arc set *is $\mathcal{NP}$-hard on Euler digraphs.*

**Proof:** We show a reduction from Feedback arc set on general digraphs. Let $D = (V, A)$ be a digraph, and let $D'$ be its Euler two-vertex extension, with added vertices $x$, $y$. We claim that $\text{fas}(D') = \text{fas}(D) + d^+(y)$. Indeed, let $Y \subseteq A(D')$ be the set of out-arcs of $y$. By the above observation, there is a minimum feedback arc set that includes $Y$, and every directed cycle in $D' - Y$ is also contained in $D$. □

Regarding properties of approximation and parameterized complexity, the simple reduction used above gives us no lower bounds, since the output parameter value is unbounded in $k$. We are also not aware of any lower bounds-preserving reduction from FEEDBACK VERTEX SET on undirected graphs to FEEDBACK ARC SET on Euler digraphs. Still, it can be interesting to compare with what is known for general digraphs and for FEEDBACK VERTEX SET on undirected graphs (recall that FEEDBACK EDGE SET can be solved in polynomial time, and therefore does not serve as a point of comparison).

For (unweighted) FEEDBACK ARC SET on general digraphs, the best approximation result is an $O(\log \tau^* \log\log \tau^*)$-approximation, where $\tau^* \leq \text{fas}(D)$ is the cost of a natural LP-relaxation of the problem [13]. The problem does not admit a constant-factor approximation under the Unique Games Conjecture (see Guruswami and Lee [20]). The problem has an FPT algorithm with a running time of $O^*(4^k k!)$ [5], and it is a famous open problem whether it has a single-exponential FPT algorithm, i.e., an FPT algorithm with a running time of $O^*(2^{O(k)})$, and whether it admits a polynomial kernel.

In contrast, FEEDBACK VERTEX SET on undirected graphs admits single-exponential FPT algorithms [11, 29], a polynomial kernel with $4k^2$ vertices [40] (recently improved to $2k^2$ vertices and linear time in [24]), and a 2-approximation [7].

Therefore, natural open questions are what the status of each of these three questions is for Euler digraphs.

**Problem 4.4.4** *Does* FEEDBACK ARC SET *on Euler digraphs allow (1) a single-exponential time FPT algorithm, (2) a polynomial kernel, and/or (3) a constant-factor approximation?*

Finally, although it does not serve to close any of the above-mentioned open questions, we note a few properties of FEEDBACK ARC SET on Euler digraphs that do not hold for the general FEEDBACK ARC SET problem. First, we note that instances with a small feedback arc set are structurally simple.

**Lemma 4.4.5** ([21]) *Every Euler digraph $D$ has undirected cutwidth at most $2fas(D)$.*

**Proof:** Let $F$ be a minimum feedback arc set of $D$, and let $\sigma = (v_1 \dots v_n)$ be an acyclic ordering of $D - F$. Let $i \in [n]$ and let $V_i = \{v_1, \dots, v_i\}$. Then for any $i$, $d^-(V_i) \leq \text{fas}(D)$ since every such arc is contained in the feedback arc set, and $d^+(V_i) = d^-(V_i)$ since $D$ is Euler. □

Recall that the undirected pathwidth is bounded by the undirected cutwidth; therefore the undirected pathwidth is also bounded. Clearly, no such statement is possible for the general FAS problem, since acyclic digraphs can have arbitrary underlying undirected graphs.

Finally, we make a remark about the iterative compression approach to FPT algorithms. This is an important method in parameterized complexity, which among many other cases is used in the algorithm for DIRECTED FEEDBACK VERTEX SET [5] and in the currently fastest algorithm for (UNDIRECTED) FEEDBACK VERTEX SET [11, 29]. In this approach, a parameterized problem is solved by iteratively solving the problem on a sequence of subgraphs of the original graph, in each step using the solution from the previous step to produce a solution for the next step. Concretely, let COMPRESSION FEEDBACK ARC SET be the FEEDBACK ARC SET problem where the input additionally includes a feedback arc set of size $k+1$. Assume that we have an FPT algorithm for COMPRESSION FEEDBACK ARC SET that either produces an output feedback arc set of size at most $k$, or concludes that no such solution exists. Then we can solve FEEDBACK ARC SET for general digraphs using $|A|$ calls to this algorithm, as follows: Let $D = (V, A)$ be a digraph, enumerate the arcs as $A = \{a_1, \ldots, a_m\}$, and define $D_i = D(\{a_1, \ldots, a_i\})$ for $i \in [m]$. Compute a trivial solution for $D_k$ (e.g., the entire arc set). Then for every $k < i \leq m$, if $F_i$ is a solution for $D_i$ with $|F_i| \leq k$ then $F_i \cup \{a_{i+1}\}$ is a solution for $D_{i+1}$ of size at most $k+1$, which can be fed into the compression algorithm. It remains only to observe that if any instance $D_i$ is concluded to be negative, i.e., not to have a feedback vertex set of size $k+1$, then the same is true for $D$.

The obstacle to the immediate application of this strategy in Euler digraphs is that the digraphs $D_1$, $D_2$, ... are in general no longer Euler. However, we observe that the strategy can be adopted by the use of the splitting off operation.

**Lemma 4.4.6** FEEDBACK ARC SET *on an Euler digraph* $D = (V, A)$ *with parameter* $k$ *can be solved using polynomial-time processing and at most* $|A|$ *calls to a solver for* COMPRESSION FEEDBACK ARC SET *on balanced digraphs, where the calls to the solver all use graphs with at most* $|V|$ *vertices, at most* $|A|$ *arcs, and parameter at most* $k+1$.

**Proof:** We use the iterative compression approach, constructing a sequence of graphs $D_i = (V, A_i)$ as follows. Begin with $D = D_m$, then for every $i \in [m-1]$ we attempt to identify a simple splitting pair $uv$, $vw$ in $A_{i+1}$. If there is one, then we construct $D_i$ from $D_{i+1}$ by splitting off $uv$ and $vw$ in $D_{i+1}$, i.e., $A_i = (A_{i+1} \setminus \{uv, vw\}) \cup \{uw\}$. If we cannot find such a pair, then by Proposition 4.1.3, $D_{i+1}$ takes the form of a disjoint union of complete digraphs. At this point, the instance $D_{i+1}$ can be solved in polynomial time, since every ordering of the vertex sets yields the same number of backward arcs. Therefore the sequence $D_m$, $D_{m-1}$, ... can be constructed, yielding a sequence $D_1$, ..., $D_t$ of gradually larger Euler digraphs, where $t \leq m$, and where we can find a minimum solution for $D_1$ in polynomial time.

It remains to show that the sequence is useful for iterative compression, i.e., that $\text{fas}(D_i) \leq \text{fas}(D_{i+1}) \leq \text{fas}(D_i)+1$ for every $1 \leq i < t$. Let $i \in [t-1]$, and let $F_i$ be a feedback arc set for $D_i$ with at most $k$ arcs. Assume that $D_i$ was created by splitting off the arcs $uv$, $vw$ in $D_{i+1}$. Define $F_i'$ by replacing $uw$ in $F_i$ by the pair $uv$, $vw$ if $uw \in F_i$, otherwise let $F_i' = F_i$. Then clearly $|F_i'| \leq k+1$, and it is easy to verify that $F_i'$ is a feedback arc set for $D_{i+1}$. In the other direction, let $F_{i+1}$ be a feedback arc set for $F_{i+1}$, and construct $F_i$ as $(F_i \setminus \{uv, vw\}) \cup \{uw\}$ if $F_{i+1} \cap \{uv, vw\} \neq \emptyset$, and otherwise $F_i = F_{i+1}$. Then again it is easy to verify that $F_i$ is a feedback arc set for $D_i$; hence if $D_i$ does not have a feedback arc set of size $k$ then neither does $D_j$ for any $j > i$ and we may reject the instance $D$. We conclude that we can solve FEEDBACK ARC SET for Euler (or balanced) digraphs using $t \leq m$ calls to a solver for COMPRESSION FEEDBACK ARC SET for balanced digraphs, without increasing the number of arcs or vertices or the value of $k$, as required. □

Finally, we note that the currently fastest algorithm for FEEDBACK VERTEX SET on undirected graphs is actually based on an algorithm with a running time of $O^*(3^w)$ where $w$ is the treewidth of the graph, which yields an FPT algorithm with a running time of $O^*(3^k)$ using iterative compression [11]. Also note that the naive state space of the treewidth dynamic programming algorithm for feedback vertex set would seem to need to enumerate either induced forests on the vertices of the bag, or at the very least all partitions of the vertices of the bag, both of which number $2^{\Theta(w \log w)}$ for $w$ vertices. Therefore, it seems worthwhile to ask the following question separately from the above.

**Problem 4.4.7** *Can* FEEDBACK ARC SET *on Euler digraphs be solved in $O^*(2^{O(w)})$ time, where $w$ is the width of a provided undirected tree decomposition?*

### 4.4.2 Arc-Disjoint Cycles

Next, we consider the ARC-DISJOINT CYCLES problem in Euler digraphs. As we saw, this problem is hard on general digraphs; Gutin, Jones, Sheng and Wahlström [21] showed the problem to be FPT on Euler digraphs. The strategy is based on a win-win approach, where they show that for every Euler digraph $D$, either $D$ contains at least $k$ pairwise arc-disjoint cycles, or $D$ has undirected pathwidth at most $f(k)$ for some function $f(k)$, in which case we can solve the problem by dynamic programming. Gutin *et al.* also solve a related problem called the DIRECTED $k$-CHINESE POSTMAN PROBLEM, but we will focus on ARC-DISJOINT CYCLES.

The basis for the strategy is the following result of Reed, Robertson, Seymour and Thomas [35], settling a conjecture by Younger [44].

**Theorem 4.4.8** ([35]) *There is a function $f : \mathbb{N} \to \mathbb{N}$ such that every digraph $D$ contains either $k$ pairwise arc-disjoint cycles or has a feedback arc set of size at most $f(k)$.*

The vertex-version of the result also holds, i.e., every digraph $D$ either contains at least $k$ pairwise vertex-disjoint cycles or has feedback vertex set number of at most $f(k)$, but the above version will be more useful to us. We also remark that the function $f(k)$ grows very rapidly; according to the authors, $f(k)$ is an iterated exponential whose height is bounded by another iterated exponential. Therefore, the resulting FPT algorithm we describe, which uses this theorem, will be a purely theoretical result, showing that the problem is FPT but without giving a running time bound that would be practically useful.

At a high level, the algorithm goes as follows. Let $D$ be an Euler digraph and $k$ an integer. If $D$ contains $k$ cycles then the instance is positive; if not, then by Theorem 4.4.8 $D$ has a feedback arc set of size at most $f(k)$, and by Lemma 4.4.5 it has undirected cutwidth, and thereby undirected pathwidth, at most $2f(k)$. Therefore, a decision algorithm could compute $f(k)$, or an upper bound on it; use the FEEDBACK ARC SET algorithm with parameter $f(k)$ to check whether $D$ has a "small" feedback arc set; and use the feedback arc set, if it exists, to compute a bounded width path decomposition of $D$. The path decomposition can then be used as the basis for a standard dynamic programming algorithm. If the FAS algorithm instead signals that $\text{fas}(D) > f(k)$, then by Theorem 4.4.8 the instance must be positive.

To turn this into a constructive algorithm, i.e., an algorithm that actually produces the cycles as an output, involves some surprising subtleties. The usual approach to this problem would be via self-reducibility: Given an algorithm that can detect the existence of an object in $D$, we can apply it repeatedly to subgraphs of $D$ until we find a subgraph $D'$ of $D$ such that $D'$ contains the object but no strict subgraph of $D'$ does, at which point finding the object is hopefully trivial.

This strategy has two obstacles in the current situation. First, the decision algorithm would only apply to Euler digraphs, and an arbitrary subgraph of $D$ would in general not be Euler. Second, even the subgraph-minimal case, when $D$ contains $k$ cycles but no strict subgraph of $D$ does, is not necessarily trivial.

Gutin *et al.* [21] solved the problem by using the FPT approximation algorithm of Grohe and Grüber [18] mentioned earlier. This is an FPT algorithm parameterized by $k$ which on an input digraph $D$ (not necessarily Euler) either concludes that $D$ does not contain $k$ disjoint cycles, or returns at least $g(k)$ disjoint cycles, for some growing, unbounded function $g(k)$. Combining this result with Theorem 4.4.8 and with the FEEDBACK ARC SET algorithm as above yields a constructive FPT algorithm (see the paper for details). Here, instead, we show a different approach, based on the splitting-off strategy as in Lemma 4.4.6. We begin by noting the dynamic programming result from

Gutin *et al.* [21]. (We note that this algorithm was developed for a more general problem than cycle-packing, and therefore it is possible that the running time can be improved; but we will not investigate the question of fastest running time.)

**Lemma 4.4.9** ([21]) *There is an algorithm that, given a digraph $D$, a vertex ordering of $D$ of undirected cutwidth $p$ and an integer $k$, finds $k$ arc-disjoint cycles in $D$ if they exist, and runs in time $O^*(2^{(p+2)k})$.*

**Theorem 4.4.10** *Let $f^*(k)$ be the smallest value such that every Euler digraph contains either at least $k$ arc-disjoint cycles or has a feedback arc set number at most $f^*(k)$. There is an FPT algorithm, parameterized by $k$, that on input $(D, k)$ in time $O^*(2^{2kf^*(k)})$ either returns $k$ arc-disjoint cycles in $D$ or concludes that no such solution exists. The algorithm does not need to know the value of $f^*(k)$.*

**Proof:** We proceed as in Lemma 4.4.6, and starting from $D_m = D$ we compute a sequence of gradually smaller digraphs, computing $D_i$ from $D_{i+1}$ using a simple splitting pair $uv, vw \in A(D_{i+1})$. Let $D'$ be the transitive digraph resulting at the end of this process. It is trivial to find a maximum cycle packing in $D'$, by Proposition 4.1.3 and since the arc set of a complete digraph on $t$ vertices decomposes into $\binom{t}{2}$ arc-disjoint directed cycles of length 2. Note that the number of cycles produced in this is also identical to the feedback arc set number of $D'$. Hence, we either find at least $k$ cycles in $D'$ or we can construct a feedback arc set of $D'$ of size less than $k$.

Now we "unroll" the splitting-off sequence above as follows. Let the current graph be $D_i$, created from $D_{i+1}$ by splitting off the arcs $uv, vw$. As a loop invariant, for every graph $D_i$ we have either located $k$ arc-disjoint cycles, or $D_i$ contains fewer than $k$ arc-disjoint cycles and we have computed a minimum feedback arc set, necessarily of size at most $f^*(k)$. If we have found $k$ arc-disjoint cycles in $D_i$, then it is clear that this cycle packing can be transformed to a cycle packing in the original graph $D$, by repeatedly undoing the splitting-off operation. Hence, we assume that the instance $D_i$ is negative, and let $F_i$ be a minimum feedback arc set of $D_i$, where $|F_i| \le f^*(k)$. As in Lemma 4.4.6, we produce a feedback arc set $F_i'$ of $D_{i+1}$ of size at most $f^*(k) + 1$, by replacing $uw$ by $uv, vw$ in $F_i$ if $uw \in F_i$; we then use the algorithm of Chen *et al.* [5] to compress $F_i'$ to a minimum feedback arc set $F_{i+1}$ of $D_{i+1}$. This gives us a vertex ordering of $D_{i+1}$ of undirected cutwidth $p \le 2|F_{i+1}| \le 2f^*(k) + 2$, and we can determine whether $D_{i+1}$ contains $k$ arc-disjoint cycles using Lemma 4.4.9, in time $O^*(2^{2kf^*(k)})$. If $D_{i+1}$ contains $k$ arc-disjoint cycles, then we can unroll the splitting-off sequence to produce $k$ arc-disjoint cycles in $D$; otherwise, $|F_{i+1}| \le f^*(k)$ and the process can be repeated. □

Finally, we note that the upper bound $f(k)$ known on the relation between the cycle-packing number and the feedback arc set number is potentially very far from being tight. In fact, even for general digraphs the best reported lower bound is $\Omega(k \log k)$ due to Alon, as reported by Reed *et al.* [35]. However, as we have seen, the structure of cycles in Euler digraphs is much simpler than in general digraphs. It seems worthwhile to pursue a tighter bound for this special class.

**Problem 4.4.11** *Can the bound on $f(k)$ be improved for Euler digraphs, possibly to polynomial or even $O(k \log k)$?*

Finally, let us consider the existence of polynomial kernels for ARC-DISJOINT CYCLES on Euler digraphs. It is tempting to once again look at the undirected version of the problem, i.e., EDGE-DISJOINT CYCLES, to heuristically indicate whether a kernel is likely. In this case, we find that EDGE-DISJOINT CYCLES does have a polynomial kernel [3], and furthermore, by the classical Erdős–Pósa result, the corresponding function $f(k)$ in undirected graphs has $f(k) = O(k \log k)$ (both in the edge- and vertex-versions) [12]. However, fundamentally these results rely upon statements about short girth in undirected graphs (e.g., a graph of minimum degree 3 has girth $O(\log n)$), which does not transfer to the Euler digraph case.

**Problem 4.4.12** *Does* ARC-DISJOINT CYCLES *have a polynomial kernel on Euler digraphs?*

### 4.4.3 Additional Topics

Finally, we review a few additional topics regarding cycle-packings in Euler digraphs.

Questions of arc-disjoint cycle-packing have been considered in extremal graph theory. Alon, McDiarmid and Molloy [1] showed that every $k$-regular digraph contains a packing of $\Omega(k^2)$ arc-disjoint cycles, and conjectured that this can be sharpened to $\binom{k+1}{2}$ arc-disjoint cycles. They also give a construction showing that this result would be tight. Let $C_n^k$, $n \geq 2k+1$, be the digraph with vertex set $\{0, \ldots, n-1\}$ and all arcs $uv$ where $v = u+i \pmod{n}$, $i \in [k]$. Then $\mathrm{fas}(C_n^k) = \binom{k+1}{2}$, witnessed by the vertex ordering $0, \ldots, n-1$, and $C_n^k$ contains $\binom{k+1}{2}$ arc-disjoint cycles, since there are $k$ arc-disjoint cycles through the vertex $n-1$, whose removal leave a graph isomorphic to $C_{n-1}^{k-1}$.

Brualdi and Shen [4] gave the following further conjectures.

**Conjecture 4.4.13** *Let $k \geq 2$ be an integer. Every bipartite Euler digraph with partition sizes $m$ and $n$ and at least $mn/(k+1)$ arcs contains a cycle of length at most $2k$.*

**Conjecture 4.4.14** *Every bipartite Euler digraph $D = (V, A)$ with partition sizes $m$ and $n$ contains a collection of $|A|^2/(4mn)$ arc-disjoint cycles.*

In particular, the latter conjecture specializes into the conjecture that every Euler bipartite tournament decomposes into arc-disjoint 4-cycles.

Another question is the following.

**Problem 4.4.15** *For which Euler digraphs is the cycle-packing number equal to the feedback arc set number?*

Let us say that a digraph $D$ **packs** if $D$ contains fas($D$) arc-disjoint cycles. To what extent can we characterize Euler digraphs which pack? One important result here is by Seymour [38], who showed that it holds for Euler digraphs which can be linklessly embedded in 3-space. This can be viewed as an "Euler generalization" of the result that all planar digraphs pack, which follows from the Lucchesi–Younger theorem [31, 38].

This result also carries over to the weighted version, as follows. Let $D = (V, A)$ be a digraph (not necessarily Euler) that can be linklessly embedded in 3-space, and let $w : A \to \mathbb{Z}_+$ be a balanced set of arc-weights, i.e., for every vertex $v \in V$, $\sum_{uv \in A} w(uv) = \sum_{vw \in A} w(vw)$. Then the arc-disjoint cycle-packing number of $D$, with arc capacities $w$, and the weighted feedback arc set number, with arc costs $w$, are the same.

But for the more general question, presumably asking which individual digraphs $D$ pack is too ambitious to expect a good answer. For general digraphs, a natural option is to restrict attention to a hereditary class of digraphs. For the non-Euler case, Guenin and Thomas [19] characterized the class of digraphs $D$ such that for every subdigraph $H$ of $D$, the feedback vertex set number and the vertex-disjoint cycle-packing number of $H$ agree. The characterization is in terms of a list of forbidden butterfly minors for the class. Naturally, if the line graph of $D$ belongs to this class, then $D$ and every subdigraph of $D$ pack in the above sense. However, this does not take into account the restriction that we are only concerned with Euler digraphs. For example, consider a digraph $D$ on six vertices $a_i, b_i$, $i \in [3]$, with arcs $a_ib_i$ and $b_ia_j$, $i \neq j$, $i, j \in [3]$. It is easily verified that this graph does not pack; it has arc-disjoint cycle-packing number 1 but feedback arc set number 2. On the other hand, consider the graph $D'$ which instead contains two copies of the arcs $a_ib_i$, $i \in [3]$. Then $D'$ is an Euler digraph, which decomposes into three arc-disjoint 4-cycles, and with feedback arc set number three (for example, the two arcs $a_1b_1$ and the arc $b_2a_3$). Thus $D'$ packs, and it can be verified that every Euler subdigraph of $D'$ also packs. Thus in particular, a question one could ask is, what is the class of Euler digraphs $D$ such that every Euler subdigraph of $D$ packs? Regarding the nature of the characterization, consider the following. Let $H$ be an Euler digraph that does not pack. If $D$ is a digraph which has $H$ as a topological minor, then it follows that $D$ has an Euler subdigraph that does not pack. The same does not appear to be true for butterfly minors (although we have no counterexample, the "minor model" of a butterfly minor is usually a non-Euler subdigraph, even when the minor itself is Euler).

Yet further options include considering richer operations than subgraphs, e.g., considering graph classes closed under splitting-off operations (and removal of loops, i.e., taking Euler subgraphs). This may be easier to answer, on account of having a more powerful containment operation.

We refrain from explicitly singling out one of these questions as a main open problem, as it is not clear to us whether any one of them will be more productive or feasible than the others.

## 4.5 Linkages and Cut Problems

We now move on to variants of linkages and multiflow-type problems. The results in this section are of mainly two variants. First, in Sections 4.5.1 and 4.5.2 we consider path-packing and unsplittable multi-commodity flow type problems in the style of WEAK $k$-LINKAGE, where the exact endpoints of the paths we are asked to pack are specified. We will in particular show that TWO-COMMODITY FLOW admits a polynomial-time algorithm, and recall the long-open question of whether WEAK $k$-LINKAGE is FPT parameterized by $k$.

Then, in Sections 4.5.3 and 4.5.4 we consider an alternative setup, where we are asked to find a maximum path-packing on a set of terminals according to some condition, but it is not specified exactly how many paths of each type we are required to pack. A main result here is the classical result that the so-called FREE MULTIFLOW problem is in $P$ on Euler digraphs. We observe (as Frank did in the 1980s [16]) that, remarkably, this implies that ARC MULTIWAY CUT is in $\mathcal{P}$ on Euler digraphs, a result that does not carry over even to undirected Euler graphs. We also review some weighted generalizations of this result.

### 4.5.1 Two-Commodity Flow

In this section, we consider problems related to TWO-COMMODITY FLOW and MULTI-COMMODITY FLOW.

We will use the following formulation. Let $D = (V, A)$ and $H = (V, F)$ be digraphs (not necessarily Euler), where $D$ is referred to as the **supply graph** and $H$ as the **demand graph**. The goal is to find a set of arc-disjoint paths in $D$ meeting the demand of $H$, or equivalently, find a collection of pairwise arc-disjoint cycles in $D + H$ such that every cycle uses exactly one arc from $F$ and all arcs of $F$ are covered by the cycles. We refer to the non-isolated vertices of $H$ as the **terminals**. Slightly abusing notation, we refer to this as the WEAK $k$-LINKAGE problem, although with input represented as a pair of directed multigraphs $(D, H)$ as above rather than in the equivalent alternative representation previously defined in Section 1.6.

In general digraphs, this problem is $\mathcal{NP}$-hard even when $H$ consists of just two arcs. In fact, for general digraphs, the only polynomial-time solvable cases of the problem reduce to standard $s$-$t$ cuts via Menger's theorem; see the end of this subsection.

In this section, we show that when the graph $D + H$ is Euler, then, effectively, we can handle the case where $H$ consists of two pairs of terminals. The same result also extends to the arc-capacitated two-commodity flow problem, in the case when the capacities of $D + H$ are balanced; see later.

These results are due to Frank [16], who extended similar results from undirected Euler graphs to the directed case. The proof below is essentially from Frank.

The central result is that given a pair of digraphs $(D, H)$ on vertex set $V$, such that $D + H$ is Euler and $H$ is acyclic and consists of two pairs of parallel arcs, the instance $(D, H)$ of WEAK $k$-LINKAGE is positive if and only if it satisfies the **directed cut criterion**:

$$d_D^+(X) \geq d_H^-(X) \text{ for all } X \subseteq V. \tag{4.1}$$

Clearly, this is a necessary condition. We show that it is also sufficient. We say that a set $X \subseteq V$ is a **tight set** if equality holds for $X$ in the statement above, i.e., $d_D^+(X) = d_H^-(X)$. For the duration of this proof, for any digraph $D$ we will define

$$d_D^*(A, B) = |(A \setminus B, B \setminus A)_D| + |(B \setminus A, A \setminus B)_D|,$$

i.e., $d_D^*(A, B)$ counts the number of arcs with one end in $A \setminus B$ and one in $B \setminus A$, regardless of orientation. We begin with some statements about tight sets.

**Lemma 4.5.1** *Let $(D = (V, A), H = (V, F))$ be an instance of* WEAK $k$-LINKAGE *such that $D + H$ is Euler and the directed cut criterion (4.1) holds for $(D, H)$. Then the following hold.*

1. *If $X \subseteq V$ is a tight set, then so is $V \setminus X$.*
2. *If $X, Y \subseteq V$ are tight sets, then $d_H^*(X, Y) \geq d_D^*(X, Y)$ and if equality holds then $X \cap Y$ and $X \cup Y$ are both tight sets.*
3. *If $X, Y \subseteq V$ are tight sets, then $d_H^*(X, V \setminus Y) \geq d_D^*(X, V \setminus Y)$ and if equality holds then $X \setminus Y$ and $Y \setminus X$ are both tight sets.*

**Proof:** *1.* Since $D + H$ is Euler, we have

$$d_D^+(X) + d_H^+(X) = d_D^-(X) + d_H^-(X)$$

for all $X \subseteq V$, from which the statement follows.

*2.* Recall that for any digraph and any vertex set $X$, we have

$$d^+(X) + d^+(Y) = d^+(X \cup Y) + d^+(X \cap Y) + d^*(X, Y),$$

and similarly for $d^-(X)$ and $d^-(Y)$. Therefore, rearranging terms we get

$$\begin{aligned}(d_D^+(X) - d_H^-(X)) + (d_D^+(Y) - d_H^-(Y)) &= \\ (d_D^+(X \cap Y) + d_D^+(X \cup Y) + d_D^*(X,Y)) &- \\ -(d_H^-(X \cap Y) + d_H^-(X \cup Y) + d_H^*(X,Y)) &= \\ (d_D^+(X \cap Y) - d_H^-(X \cap Y)) + (d_D^+(X \cup Y) - d_H^-(X \cup Y)) &+ \\ +d_D^*(X,Y) - d_H^*(X,Y) &= 0,\end{aligned}$$

where the whole expression equals 0 since $X$ and $Y$ are tight. Rewriting the last line, we have

$$(d_D^+(X\cap Y)-d_H^-(X\cap Y))+(d_D^+(X\cup Y)-d_H^-(X\cup Y)) = d_H^*(X,Y)-d_D^*(X,Y),$$

where the left-hand side is non-negative.

*3.* This statement follows by combining the two previous ones. □

**Theorem 4.5.2** *Let $(D = (V, A), H = (V, F))$ be an instance of* WEAK $k$-LINKAGE *such that $D+H$ is Euler, and $H$ is acyclic and consists of two sets of parallel arcs. Then $(D, H)$ is a 'Yes'-instance if and only if it satisfies the directed cut criterion (4.1).*

**Proof:** Let $T = \{s_1, t_1, s_2, t_2\}$ be the terminals of $H$, and assume that $F$ consists of $k_1 > 0$ arcs $t_1s_1$ and $k_2 > 0$ arcs $t_2s_2$. Hence the task is equivalent to packing $k_1 + k_2$ arc-disjoint paths in $D$ so that $k_1$ of these are from $s_1$ to $t_1$ and $k_2$ are from $s_2$ to $t_2$. We may assume that $D$ has no isolated vertices in $V \setminus T$ and every connected component of $D+H$ intersects $T$. Also observe that the result follows from Menger's theorem if $T$ intersects more than one connected component of $D + H$; hence we assume that $D + H$ is connected.

Assume for a contradiction that the theorem is false, and let $(D, H)$ be a minimum counterexample with respect to $|A(D + H)|$. Since it is clear that the directed cut criterion is a necessary condition, this implies that $(D, H)$ is a negative instance that meets the directed cut criterion, and that the theorem holds for every instance $(D', H')$ where $|A(D' + H')| < |A(D+H)|$.

In particular, consider a pair of arcs $xy, yz \in A(D)$ such that $|\{x, y, z\}| = 3$, i.e., when $x, y, z$ are distinct. Let $D'$ be the result of splitting of $xy, yz$ in $D$. Then $D' + H$ is Euler, and either $(D', H)$ fails to meet the directed cut criterion, or the instance $(D', H)$ is positive. But in the latter case, the paths packed in $D'$ also exist in $D$, by expanding one arc $xz$ (if used) into the two arcs $xy$, $yz$; hence we conclude that for every such pair of arcs $xy, yz \in A(D)$, splitting off $xy$ and $yz$ will break the directed cut criterion. Now note first that this is possible if and only if there is a tight set $X$ with either $X \cap \{x, y, z\} = \{y\}$ or $X \cap \{x, y, z\} = \{x, z\}$, and second, by Lemma 4.5.1 in fact both these tight sets would exist. Furthermore, since $d_D^+(X), d_D^+(V \setminus X) > 0$ (as evidenced by the arcs $xy$, $yz$), we also have $d_H^-(X), d_H^-(V \setminus X) > 0$. Therefore, we find that for every set $X$ such that splitting off a pair of arcs breaks the directed cut criterion at $X$, we have $X \cap T = \{s_i, t_j\}$, $i \neq j$.

We proceed with the proof. First consider the case when $V = T$. If $D$ contains an arc $s_i t_i$, $i \in \{1,2\}$, then we can remove that arc together with a copy of $t_i s_i$ from $H$, to produce a smaller instance $(D', H')$. It is easy to see that the directed cut condition holds in $(D', H')$, therefore there is a path-packing in $D'$, which can be extended by the additional arc $s_i t_i$ to a solution for $(D, H)$, and we are done. We claim that we can find a pair of arcs $su$, $uv$ in $D$ where $s \in \{s_1, s_2\}$ and $|\{s, u, v\}| = 3$. Indeed, by the cut criterion we have $d_D^+(s_i) \geq k_i$ for $i = 1, 2$, and $|(\{s_1, s_2\}, \{t_1, t_2\})_D| \geq k_1 + k_2$. The only case where you cannot find the pair $su$, $uv$ is if $A$ consists entirely of arcs $s_1 t_2$ and $s_2 t_1$, but in such a case we have $d_D^+(\{s_1, t_2\}) = 0 < d_H^-(\{s_1, t_2\}) = k_1$. Hence $su$, $uv$ exist, and as above, splitting off the pair will break the cut criterion, showing that there are tight sets $X$, $V \setminus X$ such that $X \cap \{s, u, v\} = \{u\}$ and $X \cap T = \{s_i, t_j\}$, $i \neq j$. Without loss of generality we may assume that $s = s_1$, so that $X = \{s_2, t_1\}$ and $\{s, v\} = \{s_1, t_2\}$. But then there is no way to select the arcs $su$, $uv$ without using an arc $s_i t_i$, $i \in \{1, 2\}$. Hence if there is a counterexample, it has $T \subset V$.

Now, let $u \in V \setminus T$ such that there is an arc $ux \in A$ for some $x \in T$; clearly such a vertex exists, and there is also at least one arc $vu \in A$, $v \neq x$. As before, we conclude that there must exist tight sets $X$, $V \setminus X$ such that $X \cap T = \{s_i, t_j\}$, $\{i, j\} = \{1, 2\}$, and $\{v, u, x\} \cap X = \{u\}$. Also note that since $x \in T$ in fact $X \cap T$ is fixed by the condition that $x \notin X$. For every arc $vu \in A$, let $X_v$ be a tight set proving that we cannot split off $vu$ and $ux$. Then for every pair of such sets $X_a$ and $X_b$, since $X_a \cap T = X_b \cap T$ we have $d_H^*(X_a, X_b) = 0$, so by Lemma 4.5.1 both $X_a \cup X_b$ and $X_a \cap X_b$ are tight. Let $X$ be the intersection of all such sets. We claim that there must be an arc $v'u \in A$ such that $v' \in X$, $v' \neq u$. Assume not, and consider the set $X' = X \setminus \{u\}$. Then this set has $d_H^-(X') = d_H^-(X)$ and $d_D^+(X') < d_D^+(X')$, contrary to the directed cut criterion. But now splitting off $v'u$ and $us$ cannot break the directed cut criterion, since $X \subseteq X_{v'}$ and $v' \in X$. This gives us a path packing in $(D', H)$ which can easily be converted to a path packing in $(D, H)$ of the same size. □

As Frank observes, this proof suggests an algorithm for finding such a path-packing: if the directed cut criterion does not apply, reject the instance. If there is an arc $s_i t_i$ in $D$ and $t_i s_i$ in $H$, then remove this pair of arcs and continue. Otherwise, find a pair of arcs $uv$, $vw$ that can be split off without breaking the directed cut criterion and recursively find a path packing in the resulting instance $(D', H)$; then finally, if the new arc $uw$ is used in one of the paths in $D'$, replace it by the old arcs $uv$, $vw$. This process will run in polynomial time, assuming the ability to test the directed cut criterion.

**Lemma 4.5.3** *The directed cut criterion for the instances $H$ considered here can be tested with three max-flow computations.*

**Proof:** Let $X \subseteq V$, and consider the cases for $d_H^-(X)$. If $d_H^-(X) = 0$, then the cut criterion holds for $X$. If $d_H^-(X) = k_i$ but $d_D^+(X) < k_i$, then $X$ represents an arc-cut of size less than $k_i$ from $s_i$ to $t_i$ for $i \in \{1,2\}$, which can be tested via max-flow computations. Finally, if $d_H^-(X) = k_1 + k_2$, then $X$ represents an arc-cut of size less than $k_1 + k_2$ from $\{s_1, s_2\}$ to $\{t_1, t_2\}$, which can be tested with a max-flow computation by adding a new meta-source $s$ and a meta-sink $t$. □

Finally, Frank shows a strongly polynomial time solution for the weighted version of the problem, TWO-COMMODITY FLOW. In this problem, the input is a digraph $D = (V, A)$ with an **arc capacity** function $c : A \to \mathbb{Z}_+$ as well as ordered request pairs $(s_i, t_i)$ with demand values $k_i$, $i \in \{1,2\}$, with the condition that at every vertex, the sums of incoming and outgoing capacities and demands are equal; equivalently, replacing each arc $a$ by $c(a)$ parallel copies and adding $k_i$ copies of the arc $t_i s_i$ defines an Euler digraph. We refer to this as a **balanced** two-commodity flow instance. Note that $D + H$ itself does not need to be Euler. The algorithm uses the same strategy as above, using a weighted version of the splitting operation, with some additional work required to prove that the number of steps is bounded by a polynomial. The proof, as above, is based on reasoning about the structure of tight sets, and shows that any sequence of weighted splitting operations has polynomially bounded length.

**Theorem 4.5.4** ([16]) *The* TWO-COMMODITY FLOW *problem can be solved in strongly polynomial time for balanced instances.*

We will consider the more general WEAK $k$-LINKAGE question later in this section, but for now we wrap up by recalling a characterization by Frank of when the directed cut criterion is a necessary and sufficient condition.

Let a **star** be a directed multigraph where there is either a common tail $s$ to all arcs, or a common head $t$ to all arcs. It is not hard to see that $(D, H)$-PATH PACKING can be solved via a max-flow computation if $H$ is a star, hence the problem is in $\mathcal{P}$. Fortune, Hopcroft and Wyllie [15] showed that for general digraphs, the converse is true – for any fixed digraph $H$ which is not a star, $(D, H)$-PATH PACKING is $\mathcal{NP}$-hard. For Euler digraphs, Frank showed the following.

**Theorem 4.5.5** ([16]) *Say that the directed cut criterion* **solves** *$H$ for a directed multigraph $H$ if, for every digraph $D$ such that $D + H$ is Euler, the instance $(D, H)$ of* WEAK $k$-LINKAGE *is positive if and only if the directed cut criterion is met. Then for any $H$, the directed cut criterion solves $H$ if and only if $H$ is the (arc-disjoint) union of two stars.*

**Proof:** Assume first that $H$ is the union of two stars. We convert $(D, H)$ into an equivalent instance of two-commodity flow. For each star with a common tail $t_i$ to the arcs, introduce a new vertex $s_i$, and replace every arc $t_i v$ in $H$ by an arc $t_i s_i$ in $H$ and an arc $s_i v$ in $D$. For a star with a common head $s_i$, instead introduce a new vertex $t_i$ in the same way. This reduces to the case where $H$ consists of $k_i$ arcs $t_i s_i$ for $i = 1, 2$. If the resulting graph $H$ is not acyclic, we can additionally introduce vertices $s'_i$, $t'_i$ with $k_i$ arcs $s'_i s_i$ and $t_i t'_i$ in $D$, and replace the arcs $t_i s_i$ in $H$ by $t'_i s'_i$. It is clear that this preserves the property of $D + H$ being Euler and the existence of a solution, and therefore the resulting instance can be solved by the algorithm above.

In the other direction, it is easy to observe that if $H$ is solved by the directed cut criterion, then so is every subgraph of $H$. Hence, it is sufficient to identify a constant number of digraphs $H$ which will occur as a subgraph in every graph $H$ that cannot be decomposed into two stars, and for each such $H$ show an instance where the directed cut criterion is insufficient. Such a list of instances is provided by Frank [16]. □

### 4.5.2 General Arc-Disjoint Paths Problems

We now consider the more general question of when WEAK $k$-LINKAGE is tractable for Euler digraphs. This question has several variants. One may consider the case when $D + H$ is Euler, or when $D$ is already Euler (or, indeed when both of $D$ and $H$ are Euler separately); one may consider either the basic WEAK $k$-LINKAGE problem or the weighted multi-commodity flow variant (where arcs of $D$ and $H$ have capacities, respectively demands); and one may consider $H$ to be one-time fixed or provided as problem input. There have also been several investigations into the complexity of the problem under various restrictions, including Ibaraki and Poljak [23], Vygen [43], Naves and Sebö [33], and Frank [17].

For the negative cases, we begin by noting that the three-commodity flow problem is easily seen to be $\mathcal{NP}$-hard when $D$ is Euler but $H$ is not. Let $(D, H)$ be the input of WEAK $k$-LINKAGE where $H$ consists of two arcs; as noted, this is an $\mathcal{NP}$-hard problem on general digraphs. Let $D'$ be the Euler two-vertex extension of $D$, with new vertices $x$ and $y$, and create a demand graph $H'$ from $H$ by adding the vertices $x$ and $y$ as well as $\mu_{D'}(x, y)$ copies of the arc $yx$. Then clearly, $(D, H)$ is positive if and only if $(D', H')$ is positive, and $D'$ is Euler (although $D' + H'$ is not). To reach a situation where $D' + H'$ is Euler, we can use a slight variation of this (used by Ibaraki and Poljak [23]).

**Lemma 4.5.6** WEAK $k$-LINKAGE *is $\mathcal{NP}$-hard when $D + H$ is Euler and the underlying digraph of $H$ (where all arc multiplicities are reduced to 1) has three arcs.*

**Proof:** Recall that WEAK 2-LINKAGE is $\mathcal{NP}$-hard even when $A(H) = \{st, ts\}$. We show a reduction from this problem to an instance $(D', H')$ of WEAK K-LINKAGE where $D' + H'$ is Euler. Let $(D, H)$ with $H$ as above be an instance of WEAK 2-LINKAGE, and let $D'$ be the Euler two-vertex extension of $D + H$, adding vertices $x$ and $y$. Now let $H'$ be $H$ plus all copies of the arc $xy$ in $D'$, and remove the arcs $xy$ from $D'$. We claim that $(D', H')$ is a positive instance of WEAK $k$-LINKAGE if and only if $(D, H)$ is a positive instance of WEAK 2-LINKAGE.

First, assume that $(D', H')$ is positive, and let $\mathcal{P}$ be the corresponding collection of paths. Then $\mathcal{P}$ contains an $st$-path $P_1$ and a $ts$-path $P_2$, both of which are contained in $D'$. But since $x$ and $y$ are a sink, respectively a source, in $D'$, both $P_1$ and $P_2$ must exist in $D$ as well.

In the other direction, assume that $D$ contains an arc-disjoint pair of an $st$-path $P_1$ and a $ts$-path $P_2$. Then $P_1 + P_2$ form an Euler digraph, and therefore $D'' := (D' + H') - (P_1 + P_2 + st + ts)$ is balanced. It follows that there are $\mu_{D''}(x, y)$ arc-disjoint cycles through $y$ in $D''$, and each of these cycles must use an arc $xy$ from $H'$. Removing the arc $xy$ from each of these cycles yields a collection of $yx$-paths in $D''$ that together with $P_1$ and $P_2$ forms an arc-disjoint path-packing in $D'$. □

Hence, in particular, the results on TWO-COMMODITY FLOW cannot be extended to three or more commodities. Vygen [43] shows that the problem is still $\mathcal{NP}$-hard if $D$ is additionally assumed to be acyclic.

For the case when $H$ is fixed, say $|A(H)| = k$, the complexity of the problem is notoriously open. The case when $k = 3$ was solved by Ibaraki and Poljak [23]. Specifically, they take the following approach. Let $H$ be an Euler digraph with disjoint arcs $t_i s_i$, $i = 1, 2, 3$. Then we can reduce the instance $(D, H)$ of WEAK 3-LINKAGE to an instance where $H = C_3$ as follows. Add three terminals $x$, $y$, $z$ to $D$, and arcs $xs_1$, $t_1 y$, $ys_2$, $t_2 z$, $zs_3$, $t_3 x$. Then the original instance has a weak 3-linkage if and only if the resulting graph has arc-disjoint $xy$-, $yz$- and $zx$-paths. We then find that the problem has a solution if there is an Euler trail of $D$ that, starting from $x$, visits the new terminals in the order $x$, $y$, $z$ (i.e., a trail such that the first visit to $y$ after the start at $x$ comes before the last visit to $z$). They are then able to solve the problem in polynomial time by carefully investigating the structure of minimal negative instances. Thus WEAK 3-LINKAGE is in $\mathcal{P}$ if $D + H$ is Euler.

For general $H$, as far as we know, the possibilities range from the problem being $\mathcal{NP}$-complete for $k = 4$ to the problem being FPT parameterized by $k$.

**Problem 4.5.7** *What is the status of* WEAK $k$-LINKAGE *for inputs* $(D, H)$ *where* $D + H$ *is Euler, parameterized by* $k$*? Is it FPT or in XP?*

A slight variant was considered by Frank, Ibaraki and Nagamochi [17]. They consider the problem variant where the input is an Euler digraph $D$ and an undirected graph $H$, say $E(H) = \{ab, cd\}$, and the task is to find a pair of arc-disjoint paths $P_1$, $P_2$ in $D$ where $P_1$ is either an $ab$-path or a $ba$-path, and $P_2$ either a $cd$-path or a $dc$-path. They show that this problem can be solved in polynomial time, via an extensive investigation into the structure of minimal infeasible instances.

They note that this problem generalizes the result of Ibaraki and Nagamochi, as follows. Let $(D, H)$ be an input to WEAK 3-LINKAGE where $D+H$ is Euler and $A(H) = \{t_i s_i \mid i \in [3]\}$. Create a graph $D'$ from $D$ by adding four new vertices $a$, $b$, $c$, $d$ and arcs $t_1c$, $cs_2$, $t_2d$, $da$, $as_3$, $t_3b$, $bs_1$. Observe that $D'$ is Euler. Now we may observe the following. Any $ba$-path in $D'$ will exhaust all arcs incident with $d$ in $D'$, and similarly a $dc$-path will exhaust $a$. Thus if the instance is positive, then $P_1$ is an $ab$-path and $P_2$ is a $cd$-path. Then $P_2 + da + P_1$ form a directed $cb$-path, hence since $D'$ is Euler, $D' - (P_2 + da + P_1)$ contains a directed $bc$-path. It is clear that these paths together must form a weak 3-linkage for $D$.

We end with a different question, again via Frank [16].

**Problem 4.5.8** *Is* WEAK $k$-LINKAGE *with input* $(D, H)$ *in* $\mathcal{P}$ *if* $D + H$ *is Euler and planar?*

The undirected version of this question is known to hold, i.e., EDGE-DISJOINT PACKING for undirected graphs, with input $(G, H)$, is in $\mathcal{P}$ if $G+H$ is planar. In fact, for this version, the problem is solved by the undirected version of the cut criterion (4.1). (The corresponding statement does not hold for the directed version [16].)

### 4.5.3 Free Multiflow and Arc Multiway Cut

We now turn to a slightly different model of path-packing problems, and in the process we will cover a less well known, but very interesting result due to Frank [16] on a polynomial-time solvable multicut problem on Euler digraphs.

The general setup here is as follows. We have a digraph $D = (V, A)$ and a set of terminals $T \subseteq V$, but instead of having an *exact* set of path requests (encoded as a digraph $H$ over $T$, as in the previous section), we have a notion of *allowed* or *disallowed* terminal-terminal paths, and we are looking for a maximum arc-disjoint path-packing that consists entirely of "allowed" paths. Somewhat more generally, we can also introduce *weights* for paths, depending on their type, and ask for an arc-disjoint path-packing of maximum weight.

Let us begin with the FREE MULTIFLOW problem, where every simple terminal-terminal path is "allowed" under the above notion. More concretely, the input to FREE MULTIFLOW is a digraph $D = (V, A)$ and a set of terminals $T \subseteq V$, and the task is to find a maximum arc-disjoint packing of directed paths in $D$ where each path goes between distinct terminals in $T$. Frank showed that if $D$ is Euler, then the problem has a simple min-max formula, as follows. For disjoint sets $X, Y \subset V(D)$ in a digraph $D$, let $\lambda_D(X, Y)$ denote the maximum number of arc-disjoint paths from $X$ to $Y$ in $D$. We will slightly abuse notation and use single vertices in place of singleton sets (e.g., we write $t$ rather than $\{t\}$). Then the maximum number of paths in the free multiflow packing equals

$$\sum_{t \in T} \lambda_D(t, T - t),$$

the sum of the sizes of isolating cuts in $D$.

For general digraphs, the FREE MULTIFLOW problem is $\mathcal{NP}$-hard for $|T| \geq 2$ as it generalizes the WEAK 2-LINKAGE problem, but for undirected graphs there are several classical polynomial-time results in this direction. The most immediately corresponding result is due to Lovász [30], who proved the corresponding statement for undirected Euler graphs, and on which Frank's result is based. However, multiple more general results exist, including Mader's theorem on packing internally vertex-disjoint terminal-terminal paths in general undirected graphs (see Schrijver [37]), as well as generalizations in terms of packing paths in group-labelled graphs (see Chudnovsky, Geelen, Gerards, Goddyn, Lohman, and Seymour [8]) and in general permutation-labelled graphs (see Pap [34]).

However, there is one unique feature of the Euler digraph result that is not mirrored in any of the variants of the problem on undirected graphs (even Euler graphs). For each of the above packing problems, one can define a natural dual **cut problem** (or, alternatively expressed, a **path-hitting problem**) of finding a minimum set $X$ of edges (or arcs, or vertices) such that removing every element of $X$ leaves a graph with no allowed paths of the respective type. Concretely, the dual to the FREE MULTIFLOW problem would be the classical problem ARC MULTIWAY CUT, of finding a minimum set $X$ of arcs in a digraph $D$ such that $D - X$ contains no terminal-terminal path. Whereas the undirected version, EDGE MULTIWAY CUT, is $\mathcal{NP}$-hard on undirected Euler graphs for 3 terminals, the min-max theorem for Euler digraphs directly implies that for this graph class, ARC MULTIWAY CUT is in $\mathcal{P}$.

For the rest of this section, let $D = (V, A)$ be an Euler digraph and $T \subseteq V$ a set of terminals. Let $T = \{t_1, \ldots, t_p\}$ for $p = |T|$, and recall that for each $i \in [p]$, $d_i = \lambda_D(t_i, T - t_i)$ denotes the maximum number of arc-disjoint paths from $t_i$ to the remaining terminals.

We will prove the following theorem, from which an algorithm for ARC MULTIWAY CUT will follow easily.

**Theorem 4.5.9** *Let $D = (V, A)$ be an Euler digraph and $T \subseteq V$ a set of terminals. There is an arc-disjoint packing $\mathcal{P}$ of terminal-terminal paths in $D$ such that each terminal $t_i \in T$ is the starting vertex of $d_i$ paths in $\mathcal{P}$.*

Theorem 4.5.9 will in turn follow from the following statement.

**Theorem 4.5.10** *Let $vw \in A$ be an arbitrary arc where $v \notin T$. Then there is an arc $uv \in A$ such that splitting off $uv$ and $vw$ will not change the value of $d_i$ for any $i \in [p]$.*

**Proof:** Recall that $d^+_H(S)$ denotes the number of arcs out of $S$ in a digraph $H$.

Let $uv$, $vw \in A$ be an arbitrary pair of arcs, and assume that splitting them off in $D$ decreases the value of $d_i$ for some $i \in [p]$. Let $D'$ be the result of the splitting-off operation. Then there exists a set $V_i \subseteq V$ such that $V_i \cap T = \{t_i\}$, and $d^+_{D'}(V_i) < d_i$ in $D'$, which is true if and only if $d^+_D(V_i) = d_i$ and $V_i \cap \{u, v, w\} \in \{\{v\}, \{u, w\}\}$. We say that the set $V_i$ *blocks* the splitting off $uv$, $vw$ in $D$. Therefore, similarly to the concept of tight sets in the previous section, let us refer to a set $V_i \subseteq V$ as *$i$-critical* if $V_i \cap T = \{t_i\}$ and $d^+_D(V_i) = d_i$. A set $U \subseteq V$ is *critical* if it is $i$-critical for some $i \in [p]$. We make a few observations about critical sets.

*Observation 1:* For every $i \in [p]$, the $i$-critical sets are closed under union and intersection, and for two $i$-critical sets $V_i$, $V_i'$, there is no arc between $V_i \setminus V_i'$ and $V_i' \setminus V_i$. This follows from the well-known submodularity of cuts that for any $X, Y \subseteq V$ we have

$$d^+_D(X) + d^+_D(Y) \geq d^+_D(X \cap Y) + d^+_D(X \cup Y),$$

with equality only if there are no arcs between their symmetric differences.

*Observation 2:* If $V_i$ is $i$-critical and $V_j$ is $j$-critical, $i \neq j$, then $V_i \setminus V_j$ is $i$-critical and $V_j \setminus V_i$ is $j$-critical, and there is no arc between $V_i \cap V_j$ and $V \setminus (V_i \cup V_j)$. This follows similarly as above, with the additional ingredient that for an Euler digraph $D$, we have $d^+_D(S) = d^+_D(V \setminus S)$.

The proof now proceeds in two cases. In the first case, assume that there is a critical set $V_i$, without loss of generality $i$-critical, such that $v \in V_i$ and $w \notin V_i$. Let $V_i$ be a minimal such set, and pick an arc $uv \in A$ such that $u \in V_i$. Such an arc must exist, as otherwise $d^+(V_i - v) < d^+(V_i) = d_i$. Let $U$ be a critical set that blocks the splitting off of $uv$ and $vw$. First, assume that $v \in U$; hence $u, w \notin U$. Then $U$ is not $i$-critical, since otherwise $U \cap V_i \subset V_i$, contradicting the choice of $V_i$; but if $U$ is $j$-critical, $j \neq i$, then $v \in U \cap V_i$ while $w \notin U \cup V_i$, which contradicts observation 2. Thus $U \cap \{u, v, w\} = \{u, w\}$ But $U$ cannot be $i$-critical, by observation 1 and the arc $uv$, and it cannot be $j$-critical, $i \neq j$, since $V_i \setminus U \subset V_i$, contradicting the choice of $V_i$. Thus the first case is handled.

In the second case, let $V_i$ be a maximal critical set with $w \in V_i$, $v \notin V_i$. Assume that $V_i$ is $i$-critical. Let $uv \in A$ such that $u \notin V_i$; this exists, since otherwise $d^+(V_i + v) < d^+(V_i) = d_i$. Let $U$ be a critical set that blocks the splitting of $uv$ and $vw$. Then $U \cap \{u, v, w\} = \{u, w\}$, as otherwise the set $U$ brings us back to case 1. If $U$ is $i$-critical, then $U \cup V_i$ is an $i$-critical set contradicting the choice of $V_i$, but if not, then $w \in U \cap V_i$ and $v \notin U \cup V_i$, which is a contradiction by observation 2. Therefore, in both cases we find that there exists an arc $uv$ such that $uv$ and $vw$ can be split off, and clearly if neither case applies then there cannot exist a critical set blocking the splitting off of $uv$ and $vw$ for any arc $uv$. □

The proof of the path-packing statement (Theorem 4.5.9) follows from this, by first repeatedly splitting off arcs until every vertex except $T$ is isolated, then unrolling these operations while maintaining a path-packing. Frank notes that a capacitated version can also be shown to be solvable in strongly polynomial time.

The following is an easy corollary of the path-packing result.

**Theorem 4.5.11** *The* ARC MULTIWAY CUT *problem on Euler digraphs is polynomial-time solvable.*

**Proof:** Clearly, the solution must have cardinality at least $\sum_{i=1}^{p} d_i$, by the existence of a path packing. But it is not difficult to produce a solution of exactly this size. For $i = 1, \ldots, p$, let $V_i$ be the $i$-critical set of minimum cardinality. By observation 1 of the previous proof these sets are unique, and by observation 2 the sets are also pairwise disjoint. Thus the set $\bigcup_i (V_i, V \setminus V_i)_D$ is an arc multiway cut of cardinality matching the size of a path-packing, and therefore clearly optimal. □

As noted, the remarkable aspect of this result is that no comparable statement can be found for undirected graphs. If $G$ is undirected and Euler, then even though there is a min-max result corresponding to Theorem 4.5.9 for the size of a path packing, there is no corresponding way to find a cut that intersects every path only once. In particular, the collection of closest min-cuts would hit some paths twice.

Finally, we have the following variant of Theorem 4.5.10, which generalizes a result of Lovász [30] for undirected Euler graphs, and has been shown independently by Frank [16] and by Jackson [25]. The proof is in the same spirit as Theorem 4.5.10.

**Theorem 4.5.12** *Let $D = (V, A)$ be an Euler digraph and let $vw \in A$. There exists an arc $uv \in A$ such that splitting off $uv$ and $vw$ does not affect the value of $\lambda(s, t)$ for any vertices $s, t \in V - v$.*

### 4.5.4 General Integral Weighted Path Packings

We now review a weighted generalization of FREE MULTIFLOW due to Hirai and Koichi [22]. To present the result, we need to introduce several notions.

First, we define networks. We will use a slightly different notion of a network than that given in Section 1.9, so to avoid ambiguity we introduce a different term. A **terminal network** is a triple $(D, T, c)$ consisting of a digraph $D = (V, A)$, a set of terminals $T \subseteq V$, and a set of integer arc capacities $c : A \to \mathbb{Z}_+$. We say that the network is **balanced at** $v$ for a vertex $v \in V$ if

$$\sum_{uv \in A} c(uv) = \sum_{vw \in A} c(vw).$$

A **balanced terminal network** (respectively **inner balanced terminal network**) is a network which is balanced at every vertex $v$ (respectively at every vertex $v \in V \setminus T$). This definition of a network differs from the usual one in that instead of an explicit balance vector, we have a set of terminals over which a flow is to be maximized. A **multiflow over** $T$ is a pair $(\mathcal{P}, \lambda)$ where $\mathcal{P}$ is a collection of directed paths with all endpoints in $T$, $\lambda : \mathcal{P} \to \mathbb{R}_+$ a set of **flow values** for the paths in $\mathcal{P}$, and $(\mathcal{P}, \lambda)$ satisfy the capacity constraints, i.e.,

$$\sum_{P \in \mathcal{P}: a \in P} \lambda(P) \leq c(a)$$

for every arc $a \in A$. Finally, a **directed distance** on $T$ is a function $\mu : T \times T \to \mathbb{R}_+$ such that $\mu(x, x) = 0$ for every $x \in T$. Note that the triangle inequality is not required to hold. For a directed path $P$, starting and ending at terminals $s$ and $t$ in $T$, we let $\mu(P) = \mu(s, t)$. For a directed distance $\mu$ on $T$ and a multiflow $(\mathcal{P}, \lambda)$ over $T$, the **$\mu$-weighted flow value** of $(\mathcal{P}, \lambda)$ equals

$$\sum_{P \in \mathcal{P}} \lambda \cdot \mu(P).$$

The $\mu$-WEIGHTED MAXIMUM MULTIFLOW PROBLEM ($\mu$-MFP) is then defined as the problem where the input is a terminal network $(D, T, c)$ and a directed distance $\mu$ on $T$, and the task is to find a multiflow $(\mathcal{P}, \lambda)$ over $T$ which maximizes the $\mu$-weighted flow value.

This problem will in general have a fractional optimum, but for some directed distances $\mu$, the system will have an integral optimum for every balanced terminal network – for example, if we fix a directed distance $\mu$ where $\mu(s, t) = 1$ for all $s, t \in T$, $s \neq t$, then $\mu$-MFP corresponds to the FREE MULTIFLOW problem, and the statement would follow from Theorem 4.5.9. The results of Hirai and Koichi imply a characterization of all directed distances $\mu$ such that $\mu$-MFP has an integral optimum for every balanced terminal network.

The characterization is as follows. Let $\Gamma$ be an oriented tree, and $\alpha$ a set of non-negative (real-valued) edge weights of $\Gamma$. We define a directed metric $d_{\Gamma,\alpha}$ on $V(\Gamma)$ by letting $d_{\Gamma,\alpha}(u, v)$ for $u, v \in V(\Gamma)$ be the sum of $\alpha(e)$ over all edges $e$ of $E(\Gamma)$ that are oriented from $u$ to $v$ in the path $P_{uv}$ from $u$ to $v$ in $\Gamma$. An **oriented tree realization** of a directed distance $\mu$ on $T$ is a triple $(\Gamma, \alpha, \{F_t\}_{t \in T})$ where $\Gamma$ is an oriented tree, $\alpha : E(\Gamma) \to \mathbb{R}_+$ a set of of non-negative edge lengths of $\Gamma$, and $\{F_t\}_{t \in T}$ a collection of subtrees of $\Gamma$, such that

$$\mu(s, t) = \min_{a \in F_s, b \in F_t} d_{\Gamma,\alpha}(a, b)$$

for all pairs $s, t \in T$.

**Theorem 4.5.13** ([22]) *Let $\mu$ be a directed distance on a set of terminals $T$. Then the $\mu$-MFP is integral for every balanced network $(D, T, c)$ if and only if $\mu$ has an oriented tree realization.*

Let us consider a few examples. First, let $\Gamma$ consist of a single arc $st$, with $\alpha(st) > 0$. Then $d_\Gamma(s,t) = \alpha(st) > 0$, while $d_{\Gamma,\alpha}(t,s) = 0$ since the arc $st$ is traversed in the wrong direction. Hence $\mu$-MFP reduces to the usual max-flow problem for any directed distance $\mu$ realized by $\Gamma$. For another example, the FREE MULTIFLOW problem can be realized by a unit-weighted star $\Gamma$, with all arcs oriented into the root and with the collection $\{F_t\}_{t\in T}$ being a bijection between $T$ and the leaves of $\Gamma$.

Hirai and Koichi also give a matching min-max theorem for the positive cases, in terms of packing cuts in the oriented tree that realizes $\mu$; we omit the details here.

## References

1. N. Alon, C. McDiarmid, and M. Molloy. Edge-disjoint cycles in regular directed graphs. *J. Graph Theory*, 22(3):231–237, 1996.
2. J. Bang-Jensen and G. Gutin. *Digraphs: Theory, Algorithms and Applications.* Springer-Verlag, London, 2nd edition, 2009.
3. H.L. Bodlaender, S. Thomassé, and A. Yeo. Kernel bounds for disjoint cycles and disjoint paths. *Theor. Comput. Sci.*, 412(35):4570–4578, 2011.
4. R.A. Brualdi and J. Shen. Disjoint cycles in eulerian digraphs and the diameter of interchange graphs. *J. Combin. Theory Ser. B*, 85(2):189–196, 2002.
5. J. Chen, Y. Liu, S. Lu, B. O'Sullivan, and I. Razgon. Directed Feedback Vertex Set is Fixed-Parameter Tractable. In *STOC 2008: 40th Annual ACM Symposium on Theory of Computing.* ACM Press, 2008.
6. R. Chitnis, M. Cygan, M. Hajiaghayi, M. Pilipczuk, and M. Pilipczuk. Designing FPT algorithms for cut problems using randomized contractions. *SIAM J. Comput.*, 45(4):1171–1229, 2016.
7. F.A. Chudak, M.X. Goemans, D.S. Hochbaum, and D.P. Williamson. A primal-dual interpretation of two 2-approximation algorithms for the feedback vertex set problem in undirected graphs. *Oper. Res. Lett.*, 22(4-5):111–118, 1998.
8. M. Chudnovsky, J. Geelen, B. Gerards, L. A. Goddyn, M. Lohman, and P.D. Seymour. Packing non-zero A-paths in group-labelled graphs. *Combinatorica*, 26(5):521–532, 2006.
9. B. Courcelle. The monadic second-order logic of graphs. I: Recognizable sets of finite graphs. *Inform. Comput.*, 85:12–75, 1990.
10. M. Cygan, D. Lokshtanov, M. Pilipczuk, M. Pilipczuk, and S. Saurabh. Minimum bisection is fixed parameter tractable. In *STOC 2014: 46th annual ACM Symposium on Theory of Computing*, pages 323–332. ACM, 2014.
11. M. Cygan, J. Nederlof, M. Pilipczuk, M. Pilipczuk, J.M.M. van Rooij, and J.O. Wojtaszczyk. Solving connectivity problems parameterized by treewidth in single exponential time. In R. Ostrovsky, editor, *FOCS 2011: 52nd Annual Symposium on Foundations of Computer Science*, pages 150–159. IEEE Computer Society, 2011.
12. R. Diestel. *Graph Theory, 4th Edition*, volume 173 of *Graduate texts in mathematics.* Springer, 2012.
13. G. Even, J. Naor, B. Schieber, and M. Sudan. Approximating minimum feedback sets and multicuts in directed graphs. *Algorithmica*, 20(2):151–174, 1998.

14. H. Fleischner and B. Jackson. Compatible Euler tours in Eulerian digraphs. In *Cycles and rays (Montreal, PQ, 1987)*, volume 301 of *NATO Adv. Sci. Inst. Ser. C: Math. Phys. Sci.*, pages 95–100. Kluwer, Dordrecht, 1990.
15. S. Fortune, J.E. Hopcroft, and J. Wyllie. The directed subgraph homeomorphism problem. *Theor. Comput. Sci.*, 10:111–121, 1980.
16. A. Frank. On connectivity properties of Eulerian digraphs. In *Graph theory in memory of G. A. Dirac (Sandbjerg, 1985)*, volume 41 of *Ann. Discrete Math.*, pages 179–194. North-Holland, 1989.
17. A. Frank, T. Ibaraki, and H. Nagamochi. Two arc-disjoint paths in Eulerian digraphs. *SIAM J. Discrete Math.*, 11(4):557–589 (electronic), 1998.
18. M. Grohe and M. Grüber. Parameterized approximability of the disjoint cycle problem. In *ICALP 2007: 34th Int. Colloquium on Automata, Languages and Programming*, volume 4596 of *Lect. Notes Comput. Sci.*, pages 363–374. Springer-Verlag, 2007.
19. B. Guenin and R. Thomas. Packing directed circuits exactly. *Combinatorica*, 31(4):397–421, 2011.
20. V. Guruswami and E. Lee. Simple proof of hardness of feedback vertex set. *Theory Comput.*, 12(6):1–11, 2016.
21. G. Gutin, M. Jones, B. Sheng, and M. Wahlström. Parameterized directed $k$-chinese postman problem and $k$ arc-disjoint cycles problem on euler digraphs. In *WG 2014: 40th International Workshop on Graph-Theoretic Concepts in Computer Science*, volume 8747 of *Lect. Notes Comput. Sci.*, pages 250–262, 2014.
22. H. Hirai and S. Koichi. On duality and fractionality of multicommodity flows in directed networks. *Discrete Optim.*, 8(3):428–445, 2011.
23. T. Ibaraki and S. Poljak. Weak three-linking in Eulerian digraphs. *SIAM J. Discrete Math.*, 4(1):84–98, 1991.
24. Y. Iwata. Linear-time kernelization for feedback vertex set. *CoRR*, abs/1608.01463, 2016.
25. B. Jackson. Some remarks on arc-connectivity, vertex splitting, and orientation in graphs and digraphs. *J. Graph Theory*, 12(3):429–436, 1988.
26. T. Johnson, N. Robertson, P.D. Seymour, and R. Thomas. Directed Tree-Width. *J. Combin. Theory Ser. B*, 82(1):138–154, 2001.
27. K. Kawarabayashi and S. Kreutzer. The directed grid theorem. In *STOC 2015: 47th Annual ACM on Symposium on Theory of Computing*, pages 655–664. ACM, 2015.
28. K. Kawarabayashi and M. Thorup. The minimum $k$-way cut of bounded size is fixed-parameter tractable. In R. Ostrovsky, editor, *FOCS 2011: 52nd Annual Symposium on Foundations of Computer Science*, pages 160–169. IEEE Computer Society, 2011.
29. T. Kociumaka and M. Pilipczuk. Faster deterministic feedback vertex set. *Inform. Process. Lett.*, 114(10):556–560, 2014.
30. L. Lovász. On some connectivity properties of Eulerian graphs. *Acta Math. Acad. Sci. Hung.*, 28(1–2):129–138, 1976.
31. C.L. Lucchesi and D.H. Younger. A minimax theorem for directed graphs. *J. London Math. Soc. (2)*, 17(3):369–374, 1978.
32. D. Marx, B. O'Sullivan, and I. Razgon. Finding small separators in linear time via treewidth reduction. *ACM Trans. Algor.*, 9(4):30:1–30:35, 2013.
33. G. Naves and A. Sebö. Multiflow feasibility: An annotated tableau. In W.J. Cook, L. Lovász, and J. Vygen, editors, *Research Trends in Combinatorial Optimization, Bonn Workshop on Combinatorial Optimization, 2008*, pages 261–283. Springer, 2008.
34. G. Pap. Packing non-returning $A$-paths. *Combinatorica*, 27(2):247–251, 2007.

35. B. Reed, N. Robertson, P.D. Seymour, and R. Thomas. Packing directed circuits. *Combinatorica*, 16(4):535–554, 1996.
36. N. Robertson and P.D. Seymour. Graph minors. XIII: The disjoint paths problem. *J. Combin. Theory Ser. B*, 63:65–110, 1995.
37. A. Schrijver. A short proof of Mader's S-paths theorem. *J. Combin. Theory Ser. B*, 82(2):319–321, 2001.
38. P.D. Seymour. Packing circuits in Eulerian digraphs. *Combinatorica*, 16(2):223–231, 1996.
39. A. Slivkins. Parameterized tractability of edge-disjoint paths on directed acyclic graphs. In *ESA 2003: 11th Annual Eur. Symp. Algorithms*, volume 2832 of *Lect. Notes Comput. Sci.*, pages 482–493, 2003.
40. S. Thomassé. A $4k^2$ kernel for feedback vertex set. *ACM Trans. Algor.*, 6(2), 2010.
41. W.T. Tutte and C.A.B. Smith. On unicursal paths in a network of degree 4. *Am. Math. Monthly*, 48(4):233–237, 1941.
42. T. van Aardenne-Ehrenfest and N.G. de Bruijn. Circuits and trees in oriented linear graphs. *Simon Stevin*, 28:203–217, 1951.
43. J. Vygen. NP-completeness of some edge-disjoint paths problems. *Discrete Appl. Math.*, 61(1):83–90, 1995.
44. D.H. Younger. Graphs with interlinked directed circuits,. In *Midwest symposium on circuit theory 2*, pages XVI 2.1–XVI 2.7, 1973.

# 5. Planar Digraphs

**Marcin Pilipczuk and Michał Pilipczuk**

In this chapter we focus on *planar* directed graphs, that is, directed graphs that can be drawn on a plane (or, equivalently, on a sphere) without arc crossings. We will alternate between the planar and spherical embeddings, picking the more convenient for the current argumentation.

A **planar embedding** of a digraph $D$ is a mapping $\pi$ that assigns a distinct point in the Euclidean plane to every vertex of $D$, and a curve without self-intersections to every arc of $D$ in such a manner that for every arc $e = (u, v)$, the curve $\pi(e)$ has endpoints $\pi(u)$ and $\pi(v)$, and the images of two arcs are disjoint (except for endpoints if the arcs in question share end vertices). A **face** in an embedding $\pi$ is a connected component of the plane minus the image of $\pi$; a face is **incident** with all vertices and arcs whose images under $\pi$ lie in the closure of the face. A **spherical embedding** is defined analogously with the target surface being a sphere instead of a plane; intuitively, the main difference between a planar and a spherical embedding is that the first distinguishes one face as an infinite one.

After this very brief introduction, we refrain here from introducing all formal definitions and notation concerning graph embeddings, assuming instead a common intuitive understanding. In case of doubt, we refer to other monographs for formal details, e.g., to the book of Mohar and Thomassen [22].

The main goal of this chapter is to show, from multiple angles, how the planarity assumption imposes structure on digraphs and how such structure, in conjunction with topological arguments, can be used algorithmically. In other words, the main focus here is to show various algorithmic techniques used to tackle planar digraphs. Thus, instead of providing a survey of the vast number of algorithmic results concerning embedded digraphs, we highlight three of them, chosen to highlight different aspects of planar digraphs.

First, in Section 5.1 we show an example of a low polynomial-time algorithm for planar graphs, namely a near-linear algorithm for single-source and single-sink maximum flow. Second, in Section 5.2, we discuss the classic problem $k$-Disjoint paths, where the topology assumption greatly improves the tractability of the problem. Finally, in Section 5.3 we discuss the Directed Grid Theorem for planar digraphs.

Marcin Pilipczuk (✉) and Michał Pilipczuk
Institute of Informatics,
University of Warsaw, Banacha 2, 02-097 Warsaw, Poland
e-mail: malcin@mimuw.edu.pl

 
J. Bang-Jensen and G. Gutin (eds.), *Classes of Directed Graphs*, Springer Monographs in Mathematics, https://doi.org/10.1007/978-3-319-71840-8_5

While we tried to make the description in every section as self-contained as possible, some technical details are missing in order to make the presentation clear and concise. In every section, we provide relevant references to full proofs and further reading.

## 5.1 Low Polynomial-Time Algorithms

Part of the importance of planar graphs stems from the fact that many problems admit much more efficient solutions when the input graph is required to be planar. One of the areas where such improvements are particularly visible are low polynomial-time algorithms, such as algorithms for shortest paths or maximum flows. Decades of research led to linear-time or near-linear-time (e.g., $\mathcal{O}(n \log n)$ or even $\mathcal{O}(n \log \log n)$) algorithms for problems requiring significantly larger running time in general graphs.

In this section, we do not aim at a full survey of these results for planar digraphs; the interested reader is referred to the free online book of Klein and Mozes [19]. Instead, we present one of the most elegant results in the area, namely the $\mathcal{O}(n \log n)$-time algorithm for finding the maximum flow between two given vertices due to Borradaile and Klein [2], with the simplified analysis due to Erickson [9]. We chose this result, as it involves a number of interesting techniques and properties of planar (di)graphs: duality of spanning trees in primal and dual graphs, duality of separators and cycles in dual graphs, as well as winding numbers analyzed via universal covers. The exposition mostly follows Chapter 10 of the book of Klein and Mozes [19], but we mainly focus on intuition, sweeping most of the technical details under the rug.

Because we will be working with residual capacities, we assume that we are given as an input a planar digraph $D$ where for every arc $e = (u, v)$ in $D$ its reversed twin $\mathrm{rev}(e) = (v, u)$ is also in $D$. The input also specifies two distinguished vertices $s$ and $t$, called the **source** and **sink**, and a capacity function $u\colon A(D) \to \mathbb{Z}_{\geq 0}$. If we replace every pair of arcs $\{e = (u, v), \mathrm{rev}(e)\}$ by an undirected edge $uv$, we obtain a planar undirected graph $G$. Without loss of generality, we can assume that $G$ is connected. Let us fix some planar embedding of $G$ where $t$ lies on the outer face, denoted $f^t$.

In what follows, we will work with the assumed embedding of $G$, but also implicitly treat every undirected edge $uv$ of $G$ as two arcs $(u, v)$ and $(v, u)$ of $D$. Thus, for an arc $e$ of $D$, we will speak about the face $f^-(e)$ to the right (clockwise) of $e$ and the face $f^+(e)$ to the left (counter-clockwise) of $e$. Note that these notions formally refer to the faces of the embedding of $G$. We refer to Figure 5.1 for the basic notation of the dual graphs used in this proof.

For this fixed embedding, a dual of the graph $G$ is a graph $G^*$ whose vertex set is the set of faces of the embedding, and where an edge $uv \in E(G)$ corresponds to an edge joining the two faces incident to $uv$ in the embedding of $G$. Clearly, $G^*$ is a planar graph with a natural embedding induced by the embedding of $G$. As in the case of $D$ and $G$, if we replace every edge of $G^*$ with two arcs in both directions, we obtain a digraph $D^*$. If $e = (u, v)$ is an

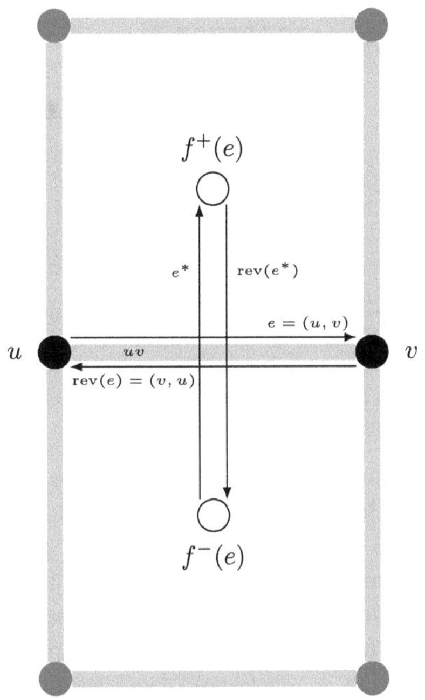


**Figure 5.1** Notation of the dual graphs, that is, graphs $D$, $G$, $G^*$, and $D^*$.

arc of $D$, then by $e^* = (f^-(e), f^+(e))$ we denote the corresponding arc of $D^*$. We translate the capacities in $D$ to *lengths* or *distances* in $D^*$: for an arc $e \in A(D)$, we assign in $D^*$ distances $w(e^*) = u(e)$ and $w(\text{rev}(e^*)) = u(\text{rev}(e))$.

Furthermore, in this section we assume that every multiset of arcs of $D^*$ of polynomial size has a distinct sum of capacities. This property will turn out to be very helpful in the analysis. In general, this can be obtained by slightly perturbing every capacity; however, such a step would require some technical analysis of the required precision. Luckily, as we will discuss later, in our algorithm we can mimick such a property by a number of carefully chosen tie-breaking rules.

### 5.1.1 Warm-Up: Source also Lying on the Outer Face

As a warm-up, let us consider the case when the source $s$ also lies on the outer face $f^t$. Draw a curve from $s$ to $t$ inside $f^t$: the curve partitions the arcs incident to $f^t$ in $D^*$ into two sets, $A_l^*$ and $A_r^*$, to the left and to the right of the curve, respectively. Consider a graph $D_{lr}^*$, constructed from $D^*$ by splitting $f^t$ into two vertices $f_l^t$ and $f_r^t$; the first one is incident with arcs $A_l^*$, and the second one with $A_r^*$. The critical observation is that a minimum cut between $s$ and $t$ in $D$ corresponds to a shortest path from $f_l^t$ to $f_r^t$ in $D_{lr}^*$; see Figure 5.2. This can be found in $\mathcal{O}(n \log n)$ time using Dijkstra's

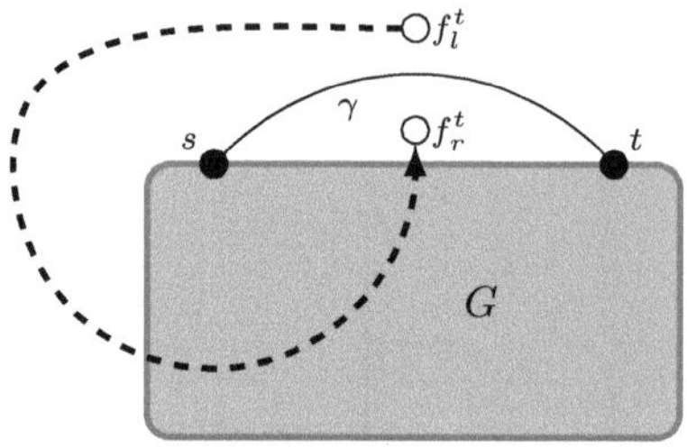


**Figure 5.2** Finding a minimum cut is equivalent to finding a shortest path in the dual in the case of $s$ and $t$ lying on a common face. The edge $\gamma$ is an auxiliary edge of infinite distance that splits the face incident with $s$ and $t$ into two faces $f_l^t$ and $f_r^t$; a shortest path in the dual graph between these faces corresponds to a minimum cut between $s$ and $t$ in the primal graph.

algorithm, or in linear time using the algorithm of Henzinger, Klein, Rao, and Subramanian [14]. Both these algorithms find not only a shortest path from $f_l^t$ to $f_r^t$, but also the minimum distances from $f_l^t$ to all the vertex of $D_{lr}^*$.

To obtain a maximum flow, we need to work a bit harder. Let $\text{dist}(f)$ be the (shortest path) distance from $f_l^t$ to $f$ in the graph $D_{lr}^*$. This distance has been computed already by the shortest path computation that identified a minimum cut. For an edge $f^-(e)f^+(e)$ of $G^*$ originating in an arc $e$ of $D$, we send a flow of size $\text{dist}(f^+(e)) - \text{dist}(f^-(e))$ along the arc $e$ (that is, if $\text{dist}(f^+(e)) < \text{dist}(f^-(e))$ we send a flow of $\text{dist}(f^-(e)) - \text{dist}(f^+(e))$ along $\text{rev}(e)$). Let $x$ be the flow defined. Observe the following:

- Since $\text{dist}(f)$ is the distance from $f_l^t$ to $f$, the flow $x$ respects capacities: $x(e) = \text{dist}(f^+(e)) - \text{dist}(f^-(e)) \leq w(e^*) = u(e)$.
- Since $G^*$ is dual to $G$, the flow $x$ respects the conservation property at every vertex except for $s$ and $t$; the latter is because in $D_{lr}^*$ the face $f^t$ has been split in two. One can view this splitting as drawing an auxiliary edge $st$, that is not present in $x$. Consequently, $x$ is an $(s,t)$-flow of value $\text{dist}(f_r^t)$.

From the above, we can obtain the following result of [13, 14]:

**Theorem 5.1.1** *Given a planar digraph $D$ with capacities and two distinguished vertices $s$ and $t$, such that $D$ can be embedded on a plane with $s$ and $t$ lying on the same face, a maximum $(s,t)$-flow and a minimum $(s,t)$-cut can be found in linear time.*

### 5.1.2 The Algorithm for the General Case

In the general case, we no longer assume that $s$ lies on the face $f^t$, and hence we cannot construct a planar digraph $D_{lr}^*$. However, we can still rely on the crucial idea of the flow construction in the previous section: a shortest paths computation from $f^t$ in $D^*$ yields a distance function $\text{dist}(\cdot)$ that can be used as a potential on faces to define a flow.

That is, similarly as in the previous case, let $\text{dist}(f)$ be the distance of $f$ from $f^t$ in $D^*$, and define a flow $x$ as before: $x(e) = \text{dist}(f^+(e)) - \text{dist}(f^-(e))$ for an arc $e$ of $D$ with $\text{dist}(f^+(e)) \geq \text{dist}(f^-(e))$. Since now $G^*$ is the actual dual of $G$ (we do not split $f^t$), with the same argument as in the previous section, $x$ is a circulation respecting capacities.

Furthermore, let $T^*$ be the computed shortest path tree in $D^*$, which is an out-branching with root $f^t$. Note that, since $T^*$ is a shortest path tree, for every arc $(f^-(e), f^+(e))$ of $T^*$ we have $\text{dist}(f^+(e)) = \text{dist}(f^-(e)) + w(e^*)$ and, consequently, the arc $e$ is saturated in the flow $x$.

We shall now treat $x$ as a flow from $s$ to $t$. Initially the amount of the flow sent from $s$ to $t$ is zero, since $x$ is a circulation at the beginning. We will gradually increase the amount of flow sent from $s$ to $t$ while maintaining the following invariant:

$$\begin{array}{c} T^* \text{ is an out-branching with root } f^t \\ \text{and all corresponding arcs of } D \text{ are saturated by } x. \end{array} \tag{5.1}$$

At every step, given $T^*$, let $T^*_G$ be the corresponding (undirected) spanning tree in $G^*$. Let $T_G$ be the set of edges of $G$ that are not crossed by the edges of $T^*_G$; then $T_G$ is a spanning tree of $G$. The tree $T_G$ contains a unique $s$-to-$t$ path $P$ in $D$. We augment $x$ by sending the maximum possible amount of flow along this path (which may be zero, if one of the arcs of $P$ is already saturated).

Then, we modify the out-branching $T^*$ as follows. Let $e$ be one of the arcs saturated on the path $P$. We would like to add the arc $e^* = (f^-(e), f^+(e))$ to $T^*$. However, then $T^*$ has one arc too many—it would no longer be an out-branching—and we need to fix it.

First, consider the case when $f^-(e)$ is a descendant of $f^+(e)$ in the out-branching $T^*$ (see Figure 5.3). Then $e^*$, together with the path from $f^+(e)$ to $f^-(e)$ in $T^*$, form a directed cycle $C^*$ in $D^*$. Note that the cycle $C^*$ has the vertex $s$ to the left and the vertex $t$ to the right. Consequently, the arcs of $D$ corresponding to the arcs of $C^*$ form an $(s,t)$-cut that, by Invariant (5.1) and the choice of $e$, consists of arcs saturated by $x$. This cut certifies that $x$ is a maximum $(s,t)$-flow and we can terminate the algorithm.

In the other case, when $f^-(e)$ is not a descendant of $f^+(e)$ in $T^*$, we replace the arc $e'$ of $T^*$ that has tail in $f^+(e)$ with the arc $e^*$; see Figure 5.4. Since $f^-(e)$ is not a descendant of $f^+(e)$, $f^-(e)$ and $f^+(e)$ lie in different connected components of $T^* \setminus \{e'\}$ and, consequently, such an operation maintains the invariant that $T^*$ is an out-branching. Furthermore, since we choose $e^*$ to be saturated, Invariant (5.1) remains satisfied.

### 5.1.3 Implementing a Single Step

It turns out that a single step of the algorithm can be implemented very efficiently, in $\mathcal{O}(\log n)$ time. However, since such an improvement belongs to the area of advanced data structures, we present here only the key ideas.

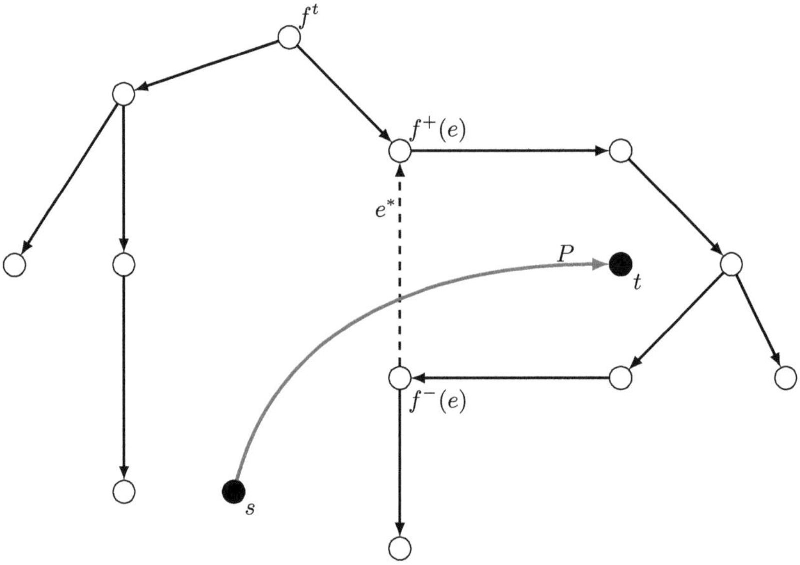


**Figure 5.3** When $f^-(e)$ is a descendant of $f^+(e)$, then the saturated arcs $e^*$ and of $T^*$ form a saturated cut certifying that the current flow is a maximum one.

Let us analyze our needs. We need to maintain the trees $T_G^*$ and $T_G$. In a single step, we first need to compute the minimum residual capacity on a single path in $T_G$, and then augment the flow $x$ by sending this capacity along the path. Then, we modify $T_G$ and $T_G^*$ by switching a constant number of edges. All these operations can be performed in amortized $\mathcal{O}(\log n)$ time per operation using one of the elaborate data structures for maintaining dynamic trees, such as the link-cut trees of Sleator and Tarjan [28]. For full details, we refer to the book of Klein and Mozes [19].

Recall that, for the sake of further analysis, we have assumed that every polynomial-size multiset of arcs of $D^*$ has unique total length. We remark here that this can be mimicked in the algorithm by careful tie-breaking in two places where the algorithm can make an arbitrary choice: when it chooses the initial shortest-path out-branching $T^*$, and when it chooses the saturated arc $e$ in each step of the algorithm.

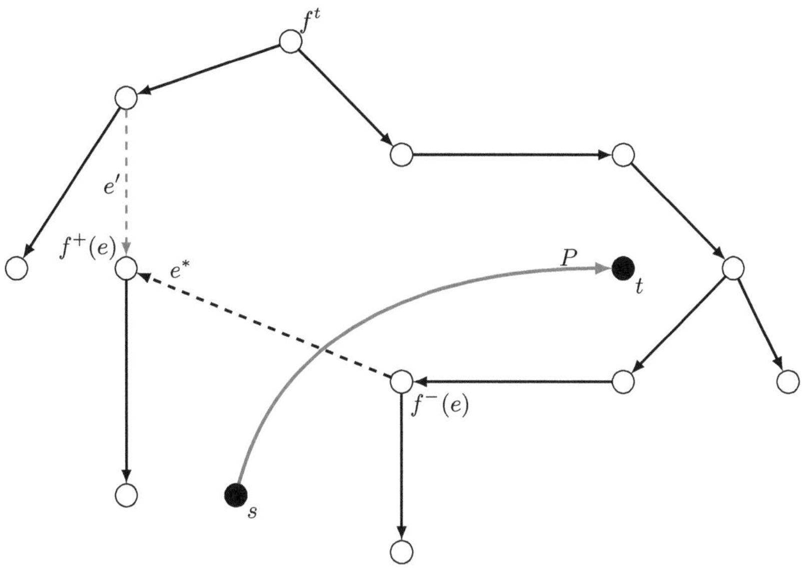


**Figure 5.4** When $f^-(e)$ is not a descendant of $f^+(e)$, we replace $e'$ with $e^*$ in the out-branching $T^*$.

### 5.1.4 Bounding the Number of Steps

In this section we focus on the following question: *how many steps can the algorithm make?* We show that every arc of $D^*$ is evicted from $T^*$ at most once, giving an $\mathcal{O}(n)$ bound on the number of steps, and, consequently, the promised $\mathcal{O}(n \log n)$ bound on the running time of the algorithm.

**Winding numbers.** For the moment, it is convenient to interpret the planar embedding of $D$ and $D^*$ as an embedding on a sphere, where $t$ is placed at the north pole and $s$ is placed at the south pole; see Figure 5.5. One can think of the choice of the initial circulation $x$ as a maximally *westbound* circulation in this embedding: we circulate as much flow as possible around the north pole in the westbound direction. Each iteration corresponds to "unwinding" some of this flow, and sending it from $s$ to $t$.

To measure this "unwinding", we need to fix some reference curve that would serve as a prime meridian between $s$ and $t$. Although any $s$-to-$t$ path $A$ in $G$ would suffice, for clarity we choose $Q$ to be the $s$-to-$t$ path in $T_G$ at the first iteration of the algorithm. In the embedding, without loss of generality we can assume that $Q$ is drawn as a straight line along the prime meridian, and we can use the notion of *west* or *east* of $Q$. To use $Q$ as a reference line, we define a *winding number* of a walk $W$ in $D^*$ as the total number of signed crossings of $Q$ by $W$. That is, we go along the walk $W$, and whenever we cross $Q$ eastbound, we add 1 to the winding number, and when

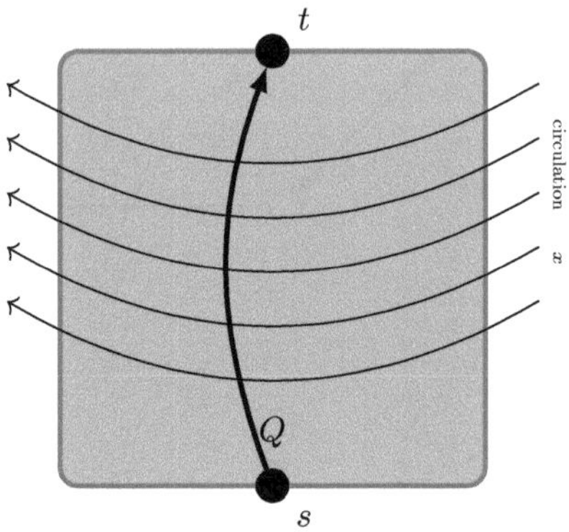


**Figure 5.5** Visualizing $t$ as the north pole, $s$ as the south pole, the reference path $Q$ as the prime meridian, and the initial circulation $x$ as a maximally westbound circulation.

we cross $Q$ westbound, we subtract 1. In the current step of the algorithm, given the current out-branching $T^*$, the **winding number** of a vertex $f$ of $D^*$ is the winding number of the unique root-to-$f$ path in $T^*$. Note that the choice of $Q$ ensures that every winding number is zero at the beginning of the algorithm. We emphasize that, although $T^*$ and $T_G$ change in the course of the algorithm, the path (meridian) $Q$ remains fixed.

The following critical observation due to Erickson [9] formalizes the "unwinding" nature of a single step of the algorithm.

**Lemma 5.1.2** *Assume that in a step of the algorithm, in an out-branching $T^*$ a new arc $e^*$ is introduced and an arc $e'$ with tail $f^+(e)$ is removed. Then, in the new out-branching, the winding number of every descendant of $f^+(e)$ is increased by one, while all other winding numbers of vertices of $D^*$ stay the same.*

**Proof:** First, note that replacing $e'$ with $e^*$ changes the root-to-$f$ paths in $T^*$ only for vertices that are descendants of $f^+(e)$ in $T^*$. Consequently, the winding number of every other vertex is not changed in the step of the algorithm.

For the affected vertices, consider the out-branching $T^*$ before the step, and let $P_-$ and $P_+$ be the root-to-$f^-(e)$ and root-to-$f^+(e)$ paths, respectively. Let $w$ be the last vertex in common of $P_-$ and $P_+$, and let $C$ be a closed walk in $D^*$ that consists of $P_-$, the arc $e^*$, and the reversed path $P_+$. Note that during the step, for every descendant $f$ of $f^+(e)$ in $T^*$ the root-to-$f$ path in $T^*$ changes in the following manner: its prefix $P_+$ is replaced by $P_-$ followed by the arc $e^*$. Consequently, the change of the winding number of the root-to-$f$ paths equals the winding number of $C$.

By the choice of $w$ and the fact that $T^*$ is an out-branching, $C$ is actually a simple cycle in $D^*$. Furthermore, by the choice of $e^*$ in the step of the algorithm, $C$ has $t$ to its left, and $s$ to its right; in other words, it is an eastbound cycle in $D^*$, and thus has winding number exactly $+1$. This finishes the proof of the claim. $\square$

Observe that Lemma 5.1.2 alone proves that the algorithm makes $\mathcal{O}(n^2)$ steps, as every winding number cannot be larger than the size of $D^*$ (every root-to-$f$ path in $T^*$ is a simple path). We now present a more elaborate argument to show a linear bound.

**Shortest paths.** Recall that the distances $\mathrm{dist}(\cdot)$ in $D^*$ have been inherited from the capacities $u(\cdot)$ in $D$ in a standard manner. Given a flow $y$ in $D$, we can consider the residual capacities $u_y := u - y$, and define accordingly the residual distances $\mathrm{dist}_y$.

If a flow $y$ respects capacities—and the flow $x$ maintained by the algorithm does respect the capacities—then no arc of $D^*$ has negative length in $\mathrm{dist}_y$. Invariant (5.1) ensures that every arc of $T^*$ has zero length in $\mathrm{dist}_x$. As a corollary, we infer that $T^*$ is a shortest-path out-branching from $f^t$ with respect to the distances $\mathrm{dist}_x$.

Consider now a flow $y$ that sends the same amount of flow from $s$ to $t$ as $x$, but sends all the flow along the path $Q$, ignoring the capacities. Although $y$ may not respect the capacities, we can still define $u_y$ and $\mathrm{dist}_y$. Readers familiar with the potential method in designing shortest path algorithms will find the following lemma immediate.

**Lemma 5.1.3** *$T^*$ is a shortest-path out-branching from $f^t$ with respect to the distances* $\mathrm{dist}_y$.

**Proof:** The crux is that a flow $y' := x - y$ (i.e., the flow $x$ that additionally sends back the flow from $t$ to $s$ along the reversed path $Q$) is a circulation (possibly not respecting the capacities).

Since $y'$ is a circulation, we can define a potential function $\zeta : V(D^*) \to \mathbb{R}$ such that $y'(e) = \zeta(f^+(e)) - \zeta(f^-(e))$ for arcs $e$ of $D$ with $\zeta(f^+(e)) \geq \zeta(f^-(e))$. Indeed, we can treat the values of $y'$ as (possibly negative) capacities of the arcs of $D$, translate them into a distance function $\mathrm{dist}'$ in $D^*$ as before, and define $\zeta(f)$ to be the minimum distance from $f^t$ to $f$ with respect to distances $\mathrm{dist}'$. A direct check shows that $\zeta$ satisfies the required properties and, since $y'$ is a circulation, *every* walk from $f^t$ to $f$ has total length exactly $\zeta(f)$.

Consequently, if a path $P$ from $f^t$ to $f$ has length $\mathrm{dist}_x(P)$ with respect to distances $\mathrm{dist}_x$, then it has length $\mathrm{dist}_x(P) - \zeta(f)$ with respect to distances $\mathrm{dist}_y$. Since $\zeta(f)$ does not depend on the path $P$, but only on the endpoint $f$, we have that $P$ is a shortest path from $f^t$ with respect to $\mathrm{dist}_x$ if and only if it is a shortest path with respect to $\mathrm{dist}_y$. The lemma follows. $\square$

However, the simplicity of the flow $y$ allows us to easily relate the distances in $\mathrm{dist}_y$ to the distances in dist that originated from the original capacities

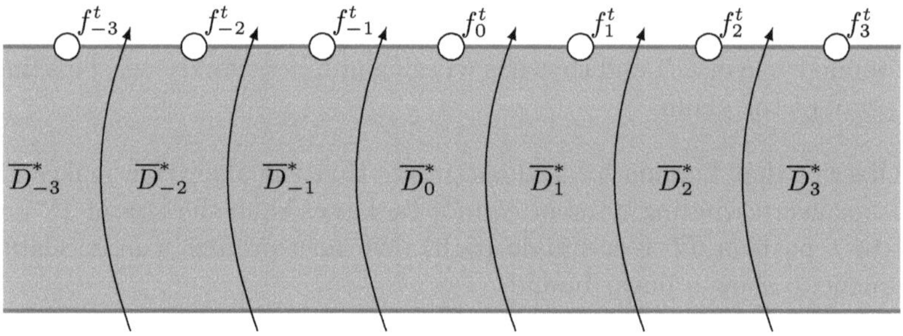


**Figure 5.6** Universal cover of $D^*$.

$u$. Indeed, if a path $P$ has winding number $i$ and the flow $y$ sends $\lambda$ amount of flow, then

$$\mathrm{dist}_y(P) = \mathrm{dist}(P) - \lambda \cdot i.$$

That is, the difference $\mathrm{dist}_y(P) - \mathrm{dist}(P)$ depends only on the winding number of $P$. Consequently, we obtain the following:

**Corollary 5.1.4** *For every vertex $f$ of $D^*$, the root-to-$f$ path in $T^*$ is the shortest $f^t$-to-$f$ path in $D^*$ among the paths that have winding numbers equal to the winding number of $f$.*

**Universal cover.** Corollary 5.1.4 speaks about a shortest path among all paths of a given winding number. A convenient way to tackle the winding number is via **universal covers.**

In our setting, consider the following *infinite* cover $\overline{D}^*$ of the graph $D^*$: we cut $D^*$ along the path $Q$ (which is a simple path in $D$, and thus corresponds to a face-edge curve of $G^*$) and glue countably many copies of $D^*$ cut along the path $Q$; see Figure 5.6. The cover $\overline{D}^*$ inherits the distances dist from $D^*$. We number the copies with integers, increasing in the eastbound direction. The $i$-th copy of $D^*$ is denoted by $\overline{D}^*_i$, the $i$-th copy of a vertex $f$ is denoted by $f_i$, etc. Since the path $Q$ leads from $s$ to $t$, the graph $\overline{D}^*$ has a single face $t^*$ corresponding to the vertex $t$ (the north pole) and a single face $s^*$ corresponding to the vertex $s$ (the south pole). As in Figure 5.6, one can view the embedding of $\overline{D}^*$ as an infinite strip, with $t^*$ and $s^*$ on its sides.

Observe that, given an integer $i$, every walk $W$ in $D^*$ can be lifted uniquely to a walk $\overline{W}_i$ in $\overline{D}^*$ that starts in the $i$-th copy of the first vertex of $W$, and then proceeds along the corresponding copies of the edges of $W$. The crux of the construction lies in the following observation: if the winding number of $W$ is $j$, then the last vertex of $\overline{W}_i$ lies in $\overline{D}^*_{i+j}$. In other words, when walking in $\overline{D}^*$, the index of the current copy reflects the winding number of the path traversed so far (when projected back to $D^*$).

Consequently, if at some iteration the root-to-$f$ path in $T^*$ has winding number $i$, then it corresponds to a path from $f^t_{-i}$ to $f_0$ and, in the other

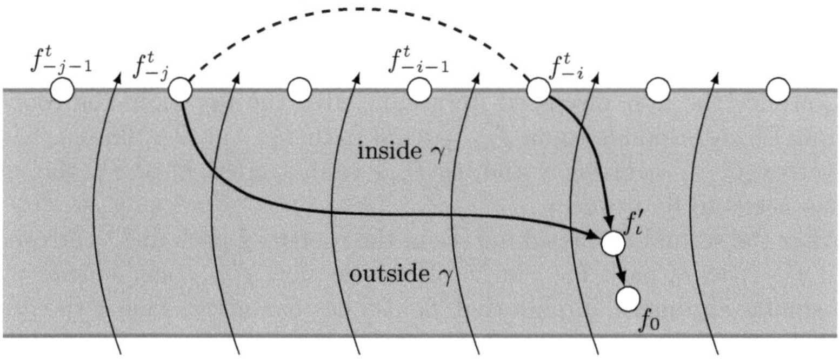


**Figure 5.7** Final argument in the proof of the linear bound on the number of steps of the algorithm: the vertex $f_0$ has to be inside and outside $\gamma$ at the same time, as it needs to be reachable both from $f^t_{-i-1}$ and $f^t_{-j-1}$ without intersecting the closed curve $\gamma$.

direction, every $f^t_{-i}$-to-$f_0$ path in $\overline{D}^*$ projects to a $f^t$-to-$f$ path in $D^*$ of winding number $i$. By Corollary 5.1.4, we have the following.

**Lemma 5.1.5** *If at some iteration the root-to-$f$ path in $T^*$ has winding number i, then it corresponds to a shortest path from $f^t_{-i}$ to $f_0$ in $\overline{D}^*$.*

Recall now that we have assumed that every nonempty multiset of arcs in $D^*$ of polynomial size has unique total cost. This implies that a shortest path from $f^t_{-i}$ to $f_0$ is unique for any vertex $f$ of $D^*$ and any $i$ bounded polynomially in the size of $D$. Furthermore, if we draw all these shortest paths for a fixed vertex $f$ and $|i| \in \mathcal{O}(n^2)$, they do not cross, that is, we obtain an in-branching in $\overline{D}^*$ with root $f_0$.

Aiming at a contradiction, consider now an arc $e$ of $D^*$ that was evicted *twice* from the tree $T^*$. Assume that the head of $e$ is $f$ and the tail is $f'$, and assume that the winding number of $f$ just before the first eviction is $i$, and before the second is $j$. Due to Lemma 5.1.2, the winding number of $f$ increased by one in both considered steps of the algorithm (when $e$ is evicted from $T^*$), which implies that $i < j$. Furthermore, it cannot hold that $i+1 = j$, as a arc from $T^*$ different than $e$ has its head in $f$ immediately after the first of the considered steps, and thus the root-to-$f$ path in $T^*$ needs to change at least once between the considered steps. Thus, we have $j - i \geq 2$.

As we discussed, the root-to-$f$ paths in $T^*$ in the two considered steps correspond to two paths in $\overline{D}^*$, one from $f^t_{-i}$ to $f_0$ (henceforth denoted $P_i$) and one from $f^t_{-j}$ to $f_0$ (henceforth denoted $P_j$). Let $P'_i$ and $P'_j$ be the paths $P_i$ and $P_j$ with the last arc removed; note that the endpoint of $P'_i$ and $P'_j$ is $f'_\iota$ for some $\iota \in \{-1, 0, 1\}$. If we connect $f^t_{-i}$ with $f^t_{-j}$ by a curve inside the face $t^*$, together with $P'_i$ and $P'_j$ we obtain a closed curve $\gamma$.

Since $P_i$ and $P_j$ are simple paths, we have that $f_0$ does not lie on $\gamma$. Since $P_i$ and $P_j$ do not intersect (by the uniqueness assumption), we can speak about vertices or arcs of $\overline{D}^*$ **inside** and **outside** the curve $\gamma$ (see Figure 5.7). The main question now is: where does the vertex $f_0$ lie: inside or outside $\gamma$?

Consider the first discussed iteration. After the iteration, the root-to-$f$ path in $T^*$ corresponds to an $f^t_{-i-1}$-to-$f_0$ path $P_{i+1}$ in $\overline{D}^*$. Since $j - i \geq 2$, the vertex $f^t_{-i-1}$ is *inside* $\gamma$ and, as $P_{i+1}$ cannot cross $P_i$ or $P_j$, the vertex $f_0$ also needs to lie *inside* $\gamma$.

After the second discussed iteration, the root-to-$f$ path in $T^*$ corresponds to an $f^t_{-j-1}$-to-$f_0$ path $P_{j+1}$ in $\overline{D}^*$. However, now $f^t_{-j-1}$ lies *outside* $\gamma$ and, by a similar argument, implies that $f_0$ also lies *outside* $\gamma$. This is the desired contradiction. Thus, every arc can be evicted from $T^*$ at most once, giving an $\mathcal{O}(n)$ bound on the number of steps and, consequently, the claimed $\mathcal{O}(n \log n)$ running time bound for the algorithm.

### 5.1.5 Perspective

We have presented an algorithm for finding maximum single-source single-sink flows in planar digraphs running in near-linear time $\mathcal{O}(n \log n)$. While this result definitely does not cover the vast literature on algorithms in planar digraphs that run in low-polynomial time, we have chosen it to present key properties of planar digraphs that allow such running times. For a more exhaustive picture of related algorithms, as well as a presentation of the above algorithm from a different angle, we refer to the free textbook of Klein and Mozes [19].

## 5.2 The Disjoint Paths Problem

Let us consider the following problem:

> $k$-Disjoint paths
> **Input:** A digraph $D$ with $k$ pairs of terminals $(s_1, t_1), \ldots, (s_k, t_k)$.
> **Question:** Does $D$ have vertex-disjoint directed paths $P_1, \ldots, P_k$ such that each $P_i$ leads from $s_i$ to $t_i$?

In the undirected setting, the fixed-parameter tractability of this problem is one of the main algorithmic corollaries of the Graph Minors project of Robertson and Seymour: they gave an algorithm for it with running time $f(k) \cdot n^3$ [24]. In directed graphs, however, the problem is completely intractable, as it is already NP-hard for $k = 2$, as shown by Fortune, Hopcroft, and Wyllie [11]. Some tractability can be retained in certain subclasses of digraphs. For instance, the problem can be solved in time $n^{k+\mathcal{O}(1)}$ in acyclic digraphs by a simple dynamic programming algorithm, but it remains W[1]-hard in this setting, as shown by Slivkins [29], which means that the existence of a fixed-parameter algorithm with running time of the form $f(k) \cdot n^{\mathcal{O}(1)}$ is

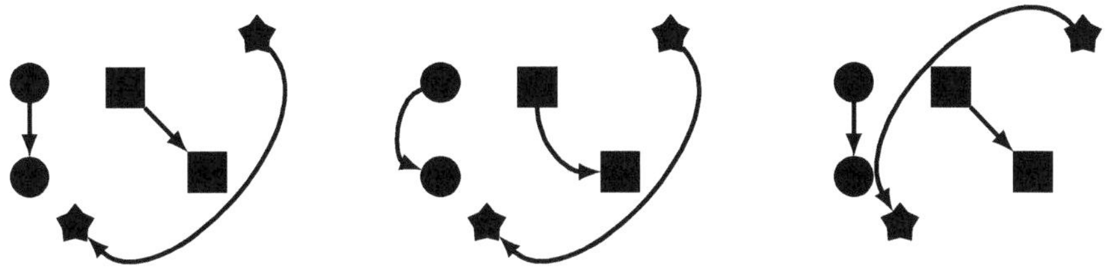

**Figure 5.8** Three solutions to $k$-DISJOINT PATHS on three terminal pairs, marked by different shapes. The first two are homotopic to each other, but not to the third.

unlikely. In this context, planar digraphs seem to be a setting where tractability is plausible, due to the inherent topological character of the $k$-DISJOINT PATHS problem. Indeed, in this section we will sketch the following result of Schrijver [26].

**Theorem 5.2.1** ([26]) *The $k$-DISJOINT PATHS problem can be solved in time $n^{\mathcal{O}(k^2)}$ when the input digraph is planar.*

Take an instance $(D, ((s_i, t_i))_{i=1,\ldots,k})$ of the problem where $D$ is planar, and suppose there is a solution $P_1, \ldots, P_k$. Imagine each path $P_i$ as a piece of string in the plane; vertex-disjointness means that the strings neither cross nor touch each other. Now abstract away the embedding of the graph and examine the picture consisting only of the strings. In the problem we do not care how long the paths are or which vertices they exactly traverse. We are content with a solution as long as the paths are vertex-disjoint and connect respective terminal pairs. Hence, we could consider two solutions as **homotopy equivalent** if one can be transformed into the other by a continuous transformation where terminals stay fixed and strings are not allowed to jump over terminals. More formally, for each $i = 1, 2 \ldots, k$, the $i$th paths in both solutions are required to be homotopic on the sphere with the other terminals pierced out; see Figure 5.8.

The intuition is that the number of such string pictures, or rather of the equivalence classes of homotopy equivalence, that can be realized in the input digraph should not be too large. If we were able to quickly search for a solution within any such class, then the whole problem could be solved efficiently. Even though this is not what will actually happen in the algorithm, as it will rely on a weaker notion than homotopy equivalence, this intuition is a good first approximation of how the problem should be attacked.

More precisely, we will consider the **homology equivalence** for solutions, because for this notion of equivalence we are able to efficiently look for a solution within a fixed equivalence class. Homotopy equivalent solutions are always homologous, but the converse direction is not necessarily true. In order to study homology equivalence, we need to introduce a certain mathematical language. In particular, we first look at the notion of **cohomology equivalence**, which intuitively is the same as homology equivalence, but in the dual digraph. While cohomology equivalence can be defined in any digraph, the translation between homology and cohomology relies on the relation between an embedded graph and its dual, and thus makes sense only for surface-embedded graphs.

### 5.2.1 Cohomology Equivalence and Feasibility

Cohomology equivalence is defined for digraphs with arcs labeled by elements of some fixed group. Let us fix $\Lambda$ to be a free group on $k$ generators $g_1, \ldots, g_k$. That is, the support of $\Lambda$ is the set of all finite words over symbols $g_1, g_1^{-1}, \ldots, g_k, g_k^{-1}$ that are **reduced**: symbols $g_i$ and $g_i^{-1}$ standing next to each other cancel out. The product of two elements $x, y$ in $\Lambda$, denoted $x \cdot y$, is defined as the concatenation of $x$ and $y$ followed by an exhaustive application of reductions as above. The neutral element of $\Lambda$ is the empty word, denoted by $\varepsilon$. For a digraph $D$, a **$\Lambda$-labeling** of $D$ is any function $\phi\colon A(D) \to \Lambda$ that assigns elements of $\Lambda$ to the arcs of $D$.

**Definition 5.2.2** *A pair of $\Lambda$-labelings $\phi$ and $\psi$ of a digraph $D$ is called* **cohomologous** *if there exists a function $\rho\colon V(D) \to \Lambda$ such that for each arc $(u, v) \in A(D)$,*

$$\psi((u, v)) = (\rho(u))^{-1} \cdot \phi((u, v)) \cdot \rho(v).$$

*We say that $\psi$ is cohomologous to $\phi$* **via** *$\rho$.*

It is clear that each $\Lambda$-labeling is cohomologous to itself by taking $\rho(u) = \varepsilon$ for each vertex $u$. Also, the relation of being cohomologous is symmetric and transitive: if $\phi$ is cohomologous to $\psi$ via $\rho$ and $\psi$ is cohomologous to $\zeta$ via $\mu$, then $\psi$ is cohomologous to $\phi$ via $\rho^{-1}$ and $\phi$ is cohomologous to $\zeta$ via $\nu$ defined as $\nu(u) = \rho(u) \cdot \mu(u)$.

Before we continue, let us discuss the intuition behind this notion. It is easy to see that a $\Lambda$-labeling $\phi$ together with $\rho\colon V(D) \to \Lambda$ uniquely define the labeling $\psi$ cohomologous to $\phi$ via $\rho$. Consider now changing the value of such $\rho$ in one vertex $u$ from $\rho(u)$ to, say, $\rho(u) \cdot g_1$, where $g_1$ is the first generator of $\Lambda$. It is easy to see that this triggers the following modification to $\psi$: for each arc $a$ with $u$ as the head, $\psi(a)$ gets right-multiplied by $g_1$, while for each arc $a'$ with $u$ as the tail, $\psi(a')$ gets left-multiplied by $g_1^{-1}$. Intuitively, this can be seen as "pulling" the group element $g_1$ over $u$ from the arcs outgoing from it to the arcs incoming to it, and $\Lambda$-labelings cohomologous to $\phi$ are exactly those that can be obtained from $\phi$ by a sequence of such "pulls". If now $D$ was the dual of some digraph $D^*$, then $u$ corresponds to some face of $D^*$, and the pull can be seen as "dragging" the generator $g_1$ over the face;

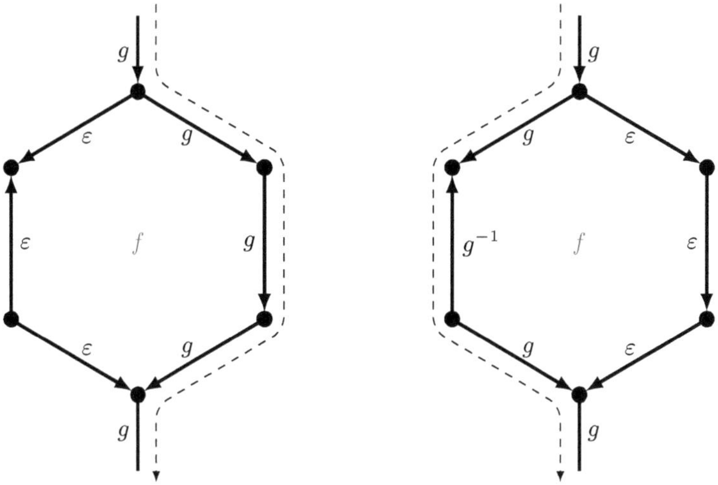


**Figure 5.9** Illustration of the "dragging" intuition. On the left panel, the values $g$ on the arcs in the dual graph correspond to a directed dashed path in the depicted primal graph. By dragging the value $g$ over the face $f$, one obtains the dashed path on the right panel; note that now the value on the middle arc is $g^{-1}$ as it is traversed in the reverse direction.

see also Figure 5.9. This models a continuous modification of a solution to the $k$-Disjoint paths problem by shifting some path by one face.

As we discussed, the main point of the approach is to show that we can efficiently search for a solution within a class of candidate solutions which are considered topologically equivalent. The main engine for this will be a polynomial-time algorithm for the Cohomology feasibility problem, defined as follows. Suppose we are given a digraph $D$ and a $\Lambda$-labeling $\phi$. Suppose further that for each arc $a \in A(D)$, we are given a set $H(a) \subseteq \Lambda$ that is **hereditary** in the following sense: if $x \in H(a)$, then every subword of the word $x$ also belongs to $H(a)$. These sets are given by an oracle, that is, we assume there is a polynomial-time algorithm that given a word $x$ and an arc $a$, checks whether $x \in H(a)$. Finally, we are also given a set $S \subseteq V(D)$ of *fixed* vertices. The goal is to determine whether there exists a $\Lambda$-labeling $\psi$ that is cohomologous to $\phi$ via $\rho$ satisfying the following conditions:

- $\psi(a) \in H(a)$ for each arc $a \in A(D)$; and
- $\rho(u) = \varepsilon$ for each vertex $u \in S$.

The intuition for the $k$-Disjoint paths problem is as follows. The digraph $D$ is the dual of the original digraph. The initial labeling $\phi$ corresponds to a crude picture of the solution, where the paths can touch or even share some subpaths, but they cannot cross in the plane. We are looking for a solution that is homologous (that is, cohomologous in the dual) and respects the disjointness conditions. By appropriately defining the dual and setting sets $H(a)$, the first property of $\psi$ will be equivalent to the disjointness of the paths. The second property will be used to ensure that the paths are not allowed to jump over terminals.

The backbone of the result of Schrijver is the following algorithmic result for COHOMOLOGY FEASIBILITY.

**Theorem 5.2.3** ([26]) *The* COHOMOLOGY FEASIBILITY *problem for free finitely generated groups is polynomial-time solvable.*

The proof of Theorem 5.2.3 is very technical, but the crux can be explained in modern terms as follows. We think of COHOMOLOGY FEASIBILITY as a constraint satisfaction problem (CSP) where vertices $u \in V(D)$ are to be labeled by elements $\rho(u)$ from the domain $\Lambda$ such that some specific constraints are satisfied. It appears that the CSP problems constructed in this way are polynomial-time solvable, because they have certain *persistence* properties. Very roughly speaking, if some part of the problem can be solved without breaking any constraint, then one can greedily fix this solution on this part; this is the same phenomenon that leads to polynomial-time solvability of the 2-SAT problem. Stating and verifying the persistence, however, requires a lot of technical work. An interesting by-product of this approach is that if the algorithm of Theorem 5.2.3 reports failure, it also provides a combinatorial certificate for the non-existence of a solution, which can be exploited algorithmically. We refer to the notes of Schrijver for details [27].

### 5.2.2 Homology Equivalence and Duals

Having understood cohomology equivalence and the COHOMOLOGY FEASIBILITY problem, we now move to homology. Suppose we are given a planar digraph $D$, say embedded on a sphere with a fixed orientation. For each arc $a \in A(D)$, let $f^-(a)$ and $f^+(a)$ be the faces incident to $a$ on the clockwise and counter-clockwise side, respectively. Similarly as in the previous section, we define the dual $D^*$ of $D$ as follows; see Fig. 5.10 for an example. The vertex set of $D^*$ is the set $F(D)$ of the faces of $D$. For each arc $a$ of $D$, we add the *dual* arc $a^* = (f^-(a), f^+(a))$ to the arc set of $D^*$. A sphere embedding of $D$ naturally gives rise to a sphere embedding of $D^*$, where each arc crosses its dual at one point.

Now homology is defined as a dual notion to cohomology, hence we are allowed to pull over faces instead of vertices.

**Definition 5.2.4** *A pair of $\Lambda$-labelings $\phi$ and $\psi$ of a sphere-embedded digraph $D$ is called* **homologous** *if there exists a function $\rho\colon F(D) \to \Lambda$ such that for each arc $a \in A(D)$,*

$$\psi(a) = (\rho(f^-(a)))^{-1} \cdot \phi(a) \cdot \rho(f^+(a)).$$

*We say that $\psi$ is homologous to $\phi$* **via** *$\rho$.*

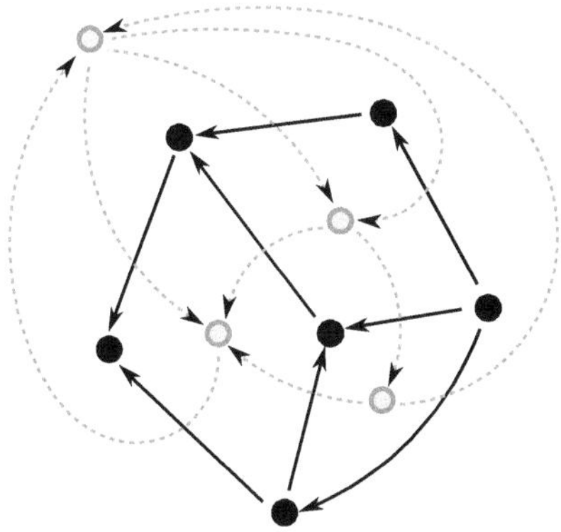

**Figure 5.10** A planar digraph (black) and its dual (grey).

Thus, the Cohomology feasibility problem in the dual $D^*$ naturally translates to the analogous problem in $D$, where we are looking for a homologous $\Lambda$-labeling satisfying certain constraints. For instance, if in the Cohomology feasibility problem on $D^*$ we put $H(a^*) = \{\varepsilon, g_1, g_2, \ldots, g_k\}$ for each arc $a \in A(D)$, then we are effectively looking for a $\Lambda$-labeling $\psi$ of $D$ homologous to the given labeling $\phi$ such that the label of each arc is either the neutral element or one of the generators. Thus, each generator $g_i$ gives rise to the arc subset $\psi^{-1}(g_i)$ such that those subsets are pairwise disjoint. By appropriately choosing $\phi$ we will be able ensure that $\psi^{-1}(g_i)$ contains a path from $s_i$ to $t_i$ and these paths are non-crossing as curves in the plane, however they may touch at vertices. To ensure real vertex-disjointness, we need to augment the dual graph slightly.

Take the dual $D^*$ of $D$. For each vertex $u \in V(D)$ and each pair of faces $f_1, f_2$ that are incident to $u$, but are not consecutive in the cyclic ordering of faces around $u$, we add arcs $(f_1, f_2)$ and $(f_2, f_1)$. These new arcs will be called **contact arcs**, and the digraph obtained from the dual by adding all contact arcs is called the **extended dual**, denoted $D^+$. Note that the extended dual is not necessarily planar, but this will not be a problem, since the algorithm for Cohomology feasibility works on any digraph.

### 5.2.3 Disjoint Paths in Directed Planar Graphs

With all the tools prepared, we are ready to encode the search for a solution within one homology type as an instance of Cohomology feasibility. We first need to describe a homology type via a representative $\Lambda$-labeling.

Let us fix an instance $(D, ((s_i, t_i))_{i=1,\ldots,k})$ of $k$-Disjoint paths. Without loss of generality we may assume that each source $s_i$ has exactly one outgoing arc and no incoming arcs, whereas each sink $t_i$ has exactly one incoming arc and no outgoing arcs. Indeed, we may add new sources and sinks as degree-one vertices adjacent only to the corresponding old sources and sinks. The following definition describes initial labelings we are interested in.

**Definition 5.2.5** *A $\Lambda$-labeling $\phi \colon V(D) \to \Lambda$ is* **consistent** *if the following conditions are satisfied:*

- *For each source $s_i$ and $t_i$, both the only arc outgoing from $s_i$ and the only arc incoming to $t_i$ are labeled by $g_i$ in $\phi$.*
- *For each non-terminal node $u$, if $a_1, \ldots, a_\ell$ are arcs incident to $u$ in the clockwise order around $u$, and $b_1, \ldots, b_\ell \in \{-1, +1\}$ are such that $a_i$ has $u$ as the head if and only if $b_i = +1$, then*

$$\phi(a_1)^{b_1} \cdot \phi(a_2)^{b_2} \cdot \ldots \cdot \phi(a_\ell)^{b_\ell} = \varepsilon.$$

Note that in the second condition it does not matter from which arc we start the enumeration of arcs incident to $u$: if the product is $\varepsilon$ for one possible starting arc, it is $\varepsilon$ for all of them.

Observe that the conditions in the above definition somewhat resemble flow conservation equations. The first condition says that each $s_i$ is a "source" of one unit of the flow of type $g_i$, and each $t_i$ is a "sink" of $g_i$. The second condition says that every nonterminal vertex satisfies a conservation property much stronger than the usual flow conservation: not only the incoming flow needs to be equal to the outgoing one, but also in some sense the flow paths cannot "cross" at a vertex.

On the other hand, the definition of a consistent labeling allows for multiple paths to be routed via the same arc, and even in the wrong direction; this corresponds to the possibility of having the label being not just a single generator. The idea is to express the requirement that this is forbidden in the language of the Cohomology feasibility problem.

Let $\phi$ be a consistent $\Lambda$-labeling of $D$. Consider now the following Cohomology feasibility instance $I(\phi)$ on the extended dual $D^+$. As the given $\Lambda$-labeling of $D^+$ we take $\phi^+$ defined as follows:

- For each arc $a$ of $D$, put $\phi^+(a^*) = \phi(a)$.
- For each contact arc $(f_1, f_2)$, say added for a vertex $u$, let $a_1, \ldots, a_p$ be the consecutive arcs incident to $u$ that we encounter when scanning the arcs around $u$ in the clockwise order, starting from $f_1$ and ending in $f_2$. Further, let $b_1, \ldots, b_p \in \{-1, +1\}$ be such that $a_i$ has $u$ as the head if and only if $b_i = +1$. Then $\phi^+((f_1, f_2)) = \prod_{i=1}^{p} \phi(a_i)^{b_i}$.

Next, we put $H(a^*) = \{\varepsilon, g_1, \ldots, g_k\}$ for each $a \in A(D)$, while for each contact arc $(f_1, f_2)$, we put $H((f_1, f_2)) = \{\varepsilon, g_1, \ldots, g_k, g_1^{-1}, \ldots, g_k^{-1}\}$. Finally, the set $S$ of forbidden vertices of $D^+$ consists of all faces of $D$ that are incident to some terminal. The following proposition, whose proof we leave as an easy exercise, explains that solving the instance $(D^+, \phi^+, H, S)$ of Cohomology feasibility immediately yields the solution to the whole problem.

**Proposition 5.2.6** *Suppose $\psi$ is a solution to the instance $(D^+, \phi^+, H, S)$. For $i = 1, 2, \ldots, k$, let $X_i$ be the set of those arcs $a$ of $D$ for which $\psi(a^*) = g_i$. Then the subgraphs induced by $X_1, \ldots, X_k$ in $D$ are pairwise vertex-disjoint and the subgraph induced by $X_i$ contains a directed path leading from $s_i$ to $t_i$.*

If now $\mathcal{P} = (P_1, \ldots, P_k)$ is a solution to the original instance, then we can define a consistent labeling $\phi_{\mathcal{P}}$ of $D$ as follows: for each arc $a \in A(D)$, put $\psi_{\mathcal{P}}(a) = g_i$ if $a$ lies on $P_i$, and put $\psi_{\mathcal{P}}(a) = \varepsilon$ if $a$ does not lie on any of the paths $P_i$. Then it is easy to see that $\psi^+_{\mathcal{P}}$ is a feasible solution to $(D^+, \phi^+, H, S)$ for any consistent labeling $\phi$ with the following property: $\phi$ is homologous to $\psi_{\mathcal{P}}$ via some $\rho$ which maps all faces of $S$ to $\varepsilon$.

Thus, we will apply the following strategy: we enumerate a small set $\Phi$ of consistent labelings of $D$ such that if there is a solution $\mathcal{P}$ to the problem, then $\Phi$ contains a labeling $\phi$ that is **well-homologous** to $\psi_{\mathcal{P}}$, that is, homologous via some $\rho$ as above. Such a set $\Phi$ will be called **exhaustive**. Then the algorithm for $k$-DISJOINT PATHS boils down to iterating through an exhaustive set $\Phi$, and for each $\phi \in \Phi$ solving the COHOMOLOGY FEASIBILITY instance $(D^+, \phi^+, H, S)$. If we obtain a solution for any of these instances, Proposition 5.2.6 gives us a way to extract a solution to the original problem. Otherwise, if none of the instances has a solution, then we can conclude that the original problem has no solution, because $\Phi$ is exhaustive.

Thus, to conclude the proof of Theorem 5.2.1, it remains to prove the following lemma. Since a complete verification requires some technical details, we give a short sketch.

**Lemma 5.2.7** *There exists an exhaustive set $\Phi$ of size $n^{\mathcal{O}(k^2)}$ which can be constructed in time $n^{\mathcal{O}(k^2)}$.*

**Proof: (Sketch)** First, we generalize the problem slightly. We will be interested in families of walks $\mathcal{P} = (P_1, \ldots, P_k)$ such that:

- Each $P_i$ is a walk connecting $s_i$ with $t_i$ in the undirected graph underlying $D$. That is, we do not require that the arcs on each $P_i$ are oriented in the direction from $s_i$ to $t_i$, and a vertex can be visited by $P_i$ multiple times.
- Walks $P_i$ are pairwise arc-disjoint and non-crossing. That is, whenever we look at two visits of a vertex $u$ by some $P_i$ and $P_j$ (possibly $i = j$), then the four arcs incident to $u$ in these two visits are not interlacing in the cyclic order of arcs around $u$.

We will call such families of walks **pre-solutions**. As before, each pre-solution $\mathcal{P}$ naturally defines a consistent labeling $\psi_{\mathcal{P}}$. We are interested in finding a small set $\Phi$ of consistent labelings of $D$ that is exhaustive for all pre-solutions: for each pre-solution $\mathcal{P}$, there exists a labeling $\phi$ in $\Phi$ that is well-homologous to $\psi_{\mathcal{P}}$ as in the definition of being exhaustive. As every solution is also a pre-solution, this suffices to prove the lemma.

The next step is to simplify the graph at hand to the case where there is exactly one vertex other than sources and sinks. However we will introduce loops (arcs with the head equal to the tail). Consider any non-loop arc $a$ such that neither the head nor the tail of $a$ is a terminal. Construct the digraph $D'$ by contracting $a$: identify the head and the tail of $a$ and remove $a$ from the graph. Every arc that is parallel to $a$, that is, has the same head and tail as $a$, or its head is the tail of $a$ and vice versa, becomes a loop at the vertex obtained by identifying the endpoints of $a$. It is easy to see that every pre-solution in $D$ can be naturally projected to a pre-solution in $D'$, and every consistent labeling $\phi'$ of $D'$ can be naturally lifted to a consistent labeling $\phi$ of $D$ so that the following holds: if $\psi_{\mathcal{P}'}$ is well-homologous to $\phi'$ in $D'$, where $\mathcal{P}'$ is the projection of $\mathcal{P}$, then $\psi_{\mathcal{P}}$ is well-homologous to $\phi$ in $D$. Thus, it suffices to find a small exhaustive set in $D'$.

Supposing the original digraph is weakly connected, we can apply this reduction exhaustively until the vertex set of $D$ consists of sources $s_i$, each with one outgoing arc, sinks $t_i$, each with one incoming arc, and one vertex $u$ that has multiple loops attached to it. The number of these loops is bounded by $m$, the number of arcs of the initial graph, which is bounded linearly in $n$.

Let $T$ be the set of all terminals. Each loop $a$ at the vertex $u$ can be associated with a partition $\{X_a, Y_a\}$ of $T$ as follows: the drawing of $a$ on the sphere divides it into two regions, and $X_a$ and $Y_a$ are the subsets of terminals contained in these regions, respectively. Two loops $a, a'$ at $u$ will be called **parallel** if the partitions $\{X_a, Y_a\}$ and $\{X_{a'}, Y_{a'}\}$ are equal; of course, being parallel is an equivalence relation. Since the drawing of the loops is non-crossing, it is not hard to convince oneself that parallel loops are homotopic in the topological space formed by the sphere on which the whole drawing is embedded, with the terminals pierced out. Therefore, the equivalence classes of the relation of being parallel really look like sets of parallel arcs: they can be ordered so that there are faces of length 2 between every two consecutive ones, as in Fig. 5.11. Each such equivalence class will be called a **bundle**. Since we do not care about the orientation of arcs in pre-solutions, we may assume that all arcs in each bundle are oriented in the same manner, as in Fig. 5.11. Formally, each 2-face between consecutive arcs of the bundle is not an oriented 2-cycle.

We may assume that there is no bundle for which the corresponding partition is $\{\emptyset, T\}$, as arcs from such a bundle can be always removed from walks of any pre-solution without any harm. Then it is not hard to prove that since the bundles are non-crossing, their number is bounded by $2|T| - 3 \leq 4k$. By somehow abusing the notation, we treat the arcs outgoing from sources and incoming to sinks also as one-arc bundles, which increases the total number of bundles to at most $6k$.

We now explain the crux of the argument. Consider any pre-solution $\mathcal{P} = (P_1, \ldots, P_k)$. Take any walk $P_i$ and let $a_1, a_2, \ldots, a_p$ be the consecutive arcs traversed by $P_i$. Further, let $B_1, B_2, \ldots, B_p$ be bundles such that $a_j \in B_j$ for each $j \in \{1, 2, \ldots, p\}$. For each $j = 1, 2, \ldots, p-1$, let us *charge* the pair $(B_j^\alpha, B_{j+1}^\beta)$, where $\alpha$ is equal to $\pm 1$ depending on whether $a_j$ is oriented in the

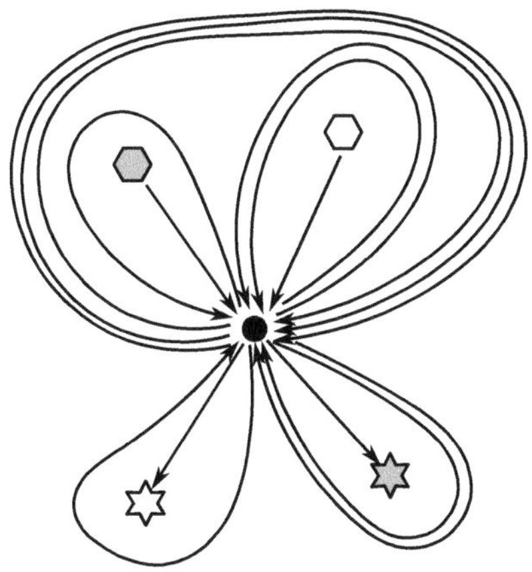

**Figure 5.11** The situation after applying the contractions. Sources are depicted as hexagons, sinks as stars, and the middle vertex is $u$. The loops at $u$ are partitioned into 5 bundles.

direction from $s_i$ to $t_i$ on $P_i$, or from $t_i$ to $s_i$; $\beta$ is defined in the same manner for $a_{j+1}$. For a pair $(A^\alpha, B^\beta)$, where $A, B$ are bundles and $\alpha, \beta \in \{-1, +1\}$, let $c(A^\alpha, B^\beta)$ be the number of times the pair $(A^\alpha, B^\beta)$ is charged; obviously $c(A^\alpha, B^\beta) \leq m$.

The following claim is now crucial: the system of numbers $c(A^\alpha, B^\beta)$ uniquely defines a pre-solution, up to being well-homologous. The proof of this fact is not hard and boils down to a careful reconstruction of a pre-solution from the numbers $c(A^\alpha, B^\beta)$, using the fact that walks in a pre-solution are pairwise non-crossing. There are at most $4 \cdot (6k)^2$ numbers $c(A^\alpha, B^\beta)$, and each of them attains a value between 0 and $m$, hence the number of pre-solutions reconstructed in this manner is bounded by $n^{\mathcal{O}(k^2)}$. □

### 5.2.4 Fixed-Parameter Algorithm: Highlights

The algorithm of Schrijver that we sketched above was later revisited by Cygan, Marx, Pilipczuk, and Pilipczuk [7], who improved the running time from the form $n^{f(k)}$ to fixed-parameter tractable. More precisely, they proved the following.

**Theorem 5.2.8** ([7]) *The $k$-Disjoint paths problem can be solved in time $2^{2^{\mathcal{O}(k^2)}} \cdot n^c$ when the input digraph is planar, where $c$ is a universal constant.*

To prove Theorem 5.2.8 it is sufficient to give an exhaustive set of size $2^{2^{\mathcal{O}(k^2)}} \cdot n^c$, as the size of an exhaustive set was the only bottleneck in the algorithm of Schrijver. Unfortunately, the number of different homology classes of solutions can be as large as $n^{\Omega(k)}$, hence we cannot hope for such a small exhaustive set in general. Therefore, Cygan *et al.* resorted to using the **irrelevant vertex technique** as follows.

Let $u$ be a non-terminal vertex of the input digraph $D$. A sequence $C_1, C_2, \ldots, C_\ell$ of vertex-disjoint cycles in $D$ is called a **concentric sequence of alternating orientation around** $u$ if the following conditions are satisfied.

- Each cycle $C_i$ separates cycles $C_j$ for $j < i$ from cycles $C_j$ for $j > i$ in the plane.
- None of the cycles passes through $u$. Moreover, for each $i = 1, 2, \ldots, k$, the region of the plane with $C_i$ cut out to which $u$ belongs does not contain any terminals.
- For even $i$, the cycle $C_i$ goes around $u$ in the clockwise direction, and for odd $i$ in the counterclockwise.

Intuitively, if a vertex $u$ can be encircled by such a concentric sequence of alternating orientation of large size, then it is "far" from terminals and not likely to be used in the solution. Cygan *et al.* formalized this intuition by proving that given the sequence is large enough, any solution can be rerouted to a solution that does not traverse $u$, and hence $u$ can be safely removed from the instance.

**Lemma 5.2.9** ([7]) *There is a function $d(k) \in 2^{\mathcal{O}(k^2)}$ such that the following holds. Suppose $u$ is a non-terminal vertex around which there exists a concentric sequence of cycles of alternating orientation of size $d(k)$. Then if there exists a solution, there is also a solution in which $u$ is not traversed by any path.*

Therefore, we can remove such vertices exhaustively from the instance. Cygan *et al.* then show that in the absence of such vertices, there is a small exhaustive set.

**Lemma 5.2.10** ([7]) *Suppose there is no vertex $u$ that satisfies the prerequisite of Lemma 5.2.9. Then there exists an exhaustive set $\Phi$ of size at most $2^{2^{\mathcal{O}(k^2)}}$ that can be constructed in time $2^{2^{\mathcal{O}(k^2)}} \cdot n^{\mathcal{O}(1)}$.*

The algorithm claimed in Theorem 5.2.8 now boils down to solving an instance of Cohomology feasibility for each labeling in $\Phi$, exactly as in the previous section. The improved bound on the size of the exhaustive set gives us the fixed-parameter tractable upper bound on the running time.

The proof of Lemma 5.2.9 in [7] is based on a complicated analysis of the interaction of a solution to the $k$-Disjoint paths with a sequence of concentric cycles of alternating orientation. It is proved that if the sequence is large enough, its cycles can be used as shortcuts for the paths in the solution, so that the paths can be rerouted simultaneously in order not to traverse vertex $u$. This argument is based on a similar analysis for the undirected case performed by Adler, Kolliopoulos, Krause, Lokshtanov, Saurabh, and Thilikos [1].

The most technically involved part of the reasoning is the proof of Lemma 5.2.10. Cygan *et al.* proved that in absence of vertices that are irrelevant in the sense of Lemma 5.2.9, the graph can be decomposed into a small number of components, each of them embedded into a disc or into a ring in the plane. The boundary of each component is well-behaved: if one travels along the boundary of, say, a disc component, then the number of times one sees an arc incoming to the component after an outgoing one, or vice versa, is bounded by a function of $k$. Having computed such a decomposition, one enumerates an exhaustive set of $\Lambda$-labelings by means of a branching procedure that "guesses" consecutive parts of a homology type. Both the depth and the degree of the search tree of this branching procedure are bounded in terms of $k$, hence the number of labelings produced by the procedure is bounded by a function of $k$.

### 5.2.5 Perspective

The fixed-parameter algorithm of [7] has double-exponential dependency on the parameter, namely $2^{2^{\mathcal{O}(k^2)}}$, which is very close to the $2^{2^{\mathcal{O}(k)}}$ dependency in the fastest known algorithm for undirected planar graphs, due to Adler, Kolliopoulos, Krause, Lokshtanov, Saurabh, and Thilikos [1]. In the undirected case, the algorithm of [1] follows a typical irrelevant vertex approach: if the treewidth of the graph is larger than $\Delta := 2^{\theta(k)}$, an irrelevant vertex inside a $\mathcal{O}(\Delta) \times \mathcal{O}(\Delta)$ grid minor is identified and deleted, whereas in the other case a standard dynamic programming routine on graphs of bounded treewidth runs in time $2^{\mathcal{O}((\Delta+k)\log\Delta)}n = 2^{2^{\mathcal{O}(k)}}n$. In [1], the authors show that this is the limit of this approach: the dependency $\Delta = 2^{\Omega(k)}$ is necessary for the irrelevant vertex rule, while multiple lower bounds for dynamic programming algorithms on graphs of bounded treewidth (see the survey of Lokshtanov, Marx, and Saurabh [20]) strongly suggest that an exponential dependency on the treewidth bound $\Delta$ is necessary for the second step of the algorithm. Hence, while there are no known lower bounds refuting the existence of an algorithm for $k$-Disjoint paths in undirected planar graphs with only single-exponential dependency on the parameter, such an algorithm would need to depart significantly from the current framework and the question of its existence remains widely open.

## 5.3 Directed Grids

In this section we discuss the Directed Grid Theorem (Theorem 9.3.14) in the context of planar digraphs. The Directed Grid Theorem is a directed

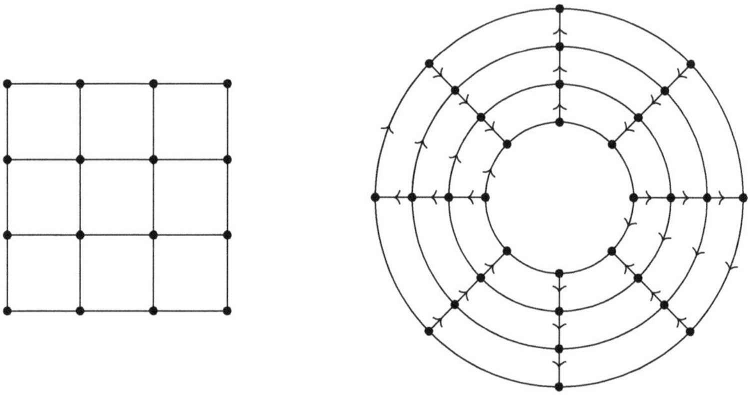

**Figure 5.12** An undirected grid and a directed cylindrical grid.

analog of the Excluded Grid Theorem for undirected graphs, asserting that any graph of sufficiently large treewidth contains a large grid as a minor.

For digraphs, we need first to replace the notion of (undirected) treewidth with **directed treewidth**, introduced by Johnson, Robertson, Seymour, and Thomas [15]. Treewidth is a graph width measure that focuses on the structure of cuts in undirected graphs; directed treewidth is a graph width measure that aims at understanding the structure of cuts in a graph — but, this time, *directed* cuts. Directed treewidth and other digraph measures will be discussed in depth in Chapter 9 and hence we refrain here from providing the (quite complex) formal definition of this measure. Instead, we will work with a dual notion of well-linked sets, introduced later in this section.

Let us move to the directed counterpart of the second ingredient of the Excluded Grid Theorem: instead of a grid, we have here the **directed cylindrical grid**. A cylindrical grid is depicted in Figure 5.12. It consists of $k$ vertex-disjoint directed cycles $C_1, C_2, \ldots, C_n$, linked by $2k$ vertex-disjoint paths $P_1, Q_1, P_2, Q_2, \ldots, P_k, Q_k$. The paths $P_i$ connect the cycles in the increasing order of indices, while the paths $Q_i$ connect the cycles in the decreasing order of indices. Along every cycle, the order of paths seen on that cycle is $P_1, Q_1, P_2, Q_2, \ldots, P_k, Q_k$. In 2001, Johnson, Robertson, Seymour, and Thomas conjectured that the cylindrical grid plays the role of the undirected grid as a canonical obstacle to small directed treewidth. This conjecture has only been recently proven by Kawarabayashi and Kreutzer [17]:

**Theorem 5.3.1** ([17]) *For every positive integer $k$ there exists an integer $f(k)$ such that every digraph of directed treewidth at least $f(k)$ contains a cylindical grid of order $k$ as a (butterfly) minor.*

A digraph $D'$ is a **butterfly minor** of $D$ if $D'$ can be obtained from $D$ by means of arc and vertex deletion, as well as contraction of arcs $e = (u, v)$ for which $e$ is the only outgoing arc of $u$ or the only ingoing arc of $v$.

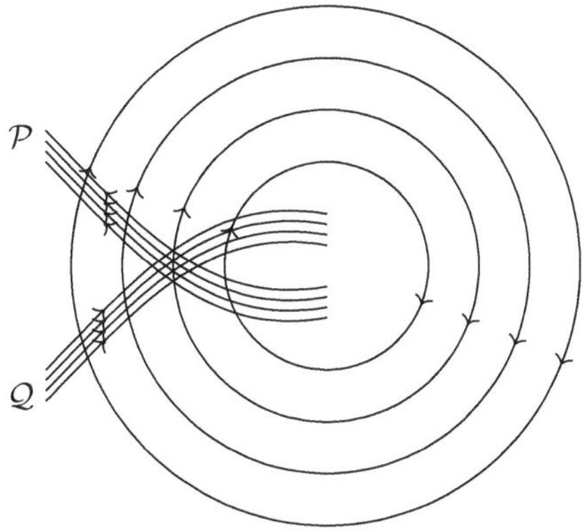


**Figure 5.13** A schematic view of a relaxed cylindrical grid of order 4. Formally, the linkages $\mathcal{P}$ and $\mathcal{Q}$ may start and end on the extreme cycles, but we will construct them as leading between the outside and inside of the concentric cycles.

In an unpublished manuscript dating back to 2001 [16], Johnson, Robertson, Seymour, and Thomas proved the theorem for planar digraphs. Our goal in this section is to sketch the proof of this theorem, following recent work of Chekuri, Ene, and Pilipczuk [5] that applied the ideas of [16] to design an approximation algorithm for the $k$-DISJOINT PATHS. We will not obtain such a rigid structure as the cylindrical grid, but a relaxed one (see Figure 5.13):

**Definition 5.3.2** *A* **relaxed cylindrical grid** *of order $k$ in a digraph $G$ embedded on a sphere consists of*

- *a sequence $C_1, C_2, \ldots, C_k$ of vertex-disjoint cycles arranged* concentrically, *that is, for every $1 \leq i < j \leq k$, the cycle $C_i$ is to the left of $C_j$;*
- *a linkage $\mathcal{P}$ of order $k$, in which every path starts at a vertex on or to the left of $C_1$, and ends at a vertex on or to the right of $C_k$;*[1]
- *a linkage $\mathcal{Q}$ of order $k$, in which every path starts at a vertex on or to the right of $C_k$ and ends at a vertex on or to the left of $C_1$.*

In other words, in a relaxed cylindrical grid we relax the requirement that the paths $P_i$ cannot intersect the paths $Q_j$ and we relax the required order in which these paths intersect every cycle $C_i$. Note that due to the spherical embedding of the graph, every path in the linkages $\mathcal{P}$ and $\mathcal{Q}$ intersects every cycle $C_i$.

Having sacrificed the rigid structure of a cylindrical grid, we will aim at a near-linear relation between the grid size and the directed treewidth. That is, our goal is to sketch the proof of the following theorem:

**Theorem 5.3.3** ([5]) *There exists a polynomial $p$ such that every planar digraph $G$ of directed treewidth $k$ contains a relaxed cylindrical grid of order at least $k/p(\log k)$.*

[1] Recall that a **linkage** is a family of pairwise vertex-disjoint paths.

In other words, the size of the obtained relaxed cylindrical grid is the same as the directed treewidth, up to polylogarithmic factors.

### 5.3.1 Well-Linked Sets

As announced at the begining of this section, instead of directed treewidth we will work with a dual notion of a **well-linked set**. To this end, let us first recall the notion of a **separation** in a digraph $D$: a pair of vertex subsets $(A, B)$ is a separation in $D$ if $A \cup B = V(D)$ and there is no arc with tail in $A \setminus B$ and head in $B \setminus A$. The **order** of the separation $(A, B)$ is $|A \cap B|$.

A set $X \subseteq V(D)$ is **node-well-linked** in $D$ if for any two disjoint subsets $A, B$ of $X$ of equal size, there exists $|A| = |B|$ vertex-disjoint paths such that every vertex of $A$ is a starting vertex of exactly one path, and every vertex of $B$ is an ending vertex of exactly one path. By relaxing vertex-disjointness to arc-disjointness we obtain the notion of an **edge-well-linked** set. By Menger's theorem, a set $X \subseteq V(D)$ is edge-well-linked if and only if for any partition $V(D) = A \uplus B$ the number of edges in $\delta^+(A)$ is at least $\min\{|X \cap A|, |X \cap B|\}$. Similarly, a set $X \subseteq V(D)$ is node-well-linked if and only if any separation $(A, B)$ of $D$ has order at least $\min\{|X \cap A|, |X \cap B|\}$. The second equivalent notion allows us to define fractional well-linkedness: for a real $\alpha \in [0, 1]$, a set $X \subseteq V(D)$ is **$\alpha$-edge-well-linked** if for every partition $V(D) = A \uplus B$ we have $|\delta^+(A)| \geq \alpha \min\{|X \cap A|, |X \cap B|\}$, while it is **$\alpha$-node-well-linked** if every separation $(A, B)$ has order at least $\alpha \min\{|X \cap A|, |X \cap B|\}$.

Observe that node-well-linkedness is stronger than edge-well-linkedness: any $\alpha$-node-well-linked set is also $\alpha$-edge-well-linked, while in the other direction we lose a factor proportial to the maximum degree: an $\alpha$-edge-well-linked set in a digraph of maximum degree $\Delta$ is $\alpha/\Delta$-node-well-linked.

Johnson, Robertson, Seymour, and Thomas [15, 16] showed that the size of the largest node-well-linked set is tightly related to directed treewidth.

**Theorem 5.3.4** ([15, 16]) *Every digraph of directed treewidth $k$ contains a node-well-linked set of size $\Omega(k)$, and, conversely, every digraph containing a node-well-linked set of size $k$ has directed treewidth $\Omega(k)$.*

A standard tool in studying well-linked sets is the following lemma that shows that one can extract an $\Omega(1)$-node-well-linked set from an $\alpha$-node-well-linked set without losing much more than necessary. This particular statement for directed graphs is due to Chekuri and Ene [4].

**Lemma 5.3.5** ([4]) *If $X$ is an $\alpha$-node-well-linked set in a digraph $D$, then there exists a set $X' \subseteq X$ of size $\Omega(\alpha|X|)$ that is $\frac{1}{32}$-node-well-linked in $D$.*

### 5.3.2 Eulerian Digraphs

A digraph is **Eulerian** if it is weakly connected and for every vertex $v$, the in-degree and the out-degree of $v$ are equal. Note that in an Eulerian digraph, the maximum in-degree is equal to the maximum out-degree. We will use the following simple "balancedness" argument in Eulerian digraphs.

**Lemma 5.3.6** *Suppose $D$ is an Eulerian digraph and $V(D) = A \uplus B$ is a partition of the vertex set of $D$. Then the number of arcs of $D$ that have tail in $A$ and head in $B$ is equal to the number of arcs of $D$ that have tail in $B$ and head in $A$.*

**Proof:** Since $D$ is Eulerian, by summing the in-degrees and the out-degrees of vertices in $A$ we infer that the number of arcs with heads in $A$ is equal to the number of arcs with tails in $A$. By subtracting the number of arcs with both heads and tails in $A$ we obtain the asserted equality. □

The critical insight of the work of Johnson, Robertson, Seymour, and Thomas [16] is that Eulerian digraphs of small maximum degree behave in some ways similarly as undirected graphs. This can be seen in the following simple lemma, used, e.g., in [5].

**Lemma 5.3.7** *Let $A, B$ be two vertex subsets in an Eulerian digraph $D$ of maximum in-degree $\Delta$, and let $k$ be a nonnegative integer. Then, if in the underlying undirected graph there exist $(\Delta+1)k+1$ vertex-disjoint undirected paths from $A$ to $B$, then in $D$ there exist $k+1$ vertex-disjoint directed paths from $A$ to $B$.*

**Proof:** If the conclusion is not true, then by Menger's theorem there exists a separation $(A', B')$ of order at most $k$ separating $A$ from $B$. That is, we have $A' \cup B' = V(D)$, $A \subseteq A'$, $B \subseteq B'$, $|A' \cap B'| \leq k$, and no arc of $D$ has its tail in $A' \setminus B'$ and its head in $B' \setminus A'$. Since there are $(\Delta+1)k+1$ undirected paths from $A$ to $B$, and only $k$ of them can pass through $A' \cap B'$, the remaining $\Delta k+1$ paths need to go via arcs connecting $A' \setminus B'$ and $B' \setminus A'$. Since there are no arcs with tail in $A' \setminus B'$ and head in $B' \setminus A'$, we infer that there are at least $\Delta k + 1$ arcs with tail in $B' \setminus A'$ and head in $A' \setminus B'$. However, $D$ contains at most $\Delta |A' \cap B'| \leq \Delta k$ arcs with tail in $A' \setminus B'$ and head in $B'$, as every such arc needs to have its head in $A' \cap B'$. This is a contradiction, as by Lemma 5.3.6, the number of arcs with tail in $A' \setminus B'$ and head in $B'$ should be equal to the number of arcs with tail in $B'$ and head in $A' \setminus B'$. □

Lemma 5.3.7 shows the surprising power of the "balancedness" argument of Lemma 5.3.6. In planar digraphs, we can exploit this argument even further, focusing on cuts represented by curves.

Let $D$ be a digraph embedded in the plane. A curve $\gamma$ on a sphere is in *general position* with respect to $D$ if $\gamma$ has a finite number of intersections with (the embedding of) $D$, and whenever $\gamma$ intersects an arc $e$ of $D$, it intersects $e$ transversally, that is, in a small neighborhood of the intersection the arc $e$ splits $\gamma$ into two parts lying on the opposite sides of $e$. Furthermore, if $\gamma$ in general position with respect to $D$ does not visit any vertex of $D$, it is called a **face-edge** curve. An *imbalance* of a curve $\gamma$ is the difference between the number of arcs of $D$ traversing $\gamma$ from left to right and the number of arcs of $D$ traversing $\gamma$ from right to left. By Lemma 5.3.6, we have the following:

**Lemma 5.3.8** *Every closed face-edge curve $\gamma$ with respect to an Eulerian digraph $D$ has zero imbalance.*

### 5.3.3 Cut-Matching Game

In Theorem 5.3.3 the given digraph $D$ may be far from being Eulerian. Quite surprisingly, we can turn $D$ into an Eulerian digraph with small maximum degree without losing much on the directed treewidth assumption. In [16], the authors obtained constant maximum degree by elaborate structural arguments, yielding a significant toll on the final relation between directed treewidth and the size of the obtained grid. The approach of [5], originating in the techniques developed in the area of routing, is conceptually cleaner, but leads only to a polylogarithmic bound on the maximum degree.

The key idea of [5] is to use the so-called **cut-matching game** to construct an embedding. To define this game, we first need to recall the notion of an *edge expansion*:

**Definition 5.3.9** *Let $G$ be an undirected multigraph. The* **edge expansion** *of a set $S \subseteq V(G)$ is defined as the ratio*

$$\frac{|\delta(S)|}{\min\{|S|, |V(G) \setminus S|\}},$$

*where $\delta(S)$ is the set of edges with exactly one endpoint in $S$. The edge expansion of a graph is the minimum edge expansion among all sets $S \subseteq V(G)$.*

*In directed (multi)graphs, the* **directed edge expansion** *is defined by replacing $\delta(S)$ with $\delta^+(S)$: the set of arcs with tails in $S$ and heads outside of $S$.*

The crucial property of digraphs with large directed edge expansion is that they contain large well-linked sets; in particular, note that the definition of edge expansion immediately implies that if $D$ has edge expansion $\alpha$, then $V(D)$ is $\alpha$-edge-well-linked.

The cut-matching game of Khandekar, Rao, and Vazirani [18] is played on an $n$-vertex multigraph $G$ for even $n$, which is initially empty. In every round, the first player, called the **Cut Player**, chooses a partition $V(G) = A \uplus B$ of the vertex set into two equal-sized sets $A$ and $B$. Then, the second player, called the **Matching Player**, chooses a perfect matching between $A$ and

$B$, which is then added to $G$ (which may lead to $G$ being a multigraph). The game ends when the graph $G$ has edge expansion at least $\alpha$, where $\alpha$ is a parameter of the game. The Cut Player wants to conclude the game as quickly as possible, while the Matching Player tries to stall the game. The main result of Khandekar, Rao, and Vazirani [18] is the following:

**Theorem 5.3.10** ([18]) *For every constant $\alpha$ there exists a randomized strategy for the Cut Player in* ***undirected*** *graphs that finishes the game in expected $\mathcal{O}(\log^2 n)$ rounds. A single move of the strategy is computable in polynomial time.*

In the directed version of the game, the matching is oriented from $A$ to $B$ (i.e., every added arc has its tail in $A$ and head in $B$), and the game ends when the directed edge expansion reaches a required threshold. This variant has been analyzed by Louis [21], who proved an analogous statement:

**Theorem 5.3.11** ([21]) *For every constant $\alpha$ there exists a randomized strategy for the Cut Player in* ***directed*** *graphs that finishes the game in expected $\mathcal{O}(\log^2 n)$ rounds. A single move of the strategy is computable in polynomial time.*

Both Theorems 5.3.10 and 5.3.11 provide a randomized strategy, with a bound on the expected number of rounds. In this description we will henceforth ignore the randomization aspect, as it is irrelevant for the purely graph theoretical existential claims.

The strength of the cut-matching game lies in the small, only polylogarithmic, number of rounds needed for the Cut Player. Consider a digraph $D$ with a node-well-linked set $X$. Without loss of generality assume that $k := |X|$ is even (we can always drop one vertex of $X$). We will play the directed version of the cut-matching game, constructing a new digraph $D_X$ with vertex set $X$. For the Matching Player, let us implement the following strategy. Given a partition $X = X_1 \uplus X_2$ into two equal-sized sets, we invoke the definition of node-well-linkedness to obtain a linkage $\mathcal{P}(X_1, X_2)$ in $D$ from $X_1$ to $X_2$. This linkage induces a directed matching between $X_1$ and $X_2$: we pair up vertices that were linked by a path in the linkage $\mathcal{P}(X_1, X_2)$. This matching is the response of the Matching Player for the partition $X = X_1 \uplus X_2$.

The result of Louis [21] shows that the Cut Player can obtain a digraph with constant directed edge expansion in $L := \mathcal{O}(\log^2 k)$ rounds. Furthermore, we can assume that whenever the Cut Player plays a partition $(X_1, X_2)$, she also immediately after plays the partition $(X_2, X_1)$. With the above behavior of the Matching Player, we obtain a final digraph $D_X$ of constant directed edge expansion and every vertex of $D_X$ has in- and out-degree $L$. This digraph $D_X$ naturally projects down to $D$, that is, we can construct a digraph $H_X$, starting from $V(H_X) = V(D)$, and for every round of the game with partition $X = X_1 \uplus X_2$ we add the linkage $\mathcal{P}(X_1, X_2)$ to $H_X$. More precisely, we add all arcs of all paths in $\mathcal{P}(X_1, X_2)$ to $H_X$, duplicating some arcs of $D$ if necessary. In this manner, every vertex of $H_X$ has equal in- and out-degree and these degrees are bounded by $2L$. Furthermore, since $D_X$ has edge expansion $\Omega(1)$, we have that $X = V(D_X)$ is $\Omega(1)$-edge-well-linked in $D_X$; by the construction of $H_X$, we have that $X$ is also $\Omega(1)$-edge-well-linked in $H_X$. By the degree bound, $X$ is $\Omega(1/L)$-node-well-linked in $H_X$. By Lemma 5.3.5, we can find a set $X' \subseteq X$ of size $\Omega(|X|/L) = \Omega(k/\log^2 k)$ that is $\frac{1}{32}$-node-well-linked in $H_X$.

The following lemma summarizes the above reasoning.

**Lemma 5.3.12** *Let $D$ be a digraph with a node-well-linked set $X$ of size $k$. Then there exists an integer $L = \mathcal{O}(\log^2 k)$ and a subgraph $H_X$ of the graph $D$ with every edge duplicated at most $L$ times, such that every vertex of $H_X$ has equal in- and out-degree, these degrees are bounded by $L$, and $X$ is $\Omega(1)$-edge-well-linked in $H_X$. Furthermore, there exists a set $X' \subseteq X$ of size $\Omega(k/\log^2 k)$ that is $1/32$-node-well-linked in $H_X$.*

Observe that if $D$ is planar, then so is the graph $H_X$ given by Lemma 5.3.12.

The final observation is that in our case it is sufficient to find a relaxed cylindrical grid in $H_X$ instead of $D$: a relaxed cylindrical grid in $H_X$ projects naturally onto $D$, and the duplicated edges do not break the structure, as we required vertex-disjointness of both the cycles $C_i$ and the linkages $\mathcal{P}$ and $\mathcal{Q}$. Thus, by losing an $\mathcal{O}(\log^2 k)$ factor in the size of the well-linked set $X$, and relaxing node-well-linkedness to $1/32$-node-well-linkedness, we can henceforth assume that the given graph $D$ is Eulerian with maximum degree $\Delta = \mathcal{O}(\log^2 k)$.

### 5.3.4 Finding a Grid in an Eulerian Digraph

In this section we show the following:

**Theorem 5.3.13** *If a planar Eulerian digraph $D$ of maximum degree $\Delta$ contains an $\alpha$-node-well-linked set $X$ of size $k$, then it also contains a relaxed cylindrical grid of order $\Omega(\alpha k/\Delta^2)$.*

As the previous section reduced us to this case with $\alpha = 1/32$ and $\Delta = \mathcal{O}(\log^2 k)$, for the proof of Theorem 5.3.3 it suffices to prove Theorem 5.3.13. We follow the exposition of [5], which builds upon the arguments of [16].

The proof of Theorem 5.3.13 heavily relies on the assumption that $D$ is Eulerian via tools introduced in Section 5.3.2. On a very high level, we start with a large *undirected* grid in $D$ and then argue about *directed* structures inside this grid using arguments relying on the assumption that $D$ is Eulerian. Let $G$ be the undirected (multi)graph underlying of $D$.

#### Obtaining an undirected grid

The first step is to obtain an undirected grid in $D$. To this end, we recall that in undirected planar graphs, a linear relation between treewidth and the largest grid minor is known [12, 25]:

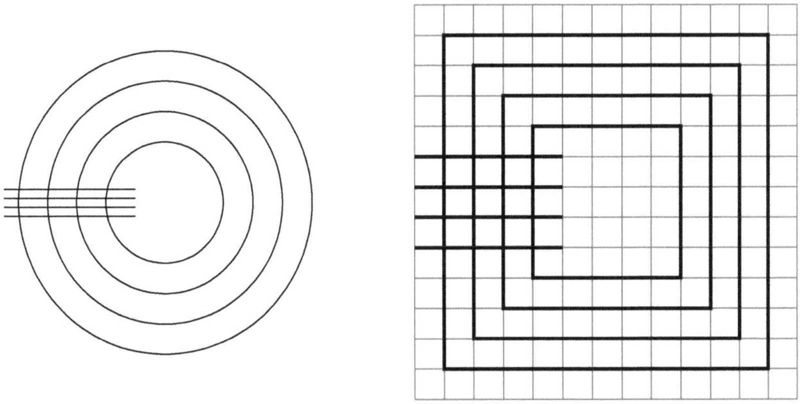

**Figure 5.14** Structure obtained by Lemma 5.3.15 and how it can be found inside a sufficiently large undirected grid minor.

**Theorem 5.3.14** ([12, 25]) *A planar undirected graph of treewidth $k$ contains a grid of sidelength $9k/2$ as a minor.*

Note that if $X$ is $\alpha$-node-well-linked in $D$, it is also $\alpha$-node-well-linked in $G$. Furthermore, a graph containing an $\alpha$-node-well-linked set of size $k$ has treewidth $\Omega(\alpha k)$. As a result we obtain the following claim; see Fig. 5.14 for a pictorial proof. Recall that in the context of undirected graphs embedded on a plane, a sequence $C_1, C_2, \ldots, C_r$ of vertex-disjoint cycles is **concentric** if each cycle $C_i$ separates the cycles $\{C_j : j < i\}$ from the cycles $\{C_j : j > i\}$.

**Lemma 5.3.15** *There exists an integer $r = \Omega(\alpha k)$ such that $G$ contains a sequence $C_1, C_2, \ldots, C_r$ of $r$ concentric cycles and a set of $r$ vertex-disjoint paths connecting $C_1$ with $C_r$.*

**Isles.** We now need the following notion. Given a vertex $v \in V(G)$, a set $Q \subseteq V(G)$ with $v \notin Q$, and an integer $\ell$, a *$(v, Q, \ell)$-isle* is a set $S \subseteq V(G)$ such that $v \in S$, $S \cap Q = \emptyset$, $G[S]$ is connected, and $|N_G(S)| \leq \ell$. In other words, $S$ is a connected part of the graph around $v$ with small boundary and separated from $Q$.

Fix $\ell = \Theta(r/\Delta) = \Theta(\alpha k/\Delta)$. The constants hidden in the $\Theta(\cdot)$ notation will be chosen in the course of the argumentation, but the reader may think that $\ell$ is a small (but constant) fraction of $r/\Delta$, in particular $\ell$ is much smaller than $r$. Pick a vertex $v_1$ on the cycle $C_1$. Since we can assume that $2\Delta < \ell = \Theta(\alpha k/\Delta)$ (as otherwise the statement of Theorem 5.3.13 is immediate), $\{v_1\}$ is a $(v_1, V(C_r), \ell)$-isle. Let $S_1$ be an inclusion-wise maximal $(v_1, V(C_r), \ell)$-isle, and let us analyze its properties.

First, since $\ell < r$ and $G$ contains $r$ vertex-disjoint paths from $C_1$ to $C_r$, the set $S_1$ cannot contain the whole cycle $C_i$ for any $i$. Since $G[S_1]$ is connected and the cycles $C_i$ are concentric, $S_1$ is disjoint from every cycle $C_i$ for $i > \ell$.

Note that $\ell$ is much smaller than $r$; the last statement shows that $S_1$ lives locally in the graph $G$, and does not go deep into the set of concentric cycles $\{C_i : 1 \leq i \leq r\}$.

Symmetrically, we pick an arbitrary vertex $v_r$ on $C_r$ and define a maximal $(v_r, V(C_1), \ell)$-isle $S_r$; we again have that $S_r$ is disjoint from cycles $C_i$ for $i \leq r - \ell$. Since we can assume that $\ell$ is much smaller than $r$, the isles $S_1$ and $S_r$ are disjoint and separated by $r - 2\ell$ cycles $C_i$.

By $N^i_G[S]$ we denote the set of vertices within distance at most $i$ from $S$ in $G$. We have that $N^2_G[S_1]$ does not intersect the cycle $C_{\ell+3}$; by the maximality of $S_1$, there are $\ell+1$ vertex-disjoint paths connecting $N^2_G[S_1]$ with $C_r$. Symmetrically, there are $\ell + 1$ vertex-disjoint paths connecting $N^2_G[S_r]$ and $C_1$. Since $\ell$ is much smaller than $r$, there are many more than $\ell$ cycles $C_i$ for $\ell + 3 \leq i \leq r - \ell - 2$; note that all these cycles are disjoint from $N^2_G[S_1 \cup S_r]$ and separate $S_1$ from $S_r$. By combining the aforementioned linkages of $\ell + 1$ paths and these cycles, we obtain that there exists a flow of size at least $\ell/3$ from $N^2_G[S_1]$ to $N^2_G[S_r]$: just treat the linkages and cycles as flow paths each carrying a flow of $1/3$ to avoid congestion, and combine the flow paths naively, following first the flow paths from $N^2_G[S_1]$ to $C_r$, then cycles $C_i$ for $\ell + 3 \leq i \leq r - \ell - 2$, and finally the flow paths from $C_1$ to $N^2_G[S_r]$. By the integrality of flows, there exists a linkage in $G$ of size at least $\ell/3$ leading from $N^2_G[S_1]$ to $N^2_G[S_r]$. A symmetric reasoning yields a linkage in $G$ of size at least $\ell/3$ leading from $N^2_G[S_r]$ to $N^2_G[S_1]$. These linkages are undirected (in $G$), but the digraph $D$ is Eulerian: by Lemma 5.3.7, in $D$, there exists a (directed) linkage $\mathcal{P}$ from $N^2_G[S_1]$ to $N^2_G[S_r]$ and a (directed) linkage $\mathcal{Q}$ from $N^2_G[S_r]$ to $N^2_G[S_1]$, both of size at least $\ell/(3(\Delta+1)) = \Theta(\alpha k/\Delta^2)$. Note that every path in $\mathcal{P}$ and $\mathcal{Q}$ intersects every cycle $C_i$ for $\ell + 3 \leq i \leq r - \ell - 2$. Figure 5.15 illustrates the structure obtained so far.

The linkages $\mathcal{P}$ and $\mathcal{Q}$ will form the desired linkages between the extreme cycles in the desired relaxed cylindrical grid. To conclude the construction, we need to show that there are $\Theta(\alpha k/\Delta^2)$ concentric directed cycles with $N^1_G[S_1]$ on one side and $N^1_G[S_r]$ on the other side, so that they intersect every path in $\mathcal{P} \cup \mathcal{Q}$. To prove their existence, we use the (undirected) cycles $C_i$.

**Cycles.** Let $D'$ be the digraph $D$ with the vertices of $N^1_G[S_1] \cup N^1_G[S_r]$ removed. Note that $D'$ is no longer Eulerian, but it is close to being Eulerian: since $S_1$ and $S_r$ are isles, we have $|N_G(S_1)|, |N_G(S_r)| \leq \ell$ and, consequently, at most $2\ell\Delta$ arcs connect $N^1_G[S_1] \cup N^1_G[S_r]$ with the vertices of $D'$.

Consider now the spherical embedding of $D$ and the naturally induced embedding of $D'$. There are two distinguished faces of the embedding of $D'$: $f_1$, which contains $S_1$ in the embedding of $D$, and $f_r$, which contains $S_r$. Let us try to find as many as possible vertex-disjoint directed cycles that have $f_1$ to the left and $f_r$ to the right.

The crucial observation is that there is a well-defined notion of a directed cycle that has $f_1$ to the left, but is as close to $f_1$ as possible, in the sense that it has as few faces of $D'$ to the left as possible. To see this, consider the

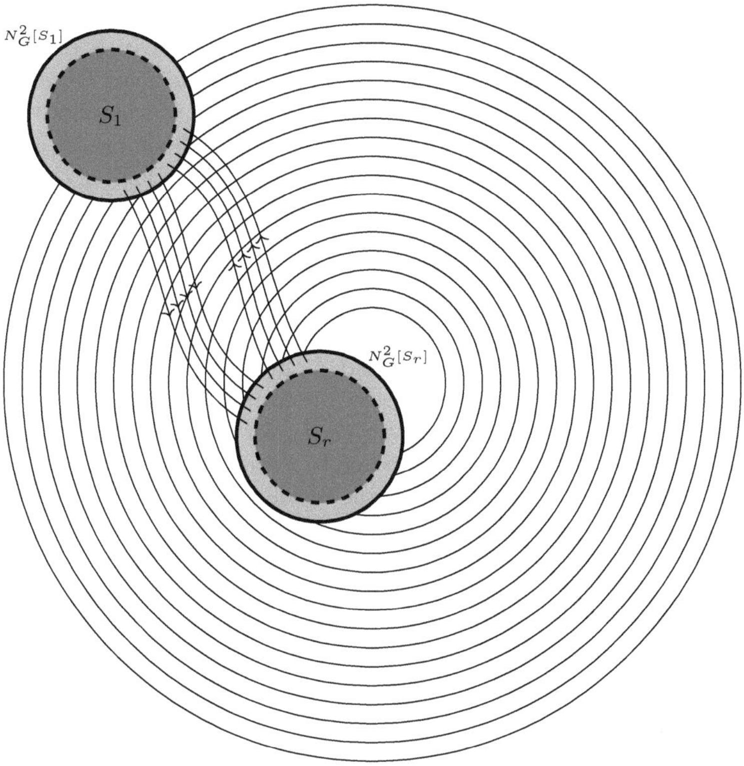


**Figure 5.15** Structure obtained from isles $S_1$ and $S_r$. To finish the construction, we lack sufficiently many concentric *directed* cycles separating $S_1$ from $S_r$, but we have many *undirected* ones.

following procedure: mark $f_1$ and every face of $D'$ that is reachable from $f_1$ via face-edge curves in $D'$ that are crossed by the arcs of $D'$ only from left to right. If such a curve $\gamma$ reaches a face $f$, then $\gamma$ certifies that $f$ needs to be to the left of any cycle in $D'$ that keeps $f_1$ to the left; in particular, if $f_r$ is marked, the corresponding curve shows that there is no cycle in $D'$ that keeps $f_1$ to the left and $f_r$ to the right. In the other direction, it is easy to see that the boundary of the region of unmarked faces that contain $f_r$ (if $f_r$ is unmarked) forms the desired directed cycle.

By iterating the above argument, we can obtain the following claim:

**Lemma 5.3.16** ([5]) *For any integer $t$, in $D'$ there exists either:*

1. *a family of vertex-disjoint cycles $D_1, D_2, \dots, D_t$, each having $f_1$ to the left and $f_r$ to the right;*
2. *a curve $\gamma$ in general position with respect to $D'$ that starts in $f_1$, ends in $f_r$, passes through at most $t$ vertices of $D'$, and such that every arc of $D'$ crossing $\gamma$ crosses it from left to right.*

We pick $t = \Theta(\alpha k/\Delta^2)$ and apply Lemma 5.3.16: once directly, and once with the roles of $f_1$ and $f_r$ swapped. If any of the application resulted in a family of $t$ directed cycles, these cycles, together with linkages $\mathcal{P}$ and $\mathcal{Q}$, form the desired relaxed grid. Thus, we are left with the case when both applications returned a curve; note that we may assume without loss of generality that each of these curves is without self-intersections. By joining these curves together inside $f_1$ and $f_r$, we obtain a closed curve $\gamma_0$ in general position with respect to $D'$ that intersects at most $2t$ vertices and every arc crossing $\gamma_0$ crosses it from left to right. We modify $\gamma_0$ slightly as follows: whenever $\gamma_0$ visits a vertex $v$ we move it a little so that it intersects a number of arcs incident with $v$ instead. In this manner, the obtained curve $\gamma$ is a closed face-edge curve in $D'$ that visits both $f_1$ and $f_r$, does not visit any vertex of $D'$, and at most $2t\Delta$ arcs intersecting $\gamma$ cross it from right to left.

However, $\gamma$ needs to cross every cycle $C_i$ for $\ell + 3 \leq i \leq r - \ell - 2$; by taking $\ell$ to be sufficiently small compared to $r$, there are at least $r/2 = \Theta(\alpha k)$ such cycles. Since $2t\Delta = \Theta(\alpha k/\Delta)$, the absolute value of the imbalance of the curve $\gamma$ can be assumed to be at least $r/4$.

Consider now a digraph $D''$, obtained similarly as $D'$ from $D$, but instead of removing $N_G^1[S_1]$, we contract it onto a single vertex $w_1$, similarly we also contract $N_G^1[S_r]$ onto a new vertex $w_r$. Any loops thus created at $w_1$ or $w_r$ are removed. Note that $D''$ remains Eulerian and the degree of $w_1$ and $w_r$ is at most $\ell\Delta$ in $D''$. Furthermore, by slight modifications of $\gamma$ inside $f_1$ and $f_r$, we may assume that $\gamma$ is in general position with respect to $D''$ as well, visits neither $w_1$ nor $w_r$, and crosses every arc incident to these two vertices at most once (they are drawn inside $f_1$ and $f_r$, where we can freely manipulate $\gamma$). However, now $\gamma$ is a closed curve in general position with respect to an Eulerian digraph $D''$, and thus has zero imbalance. Recall that $D'$ and $D''$ differ on at most $2\ell\Delta$ edges, each crossed by $\gamma$ at most once. By picking a sufficiently small constant in the definition of $\ell = \Theta(r/\Delta)$ we obtain $2\ell\Delta < r/4$, yielding a contradiction.

Thus, at least one application of Lemma 5.3.16 resulted in a family of cycles, giving the final ingredient of the relaxed cylindrical grid, and concluding the proofs of Theorems 5.3.13 and 5.3.3.

### 5.3.5 Perspective

Theorem 5.3.3 shows that if one relaxes the structure of the cylindrical grid to allow intersections of the radial linkages, we can obtain good (up to polylogarithmic factors) dependency between directed treewidth and the size of the grid. This resembles the situation from undirected graphs, where linear dependency between treewidth and the size of largest grid minor gave rise to multiple algorithmic applications through the theory of bidimensionality [10].

In the context of routing, the above theorem fits into a more general approach for designing approximation algorithms for the $k$-Disjoint paths problem, pioneered by Chekuri, Khanna, and Shepherd [3]. This approach turned out to be very successful in the context of undirected graphs, leading to a poly-logarithmic approximation with congestion 2 for the edge-disjoint version of $k$-DISJOINT PATHS by Chuzhoy and Li [6].

The first step is to decompose the input instance into a number of subinstances where in each subinstance the set of terminals is (fractionally) well-linked. This well-linkedness in turn allows us to reason about the existence of a good **crossbar**, a grid-like routing structure. The well-linkedness also implies the existence of a large flow between the terminals and the crossbar; an approximate solution is formed by these flow paths, joined together inside the crossbar in a way respecting the terminal pairs.

The crucial ingredient in this approach is to prove the existence of a crossbar in the presence of a large well-linked set; if the approximation factor is to be poly-logarithmic, the ratio between the size of the well-linked set and the size of the crossbar needs to be poly-logarithmic as well. The presented theorem serves as such an ingredient in the context of planar digraphs.

Apart from the context of routing [5], we do not know any other applications of Theorem 5.3.3. Furthermore, a number of questions regarding generalizations appear:

1. Can we reduce the upper bound on the maximum degree to constant, as opposed to poly-logarithmic, with only a poly-logarithmic loss on the directed treewidth? The cut-matching game approach has an inherent $\mathcal{O}(\log^2 k)$ factor due to the number of rounds, while the arguments of [16] lead to a maximum degree of 6, but give a much worse parameter dependency.
2. Can we conduct the final part of the proof of [16], that is, obtain a regular cylindrical grid from a relaxed one, with only a poly-logarithmic loss on the size? Such an improvement may be needed if one wants to lower the allowed congestion in the approximation algorithm of [5].
3. Can we generalize these developments to other sparse graph classes? In undirected graphs, many results in the theory of bidimensionality hold in apex-minor-free or general proper minor-closed graph classes.

We remark here that the first part of the proof, which leads to an Eulerian digraph with a poly-logarithmic maximum degree and is based on the cut-matching game, works in general graphs; that is, this part does not require the planarity assumption. On the other hand, the second part of the reasoning seems to crucially depend on the topological structure of the digraph.

The existence of a large (relaxed) directed grid in the presence of a large well-linked set is also related to the Erdős–Pósa property of cycles. In undirected graphs, the classic result of Erdős and Pósa [8] asserts that if a graph does not contain $k$ vertex-disjoint cycles, it admits a set of $\mathcal{O}(k \log k)$ vertices that intersect every cycle. For directed graphs, a similar relation has been conjectured by Younger [30]; the conjecture was confirmed in 1996 by Reed, Robertson, Seymour, and Thomas [23]. However, the relation between the number of vertex-disjoint cycles and the size of the hitting set is not explicit in [23] and at least exponential. Improving this relation to, say, polynomial remains widely open. More discussion on various aspects of the Erdős–Pósa property in directed graphs can be found in Section 9.5.3

Apart from the above questions, a number of very important questions remain regarding the Directed Grid Theorem in the general setting, where the proof of Kawarabayashi and Kreutzer [17] gives only a very weak parameter dependency. A discussion on these issues can be found in Chapter 9.

## References


1. I. Adler, S.G. Kolliopoulos, P.K. Krause, D. Lokshtanov, S. Saurabh, and D. M. Thilikos. Irrelevant vertices for the planar Disjoint Paths problem. *J. Combin. Theory Ser. B*, 122:815–843, 2017.
2. G. Borradaile and P.N. Klein. An $O(n \log n)$ algorithm for maximum $st$-flow in a directed planar graph. *J. Assoc. Comput. Mach.*, 56(2), 2009.
3. C. Chekuri and A. Ene. The all-or-nothing flow problem in directed graphs with symmetric demand pairs. *Math. Program.*, 154(1-2):249–272, 2015.
4. C. Chekuri, A. Ene, and M. Pilipczuk. Constant congestion routing of symmetric demands in planar directed graphs. In I. Chatzigiannakis, M. Mitzenmacher, Y. Rabani, and D. Sangiorgi, editors, *ICALP 2016: The 43rd International Colloquium on Automata, Languages, and Programming*, volume 55 of *LIPIcs*, pages 7:1–7:14, 2016.
5. C. Chekuri, S. Khanna, and F.B. Shepherd. Multicommodity flow, well-linked terminals, and routing problems. In H. N. Gabow and R. Fagin, editors, *STOC 2005: 37th Annual ACM Symposium on Theory of Computing*, pages 183–192. ACM, 2005.
6. J. Chuzhoy and S. Li. A polylogarithmic approximation algorithm for edge-disjoint paths with congestion 2. *J. Assoc. Comput. Mach.*, 63(5):45:1–45:51, 2016.
7. M. Cygan, D. Marx, M. Pilipczuk, and M. Pilipczuk. The planar directed $k$-Vertex-Disjoint Paths problem is fixed-parameter tractable. In *FOCS 2013: 54th Annual IEEE Symposium on Foundations of Computer Science*, pages 197–206. IEEE Computer Society, 2013.
8. P. Erdős and L. Pósa. On independent circuits contained in a graph. *Canadian J. Math.*, 17:347–352, 1965.
9. J. Erickson. Maximum flows and parametric shortest paths in planar graphs. In M. Charikar, editor, *SODA 2010: 21st Annual ACM-SIAM Symposium on Discrete Algorithms*, pages 794–804. SIAM, 2010.
10. F.V. Fomin, E.D. Demaine, M. Hajiaghayi, and D.M. Thilikos. Bidimensionality. In *Encyclopedia of Algorithms*, pages 203–207. Springer, 2016.
11. S. Fortune, J.E. Hopcroft, and J. Wyllie. The directed subgraph homeomorphism problem. *Theor. Comput. Sci.*, 10:111–121, 1980.
12. Q.-P. Gu and H. Tamaki. Improved bounds on the planar branchwidth with respect to the largest grid minor size. *Algorithmica*, 64(3):416–453, 2012.
13. R. Hassin. Maximum flow in $(s,t)$ planar networks. *Inform. Process. Lett.*, 13(3):107, 1981.
14. M.R. Henzinger, P.N. Klein, S. Rao, and S. Subramanian. Faster shortest-path algorithms for planar graphs. *J. Comput. Syst. Sci.*, 55(1):3–23, 1997.
15. T. Johnson, N. Robertson, P.D. Seymour, and R. Thomas. Directed Tree-Width. *J. Combin. Theory Ser. B*, 82(1):138–154, 2001.

16. T. Johnson, N. Robertson, P. D. Seymour, and R. Thomas. Excluding a grid minor in planar digraphs. *CoRR*, arXiv:1510.00473, 2015.
17. K. Kawarabayashi and S. Kreutzer. The directed grid theorem. In *STOC 2015: 47th Annual ACM on Symposium on Theory of Computing*, pages 655–664. ACM, 2015.
18. R. Khandekar, S. Rao, and U.V. Vazirani. Graph partitioning using single commodity flows. *J. ACM*, 56(4), 2009.
19. P.N. Klein and S. Mozes. *Optimization Algorithms for Planar Graphs*. Available at http://planarity.org, 2016.
20. D. Lokshtanov, D. Marx, and S. Saurabh. Lower bounds based on the exponential time hypothesis. *Bulletin of the EATCS*, 105:41–72, 2011.
21. A. Louis. Cut-matching games on directed graphs. *CoRR*, arXiv:1010.1047, 2010.
22. B. Mohar and C. Thomassen. *Graphs on Surfaces*. Johns Hopkins University Press, 2001.
23. B. Reed, N. Robertson, P.D. Seymour, and R. Thomas. Packing directed circuits. *Combinatorica*, 16(4):535–554, 1996.
24. N. Robertson and P.D. Seymour. Graph minors. XIII: The disjoint paths problem. *J. Combin. Theory Ser. B*, 63:65–110, 1995.
25. N. Robertson, P.D. Seymour, and R. Thomas. Quickly excluding a planar graph. *J. Combin. Theory Ser. B*, 62(2):323–348, 1994.
26. A. Schrijver. Finding $k$ disjoint paths in a directed planar graph. *SIAM J. Comput.*, 23(4):780–788, 1994.
27. A. Schrijver. Free partially commutative groups, cohomology, and paths and circuits in directed graphs on surfaces. Preprint, available at http://homepages.cwi.nl/lex/files/dppg.pdf.
28. D.D. Sleator and R.E. Tarjan. A data structure for dynamic trees. *J. Comput. Syst. Sci.*, 26(3):362–391, 1983.
29. A. Slivkins. Parameterized tractability of edge-disjoint paths on directed acyclic graphs. In *ESA 2003: 11th Annual Eur. Symp. Algorithms*, volume 2832 of *Lect. Notes Comput. Sci.*, pages 482–493, 2003.
30. D.H. Younger. Graphs with interlinked directed circuits,. In *Midwest symposium on circuit theory 2*, pages XVI 2.1–XVI 2.7, 1973.

# 6. Locally Semicomplete Digraphs and Generalizations

**Jørgen Bang-Jensen**

Locally semicomplete digraphs were introduced by Bang-Jensen [9], who discovered them by coincidence in 1988 while working on homomorphism problems for semicomplete digraphs. He wanted to find another class of digraphs for which the so-called sub-indicator construction (see the paper [21] by Bang-Jensen, MacGillivray and Hell) would be useful. This prompted him to look at digraphs where the out-neighbourhoods and the in-neighbourhoods were semicomplete. From then it took only curiosity to start the research on locally semicomplete digraphs. The research on these digraphs and some of their generalizations has been very successful, leading to several doctoral theses such as those by Guo, Huang and Tewes [43, 55, 66] and more than 100 research papers on these topics.

The research, which we will shed some light on in this chapter, has revealed that locally semicomplete digraphs share a large number of properties with semicomplete digraphs. Sometimes the proofs of results for locally semicomplete digraphs are almost identical to those for semicomplete digraphs (e.g. for existence of a Hamiltonian path or cycle) but, not surprisingly, they are often more complicated. In that case one may often benefit from a classification theorem that we describe in Section 6.6. This allows one to concentrate only on those locally semicomplete digraphs that are not semicomplete (when trying to extend a result from that class). The classification theorem which we prove in Section 6.6 implies that locally semicomplete digraphs that are not semicomplete can be divided into two classes: round decomposable locally semicomplete digraphs, which are very well structured and allow for easy solution of several hard problems, and the so-called evil locally semicomplete digraphs , which are much closer to semicomplete digraphs when it comes to the difficulty of many problems. The chapter is organized as follows: In Section 6.1 we give some necessary new definitions. In Section 6.2 we introduce an important property, called the path merging property, which is already sufficient to imply hamiltonicity for strong digraphs with no cut-vertex. In Section 6.3 we describe the structure of non-strong locally (in-)semicomplete digraphs. In Section 6.4 we consider the existence of Hamiltonian paths and cycles in locally semicomplete digraphs and some generalizations. Section 6.5 intro-

J. Bang-Jensen (✉)
Department of Mathematics and Computer Science,
University of Southern Denmark, Campusvej 55, 5230 Odense M, Denmark
e-mail: jbj@imada.sdu.dk

 
J. Bang-Jensen and G. Gutin (eds.), *Classes of Directed Graphs*, Springer Monographs in Mathematics, https://doi.org/10.1007/978-3-319-71840-8_6

duces round (decomposable) digraphs which form an important subclass of locally semicomplete digraphs. In Section 6.6 we give a classification of locally semicomplete digraphs that has been very useful in proofs of a number of results for the class. Section 6.7 deals with Hamiltonian connectivity and Section 6.8 with pancyclicity of locally semicomplete digraphs. In Section 6.9 we discuss various results on cycle factors with a given number of cycles. Then in Sections 6.10 and 6.11 we describe a number of results on weak linkages, respectively linkages in locally semicomplete digraphs. Section 6.12 deals with results concerning arc-disjoint spanning subdigraphs, Section 6.13 deals with kernels and Section 6.14 with feedback sets in locally semicomplete digraphs. Section 6.15 is about orientations of locally semicomplete digraphs, that is, oriented graphs that we obtain by deleting one arc from every 2-cycle. In Section 6.16 we present a generalization of round digraphs and finally in Section 6.17 we cover a few more topics on locally semicomplete digraphs.

## 6.1 New Definitions

For notational simplicity we will use $P_n$ and $C_n$ for the path, respectively cycle on $n$ vertices instead of $\overrightarrow{P}_n$, $\overrightarrow{C}_n$. Also every path and cycle in this chapter will be directed.

A digraph $D$ is **locally in-semicomplete** (**locally out-semicomplete**) if the induced subdigraph $D[N^-(v)]$ $(D[N^+(v)])$ is semicomplete for every vertex $v \in V(D)$. Clearly, the converse of a locally in-semicomplete digraph is a locally out-semicomplete digraph and vice versa. A digraph $D$ is **locally semicomplete** if it is both locally in- and locally out-semicomplete. Clearly every semicomplete digraph is locally semicomplete.

A locally in-semicomplete digraph with no cycle of length 2 is a **locally in-tournament digraph**. Similarly, one can define **locally out-tournament digraphs** and **locally tournament digraphs**. For convenience, we will often refer to locally tournament digraphs as **local tournaments** and to locally in-tournament (out-tournament) digraphs as **local in-tournaments** (**local out-tournaments**). As every result for local in-tournaments has an analogue for local out-tournaments by considering the converse digraphs, we will often just state a result for one of these classes.

Recall that an **extension** of a digraph $D$ is any digraph $D'$ which can be obtained by substituting an independent set $I_v$ for each vertex $v \in V(D)$. Thus $D' = D[I_{v_1}, I_{v_2}, \ldots, I_{v_n}]$, where $n = |V(D)|$. This definition also applies to classes of digraphs, e.g. an **extended locally in-semicomplete digraph** is any digraph $D'$ which is an extension of some locally in-semicomplete digraph $D$.

The **second power**[1] **of a cycle** $C_n$, denoted by $C_n^2$, is the digraph obtained from $C_n$ by adding the arcs $\{v_iv_{i+2} : 1 \leq i \leq n\}$, where $C_n = v_1v_2 \ldots v_{n-1}v_nv_1$ and the subscripts are modulo $n$.

[1] Also called the **square**.

## 6.2 The Path Merging Property

The sorting problem for a set $S = \{x_1, x_2, \dots, x_n\}$ of distinct integers[2] is a special case of the Hamiltonian path problem for tournaments: Define $T(S)$ to be the tournament on $n$ vertices $v_1, \dots, v_n$ and arcs $\{v_iv_j | i \neq j \text{ and } v_i < v_j\}$. Clearly $T(S)$ is a transitive tournament. Now the unique Hamiltonian path $P = v_{i_1}v_{i_2}\dots v_{i_n}$ in $T(S)$ corresponds to the sorted order $x_{i_1} < x_{i_2} < \dots < x_{i_n}$ on $S$. Conversely, one can generalize several sorting algorithms, including the merge-sort algorithm, to algorithms for finding Hamiltonian paths in arbitrary tournaments. This inspired Bang-Jensen [4] to introduce the following notion for digraphs.

A digraph $D$ is **path-mergeable** if for every choice of vertices $x, y \in V(D)$ and every pair of internally disjoint $(x, y)$-paths $P, P'$ there exists an $(x, y)$-path $P^*$ in $D$ such that $V(P^*) = V(P) \cup V(P')$.

**Theorem 6.2.1** ([4]) *A digraph $D = (V, A)$ is path-mergeable if and only if for every pair of distinct vertices $x, y \in V(D)$ and every pair of internally disjoint $(x, y)$-paths $P = xx_1 \dots x_ry$, $P' = xy_1 \dots y_sy$, $r, s \geq 1$ in $D$, either there exists an $i \in \{1, \dots, r\}$ such that $x_iy_1 \in A$, or there exists a $j \in [s]$ such that $y_jx_1 \in A$.*

**Proof:** We prove the 'only if' statement by induction on $r+s$. It is obvious for $r = s = 1$, so suppose that $r + s \geq 3$. If there is no arc between $\{x_1, \dots, x_r\}$ and $\{y_1, \dots, y_s\}$, then clearly $P, P'$ cannot be merged into one path. Hence we may assume without loss of generality that there is an arc $x_iy_j$ for some $i, j$, $1 \leq i \leq r, 1 \leq j \leq s$. If $j = 1$, then the claim follows. Otherwise apply induction to the paths $P[x, x_i]y_j$, $xP'[y_1, y_j]$.

The proof of the 'if' statement is left to the reader. It is similar to the proof of Proposition 6.2.3 below. □

Bang-Jensen showed how to use Theorem 6.2.1 to obtain the following.

**Theorem 6.2.2** ([4]) *Path-mergeable digraphs can be recognized in polynomial time.* ◇

The next result shows that if a digraph is path-mergeable, then the merging of paths can always be done in a particularly nice way. This illustrates the similarity between path-merging and the merging subroutine in the well-known **merge sort** algorithm, see e.g. [36, Chapter 2].

**Proposition 6.2.3** *Let $D = (V, A)$ be a digraph which is path-mergeable and let $P = xx_1 \dots x_ry$, $P' = xy_1 \dots y_sy$, $r, s \geq 0$ be internally disjoint $(x, y)$-paths in $D$. The paths $P$ and $P'$ can be merged into one $(x, y)$-path $P^*$ such*

---

[2] The reduction also works when some numbers may be equal, in which case we obtain a semicomplete digraph instead.

*that vertices from P (respectively, P′) remain in the same order as on that path. Furthermore, the merging can be done in at most* $2(r+s)$ *steps.*

**Proof:** We prove the result by induction on $r+s$. It is obvious if $r=0$ or $s=0$, so suppose that $r,s \geq 1$. By Theorem 6.2.1 there exists an $i$ such that either $x_iy_1$ or $y_ix_1$ is an arc. By scanning both paths forward one arc at a time, we can find $i$ in at most $2i$ steps; suppose without loss of generality $x_iy_1 \in A$. By applying the induction hypothesis to the paths $P[x_i,x_r]y$ and $x_iP'[y_1,y_s]y$, we see that we can merge them into a single path $Q$ in the required order-preserving way in at most $2(r+s-i)$ steps. The required path $P^*$ is obtained by concatenating the paths $xP[x_1,x_i]$ and $Q$, and we have found it in at most $2(r+s)$ steps, as required. □

The path-mergeability can be generalized in a natural way as follows. A digraph $D$ is **in-path-mergeable (out-path-mergeable)** if, for every vertex $y \in V(D)$ and every pair $P,Q$ of internally disjoint paths with common terminal (initial) vertex $y$, there is a path $R$ such that $V(R)=V(P)\cup V(Q)$, the path $R$ terminates (starts) at $y$ and starts (terminates) at a vertex which is the initial (terminal) vertex of either $P$ or $Q$ (or, possibly, both). Observe that, in this definition, the initial vertices of the paths $P$ and $Q$ may coincide. Therefore, every in-path-mergeable (out-path-mergeable) digraph is path-mergeable. However, it is easy to see that not every path-mergeable digraph is in-path-mergeable. Clearly, every in-path-mergeable (out-path-mergeable) digraph is locally in-semicomplete (locally out-semicomplete). The converse is also true (hence this is another way of characterizing locally in-semicomplete digraphs).

**Proposition 6.2.4** ([16]) *Every locally in-semicomplete (out-semicomplete) digraph is in-path-mergeable (out-path-mergeable).* ◇

**Proof:** Let $D$ be a locally in-semicomplete digraph and let $P=y_1y_2\dots y_k$, $Q=z_1z_2\dots z_t$ be a pair of internally disjoint paths such that $y_k=z_t$. We show that there exists a path $R$ in $D$ such that $V(R)=V(P)\cup V(Q)$ and $R$ starts in either $y_1$ or $z_1$ and ends in $y_k=z_t$. When $|A(P)|+|A(Q)|=2$ our claim follows from the definition of an in-semicomplete digraph. Assume now that $|A(P)|+|A(Q)|\geq 3$. Since $D$ is in-semicomplete, either $y_{k-1}z_{t-1}$ or $z_{t-1}y_{k-1}$ (or both) is an arc. Now the claim follows by induction applied to either the paths $P[y_1,y_{k-1}]z_{t-1}$ and $Q[z_1,z_{t-1}]$ or the paths $P[y_1,y_{k-1}]$, $Q[z_1,z_{t-1}]y_{k-1}$. The claim for locally out-semicomplete digraphs holds as they are the converses of locally in-semicomplete digraphs. □

**Corollary 6.2.5** ([4]) *Every locally in-semicomplete (out-semicomplete) digraph and hence every locally semicomplete digraph is path-mergeable.*

## 6.3 The Structure of Non-strong Locally (In-)Semicomplete Digraphs

In this section we describe basic results on the structure of non-strong locally in-semicomplete digraphs and locally semicomplete digraphs. We start with the following results due to Bang-Jensen, Huang and Prisner.

**Lemma 6.3.1** ([24]) *Every connected locally in-semicomplete digraph $D$ has an out-branching.* □

**Proof:** Consider an out-tree $T_s^+$ which has the maximum number of arcs among all out-trees in $D$. We show that $T_s^+$ must be an out-branching. Suppose not and let $y \in V(D) - V(T_s^+)$ be a vertex such that there is an arc $yz$ from $y$ to $V(T_s^+)$. Considering the directed path from $s$ to $z$ in $T_s^+$ and using the maximality of $T_s^+$ we conclude that $ys$ is an arc of $A(D)$, contradicting the maximality of $T_s^+$ as we can get a better out-tree rooted at $y$. □

**Theorem 6.3.2** ([24]) *Let $D$ be a locally in-semicomplete digraph.*

*(i) Let $X$ and $Y$ be distinct strong components of $D$. If a vertex $x \in X$ dominates some vertex in $Y$, then $x \mapsto Y$.*
*(ii) If $D$ is connected, then the strong component digraph $SC(D)$ has an out-branching.*

**Proof:** Let $X$ and $Y$ be strong components of $D$ for which there is an arc $xy$ from $X$ to $Y$. Since $Y$ is strong, there is a $(y', y)$-path in $Y$ for every $y' \in V(Y)$. By the definition of locally in-semicomplete digraphs and the fact that there is no arc from $Y$ to $X$, we can conclude that $xy' \in A$. This proves (i).

Part (ii) follows from the fact that, by (i), $SC(D)$ is itself a locally in-tournament digraph and Lemma 6.3.1. □

The most basic properties of strong components of a connected non-strong locally semicomplete digraph are given in the following result, due to Bang-Jensen.

**Theorem 6.3.3** ([9]) *Let $D$ be a connected locally semicomplete digraph that is not strong. Then the following holds for $D$.*

*(a) If $X$ and $Y$ are distinct strong components of $D$ with at least one arc between them, then either $X \mapsto Y$ or $Y \mapsto X$.*
*(b) If $X$ and $Y$ are strong components of $D$ such that $X \mapsto Y$, then $X$ and $Y$ are semicomplete digraphs.*
*(c) The strong components of $D$ can be ordered uniquely as $D_1, D_2, \ldots, D_p$ such that there is no arc from $D_j$ to $D_i$ when $j > i$, and $D_i$ dominates $D_{i+1}$ for $i \in [p-1]$.*

**Proof:** Recall that a locally semicomplete digraph is a locally in-semicomplete digraph as well as a locally out-semicomplete digraph. Part (a) of this theorem follows immediately from Part (i) of Theorem 6.3.2 and its analogue for locally out-semicomplete digraphs. Part (b) can be easily obtained from the definition of a locally semicomplete digraph. Finally, to prove (c) first observe that by Theorem 6.3.2 (and its analogue for locally out-semicomplete digraphs), $SC(D)$ has an out-branching and an in-branching. This implies that $D$ has a unique initial component $D_1$ and a unique terminal component $D_p$ and thus every internal strong component $D_i$ is reachable for $D_1$ and can reach $D_p$. Combining this with the fact that $D$ is path-mergeable proves the claim. □

Guo and Volkmann introduced the following very useful way of decomposing non-strong locally semicomplete digraphs.

**Theorem 6.3.4** ([47, 48]) *Let $D$ be a connected locally semicomplete digraph that is not strong and let $D_1, \dots, D_p$ be the acyclic ordering of strong components of $D$. Then $D$ can be decomposed into $r \geq 2$ induced semicomplete subdigraphs $D'_1, D'_2, \dots, D'_r$ as follows:*

- $D'_1 = D_p, \quad \lambda_1 = p,$
- $\lambda_{i+1} = \min\{\, j \mid N^+(D_j) \cap V(D'_i) \neq \emptyset\}$, *for each* $i \in [r-1]$,
- $D'_{i+1} = D\langle V(D_{\lambda_{i+1}}) \cup V(D_{\lambda_{i+1}+1}) \cup \cdots \cup V(D_{\lambda_i - 1})\rangle$, *for each* $i \in [r-1]$.

*The subdigraphs $D'_1, D'_2, \dots, D'_r$ satisfy the properties below:*

*(a) $D'_i$ consists of some strong components of $D$ and is semicomplete for each $i \in [r]$;*
*(b) $D'_{i+1}$ dominates the initial component of $D'_i$ and there exists no arc from $D'_i$ to $D'_{i+1}$ for any $i \in [r-1]$;*
*(c) if $r \geq 3$, then there is no arc between $D'_i$ and $D'_j$ for $i, j$ satisfying $|j-i| \geq 2$.* ◇

For a connected, but not strongly connected locally semicomplete digraph $D$, the unique sequence $D'_1, D'_2, \dots, D'_r$ defined in Theorem 6.3.4 is called the **semicomplete decomposition** of $D$.

## 6.4 Hamiltonian Paths and Cycles

Recall that we call a digraph **traceable** if it has a Hamiltonian path. Below we shall characterize traceable locally in-semicomplete digraphs and show that a locally in-semicomplete digraph has a Hamiltonian cycle whenever it satisfies the obviously necessary condition of being strongly connected. In order to save some space below, we shall state and prove the relevant result for the largest class among locally semicomplete, locally in-semicomplete and path-mergeable digraphs for which it holds. See [9] and [24] for short proofs of hamiltonicity in strong locally semicomplete digraphs and strong locally in-semicomplete digraphs.

**Theorem 6.4.1** *A locally in-semicomplete digraph has a Hamiltonian path ending at a vertex s if and only it has an in-branching rooted at s.*

**Proof:** Necessity is obvious. We prove the sufficiency by induction on the number $n$ of vertices. If $n \leq 2$, then every in-branching is also a Hamiltonian path so we can proceed to the induction step. Let $B_s^-$ be an in-branching of $D$, let $u$ be a leaf of $B_s^-$ and let $v$ be the out-neighbour of $u$ in $B_s^-$. Then $B_s^- - u$ is an in-branching in $D - u$ so by induction $D - u$ has a Hamiltonian path $v_1 v_2 \ldots v_{n-1}$. Note that $v = v_j$ for some $j \in [n-1]$. Let $i \in [j]$ be the minimum index such that $u$ dominates $v_i$. Then $v_1 \ldots v_{i-1} u v_i \ldots v_{n-1}$ is the desired Hamiltonian path. □

Detecting whether an arbitrary digraph has an in-branching can be done in time $O(n+m)$ so the proof above can be turned into an $O(n^2)$ algorithm for finding a Hamiltonian path in an in-semicomplete digraph or detecting that it has no branching. Bang-Jensen and Hell [20] gave a faster algorithm.

**Theorem 6.4.2** ([20]) *Suppose we are given an in-semicomplete digraph $D$ with an in-branching $B_s^-$. Then in time $O(n \log n)$ one can construct a Hamiltonian path of $D$.*

Bang-Jensen and Hell also proved the following.

**Theorem 6.4.3** ([20]) *There is an $O(m + n \log n)$ algorithm for finding a longest path in an in-semicomplete digraph.*

**Corollary 6.4.4** ([9]) *Every connected locally semicomplete digraph has a Hamiltonian path. Furthermore such a path ending (starting) in $v$ exists if and only if $v$ belongs to the terminal (initial) strong component of $D$.*

The complexity of the Hamiltonian path problem for path-mergeable digraphs is open.

**Problem 6.4.5** ([17]) *Characterize traceable path-mergeable digraphs. Is there a polynomial algorithm to decide whether a path-mergeable digraph is traceable?*

### 6.4.1 Hamilton Cycles in Path-Mergeable Digraphs

We now show that the property of being path-mergeable and strongly connected is already sufficient to guarantee a Hamiltonian cycle, provided that $D$ satisfies the obviously necessary condition that the underlying digraph has no cut-vertex. A corollary of this result is that every strongly connected locally in-semicomplete digraph is Hamiltonian.

We begin with a simple lemma which forms the basis for the proof of Theorem 6.4.7. For a cycle $C$, a $C$- **bypass** is a path of length at least two with both end-vertices on $C$ and no other vertices on $C$.

**Lemma 6.4.6** ([4]) *Let $D$ be a path-mergeable digraph and let $C$ be a cycle in $D$. If $D$ has a $C$-bypass $P$, then there exists a cycle in $D$ containing precisely the vertices $V(C) \cup V(P)$.*

**Proof:** Let $P$ be an $(x, y)$-path. Then the paths $P$ and $C[x, y]$ can be merged into one $(x, y)$-path $R$, which together with $C[y, x]$ forms the desired cycle. □

**Theorem 6.4.7** ([4]) *A path-mergeable digraph $D$ of order $n \geq 2$ is Hamiltonian if and only if $D$ is strong and $UG(D)$ is 2-connected.*

**Proof:** 'Only if' is obvious; we prove 'if'. Suppose that $D$ is strong, $UG(D)$ is 2-connected and $D$ is not Hamiltonian. Let $C = u_1u_2 \dots u_pu_1$ be a longest cycle in $D$. Observe that, by Lemma 6.4.6, there is no $C$-bypass. For each $i \in [p]$ let $X_i$ (respectively, $Y_i$) be the set of vertices of $D - V(C)$ that can be reached from $u_i$ (respectively, from which $u_i$ can be reached) by a path in $D - (V(C) - u_i)$. Since $D$ is strong,

$$X_1 \cup \dots \cup X_p = Y_1 \cup \dots \cup Y_p = V(D) - V(C).$$

Since there is no $C$-bypass, every path starting at a vertex in $X_i$ and ending at a vertex in $C$ must end at $u_i$. Thus, $X_i \subseteq Y_i$. Similarly, $Y_i \subseteq X_i$ and, hence, $X_i = Y_i$. Since there is no $C$-bypass, the sets $X_i$ are disjoint. Since we assumed that $D$ is not Hamiltonian, at least one of these sets, say $X_1$, is non-empty. Since $UG(D)$ is 2-connected, there is an arc with one end-vertex in $X_1$ and the other in $V(D) - (X_1 \cup u_1)$, and no matter what its orientation is, this implies that there is a $C$-bypass, a contradiction. □

Using the proof of this theorem, Lemma 6.4.6 and Proposition 6.2.3, it is not difficult to show the following:

**Corollary 6.4.8** ([4]) *There is an $O(nm)$ algorithm to decide whether a given strong path-mergeable digraph has a Hamiltonian cycle and find one if it exists.* ◇

Clearly, Theorem 6.4.7 and Corollary 6.4.8 imply an obvious characterization of longest cycles in path-mergeable digraphs and a polynomial algorithm to find a longest cycle.

It is easy to show that a strong locally in-semicomplete digraph cannot have a cut-vertex and hence we get the following.

**Theorem 6.4.9** ([24]) *A locally in-semicomplete digraph is Hamiltonian if and only if it is strongly connected.*

Bang-Jensen and Hell [20] showed that one can find a Hamiltonian cycle in a strong locally in-semicomplete digraph in time $O(m + n \log n)$.

## 6.5 Round Decomposable Digraphs

We now study a subclass of locally semicomplete digraphs with a particularly nice structure which implies a lot of useful properties, as we shall see later in this chapter. A digraph on $n$ vertices is **round** if we can label its vertices $v_1, v_2, \ldots, v_n$ so that for each $i$, we have $N^+(v_i) = \{v_{i+1}, \ldots, v_{i+d^+(v_i)}\}$ and $N^-(v_i) = \{v_{i-d^-(v_i)}, \ldots, v_{i-1}\}$ (all subscripts are taken modulo $n$). We will refer to the labelling $v_1, v_2, \ldots, v_n$ as a **round labelling** of $D$. See Figure 6.1 for an example of a round digraph. Observe that every strong round digraph $D$ is Hamiltonian, since $v_1 v_2 \ldots v_n v_1$ form a Hamiltonian cycle, whenever $v_1, v_2, \ldots, v_n$ is a round labelling. Round digraphs form a subclass of locally semicomplete digraphs.

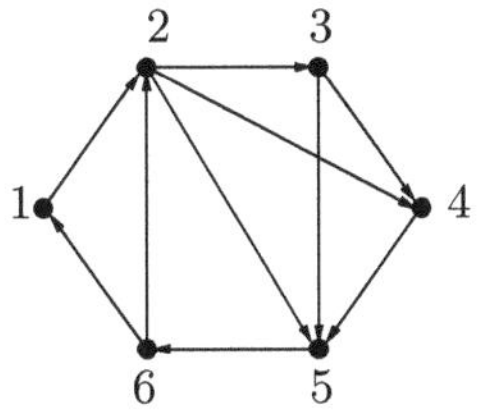


**Figure 6.1** A round digraph with a round labelling.

**Proposition 6.5.1** ([56]) *Every round digraph is locally semicomplete.*

**Proof:** Let $D$ be a round digraph and let $v_1, v_2, \ldots, v_n$ be a round labelling of $D$. Consider an arbitrary vertex, say $v_i$. Let $x, y$ be a pair of out-neighbours of $v_i$. We show that $x$ and $y$ are adjacent. Assume without loss of generality that $v_i, x, y$ appear in that circular order in the round labelling. Since $v_i \to y$ and the in-neighbours of $y$ appear consecutively preceding $y$, we must have $x \to y$. Thus the out-neighbours of $v_i$ are pairwise adjacent. Similarly, we can show that the in-neighbours of $v_i$ are also pairwise adjacent. Therefore, $D$ is locally semicomplete. $\square$

The main result of this subsection is Theorem 6.5.2, due to Huang [56], which gives a characterization of round locally semicomplete digraphs. This characterization generalizes the corresponding characterizations of round local tournaments and tournaments, due to Bang-Jensen [9] and Alspach and Tabib [2], respectively.

An arc $xy$ of a digraph $D$ is **ordinary** if $yx$ is not in $D$. A cycle or path $Q$ of a digraph $D$ is **ordinary** if all arcs of $Q$ are ordinary.

**Theorem 6.5.2 (Huang** [56]) *A connected locally semicomplete digraph $D$ is round if and only if the following holds for each vertex $x$ of $D$:*

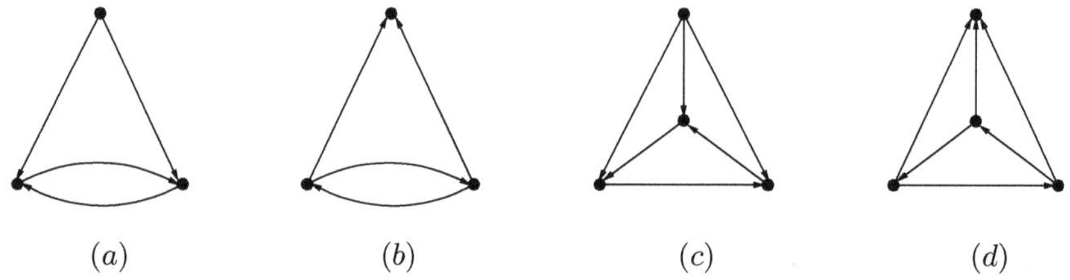


**Figure 6.2** Some forbidden digraphs in Huang's characterization.

*(a)* $N^+(x) \setminus N^-(x)$ *and* $N^-(x) \setminus N^+(x)$ *induce transitive tournaments; and*
*(b)* $N^+(x) \cap N^-(x)$ *induces a (semicomplete) subdigraph containing no ordinary cycle.* ◇

The proof of sufficiency of the conditions of this theorem in [15, 56] can be transformed into a polynomial time algorithm to decide whether a digraph $D$ is round and to find a round labelling of $D$ (if $D$ is round).

**Corollary 6.5.3** ([9]) *A connected local tournament $D$ is round if and only if, for each vertex $x$ of $D$, $N^+(x)$ and $N^-(x)$ induce transitive tournaments.* □

We now turn to locally semicomplete digraphs that are not round but which can be obtained from such a digraph by substituting sets of vertices for each vertex. A locally semicomplete digraph $D$ is **round decomposable** if there exists a round local tournament $R$ on $r \geq 2$ vertices such that $D = R[S_1, \ldots, S_r]$, where each $S_i$ is a strong semicomplete digraph. We call $R[S_1, \ldots, S_r]$ a **round decomposition** of $D$.

**Corollary 6.5.4** ([9]) *Every connected non-strong locally semicomplete digraph $D$ has a unique round decomposition given by $R[D_1, D_2, \ldots, D_p]$, where $D_1, D_2, \ldots, D_p$ is the acyclic ordering of strong components of $D$ and $R$ is the acyclic round local tournament which one obtains by taking an arbitrary vertex from each $D_i$.* ◇

### 6.5.1 Strong Round Decomposable Locally Semicomplete Digraphs

In the previous subsection we saw that every connected non-strong locally semicomplete digraph is round decomposable. This property does not hold for strong locally semicomplete digraphs (see Lemma 6.6.4).

The following assertions, due to Bang-Jensen, Guo, Gutin and Volkmann, provides some important properties concerning round decompositions of strong locally semicomplete digraphs.

**Proposition 6.5.5** ([13]) *Let $R[H_1, H_2, \ldots, H_\alpha]$ be a round decomposition of a strong locally semicomplete digraph $D$. Then, for every inclusion-wise minimal separating set $S$, there are two integers $i$ and $k \geq 0$ such that $S = V(H_i) \cup \ldots \cup V(H_{i+k})$.*

**Corollary 6.5.6** ([13]) *If a locally semicomplete digraph $D$ is round decomposable, then it has a unique round decomposition $D = R[D_1, D_2, \ldots, D_\alpha]$.*

**Proof:** Suppose that $D$ has two different round decompositions: $D = R[D_1, \ldots, D_\alpha]$ and $D = R'[H_1, \ldots, H_\beta]$.

By Corollary 6.5.4, we may assume that $D$ is strong. By the definition of a round decomposition, this implies that $\alpha, \beta \geq 3$. Let $S$ be a minimal separating set of $D$. By Proposition 6.5.5, we may assume without loss of generality that $S = V(D_1 \cup \ldots \cup D_i) = V(H_1 \cup \ldots \cup H_j)$ for some $i$ and $j$. Since $D - S$ is non-strong, by Corollary 6.5.4, $D_{i+1} = H_{j+1}, \ldots, D_\alpha = H_\beta$ (in particular, $\alpha - i = \beta - j$). Now it suffices to prove that

$$D_1 = H_1, \ldots, D_i = H_j (\text{in particular}, i = j). \tag{6.1}$$

If $D[S]$ is non-strong, then (6.1) follows by Corollary 6.5.4. If $D[S]$ is strong, then first consider the case $\alpha = 3$. Then $S = V(D_1)$, because $D - S$ is non-strong and $\alpha = 3$. Assuming that $j > 1$, we obtain that the subdigraph of $D$ induced by $S$ has a strong round decomposition. This contradicts the fact that $R'$ is a local tournament, since the in-neighbourhood of the vertex $r'_{j+1}$ in $R'$ contains a cycle (where $r'_p$ corresponds to $H_p$, $p = 1, \ldots, \beta$). Therefore, (6.1) is true for $\alpha = 3$. If $\alpha > 3$, then we can find a separating set in $D\langle S \rangle$ and conclude by induction that (6.1) holds. □

Proposition 6.5.5 allows us to construct a polynomial algorithm for checking whether a locally semicomplete digraph is round decomposable.

**Proposition 6.5.7** ([13]) *There exists a polynomial algorithm for deciding whether a given locally semicomplete digraph $D$ has a round decomposition and to find this decomposition if it exists.*

**Proof:** We only give a sketch of such an algorithm. Find a minimal separating set $S$ in $D$ starting with $S' = N^+(x)$ for a vertex $x \in V(D)$ and deleting vertices from $S'$ until a minimal separating set is obtained. Construct the strong components of $D\langle S \rangle$ and $D - S$ and label these $D_1, D_2, \ldots, D_\alpha$, where $D_1, \ldots, D_p$, $p \geq 1$, form an acyclic ordering of the strong components of $D[S]$ and $D_{p+1}, \ldots, D_\alpha$ form an acyclic ordering of the strong components of $D - S$. For every pair $D_i$ and $D_j$ $(1 \leq i \neq j \leq \alpha)$, we check the following: if there exist some arcs between $D_i$ and $D_j$, then either $D_i \mapsto D_j$ or $D_j \mapsto D_i$. If we find a pair for which the above condition is false, then $D$ is not round decomposable. Otherwise, we form a digraph $R = D\langle \{x_1, x_2, \ldots, x_\alpha\} \rangle$, where $x_i \in V(D_i)$ for each $i \in [\alpha]$. We check whether $R$ is round using Corollary 6.5.3.

If $R$ is not round, then $D$ is not round decomposable. Otherwise, $D$ is round decomposable and $D = R[D_1, \ldots, D_\alpha]$.

It is not difficult to verify that our algorithm is correct and polynomial. □

## 6.6 Classification of Locally Semicomplete Digraphs

Based on the results from the previous sections we are now ready to take the final step towards a full classification of locally semicomplete digraphs. We start this subsection with an important lemma on minimal separating sets of locally semicomplete digraphs.

**Lemma 6.6.1** ([9]) *Let $D$ be a strong locally semicomplete digraph and $S$ a minimal separating set in $D$, then $D - S$ is connected.*

**Lemma 6.6.2** ([13]) *If a strong locally semicomplete digraph $D$ is not semicomplete, then there exists a minimal separating set $S \subset V(D)$ such that $D - S$ is connected but not semicomplete. Furthermore, if $D_1, D_2, \ldots, D_p$ is the acyclic ordering of the strong components of $D$ and $D'_1, D'_2, \ldots, D'_r$ is the semicomplete decomposition of $D - S$, then $r \geq 3$, $D\langle S\rangle$ is semicomplete and we have $D_p \mapsto S \mapsto D_1$.*

**Proof:** Suppose $D - S$ is semicomplete for every minimal separating set $S$. Then $D - S$ is semicomplete for all separating sets $S$. Hence $D$ is semicomplete, because any pair of non-adjacent vertices can be separated by some separating set $S$. Together with Lemma 6.6.1 this proves the first claim of the lemma.

Let $S$ be a minimal separating set such that $D - S$ is not semicomplete. Clearly, if $r = 2$ (in Theorem 6.3.4), then $D - S$ would be semicomplete. Thus, $r \geq 3$. By the minimality of $S$ every vertex $s \in S$ dominates a vertex in $D_1$ and is dominated by a vertex in $D_p$. Thus if some $x \in D_p$ was dominated by $s \in S$, then, by the definition of a locally semicomplete digraph, we would have $D_1 \mapsto D_p$, contradicting the fact that $r \geq 3$. Hence (using that $D_p$ is strongly connected) we get that $D_p \mapsto S$ and similarly $S \mapsto D_1$. From the last observation it follows that $S$ is semicomplete. □

Now we consider strongly connected locally semicomplete digraphs which are not semicomplete and not round decomposable. We first show that the semicomplete decomposition of $D - S$ has exactly three components, whenever $S$ is a minimal separating set such that $D - S$ is not semicomplete.

**Lemma 6.6.3** ([13]) *Let $D$ be a strong locally semicomplete digraph which is not semicomplete. Either $D$ is round decomposable, or $D$ has a minimal separating set $S$ such that the semicomplete decomposition of $D - S$ has exactly three components $D'_1, D'_2, D'_3$.*

**Proof:** By Lemma 6.6.2, $D$ has a minimal separating set $S$ such that the semicomplete decomposition of $D-S$ has at least three components.

Assume now that the semicomplete decomposition of $D-S$ has more than three components $D'_1,\dots,D'_r$ $(r\geq 4)$. Let $D_1,D_2,\dots,D_p$ be the acyclic ordering of strong components of $D-S$. According to Theorem 6.3.4 (c), there is no arc between $D'_i$ and $D'_j$ if $|i-j|\geq 2$. It follows from the definition of a locally semicomplete digraph that

$$N^+(D'_i)\cap S=\emptyset \text{ for } i\geq 3 \text{ and } N^-(D'_j)\cap S=\emptyset \text{ for } j\leq r-2. \qquad (6.2)$$

By Lemma 6.6.2, $D\langle S\rangle$ is semicomplete and $S=N^+(D_p)$. Let $D_{p+1},\dots,$ $D_{p+q}$ be the acyclic ordering of the strong components of $D\langle S\rangle$. Using (6.2) and the assumption $r\geq 4$, it is easy to check that if there is an arc between $D_i$ and $D_j$ $(1\leq i\neq j\leq p+q)$, then $D_i{\mapsto}D_j$ or $D_j{\mapsto}D_i$. Let $R=D\langle\{x_1,x_2,\dots,x_{p+q}\}\rangle$ with $x_i\in V(D_i)$ for each $i\in[p+q]$. Now it suffices to prove that $R$ is a round local tournament.

Since $R$ is a subdigraph of $D$ and no pair $D_i$, $D_j$ induces a strong digraph, we see that $R$ is a local tournament. By Corollary 6.5.4 each of the subdigraphs $R'=R-\{x_{p+1},\dots,x_{p+q}\}$, $R''=R-V(R)\cap V(D'_{r-1})$ and $R'''=R-V(R)\cap V(D'_2)$ is round. Since $N^+(v)\cap V(R)$ (as well as $N^-(v)\cap V(R)$) is completely contained in one of the sets $V(R'),V(R'')$ and $V(R''')$ for every $v\in V(R)$, we see that $R$ is round.

Thus if $r\geq 4$, then $D$ is round decomposable. □

Now we are ready to give a characterization of locally semicomplete digraphs which are not semicomplete and not round decomposable. This characterization was proved for the first time by Guo in [43]. A weaker form was obtained earlier by Bang-Jensen in [3]. Here we give the proof of this result from [13].

**Lemma 6.6.4** *Let $D$ be a strong locally semicomplete digraph which is not semicomplete. Then $D$ is not round decomposable if and only if the following conditions are satisfied:*

*(a) There is a minimal separating set $S$ such that $D-S$ is not semicomplete and for each such $S$, $D\langle S\rangle$ is semicomplete and the semicomplete decomposition of $D-S$ has exactly three components $D'_1,D'_2,D'_3$,*
*(b) There are integers $\alpha,\beta,\mu,\nu$ with $\lambda_2\leq\alpha\leq\beta\leq p-1$ and $p+1\leq\mu\leq\nu\leq p+q$ such that*

$$N^-(D_\alpha)\cap V(D_\mu)\neq\emptyset \text{ and } N^+(D_\alpha)\cap V(D_\nu)\neq\emptyset,$$

$$\text{or}\quad N^-(D_\mu)\cap V(D_\alpha)\neq\emptyset \text{ and } N^+(D_\mu)\cap V(D_\beta)\neq\emptyset,$$

*where $D_1, D_2, \ldots, D_p$ and $D_{p+1}, \ldots, D_{p+q}$ are the acyclic orderings of the strong components of $D-S$ and $D\langle S\rangle$, respectively, and $D_{\lambda_2}$ is the initial component of $D_2'$.*

**Proof:** If $D$ is round decomposable and satisfies (a), then we must have $D = R[D_1, D_2, \ldots, D_{p+q}]$, where $R$ is the digraph obtained from $D$ by contracting each $D_i$ into one vertex. This follows from Corollary 6.5.4 and the fact that each of the digraphs $D-S$ and $D-V(D_2')$ has a round decomposition that agrees with this structure. Now it is easy to see that $D$ does not satisfy (b).

Suppose now that $D$ is not round decomposable. By Lemmas 6.6.2 and 6.6.3, $D$ satisfies (a), so we only have to prove that it also satisfies (b).

If there are no arcs from $S$ to $D_2'$, then it is easy to see that $D$ has a round decomposition. If there exist components $D_{p+i}$ and $D_j$ with $V(D_j) \subseteq V(D_2')$ such that there are arcs in both directions between $D_{p+i}$ and $D_j$, then $D$ satisfies (b). So we can assume that for every pair of sets from the collection $D_1, D_2, \ldots, D_{p+q}$, either there are no arcs between these sets, or one set completely dominates the other. Then, by Corollary 6.5.3, $D$ is round decomposable, with round decomposition $D = R[D_1, D_2, \ldots, D_{p+q}]$ as above, unless we have three subdigraphs $X, Y, Z \in \{D_1, D_2, \ldots, D_{p+q}\}$ such that $X \mapsto Y \mapsto Z \mapsto X$ and there exists a subdigraph $W \in \{D_1, D_2, \ldots, D_{p+q}\} - \{X, Y, Z\}$ such that either $W \mapsto X, Y, Z$ or $X, Y, Z \mapsto W$.

One of the subdigraphs $X, Y, Z$, say without loss of generality $X$, is a strong component of $D\langle S\rangle$. If we also have $V(Y) \subseteq S$, then $V(Z) \subseteq V(D_2')$ and $W$ is either in $D\langle S\rangle$ or in $D_2'$ (there are four possible positions for $W$ satisfying either $W \mapsto X, Y, Z$ or $X, Y, Z \mapsto W$). In each of these cases it is easy to see that $D$ satisfies (b). For example, if $W$ is in $D\langle S\rangle$ and $W \mapsto X, Y, Z$, then any arc from $W$ to $Z$ and from $Z$ to $X$ satisfies the first part of (b). The proof is similar when $V(Y) \subseteq V(D_3')$. Hence we can assume that $V(Y) \subseteq V(D_2')$. If $Z = D_p$, then $W$ must be either in $D\langle S\rangle$ and $X, Y, Z \mapsto W$, or $V(W) \subseteq V(D_2')$ and $W \mapsto X, Y, Z$ (which means that $W = D_i$ and $Y = D_j$ for some $\lambda_2 \leq i < j < p$). In both cases it is easy to see that $D$ satisfies (b). The last case $V(Y), V(Z) \subseteq V(D_2')$ can be treated similarly. □

We can now state the classification of locally semicomplete digraphs due to Bang-Jensen, Guo, Gutin and Volkmann.

**Theorem 6.6.5 (Bang-Jensen, Guo, Gutin, Volkmann** [13]**)** *Let $D$ be a connected locally semicomplete digraph. Then exactly one of the following possibilities holds.*

*(a) $D$ is round decomposable with a unique round decomposition given by $D = R[D_1, D_2, \ldots, D_\alpha]$, where $R$ is a round local tournament on $\alpha \geq 2$ vertices and $D_i$ is a strong semicomplete digraph for each $i \in [\alpha]$,*
*(b) $D$ is not round decomposable and not semicomplete and it has the structure as described in Lemma 6.6.4 (so $D$ is evil),*
*(c) $D$ is a semicomplete digraph which is not round decomposable.* □

Locally semicomplete digraphs which are not semicomplete and not round decomposable are also called **evil** locally semicomplete digraphs [12], since this is by far the most complicated class among the two non-semicomplete cases classes (a) and (b) above.

We finish this section with the following useful result which has been used in many proofs on locally semicomplete digraphs (see e.g. the proof of Lemma 6.8.5 below).

**Proposition 6.6.6** ([13]) *Let $D$ be a strong evil locally semicomplete digraph and let $S$ be a minimal separating set of $D$ such that $D-S$ is not semicomplete. Let $D_1,\dots,D_p$ be the acyclic ordering of the strong components of $D-S$ and $D_{p+1},\dots,D_{p+q}$ be the acyclic ordering of the strong components of $D\langle S\rangle$. Suppose that there is an arc $s\to v$ from $S$ to $D_2'$ with $s\in V(D_i)$ and $v\in V(D_j)$, then*

$$D_i \cup D_{i+1} \cup \ldots \cup D_{p+q} \mapsto D_3' \mapsto D_{\lambda_2} \cup \ldots \cup D_j. \qquad \square$$

**Problem 6.6.7** *Does there exist a nice structural characterization of those locally in-semicomplete digraphs that are not locally semicomplete?*

## 6.7 Hamiltonian Connectivity

Recall that an $[x,y]$-path in a digraph $D=(V,A)$ is a path which either starts at $x$ and ends at $y$ or oppositely. We say that $D$ is **weakly Hamiltonian-connected** if it has a Hamiltonian $[x,y]$-path (also called an $[x,y]$-**Hamiltonian path**) for every choice of distinct vertices $x,y\in V$. Our next goal is to describe the solution of the $[x,y]$-Hamiltonian path problem for locally semicomplete digraphs. Notice that this solution also covers the case of semicomplete digraphs and so, in particular, it generalizes Theorem 2.6.3 to semicomplete digraphs.

We start by establishing the notation for some special locally semicomplete digraphs. Up to isomorphism there is a unique strong tournament with four vertices. We denote this by $T_4^1$. It has the following vertices and arcs:

$$V(T_4^1)=\{a_1,a_2,a_3,a_4\}, A(T_4^1)=\{a_1a_2,a_2a_3,a_3a_4,a_4a_1,a_1a_3,a_2a_4\}.$$

The semicomplete digraphs $T_4^2$, $T_4^3$, and $T_4^4$ are obtained from $T_4^1$ by adding some arcs, namely:

$$A(T_4^2)=A(T_4^1)\cup\{a_3a_1,a_4a_2\},$$

$$A(T_4^3)=A(T_4^1)\cup\{a_3a_1\}, A(T_4^4)=A(T_4^1)\cup\{a_1a_4\}.$$

Let $\mathcal{T}_4 = \{T_4^1, T_4^2, T_4^3, T_4^4\}$. It is easy to check that every digraph of $\mathcal{T}_4$ has a unique Hamiltonian cycle and has no Hamiltonian path between two vertices which are not consecutive on this Hamiltonian cycle (two such vertices are called **opposite**).

Let $\mathcal{T}_6$ be the set of semicomplete digraphs with the vertex set $\{x_1, x_2, a_1, a_2, a_3, a_4\}$ such that each member $D$ of $\mathcal{T}_6$ has a cycle $a_1a_2a_3a_4a_1$ and the digraph $D\langle\{a_1, a_2, a_3, a_4\}\rangle$ is isomorphic to one member of $\mathcal{T}_4$, in addition, $x_i \to \{a_1, a_3\} \to x_{3-i} \to \{a_2, a_4\} \to x_i$ for $i = 1$ or $i = 2$. It is straightforward to verify that $\mathcal{T}_6$ contains only two tournaments (denoted by $T_6'$ and $T_6''$), namely, the ones shown in Fig. 2.2, and that $|\mathcal{T}_6| = 11$. Since none of the digraphs of $\mathcal{T}_4$ has a Hamiltonian path connecting any two opposite vertices, no digraph of $\mathcal{T}_6$ has a Hamiltonian path between $x_1$ and $x_2$.

For every even integer $n \geq 4$ there is only one 2-strong, 2-regular locally semicomplete digraph on $n$ vertices, namely, the second power $\vec{C}_n^2$ of an $n$-cycle. We define

$$\mathcal{T}^* = \{C_n^2 \mid n \text{ is even and } n \geq 4\}.$$

It is not difficult to prove that every digraph of $\mathcal{T}^*$ has a unique Hamiltonian cycle and is not weakly Hamiltonian-connected (see [10]). For instance, if the unique Hamiltonian cycle of $C_6^2$ is denoted by $u_1u_2u_3u_4u_5u_6u_1$, then $u_1u_3u_5u_1$ and $u_2u_4u_6u_2$ are two cycles of $C_6^2$ and there is no Hamiltonian path between any two vertices of $\{u_1, u_3, u_5\}$ or of $\{u_2, u_4, u_6\}$.

Let $T_8^1$ be the digraph consisting of $C_6^2$ together with two new vertices $x_1$ and $x_2$ such that $x_1 \to \{u_1, u_3, u_5\} \to x_2 \to \{u_2, u_4, u_6\} \to x_1$. Furthermore, $T_8^2$ ($T_8^3$, respectively) is defined as the digraph obtained from $T_8^1$ by adding the arc $x_1x_2$ (the arcs $x_1x_2$ and $x_2x_1$, respectively). Let $\mathcal{T}_8 = \{T_8^1, T_8^2, T_8^3\}$. It is easy to see that every element of $\mathcal{T}_8$ is a 3-strong locally semicomplete digraph and has no Hamiltonian path between $x_1$ and $x_2$.

Before we present the main result, we state the following two lemmas that were used in the proof of Theorem 6.7.3 by Bang-Jensen, Guo and Volkmann in [14]. The first lemma generalizes the structure found in the last part of the proof of Theorem 2.6.3.

**Lemma 6.7.1** ([14]) *Let $D$ be a strong locally semicomplete digraph on $n \geq 4$ vertices and $x_1, x_2$ two distinct vertices of $D$. If $D - \{x_1, x_2\}$ is strong, and $N^+(x_1) \cap N^+(x_2) \neq \emptyset$ or $N^-(x_1) \cap N^-(x_2) \neq \emptyset$, then $D$ has a Hamiltonian path connecting $x_1$ and $x_2$.*

Another useful ingredient in the proof of Theorem 6.7.3 is the following linking result. An **odd chain** is the second power, $P_{2k+1}^2$ for some $k \geq 1$, of a path on an odd number of vertices.

**Lemma 6.7.2** ([14]) *Let $D$ be a connected, locally semicomplete digraph with $p \geq 4$ strong components and an acyclic ordering $D_1, D_2, ..., D_p$ of these. Suppose that $V(D_1) = \{u_1\}$ and $V(D_p) = \{v_1\}$ and that $D - x$ is connected for every vertex $x$. Then, for every choice of $u_2 \in V(D_2)$ and $v_2 \in V(D_{p-1})$, $D$ has two vertex disjoint paths $P_1$ from $u_2$ to $v_1$ and $P_2$ from $u_1$ to $v_2$ with $V(P_1) \cup V(P_2) = V(D)$ if and only if $D$ is not an odd chain from $u_1$ to $v_1$.*

**Proof:** If $D$ is an odd chain, it is easy to see that $D$ has no two vertex-disjoint $(u_i, v_{3-i})$-paths, for $i = 1, 2$. We prove by induction on $p$ that the converse is true as well. Suppose that $D$ is not an odd chain from $u_1$ to $v_1$. Since the subdigraph $D - x$ is connected for every vertex $x$, $|N^+(D_i)| \geq 2$ for all $i \leq p-2$ and $|N^-(D_j)| \geq 2$ for all $j \geq 3$. If $p = 4$, then it is not difficult to see that $D$ has two vertex-disjoint paths $P_1$ from $u_2$ to $v_1$ and $P_2$ from $u_1$ to $v_2$ with $V(P_1) \cup V(P_2) = V(D)$. If $p = 5$, it is also not difficult to check that $D$ has the desired paths, unless $D$ is a chain on five vertices. So we assume that $p \geq 6$. Now we consider the digraph $D'$, which is obtained from $D$ by deleting the vertex sets $\{u_1, v_1\}$, $V(D_2 - u_2)$ and $V(D_{p-1} - v_2)$.

Using the assumption on $D$, it is not difficult to show that $D'$ is a connected, but not strongly connected locally semicomplete digraph with the acyclic ordering $\{u_2\}, D_3, D_4, \ldots, D_{p-2}, \{v_2\}$ of its strong components. Furthermore, for every vertex $y$ of $D'$, the subdigraph $D' - y$ is still connected. Let $u$ be an arbitrary vertex of $D_3$ and $v$ an arbitrary vertex of $D_{p-2}$. Note that there is a $(u_1, u)$-Hamiltonian path $P$ in $D[\{u_1, u\} \cup V(D_2 - u_2)]$ and similarly there is a $(v, v_1)$-Hamiltonian path $Q$ in $D[\{v, v_1\} \cup V(D_{p-1} - v_2)]$. Hence if $D'$ has disjoint $(u_2, v)$-, $(u, v_2)$-paths which cover all vertices of $D'$, then $D$ has the desired paths. So we can assume $D'$ has no such paths. By induction, $D'$ is an odd chain from $u_2$ to $v_2$. Now using that $D$ is not an odd chain from $u_1$ to $v_1$ it is easy to see that $D$ has the desired paths. We leave the details to the reader. □

A weaker version of Lemma 6.7.2 was proved in [10, Theorem 4.5].

Below we give a characterization, due to Bang-Jensen, Guo and Volkmann, for the existence of an $[x, y]$-Hamiltonian path in a locally semicomplete digraph. Note again the similarity to Theorem 2.6.3.

**Theorem 6.7.3** ([14]) *Let $D$ be a connected locally semicomplete digraph on $n$ vertices and $x_1$ and $x_2$ be two distinct vertices of $D$. Then $D$ has no Hamiltonian $[x_1, x_2]$-path if and only if one of the following conditions is satisfied:*

*(1) $D$ is not strong and either the initial or the terminal component of $D$ (or both) contains none of $x_1, x_2$.*

*(2) $D$ is strongly connected, but not 2-strong,*

*(2.1) there is an $i \in \{1, 2\}$ such that $D - x_i$ is not strong and $x_{3-i}$ belongs to neither the initial nor the terminal component of $D - x_i$;*

*(2.2)* $D - x_1$ *and* $D - x_2$ *are strong,* $s$ *is a separating vertex of* $D$, $D_1, D_2, ..., D_p$ *is the acyclic ordering of the strong components of* $D - s$, $x_i \in V(D_\alpha)$ *and* $x_{3-i} \in V(D_\beta)$ *with* $\alpha \leq \beta - 2$. *Furthermore,* $V(D_{\alpha+1}) \cup V(D_{\alpha+2}) \cup ... \cup V(D_{\beta-1})$ *contains a separating vertex of* $D$, *or* $D' = D\langle V(D_\alpha) \cup V(D_{\alpha+1}) \cup ... \cup V(D_\beta)\rangle$ *is an odd chain from* $x_i$ *to* $x_{3-i}$ *with* $N^-(D_{\alpha+2}) \cap V(D - V(D')) = \emptyset$ *and* $N^+(D_{\beta-2}) \cap V(D - V(D')) = \emptyset$.

*(3)* $D$ *is 2-strong and is isomorphic to* $T_4^2$ *or to one member of* $\mathcal{T}_6 \cup \mathcal{T}_8 \cup \mathcal{T}^*$ *and* $x_1, x_2$ *are the corresponding vertices in the definitions.* ◇

As an easy consequence of Theorem 6.7.3, we obtain a characterization of weakly Hamiltonian-connected locally semicomplete digraphs. The proof is left to the interested reader.

**Theorem 6.7.4** ([14]) *A locally semicomplete digraph* $D$ *with at least three vertices is weakly Hamiltonian–connected if and only if it satisfies (a), (b) and (c) below:*

*(a)* $D$ *is strong,*
*(b) For every* $x \in V(D)$, $D - x$ *has at most two strong components,*
*(c)* $D$ *is not isomorphic to any member of* $\mathcal{T}_6 \cup \mathcal{T}_8 \cup \mathcal{T}^*$. ◇

The following is an easy corollary of Theorem 6.7.3.

**Corollary 6.7.5** *There exists a polynomial algorithm for deciding whether a given locally semicomplete digraph* $D$ *has an* $[x, y]$*-Hamiltonian path for two specified vertices* $x, y$.

We conjecture that this can be extended to locally in-semicomplete digraphs.

**Conjecture 6.7.6** *There exists a polynomial algorithm for deciding whether a given locally in-semicomplete digraph* $D$ *has an* $[x, y]$*-Hamiltonian path for two specified vertices* $x, y$.

Guo [45] extended Theorem 2.6.7 to locally semicomplete digraphs.

**Theorem 6.7.7 (Guo [45])** *Let* $D$ *be a 2-strong locally semicomplete digraph and let* $x, y$ *be two distinct vertices of* $D$. *Then* $D$ *contains a Hamiltonian path from* $x$ *to* $y$ *if (a) or (b) below is satisfied.*

*(a) There are three internally disjoint* $(x, y)$*-paths in* $D$, *each of which is of length at least 2 and* $D$ *is not isomorphic to any of the digraphs* $T_8^1$ *and* $T_8^2$ *(see the definition in the preceding section).*
*(b) The digraph* $D$ *has two internally disjoint* $(x, y)$*-paths* $P_1, P_2$, *each of which is of length at least 2 and a path* $P$ *which either starts at* $x$ *or ends at* $y$ *and has only* $x$ *or* $y$ *in common with* $P_1, P_2$ *such that* $V(D) = V(P_1) \cup V(P_2) \cup V(P)$. *Furthermore, for any vertex* $z \notin V(P_1) \cup V(P_2)$, $z$ *has a neighbour on* $P_1 - \{x, y\}$ *if and only if it has a neighbour on* $P_2 - \{x, y\}$. ◇

Since neither of the two exceptions in (a) is 4-strong, Theorem 6.7.7 implies the following generalization of Theorem 2.6.7:

**Corollary 6.7.8** ([45]) *If a locally semicomplete digraph is 4-strong, then it is Hamiltonian-connected.* ◇

In [43] Guo used Theorem 6.7.7 to give a complete characterization of those 3-strongly connected arc-3-cyclic (that is, every arc is in a 3-cycle) locally tournament digraphs with no Hamiltonian path from $x$ to $y$ for specified vertices $x$ and $y$. In particular, this characterization shows that there exist infinitely many 3-strongly connected digraphs which are locally tournament digraphs (but not semicomplete digraphs) and are not Hamiltonian-connected. Thus, as far as this problem is concerned, it is not only the subclass of semicomplete digraphs which contain difficult instances within the class of locally semicomplete digraphs. It should be noted that Guo's proof does not rely on Theorem 2.6.7. However, due to the non-semicomplete exceptions mentioned above, it seems unlikely that a much simpler proof of Corollary 6.7.8 can be found using Theorems 2.6.7 and 6.6.5.

We conjecture the following generalization of Theorem 2.6.9.

**Conjecture 6.7.9** *There exists a polynomial algorithm for deciding whether a given locally semicomplete digraph $D$ has an $(x,y)$-Hamiltonian path for two specified vertices $x, y$.*

## 6.8 Pancyclicity

Recall that by Moon's theorem (Theorem 2.2.7) and Theorem 2.2.9 every strong semicomplete digraph is vertex-pancyclic, that is, every vertex is contained in a $k$-cycle for every $k = 3, 4, \ldots, n$. Below we use the structure theorem for locally semicomplete digraphs (Theorem 6.6.5) to find a characterization of those locally semicomplete digraphs which are pancyclic (vertex-pancyclic). By the remark above, we need only consider locally semicomplete digraphs that are not semicomplete, corresponding to cases (a) and (b) of Theorem 6.6.5.

Our first goal (Corollary 6.8.4) is a characterization of those round decomposable locally semicomplete digraphs which are (vertex-)pancyclic.

**Lemma 6.8.1** *Let $R$ be a strong round local tournament and let $C$ be a shortest cycle of $R$ and suppose $C$ has $k \geq 3$ vertices. Then for every round labelling $v_0, v_1, \ldots, v_{n-1}$ of $R$ such that $v_0 \in V(C)$ there exist indices $0 < a_1 < a_2 < \ldots < a_{k-1} < n$ such that $C = v_0 v_{a_1} v_{a_2} \ldots v_{a_{k-1}} v_0$.*

**Proof:** Let $C$ be a shortest cycle and let $\mathcal{L} = v_0, v_1, \ldots, v_{n-1}$ be a round labelling of $R$ such that $v_0 \in V(C)$. If the claim is not true, then there exists a number $2 \leq l < k-1$ such that $C = v_0 v_{a_1} v_{a_2} \ldots v_{a_{k-1}} v_0$, where

$0 < a_1 < \ldots < a_{l-1}$ and $a_l < a_{l-1}$. Now the fact that $\mathcal{L}$ is a round labelling of $R$ implies that $v_{l-1} \to v_0$, contradicting the fact that $C$ is a shortest cycle. □

Recall that the **girth** $g(D)$ of a digraph is the length of a shortest cycle in $D = (V, A)$. For a vertex $v \in V$ we let $g_v(D)$ denote the length of a shortest cycle in $D$ through $v$. The next lemma shows that every round local tournament $R$ is $g(R)$-pancyclic.

**Lemma 6.8.2** *A strong round local tournament digraph $R$ on $r$ vertices has cycles of length $k, k+1, ..., r$, where $k = g(R)$.*

**Proof:** By Lemma 6.8.1 we may assume that $R$ contains a cycle of the form $v_{i_1}v_{i_2}...v_{i_k}v_{i_1}$, where $0 = i_1 < i_2 < ... < i_k < r$. Because $D$ is strong, $v_{i_m}$ dominates all the vertices $v_{i_m+1}, \ldots, v_{i_{m+1}}$ for $m = 1, 2, ..., k$. Now it is easy to see that $D$ has cycles of lengths $k, k+1, ..., r$ through the vertices $v_{i_1}, v_{i_2}, ..., v_{i_k}$. □

There is also a very nice structure on cycles through a given vertex in a round local tournament digraph. The proof is very similar to the proof of Lemma 6.8.2.

**Lemma 6.8.3** ([13]) *If a strong round local tournament digraph with $r$ vertices has a cycle of length $k$ through a vertex $v$, then it has cycles of all lengths $k, k+1, \ldots, r$ through $v$.* ◇

**Corollary 6.8.4** ([13]) *Let $D$ be a strongly connected round decomposable locally semicomplete digraph with round decomposition $D = R[S_1, \ldots, S_p]$. Let $V(R) = \{r_1, r_2, \ldots, r_p\}$, where $r_i$ is the vertex of $R$ corresponding to $S_i$. Then*

*(1) $D$ is pancyclic if and only if either the girth of $R$ is 3 or $g(R) \leq \max_{1 \leq i \leq p} |V(S_i)| + 1$.*
*(2) $D$ is vertex-pancyclic if and only if, for each $i = 1, ..., p$, either $g_{r_i}(R) = 3$ or $g_{r_i}(R) \leq |V(S_i)| + 1$.*

**Proof:** As each $S_i$ is semicomplete, it has a Hamiltonian path $P_i$. Furthermore, since $R$ is a strong locally semicomplete digraph, it is Hamiltonian by Theorem 6.4.9. Thus, starting from a $p$-cycle with one vertex from each $S_i$, we can get cycles of all lengths $p+1, p+2, \ldots, n$, by taking appropriate pieces of Hamiltonian paths $P_1, P_2, \ldots, P_p$ in $S_1, \ldots, S_p$. Thus, if $g(R) = 3$, then $D$ is pancyclic by Lemma 6.8.2. If $g(R) \leq \max_{1 \leq i \leq r} |V(S_i)| + 1$, then $D$ is pancyclic by Lemma 6.8.2 and the fact that (by Theorem 2.2.9) every $S_i$ has cycles of lengths $3, 4, \ldots, |V(S_i)|$. If $g(R) > 3$ and, for every $i = 1, ..., r$, $g(R) > |V(S_i)| + 1$, then $D$ is not pancyclic since it has no $(g(R)-1)$-cycle. The second part of the lemma can be proved analogously by first proving that for each $i = 1, 2, \ldots, p$, every vertex in $S_i$ is on cycles of all lengths

$g_{r_i}(R), g_{r_i}(R)+1,\dots,n$ (using Lemma 6.8.3) and then applying Theorem 2.2.9. □

To complete the characterization of (vertex-)pancyclic locally semicomplete digraphs it suffices to prove the following lemma (recall Theorem 6.6.5).

**Lemma 6.8.5** ([13]) *Let $D$ be a strong locally semicomplete digraph on $n$ vertices which is not round decomposable. Then $D$ is vertex-pancyclic.*

**Proof:** If $D$ is semicomplete, then the claim follows from Theorem 2.2.9. So we assume that $D$ is not semicomplete. Thus, $D$ has the structure described in Lemma 6.6.4.

Let $S$ be a minimal separating set of $D$ such that $D-S$ is not semicomplete and let $D_1, D_2, ..., D_p$ be the acyclic ordering of the strong components of $D-S$. Since the subdigraph $D\langle S\rangle$ is semicomplete, it has a unique acyclic ordering $D_{p+1}, ..., D_{p+q}$ with $q \geq 1$ of its strong components. Recalling Lemma 6.6.4(a), the semicomplete decomposition of $D-S$ contains exactly three components $D'_1, D'_2, D'_3$. Recall that the index of the initial component of $D'_2$ is $\lambda_2$. From Theorem 6.3.4 and Lemma 6.6.2, we see that $D'_2 \Rightarrow D'_1 \Rightarrow S \Rightarrow D_1$ and there is no arc between $D'_1$ and $D'_3$.

We first consider the spanning subdigraph $D^*$ of $D$ which is obtained by deleting all the arcs between $S$ and $D'_2$. By Lemma 6.6.4, $D^*$ is a round decomposable locally semicomplete digraph and $D^* = R^*[D_1, D_2, \dots, D_{p+q}]$, where $R^*$ is the round locally semicomplete digraph obtained from $D^*$ by contracting each $D_i$ to one vertex (or, equivalently, $R^*$ is the digraph obtained by keeping an arbitrary vertex from each $D_i$ and deleting the rest). It can be checked easily that $g_v(R^*) \leq 5$ for every $v \in V(R^*)$. Thus $D^*$ is vertex 5-pancyclic by the remark in the proof of Corollary 6.8.4 (in the case when $n=4$, $D$ is easily seen to be vertex-pancyclic so we may assume $n \geq 5$). Thus, it remains to show that every vertex of $D$ lies on a 3-cycle and a 4-cycle.

We define

$$t = \max\{\, i \mid N^+(S) \cap V(D_i) \neq \emptyset, \lambda_2 \leq i < p\},$$

$$A = V(D_{\lambda_2}) \cup \dots \cup V(D_t),$$

$$t' = \min\{\, j \mid N^+(D_j) \cap V(D'_2) \neq \emptyset, p+1 \leq j \leq p+q\}$$

$$\text{and } B = V(D_{t'}) \cup \dots \cup V(D_{p+q}).$$

It follows from Proposition 6.6.6 that $B \mapsto D'_3 \mapsto A$.

Since we have $S \mapsto D_1 \mapsto D_{\lambda_2} \mapsto D'_1 \mapsto S$, every vertex of $S$ is in a 4-cycle and since we have $B \mapsto D'_3 \mapsto A \mapsto D'_1 \mapsto S$, each vertex of $V(D'_3) \cup A \cup V(D'_1)$ is contained in a 4-cycle.

By the definition of $t'$ and $A$, there is an arc $sa$ from $D_{t'}$ to $A$. It follows from Lemma 6.6.4(b) that there is an arc $a's'$ from $A$ to $B$. Let $v \in V(D'_1)$ and $w \in V(D'_3)$ be arbitrarily chosen. Then $savs$ and $s'wa's'$ are 3-cycles.

Suppose $D'_2$ contains a vertex $x$ that is not in $A$, then $A \mapsto x$. We also have $x, s' \in N^+(a')$ and this implies that $x \to s'$. From this we get that $x \mapsto D_{t'}$, in particular, $x \to s$. Hence $xsax$ is a 3-cycle and $xvsax$ is a 4-cycle. Thus, it only remains to show that every vertex of $S \cup A$ is contained in a 3-cycle.

Let $u$ be a vertex of $S$ and let $D_\ell$ be the strong component containing $u$. If $D_\ell$ has at least three vertices, then $u$ lies on a 3-cycle by Theorem 2.2.7. So we assume $|V(D_\ell)| \leq 2$. If $\ell < t'$, then $u$ and $a'$ are adjacent because $D_\ell$ dominates the vertex $s'$ of $B$. If $\ell \geq t'$, then either $u = s$ or $s \to u$ (if $V(D_\ell) = \{s, u\}$, then $usu$ is a 2-cycle) and hence $u, a$ are adjacent. Therefore, in any case, $u$ is adjacent to one of $\{a, a'\}$. Assume without loss of generality that $a$ and $u$ are adjacent. If $u \to a$, then $uavu$ is a 3-cycle. If $a \to u$, then $uwau$ is a 3-cycle because $D'_3 \to A$. Hence, every vertex of $S$ has the desired property.

Finally, we note that $S' = N^+(D'_3)$ is a subset of $V(D'_2)$ and it is also a minimal separating set of $D$. Furthermore, $D - S'$ is not semicomplete. From the proof above, every vertex of $S'$ is also in a 3-cycle. So the proof of the theorem is completed by the fact that $A \subseteq S'$. □

Combining Corollary 6.8.4 and Lemma 6.8.5 we have the following characterization of pancyclic and vertex-pancyclic locally semicomplete digraphs due to Bang-Jensen, Guo, Gutin and Volkmann:

**Theorem 6.8.6** ([13]) *A strong locally semicomplete digraph $D$ is pancyclic if and only if it is not of the form $D = R[S_1, \ldots, S_p]$, where $R$ is a round local tournament digraph on $p$ vertices with $g(R) > \max\{2, |V(S_1)|, \ldots, |V(S_p)|\} + 1$. Furthermore, $D$ is vertex-pancyclic if and only if $D$ is not of the form $D = R[S_1, \ldots, S_p]$, where $R$ is a round local tournament digraph with $g_{r_i}(R) > \max\{2, |V(S_i)|\} + 1$ for some $i \in \{1, \ldots, p\}$, where $r_i$ is the vertex of $R$ corresponding to $S_i$.* ◇

Tewes studied pancyclicity for locally in-tournament digraphs and obtained several bounds on the minimum in-degree which guarantees the existence of a cycle of a given length $k$.

**Theorem 6.8.7** ([67]) *Let $D$ be a strong locally in-tournament digraph on $n$ vertices and let $3 \leq k \leq n$ be an integer such that $\delta^-(D) \geq 3n/(2k+2) - \frac{1}{2}$. Then $D$ has a cycle of length $k$.*

Tewes showed that the bound on the in-degree is the best possible when $k \geq \sqrt{n+1}$. For values of $k$ with $3 \leq k \leq \sqrt{n+1}$ Tewes proved the existence of a function $f(k)$ such that $\delta^-(D) > f(k)$ implies $k$-pancyclicity and proved that the function, which we will not define here (see [67, Definition 4.2]), is the best possible.

Theorem 6.8.7 immediately implies the following.

**Corollary 6.8.8** ([67]) *Let $D$ be a strong locally in-tournament digraph on $n$ vertices such that $\delta^-(D) > 3n/(2k+2) - \frac{1}{2}$ for some integer $3 \leq k \leq n$. Then $D$ is $k$-pancyclic.*

For further work on pancyclicity of locally in-tournament digraphs, see the papers [65] by Peters and Volkmann on vertex 6-pancyclic locally in-semicomplete digraphs and [68] by Tewes on pancyclic orderings of locally in-semicomplete digraphs.

## 6.9 Cycle Factors with a Fixed Number of Cycles

An obvious necessary condition for a digraph $D$ on $n$ vertices to contain a 2-cycle factor is that the girth of $D$ is at most $n/2$. The second power $D = C^2_{2k+1}$ of an odd cycle has girth $k+1$ and $D$ is a 2-strong locally semicomplete digraph. This shows that Theorem 2.8.1 cannot be extended to locally semicomplete digraphs. Confirming a conjecture by Bang-Jensen [10], Guo and Volkmann proved that powers of odd cycles are the only exceptions when $n \geq 8$.

**Theorem 6.9.1** ([47]) *Let $D$ be a 2-strong locally semicomplete digraph on $n \geq 8$ vertices. Then $D$ has a 2-cycle factor such that both cycles have length at least 3 if and only if $D$ is not the second power of an odd cycle.* ◇

Guo and Volkmann have shown that, although Theorem 2.8.1 cannot be extended to locally semicomplete digraphs, there is still enough structure to allow 2-cycle factors with many different lengths.

**Theorem 6.9.2** ([46]) *Let $D$ be a 2-strong locally semicomplete digraph on $n \geq 8$ vertices. If $D$ has no induced cycle of length at least 4, then for every $3 \leq k \leq n-3$, $D$ has a pair of disjoint cycles of lengths $k$ and $n-k$, respectively.*

Meierling and Volkmann proved the following extension of Theorem 6.9.1 in the case of oriented graphs.

**Theorem 6.9.3** ([61]) *Every 2-strong locally in-tournament digraph $D$ on $n \geq 8$ vertices has a pair of complementary cycles if and only if $D$ is not the second power of an odd cycle.* □

Bang-Jensen and Nielsen [28] gave a polynomial algorithm for checking whether a given locally semicomplete digraph has a 2-cycle factor.

Guo posed the following problem.

**Problem 6.9.4** ([43]) *Does there exists an analogue of Theorem 2.8.6 for locally semicomplete digraphs?*

Denote by $g^i(D)$ the length of a longest induced cycle in $D$. If $D$ is locally semicomplete, then $g^i(D) \leq 4$ unless $D$ is round decomposable. This is due to the following easy consequence of Theorem 6.6.5.

**Theorem 6.9.5** *A locally semicomplete digraph $D$ has the property that either $D$ is round decomposable with a unique round decomposition, or the vertex set of every cycle induces a digraph containing a cycle on at most 4 vertices.*

Gould and Guo found an analogue of Theorem 2.8.6 for the case of round decomposable locally semicomplete digraphs.

**Theorem 6.9.6** ([42]) *Let $D = R[S_1, S_2, \dots, S_r]$ be a $k$-strong round decomposable locally semicomplete digraph on $n \geq 2(k-1)g(R)$ vertices. Then $D$ contains a $g(R)$-cycle $C$ such that $D - V(C)$ is $(k-1)$-strong.*

**Corollary 6.9.7** ([42]) *Let $D$ be a round decomposable $k$-strong locally semicomplete digraph on $n \geq 2(k-1)g^i(D)$ vertices. Then for every choice of $k$ integers $n_1, n_2, \dots, n_k \geq g^i(D)$ such that $n_1 + n_2 + \dots + n_k = n$, $D$ has a $k$-cycle factor $C_1, C_2, \dots, C_k$ such that $C_i$ has length $n_i$, $i \in [k]$.*

**Proof:** Let $D = R_1[S_1^1, S_2^1, \dots, S_{|V(R_1)|}^1]$ be the round decomposition of $D$. By Theorem 6.9.6, the induced subdigraph $R_1$ of $D_1$ has a shortest cycle $C_1'$ such that $D_2 = D - V(C_1')$ is $(k-1)$-strong. It is easy to check that $D_2$ is also round decomposable with unique round decomposition $R_2[S_1^2, \dots, S_{|V(R_2)|}^2]$. Hence we can repeat the process $k-1$ more times, always picking a shortest cycle in the current round digraph $R_i$, to obtain disjoint cycles $C_1', C_2' \dots, C_k'$ each of which are cycles in $R$ and have length at most $g^i(D)$. Note that, the choice of the cycles $C_1', C_2' \dots, C_k'$ (each going once around in the round ordering of $R_1$) implies that every vertex of $D$ has at least one in-neighbour and at least one out-neighbour on $C_j'$ for $j \in [k]$. This implies that for each $i \in [k]$ we can insert exactly $n_i - |V(C_i')|$ vertices among the vertices of $V(D) - (V(C_1') \cup \dots \cup V(C_k'))$ into $C_i'$ such that the resulting subdigraph $H_i$ is a strong locally semicomplete digraph on $n_i$ vertices. Thus if $C_i$ denotes a Hamiltonian cycle of $H_i$, $i \in [k]$, we see that $C_1, C_2, \dots, C_k$ is the desired $k$-factor. □

Gould and Guo showed that in the case of $k$-strong evil locally semicomplete digraphs one can always obtain a cycle factor on $k$ cycles such that almost all of the cycles are short.

**Theorem 6.9.8** ([42]) *Let $D$ be an evil locally semicomplete digraph on $n \geq 20(k-1)$ vertices. Then $D$ has a cycle factor consisting of $k$ cycles such that at least $(k-2)$ of these have length at most 4.*

## 6.10 Arc-Disjoint Paths

The next topic we consider is the WEAK-$k$-LINKAGE problem.

> WEAK-$k$-LINKAGE
> **Input:** A digraph $D = (V, A)$ and not necessarily distinct vertices $s_1, \dots, s_k, t_1, \dots, t_k$.
> **Question:** Does $D$ contain arc disjoint paths $P_1, \dots, P_k$ such that $P_i$ is an $(s_i, t_i)$-path for $i \in [k]$?

Recall from Theorem 2.5.1 that Fradkin and Seymour gave a polynomial algorithm for the WEAK-$k$-LINKAGE problem in digraphs of bounded independence number. Using that result, Theorem 6.6.5 as well as some new results concerning the structure of arc-disjoint paths in round decomposable digraphs, which are described in Section 8.5.2, Bang-Jensen and Maddaloni obtained the following.

**Theorem 6.10.1** ([26]) *For every fixed natural number $k$ there exists a polynomial algorithm for the* WEAK-$k$-LINKAGE *problem for locally semicomplete digraphs.*

**Conjecture 6.10.2** *For every fixed natural number $k$ the* WEAK-$k$-LINKAGE *problem is polynomially solvable for locally in-semicomplete digraphs.*

The conjecture is open even for $k = 2$. Below we illustrate a solution, due to Bang-Jensen, for the special case of the WEAK 2-LINKAGE problem where we are seeking arc-disjoint $(x, y)$- and $(y, z)$-paths for distinct vertices $x, y, z$. In this case it is possible to find a complete charaterization even for extended locally in-semicomplete digraphs

Two vertices are called **similar** if and only if they are non-adjacent and have the same in- and out-neighbours. Note that if $x, y$ are non-adjacent vertices with a common out-neighbour $w$ in an extended locally in-semicomplete digraph, then $x$ and $y$ are similar vertices, by the definition of an extension and the definition of a locally in-semicomplete digraph.

The following lemma can be proved along the same lines as Lemma 6.10.4.

**Lemma 6.10.3** ([16]) *Let $D$ be a strong extended locally in-semicomplete digraph and let $x, y$ be distinct vertices of $D$. Then $D$ has arc-disjoint paths $P, Q$ such that $P$ is an $(x, y)$-path and $Q$ is a $(y, x)$-path if and only if there is no arc $a$ such that $D - a$ contains no $(x, y)$-path and no $(y, x)$-path.* ◇

**Lemma 6.10.4** ([5]) *Let $D = (V, A)$ be an extended locally in-semicomplete digraph and $x, y, z$ vertices of $D$ such that $x \neq z$ and $D$ contains a path from $y$ to $z$. If $D$ has arc-disjoint $(x, y)$-, $(x, z)$-paths, then $D$ contains arc-disjoint $(x, y)$-, $(y, z)$-paths.*

**Proof:** Let $P_1$ and $P_2$ be arc-disjoint paths such that $P_2$ is an $(x,z)$-path and $P_1$ is a minimal $(x,y)$-path. If $y \in V(P_2)$, or $yx \in A$, then the claim is trivial, so we assume that none of these hold. We can also assume that $x$ and $y$ are not similar vertices, because if they are, then $y$ dominates the successor of $x$ on $P_2$ and again the claim is trivial.

If $D$ has a $(y,z)$-path whose first intersection with $V(P_1) \cup V(P_2)$ (starting from $y$) is on $P_2$, then the desired paths clearly exist. Hence we may assume that $D$ contains a path from $y$ to $V(P_1) \cup V(P_2) - y$ whose only vertex $w$ from $V(P_1) \cup V(P_2) - y$ is in $V(P_1) - V(P_2)$. Now choose $P$ among all such paths so that $w$ is as close as possible to $x$ on $P_1$. By the assumption above $w \neq x$. Let $u$ ($v$) denote the predecessor of $w$ on $P_1$ ($P$), i.e., $u = w^-_{P_1}$ and $v = w^-_P$.

Suppose first that $u$ and $v$ are not adjacent. Then, by the remark just before Lemma 6.10.3, $u$ and $v$ are similar. Now the choice of $P$ implies that $v = y$ (otherwise the predecessor of $v$ on $P$ dominates $u$, contradicting the choice of $P$). By the assumption that $x$ and $y$ are not similar we conclude that $u \neq x$, but then $u^-_{P_1} y \in A$, contradicting the minimality of $P_1$.

Thus we may assume that $u$ and $v$ are adjacent. By the choice of $P$, this implies that $uv \in A$. Choose $r$ as the first vertex on $P$ which is dominated by $u$. By the minimality of $P_1$, $r \neq y$. Let $s$ be the predecessor of $r$ on $P$. The choice of $r$ and $P$ implies that $u$ and $s$ are similar. Thus as above, we must have $s = y$, and since $u \neq x$ we reach a contradiction as before. □

The digraph $D = (V, A)$ with vertex set $V = \{x, u, v, y, z\}$ and arc set $A = \{xu, uv, vy, yu, vz, xz\}$ shows that the conclusion of Lemma 6.10.4 does not hold for general digraphs.

Using Lemma 6.10.4 we can now characterize those extended locally in-semicomplete digraphs which do not have arc-disjoint $(x,y)$-, $(y,z)$-paths.

**Theorem 6.10.5** ([5]) *An extended locally in-semicomplete digraph $D$ has arc-disjoint $(x,y)$-, $(y,z)$-paths if and only if it has an $(x,y)$-path and a $(y,z)$-path and $D$ has no arc $e$ such that $D-e$ has no $(x,y)$-path and no $(y,z)$-path.*

**Proof:** Clearly if $D$ has an arc $e$ whose removal separates $x$ from $y$ and $y$ from $z$, then the paths cannot exist. Now assume that $D$ has no such arc and that $D$ has an $(x,y)$-path and a $(y,z)$-path. We prove that $D$ has the desired paths. By Lemma 6.10.3 we may assume $x \neq z$.

By Lemma 6.10.4, we may assume that $D$ contains no pair of arc-disjoint $(x,y)$-,$(x,z)$-paths. Thus, by Menger's theorem, there exists an arc $e = uv$ such that $D - e$ has no path from $x$ to $\{y,z\}$. Let $X = \{w : \exists (x,w) -$ path in $D - e\}$ and $B = V(D) - X$. Then $x \in X$, $y, z \in B$ and the only arc from $X$ to $B$ is $e$.

Since $D$ contains an $(x,y)$-path, $D[X]$ has an $(x,u)$-path and $D[B]$ has a $(v,y)$-path. $D[B]$ also has a $(y,z)$-path, since $e$ does not destroy all paths from $y$ to $z$.

If $v = y$, the desired paths clearly exist (and can in fact be chosen vertex disjoint). If $v = z$, then it follows from our assumption that there is no arc $a$ in $D[B]$ which separates $y$ from $z$ and also $z$ from $y$. Now it follows from Lemma 6.10.3 that $D[B]$ contains arc-disjoint $(z, y)$-, $(y, z)$-paths and hence $D$ contains the desired paths. Thus we may assume $v \neq y, z$.

Now it is clear that the desired paths exist if and only if $D[B]$ has arc-disjoint $(v, y)$-, $(y, z)$-paths. By induction this is the case unless there exists an arc $e' = ab$ in $D[B]$ such that $D[B] - e'$ has no path from $v$ to $y$ and no path from $y$ to $z$, but then $e'$ separates $x$ from $y$ and $y$ from $z$ in $D$, contradicting the assumption that $D$ has no such arc. □

The proof above is constructive and hence we have the following

**Corollary 6.10.6** ([6]) *There exists a polynomial algorithm which, given an extended in-semicomplete digraph $D$ and distinct vertices $x, y, z$, either returns a pair of arc-disjoint $(x, y)$-, $(y, z)$-paths or an arc $a$ such that $D - a$ has no $(x, y)$-path and no$(y, z)$-path.* □

## 6.11 Vertex-Disjoint Paths

We now turn to vertex-disjoint paths with prescribed end vertices.

| |
|---|
| $k$-DISJOINT PATHS<br>**Input:** A digraph $D = (V, A)$ and distinct vertices $s_1, \ldots, s_k, t_1, \ldots, t_k$.<br>**Question:** Does $D$ contain vertex disjoint paths $P_1, \ldots, P_k$ such that $P_i$ is an $(s_i, t_i)$-path for $i \in [k]$? |

Note that this is the same as the $k$-LINKAGE problem defined in Chapter 1, except that now we insist that the paths are completely disjoint.

### 6.11.1 Algorithmic Results

Recall from Theorem 2.5.11 that for every fixed pair of integers $p, k$ the $k$-DISJOINT PATHS problem is polynomially solvable for all digraphs whose vertex sets can be partitioned into $p$ disjoint sets, each of which induces a semicomplete digraph.

Bang-Jensen, Christiansen and Maddaloni used this result, Theorem 6.6.5 and a number of new results, primarily on linkings in round decomposable digraphs to obtain a polynomial algorithm for the $k$-DISJOINT PATHS problem in locally semicomplete digraphs.

We show below that the case of round digraphs can be solved using the algorithms for the $k$-DISJOINT PATHS problem on acyclic digraphs.

**Theorem 6.11.1** ([11]) *For every fixed k, there exists a polynomial algorithm which solves the k-*DISJOINT PATHS *problem on round digraphs.*

**Proof:** Let $D$ be a round digraph with round ordering $v_1, ..., v_n$ and let $\Pi = \{(s_1, t_1), ..., (s_k, t_k)\}$ be a set of pairs of vertices of $D$ for which we seek a $\Pi$-linkage. Given $j \in [n-1]$, we say that an arc $v_a v_b \in A(D)$ is **over** $v_j v_{j+1}$ if $v_b \in \{v_{j+1}, v_{j+2}, ..., v_{a-1}\}$. Note that the removal of all the arcs over $v_j v_{j+1}$ from $D$ leaves an acyclic digraph. We show that if $(D, \Pi)$ has a $\Pi$-linkage, then there exists a linkage such that each of the paths uses at most one arc over any $v_j v_{j+1}$, namely the linkage that minimizes the total number of used vertices.

Suppose, by contradiction, that an $(s_i, t_i)$-path $P$ uses two arcs over $v_j v_{j+1}$ and call them $u_1 w_1$ and $u_2 w_2$. Assume without loss of generality that the arc $u_1 w_1$ precedes $u_2 w_2$ on the path $P$. There are four possibilities for the relative positions of the four vertices in the round ordering:

$$(u_1, u_2, w_1, w_2),\ (u_2, u_1, w_1, w_2),\ (u_1, u_2, w_2, w_1),\ (u_2, u_1, w_2, w_1).$$

In all these cases the path $P$ can be shortened by using, for instance, the arc $u_1 u_2$ in the first case and $u_1 w_2$ in the other cases (such arcs exist by the round property). It follows that $P$ uses at most one arc over $v_j v_{j+1}$.

A polynomial algorithm is obtained by selecting a $j \in [n-1]$, then for every choice of an ordered $h$-tuple of pairs $((s_{i_1}, t_{i_1}), ..., (s_{i_h}, t_{i_h}))$ (with $0 \leq h \leq k$) and every choice of arcs $u_1 w_1, ..., u_h w_h$ over $v_j v_{j+1}$ we do the following: construct the digraph $D'$ by deleting all the arcs over $v_j v_{j+1}$ from $D$ and run the algorithm for $k$-linkage on acyclic digraphs (from Theorem 3.4.1) with input $D'$ and terminals $(s_{i_1}, u_1), (w_1, t_{i_1}), ..., (s_{i_h}, u_h), (w_h, t_{i_h})$ plus the remaining original pairs. If a solution is found, construct a solution for the original instance by using the selected arcs $u_1 w_1, ..., u_h w_h$. If there is no solution for each of the possible choices, it means there is no linkage using at most $k$ arcs over $v_j v_{j+1}$, and hence no linkage at all.

The above algorithm involves running a polynomial number of times the polynomial algorithm from Theorem 3.4.1 and hence is polynomial. □

**Theorem 6.11.2** ([11]) *For every fixed k, there exists a polynomial algorithm to solve the k-*DISJOINT PATHS *problem on round decomposable digraphs.*

Lemma 6.6.4 implies that if $D$ is an evil locally semicomplete digraph, then $D$ can be covered by 3 disjoint semicomplete subdigraphs of $D$ (e.g. the digraphs $D'_3, D'_2, D\langle V(S) \cup V(D'_1)\rangle$). In fact two semicomplete digraphs always suffice [54] but we only need the weaker version below. By Corollary 6.5.6 it is possible to decide in polynomial time whether a locally semicomplete digraph $D$ is round decomposable. By running the algorithm from Theorem 6.11.2 if $D$ is round-decomposable and the algorithm from Theorem 2.5.11 if $D$ is not round-decomposable, we get the following theorem.

**Theorem 6.11.3** ([11]) *For every fixed $k$, there exists a polynomial algorithm to solve the $k$-*DISJOINT PATHS *problem on locally semicomplete digraphs.*

We believe that the following much stronger assertion holds. The conjecture is open already for $k = 2$.

**Conjecture 6.11.4** *For every fixed $k$, there exists a polynomial algorithm to solve the $k$-*DISJOINT PATHS *problem on locally in-semicomplete digraphs.*

### 6.11.2 Sufficient Conditions for Being $k$-Linked

Recall that by Theorem 2.5.15, every $452k$-strong semicomplete digraph is $k$-linked. Our first result is a characterization, due to Bang-Jensen, Christiansen and Maddaloni, of $k$-linked round-decomposable digraphs. We need the following lemma.

**Lemma 6.11.5** ([11]) *If $D$ is digraph which has a decomposition as $D = R[M_1, ..., M_r]$, with $R$ round, such that $d^+(M_i) \geq 2k - 1$ for $i = 1, ..., r$, then $D$ is $k$-linked.*

**Proof:** We use induction on $k$.

For $k = 1$, the above condition, together with the round property of $R$, implies strong connectivity for $D$, so there is a path between each pair of vertices.

Assume that the statement is true for $k$, we prove that every digraph decomposable as $D = R[M_1, ..., M_r]$, with $R$ round such that $d^+(M_i) \geq 2k + 1$ for $i = 1, ..., q$ is $k + 1$-linked. Suppose that we want a linking between $s_1, ..., s_{k+1} \in V(D)$ and $t_1, ..., t_{k+1} \in V(D)$, respectively. We construct an $(s_1, t_1)$-path $P$ whose removal leaves a digraph $D' = Q[M'_1, ..., M'_q]$, with $Q$ round and $d^+_{D'}(M'_i) \geq 2k-1$ for $i = 1, ..., q$. Thus, by the induction hypothesis, $D'$ is $k$-linked, so we are done.

The path $P$ starts from $s_1 \in M_i$ and uses an available widest arc: an arc $s_1v$ such that $v \notin \{s_2, ..., s_{k+1}, t_2, ..., t_{k+1}\}$ and $v \in M_j$, with $M_j$ maximizing the distance from $M_i$ in the round ordering of $R$, namely for every $l$ such that $M_i < M_j < M_l$ in the round ordering $s_1$ has no arc to $M_l$; the path $P$ keeps using widest available arcs until a vertex adjacent to $t_1$ is reached, in which case the path continues to $t_1$. Now for $i = 1, ..., r$, define $M'_i := M_i - V(P)$, let $r'$ be the number of nonempty sets of the form $M'_i$, and $R'$ be the round digraph obtained from $R$ by removing the vertices $v_i$ such that $M'_i = \emptyset$. The digraph $D' = R'[M'_1, ...M'_{r'}]$ is as desired. Indeed, for every $i$ there do not exist three vertices $x, y, z$ of $P$ inside $N^+_D(M_i)$, since, by the fact that $x, y, z \in N^+_D(M_i)$ and by the round property of $R$, one of the vertices dominates the other two or the three vertices belong to the same module in the decomposition. In both cases one of the arcs of $P$ would not be the widest available or will not be directed to the target. It follows that for every $i$, $N^+_{D'}(M'_i)$ has size at least $2k-1$, so $D'$ has the desired property. □

**Corollary 6.11.6** ([11]) *Let $D$ be a digraph on $n \geq 2k$ vertices that is not semicomplete and is decomposable as $D = R[M_1, ..., M_r]$, with $R$ round and $M_1, ..., M_r$ semicomplete. The digraph $D$ is $k$-linked if and only if it is $(2k-1)$-strong.*

**Proof:** Suppose that $D$ is $(2k-1)$-strong. Given that $D = R[M_1, ..., M_r]$ is not semicomplete, we have $r \geq 3$ and for every $i$, $D - N^+(M_i)$ is non-empty. It follows that for every $i$, $N^+(M_i)$ is a separator and hence must be of size at least $2k-1$. Therefore $D$ satisfies the hypothesis of Lemma 6.11.5 and thus $D$ is $k$-linked.

Vice versa: a $k$-linked digraph on $n \geq 2k$ vertices must necessarily be $(2k-1)$-strong, otherwise a set of size at most $2k-2$ would separate two vertices $s, t$ of the digraph, so if these vertices formed the first $k-1$ pairs and $s, t$ the $k$-th pair, there is no good linkage. □

Corollary 6.11.6 immediately applies to round decomposable digraphs.

**Theorem 6.11.7** ([11]) *Let $D$ be a round decomposable digraph on $n \geq 2k$ vertices that is not semicomplete. Then $D$ is $k$-linked if and only if it is $(2k-1)$-strong.*

Note that the decomposition of Lemma 6.11.5, $D = R[M_1, ..., M_r]$, need not be a proper decomposition (that is, $R \neq D$), indeed even if $|M_i| = 1$ for every $i$, the proof holds. Therefore the previous results hold for round digraphs too.

**Theorem 6.11.8** ([11]) *Let $k$ be an integer. A round digraph on $n \geq 2k$ vertices is $(2k-1)$-strong if and only if it is $k$-linked.*

Bang-Jensen proved the following result, which generalizes a previous result by Thomassen for semicomplete digraphs.

**Theorem 6.11.9** ([8]) *There exists, for each natural number $k$, a natural number $f(k)$ such that every $f(k)$-strong locally semicomplete digraph is $k$-linked.*

We believe that Pokrovskiey's result on $k$-linked semicomplete digraphs (Theorem 2.5.15) can be extended to locally semicomplete digraphs.

**Conjecture 6.11.10** *There exists a constant $B$ such that every $Bk$-strong locally semicomplete digraph is $k$-linked.*

By Theorems 6.11.7 and 2.5.15 the conjecture holds for locally semicomplete digraphs that are not evil.

Bang-Jensen, Christiansen and Maddaloni extended Theorem 2.5.12 to locally semicomplete digraphs. The proof of Theorem 6.11.11 makes extensive use of Theorem 6.6.5 and several refinements of this concerning the structure of evil locally semicomplete digraphs.

**Theorem 6.11.11** ([11]) *Every 5-strong locally semicomplete digraph is 2-linked.*

## 6.12 Arc-Disjoint Spanning Subdigraphs

Recall from Theorem 2.12.14 that every 2-arc-strong semicomplete digraph which is not the second power of a 4-cycle has an arc-decomposition into two spanning strong subdigraphs. It is easy to see that the second power $C_n^2$ of an $n$-cycle is a 2-strong locally semicomplete digraph and that it has an arc-decomposition into two spanning strong subdigraphs if and only if $n$ is odd. Hence there are infinitely many 2-arc-strong locally semicomplete digraphs, which do not have an arc-decomposition into two spanning strong subdigraphs. Bang-Jensen and Huang proved that second powers of odd cycles are the only exceptions.

**Theorem 6.12.1** ([23]) *A 2-arc-strong locally semicomplete digraph $D = (V, A)$ has an arc-decomposition into two spanning strong subdigraphs $D_1 = (V, A_1)$, $D_2 = (V, A_2)$ if and only if $D$ is not the second power of an even cycle. Furthermore, there exists a polynomial algorithm for constructing an arc-decomposition into two spanning strong subdigraphs whenever the input is a 2-arc-strong locally semicomplete digraph which is not the second power of an even cycle.*

The complicated proof makes heavy use of Theorem 6.6.5 as well as several other tools.

Since no second power of a cycle is 3-arc-strong, Theorem 6.12.1 implies the following:

**Corollary 6.12.2** *Every 3-arc-strong locally semicomplete digraph has an arc-decomposition into two spanning strong subdigraphs.* □

Conjecture 2.12.12 does not hold if we replace 'semicomplete digraph' by 'locally semicomplete digraph' as by Theorem 6.12.1 the second power of an even cycle cannot be decomposed into two arc-disjoint Hamiltonian cycles. If $n$ is relatively prime to both 2 and 3, then it is easy to see that $C_n^3$ can be decomposed into three arc-disjoint Hamiltonian cycles. In fact, such a decomposition does not exist for any other $n$.

**Proposition 6.12.3** ([23]) *If $n$ is not relatively prime to 2 or 3, then $C_n^3$ cannot be decomposed into arc-disjoint Hamiltonian cycles.*

**Conjecture 6.12.4** ([23]) *A $k$-arc-strong locally semicomplete digraph $D$ can be decomposed into $k$ arc-disjoint strong spanning subdigraphs if and only if $D$ is not the $k$-th power $C_n^k$ of an $n$-cycle $C_n$ where $n$ is divisible by some $i = 2, 3, \ldots, k$.*

If true the above conjecture would imply the following:

**Conjecture 6.12.5** ([23]) *For every natural number $k$ there exists a natural number $f(k)$ such that every $k$-arc-strong locally semicomplete digraph $D = (V, A)$ with $\delta(D) \geq f(k)$ has an arc-decomposition $A = A_1 \cup A_2 \cup \ldots \cup A_k$ such that each of the spanning subdigraphs $D_i = (V, A_i)$, $i = 1, 2, \ldots, k$ are strong.*

The following result due to Bang-Jensen and Huang shows that for round decomposable locally semicomplete digraphs that are not round we can choose one of the digraphs $D_1, D_2$ in the decomposition into arc-disjoint spanning subdigraphs to be a Hamiltonian cycle.

**Lemma 6.12.6** ([23]) *Let $D = R[D_1, D_2, \ldots, D_r]$, $r \geq 3$ be a 2-arc-strong round decomposable locally semicomplete digraph for which at least one $D_i$ has more than one vertex. Then $D$ contains a Hamiltonian cycle $C$ such that $D - A(C)$ is strong. Furthermore, we can find $C$ in polynomial time.*

**Proof:** First note that if $|V(D_i)| > 1$ then $D_i$ contains a Hamiltonian cycle as $D_i$ is semicomplete. Denote by $C_i$ a Hamiltonian cycle in $D_i$ for each $i$ with $|V(D_i)| > 1$. Let $s_1, s_2, \ldots, s_r$ be a round labeling of $R$. Form a Hamiltonian cycle $C$ of $D$ by replacing each $s_i$ for which $|V(D_i)| > 1$ by $C_i$ minus one arc $a_i = u_iv_i$. Let $u_i$ be the only vertex in $D_i$ when $|V(D_i)| = 1$. To show that $D - A(C)$ is strong, form a subdigraph $D'$ of $D$ as follows: First of all $D'$ contains all arcs $a_i$. For each $i, j$ such that $|V(D_i)| > 1$, $|V(D_j)| > 1$ and $|V(D_k)| = 1$ with $i < k < j$, $D'$ contains all arcs of the two paths $v_iu_{i+1}u_{i+3} \ldots u_{j-2}u_j$, $v_iu_{i+2}u_{i+4} \ldots u_{j-1}u_j$ when $j = i + 2p + 1$ modulo $n$ for some $p$ and all arcs of the two paths $v_iu_{i+1}u_{i+3} \ldots u_{j-1}u_j$, $v_iu_{i+2}u_{i+4} \ldots u_{j-2}u_j$ when $j = i + 2p$ modulo $n$ for some $p$. It is easy to see that $D'$ is strong and contains no arc from $C$. Every vertex of $D$ which is not in $D'$ (they all belong to $D_i$'s with at least 3 vertices) can be added as an ear via a path of length 2 without using any arc from $C$ and hence $D'$ is strong. Hence $D - A(C)$ is strong. □

As mentioned in Section 2.12, Thomassen conjectured that every 3-strong tournament has a pair of arc-disjoint Hamiltonian cycles. Bang-Jensen and Huang posed the following extension of Thomassen's conjecture.

**Conjecture 6.12.7** ([23]) *Every 3-strong arc-local tournament digraph has a pair of arc-disjoint Hamiltonian cycles.*

Li and Han [58] have verified the conjecture for round decomposable locally semicomplete digraphs.

We now consider arc-disjoint in-and out-branchings and start with the following consequence of Theorem 6.12.1.

**Corollary 6.12.8** *Every 2-arc-strong locally semicomplete digraph* $D = (V, A)$ *contains arc-disjoint in- and out-branchings* $B_u^-, B_v^+$ *for every choice of* $u, v \in V$. *Furthermore, there is a polynomial algorithm which constructs such a pair of branchings in a given 2-arc-strong locally semicomplete digraph.*

**Proof:** If $D$ is not the second power of an even cycle, then by Theorem 6.12.1, $D$ has an arc-decomposition into strong spanning subdigraphs $D_1, D_2$ and such a decomposition can be found in polynomial time. Now we can take $B_u^-$ to be any in-branching rooted at $u$ in $D_1$ and $B_v^+$ to be any out-branching rooted at $v$ in $D_2$. Suppose that $D$ is the second power of an even cycle $v_1v_2 \dots v_{2k}v_1$. Assume without loss of generality that $v = v_1$. It is easy to verify that $D' = D - A(P)$ is strong where $P$ is the Hamiltonian path $v_1v_3v_4 \dots v_{2k}v_2$. Thus we can take any in-branching rooted at $u$ in $D'$. Clearly this implies the complexity claim. □

No characterization is known of those locally semicomplete digraphs that do not have a pair of disjoint branchings $B_s^+, B_t^-$ for given distinct vertices $s, t$. When $s = t$, that is, the two branchings are rooted at the same vertex, Bang-Jensen and Huang found such a characterization [22]. The complicated proof makes heavy use of Theorem 6.6.5 as well as several other tools such as arc-contractions, etc. The characterization in [22] implies the following.

**Theorem 6.12.9** ([22]) *There exists a polynomial algorithm for deciding if a given locally semicomplete digraph* $D$ *with a special vertex* $s$ *has a pair of arc-disjoint in- and out-branchings* $B_s^-, B_s^+$ *rooted at* $s$.

## 6.13 Kernels and Quasi-Kernels

A set $K$ of vertices in a digraph $D = (V, A)$ is a **kernel** if $K$ is independent and the first closed neighbourhood of $K$, $N^-[K]$, is equal to $V$. This notion was introduced by von Neumann in [71]; kernels have found many applications, for instance in game theory (a kernel represents a set of winning positions, cf. [71] and Chapter 14 in the book by Berge [31]), in logic [32] and in list edge-colouring of graphs (see [16, Section 17.9]). The problem KERNEL is to decide whether a given digraph $D$ has a kernel. Chvátal (see [41], p. 204) proved that KERNEL is $\mathcal{NP}$-complete.

Using Theorem 6.6.5 Bang-Jensen, Guo, Gutin and Volkmann showed that KERNEL is polynomial for locally semicomplete digraphs.

**Theorem 6.13.1** ([13]) KERNEL *is polynomially solvable for locally semicomplete digraphs.*

The key lemma used in the proof of Theorem 6.13.1 is the following. Note that for the kernel problem, the difficulty lies in handling locally semicomplete digraphs that are neither semicomplete nor evil as the independence number of the later is at most 2.

**Lemma 6.13.2** ([13]) *There exists a polynomial algorithm to decide if a round local tournament has a kernel.*

**Proof:** Let $R$ be a round local tournament with vertex set $\{v_0, v_1, \ldots, v_{r-1}\}$. Let $T_R$ be a clock with a dial on $r$ hours $v_0, v_1, \ldots, v_{r-1}$ corresponding to the vertices of $R$, and define for each $v_i$ the time interval $T_i = [v_i, v_{i+d^+(v_i)}]$. It is easy to see that $R$ has a kernel if and only if the dial of the time clock $T_R$ can be covered by pairwise non-overlapping time intervals. This can be checked in time $O(r^2)$. Note that if $R$ is not strong and $R$ has a kernel, then it is unique (this corresponds to a unique way to cover the dial of $T_R$). □

**Conjecture 6.13.3** ([17]) Kernel *is polynomially solvable for locally in-semicomplete digraphs.*

**Problem 6.13.4** ([17]) *What is the complexity of* Kernel *for path-mergeable digraphs?*

A digraph $D$ is **critically kernel imperfect** if $D$ has no kernel but every proper induced subdigraph of $D$ has a kernel. Galeana-Sánchez and Olsen [40] characterized locally semicomplete digraphs that are critically kernel imperfect. For each integer $m \geq 4$ the semicomplete digraph $Q_m$ is obtained from the complete digraph $\overleftrightarrow{K}_m$ by deleting the arcs of a Hamiltonian cycle.

**Theorem 6.13.5** ([40]) *A locally semicomplete digraph is critically kernel imperfect if and only if it is either the second power $C_7^2$ of a 7-cycle, an odd cycle or one of the semicomplete digraphs $Q_m$ defined above, where $m \geq 4$.*

A **quasi-kernel** in a digraph $D = (V, A)$ is an independent set $Q \subseteq V$ with the property that for every vertex $x \in V - Q$, either $x$ dominates some vertex in $Q$ or $x$ dominates a vertex $y \in V - Q$ which has an out-neighbour in $Q$. In other words every vertex has distance at most 2 to $Q$. Chvátal and Lovász [35] proved that every digraph has a quasi-kernel. For a beautiful proof of this, due to Thomassé see [34].

Clearly a digraph containing a vertex $x$ of out-degree zero cannot have two disjoint quasi-kernels as $x$ must belong to every quasi-kernel. Heard and Huang showed that not having a sink (i.e. a vertex of out-degree zero) is a sufficient condition for a locally semicomplete digraphs to have two disjoint quasi-kernels.

**Theorem 6.13.6** ([51]) *Every locally semicomplete digraph $D$ with $\delta^+(D) \geq 1$ has a pair of disjoint quasi-kernels.*

In [49] an example of a locally out-tournament digraph of minimum out-degree 1 and no pair of disjoint quasi-kernels is given.

**Problem 6.13.7** *Characterize locally out-semicomplete digraphs with a pair of disjoint quasi-kernels.*

**Problem 6.13.8** *Does every locally in-semicomplete digraph $D$ which has $\delta^+(D) \geq 1$ have a pair of disjoint quasi-kernels?*

## 6.14 Feedback Sets in Locally Semicomplete Digraphs

Most of the results we discuss in this section are due to Bang-Jensen, Maddaloni and Saurabh [27]. Recall that a set of arcs (vertices) $X$ is a **feedback arc set** (**feedback vertex set**) in a digraph $D$ if $D - X$ is acyclic. The problems we focus on in this section are the decision versions of the feedback arc set and feedback vertex set problems, namely.

| FEEDBACK ARC SET ($k$-FAS) | **Parameter:** $k$ |
|---|---|
| **Input:** A digraph $D$ and a positive integer $k$. | |
| **Question:** Does there exist a feedback arc set of size at most $k$? | |

| FEEDBACK VERTEX SET ($k$-FVS) | **Parameter:** $k$ |
|---|---|
| **Input:** A digraph $D$ and a positive integer $k$. | |
| **Question:** Does there exist a feedback vertex set of size at most $k$? | |

On general digraphs both $k$-FAS and $k$-FVS are equivalent [38]. Indeed, it is not difficult to see that given an instance $(D, k)$ of $k$-FAS one can produce an equivalent instance of $k$-FVS by taking the line digraph of $D$. On the other hand, to obtain a reduction from $k$-FVS to $k$-FAS consider the following: given an instance $(D, k)$ of $k$-FVS one can produce an equivalent instance of $k$-FAS by performing the vertex splitting procedure on $D$. The vertex splitting procedure consists in substituting each vertex $v$ by two new vertices $v^+, v^-$, deleting $v$, adding the arc $v^-v^+$, making all out-neighbours of $v$ in $D$ out-neighbours of $v^+$ and making all in-neighbours of $v$ in $D$ in-neighbours of $v^-$.

### 6.14.1 Feedback Vertex Sets

**Lemma 6.14.1** ([27]) *Let $D$ be a round digraph. For all $v \in V(D)$ the digraphs $D - N^+(v)$ and $D - N^-(v)$ are both acyclic and the minimum feedback vertex sets of $D$ are exactly the minimum neighbourhoods of $D$.*

**Proof:** The first claim follows from the definition of a round digraph: Let $v_1, v_2, \ldots, v_n$ be a round labelling of $V(D)$. If $v_iv_j$ is an arc then $v_{i+1}v_j, \ldots, v_{j-1}v_j$ and $v_iv_{i+1}, \ldots, v_iv_{j-1}$ are also arcs and every cycle $C$ contains an arc which is over $v_iv_{i+1}$ for every $i \in [n]$. So deleting any in- or out-neighbourhood kills all cycles. To prove the second part, let $F$ be a feedback vertex set of $D$. There exists a $v \in V$ such that $\min(d^+_{D-F}(v), d^-_{D-F}(v)) = 0$, in particular $F$ contains a set of the form $N^+(v)$ or $N^-(v)$.

Let $N = \{v_a, v_{a+1}..., v_b\}$ be an out-neighbourhood and suppose there exist a cycle $C$ in $D - N$. Then $C$ contains an arc $v_iv_j$, which is over $v_av_{a+1}$, implying $v_i \in N^-(v_{b+1})$, $v_j \in N^+(v_{a-1})$. But then $N$ cannot be a neighbourhood (of $v_{a-1}$ or $v_{b+1}$), contradiction. It follows that $N$ is a feedback vertex set.

Now given a minimum fvs $F$, there exists a neighbourhood, and thus fvs, $N$ such that $F \supseteq N$, but $|F| \leq |N|$, thus $F = N$, moreover $N$ is a minimum neighbourhood, otherwise if there existed a neighbourhood $N'$ with $|N'| < |N|$, it would be a feedback vertex set of size smaller than $|F|$. Similarly every minimum neighbourhood is a minimum fvs. □

From Lemma 6.14.1 we can get the following:

**Corollary 6.14.2** ([27]) *There exists a linear time algorithm for FVS on round digraphs.*

For the full class of locally semicomplete digraphs it is also possible to obtain a polynomial size problem-kernel. The proof of this is considerably more complicated than for the round case.

**Theorem 6.14.3** ([27]) *There is an $O(k^4)$ problem-kernel for k-FVS on locally semicomplete digraphs*

### 6.14.2 Feedback Arc Sets

Bessy et al., proved the following.

**Theorem 6.14.4** ([37]) *The problem k-FAS on tournaments has a kernel with $O(k)$ vertices.*

Below we use the abbreviation 'fas' for 'feedback arc set'.

**Lemma 6.14.5** ([27]) *Let $R$ be a round directed multigraph without 2-cycles and with round labelling $O = v_1, ..., v_n$, $n \geq 2$. There exists a minimum fas of $R$ consisting of all the arcs over $v_av_{a+1}$, for some $a \in [n]$.*

**Proof:** We use induction on the number $n$ of vertices. If $n = 2$, then $R$ is acyclic since $v_1v_2 \notin A(R)$ or $v_2v_1 \notin A(R)$ and the statement vacuously holds. Therefore assume that $n > 2$ and that $R$ is strong (otherwise it is acyclic). Then, by the definition of a round digraph, we have $v_av_{a+1} \in A(R)$, for $a = 1, ..., n$ (where $v_{n+1} = v_1$). Let $F$ be a minimum fas of $R$. There must exist a vertex $v_i$ such that $d^-_{R-F}(v_i) = 0$.

Consider the round digraph $R' := R - v_i$, with the round labelling induced by the round labelling of $R$. By induction hypothesis, there exists a minimum fas $F'$ of $R'$ consisting of the arcs over $wz$, for some consecutive vertices $w, z$ with respect to the labelling of $R'$ (it could be that $w$ and $z$ are not consecutive in the labelling of $R$, when $w = v_{i-1}$, $z = v_{i+1}$). We can assume $F' \neq \emptyset$, otherwise $F$ consists of the arcs over $v_{i-1}v_i$. Note that if $F'$ is a minimum fas of $R$ we are done; if not, then there is a cycle and thus an arc over $wz$ in $R - F'$, but the only arcs in $R - F'$ over $wz$ are those incident with $v_i$, hence either $z \in N^+(v_i)$ or $v_i \in N^+(w)$. Assume that the first case holds. Let $X := N^+(v_i) \cap N^-(z)$ and $Y := N^+(v_i) \cap N^+(w)$ and note that $X$ and $Y$ are proper segments of the round labelling starting at $v_{i+1}$ and $z$, respectively (this follows from the fact that $R$ has no 2-cycles: if $Z$ and $X$ overlapped in some vertex $x$ then $xwx$ would be a 2-cycle). Let $\phi$ be the number of arcs from $N^-(v_i)$ to $Y$ (they are all over $wz$). By the minimality of $F'$, we have that $|F| - d^-(v_i) \geq |F'|$. We also have that $|F| \leq |F'| + |A(v_i, Y)|$, given that $F' \cup A(v_i, Y)$ is a fas[3] of $R$. These two inequalities imply $d^-(v_i) \leq |A(v_i, Y)|$.

Now let $F''$ be the set of arcs over $v_{i-1}v_i$ (in $R$), we have:

$$\begin{aligned} |F''| &\leq d^-(v_i) \cdot |X| + d^-(v_i) + \phi \\ &\leq |A(v_i, Y)| \cdot |X| + d^-(v_i) + \phi \\ &\leq |A(X, Y)| + d^-(v_i) + \phi. \end{aligned}$$

On the other hand

$$|F| \geq |F'| + d^-(v_i) \geq |A(X, Y)| + d^-(v_i) + \phi.$$

We deduce that $|F''| \leq |F|$ and thus $F''$ is a minimum fas as required.

Similarly, if $v_i \in N^+(w)$, let $X := N^+(v_i) \cap N^+(w)$ and $Y := N^-(v_i) \cap N^-(z)$. Again we have $|F| - d^-(v_i) \geq |F'|$ and $|F| \leq |F'| + |A(Y, v_i)|$, implying that $d^-(v_i) \leq |A(Y, v_i)|$. (In fact $d^-(v_i) = |A(Y, v_i)|$ and we are in the case $w = v_{i-1}$, $z = v_{i+1}$.) So letting $F''$ be the set of arcs over $v_{i-1}v_i$ in $R$, we have: $|F''| \leq d^-(v_i) \cdot |X| + d^-(v_i) = |A(Y, v_i)| \cdot |X| + d^-(v_i) \leq |A(Y, X)| + d^-(v_i) \leq |F'| + d^-(v_i) \leq |F|$, showing that $F''$ is a minimum fas as required. □

By finding the minimum size set of arcs over $v_a v_{a+1}$, for $1 \leq a \leq n$ we get the following

**Corollary 6.14.6** ([27]) *There is a linear algorithm to find a minimum fas on round directed multigraphs without 2-cycles.*

There is also a problem kernel for $k$-FAS on round decomposable digraphs.

[3] This is not necessarily true if there are 2-cycles.

**Theorem 6.14.7** ([27]) *There exists an $O(k)$ problem-kernel for $k$-FAS on round decomposable digraphs.*

Using probabilistic techniques combined with ideas from [1] the authors of [27] obtained the following two results.

**Theorem 6.14.8** ([27]) *For locally semicomplete digraphs that are neither semicomplete nor round decomposable $k$-FAS can be solved in expected time $O(n^3 \cdot 2^{O(\sqrt{k}\log k)} + n^2 M(n)\log^2(n))$, where $M(n)$ is the complexity of matrix multiplication [38].*

**Theorem 6.14.9** *There exists an algorithm with running time $O(n^3 \log n \cdot 2^{O(\sqrt{k}(\log k)^{O(1)})} + n^2 M(n)\log^2(n))$ for solving $k$-fas on the class of digraphs whose vertex set can be partitioned into two sets inducing semicomplete digraphs and, in particular, for locally semicomplete digraphs that are not round decomposable.*

Combining the results above with Theorem 6.6.5 and the solution for semicomplete digraphs in [1] Bang-Jensen, Maddaloni and Saurabh obtained the following.

**Theorem 6.14.10** ([27]) *For locally semicomplete digraphs the $k$-FAS problem can be solved in $O(n^3 \log n \cdot 2^{O(\sqrt{k}(\log k)^{O(1)})} + n^2 M(n)\log^2(n))$ time.*

## 6.15 Orientations of Locally Semicomplete Digraphs

Recall that an **orientation** of a digraph $D$ is any spanning oriented graph that can be obtained by deleting exactly one arc from every 2-cycle in $D$ and keeping all ordinary arcs. It is easy to see that every orientation of a locally semicomplete digraph is a local tournament.

### 6.15.1 Diameter Preserving Orientations

Denote by $\text{diam}_{\min}(D)$ the minimum diameter among all orientations of $D$. The following example, due to Gutin and Yeo, shows that for general locally semicomplete digraphs we cannot guarantee the existence of an orientation whose diameter is the same as that of the original digraph, no matter how large the diameter of $D$ is. Consider the following digraph $D_k = (V, A)$:

$$V = \{x_1, x_2, ..., x_k\},\ A = \{x_i x_{i+1} :\ i \in [k-1]\} \cup \{x_k x_1, x_k x_2, x_1 x_3, x_2 x_1\}.$$

It is easy to check that $\text{diam}(D_k) = k-2$ and $\text{diam}(D_k - x_1x_2) = \text{diam}(D_k - x_2x_1) = k-1$. The digraph $D_k$ contains a pair of vertices that are twins, namely $x_1$ and $x_2$. Two vertices $x$ and $y$ of a digraph $D$ are **twins** if $N^+[x] = N^+[y]$ and $N^-[x] = N^-[y]$. Observe that if $x$ and $y$ are twins, then the 2-cycle $xyx$ is in $D$.

Gutin and Yeo proved that we can almost preserve the diameter of the original locally semicomplete digraph $D$ in some orientation of $D$.

**Theorem 6.15.1** ([50]) *If $D$ is a strong locally semicomplete digraph then*

$$\mathrm{diam}_{\min}(D) \leq \max\{5, \mathrm{diam}(D)+1\}.$$

*If $D$ has no pair of twin vertices then* $\mathrm{diam}_{\min}(D) \leq \max\{4, \mathrm{diam}(D)\}$.

### 6.15.2 Highly Connected Orientations of Locally Semicomplete Digraphs

We now turn to highly connected orientations of locally semicomplete digraphs. Note that this is the same as studying highly connected spanning local tournaments in locally semicomplete digraphs. By Theorem 6.4.9 every strong locally semicomplete digraph has a Hamiltonian cycle and this implies that every strong locally semicomplete digraph has a spanning strong local tournament. Below we study which degree of strong connectivity suffices to guarantee a spanning $k$-strong local tournament when $k \geq 2$.

Guo [44] proved that the bound of Theorem 2.3.2 also holds for locally semicomplete digraphs.

**Theorem 6.15.2** ([44]) *For every positive integer $k$, every $(3k-2)$-strong locally semicomplete digraph contains a spanning local tournament.*

Huang [53] showed that we may delete one arc from every 2-cycle in any round decomposable locally semicomplete digraph and still maintain the same degree of strong connectivity.

**Theorem 6.15.3** ([53]) *Every $k$-strong round-decomposable locally semicomplete digraph contains a spanning $k$-strong local tournament.*

**Proposition 6.15.4** ([7]) *For every $k \geq 4$ there exists a $(2k-4)$-strong locally semicomplete digraph which is not semicomplete and which has no spanning $k$-strong local tournament.*

**Proof:** Consider the locally semicomplete digraph $D_r$ in Figure 6.3 and take $r = 2k-4$. Notice that $D'_{2k-4} = D_{2k-4} - \{u, v\}$ is a $(2k-4)$-strong semicomplete digraph which does not contain a spanning $(k-1)$-strong tournament, because in every spanning tournament of $D'_{2k-4}$ at least one vertex in $U$ $(W)$ will have out-degree (in-degree) at most $k-2$. This implies that $D_{2k-4}$ has no spanning $k$-strong local tournament since each vertex in $U$ has only one new out-neighbour $(u)$ in $D_{2k-4}$. It is easy to check (see e.g. [16, Proposition 5.8.5]) that $D_{2k-4}$ is $(2k-4)$-strong. □

**Lemma 6.15.5** ([16]) *Let $D = (V, A)$ be a $k$-strong digraph and let $D'$ be obtained from $D$ by adding a new vertex $x$ and joining it to $V$ in such a way that $x$ has at least $k$ in-neighbours in $V$ and at least $k$ out-neighbours in $V$. Then $D'$ is $k$-strong and if it is not also $(k+1)$-strong, then either $\min\{d^+_{D'}(x), d^-_{D'}(x)\} = k$ or every minimum separating set in $D'$ is also a minimum separating set in $D$.*

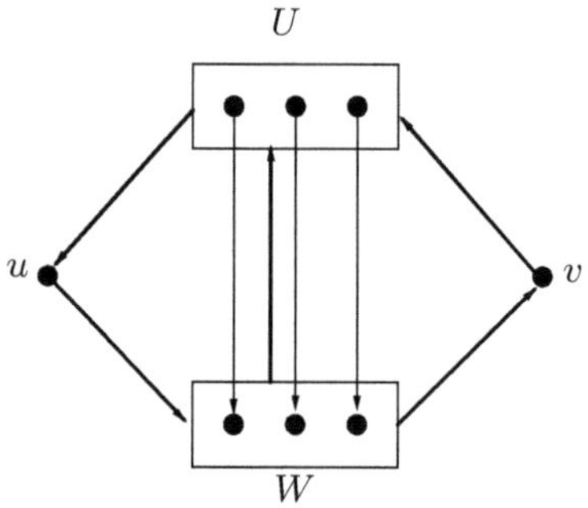


**Figure 6.3** A locally semicomplete digraph $D_r$. Each of the digraphs $U, W$ is a copy of $\overleftrightarrow{K}_r$, the complete digraph on $r \geq 2$ vertices. Bold arcs between two sets indicate that all arcs have the direction shown, except between $U$ and $W$ where the arcs of a matching go from $U$ to $W$ and all other arcs go from $W$ to $U$. Thus $D_r - \{u, v\}$ is the semicomplete digraph $H'$ described just after Conjecture 2.3.4. The only pair of non-adjacent vertices in $D_r$ is $u, v$. The figure is from [7].

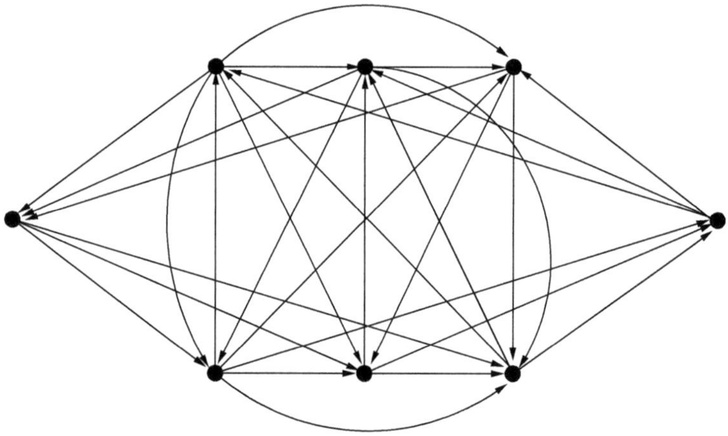

**Figure 6.4** A 3-strong non-semicomplete locally semicomplete digraph containing no 3-strong spanning local tournament. The figure is from [7].

**Theorem 6.15.6** ([7]) *Suppose $g(k)$ is an integer-valued function such that $g(1) \geq 1$ and $g(k) \geq g(k-1)+2$ and for every $k \geq 1$ every $g(k)$-strong semicomplete digraph contains a spanning $k$-strong tournament. Then every $g(k)$-strong locally semicomplete digraph contains a spanning $k$-strong local tournament which contains an arc from every 2-cycle of $D$.*

**Proof:** The proof is by induction on $k$ with the base case $k = 1$ following from Theorem 6.4.9. If $D$ is semicomplete there is nothing to prove by the choice of $g(k)$, so assume below that $D$ is not a semicomplete digraph. By Lemma 6.6.2, $D$ has a minimal separating set $S$ such that $D - S$ is connected but not semicomplete. Let $D_1, D_2, \ldots, D_p$ be the strong decomposition of

$D - S$ and let $D'_1, D'_2, \ldots, D'_h$ be the semicomplete decomposition of $D - S$ (according to Theorem 6.3.4). As $D$ is $g(k)$-strong and not semicomplete we have $|S|, |V(D'_2)| \geq g(k) \geq 2k-1$ and $h \geq 3$. Using the definition of a locally semicomplete digraph it is easy to show that $D'_2 \Rightarrow D'_1 \Rightarrow S \Rightarrow D_1$.

Let $x_1 \in V(D_1)$, $x_2 \in V(D_p)$ and $D^* = D - \{x_1, x_2\}$. Since $g(k) \geq g(k-1)+2$, $D^*$ is a $g(k-1)$-strong locally semicomplete digraph. By the induction hypothesis, $D^*$ contains a $(k-1)$-strong spanning local tournament $T^*$ which contains an arc of every 2-cycle in $D^*$. Since we have $D'_2 \Rightarrow D'_1 \Rightarrow S \Rightarrow D_1$, each of the sets $\{x_2 s | s \in S\}$, $\{y x_2 | y \in V(D'_2)\}$, $\{s x_1 | s \in S\}$ and $\{y x_2 | y \in V(D'_2)\}$ contain at least $g(k)$ arcs and none of the latter arcs are in 2-cycles (implying that $x_1$ has at least $g(k)$ out-neighbours outside $S$). This shows that we may add $x_1, x_2$ to $T^*$ together with all arcs from $D$ between $\{x_1, x_2\}$ and $V - \{x_1, x_2\}$ and delete one arc from each 2-cycle incident with $x_1$ or $x_2$ in such a way that in the resulting local tournament $T$ the vertices $x_1, x_2$ both have in- and out-degree at least $g(k)$.

By Lemma 6.15.5, $T$ is $(k-1)$-strong. Suppose it is not $k$-strong and let $S'$ be a separating set of size $k-1$. By Lemma 6.15.5, $S'$ is also a separating set of $T'$ and none of $x_1, x_2$ are in $S'$. Let $T_1, T_2, \ldots, T_q$ be the strong decomposition of $T - S'$ and let $T'_1 = T_q, T'_2, \ldots, T'_t$ be the semicomplete decomposition of $T - S'$. As $x_1$ and $x_2$ are not adjacent (there is no arc in $D$ from $V(D'_h)$ to $V(D'_1) = V(D_p)$), $T - S'$ is not semicomplete, so we must have $t \geq 3$. But now in $T$ all out-neighbours of $V(T_q)$ are in $S'$. As $T$ is a spanning local tournament of the $g(k)$-strong digraph $D$ and $g(k) \geq 2k-1 > k-1$, there must be arcs from $V(T_q)$ to $V - S' - V(T_q)$ in $D$. Since there are no arcs in $T$ between $V(T'_3)$ and $V(T'_1) = V(T_q)$ and $T$ contains an arc from every 2-cycle in $D$, there is also no arc in $D$ between $V(T'_3)$ and $V(T'_1) = V(T_q)$. On the other hand, since $D - S'$ is strong, it contains a path from $V(T_q)$ to $V(T'_3)$ so there is some 2-cycle $xyx$ in $D$ with $x \in V(T'_2)$ and $y \in V(T'_3)$. However, this and the fact that $V(T'_2) \Rightarrow V(T_q)$ implies that $x$ is adjacent to all vertices of $V(T_q)$, a contradiction. This shows that $T$ must be $k$-strong and the proof is complete. □

**Corollary 6.15.7** *Every 3-strong locally semicomplete digraph $D$ contains a spanning 2-strong local tournament and every 5-strong locally semicomplete digraph which is not semicomplete contains a spanning 3-strong local tournament.*

**Proof:** This follows from Theorem 2.3.3 and the proof of Theorem 6.15.6. When $D$ is 5-strong we need Theorem 2.3.3 to guarantee that the digraph $D^*$ in the proof of Theorem 6.15.6 has a 2-strong spanning strong local tournament $T^*$ because $D^*$ may be semicomplete. □

Bang-Jensen conjectured the following extension of Conjecture 2.3.4.

**Conjecture 6.15.8** ([7]) *For every integer $k \geq 1$ every $(2k-1)$-strong locally semicomplete digraph contains a spanning $k$-strong local tournament.*

Theorem 6.15.3 shows that the conjecture holds for round decomposable locally semicomplete digraphs. By Theorem 6.15.6 the truth of Conjecture 2.3.4 would imply that Conjecture 6.15.8 is true.

The example in Figure 6.3 shows that the connectivity bound in this conjecture cannot be significantly better than $2k-1$, namely it cannot be less than $2k-3$.

## 6.16 Out-Round Digraphs

Li, Zhang and Meng [59] (see also [57]) introduced the following generalization of round digraphs.[4] A digraph is **out-round (in-round)** if we can label its vertices $v_1, \ldots, v_n$ so that for each $i$, we have $N^+(v_i) = \{v_{i+1}, \ldots, v_{i+d^+(v_i)}\}$ ($N^-(v_i) = \{v_{i-d^-(v_i)}, \ldots, v_{i-1}\}$), where all subscripts are taken modulo $n$. The labelling $v_1, \ldots, v_n$ is referred as an **out-round labelling (in-round labelling)** of $D$.

See Figure 6.5 for examples of out-round digraphs that are not locally semicomplete and an example of a locally in-semicomplete digraph which is not out-round.

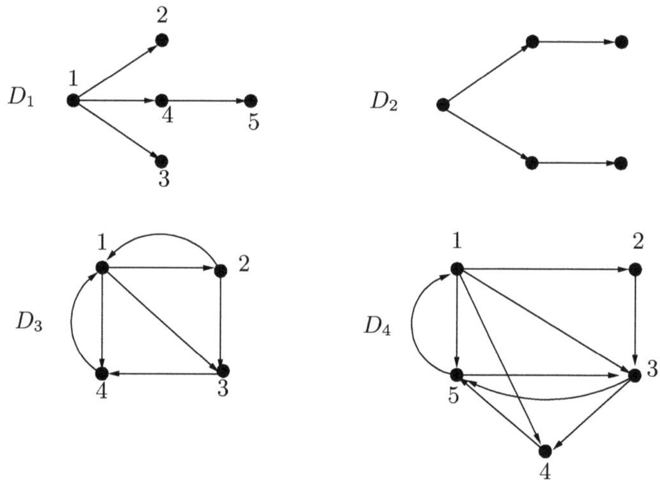


**Figure 6.5** The digraphs $D_1, D_3, D_4$ are out-round with the given labelling. The digraph $D_2$ is not out-round.

**Proposition 6.16.1** ([59]) *If a digraph is out-round, then it is also locally in-semicomplete.*

[4] They used the name **positive round** instead of out-round.

**Proof:** Let $D$ be an out-round digraph and let $v_1, \ldots, v_n$ be an out-round labelling of $D$. Consider an arbitrary vertex, say $v_i$. Let $x, y$ be a pair of in-neighbours of $v_i$. We show that $x$ and $y$ are adjacent. Without loss of generality, assume that $v_i, x, y$ appear in that circular order in the positive-round labelling. Since $x \rightarrow v_i$ and the out-neighbours of $x$ appear consecutively succeeding $x$, we must have $x \rightarrow y$. Therefore, $D$ is locally in-semicomplete. □

Li, Zhang and Meng obtained the following sufficient condition for a strong locally in-semicomplete digraph to be out-round.

**Theorem 6.16.2** ([59]) *Let $D$ be a strong locally in-semicomplete. Then $D$ is out-round if the following holds for each vertex $x$ of $D$:*

*(a) $N^+(x) \cap N^-(x)$ induces a (semicomplete) subdigraph containing no ordinary cycle;*
*(b) $N^-(x) \setminus N^+(x)$ induces a transitive tournament; and*
*(c) $N^+(x) \setminus N^-(x)$ induces an acyclic digraph with a Hamiltonian path.*

*Furthermore, there exists a polynomial algorithm for producing an out-round labelling of any digraph which satisfies the conditions above.*

The following gives a necessary condition for a digraph to be out-round.

**Lemma 6.16.3** *Let $D$ be an out-round digraph, then the following is true:*

*(a) Every induced subdigraph of $D$ is out-round.*
*(b) For each $x \in V(D)$, the subdigraphs induced by $N^-(x) \setminus N^+(x)$, $N^+(x) \setminus N^-(x)$ and $N^+(x) \cap N^-(x)$ contain no ordinary cycles.*

**Proof:** The statement (a) follows directly from the definition of an out-round digraph.

Suppose $D[N^-(x) \setminus N^+(x)]$ contains an ordinary cycle $C$ for some vertex $x$. Let $v_1, v_2, \ldots, v_n$ be an out-round labelling of $D$. Without loss of generality, assume that $x = v_1$. The cycle $C$ must contain an arc $v_iv_j$ such that $i > j$. As $C$ is ordinary, $v_jv_i \notin A$ but then we have $v_1 \in N^+(v_j)$ and $v_i \notin N^+(v_j)$, contradicting the assumption that $v_1, v_2, \ldots, v_n$ is an out-round labelling of $D$. Similarly, it is easy to show that for each vertex $x$ the subdigraphs induced by $D[N^+(x) \setminus N^-(x)]$ and $D[N^+(x) \cap N^-(x)]$ contain no ordinary cycles. □

It follows from the definition that if $D$ is a strong out-round digraph, then the subdigraph induced by $N^+(x)$ contains a Hamiltonian path for each vertex $x$ of $D$. Thus Theorem 6.16.2 and Lemma 6.16.3 imply the following characterization of strong out-round locally in-tournament digraphs.

**Theorem 6.16.4** ([59]) *A strong locally in-tournament digraph $D$ is out-round if and only if, for each vertex $x$ of $D$, $N^-(x)$ induces a transitive tournament and $N^+(x)$ induces an acyclic digraph with a Hamiltonian path.*

As the proof of Theorem 6.16.2 in [59] is algorithmic, we obtain the following.

**Corollary 6.16.5** *There exists a polynomial algorithm for recognizing out-round locally in-tournament digraphs and producing an out-round labelling of such a digraph.*

As can be seen from the digraph $D_4$ in Figure 6.5, Theorem 6.16.4 does not extend to non-strong out-round locally in-semicomplete digraphs.

**Problem 6.16.6** ([59]) *Characterize out-round digraphs.*

**Problem 6.16.7** *Find a polynomial algorithm for recognizing out-round digraphs.*

## 6.17 Miscellaneous Topics

### 6.17.1 Kings

A **k-king** in a digraph is a vertex that can reach every other vertex by a path of length at most $k$. We saw in Theorem 2.2.12 that every tournament has a 2-king. Since directed cycles are locally semicomplete digraphs, there is no integer $k$ such that every locally semicomplete digraph has a $k$-king. However, as shown by Wang, Yang and Wang [72], evil locally semicomplete digraphs all have a 2-king. More precisely, using the classification of locally semicomplete digraphs, they obtained the following:

**Theorem 6.17.1** ([72]) *Let $D$ be a connected locally semicomplete digraph. Then the following holds:*

*(a) If $D$ is not strong and $D'_1, D'_2, \ldots, D'_r$ is the semicomplete decomposition of $D$, then $D$ has a 2-king if $r = 2$ and if $r \geq 3$, then $D$ has an $(r-1)$-king.*
*(b) If $D$ is an evil locally semicomplete digraph, then it has a 2-king.*
*(c) If $D$ is strongly connected and round decomposable with round decomposition $D = R[S_1, S_2, \ldots, S_r]$, then $D$ has a $g(R)$-king, where $g(R)$ is the girth of the round digraph $R$.*

**Problem 6.17.2** *Characterize those locally in-semicomplete digraphs that have a $k$-king for some finite integer $k$.*

### 6.17.2 Cycle Extendability

The following definitions are due to Hendry [52]. A non-Hamiltonian cycle $C$ in a digraph $D$ is **extendable** if $D$ has a cycle $C'$ with $V(C') = V(C) \cup \{y\}$ for some vertex $y \in V - V(C)$. A digraph $D$ which has at least one cycle is **cycle extendable** if every non-Hamiltonian cycle of $D$ is extendable. Clearly a cycle extendable digraph is pancyclic if and only if it contains a 3-cycle and vertex-pancyclic if and only if every vertex is in a 3-cycle.

The following is an easy consequence of the proof of Theorem 2.2.7:

**Theorem 6.17.3** ([64]) *A strong tournament $T = (V, A)$ is cycle extendable unless $V$ can be partitioned into sets $U, W, Z$ such that $W \mapsto U \mapsto Z$ and $T[U]$ is strong.* $\diamond$

Tewes and Volkmann [70] generalized Hendry's definition as follows: Let $D$ be a digraph on $n$ vertices and let $3 \leq k < n$ be an integer. Then $D$ is **k-extendable** if every non-Hamiltonian cycle $C$ on at least $k$ vertices is extendable. If $D$ is $k$-extendable and every vertex is in a $k$-cycle, then $D$ is **fully k-extendable**. Tewes and Volkmann studied cycle extendability for locally in-tournament digraphs and obtained several results, including the following two.

**Theorem 6.17.4** ([69]) *Every connected locally in-tournament digraph $D$ on $n$ and minimum semi-degree $\delta^0(D) \geq 1$ vertices is $(n - \lfloor \frac{4\delta^0(D)}{3} \rfloor)$-extendable.*

**Theorem 6.17.5** ([70]) *Every strong locally in-tournament digraph $D$ on $n$ vertices and $\delta^0(D) = 2$ or $\delta^0(D) > \frac{8n-17}{31}$ is fully $(n - \lfloor \frac{4\delta^0(D)}{3} \rfloor)$-extendable.*

When $\delta^0(D) = 1$ we have $(n - \lfloor \frac{4\delta^0(D)}{3} \rfloor) = n - 1$ and there are many locally in-tournament digraphs with minimum semi-degree 1 which are not fully $(n-1)$-extendable.

Meierling proved the following result.

**Theorem 6.17.6** ([60]) *Let $D$ be a strong locally in-tournament digraph on $n$ vertices such that $3 \leq \delta^0(D) \leq \frac{3n-2}{8}$. Then every vertex of $D$ is in a cycle of length $n - \lfloor \frac{4\delta^0(D)+1}{3} \rfloor$.*

Combining this with Theorem 6.17.5, Meierling obtained the following result, which was conjectured by Tewes and Volkmann in [70].

**Theorem 6.17.7** ([60]) *Let $D$ be a strong locally in-tournament digraph on $n$ vertices such that $3 \leq \delta^0(D) \leq \frac{8n-17}{31}$, then $D$ is fully $(n - \frac{4\delta^0+1}{3})$-extendable.*

### 6.17.3 Pancyclic Arcs and Arc-Traceability

Recall that a vertex $u$ in a digraph $D$ is **out-pancyclic** if every arc $uv$ with tail $u$ is contained in a $k$-cycle for every $k = 3, \ldots, n$.

By Theorem 2.14.7 every strong tournament has an out-pancyclic vertex. The round local tournament $D = C_6^2$ shows that this result does not extend directly to local tournaments.

Meng, Li, Guo and Xu define the **pseudo-girth**, $g^s(D)$, of a local tournament $D$ as follows: If $D$ is round-decomposable with round decomposition $D = R[S_1, S_2, \ldots, S_r], r = |V(R)|$, then

$$g^s(D) = \min\{n, \max_{1 \le i \le r} \{g_{r_i}(R)\} + 1\},$$

where $r_i$ is the $i$th vertex of $R$ (the one corresponding to $S_i$ in $D$) and $g_{r_i}(R)$ is the length of a shortest cycle containing $r_i$ in $R$. If $D$ is not round decomposable, then we let $g^s(D) = 3$.

**Theorem 6.17.8** ([62]) *Let $D$ be a strong local tournament on $n$ vertices. Then $D$ has a vertex $u$ such that all out-arcs of $u$ are pseudo-girth-pancyclic, that is, they are in cycles of all lengths from $g^s(D)$ and up, if and only if $D$ is not the second power of an even cycle of length at least 6.*

To see that this result is tight, consider the round decomposable local tournament $D = C_4[C_3, C_3, C_3, C_3]$ that we obtain by substituting a 3-cycle for each vertex of a 4-cycle. The pseudo-girth of $D$ is 5 and it is easy to check that no arc between two vertices of the same $C_3$ is contained in a 4-cycle. Hence for every vertex $u$, at least one of its out-arcs is not 4-pancyclic.

### 6.17.4 Hamiltonicity of Digraphs with Degree Bounds on Certain Vertices

There are several well-known sufficient conditions for hamiltonicity in digraphs in terms of degrees of certain pairs of vertices, e.g. pairs of non-adjacent vertices. Examples are the sufficient conditions of Woodall [73] and Meyniel [63].

We saw in Section 6.4 that every strong locally semicomplete digraph is Hamiltonian. One could ask whether it is possible to obtain a sufficient condition for hamiltonicity that only applies to pairs of vertices that should be adjacent, had the digraph in question been locally semicomplete. The results below, due to Bang-Jensen, Gutin and Li, show that this is indeed the case. We say that a pair of vertices $x, y$ is a **dominated pair** (**dominating pair**) if they have a common in-neighbour (out-neighbour). Note that in a locally semicomplete digraph the vertices of any dominated (dominating) pair is adjacent.

**Theorem 6.17.9** ([18]) *Let $D$ be a strong digraph on $n$ vertices such that for every dominated pair of non-adjacent vertices $x_1, x_2$ we have that $d(x_i) \geq n$ and $d(x_{3-i}) \geq n-1$ for $i = 1$ or $i = 2$. Then $D$ is Hamiltonian.*

**Theorem 6.17.10** ([18]) *Let $D$ be a strong digraph on $n$ vertices such that for every pair of non-adjacent vertices $x_1, x_2$ which form either a dominating pair or a dominated pair we have $d^+(x_i) + d^-(x_{3-i}) \geq n$ for $i = 1, 2$. Then $D$ is Hamiltonian.*

Both of the theorems above are sharp [18]. Bang-Jensen, Gutin and Li [18] proposed the following generalization of Meyniel's theorem.

**Conjecture 6.17.11** ([18]) *Let $D$ be a strong digraph on $n$ vertices such that $d(x) + d(y) \geq 2n - 1$ for every pair of non-adjacent vertices $x, y$ which form either a dominating pair or a dominated pair. Then $D$ is Hamiltonian.*

### 6.17.5 The Directed Steiner Problem

Consider the following three problems

| |
|---|
| DIRECTED STEINER PROBLEM<br>**Input:** A strong digraph $D$ and a non-empty subset $X$ of its vertices.<br>**Find:** A strong subdigraph $D'$ of $D$ which contains all vertices of $X$ and has as few arcs as possible. |

| |
|---|
| MINIMUM COST STRONG SUBDIGRAPH<br>**Input:** A strong digraph $D$ with real-valued costs on the vertices.<br>**Find:** A strong subdigraph of $D$ of minimum cost. |

| |
|---|
| MINIMUM COST CYCLE<br>**Input:** A strong digraph $D$ with real-valued costs on the vertices.<br>**Find:** A cycle of minimum cost in $D$. |

All three problems are $\mathcal{NP}$-hard as they contain the Hamiltonian cycle problem as a special case. Feldman and Ruhl [39] proved that if $|X|$ is bounded by a constant $k$, then the DIRECTED STEINER PROBLEM is solvable in polynomial time (for fixed $k$). Their solution is quite complicated, even in the case when $|X| = 2$. The special case $X = V(D)$ of DIRECTED STEINER PROBLEM is also known as the MINIMUM SPANNING STRONG SUBDIGRAPH problem (MSSS) and has been considered in several papers, including [25, 29, 30, 33].

The result below implies that for locally in-semicomplete digraphs the DIRECTED STEINER PROBLEM is a special case of the MINIMUM COST CYCLE problem.

**Lemma 6.17.12** ([19]) *If $D = (V, A)$ is a strong locally in-semicomplete digraph and $X \subseteq V$, then the solution to the* DIRECTED STEINER PROBLEM *is always a shortest cycle (measured in number of arcs) covering $X$.*

**Proof:** Suppose $A'$ is an optimal solution to the DIRECTED STEINER PROBLEM (for $X$). Let $V'$ be the set of vertices incident with the arcs in $A'$. Then $D' = D[A']$ is a spanning subdigraph of the locally in-semicomplete digraph $D^* = D[V']$, implying that $D^*$ is strong. By Theorem 6.4.9, $D^*$ has a Hamiltonian cycle $C$. By the minimality of $A'$, we get $|A'| = |A(C)|$. □

Lemma 6.17.12 implies that the DIRECTED STEINER PROBLEM for locally in-semicomplete digraphs reduces to the MINIMUM COST CYCLE problem for locally in-semicomplete digraphs in linear time.

**Theorem 6.17.13** ([19]) *The* DIRECTED STEINER PROBLEM *is solvable in time $O(n^2)$ for locally semicomplete digraphs.* □

In the rest of this section concentrate on the MINIMUM COST CYCLE problem for locally semicomplete digraphs.

The case of round digraphs has a particularly nice solution.

**Lemma 6.17.14** ([19]) *Let $D$ be a strong round local tournament with round enumeration $v_1, v_2, \ldots, v_n$ and with real-valued costs on the vertices. Then every minimum cost cycle includes all the vertices of negative costs and $D$ has such a cycle of the form $v_{a_1} v_{a_2} \ldots v_{a_k} v_{a_1}$ where $1 \leq a_1 < a_2 < \ldots < a_k \leq n$. Furthermore, given a round enumeration of $D$, a minimum cost cycle can be found in time $O(n^2)$.*

**Lemma 6.17.15** ([19]) *Let $D$ be a strong semicomplete digraph with real-valued costs on the vertices. In time $O(n(m+n \log n))$ we can find a minimum cost cycle of $D$.*

Combining the two lemmas above with solutions for the round decomposable case (which uses the solution for round digraphs) and for evil locally semicomplete digraphs Bang-Jensen, Gutin and Yeo obtained the following result.

**Theorem 6.17.16** ([19]) *The* MINIMUM COST CYCLE *problem is solvable in time $O(nm + n^2 \log n)$ for locally semicomplete digraphs.*

We can now show that the MINIMUM COST STRONG SUBDIGRAPH problem is polynomially solvable for locally semicomplete digraphs.

**Theorem 6.17.17** ([19]) *A minimum cost strong subdigraph of a locally semicomplete digraph $D$ can be found in time $O(nm + n^2 \log n)$.*

**Proof:** Since $D$ is locally semicomplete, every strong subdigraph of $D$ with at least 2 vertices contains a spanning cycle by Theorem 6.4.9. Hence a minimum cost strong subdigraph $D'$ of $D$ can be found in time $O(nm + n^2 \log n)$ by finding a minimum cost cycle and a minimum cost vertex of $D$ and taking the cheapest of these two as $D'$. □

## References


1. N. Alon, D. Lokshtanov, and S. Saurabh. Fast FAST. In *ICALP 2009: 36th International Colloquium on Automata, Languages and Programming, Part I*, volume 5555 of *Lect. Notes Comput. Sci.*, pages 49–58, 2009.
2. B. Alspach and C. Tabib. A note on the number of 4-circuits in a tournament. In *Theory and practice of combinatorics*, volume 60 of *North-Holland Math. Stud.*, pages 13–19. North-Holland, Amsterdam, 1982.
3. J. Bang-Jensen. A reformulation of Huang's structure theorem for local tournaments with some consequences. Technical report 13, Department of Mathematics and Computer Science, Odense University, Denmark, 1994.
4. J. Bang-Jensen. Digraphs with the path-merging property. *J. Graph Theory*, 20(2):255–265, 1995.
5. J. Bang-Jensen. Disjoint paths with prescribed ends and cycles through given arcs in locally semicomplete digraphs and quasi-transitive digraphs. Technical Report 22, Department of Mathematics and Computer Science, Odense University, Denmark, 1996.
6. J. Bang-Jensen. Edge-disjoint in- and out-branchings in tournaments and related path problems. *J. Combin. Theory Ser. B*, 51(1):1–23, 1991.
7. J. Bang-Jensen. $k$-strong spanning local tournaments in locally semicomplete digraphs. *Discrete Appl. Math.*, 157(11):2536–2540, 2009.
8. J. Bang-Jensen. Linkages in locally semicomplete digraphs and quasi-transitive digraphs. *Discrete Math.*, 196(1-3):13–27, 1999.
9. J. Bang-Jensen. Locally semicomplete digraphs: a generalization of tournaments. *J. Graph Theory*, 14(3):371–390, 1990.
10. J. Bang-Jensen. On the structure of locally semicomplete digraphs. *Discrete Math.*, 100(1-3):243–265, 1992.
11. J. Bang-Jensen, T. M. Christiansen, and A. Maddaloni. Disjoint paths in decomposable digraphs. *J. Graph Theory*, 85:545–567, 2017.
12. J. Bang-Jensen and Y. Guo. A note on vertex pancyclic oriented graphs. *J. Graph Theory*, 31:313–318, 1999.
13. J. Bang-Jensen, Y. Guo, G. Gutin, and L. Volkmann. A classification of locally semicomplete digraphs. *Discrete Math.*, 167/168:101–114, 1997.
14. J. Bang-Jensen, Y. Guo, and L. Volkmann. Weakly Hamiltonian-connected locally semicomplete digraphs. *J. Graph Theory*, 21(2):163–172, 1996.
15. J. Bang-Jensen and G. Gutin. *Digraphs: Theory, Algorithms and Applications.* Springer-Verlag, London, 2000.
16. J. Bang-Jensen and G. Gutin. *Digraphs: Theory, Algorithms and Applications.* Springer-Verlag, London, 2nd edition, 2009.
17. J. Bang-Jensen and G. Gutin. Generalizations of tournaments: A survey. *J. Graph Theory*, 28:171–202, 1998.
18. J. Bang-Jensen, G. Gutin, and H. Li. Sufficient conditions for a digraph to be Hamiltonian. *J. Graph Theory*, 22(2):181–187, 1996.

19. J. Bang-Jensen, G. Gutin, and A. Yeo. Steiner type problems for digraphs that are locally semicomplete or extended semicomplete. *J. Graph Theory*, 44(3):193–207, 2003.
20. J. Bang-Jensen and P. Hell. Fast algorithms for finding Hamiltonian paths and cycles in in-tournament digraphs. *Discrete Appl. Math.*, 41(1):75–79, 1993.
21. J. Bang-Jensen, P. Hell, and G. MacGillivray. The complexity of colouring by semicomplete digraphs. *SIAM J. Discrete Math.*, 1(3):281–298, 1988.
22. J. Bang-Jensen and J. Huang. Arc-disjoint in- and out-branchings with the same root in locally semicomplete digraphs. *J. Graph Theory*, 77(4):278–298, 2014.
23. J. Bang-Jensen and J. Huang. Decomposing locally semicomplete digraphs into strong spanning subdigraphs. *J. Combin. Theory Ser. B*, 102:701–714, 2010.
24. J. Bang-Jensen, J. Huang, and E. Prisner. In-tournament digraphs. *J. Combin. Theory Ser. B*, 59(2):267–287, 1993.
25. J. Bang-Jensen, J. Huang, and A. Yeo. Strongly connected spanning subgraphs with the minimum number of arcs in quasi-transitive digraphs. *SIAM J. Discrete Math.*, 16:335–343, 2003.
26. J. Bang-Jensen and A. Maddaloni. Arc-disjoint paths in decomposable digraphs. *J. Graph Theory*, 77(2):89–110, 2014.
27. J. Bang-Jensen, A. Maddaloni, and S. Saurabh. Algorithms and kernels for feedback set problems in generalizations of tournaments. *Algorithmica*, 76(2):320–343, 2016.
28. J. Bang-Jensen and M.H. Nielsen. Finding complementary cycles in locally semicomplete digraphs. *Discrete Appl. Math.*, 146(3):245–256, 2005.
29. J. Bang-Jensen and A. Yeo. The minimum spanning strong subdigraph problem for extended semicomplete digraphs and semicomplete bipartite digraphs. *J. Algor.*, 41(1):1–19, 2001.
30. J. Bang-Jensen and A. Yeo. The minimum spanning strong subdigraph problem is fixed parameter tractable. *Discrete Appl. Math.*, 156:2924–2929, 2008.
31. C. Berge. *Graphs*. North-Holland, Amsterdam, 1985. Second revised edition of part 1 of the 1973 English version.
32. C. Berge and A.R. Rao. A combinatorial problem in logic. *Discrete Math.*, 17:23–26, 1977.
33. S. Bessy and S. Thomassé. Every strong digraph has a spanning strong subgraph with at most $n + 2\alpha - 2$ arcs. *J. Combin. Theory Ser. B*, 87(2):289–299, 2003.
34. J.A. Bondy. Short proofs of classical theorems. *J. Graph Theory*, 44(3):159–165, 2003.
35. V. Chvátal and L. Lovász. Every directed graph has a semi-kernel. *Lect. Notes Math.*, 411:175, 1974.
36. T.H. Cormen, C.E. Leiserson, R.L. Rivest, and C. Stein. *Introduction to Algorithms*. MIT Press, Cambridge, MA, 2nd edition, 2001.
37. A. Cuzzocrea, D. Taniar, S. Bessy, F. Fomin, S. Gaspers, C. Paul, A. Perez, S. Saurabh, and S. Thomassé. Kernels for feedback arc set in tournaments. *J. Comput. Syst. Sci.*, 77(6):1071–1078, 2011.
38. G. Even, J. Naor, B. Schieber, and M. Sudan. Approximating minimum feedback sets and multicuts in directed graphs. *Algorithmica*, 20(2):151–174, 1998.
39. J. Feldman and M. Ruhl. The directed Steiner network problem is tractable for a constant number of terminals. In *FOCS 1999: 40th Annual Symposium on Foundations of Computer Science*, pages 299–308. IEEE Computer Society Press, 1999.
40. H. Galeana-Sánchez and M. Olsen. A characterization of locally semicomplete CKI-digraphs. *Graphs Combin.*, 32(5):1873–1879, 2016.

41. M.R. Garey and D.S. Johnson. *Computers and intractability.* W. H. Freeman, San Francisco, 1979.
42. R. Gould and Y. Guo. Locally semicomplete digraphs with a factor composed of $k$ cycles. *J. Korean Math. Soc.*, 41:895–912, 2004.
43. Y. Guo. *Locally semicomplete digraphs.* PhD thesis, RWTH Aachen, Germany, 1995.
44. Y. Guo. Spanning local tournaments in locally semicomplete digraphs. *Discrete Appl. Math.*, 79(1-3):119–125, 1997.
45. Y. Guo. Strongly Hamiltonian-connected locally semicomplete digraphs. *J. Graph Theory*, 22(1):65–73, 1996.
46. Y. Guo and L. Volkmann. Locally semicomplete digraphs that are complementary $m$-pancyclic. *J. Graph Theory*, 21(2):121–136, 1996.
47. Y. Guo and L. Volkmann. On complementary cycles in locally semicomplete digraphs. *Discrete Math.*, 135(1-3):121–127, 1994.
48. Y. Guo and L. Volkmann. Pancyclic locally semicomplete digraphs. Unpublished manuscript, 1992.
49. G. Gutin, K.M. Koh, E.G. Tay, and A. Yeo. On the number of quasi-kernels in digraphs. *J. Graph Theory*, 46(1):48–56, 2004.
50. G. Gutin and A. Yeo. Orientations of digraphs almost preserving diameter. *Discrete Appl. Math.*, 121(1-3):129–138, 2002.
51. S. Heard and J. Huang. Disjoint quasi-kernels in digraphs. *J. Graph Theory*, 58(3):251–260, 2008.
52. G.R.T. Hendry. Extending cycles in directed graphs. *J. Combin. Theory Ser. B*, 46(2):162–172, 1989.
53. J. Huang. A note on spanning local tournaments in locally semicomplete digraphs. *Discrete Appl. Math.*, 89:277–279, 1998.
54. J. Huang. On the structure of local tournaments. *J. Combin. Theory, Ser. B*, 63(2):200–221, 1995.
55. J. Huang. *Tournament-like oriented graphs.* PhD thesis, School of Computing Science, Simon Fraser University, Canada, 1992.
56. J. Huang. Which digraphs are round? *Australas. J. Combin.*, 19:203–208, 1999.
57. R. Li. *k-ordered Graphs and Out-arc Pancyclicity of Digraphs.* PhD thesis, Fakultät Für Mathematik, Informatik und Naturwissenschaften, RWTH Aachen, Germany, 2009.
58. R. Li and T. Han. Arc-disjoint hamiltonian cycles in round decomposable locally semicomplete digraphs. *Discuss. Math. Graph Theory*, 340:2916–2924, 2017.
59. R. Li, X. Zhang, and W. Meng. A sufficient condition for a digraph to be positive round. *Optimization*, 57:345–352, 2008.
60. D. Meierling. Solution of a conjecture of Tewes and Volkmann regarding extendable cycles in in-tournaments. *J. Graph Theory*, 63(1):82–92, 2010.
61. D. Meierling and L. Volkmann. All 2-connected in-tournaments that are cycle complementary. *Discrete Math.*, 308:2115–2133, 2008.
62. W. Meng, S. Li, Y. Guo, and G. Xu. A local tournament contains a vertex whose out-arcs are pseudo-girth-pancyclic. *J. Graph Theory*, 62(4):346–361, 2009.
63. H. Meyniel. Une condition suffisante d'existence d'un circuit hamiltonien dans un graphe orienté. *J. Combin. Theory Ser. B*, 14:137–147, 1973.
64. J.W. Moon. *Topics on tournaments.* Holt, Rinehart & Winston,, New York, 1968.
65. C. Peters and L. Volkmann. Vertex 6-pancyclic in-tournaments. *Discrete Math.*, 285(1-3):227–238, 2004.
66. M. Tewes. *In-tournaments and Semicomplete Multipartite Digraphs.* PhD thesis, Lehrstuhl II für Mathematik, RWTH Aachen, 1998.

67. M. Tewes. Pancyclic in-tournaments. *Discrete Math.*, 233(1-3):193–204, 2001.
68. M. Tewes. Pancyclic orderings of in-tournaments. *Discrete Applied Math.*, 120(1-3):239–249, 2002.
69. M. Tewes and L. Volkmann. On the cycle structure of in-tournaments. *Australas. J. Combin.*, 18:293–301, 1998.
70. M. Tewes and L. Volkmann. Vertex pancyclic in-tournaments. *J. Graph Theory*, 36(2):84–104, 2001.
71. J. von Neumann and O. Morgenstern. *Theory of Games and Economic Behavior.* Princeton University Press, Princeton, NJ, 1944.
72. R. Wang, A. Yang, and S. Wang. Kings in locally semicomplete digraphs. *J. Graph Theory*, 63(4):279–287, 2010.
73. D.R. Woodall. Sufficient conditions for cycles in digraphs. *Proc. London Math. Soc.*, 24:739–755, 1972.

# 7. Semicomplete Multipartite Digraphs

**Anders Yeo**

In this chapter we will consider the class of semicomplete multipartite digraphs (SMD). A digraph is **semicomplete multipartite** if it is obtained from a complete multipartite graph by replacing every edge by an arc or a pair of opposite arcs. In other words, the vertex set of a semicomplete multipartite digraph can be partitioned into sets $V_1, V_2, \ldots, V_k$ for some $k$ such that vertices within the same set are non-adjacent and vertices between different sets are adjacent (there is at least one arc between them). The sets $V_1, V_2, \ldots, V_k$ are called the **partite sets**, or **colour classes**, of the digraph. See Figure 7.1 for examples of semicomplete multipartite digraphs. All cycles and paths in this chapter are *directed.*

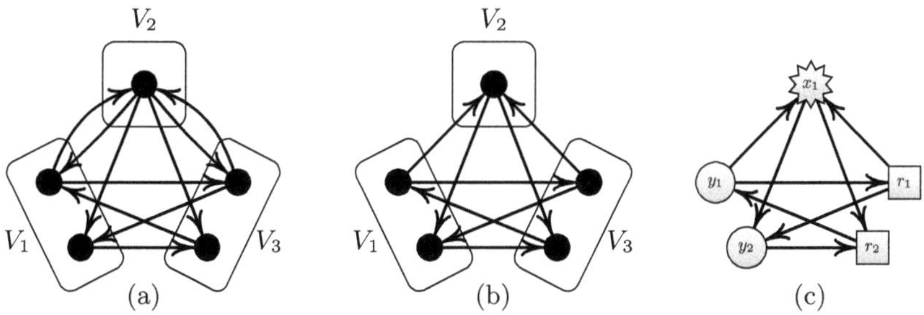


**Figure 7.1** Examples of semicomplete multipartite digraphs. Figure (a) is not a multipartite tournament as it contains 2-cycles. Figure (b) is also a multipartite tournament. Figure (c) is the same semicomplete multipartite digraph as (b), but uses an alternative way of illustrating the partite sets (colour classes) $V_1$, $V_2$ and $V_3$.

Multipartite tournaments were already considered in the book [52] (1968) on tournaments by Moon. The first study of cycles in multipartite tournaments was by Bondy [15] in 1976. Bipartite tournaments were then considered in 1981 by Beineke [13]. In the following years more and more people studied multipartite tournaments and semicomplete multipartite digraphs. In particular, these digraph classes were studied in the Ph.D. theses of Gutin, [32] (1993), Yeo, [77] (1998), Tewes, [58] (1999), Winzen, [70] (2004), and the

A. Yeo (✉)
Department of Mathematics and Computer Science,
University of Southern Denmark, Campusvej 55, 5230 Odense M, Denmark
e-mail: yeo@imada.sdu.dk

 
J. Bang-Jensen and G. Gutin (eds.), *Classes of Directed Graphs*, Springer Monographs in Mathematics, https://doi.org/10.1007/978-3-319-71840-8_7

habilitation thesis of Guo, [24] (1998). This research has continued into the twenty-first century and is still ongoing.

## 7.1 Overview of Chapter 7

Section 7.3 will begin with a structural result on cycle factors and cycle subgraphs in semicomplete multipartite digraphs that allowed first Yeo and then others to solve several conjectures in the area.

Before considering paths and cycles in general semicomplete multipartite digraphs we will, in Section 7.4, consider the class of semicomplete bipartite digraphs, which is the class of semicomplete multipartite digraphs with only two partite sets. This is a widely studied class of digraphs. In Section 7.4 we will mainly focus on paths and cycles.

In Section 7.5 we then consider results on paths in semicomplete multipartite digraphs. Many of these generalize Redei's Theorem which states that every tournament contains a Hamilton path. Hamilton paths containing given arcs are also considered in this section.

In Section 7.6 we then consider results on cycles that hold for semicomplete multipartite digraphs. Some of these generalize Camion's Theorem which states that every strong tournament contains a Hamilton cycle (Theorem 2.2.6). However the class of semicomplete multipartite digraphs is much more complex than that of tournaments and so several results for tournaments are not easily extendable to semicomplete multipartite digraphs. In general this section is devoted to long cycles and pancyclicity. In Section 7.7 we consider short cycles in semicomplete multipartite digraphs.

Section 7.8 is devoted to regular and close to regular semicomplete multipartite digraphs. It turns out that many (but not all) results that hold for regular semicomplete multipartite digraphs also hold for (large) semicomplete multipartite digraphs that are close to being regular.

In Section 7.9 we consider semicomplete multipartite digraphs with given connectivity. In fact we mainly consider which kind of cycles exist in $k$-strong semicomplete multipartite digraphs.

In Section 7.10 we study extended semicomplete digraphs, which is the class of semicomplete multipartite digraphs where if there is an arc from partite set $V_i$ to partite set $V_j$ then all arcs exist from $V_i$ to $V_j$. In other words, one can think of extended semicomplete digraphs as semicomplete digraphs where each vertex is blown up to an independent set. It turns out that many of the results that hold semicomplete digraphs also hold for extended semicomplete digraphs.

In Section 7.11 we consider orientations of semicomplete multipartite digraphs, which are the spanning subgraphs obtained by deleting one arc from every 2-cycle in the semicomplete multipartite digraph.

Section 7.12 is devoted to $r$-kings in semicomplete multipartite digraphs. An $r$-**king** is a vertex that can reach all other vertices with a path of length at most $r$. For semicomplete multipartite digraphs the notion of 4-kings is especially interesting.

In Section 7.13 we consider out-paths in semicomplete multipartite digraphs. Out-paths in semicomplete multipartite digraphs generalize cycles in tournaments and it turns out that several results on cycles in tournaments can be extended to out-paths in semicomplete multipartite digraphs. In Section 7.14 we then study a generalization of paths in tournaments.

In Section 7.15 we look for strongly connected spanning subgraphs with the minimum number of arcs. Results are given for semicomplete bipartite digraphs and extended semicomplete digraphs and an open problem for semicomplete multipartite digraphs is stated.

In Section 7.16 we consider $k$-coloured kernels in arc-coloured semicomplete multipartite digraphs. This extends the notion of kernels for digraphs.

Section 7.17 is devoted to complementary cycles in semicomplete multipartite digraphs. These results are extensions of the known result that all 2-strong tournaments of order $n \geq 8$ contain vertex disjoint cycles of length 3 and $n-3$ (see Theorem 2.8.1).

In Section 7.18 we mention how results for semicomplete multipartite digraphs can be used to prove results for tournaments. In particular, when a tournament contains a Hamilton cycle avoiding prescribed arcs.

Finally, in Section 7.19 we list a number of conjectures.

Many of the results given below for strong multipartite tournaments also hold for strong semicomplete multipartite digraphs, due to the following result of Volkmann.

**Theorem 7.1.1** ([62]) *Every strong semicomplete $c$-partite digraph with $c \geq 3$ contains a spanning strong oriented subdigraph.*

## 7.2 Further Notation

In Section 1.3 some of the following terms are defined. We recall their definitions as they will be frequently used in this chapter.

A multipartite tournament (MT) is a semicomplete multipartite digraph without 2-cycles. In other words, vertices in different partite sets have exactly one arc between them. See Figure 7.1 (b) (and (c)) for an example of a multipartite tournament. Note that Example (a) in Figure 7.1 is not a multipartite tournament as it contains at least one 2-cycle.

Let $D$ be a digraph. A $q$-**path-cycle factor** of $D$ is a spanning collection of $q$ vertex disjoint paths and any number of cycles of $D$. That is, every vertex in $D$ belongs to exactly one path or cycle of the path-cycle factor. The **path covering number** of $D$ is the minimum number of vertex disjoint paths needed to cover all the vertices of $D$, and is denoted by $pc(D)$. Note that the path covering number can also be thought of as the minimum $q$ such that $D$ contains a $q$-path-cycle factor with no cycles, which is also called a $q$-**path**

**factor**. The **path cycle covering number** of $D$ is the minimum number, $q$, such that $D$ contains a $q$-path-cycle factor and is denoted by $pcc(D)$. A **cycle factor** of $D$ is a collection vertex disjoint cycles of $D$.

## 7.3 The Irreducible Cycle Subgraph Theorem

Several results on paths and cycles are proved using Theorem 7.3.2 below, which was obtained by Yeo in [75]. Before giving the statement of this theorem we need the following notation and definitions.

Recall that if $v$ belongs to a cycle then we denote the successor of $v$ on the cycle by $v^+$ and the predecessor by $v^-$. Let $C_1$ and $C_2$ be two disjoint cycles in a semicomplete multipartite digraph $D$. Suppose that there exists some partite set $V_i$ such that the following holds.

*For every arc $u_2v_1$ from $C_2$ to $C_1$ we have $\{u_2^+, v_1^-\} \subseteq V_i$, where $u_2^+$ is the successor of $u_2$ on $C_2$ and $v_1^-$ is the predecessor of $v_1$ on $C_1$.*

In this case we say that $C_1$ $V_i$-**weakly-dominates** $C_2$ and denote this by $C_1 \rightsquigarrow_{V_i} C_2$. If $C_1 \rightsquigarrow_{V_i} C_2$ for some $i$ then we also say that $C_1$ **weakly-dominates** $C_2$ and denote this simply by $C_1 \rightsquigarrow C_2$.

See Figure 7.2 for an illustration of this definition. For example, in Figure 7.2 $w_1y_2$ is the only arc from $C_3$ to $C_2$ and $\{w_1^+, y_2^-\} \in V_3$ (as $w_1^+ = w_2$ and $y_2^- = y_1$). Therefore $C_2 \rightsquigarrow_{V_3} C_3$.

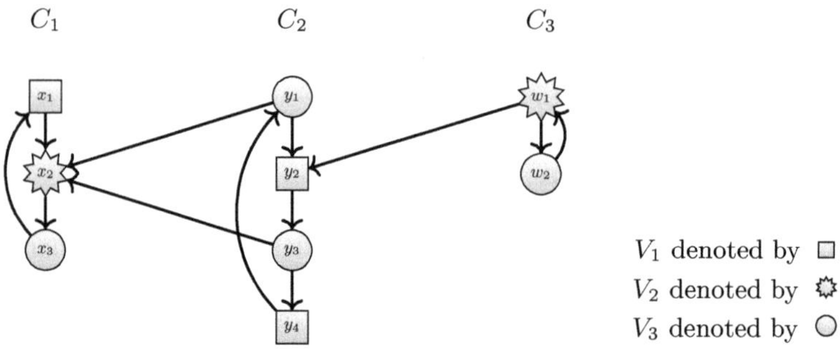


**Figure 7.2** Arcs from cycle $C_i$ to $C_j$, for $1 \leq i < j \leq 3$, are not shown. Note that $C_1 \rightsquigarrow_{V_1} C_2$ (which we also write as $C_1 \overset{\rightsquigarrow}{\square} C_2$) and $C_2 \rightsquigarrow_{V_3} C_3$ (which we also write as $C_2 \overset{\rightsquigarrow}{\bigcirc} C_3$). As there are no arcs from $C_3$ to $C_1$ we have $C_1 \rightsquigarrow_{V_i} C_3$ for all $i = 1, 2, 3$.

We are now in a position to give the simplest version of the main result of this section. Yeo proved the following result.

**Theorem 7.3.1 (Yeo** [75]**)** *Let $D$ be a semicomplete multipartite digraph with partite sets $V_1, V_2, \ldots, V_c$. Let $\mathcal{F}$ be a cycle factor of $D$ consisting of*

*$t$ cycles such that $t$ is minimum. If $t \geq 2$, then there exists an ordering, $C_1, C_2, \ldots, C_t$, of the cycles of $\mathcal{F}$ such that $C_i \leadsto C_j$ for all $1 \leq i < j \leq t$ and the following also holds.*

*(1): There exists indices $1 = i_0 < i_1 < i_2 < \cdots < i_l = t$ and a set of numbers $j_1, j_2, \ldots, j_l \in [c]$ such that if $v_j u_i$ is an arc from $C_j$ to $C_i$ with $j > i$ then $i_{a-1} \leq i < j \leq i_a$ for some $a \in [l]$ and $C_i \leadsto_{V_{j_a}} C_j$.*

Consider Figure 7.2 as an illustration of Theorem 7.3.1. Here there exist indices $(i_0, i_1, i_2) = (1, 2, 3)$ and $(j_1, j_2) = (1, 3)$. We note that there are three arcs from a cycle with higher index to one with lower index. The arc $y_1x_2$ from $C_2$ to $C_1$ satisfies statement (1) in Theorem 7.3.1 as $1 = i_0 \leq 1 < 2 \leq i_1$ and $C_1 \leadsto_{V_{j_1}} C_2$. It can easily be checked that the arcs $y_3x_2$ and $w_1y_2$ also satisfy (1). Note that as $i_1 = 2$ there can be no arcs from $C_3$ to $C_1$. In general, there can be no arcs from a $C_j$ to a $C_i$ if $i < i_a < j$ for some $a$.

It is shown in [75] that Theorem 7.3.1 is close to being best possible in the sense that if a cycle factor satisfies the structural conditions of Theorem 7.3.1 then any cycle factor with fewer than $t$ cycles would have to use internal arcs of the cycles $C_1, C_2, \ldots, C_t$. Theorem 7.3.1 turns out to be very useful when studying cycle factors in semicomplete multipartite digraphs. If we want to consider cycle subgraphs instead then we need a stronger version of this theorem, which we will now state.

**Theorem 7.3.2** ([75]) *Let $D$ be a semicomplete multipartite digraph with partite sets $V_1, V_2, \ldots, V_c$ and let $X \subseteq V(D)$. Let $\mathcal{F}$ be a cycle subgraph of $D$ consisting of $t$ cycles and covering $X$, such that $t$ is minimum. If $t \geq 2$, then there exists an ordering, $C_1, C_2, \ldots, C_t$, of the cycles of $\mathcal{F}$, such that $C_i \leadsto C_j$ for all $1 \leq i < j \leq t$. Furthermore,*

*(1): There exist indices $1 = i_0 < i_1 < i_2 < \cdots < i_l = t$ and a set of numbers $j_1, j_2, \ldots, j_l \in [c]$, such that the following holds. Let $P$ be a path from $v_j \in V(C_j)$ to $u_i \in V(C_i)$ such that $V(P) \cap V(\mathcal{F}) = \{u_i, v_j\}$ and $j > i$. Then $i_{a-1} \leq i < j \leq i_a$ for some $a \in [l]$ and $\{v_j^+, u_i^-\} \subseteq V_{j_a}$.*

Note that Theorem 7.3.1 follows from Theorem 7.3.2, by letting $X = V(D)$. This completes the description of Theorem 7.3.1 and Theorem 7.3.2 which were the main theorems in this section.

### 7.3.1 An Outline of the Proof of Theorem 7.3.1

For the sake of simplicity we will only outline the proof of Theorem 7.3.1. We refer the reader to [75] for a full proof of this theorem and for a proof of Theorem 7.3.2.

Let $P$ be an $(x, y)$-path in a digraph $D$ and let $Q = v_1v_2 \ldots v_l$ be a path or cycle in $D - V(P)$. Then we say that $P$ has a **partner** on $Q$ if there is an arc (the partner of $P$) $v_iv_{i+1}$ on $Q$ such that $v_ix, yv_{i+1} \in A(D)$. In this case the path, $P$, can be inserted in $Q$ to give a new path (or cycle) $Q[v_1, v_i]PQ[v_{i+1}, v_l]$.

Let $D$ be a digraph and $X, Y \subseteq V(D)$. Then $X \Rightarrow Y$ means that there is no arc from $Y$ to $X$. Let $D$ be a digraph with two disjoint cycles $C_1$ and $C_2$. We will write $C_1 \cong> C_2$ when the following holds. There is a vertex $x_1 \in V(C_1)$ such that $x_1 \Rightarrow V(C_2)$ and there is no vertex $y_1 \in V(C_1)$ such that $V(C_2) \Rightarrow y_1$. Furthermore, there is a vertex $x_2 \in V(C_2)$ such that $V(C_1) \Rightarrow x_2$ and there is no vertex $y_2 \in V(C_2)$ such that $y_2 \Rightarrow V(C_1)$.

The following two results were proved by Bang-Jensen, Gutin and Huang in [4].

**Theorem 7.3.3** ([4]) *Let $D$ be a digraph and suppose that $P = p_1p_2 \ldots p_l$ is a path in $D$ and $C$ is a cycle in $D - V(P)$. Suppose that for each odd $i$ the arc $p_ip_{i+1}$ has a partner on $C$ and if $l$ is odd then $p_l$ has a partner on $C$. Then $D$ contains a cycle with vertex set $V(P) \cup V(C)$.*

Bang-Jensen, Gutin and Huang used the above theorem to prove the following result.

**Theorem 7.3.4** ([4]) *Let $D$ be a semicomplete multipartite digraph containing a cycle factor $F = C_1 \cup C_2 \cup \ldots \cup C_t$ such that $t$ is minimum possible. Then for all $i$ and $j$, with $1 \leq i \neq j \leq t$, we either have $C_i \cong> C_j$ or $C_j \cong> C_i$ (but not both).*

The first part of the proof of Theorem 7.3.1 is to prove the following lemma.

**Lemma 7.3.5** ([75]) *Let $D$ be a semicomplete multipartite digraph, and let $C_1$ and $C_2$ be two disjoint cycles in $D$ such that $C_1 \cong> C_2$ and there is some arc from $C_2$ to $C_1$. Assume that there is no cycle in $D$ with vertex set $V(C_1) \cup V(C_2)$. Then there exists a unique partite set $V_i$ of $D$ such that for any $(C_2, C_1)$-path, $P = p_1p_2 \ldots p_l$, in $D$, either $\{p_1^+, p_l^-\} \subseteq V_i$ or there exists a cycle $C^*$ in $D$ with $V(C^*) = V(C_1) \cup V(C_2) \cup V(P)$.*

**Proof:** Since $C_1 \cong> C_2$ and there is some arc from $C_2$ to $C_1$, we can find an $x \in V(C_1)$ such that $x \Rightarrow C_2$ and $y^-x^+ \in A(D)$, for some $y \in V(C_2)$. Let $V_i$ be the partite set containing $x$.

We must have $y \in V_i$ as otherwise $C = C_2[y, y^-]C_1[x^+, x]y$ is a cycle with $V(C) = V(C_1) \cup V(C_2)$, a contradiction. We will now prove the following three claims, where Claim (B) and (C) (where $P = p_1p_2 \ldots p_l$ is an arbitrary $(C_2, C_1)$-path) imply the lemma.

**Claim A.** $C_1 \Rightarrow y$.

*Proof of Claim A.* Label the vertices in $C_2$ such that $C_2 = y_1y_2 \ldots y_my_1$, where $y_1 = y$, and assume that Claim A is not true. That is, there is an arc from $y_1$ to $C_1$. We define the statements $\alpha_k$ and $\beta_k$ for all odd $k$ as follows.

$\alpha_k$: The vertex $y_l \in V_i$ and $y_l$ has an arc to $C_1$ for all $l = 1, 3, 5, \ldots, k$.
$\beta_k$: The arc $y_l y_{l+1}$ has a partner in $C_1[x^+, x]$ for all $l = 1, 3, 5, \ldots, k$.

We will now prove that $\alpha_k$ and $\beta_k$ are true for all odd $k$, with $1 \leq k < m$. Clearly $\alpha_1$ holds, so if we prove the following two operations we are done by induction.

$\alpha_k$ and $\beta_{k-2}$ imply $\beta_k$ (when $k = 1$, $\alpha_k$ implies $\beta_k$): First consider the case when $y_k$ has a partner in $C_1$. As $x$ and $y_k$ both belong to $V_i$ (by $\alpha_k$) we note that the partner is not $xx^+$ and therefore the partner is on $C_1[x^+, x]$. By $\beta_{k-2}$ and Theorem 7.3.3 we can insert the path $C_2[y_1, y_k]$ into the cycle $C_1[x^+, x]C_2[y_{k+1}, y_m]x^+$, a contradiction.
So we may assume that $y_k$ has no partner in $C_1$. Since $y_k$ has an arc to $C_1$ (by $\alpha_k$) and an arc from $C_1$ (as $C_1 \not\Rightarrow C_2$), this implies that there exists a $z_k \in V(C_1)$ such that $z_k \in V_i$ and $z_k^- \to y_k \to z_k^+$. Therefore $y_{k+1} \to z_k$, since otherwise there would be a cycle, $C = C_2[y_{k+1}, y_k]C_1[z_k^+, z_k]y_{k+1}$, in $D$ with $V(C) = V(C_1) \cup V(C_2)$. Thus $y_k y_{k+1}$ has the partner $z_k^- z_k$ in $C_1$, which implies that $y_k y_{k+1}$ has a partner in $C_1[x^+, x]$, as $z_k^- \neq x$ ($z_k^- \notin V_i$).

$\alpha_{k-2}$ and $\beta_{k-2}$ imply $\alpha_k$: $y_k \in V_i$, since if not, then by $\beta_{k-2}$ and Theorem 7.3.3, we can obtain a cycle in $D$ with vertex set $V(C_1) \cup V(C_2)$ by inserting the path $P = y_1 y_2 \ldots y_{k-1}$ into the cycle $C_1[x^+, x]C_1[y_k, y_m]x^+$. If $C_1 \Rightarrow y_k$ then $z_{k-2}^- \to y_k$, where $z_{k-2}$ was defined when we proved $\beta_{k-2}$. When we defined $z_{k-2}$, we found that $y_{k-1} \to z_{k-2}$ and therefore $C = C_1[z_{k-2}, z_{k-2}^-]C_2[y_k, y_{k-1}]z_{k-2}$ is a cycle with $V(C) = V(C_1) \cup V(C_2)$, a contradiction. Therefore $\alpha_k$ holds.

Since $y_m$ has a partner in $C_1$ (namely $xx^+$), Theorem 7.3.3 implies that we can insert the path $C_2[y_1, y_m]$ into $C_2$ such that we obtain a new cycle in $D$ with vertex set $V(C_1) \cup V(C_2)$. This contradiction implies that $C_1 \Rightarrow y$.

**Claim B.** $\{p_1^+, p_l^-\} \cap V_i \neq \emptyset$, *otherwise we are done.*

*Proof of Claim B.* If $\{p_1^+, p_l^-\} \cap V_i = \emptyset$, then the following cycle has $V(C^*) = V(C_1) \cup V(C_2) \cup V(P)$ and we are done.

$$C^* = C_1[x^+, p_l^-]C_2[y, p_1]PC_1[p_l, x]C_2[p_1^+, y^-]x^+.$$

**Claim C.** $p_1^+$ *and* $p_l^-$ *belong to the same partite set, otherwise we are done.*

*Proof of Claim C.* Assume that $p_1^+$ and $p_l^-$ do not belong to the same partite set. Claim (B) implies that either $p_1^+ \in V_i$ or $p_l^- \in V_i$ (but not both by the assumption of Claim C). We may assume that $p_1^+ \to p_l^-$, since otherwise $C^* = C_1[p_l, p_l^-]C_2[p_1^+, p_1]P$ would be a cycle with $V(C^*) = V(P) \cup V(C_1) \cup V(C_2)$ and we would be done.

Now Claim (B), used for the path $P' = p_1^+ p_l^-$, implies that either $p_1^{++} \in V_i$ or $p_l^{--} \in V_i$, but not both, since either $p_1^+ \in V_i$ or $p_l^- \in V_i$. Continuing this process we obtain that $p_1^+ \to p_l^-$, $p_1^{++} \to p_l^{--}, \ldots$ which clearly is impossible since $C_1$ has a vertex which strongly dominates $C_2$. This is a contradiction and hence $p_1^+$ and $p_l^-$ belong to the same partite set.

As mentioned earlier Claim (B) and (C) imply Lemma 7.3.5. □

The next theorem is now needed.

**Theorem 7.3.6** ([75]) *Let $F$ be a cycle factor in a semicomplete multipartite digraph with the minimum possible number of cycles. Then the cycles in $F$ can be labeled in a unique way $C_1, C_2, \ldots, C_t$ such that $C_i \Rightarrow C_j$ for all $i, j$ satisfying $1 \leq i < j \leq t$.*

**Proof:** One can show that if such an ordering is not possible then there must be 3 distinct cycles $C_i$, $C_j$ and $C_k$ such that $C_i \Rightarrow C_j \Rightarrow C_k \Rightarrow C_i$ (as a non-transitive tournament contains a 3-cycle). First assume that there is some arc from $C_j$ to $C_i$. Assume $V^*$ is the partite set found in Lemma 7.3.5 and let $x \in V(C_j) \cap V^*$ be arbitrary. As $C_j \Rightarrow C_k$ there is an arc, say $xy$, from $x$ to $C_k$. Similarly there is an arc, say $y^- z$ from $y^-$ to $C_i$. Now the path $xC_k[y, y^-]z$ is a path from $C_j$ to $C_i$ starting in $V^*$, which by Lemma 7.3.5 is a contradiction.

Therefore we may assume that there is no arc from $C_j$ to $C_i$. Analogously we may assume that there is no arc from $C_k$ to $C_j$ or from $C_i$ to $C_k$. However in this case it was shown in [4] that there is a cycle $C^*$ in $D$ with vertex set $V(C_i) \cup V(C_j) \cup V(C_k)$, contradicting the minimality of $t$. □

These were the main ideas of the proof in [75]. We will now give a slightly different proof for the remaining part of the proof of Theorem 7.3.1.

**Lemma 7.3.7** *Let $F$ be a cycle factor in a semicomplete multipartite digraph with the minimum possible number of cycles. Let $C_1, C_2, \ldots, C_t$ be the ordering of the cycles in $F$ given by Theorem 7.3.6. Assume there is a $(C_j, C_i)$-arc, $x_j x_i$, and a $(C_k, C_i)$-arc, $y_k y_i$, where $i < j < k$. Let $V_{ij}$ be the partite set found in Lemma 7.3.5 when considering $C_i$ and $C_j$ and let $V_{ik}$ be the partite set found in Lemma 7.3.5 when considering $C_i$ and $C_k$. Then $V_{ij} = V_{ik}$.*

**Proof:** Define the ordering $C_1, C_2, \ldots, C_t$, the arcs $x_j x_i$ and $y_k y_i$ and the partite sets $V_{ij}$ and $V_{ik}$ as in the statement of the lemma. For the sake of contradiction assume that $V_{ij} \neq V_{ik}$. This implies that $y_i^- \to x_j^+$ (as $x_j^{++} \notin V_{ij}$ and $y_i^- \in V_{ik}$ and $x_j^+ \in V_{ij}$) and $x_i^- \to y_k^+$ (as $y_k^{++} \notin V_{ik}$ and $x_i^- \in V_{ij}$ and $y_k^+ \in V_{ik}$). We can now define two new cycles $C_1^* = C_i[x_i, y_i^-]C_j[x_j^+, x_j]x_i$ and $C_2^* = C_i[y_i, x_i^-]C_k[y_k^+, y_k]y_i$ such that $V(C_1^*) \cup V(C_2^*) = V(C_i) \cup V(C_j) \cup V(C_k)$, contradicting the minimality of $t$. □

By Lemma 7.3.7 we can define a partite set $V_i^*$ such that all $(C_j, C_i)$-arcs, $xy$, with $j > i$ have $y^+, x^- \in V_i^*$ (if there are no $(C_j, C_i)$-arcs with $j > i$ then let $V_i^* = V_{i-1}^*$ if $i > 1$ and let $V_i^*$ be arbitrary if $i = 1$).

**Lemma 7.3.8** *Let $F$ be a cycle factor in a semicomplete multipartite digraph with the minimum possible number of cycles. Let $C_1, C_2, \ldots, C_t$ be the ordering of the cycles in $F$ given by Theorem 7.3.6. Assume there is a $(C_j, C_i)$-arc in $D$ where $i < j$. Then $V_i^* = V_{i+1}^* = V_{i+2}^* = \cdots = V_{j-1}^*$.*

**Proof:** Define the ordering $C_1, C_2, \ldots, C_t$ as in the statement of the lemma and assume that $xy$ is a $(C_j, C_i)$-arc with $i < j$. For the sake of contradiction, assume that $V_k^* \neq V_i^*$ for some $i < k < j$. Note that there is an arc, say $y^- z$, from $y^-$ to $C_k$ (as $C_i \Rightarrow C_k$). Now $P = xC_i[y, y^-]z$ is a path from $C_j$ to $C_k$ where $x^+ \notin V_k^*$ (as $x^+ \in V_i^*$), contradicting Lemma 7.3.5. □

Theorem 7.3.1 follows from the above by the following argument. Let $i_0 = 1$ and then let $i_1$ be the smallest value greater than $i_0$ such that $V_{i_1}^* \neq V_{i_1-1}^*$. Then let $i_2$ be the smallest integer greater than $i_1$ such that $V_{i_2}^* \neq V_{i_2-1}^*$. Continuing this procedure gives us the correct $i$-values in Theorem 7.3.1. Now it is not difficult to see that Theorem 7.3.1 follows from the above results.

## 7.4 Semicomplete Bipartite Digraphs

A semicomplete bipartite digraph is a semicomplete multipartite digraph with only 2 partite sets. This graph class has been widely studied and some results are much nicer and/or easier for semicomplete bipartite digraphs than for semicomplete multipartite digraphs in general. We consider cycles and paths in semicomplete bipartite digraphs separately.

### 7.4.1 Cycles in Semicomplete Bipartite Digraphs

25 years after Rédei proved Theorem 2.2.4 Camion proved Theorem 2.2.6, which was the next major result for tournaments.

**Theorem** 2.2.6 **(Camion's Theorem** [17], **1959)** *Every strong tournament contains a Hamiltonian cycle.*

Unfortunately such a nice and simple theorem does not hold for semicomplete bipartite digraphs or semicomplete multipartite digraphs. However, it is still easy to determine when a semicomplete bipartite digraph has a Hamilton cycle. The following characterization was obtained independently by Gutin [31] and Häggkvist and Manoussakis [43]. The original proof of this theorem was longer, but using Theorem 7.3.1 we will give a shorter proof of the structural part.

**Theorem 7.4.1** ([31, 43]) *A semicomplete bipartite digraph has a Hamilton cycle if and only if it contains a cycle factor and is strongly connected. Furthermore, one can verify whether $D$ contains a Hamilton cycle (and find one if it does) in time $O(|V(D)|^{2.5})$.*

**Proof:** Clearly if a digraph contains a Hamilton cycle then it is strong and has a cycle factor (as the Hamilton cycle is a cycle factor). So let $D$ be a strong semicomplete bipartite digraph with a cycle factor $F = C_1 \cup C_2 \cup \cdots \cup C_t$ where $t$ is minimum. If $t = 1$ then $D$ is Hamiltonian and we are done, so assume for the sake of contradiction that $t \geq 2$. As $D$ is strong, there exists an arc $xy$ from $V(C_t)$ to $V(C_i)$ for some $i < t$. By Theorem 7.3.1 the successor of $x$ on $C_t$ and the predecessor of $y$ on $C_i$ belong to the same partite set. However this implies that $x$ and $y$ belong to the same partite set, contradicting the fact that the arc $xy$ exists.

This completes the first part of the proof. We refer the reader to [31, 43] for the complexity part. □

Theorem 7.4.1 can also be used to find longest cycles in semicomplete bipartite digraphs, as we will see below. This was first proved by Gutin in [32, 34].

**Theorem 7.4.2** ([32, 34]) *Let $D$ be a strong semicomplete bipartite digraph of order $n$. The length of a longest cycle in $D$ is equal to the maximum order of a cycle subgraph in $D$.*

*This implies that one can find a longest cycle in $D$ in time $O(n^3)$.*

**Proof:** We will prove the structural part of the theorem and refer the reader to [32, 34] for the complexity proof.

Let $D$ be a strong semicomplete bipartite digraph. Let $F = C_1 \cup C_2 \cup \cdots \cup C_t$ be a cycle subgraph of maximum order. Of all such cycle subgraphs choose $t$ to be the smallest possible. If $t = 1$ then we are done, so assume that $t \geq 2$. If any subset of at least two cycles in $F$ induce a strong subgraph of $D$ then they can be merged to one cycle due to Theorem 7.4.1, contradicting the minimality of $t$. Therefore this is not the case, which implies that we can, without loss of generality, assume that the cycles are numbered such that there is no arc from $C_j$ to $C_i$ for any $j > i$.

As $D$ is strong, there must be a path, $P$, from $V(C_t)$ to $V(C_1) \cup V(C_2) \cup \cdots \cup V(C_{t-1})$. Assume $P$ is a $(x, y)$-path and $y \in V(C_i)$, where $1 \leq i < t$. Let $y^+$ be the successor of $y$ on $C_t$ and let $x^-$ be the predecessor of $x$ on $C_i$. If $x^-$ and $y^+$ are adjacent then we can merge $C_i$, $C_t$ and $P$ into a cycle (using the arc $x^-y^+$). This cycle has more vertices than $V(C_i) \cup V(C_t)$ (as $P$ is not just an arc), a contradiction. Therefore $x^-$ and $y^+$ belong to the same partite set. However, then the predecessor, $x^{--}$, of $x^-$ on $C_i$ lies in a different partite set to $y^+$. Therefore $x^{--}y^+$ is an arc in $D$ and adding $P$ and $x^{--}y^+$ to $C_i$ and $C_j$ and deleting $x^{--}x^-$, $x^-x$ and $yy^+$ we obtain a cycle containing $V(C_i) \cup V(C_t) \cup V(P) \setminus \{x^-\}$ of at least the same order as $V(C_i) \cup V(C_t)$, contradicting the minimality of $t$. □

If $F$ is a cycle subgraph of maximum order in a strong semicomplete bipartite digraph then there is not necessarily a cycle on the same set of vertices. A similar result as that given in Theorem 7.4.2 holds for extended semicomplete digraphs (see Theorem 7.10.2), but for extended semicomplete digraphs one can always find a longest cycle with the same vertex set as any cycle subgraph of maximum order. So even though Theorem 7.4.2 and Theorem 7.10.2 look very similar the proofs are somewhat different.

The following result of Yeo holds when the desired cycle has to cover a given set of vertices.

**Theorem 7.4.3** ([71]) *Let $D$ be a strong semicomplete bipartite digraph of order $n$ and let $X \subseteq V(D)$. One can decide if there is a cycle covering $X$ in $D$ in time $O(n^5)$. Furthermore, if it exists, then we can find such a cycle in time $O(n^5)$.*

Yeo conjectured that the above also holds for semicomplete multipartite digraphs.

**Conjecture 7.4.4** ([71]) *Let $D$ be a semicomplete multipartite digraph and $X \subseteq V(D)$. There is a polynomial time algorithm for finding a cycle covering $X$ (if it exists), and which is the longest among all suchcycles.*

### 7.4.2 Even Pancyclic Bipartite Tournaments

A digraph $D$ is **even pancyclic** if it contains all cycles of even length from 4 to $2\lfloor |V(D)|/2 \rfloor$. Similarly we can define **vertex-even-pancyclic** (**arc even pancyclic**) if for every vertex (arc) there exists cycles of all even lengths from 4 to $2\lfloor |V(D)|/2 \rfloor$ containing this vertex (arc). Let $B(r,r,r,r)$ denote the bipartite tournament with partite sets $V_1 \cup V_1'$ and $V_2 \cup V_2'$, where $|V_1| = |V_1'| = |V_2| = |V_2'| = r$ and $V_1 \to V_2 \to V_1' \to V_2' \to V_1$. See Figure 7.3 for an illustration of $B(r,r,r,r)$.

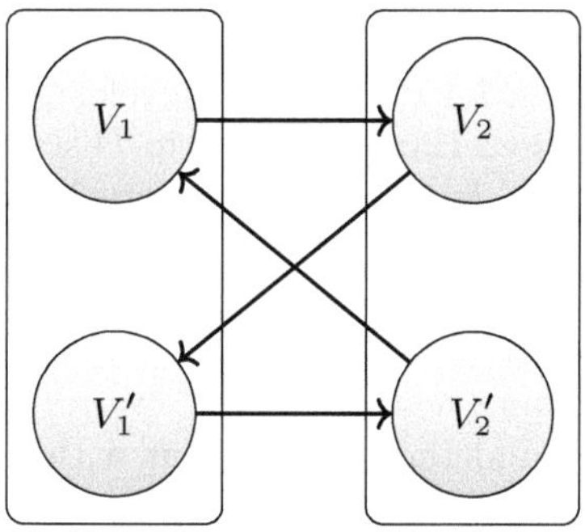


**Figure 7.3** The bipartite tournament $B(r,r,r,r)$. Note that $|V_1| = |V_1'| = |V_2| = |V_2'| = r$.

The following characterizations of even pancyclic and vertex-even-pan-breakcyclic bipartite tournaments were derived by Beineke and Little in [14] and by Zhang [78], respectively. Note that the last characterization was obtained independently by Häggkvist and Manoussakis in [43] as well.

**Theorem 7.4.5** ([14, 43, 78]) *A bipartite tournament is even pancyclic as well as vertex-even-pancyclic if and only if it is Hamiltonian and is not isomorphic to the bipartite tournament $B(r, r, r, r)$ $(r = 2, 3, \ldots)$.*

Considering regular bipartite tournaments, Amar and Manoussakis [1] and, independently, Wang [69] showed the following.

**Theorem 7.4.6** ([1, 69]) *An $r$-regular bipartite tournament is arc even pancyclic unless it is isomorphic to $B(r, r, r, r)$.*

### 7.4.3 Paths in Semicomplete Bipartite Digraphs

Theorem 7.4.1 can be used to prove the following theorem of Bang-Jensen and Gutin.

**Theorem 7.4.7** ([3]) *A semicomplete bipartite digraph, $D$, has a Hamilton path starting at the vertex $x$ if and only if $D$ contains a 1-path-cycle factor where the path starts at $x$ and $x$ can reach every other vertex of $D$.*

*Furthermore, given a 1-path-cycle factor where the path starts at $x$ and $x$ can reach all vertices in $V(D) \setminus \{x\}$, we can find a Hamilton path in $D$ starting at $x$ in time $O(|V(D)|^2)$.*

**Proof:** Let $D$ be a semicomplete bipartite digraph with a 1-path-cycle factor $F$ with cycles $C_1, C_2, \ldots, C_t$ and path $P$. Assume that $P$ is a $(x, y)$-path and $x$ can reach all vertices in $D$. Let $V_1$ and $V_2$ be the partite sets of $D$ and, without loss of generality, assume that $x \in V_1$.

If $y \in V_2$, then let $D_1$ be the semicomplete bipartite digraph obtained from $D$ by adding all arcs from $V_2$ to $x$. As $x$ can reach all vertices in $D$ we note that $D_1$ is strong. As $yx \in A(D_1)$ we note that $D_1$ has a cycle factor and therefore by Theorem 7.4.1 a Hamilton cycle, $C_1$. Removing the arc into $x$ in $C_1$ gives us the desired Hamilton path in $D$ starting in $x$.

If $y \in V_1$ then let $D_2$ be the semicomplete bipartite digraph obtained from $D$ by adding a new vertex $z$ to $V_2$ and all arcs from $V_1$ to $z$ as well as the arc $zx$. As $x$ can reach all vertices in $D$ we note that $D_2$ is strong. As $yzx$ is a path in $D_2$ we note that $D_2$ has a cycle factor and therefore by Theorem 7.4.1 a Hamilton cycle, $C_2$. As $N^+(z) = \{x\}$ we note that $zx$ is an arc in $C_2$. Removing $z$ and the arcs incident with $z$ we obtain the desired Hamilton path in $D$ starting in $x$.

This completes the first part of the proof. We refer the reader to [3] for the complexity part. □

The following result was proved by Amar, Manoussakis and Wang in [1, 69] and also follows from Theorem 7.4.6. Recall that a digraph is $r$-regular if all out-degrees and in-degrees have the value $r$ and it is regular if it is $r$-regular for some $r$.

**Theorem 7.4.8** ([1, 69]) *Every arc of a regular bipartite tournament is contained in a Hamiltonian cycle.*

In Section 7.8 we say that a digraph $D$ is **almost regular** if the maximum of all out- and in-degrees is at most one larger than the minimum of all out- and in-degrees (this will also be denoted by $i_g(D) \leq 1$). See also Section 7.8 for more results on semicomplete multipartite digraphs that are close to regular.

If $D$ is an almost regular bipartite tournament then the partite sets differ by at most two in size. However if the size of the partite sets differ by two then there is no Hamilton path in $D$. Therefore in the following theorem of Volkmann we consider the case when the size of the partite sets differ by at most one.

**Theorem 7.4.9** ([63]) *Let $T$ be an almost regular bipartite tournament with the partite sets $X$ and $Y$ such that $1 \leq |X| \leq |Y| \leq |X| + 1$. Every arc of $T$ is contained in a Hamiltonian path if and only if $T$ is not isomorphic to $T_{3,3}$ shown in Figure 7.4.*

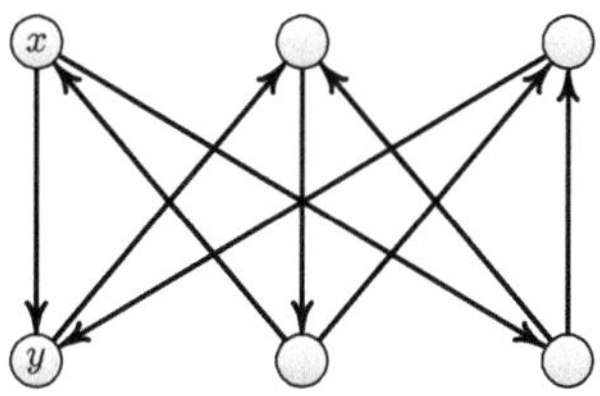


**Figure 7.4** The strong semicomplete bipartite digraph $T_{3,3}$, with no Hamilton path containing the arc $xy$.

## 7.5 Paths in Semicomplete Multipartite Digraphs

One of the first results on tournaments is Theorem 2.2.4, due to Rédei. We recall its statement here.

**Theorem** 2.2.4 (**Rédei's Theorem** [55]) *Every tournament contains a Hamiltonian path.*

Theorem 2.2.4 is not true in general for semicomplete multipartite digraphs. However the following characterization was proved by Gutin in [29]. We give a different proof below in order to illustrate the usefulness of Theorem 7.3.1.

**Theorem 7.5.1** ([29]) *A semicomplete multipartite digraph, $D$, has a Hamilton path if and only if it has a 1-path-cycle factor.*

*Furthermore, one can verify whether $D$ contains a Hamilton path (and find one if it does) in time $O(|V(D)|^{2.5})$.*

**Proof:** Let $D$ be a semicomplete multipartite digraph. If $D$ has a hamiltoninan path then clearly it has a 1-path-cycle factor as the Hamilton path is a 1-path-cycle factor. Now assume that $D$ has a 1-path-cycle factor. Add a new vertex $x$ to $D$ that is connected to all other vertices in $D$ with a 2-cycle and let the resulting digraph be $D'$. Clearly $D'$ is a semicomplete multipartite digraph with $x$ being in a partite set by itself. Note that $D'$ contains a cycle factor as the path, $P$, in the 1-path-cycle factor of $D$ can be made into a cycle in $D'$ by adding the vertex $x$ and arcs from the end of $P$ to $x$ and from $x$ to the start of $P$.

Let $\mathcal{F}$ be a cycle factor of $D'$ with the minimum number of cycles, $t$. By Theorem 7.3.1 we have $t = 1$, as the cycle containing $x$ cannot weakly-dominate or be weakly-dominated by any other cycle. Therefore $D'$ contains a Hamilton cycle. Removing $x$ from this Hamilton cycle gives us the desired Hamilton path.

We have now proved the structural part of the theorem. We refer the reader to [29] for a proof of the complexity. □

Theorem 7.5.1 can also be formulated as: a semicomplete multipartite digraph $D$ has $pc(D) = 1$ if and only if $pcc(D) = 1$. In fact, the following was proved by Gutin and easily follows from Theorem 7.5.1.

**Theorem 7.5.2** ([32]) *$pc(D) = pcc(D)$ for all semicomplete multipartite digraph $D$. Furthermore, $pc(D)$ can be calculated in time $O(|V(D)|^{2.5})$.*

Theorem 7.5.1 also implies that it is easy to find a longest path in a semicomplete multipartite digraph. The following result was first proved by Gutin.

**Theorem 7.5.3** ([33]) *Let $D$ be a semicomplete multipartite digraph of order $n$ and let $\mathcal{F}$ be a 1-path-cycle subgraph in $D$. Then there exists a path, $P$, in $D$ with $V(P) = V(\mathcal{F})$.*

*As we can find a 1-path-cycle subgraph in $D$ of maximum order in $O(n^3)$ time (see Chapter 4 in [2]) we can find a longest path in $O(n^3)$ time.*

In fact Theorem 7.5.1 can be generalised to semicomplete multipartite digraphs with costs on its vertices. If $D$ is a digraph with costs on its vertices $mp_i(D)$ is defined to be the minimum cost of an $i$-path subgraph (that is, collection of $i$ paths) of $D$. By definition $mp_0(D) = 0$. Furthermore, $mpc_i(D)$ is defined to be the minimum cost of an $i$-path-cycle subgraph (that is, a collection of $i$ paths and any number of cycles) of $D$. Theorem 7.5.1 implies the following result of Gutin.

**Theorem 7.5.4** ([32]) *$mp_i(D) = mpc_i(D)$ for all semicomplete multipartite digraphs $D$ and all $i \in [n]$.*

### 7.5.1 Hamilton Paths Containing Arcs

In Corollary 7.9.4 below we will see that all $k$-strong semicomplete multipartite digraphs with at most $k$ vertices in each partite set have a Hamilton cycle. One could imagine that all $k$-strong semicomplete multipartite digraphs with at most $k$ vertices in each partite set might therefore also have a Hamilton path through any given arc. However this is false even for tournaments, by Figure 7.5. Nevertheless Volkmann proved the following.

**Theorem 7.5.5** ([64]) *If a $(k+1)$-strong semicomplete multipartite digraph $D$ has at most $k$ vertices in each partite set then $D$ contains a Hamilton path through any given arc.*

**Proof:** Let $D$ be a $(k+1)$-strong semicomplete multipartite digraph with at most $k$ vertices in each partite set and let $xy \in A(D)$ be arbitrary. Clearly the semicomplete multipartite digraph $D - x$ is $k$-strong and has at most $k$ vertices in each partite set and therefore contains a Hamilton cycle $C$ by Corollary 7.9.4 (which we will see in Section 7.9). Deleting the arc into $y$ on $C$ and adding the arc $xy$ gives us the desired Hamilton path. □

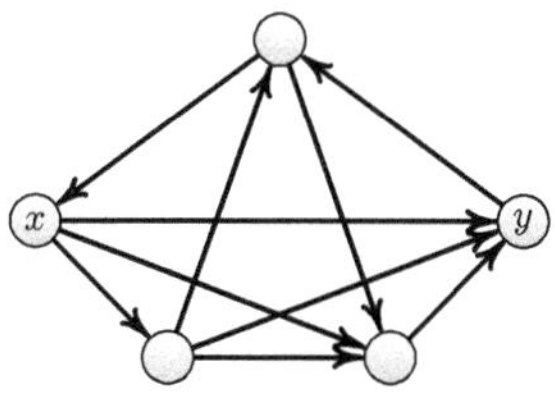


**Figure 7.5** A strong tournament with no Hamilton path containing the arc $xy$.

Observe that the proof above actually implies that in Theorem 7.5.5 we could have required the Hamilton path to start (or end) with the desired arc. Clearly Theorem 7.5.5 is best possible as in Figure 7.5 (see also [16]) we have a 1-strong semicomplete multipartite digraph with at most 1 vertex in each partite set and there is no Hamilton path containing the arc $xy$.

Meng and Li [50] proved the following result (where $\alpha(D)$ is the size of a maximum independent set in $D$, which for semicomplete multipartite digraphs corresponds to the maximum size of a partite set).

**Theorem 7.5.6** ([50]) *Let $D$ be a semicomplete multipartite digraph of order $n$ and with partite sets $V_1, V_2, \ldots, V_c$. If $d^+(x_i, V_j), d^-(x_i, V_j) \geq (|V_j| + 1)/2$ for all $x_i \in V_i$ and $j \neq i$ and $\alpha(D) \leq \frac{n-1}{2}$, then every arc of $D$ is contained in a Hamilton path of $D$.*

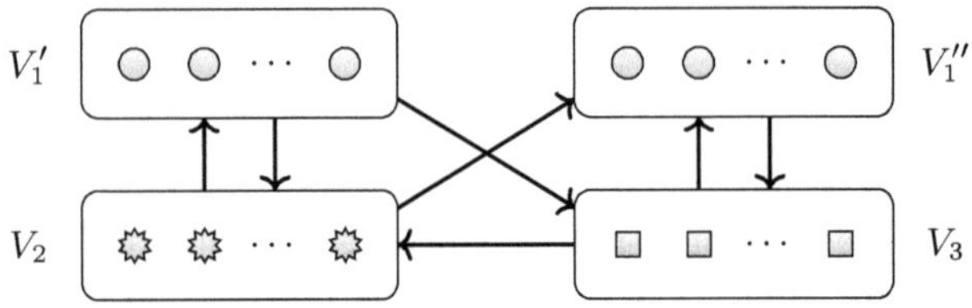


**Figure 7.6** A $k$-strong semicomplete multipartite digraph with a cycle factor but no Hamilton cycle (when $|V_1'| = |V_1''| = |V_2| = |V_3| = k$).

## 7.6 Cycles in Semicomplete Multipartite Digraphs

Unfortunately Theorem 7.4.1, which states that every strong semicomplete bipartite digraph with a cycle factor contains a Hamilton cycle, doesn't easily extend to semicomplete multipartite digraphs. In fact, no degree of strong connectivity and a cycle factor guarantees a Hamilton cycle in a semicomplete multipartite digraph, as can be seen in Figure 7.6. In Figure 7.6 we note that there is a cycle containing $V_1' \cup V_2$ and a cycle containing $V_1'' \cup V_3$, so there is a cycle factor. However, to get from $V_1'' \cup V_3$ to $V_1' \cup V_2$ we need to use an arc from $V_3$ to $V_2$ and since half the vertices belong to the partite set $V_1' \cup V_1''$ no Hamilton cycle contains an arc from $V_3$ to $V_2$. Therefore the digraph is not Hamiltonian.

Despite the fact that no degree of strong connectivity and a cycle factor implies hamiltonicity we can still decide if a semicomplete multipartite digraph has a Hamilton cycle in polynomial time. This was proved by Bang-Jensen, Gutin and Yeo in [6].

**Theorem 7.6.1** ([6]) *One can decide if a semicomplete multipartite digraph of order $n$ has a Hamilton cycle (and find one if it exists) in time $O(n^7)$.*

The proof of Theorem 7.6.1 is very complicated and uses the structural result given in Theorem 7.3.1. In [71] Yeo generalised Theorem 7.6.1 to cycles covering a given set of vertices.

**Theorem 7.6.2** ([71]) *Let $D$ be a strong semicomplete multipartite digraph of order $n$ and let $X \subseteq V(D)$. One can decide if there is a cycle covering $X$ in $D$ (and find one if it exists) in time $O(n^5)$.*

Note that Theorem 7.6.2 not only generalizes Theorem 7.6.1 but also improves the complexity from $O(n^7)$ to $O(n^5)$. For the case when $D$ is a semicomplete bipartite digraph Theorem 7.6.2 was strengthened in Theorem 7.4.3. That is, we showed that if $D$ is a strong semicomplete bipartite digraph and $X \subseteq V(D)$ then we can decide if there is a cycle covering $X$ in $D$ and if there is then we can find a *longest* such cycle in polynomial time.

### 7.6.1 Pancyclicity

Recall the following definitions from Section 1.3. A digraph $D$ is **vertex-$k$-cyclic** (**arc-$k$-cyclic**, respectively) if every vertex (arc, respectively) of $D$ is contained in a $k$-cycle. A digraph $D$ of order $n$ is **pancyclic** if it has a $k$-cycle for every $k \in \{3, 4, \ldots, n\}$. Furthermore, $D$ is **vertex-pancyclic** (**arc-pancyclic**, respectively) if $D$ is vertex-$k$-cyclic (arc-$k$-cyclic, respectively) for every $k \in \{3, 4, \ldots, n\}$.

In 1966 Moon extended Camion's Theorem (Theorem 2.2.6) to Theorem 1.5.1. Recall this theorem.

**Theorem** 1.5.1 (**Moon's Theorem** [51]) *Every strong tournament is vertex-pancyclic.*

Moon's Theorem has been extended to multipartite tournaments and semicomplete multipartite digraphs in several ways. For example, Volkmann and Guo proved the following result in [28].

**Theorem 7.6.3** ([28]) *Let $D$ be a strongly connected $c$-partite tournament. Then every partite set of $D$ has at least one vertex which belongs to cycles $C_3, C_4, \ldots, C_c$ such that $|V(C_i)| = i$ for all $i \in \{3, 4, \ldots, c\}$ and $V(C_3) \subset V(C_4) \subset \cdots \subset V(C_c)$.*

The result in Theorem 7.6.3 cannot be extended to every vertex of a partite set, as can be seen in the multipartite tournament in Figure 7.7.(c), where $x_2$ does not belong to a 4-cycle in the strong 4-partite tournament. It is not difficult to see that Theorem 7.6.3 extends Moon's Theorem, as a strong tournament of order $n$ is also a strong $n$-partite tournament where all partite sets have size one.

Another generalization of Moon's Theorem was given by Goddard and Oellermann in [21].

**Theorem 7.6.4** ([21]) *Every vertex of a strongly connected $c$-partite tournament $D$ belongs to a cycle that contains vertices from exactly $q$ partite sets for each $q \in \{3, 4, \ldots, c\}$.*

Again it is not difficult to see that this theorem extends Moon's Theorem.

## 7.7 Short Cycles in Semicomplete Multipartite Digraphs

Most of the results in this section will be stated for multipartite tournaments. However, due to Theorem 7.1.1, many of them also hold for semicomplete multipartite digraphs. Recall the following theorem of Volkmann.

**Theorem** 7.1.1 ([62]) *Every strong semicomplete $c$-partite digraph with $c \geq 3$ contains a spanning strong oriented subdigraph.*

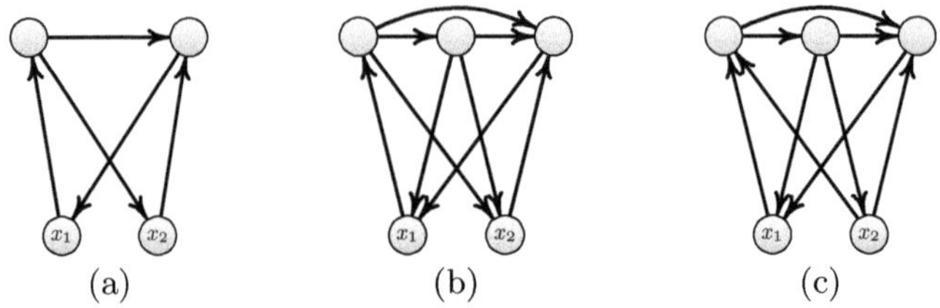


**Figure 7.7** (a) A strong 3-partite tournament with no 3-cycle containing $x_2$. (b) A strong 4-partite tournament with no 3-cycle containing $x_2$. (c) A strong 4-partite tournament with no 4-cycle containing $x_2$.

Recall Theorem 7.6.3, which generalizes many of the known results in this area.

**Theorem** 7.6.3 *Let $D$ be a strong $c$-partite tournament with $c \geq 3$ and with partite sets $V_1, V_2, \ldots, V_c$. For each $i \in [c]$, there exists a vertex $v \in V_i$ that belongs to a $k$-cycle, $C_k$, for all $k \in \{3, 4, \ldots, c\}$ such that $V(C_3) \subset V(C_4) \subset \cdots \subset V(C_c)$.*

Note that Theorem 7.6.3 also holds for semicomplete multipartite digraphs due to Theorem 7.1.1. An immediate corollary of this result is the following well known result, also due to Volkmann and Guo.

**Corollary 7.7.1** ([27]) *Let $D$ be a strong $c$-partite tournament with $c \geq 3$ and with partite sets $V_1, V_2, \ldots, V_c$. For each $i \in [c]$, there exists a vertex $v \in V_i$ that belongs to a $k$-cycle for all $k \in \{3, 4, \ldots, c\}$.*

Clearly Theorem 7.6.3 and Corollary 7.7.1 both generalize the fact that a strong tournament is vertex-pancyclic. A different way to generalize this was seen in Theorem 7.6.4, where it was shown that every vertex of a strong $c$-partite tournament $D$ belongs to a cycle containing vertices from exactly $t$ partite sets of $D$ for each $t \in \{3, 4, \ldots, c\}$.

Theorem 7.6.3 cannot be extended to all vertices in a partite set, as can be seen from the examples in Figure 7.7. Even though there are vertices in the $c$-partite digraphs in Figure 7.7 which do not belong to cycles of lengths $3, 4, \ldots, c$, clearly all vertices belong to cycles that contain vertices from $t$ partite sets for all $t = 3, 4, \ldots, c$ (as promised by Theorem 7.6.4).

In fact, every vertex of a strong multipartite tournament belongs to cycles of length $k$ or $k+1$, as proved in the following theorem of Guo, Pinkernell and Volkmann.

**Theorem 7.7.2** ([25]) *Let $D$ be a strong $c$-partite tournament and let $v \in V(D)$ be arbitrary. Then $v$ belongs to a $k$-cycle or a $(k+1)$-cycle in $D$ for every $k \in \{3, 4, \ldots, c\}$.*

In [38] Gutin, Rafiey and Yeo characterize the strong $c$-partite tournaments with a unique $c$-cycle. The characterization is quite complex, so it is not given here. By Theorem 7.6.3 we observe that a $c$-cycle always exists in a strong $c$-partite tournament $D$. Any vertex not in a $c$-cycle belongs to a $(c+1)$-cycle by Theorem 7.7.2. Therefore if $D$ is a $c$-partite tournament, but not a tournament, and it has a unique $c$-cycle, then it will contain cycles of length more than $c$.

It is not difficult to construct strong $c$-partite multipartite tournaments where the longest cycle has length $c$. Examples of such multipartite tournaments can be seen in Figure 7.8. In fact, the strong $c$-partite tournaments for which the longest cycle has length $c$ were characterized by Gutin in [37].

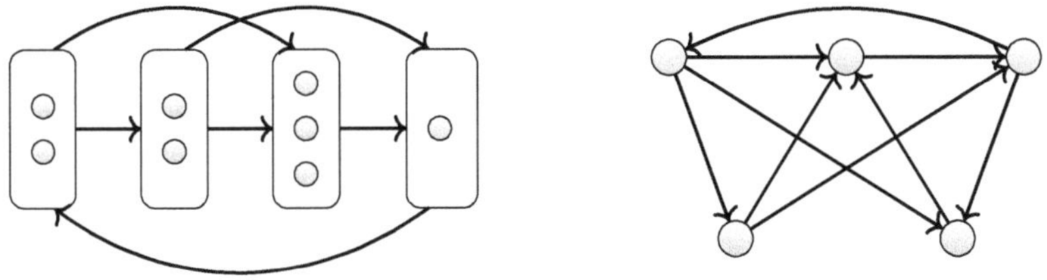

**Figure 7.8** Examples of strong 4-partite tournaments whose longest cycle has length 4.

If we not only know that the multipartite tournament is strong, but also regular, then the following generalization of Alspach's theorem was proved by Guo.

**Theorem 7.7.3** ([24]) *Let $D$ be a regular $c$-partite tournament. If every arc of $D$ belongs to a 3-cycle, then every arc of $D$ is on a $k$-cycle for all $k \in \{3, 4, \ldots, c\}$.*

The following theorem was proved by Zhou and Zhang.

**Theorem 7.7.4** ([80]) *If $D$ is a regular $c$-partite tournament with $c \geq 6$, then every arc of $D$ belongs to a $k$-cycle for all $k \in \{4, 5, \ldots, c\}$.*

## 7.8 Regular and Close to Regular Semicomplete Multipartite Digraphs

In Section 1.2 an $r$-**regular digraph** is defined as a digraph where all out- and in-degrees are equal to $r$. There are several measures of how much a digraph differs from being regular. In [74] the **local irregularity** is defined as $i_l(D) = \max\{|d^+(x) - d^-(x)|\}$ over all vertices $x$ of $D$ and the **global irregularity** as $i_g(D) = \max\{\Delta^+(D), \Delta^-(D)\} - \min\{\delta^+(D), \delta^-(D)\}$. That is, the global irregularity is the difference between the maximum out- or in-degree and the minimum out- or in-degree. Note that $i_l(D) \leq i_g(D)$ and that if $i_g(D) = 0$ then $D$ is regular.

Alspach proved the following result, which was stated previously as Corollary 2.14.4.

**Corollary** 2.14.4 *All regular tournaments are arc-pancyclic.*

We will refer to the above result as Alspach's Theorem. This theorem does not extend to multipartite tournaments. In fact C.Q. Zhang conjectured that all regular semicomplete multipartite digraphs contain a Hamilton cycle and this conjecture was open for several years in the 1990s. The conjecture was eventually proved by Yeo in [75] using Theorem 7.3.1. We give the proof here as another example of how to use Theorem 7.3.1.

**Theorem 7.8.1** ([75]) *Every regular semicomplete multipartite digraph contains a Hamilton cycle.*

**Proof:** Let $D$ be a diregular semicomplete multipartite digraph. Ore proved (in [53]) that every regular digraph contains a cycle factor, so let $F = C_1 \cup C_2 \cup \ldots \cup C_t$ be a cycle factor of $D$. We may assume that $F$ is chosen such that $t$ is minimum. If $t = 1$ then $D$ is Hamiltonian, so assume for the sake of contradiction that $t > 1$.

By Theorem 7.3.1 we may assume that $C_i \rightsquigarrow C_j$ for all $1 \leq i < j \leq t$. Furthermore, by Theorem 7.3.1, there exists an index $j_1$, such that $C_1 \rightsquigarrow_{V_{j_1}} C_k$ for all $2 \leq k \leq t$.

Let $yx \in A(D)$ be an arc from $y \in V(C_k)$, with $k \in \{2, 3, \ldots, t\}$, to $x \in V(C_1)$. By Theorem 7.3.1 we have $x^-, y^+ \in V_{j_1}$. Now we define the two distinct arcs $a_1(yx) = xy^+$ and $a_2(yx) = x^-y$. By Theorem 7.3.1, $a_1(yx)$ and $a_2(yx)$ are arcs in $D$.

If $y'x'$ and $yx$ are distinct arcs from $V(D) \setminus V(C_1)$ to $V(C_1)$, then we see that $a_1(yx)$, $a_2(yx)$, $a_1(y'x')$ and $a_2(y'x')$ are four distinct arcs from $V(C_1)$ to $V(D) \setminus V(C_1)$. Therefore the number of arcs leaving $V(C_1)$ is at least double as large as the number of arcs entering $V(C_1)$. However this contradicts the fact that $D$ is an Eulerian digraph (and therefore has equally many arcs from $V(C_1)$ to $V(D) \setminus V(C_1)$ as from $V(D) \setminus V(C_1)$ to $V(C_1)$). □

Much research has gone into semicomplete multipartite digraphs that are close to regular. For example, the following result of Yeo.

**Theorem 7.8.2** ([74]) *Let $D$ be a semicomplete multipartite digraph of order $n$. Let $n_1$ denote the size of the largest partite set in $D$ and let $n_2$ denote the size of the second largest partite set ($n_1 = n_2$ is possible). If either of the following holds then $D$ is Hamiltonian.*

- $i_g(D) \leq \frac{n-2n_1-n_2}{2} + 1$.
- $i_l(D) \leq \min\left\{n - 3n_1 + 1, \frac{n-2n_1-n_2}{2} + 1\right\}$.

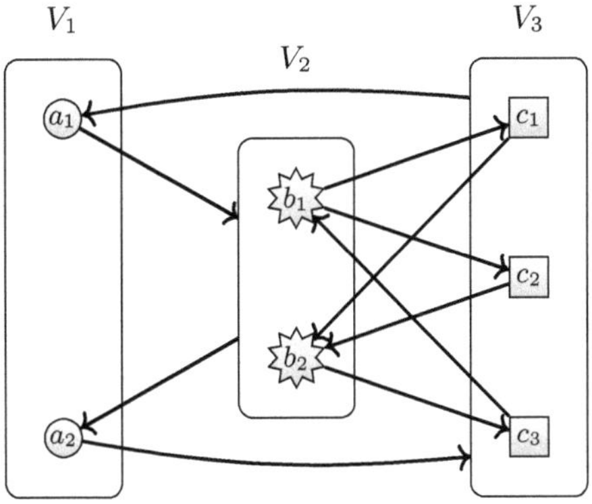


**Figure 7.9** A semicomplete multipartite digraph, $D$, with no Hamilton cycle (or cycle factor) and with $i_g(D) = i_l(D) = 1$. Note that there is no cycle factor in $D$ as if there was then between any two vertices from $V_3$ we must have a vertex from $V_2$ and $|V_2| < |V_3|$.

Theorem 7.8.2 is best possible, as in [74] there are constructed infinitely many semicomplete multipartite digraphs, $D$, with $i_l(D) = i_g(D) = \frac{n-2n_1-n_2}{2} + \frac{3}{2} \leq n - 3n_1 + 2$, which are not Hamiltonian. The smallest such example given in [74] contains 11 vertices and is a bit to big to illustrate here. However, consider the semicomplete multipartite digraph, $D$, given in Figure 7.9. Note that for this digraph we have $|V(D)| = 7$ and all out- and in-degrees are either 2 or 3 and therefore $i_g(D) = i_l(D) = 1$. Furthermore, the size of the largest partite set is $n_1 = |V_3| = 3$ and the size of the second largest is $n_2 = |V_2| = 2$. Therefore $i_g(D) = \frac{|V(D)|-2n_1-n_2}{2} + \frac{3}{2}$, which shows that one part of Theorem 7.8.2 is tight.

Another way of generalizing Theorem 7.8.1 is the following. For a digraph $D$ with vertex set $V$ and a positive integer $k$ define $f(D,k)$ as follows

$$f(D,k) = \sum_{x \in V, d^+(x) > k} (d^+(x) - k) + \sum_{x \in V, d^-(x) < k} (k - d^-(x)).$$

Note that $f(D,k) \geq 0$ and if $f(D,k) = 0$ then $D$ is $k$-regular. This is the case as if $d^+(x) \leq k$ for all $x$ and $d^-(x) \geq k$ for all $x$ then $d^+(x) = d^-(x) = k$ for all $x$ (as $\sum_{x \in V} d^+(x) = |A(D)| = \sum_{x \in V} d^-(x)$). One can show that if $f(D,k) \leq k-1$ then $D$ has a cycle factor. We can in fact show that, except for a special class of semicomplete multipartite digraphs, $f(D,k) \leq k-1$ also implies hamiltonicity for semicomplete multipartite digraphs.

For every $k \geq 2$ define $G'_k$ to be the 3-partite digraph with partite sets $V_1 = \{x\}$, $V_2 = \{y_2, y_3, \ldots, y_k\}$ and $V_3 = \{z_1, z_2, \ldots, z_k\}$ and the following arc set

$$\{yx, xz, zy, yv \mid y \in V_2,\, z \in V_3,\, v \in V_3 \setminus \{z_1\}\} \cup \{z_1 x\}.$$

See Figure 7.10 for an illustration of $G'_k$. For the sake of contradiction assume that $H$ is a Hamiltonian cycle in $G'_k$. As $d^-(z_1) = 1$, $H$ has to include the arc $xz_1$. As $xz_1$ belongs to $H$ we note that $z_1x$ does not belong to $H$ and therefore some arc from $V_2$ to $x$ must belong to $H$ (as some arc has to enter $x$). As half the vertices of $G'_k$ belong to $V_3$ no arc from $V_2$ to $x$ can belong to $H$, a contradiction. Therefore $G'_k$ is not Hamiltonian. Furthermore $f(G'_k, k) = k - 1$.

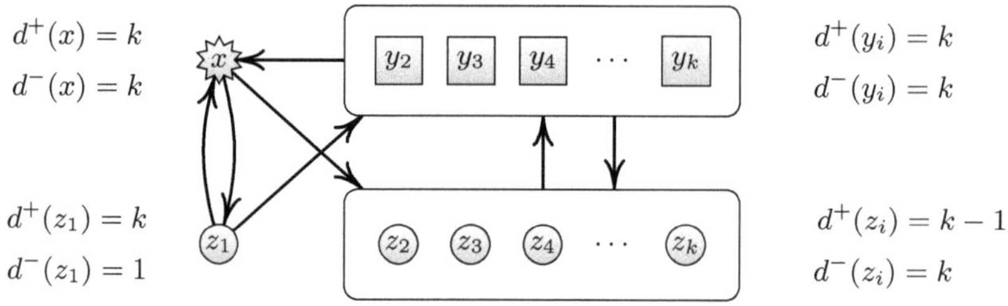


**Figure 7.10** The graph class $G'_k$.

Let $G''_k$ be the converse[1] of $G'_k$ for all $k \geq 2$. The following theorem was proved by Guo, Tewes, Volkmann and Yeo.

**Theorem 7.8.3** ([26]) *Let $D$ be a semicomplete multipartite digraph such that $f(D, k) \leq k - 1$. If $D$ is not isomorphic to $G'_k$ or $G''_k$, then $D$ contains a Hamilton cycle.*

In [26] Guo, Tewes, Volkmann and Yeo defined the notion of a **semi-partitioncomplete digraph** and proved the theorem below. A semicomplete multipartite digraph is called semi-partitioncomplete if both the out-neighbourhood and the in-neighbourhood of every vertex contains at least half the vertices from every partite set, except the partite set that the vertex belongs to.

**Theorem 7.8.4** ([26]) *Let $D$ be a strong semi-partitioncomplete semicomplete multipartite digraph with no partite set containing more than half the vertices. Then $D$ contains a Hamilton cycle.*

It is also interesting to consider Hamiltonian paths in regular or close to regular multipartite tournaments. The following result was first proved by Volkmann and Yeo in [68].

**Theorem 7.8.5** ([68]) *Every arc of a regular multipartite tournament is contained in a Hamilton path.*

[1] Recall that the converse of a digraph $D$ is the digraph $H$ which one obtains from $D$ by reversing all arcs.

A similar result almost holds for almost regular multipartite tournaments, as can be seen in the following theorem of Volkmann.

**Theorem 7.8.6** ([63]) *Let $D$ be an almost regular $c$-partite tournament with the partite sets $V_1, V_2, \ldots, V_c$ such that $|V_1| = |V_2| = \cdots = |V_c|$. Then each arc of $D$ is contained in a Hamiltonian path if and only if $D$ is not isomorphic to $T_{3,3}$ (see Figure 7.4).*

Note that Theorem 7.8.6 generalizes Theorem 7.8.5 as any regular multipartite tournament has equally many vertices in each partite set.

Considering Theorem 7.8.6, it is interesting to determine a smallest possible value $g(i)$ such that all $c$-partite tournaments, $D$, with $i_g(D) \leq i$ and $c \geq g(i)$ have a Hamiltonian path through any given arc. In the following theorem of Volkmann and Winzen we note that $g(i) \leq 4i + 4$ and $g(1) = 5$.

**Theorem 7.8.7** ([66]) *For all $i \geq 0$ the following holds. All $c$-partite tournaments, $D$, with $i_g(D) \leq i$ and $c \geq 4i + 4$ have a Hamiltonian path through any given arc. Furthermore, the following two statements also hold.*

- *All $c$-partite tournaments, $D$, with $i_g(D) \leq 1$ and $c \geq 5$ have a Hamiltonian path through any given arc.*
- *All regular $c$-partite tournaments with $c \geq 2$ have a Hamiltonian path through any given arc (see Theorem 7.8.5).*

If we are interested in Hamilton paths through paths instead of just arcs, then the following theorem of Volkmann and Yeo is useful.

**Theorem 7.8.8** ([68]) *Let $D$ be a $c$-partite tournament with partite sets $V_1, V_2, \ldots, V_c$ such that $|V_1| \leq |V_2| \leq \cdots \leq |V_c|$ and let $P$ be a path of length $q$ in $D$. If $|V(D)| \geq 2i_g(D) + 3q + 2|V_c| + |V_{c-1}| - 2$, then there exists a Hamiltonian path in $D$, starting with the path $P$.*

See also Theorems 7.4.8 and 7.4.9 in Section 7.4.3 for results on regular and close to regular semicomplete bipartite digraphs. For example, Theorem 7.4.8 is a fundamental theorem showing that every arc of a regular bipartite tournament is contained in a Hamiltonian cycle.

### 7.8.1 Connectivity in Close to Regular Semicomplete Multipartite Digraphs

In the proofs of several of the other results in this section the following results due to Yeo have been used.

**Theorem 7.8.9** ([77]) *If $D$ is a semicomplete multipartite digraph and the largest partite set has size $v^*$ then $D$ is $\left\lceil \frac{|V(D)| - 2i_l(D) - v^*}{3} \right\rceil$-strong.*

An easy corollary of the above theorem is the following.

**Theorem 7.8.10** ([72]) *If $D$ is a regular multipartite tournament with $v^*$ vertices in each partite set then $D$ is $\left\lceil \frac{|V(D)|-v^*}{3} \right\rceil$-strong.*

In fact, Theorem 7.8.9 can be slightly improved in the sense that Volkmann and Winzen, [67], characterized when equality was obtained for multipartite tournaments in Theorem 7.8.9. This characterization led to an improved bound of $\left\lceil \frac{|V(D)|-2i_l(D)-v^*+1}{3} \right\rceil$ when $v^*$ is odd.

### 7.8.2 Pancyclicity in Close to Regular Semicomplete Multipartite Digraphs

By Theorem 7.8.1 every regular semicomplete multipartite digraph contains a Hamilton cycle. This result can in many cases be extended to pancyclicity. The following results were proved by Yeo.

**Theorem 7.8.11** ([72]) *Every regular multipartite tournament with at least* 5 *partite sets is vertex-pancyclic.*

For large multipartite tournaments the following holds.

**Theorem 7.8.12** ([76]) *If $D$ is a $c$-partite tournament with $c \geq 4$ and $|V(G)| > 476i_g(D) + 13917$ then there exists a path of length $l$ between any two vertices in $D$ for all $42 \leq l < |V(D)|$.*

In fact, the lower bound of 42 can be improved as follows.

**Theorem 7.8.13** ([76]) *If $D$ is a $c$-partite tournament with $c \geq 4$ and $|V(G)| > 715i_g(D) + 13917$ and $e \in A(D)$ is arbitrary, then there exists a cycle of length $l$ containing $e$ for all $5 \leq l < |V(D)|$.*

Theorem 7.8.13 can be used to prove the following corollaries.

**Corollary 7.8.14** ([76]) *Every $c$-partite tournament with $c \geq 5$ and $|V(G)| > 715i_g(D) + 13917$ is vertex-pancyclic.*

Note that Corollary 7.8.14 compliments Theorem 7.8.11, which states that every $c$-partite tournament with $c \geq 5$ is vertex-pancyclic.

**Corollary 7.8.15** ([76]) *Every regular* 4*-partite tournament of order at least* 13918 *is vertex-pancyclic.*

The above results of Yeo give strong support for the following conjecture of Volkmann.

**Conjecture 7.8.16** ([62]) *Every regular* 4*-partite tournament is vertex-pancyclic.*

Since it is very easy to see that every arc in a regular tournament is contained in a 3-cycle, the next result of Guo is an extension of Alspach's Theorem (see Corollary 2.14.4).

**Theorem 7.8.17** ([22]) *Let $D$ be a regular $c$-partite tournament. If every arc of $D$ is contained in a 3-cycle, then every arc of $D$ belongs to an $m$-cycle for each $m \in \{4, 5, \ldots, c\}$.*

If we consider almost regular semicomplete multipartite digraphs instead of regular semicomplete multipartite digraphs, then the following result was proved by Tewes, Volkmann and Yeo.

**Theorem 7.8.18** ([59]) *Let $D$ be an almost regular $c$-partite tournament. If $c \geq 8$, then $D$ is vertex-pancyclic. If $5 \leq c \leq 7$, then $D$ is vertex-pancyclic, except for possibly a finite number of counterexamples, whose partite sets have different cardinalities.*

Tewes, Volkmann and Yeo constructed some infinite families of almost regular 4-partite tournaments that are not vertex-pancyclic. Thus, Theorem 7.8.18 as well as the following conjecture cannot be extended to almost regular 4-partite tournaments.

**Conjecture 7.8.19** ([59]) *An almost regular $c$-partite tournament with $5 \leq c \leq 7$ is vertex-pancyclic.*

## 7.9 $k$-Strong Semicomplete Multipartite Digraphs

Recall that in Section 1.5 a $k$-**strong** digraph is defined as a digraph on at least $k+1$ vertices which is still strong after the removal of any $k-1$ vertices. The following result of Bang-Jensen, Gutin and Yeo turns out to be very useful.

**Theorem 7.9.1** ([9]) *If $D$ is a $k$-strong digraph, and $X \subseteq V(D)$, has independence number $\alpha(D[X]) \leq k$, then there is a cycle subgraph in $D$ that covers $X$.*

Using Theorem 7.3.2 the following result was proved by Yeo in [75].

**Theorem 7.9.2** ([75]) *Let $D$ be a $(\lfloor k/2 \rfloor + 1)$-strong semicomplete multipartite digraph, and let $X$ be an arbitrary set of vertices in $D$ such that $X$ includes at most $k$ vertices from any partite set of $D$. If there exists a cycle subgraph in $D$ which covers $X$, then there is a cycle $C$ in $D$ such that $X \subseteq V(C)$.*

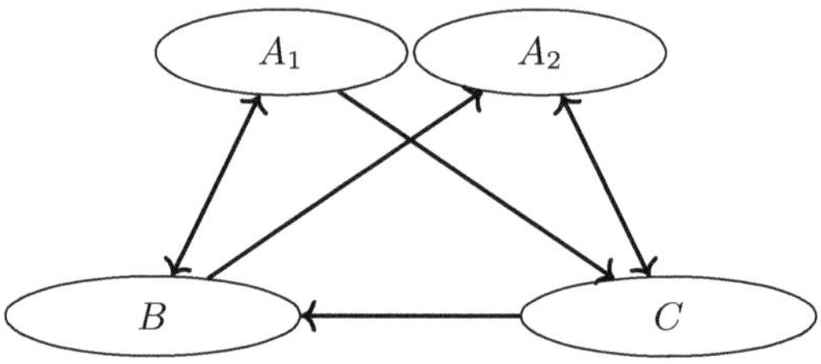


**Figure 7.11** A non-Hamiltonian $\lfloor k/2 \rfloor$-strong semicomplete multipartite digraph, with a cycle factor, where $|A_1| = |B| = \lfloor k/2 \rfloor$ and $|A_2| = |C| = \lceil k/2 \rceil$.

Theorem 7.9.2 is best possible, as can be seen by the example in Figure 7.11, which we will denote by $D^*$. Note that $D^*$ is a semicomplete multipartite digraph with partite sets $A_1 \cup A_2$, $B$ and $C$ and is $\lfloor k/2 \rfloor$-strong, as all sets $A_1$, $A_2$, $B$ and $C$ have size at least $\lfloor k/2 \rfloor$ and $C$ is a separating set of size $\lfloor k/2 \rfloor$. Furthermore, $D^*$ contains a cycle factor as there is a cycle covering $A_1 \cup B$ and another cycle covering $A_2 \cup C$. However, $D^*$ does not contain a Hamilton cycle as if $H$ was a Hamilton cycle then the vertices on $H$ would have to alternate between $A_1 \cup A_2$ and $B \cup C$ (as $A_1 \cup A_2$ is independent and $|A_1 \cup A_2| = |B \cup C|$). So no arc from $C$ to $B$ lies on $H$, which contradicts the fact that some arc has to go from $A_2 \cup C$ to $A_1 \cup B$ on $H$. Therefore $D^*$ is a non-Hamiltonian $\lfloor k/2 \rfloor$-strong semicomplete multipartite digraph with at most $k$ vertices in any partite set and which contains a cycle factor.

There are several consequences of Theorem 7.9.2. For example, the following results of Yeo.

**Corollary 7.9.3** ([75]) *Let $D$ be a $k$-strong semicomplete multipartite digraph, and let $X \subseteq V(D)$ be an arbitrary set of vertices in $D$ with at most $k$ vertices from each partite set. Then there exists a cycle $C$ with $X \subseteq V(C)$.*

**Proof:** By Theorem 7.9.1 there is a cycle subgraph in $D$ covering $X$. Theorem 7.9.2 now implies the corollary. □

Corollary 7.9.3 furthermore implies the following.

**Corollary 7.9.4** ([75]) *If a $k$-strong semicomplete multipartite digraph $D$ has at most $k$ vertices in each partite set then $D$ contains a Hamilton cycle.*

**Corollary 7.9.5** ([75]) *A $k$-strong semicomplete multipartite digraph has a cycle through any set of $k$ vertices.*

The above corollaries were originally conjectured by Guo and Volkmann, [60], and Bang-Jensen, Gutin and Yeo, [9], respectively.

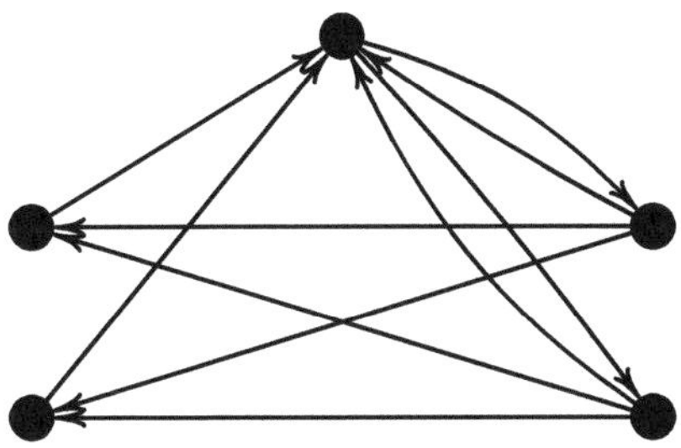

**Figure 7.12** An example of an extended semicomplete digraph.

## 7.10 Extended Semicomplete Digraphs

An extended semicomplete digraph is a semicomplete multipartite digraph such that if there is one arc between two partite sets, say from $V_i$ to $V_j$, then there is an arc from every vertex in $V_i$ to every vertex in $V_j$. See Figure 7.12 for an example of an extended semicomplete digraph. One can also think of an extended semicomplete digraph as a semicomplete digraph where every vertex has been blown up in the natural way to an independent set. For example, the extended semicomplete digraph in Figure 7.12 would be obtained from the semicomplete digraph on 3 vertices consisting of a 3-cycle and an extra edge (creating a 2-cycle with one of the existing edges). Extended semicomplete digraphs have been extensively studied. We will give some of the main results on this class of digraphs below. The main results in this area include the following theorems of Gutin.

**Theorem 7.10.1** ([30]) *An extended semicomplete digraph, $D$, has a Hamilton cycle if and only if it contains a cycle factor and is strongly connected. Furthermore, one can verify whether $D$ contains a Hamilton cycle (and find one if it does) in time $O(|V(D)|^{2.5})$.*

**Proof:** Let $D$ be a strong extended semicomplete digraph and let $F = C_1 \cup C_2 \cup \cdots \cup C_t$ be a cycle factor in $D$ where $t$ is minimum. If $t = 1$ then $D$ is Hamiltonian and we are done, so assume for the sake of contradiction that $t \geq 2$.

If some partite set $V_i$ intersects two distinct cycles $C_a$ and $C_b$ then let $x_a \in V(C_a) \cap V_i$ and $x_b \in V(C_b) \cap V_i$ be arbitrary. Let $x_a^-$ be the predecessor of $x_a$ on $C_a$ and let $x_b^-$ be the predecessor of $x_b$ on $C_b$. As $x_a^- x_a \in A(D)$ and $x_b^- x_b \in A(D)$ and $D$ is an extended semicomplete digraph we also have $x_a^- x_b \in A(D)$ and $x_b^- x_a \in A(D)$. Removing $x_a^- x_a$ and $x_b^- x_b$ from $C_a$ and $C_b$ and adding $x_a^- x_b$ and $x_b^- x_a$ we get one cycle in $D$ with vertex set $V(C_a) \cup V(C_b)$, a contradiction to the minimality of $t$. Therefore no partite set intersects more than one cycle.

However, by Theorem 7.3.1 there exists an ordering of the cycles of $F$ such that any arc, $xy$, from $V(C_t)$ to $V(C_1) \cup V(C_2) \cup \cdots \cup V(C_{t-1})$ has the property that the successor of $x$ on $C_t$ and the predecessor of $y$ in $F$ lie in the same partite set. This contradicts the fact that no partite set intersects two distinct cycles. Therefore there is no arc out of $C_t$, contradicting the fact that $D$ was strong.

This completes the first part of the proof. We refer the reader to [30] for the complexity part. □

**Theorem 7.10.2** ([32]) *Let $D$ be a strong extended semicomplete digraph and let $\mathcal{F}$ be a cycle subgraph of $D$. Then $D$ has a cycle $C$ which contains all vertices of $\mathcal{F}$. The cycle $C$ can be found in time $O(|V(D)|^3)$.*

*In particular, if $|V(\mathcal{F})|$ is maximum then $V(C) = V(\mathcal{F})$ and $C$ is a longest cycle of $D$.*

In fact, Theorem 7.10.2 can be extended in the following way, which was first done by Bang-Jensen, Huang and Yeo.

**Theorem 7.10.3** ([10]) *Let $D$ be a strong extended semicomplete digraph with partite sets $V_1, V_2, \ldots, V_s$ and assume that some longest cycle contains $m_i$ vertices from $V_i$ for all $i = 1, 2, \ldots, s$. Then every longest cycle contains $m_i$ vertices from $V_i$ and no cycle subdigraph in $D$ can contain more than $m_i$ vertices from $V_i$.*

The following result on an extended semicomplete digraph with costs on its vertices was proved by Bang-Jensen, Gutin and Yeo.

**Theorem 7.10.4** ([7]) *Let $D$ be an extended semicomplete digraph with real-valued costs on its vertices. In time $O(n^3m + n^4\log(n))$ we can find a minimum cost cycle in $D$ (or determine that no cycle exists).*

### 7.10.1 Paths in Extended Semicomplete Digraphs

Recall that an $[x, y]$-**path** is a path that starts in $x$ and ends in $y$ or starts in $y$ and ends in $x$. If a digraph, $D$, contains a Hamiltonian $[x, y]$-path for all distinct vertices $x$ and $y$ in $D$, then $D$ is **weakly Hamiltonian connected**.

The following theorem of Bang-Jensen, Gutin and Huang gives a characterization of when there exists a Hamiltonian $[x, y]$-path in an extended tournament. Note the strong similarity with Theorem 2.6.3.

**Theorem 7.10.5** ([5]) *Let $D$ be an extended tournament with distinct vertices $x_1$ and $x_2$. Then $D$ has a Hamiltonian $[x_1, x_2]$-path if and only if $D$ contains a 1-path-cycle factor where the path, $P$, is a $[x_1, x_2]$-path and $D$ does not satisfy any of the below.*

- *$D$ is not strong and either the initial or terminal component of $D$ (or both) contains neither $x_1$ or $x_2$.*
- *$D$ is strong and the following holds for some $i \in \{1, 2\}$: $D - x_i$ is not strong and one of the following holds.*

– $x_{3-i}$ *belongs to neither the initial nor the terminal component of* $D - x_i$.
– $x_{3-i}$ *belongs to the initial component of* $D - x_i$, *but there is no* $(x_{3-i}, x_i)$-*path,* $P'$ *in* $D$ *such that* $D - P'$ *contains a cycle factor.*
– $x_{3-i}$ *belongs to the terminal component of* $D - x_i$, *but there is no* $(x_i, x_{3-i})$-*path,* $P''$ *in* $D$ *such that* $D - P''$ *contains a cycle factor.*

- $D$, $D - x_1$ *and* $D - x_2$ *are all strong and* $D$ *is isomorphic to one of the tournaments in Figure 7.13.*

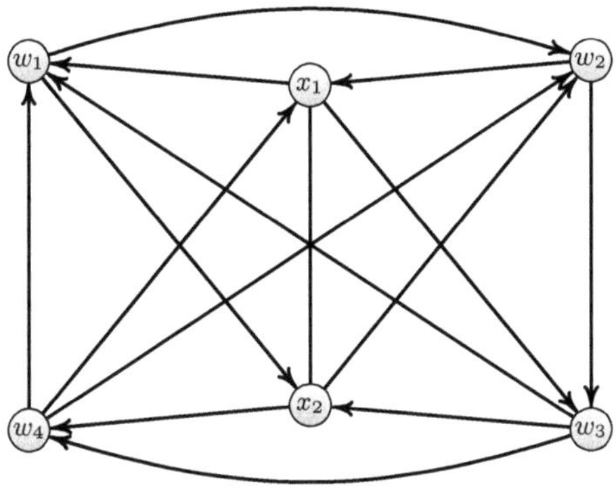


**Figure 7.13** The exceptional tournaments in Theorem 7.10.5. The edge between $x_1$ and $x_2$ can be oriented arbitrarily.

Let $T$ denote one of the two tournaments depicted in Figure 7.13 (depending on the orientation of the arc between $x_1$ and $x_2$). $T - \{x_1, x_2\}$ contains a 4-cycle, $w_1w_2w_3w_4w_1$, so $T$ contains a 1-path-cycle factor where the path is a $[x_1, x_2]$-path (containing the arc between $x_1$ and $x_2$). Furthermore, it is not difficult to see that $T$, $T - x_1$ and $T - x_2$ are all strong. We will now show that there is no Hamiltonian $[x_1, x_2]$-path in $T$. For the sake of contradiction assume that $P$ is a Hamiltonian $[x_1, x_2]$-path in $T$. First consider the case when $P$ starts in $x_1$. The second vertex on $P$ is either $w_1$ or $w_3$. If it is $w_1$ then the second last vertex on $P$ has to be $w_3$ (considering the arcs into $x_2$). However, there is no Hamiltonian path in $P - \{x_1, x_2\}$ from $w_1$ to $w_3$. Analogously if the second vertex on $P$ is $w_3$, then the second last vertex on $P$ is $w_1$, but again there is no Hamiltonian path from $w_3$ ot $w_1$ in $T - \{x_1, x_2\}$. So, there is no Hamiltonian path from $x_1$ to $x_2$ in $T$. Analogously one can show that there is no Hamiltonian path from $x_2$ to $x_1$ in $T$. Therefore there is no Hamiltonian $[x_1, x_2]$-path in $T$.

The proof of Theorem 7.10.5 is constructive and implies the following result, also from [5].

**Theorem 7.10.6** ([5]) *Let* $D$ *be an extended tournament of order* $n$ *and size* $m$ *and with distinct vertices* $x_1$ *and* $x_2$. *There exists an* $O(\sqrt{n}m)$

*algorithm to decide if $D$ contains a Hamiltonian path connecting $x_1$ and $x_2$. Furthermore, within the same time bound a Hamiltonian $[x_1, x_2]$-path can be found if it exists.*

Using Theorem 7.10.5 one can also give a characterization of which extended semicomplete digraphs are weakly Hamiltonian connected. As the characterization is basically identical to using Theorem 7.10.5 for all pair of vertices in the extended semicomplete digraph, we will omit the characterization and refer the reader to [5].

We will end this section with the following result of Bang-Jensen, Huang and Gutin.

**Theorem 7.10.7** ([5]) *Let $\{x_1, x_2, x_3\}$ be a set of three distinct vertices in a strong extended tournament $D$. Suppose that for every choice of distinct $x_i, x_j \in \{x_1, x_2, x_3\}$ there exists a 1-path-cycle factor in $D$ where the path is a $[x_i, x_j]$-path. Then there exists a Hamilton path in $D$ connecting two of the vertices in $\{x_1, x_2, x_3\}$.*

### 7.10.2 Pancyclicity in Extended Semicomplete Digraphs

A **zigzag digraph** is an extended semicomplete digraph with partite sets $V_1, V_2, \ldots, V_c$, where $V^* = V_3 \cup V_4 \cup \cdots \cup V_c$ and $c \geq 3$ and $|V_1| = |V_2| = |V^*|$. Furthermore, $V_1 \rightarrow V_2 \rightarrow V^* \rightarrow V_1$ and no vertex of $V_1 \cup V_2$ is contained in a 2-cycle. See Figure 7.14 for an example of a zigzag digraph. Note that a zigzag digraph of order $n > 3$ has no cycle of length $n-1$. Also note that a 4-partite tournament on at least 5 vertices does not contain a 5-cycle, as if $C$ is a 5-cycle then at least 2 vertices, say $x$ and $y$, on $C$ belong to the same partite set and any path from $x$ to $y$ (or $y$ to $x$) must contain at least 3 vertices, so $C$ contains at least 6 vertices, a contradiction. In [30] Gutin characterized pancyclic and vertex-pancyclic extended semicomplete digraphs as follows.

**Theorem 7.10.8** ([30]) *Let $D$ be a Hamiltonian extended semicomplete digraph of order $n \geq 5$ with $c$ partite sets ($c \geq 3$). Then $D$ is pancyclic if and only if $D$ is not a zigzag digraph and not a 4-partite tournament.*

*Furthermore, $D$ is vertex-pancyclic if and only if it is pancyclic and either $c > 3$ or $c = 3$ and $D$ contains two cycles $Z$, $Z'$ of length 2 such that $Z \cup Z'$ has vertices in the three partite sets.*

## 7.11 Orientations

An **orientation** of a digraph $D$ is an oriented graph $H$ obtained from $D$ by deleting one arc in every 2-cycle of $D$. It is often desirable to find orientations with the minimum possible diameter, as this in some sense makes for more

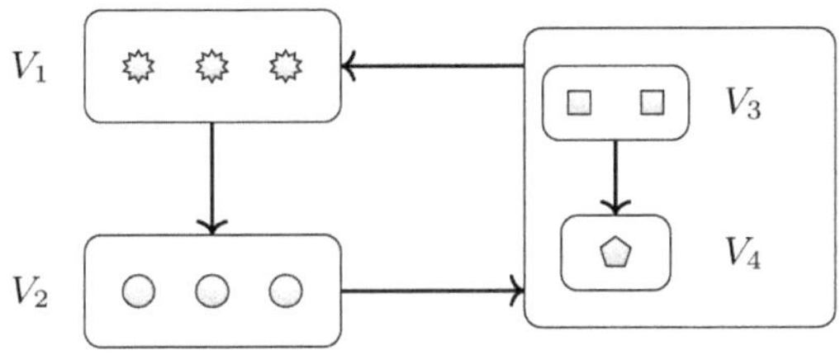


**Figure 7.14** An example of a zigzag digraph.

reliable systems in real-world applications. A digraph is called **bridgeless** if its underlying graph has no bridge. By Theorem 1.7.3 any strong bridgeless digraph has an orientation with finite diameter. Therefore we can define $\mathrm{diam}_{min}(D)$ to be the minimum possible diameter taken over all orientations of $D$ whenever $D$ is strong and bridgeless.

The following result is proved in [41] by Gutin and Yeo.

**Theorem 7.11.1** ([41]) *If $D$ is a strong semicomplete bipartite digraph not isomorphic to $\overleftrightarrow{K}_{1,n-1}$, then* $\mathrm{diam}_{min}(D) \leq \max\{5, \mathrm{diam}(D)\}$.

It was furthermore shown in [41] that Theorem 7.11.1 is best possible. If $\mathrm{diam}(D) \geq 5$ it implies that we can find an orientation of $D$ that does not increase the diameter. Theorem 7.11.1 also has a similar flavour to the fact that for all strong semicomplete digraphs, $D$, (of order at least 3) we have $\mathrm{diam}_{min}(D) \leq \max\{3, \mathrm{diam}(D)\}$.

It is not true that $\mathrm{diam}_{min}(D) \leq \max\{k, \mathrm{diam}(D)\}$ for all strong bridgeless semicomplete multipartite digraphs for any $k$, as can be seen by the example in Figure 7.15. However the following is conjectured by Gutin, Koh, Tay and Yeo in [36].

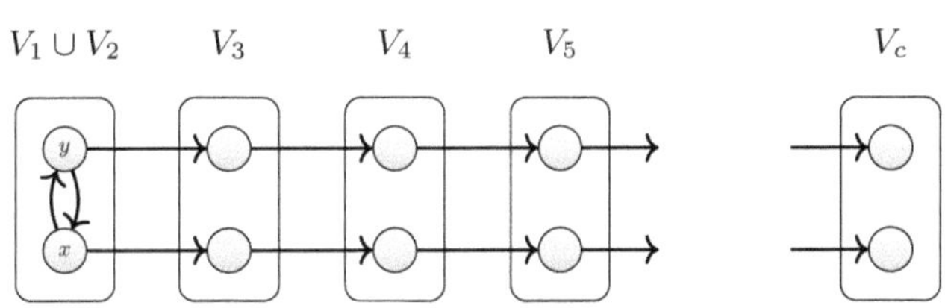


**Figure 7.15** An example of a strong semicomplete $c$-partite digraph, $D$, with $\mathrm{diam}(D) = c$, where all arcs not shown go from right-to-left. Note that the only 2-cycle, $xyx$, goes between $V_1$ and $V_2$. Furthermore, $\mathrm{diam}(D - xy) = c + 1$ and $\mathrm{diam}(D - yx) = c + 1$, so $\mathrm{diam}_{min}(D) = c + 1$.

**Conjecture 7.11.2** ([36]) *There is an absolute constant $k$ such that for every strong semicomplete multipartite digraph $D$, different from $\overleftrightarrow{K}_{1,n-1}$, we have* $\mathrm{diam}_{\min}(D) \leq \mathrm{diam}(D) + k$.

In [36] the following results were proved by Gutin, Koh, Tay and Yeo for extended semicomplete digraphs.

**Theorem 7.11.3** ([36]) *Let $D$ be a strong extended $c$-partite digraph with $c \geq 3$ and where all partite sets have size at least two and* $\mathrm{diam}(D) \geq 3$. *Then* $\mathrm{diam}_{\min}(D) \leq \mathrm{diam}(D) + 2$.

In [36] it was conjectured that the bound should be $\mathrm{diam}(D) + 1$ instead of $\mathrm{diam}(D) + 2$.

**Conjecture 7.11.4** ([36]) *Let $D$ be a strong extended $c$-partite digraph with $c \geq 3$ and where all partite sets have size at least two and* $\mathrm{diam}(D) \geq 3$. *Then* $\mathrm{diam}_{\min}(D) \leq \mathrm{diam}(D) + 1$.

In the case when $\mathrm{diam}(D) \geq 4$ and each partite set contains at least four vertices the following improvement is known.

**Theorem 7.11.5** ([36]) *Let $D$ be a strong extended $c$-partite digraph with $c \geq 3$ and where all partite sets have size at least four and* $\mathrm{diam}(D) \geq 4$. *Then* $\mathrm{diam}_{\min}(D) = \mathrm{diam}(D)$.

A different problem is to find orientations of semicomplete multipartite digraphs where the length of the longest path does not change. It turns out that this is not always possible. However the following was proved by Gutin, Tewes and Yeo.

**Theorem 7.11.6** ([39]) *Let $D$ be a strong semicomplete multipartite digraph of order $n$ which is not isomorphic to $\overleftrightarrow{K}_{1,n-1}$ and with a longest path of length $l$. Then there exists a strong spanning orientation of $D$ containing a path of length $l - 2$.*

Theorem 7.11.6 is shown in [39] to be best possible in the sense that the bound $l - 2$ cannot be improved.

One can also bound the length of a longest path in a semicomplete multipartite digraph by the length of a longest cycle, as was done by Gutin and Yeo in the following theorem.

**Theorem 7.11.7** ([42]) *Let $D$ be strong semicomplete multipartite digraph with a longest path of length $l$ and a longest cycle of length $c$. Then $l \leq 2c-1$. Furthermore, this bound is sharp.*

## 7.12 Kings in Semicomplete Multipartite Digraphs

In a digraph $D$, an $r$-**king** is a vertex $q$ such that every vertex in $D$ can be reached from $q$ by a path of length at most $r$.

It is well known that every tournament $T$ has a 2-king (see Theorem 2.2.12). In fact, every vertex of maximum out-degree in $T$ is a 2-king. Multipartite tournaments may have two or more vertices of in-degree zero, and, thus, no $r$-king for any integer $r$. However, Gutin [35] (and, independently, Petrovic and Thomassen [54]) proved that every multipartite tournament with at most one vertex of in-degree zero contains a 4-king. Moreover, it is easy to construct infinite families of $c$-partite tournaments (for every $c \geq 2$) which contain 4-kings but have no 3-kings (see the papers [35, 45] by Gutin and Koh and Tan, respectively). Therefore, in the study of multipartite tournaments, 4-kings are of special interest. Notice that while in a bipartite tournament every vertex of maximum out-degree is a 4-king, the obvious extension of this result to $c$-partite tournaments for $c \geq 3$ is not valid [20].

The latest result on 4-kings is the following of Gutin and Yeo.

**Theorem 7.12.1** ([40]) *Let $D = (V, A)$ be a semicomplete multipartite digraph and let $k$ be the number of 4-kings in $D$. Then,*

*(a): $k = 1$ if and only if $D$ has exactly one vertex of in-degree zero.*
*(b): $k = 2, 3$ or 4 if and only if the initial strong component of $D$ has $k$ vertices.*
*(c): $k = 5$ if and only if either the initial strong component $Q$ of $D$ has five vertices or $Q$ contains at least six vertices and possesses a path $P = p_0p_1p_2p_3p_4$ such that $\mathrm{dist}(p_0, p_4) = 4$ and there is no arc from $V - V(P)$ to $\{p_1, p_2, p_3, p_4\}$) (see Figure 7.16).*

For multipartite tournaments the following was shown by Koh and Tan in [46] (for bipartite tournaments) and [47] (for $c$-partite tournaments when $c \geq 3$).

**Theorem 7.12.2** ([46, 47]) *If a multipartite tournament has a unique initial strong component and no 3-king then it has at least eight 4-kings.*

## 7.13 Out-Paths in Semicomplete Multipartite Digraphs

An **out-path** of a vertex $x$ (an arc $xy$, respectively) in a digraph is a directed path starting at $x$ ($xy$, respectively) such that either the end-vertex dominates $x$ or the end-vertex and $x$ are not adjacent. Using out-paths, both Moon's and Alspach's famous theorems for tournaments have been extended by Guo as follows.

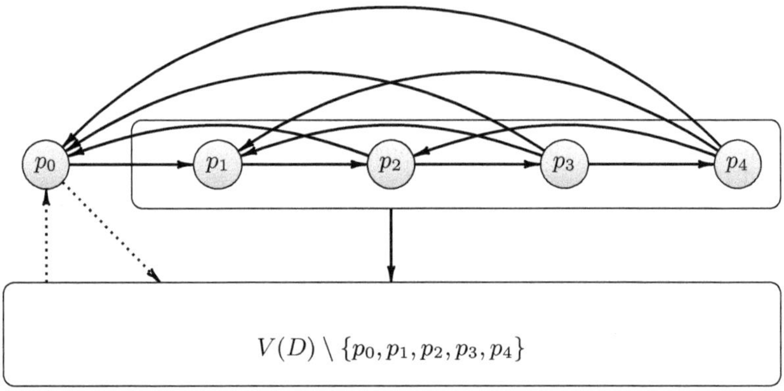


**Figure 7.16** A strong semicomplete multipartite digraph, $D$, with five 4-kings ($\{p_0, p_1, p_2, p_3, p_4\}$). The dotted lines may or may not exist. That is, $p_0$ is non-adjacent to all vertices that belong to the same partite set as $p_0$ itself and all other vertices in $V(D) \setminus \{p_0, p_1, p_2, p_3, p_4\}$ have either one arc to or from $p_0$ or belong to a 2-cycle with $p_0$.

**Theorem 7.13.1** ([23]) *If $D$ is a strongly connected semicomplete $c$-partite digraph ($c \geq 3$), then every vertex $v$ of $D$ has an out-path of length $k-1$ for all $k \in \{3, 4, \ldots, n\}$.*

Theorem 7.13.1 generalizes Moon's Theorem (Theorem 1.5.1), which states that a strong tournament is vertex-pancyclic.

**Theorem 7.13.2** ([23]) *if $D$ is a regular $c$-partite tournament $c \geq 3$, then every arc of $D$ has an out-path of length $k-1$ for all $k$ satisfying $3 \leq k \leq n$.*

This result extends Alspach's Theorem (see Corollary 2.14.4), which states that all regular tournaments are arc-pancyclic.

## 7.14 Quasi-Hamiltonian Paths in Semicomplete Multipartite Digraphs

A **quasi-Hamiltonian path** in a semicomplete multipartite digraph, $D$, is a path which visits each partite set of $D$ at least once. This is a generalization of a Hamiltonian path in a tournament.

In [11] the following results are proved by Bang-Jensen, Maddaloni and Simonsen.

**Theorem 7.14.1** ([11]) *It is $NP$-complete to decide if a given semicomplete multipartite digraph $D$ has a quasi-Hamiltonian $(x, y)$-path, where $x, y \in V(D)$.*

**Theorem 7.14.2** ([11]) *It is polynomial time solvable to decide if there is a quasi-Hamiltonian path between $x$ and $y$ in a semicomplete multipartite digraph, $D$, where $x, y \in V(D)$.*

The results in Theorems 7.14.1 and 7.14.2 are somewhat surprising as the two problems considered seem quite similar. The fact that in Theorem 7.14.1 we require the path to start in $x$ and end in $y$, as opposed to allowing either direction in Theorem 7.14.2, changes the complexity of the problem.

If the connectivity is high enough then a quasi-Hamiltonian path always exists in a multipartite tournament, as can be seen by the following theorem of Lu, Guo and Surmacs.

**Theorem 7.14.3** ([49]) *If $D$ is a 4-strong multipartite tournament then $D$ is quasi-Hamiltonian-connected (that is, there exists a quasi-Hamiltonian path starting in $x$ and ending in $y$ for all choices of $x$ and $y$).*

Theorem 7.14.3 generalizes Thomassen's result that every 4-strong tournament is strongly Hamiltonian-connected, which was stated earlier as Theorem 2.6.7. Since Thomassen proved the existence of an infinite number of 3-strong tournaments which are not strongly Hamiltonian-connected as well, Theorem 7.14.3 is, in a sense, best possible.

## 7.15 Strongly Connected Spanning Subgraphs with Minimum Number of Arcs

Let $D$ be a strong digraph. Finding a strongly connected spanning subgraph with minimum number of arcs is NP-hard, as it generalizes the Hamiltonian cycle problem. However it is polynomial time solvable for semicomplete bipartite digraphs and for extended semicomplete digraphs, as can be seen in the following theorems of Bang-Jensen and Yeo.

**Theorem 7.15.1** ([12]) *One can find a strongly connected spanning subgraph with the minimum number of arcs in a strong digraph $D$ which is either a semicomplete bipartite digraph or an extended semicomplete digraph in polynomial time.*

Recall that $pcc(D)$ is smallest positive integer $k$ such that $D$ contains a $k$-path-cycle factor. Define the number $pcc^*(D)$ as follows: If $D$ contains a cycle factor, then $pcc^*(D) = 0$ and otherwise $pcc^*(D) = pcc(D)$. Bang-Jensen and Yeo proved in [12] that every strong spanning subdigraph of a strong digraph $D$ has at least $|V(D)| + pcc^*(D)$ arcs. The following result shows that for strong digraphs that are either semicomplete bipartite or extended semicomplete, one can always find a spanning strong subdigraph with this number of arcs.

**Theorem 7.15.2** ([12]) *If $D$ is a strong semicomplete bipartite digraph or a strong extended semicomplete digraph then there exists a strong spanning strong subgraph of $D$ with $|V(D)| + pcc^*(D)$ arcs.*

The class of semicomplete bipartite digraphs and extended semicomplete digraphs are subclasses of semicomplete multipartite digraphs. However, Theorem 7.15.2 cannot be extended to semicomplete multipartite digraphs. For example, there exist strong semicomplete multipartite digraphs, $D$, with $pcc(D) = 0$ (that is, they contain a cycle factor) that are not Hamiltonian. It was, however, conjectured by Bang-Jensen and Yeo in [12] that Theorem 7.15.1 can be extended to semicomplete multipartite digraphs.

**Conjecture 7.15.3** ([12]) *One can find a strongly connected spanning subgraphs with minimum number of arcs in a strong semicomplete multipartite digraph in polynomial time.*

Clearly Conjecture 7.15.3 is true when the semicomplete multipartite digraph has a Hamiltonian cycle, by Theorem 7.6.1. However, in general, Conjecture 7.15.3 is still open.

## 7.16 *k*-Coloured Kernels in Arc-Coloured Semicomplete Multipartite Digraphs

A digraph $D$ is said to be $m$-coloured if the arcs of $D$ are coloured with $m$ colours. Given $u, v \in V(D)$, a directed path from $u$ to $v$ in $D$ is $j$-coloured if its arc set uses exactly $j$ colours.

A set $S$ is called a $k$-**coloured kernel** if the following holds.

- Every vertex not in $S$ has a $j$-coloured path to a vertex in $S$ with $j \leq k$.
- There are no two distinct vertices in $S$ that are connected by a $j$-coloured path where $j \leq k$.

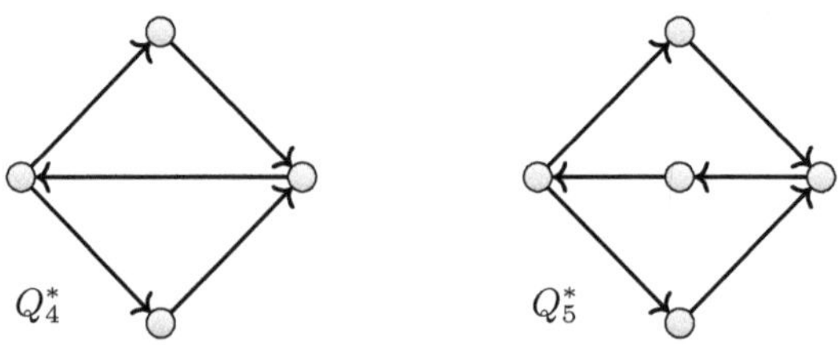


**Figure 7.17** The digraphs $Q_4^*$ and $Q_5^*$.

The main result in this area was proved by Galeana-Sánchez, Llano and Montellano-Ballesteros in [19].

**Theorem 7.16.1** ([19]) *An $m$-coloured semicomplete $c$-partite digraph $D$ has a $k$-coloured kernel provided that $c \geq 3$ and one of the following hold.*

*(i): $k \geq 4$.*
*(ii): $k = 3$ and every 4-cycle in $D$ is at most 2-coloured and, either every 5-cycle in $D$ is at most 3-coloured or every $Q_4^*$ (see Figure 7.17) contained in $D$ is at most 2-coloured.*
*(iii): $k = 2$ and every 3-cycle and every 4-cycle in $D$ is monochromatic (that is, 1-coloured).*

*Furthermore, if $D$ is an $m$-coloured semicomplete bipartite digraph and $k = 2$ (resp. $k = 3$) and every $Q_5^*$ (see Figure 7.17) contained in $D$ is at most 2-coloured (resp. 3-coloured), then $D$ has a 2-coloured (resp. 3-coloured) kernel.*

## 7.17 Complementary Cycles in Semicomplete Multipartite Digraphs

Reid proved in Theorem 2.8.1 that a 2-strong tournament of order $n \geq 8$ has 2 disjoint cycles of length 3 and $n-3$, respectively. Song (see [57]) improved this to cycles of length $t$ and $n-t$ for all $3 \leq t \leq n-3$. Reid's result was generalized to multipartite tournaments by Li, Meng and Guo in the following two theorems (recall that for semicomplete multipartite digraphs $\alpha(D)$ is the size of the largest partite set).

**Theorem 7.17.1** ([48]) *If $D$ is a $(\alpha(D)+1)$-strong $c$-partite tournament with $c \geq 6$ then $D$ contains two disjoint cycles of length 3 and $c-3$ respectively, unless $D$ is isomorphic to $T_7^1$ (see Figure 7.18).*

**Theorem 7.17.2** ([48]) *If $D$ is a $(\alpha(D)+1)$-strong $c$-partite tournament with $c \geq 6$ then $D$ contains two disjoint cycles that pass through exactly 3 and $c-3$ partite sets, respectively, unless $D$ is isomorphic to $T_7^1$ (see Figure 7.18).*

A digraph is **cycle complementary** if it contains a cycle factor with exactly two cycles. For regular semicomplete multipartite digraphs Volkmann proved the following result.

**Theorem 7.17.3** ([61]) *If $D$ is a regular semicomplete multipartite digraph with $|V(D)| \geq 8$, then $D$ is cycle complementary.*

For bipartite tournaments the above result was proved by Song [56] and Zhang and Song [79]. In [62] Volkmann conjectures the following.

**Conjecture 7.17.4** ([62]) *Let $D$ be a multipartite tournament. If $D$ is $(\alpha(D)+1)$-strong then $D$ is cycle complementary, unless $D$ is a member of a finite family of multipartite tournaments.*

More information on complementary cycles in multipartite tournaments can be found in [65], where Volkmann gives a survey on this topic.

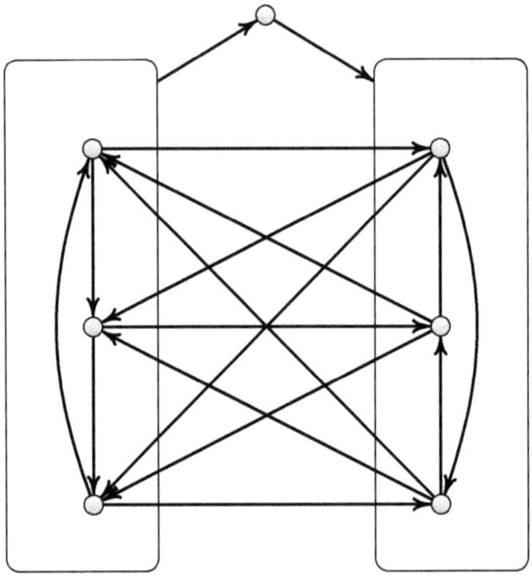

**Figure 7.18** The tournament $T_7^1$.

## 7.18 Applications of Semicomplete Multipartite Digraphs

The following result is proved by Fraisse and Thomassen in [18].

**Theorem 7.18.1** ([18]) *Let $T$ be a $k$-strong tournament and $A' \subset A(T)$ such that $|A'| \leq k-1$. Then there exists a Hamilton cycle in $T - A'$.*

This was extended by Bang-Jensen, Gutin and Yeo to the theorem below (using Theorem 7.3.1 on semicomplete multipartite digraphs), which was previously mentioned as Theorem 2.6.20. Note that in the theorem below $D$ becomes a multipartite tournament with partite sets $V_1, V_2, \ldots, V_c$.

**Theorem 7.18.2** ([8]) *Let $T = (V, A)$ be a $k$-strong tournament of order $n$ and let $V_1, V_2, \ldots, V_c$ be a partition of $V$ such that $n/2 \geq |V_1| \geq |V_2| \geq \cdots \geq |V_c| \geq 1$. Let $D$ be the digraph obtained from $T$ by deleting all arcs with endpoints in the same set $V_i$, for all $i$.*

*If $k \geq |V_1| + \sum_{i=2}^{c} \lfloor |V_i|/2 \rfloor$, then $D$ is Hamiltonian. In other words, $T$ contains a Hamiltonian cycle avoiding all arcs with endpoints in the same set $V_i$.*

Note that Theorem 7.18.2 generalizes Theorem 7.18.1, due to the following. Let $V_1, V_2, \ldots, V_c$ be the vertex sets of the connected components induced by $A'$ (where $A'$ was a set of at most $k-1$ arcs in the tournament $T$). It is not difficult to see that using these sets in Theorem 7.18.2 implies that $T$ contains a Hamilton cycle avoiding the arcs of $A'$. Clearly in Theorem 7.18.2 we can avoid a lot more than $k-1$ arcs, so Theorem 7.18.2 is stronger than Theorem 7.18.1. It was also shown in [8] that Theorem 7.18.2 is best possible.

If we consider close to regular tournaments instead of $k$-strong tournaments then the following was proved by Yeo.

**Theorem 7.18.3** ([73]) *Let $T = (V, A)$ be a tournament of order $n$ and let $n/2 \geq |V_1| \geq |V_2| \geq \cdots \geq |V_c| \geq 1$. Let $D$ be the digraph obtained from $T$ by deleting all arcs with endpoints in the same set $V_i$, for all $i$. Note that $D$ is a multipartite tournament with partite sets $V_1, V_2, \ldots, V_c$.*

*If $|V(T)| \geq \max\{2i_l(T) + 2|V_1| + 2|V_2| - 2, i_l(T) + 3|V_1| - 1\}$ then $D$ is Hamiltonian.*

*Furthermore, if $T$ is regular and the following holds, then $D$ is Hamiltonian (ignore the term $2|V_1|+2|V_2|-2\sqrt{2|V_2|-|V_1|+2}$ when $2|V_2|-|V_1|+2 < 0$)*

$$\begin{aligned}|V(T)| \geq \max\{\ & 2|V_1| + 2|V_2| - 2\sqrt{|V_2|},\\ & 2|V_1| + 2|V_2| - 2\sqrt{2|V_2| - |V_1| + 2},\\ & 3|V_1| - 2,\\ & 2|V_1| + 2\}.\end{aligned}$$

## 7.19 Conjectures

In this section we will state the main conjectures in the areas covered in this chapter. There are many other interesting problems and conjectures in the class of semicomplete multipartite digraphs, but due to space restrictions it is impossible to mention them all. The following conjectures were stated by Gutin, Bang-Jensen and Yeo in [6].

**Conjecture 7.19.1 (Bang-Jensen, Gutin, Yeo [6], 1998)** *There is a polynomial algorithm to find a longest cycle in a semicomplete multipartite digraph.*

By Theorem 7.4.2 we note that Conjecture 7.19.1 is true for semicomplete bipartite digraphs, but it is open for semicomplete multipartite digraphs.

**Conjecture 7.19.2 (Bang-Jensen, Gutin, Yeo [6], 1998)** *Let $D$ be a semicomplete multipartite digraph and $X \subseteq V(D)$. There is a polynomial time algorithm for finding a cycle containing the maximum possible number of vertices from $X$.*

Conjecture 7.19.2 is even open for the restricted class of semicomplete bipartite digraphs. Recall Conjecture 7.4.4.

**Conjecture** 7.4.4 (**Yeo** [71], **1999**) *Let $D$ be a* semicomplete multipartite digraph *and $X \subseteq V(D)$. There is a polynomial time algorithm for finding a cycle covering $X$ (if it exists), and which is the longest of all such cycles.*

By Theorem 7.4.3, Conjecture 7.4.4 is true for semicomplete bipartite digraphs. However, Conjecture 7.4.4 is still open for semicomplete multipartite digraphs in general. The following conjecture would imply that Conjecture 7.4.4 is true (by setting $Y = V(D)$).

**Conjecture 7.19.3** (**Yeo** [71], **1999**) *Let $D$ be a semicomplete multipartite digraph and $X \subseteq Y \subseteq V(D)$. There is a polynomial time algorithm for finding a cycle covering $X$ (if it exists), and containing the maximum number of vertices of $Y$ of all such cycles.*

Recall Conjecture 7.8.16.

**Conjecture** 7.8.16 (**Volkmann** [62], **2002**) *Every regular 4-partite tournament is vertex-pancyclic.*

Recall Conjectures 7.11.2 and 7.11.4.

**Conjecture** 7.11.2 (**Gutin, Koh, Tay, Yeo** [36], **2002**) *There is an absolute constant $k$ such that for every strong semicomplete multipartite digraph $D$, different from $\overleftrightarrow{K}_{1,n-1}$, we have* $\mathrm{diam}_{\min}(D) \le \mathrm{diam}(D) + k$.

**Conjecture** 7.11.4 (**Gutin, Koh, Tay, Yeo** [36], **2002**) *Let $D$ be a strong extended $c$-partite digraph with $c \geq 3$ and where all partite sets have size at least two and* $\mathrm{diam}(D) \geq 3$. *Then* $\mathrm{diam}_{\min}(D) \le \mathrm{diam}(D) + 1$.

Recall Conjecture 7.15.3.

**Conjecture** 7.15.3 (**Bang-Jensen, Yeo** [12], **2001**) *One can find a strongly connected spanning subgraph with minimum number of arcs in a strong semicomplete multipartite digraph in polynomial time.*

**Conjecture 7.19.4** (**Bang-Jensen, Maddaloni, Simonsen** [11], **2013**) *There exists a polynomial algorithm for finding the longest $[x, y]$-path in a semicomplete multipartite digraph.*

**Problem 7.19.5** (**Bang-Jensen, Maddaloni, Simonsen** [11], **2013**) *Is there a polynomial algorithm that, given a semicomplete multipartite digraph $D$ and $x, y \in V(D)$, finds a longest $[x, y]$-quasi-Hamiltonian path?*

It is possible to show that there always exists an $[x, y]$-path intersecting at most 3 different partite sets in a semicomplete multipartite digraph. This implies a polynomial algorithm to find an $[x, y]$-path that minimizes the number of partite sets intersected. The shortest path version of the above conjecture seems harder, though.

**Problem 7.19.6** (**Bang-Jensen, Maddaloni, Simonsen** [11], **2013**) *Is there a polynomial algorithm that, given a semicomplete multipartite digraph $D$ and $x, y \in V(D)$, finds a shortest $[x, y]$-quasi-Hamiltonian path?*

**Problem 7.19.7** (**Bang-Jensen, Maddaloni, Simonsen** [11], **2013**) *Is there a polynomial algorithm, that given an integer $k$ and a semicomplete multipartite digraph $D$ with partite sets $V_1, V_2, \dots, V_c$, decides whether $D$ has a path covering at least $\min\{k, |V_i|\}$ vertices from each $V_i$, $i = 1, 2, \dots, c$?*

The following conjecture of Jackson has the same flavour as the famous Kelly Conjecture, which was previously mentioned as Conjecture 2.12.8.

**Conjecture 7.19.8** (**Jackson** [44], **2013**) *Every regular bipartite tournament is decomposable into Hamilton cycles.*

Recall Conjecture 7.8.19.

**Conjecture** 7.8.19 ([59]) *An almost regular c-partite tournament with $5 \leq c \leq 7$ is vertex-pancyclic.*

Recall Conjecture 7.17.4.

**Conjecture** 7.17.4 (**Volkmann** [62], **2002**) *Let $D$ be a multipartite tournament. If $D$ is $(\alpha(D) + 1)$-strong then $D$ is cycle complementary, unless $D$ is a member of a finite family of multipartite tournaments.*

## References


1. D. Amar and Y. Manoussakis. Cycles and paths of many lengths in bipartite digraphs. *J. Combin. Theory Ser. B*, 50(2):254–264, 1990.
2. J. Bang-Jensen and G. Gutin. *Digraphs: Theory, Algorithms and Applications.* Springer-Verlag, London, 2nd edition, 2009.
3. J. Bang-Jensen and G. Gutin. On the complexity of hamiltonian path and cycle problems in certain classes of digraphs. *Discrete Appl. Math.*, 95:41–60, 1999.
4. J. Bang-Jensen, G. Gutin, and J. Huang. A sufficient condition for a semicomplete multipartite digraph to be Hamiltonian. *Discrete Math.*, 161(1-3):1–12, 1996.
5. J. Bang-Jensen, G. Gutin, and J. Huang. Weakly Hamiltonian-connected ordinary multipartite tournaments. *Discrete Math.*, 138(1-3):63–74, 1995.
6. J. Bang-Jensen, G. Gutin, and A. Yeo. A polynomial algorithm for the Hamiltonian cycle problem in semicomplete multipartite digraphs. *J. Graph Theory*, 29:111–132, 1998.
7. J. Bang-Jensen, G. Gutin, and A. Yeo. Finding a cheapest cycle in a quasi-transitive digraph with real-valued vertex costs. *Discrete Optim.*, 3:86–94, 2006.
8. J. Bang-Jensen, G. Gutin, and A. Yeo. Hamiltonian cycles avoiding prescribed arcs in tournaments. *Combin. Prob. Comput.*, 6(3):255–261, 1997.
9. J. Bang-Jensen, G. Gutin, and A. Yeo. On $k$-strong and $k$-cyclic digraphs. *Discrete Math.*, 162(1-3):1–11, 1996.
10. J. Bang-Jensen, J. Huang, and A. Yeo. Strongly connected spanning subgraphs with the minimum number of arcs in quasi-transitive digraphs. *SIAM J. Discrete Math.*, 16:335–343, 2003.

11. J. Bang-Jensen, A. Maddaloni, and S. Simonsen. Quasi-hamiltonian paths in semicomplete multipartite digraphs. *Discrete Appl. Math.*, 161:889–898, 2013.
12. J. Bang-Jensen and A. Yeo. The minimum spanning strong subdigraph problem for extended semicomplete digraphs and semicomplete bipartite digraphs. *J. Algor.*, 41(1):1–19, 2001.
13. L.W. Beineke. A tour through tournaments or bipartite and ordinary tournaments: a comparative survey. In *Combinatorics (Swansea, 1981)*, volume 52 of *London Math. Soc. Lect. Note Ser.*, pages 41–55. Cambridge Univ. Press, Cambridge, 1981.
14. L.W. Beineke and C.H.C. Little. Cycles in bipartite tournaments. *J. Combin. Theory Ser. B*, 32(2):140–145, 1982.
15. J.A. Bondy. Diconnected orientations and a conjecture of Las Vergnas. *J. London Math. Soc. (2)*, 14(2):277–282, 1976.
16. A.H. Busch, M.S. Jacobson, and K.B. Reid. On arc-traceable tournaments. *J. Graph Theory*, 53:157–166, 2006.
17. P. Camion. Chemins et circuits hamiltoniens des graphes complets. *C. R. Acad. Sci. Paris*, 249:2151–2152, 1959.
18. P. Fraisse and C. Thomassen. Hamiltonian dicycles avoiding prescribed arcs in tournaments. *Graphs Combin.*, 3(3):239–250, 1987.
19. H. Galeana-Sanchez, B. Llano, and J.J. Montellano-Ballesteros. $k$-colored kernels in semicomplete multipartite digraphs, manuscript.
20. W.D. Goddard, G. Kubicki, O.R. Oellermann, and S.L. Tian. On multipartite tournaments. *J. Combin. Theory Ser. B*, 52(2):284–300, 1991.
21. W.D. Goddard and O.R. Oellermann. On the cycle structure of multipartite tournaments. In *Graph Theory Combin. Appl. 1*, pages 525–533. Wiley, New York, 1991.
22. Y. Guo. *Locally semicomplete digraphs.* PhD thesis, RWTH Aachen, Germany, 1995.
23. Y. Guo. Outpaths in semicomplete multipartite digraphs. *Discrete Appl. Math.*, 95:273–277, 1999.
24. Y. Guo. Semicomplete Multipartite Digraphs: A Generalization of Tournaments. German Habilitation Thesis, RWTH Aachen, Germany, 1998.
25. Y. Guo, A. Pinkernell, and L. Volkmann. On cycles through a given vertex in multipartite tournaments. *Discrete Math.*, 164(1-3):165–170, 1997.
26. Y. Guo, M. Tewes, L. Volkmann, and A. Yeo. Sufficient conditions for semicomplete multipartite digraphs to be hamiltonian. *Discrete Math.*, 212(1-2):91–100, 2002.
27. Y. Guo and L. Volkmann. Cycles in multipartite tournaments. *J. Combin. Theory Ser. B*, 62(2):363–366, 1994.
28. Y. Guo and L. Volkmann. Extendable cycles in multipartite tournaments. *Graphs Combin.*, 20(2):185–190, 2004.
29. G. Gutin. Characterization of complete $n$-partite digraphs that have a Hamiltonian path. *Kibernetika (Kiev)*, no. 1:107–108, 136, 1988.
30. G. Gutin. Characterizations of vertex pancyclic and pancyclic ordinary complete multipartite digraphs. *Discrete Math.*, 141(1-3):153–162, 1995.
31. G. Gutin. Criterion for complete bipartite digraphs to be Hamiltonian. *Vestsī Akad. Navuk BSSR Ser. Fīz.-Mat. Navuk*, no. 1:109–110, 1984.
32. G. Gutin. *Cycles and paths in directed graphs.* PhD thesis, School of Mathematics, Tel Aviv University, 1993.
33. G. Gutin. Finding a longest path in a complete multipartite digraph. *SIAM J. Discrete Math.*, 6:270–273, 1993.
34. G. Gutin. Finding the largest contour in a complete bipartite digraph. *Kibernetika*, no. 2:117–118, 1987.

35. G. Gutin. The radii of $n$-partite tournaments. *Mat. Zametki*, 40(3):414–417, 430, 1986.
36. G. Gutin, K.M. Koh, E.G. Tay, and A. Yeo. Almost minimum diameter orientations of semicomplete multipartite and extended digraphs. *Graphs Combin.*, 18(3):499–506, 2002.
37. G. Gutin and A. Rafiey. When $n$-cycles in $n$-partite tournaments are longest cycles. *Discrete Math.*, 289(1-3):163–168, 2004.
38. G. Gutin, A. Rafiey, and A. Yeo. On $n$-partite tournaments with unique $n$-cycle. *Graphs Combin.*, 22(2):241–249, 2006.
39. G. Gutin, M. Tewes, and A. Yeo. Longest paths in strong spanning oriented subgraphs of strong semicomplete multipartite digraphs. *Discrete Math.*, 222:269–274, 2000.
40. G. Gutin and A. Yeo. Kings in semicomplete multipartite digraphs. *J. Graph Theory*, 33:177–183, 2000.
41. G. Gutin and A. Yeo. Orientations of digraphs almost preserving diameter. *Discrete Appl. Math.*, 121(1-3):129–138, 2002.
42. G. Gutin and A. Yeo. Solution of a conjecture of Volkmann on the number of vertices in longest paths and cycles of strong semicomplete multipartite digraphs. *Graphs Combin.*, 17:473–477, 2001.
43. R. Häggkvist and Y. Manoussakis. Cycles and paths in bipartite tournaments with spanning configurations. *Combinatorica*, 9(1):33–38, 1989.
44. B. Jackson. Long paths and cycles in oriented graphs. *J. Graph Theory*, 5(2):145–157, 1981.
45. K.M. Koh and B.P. Tan. Kings in multipartite tournaments. *Discrete Math.*, 147:171–183, 1995.
46. K.M. Koh and B.P. Tan. Number of 4-kings in bipartite tournaments with no 3-kings. *Discrete Math.*, 154(1-3):281–287, 1996.
47. K.M. Koh and B.P. Tan. The number of kings in a multipartite tournament. *Discrete Math.*, 167/168:411–418, 1997.
48. S. Li, W. Meng, and Y Guo. Three supplements to Reid's theorem in multipartite tournaments. *Discrete Applied Math.*, 158:340–348, 2010.
49. M. Lu, Y. Guo, and M. Surmacs. Strongly quasi-Hamiltonian-connected semicomplete multipartite digraphs. *Discrete Math.*, 313:2667–2672, 2013.
50. W. Meng and S. Li. Semicomplete multipartite digraphs whose every arc is contained in a Hamiltonian path. In *The 7th International Conference on Computational Science, Part III*, volume 4489 of *Lect. Notes Comput. Sci.*, pages 299–304. Springer, 2007.
51. J.W. Moon. On subtournaments of a tournament. *Can. Math. Bull.*, 9:297–301, 1966.
52. J.W. Moon. *Topics on tournaments.* Holt, Rinehart & Winston, New York, 1968.
53. O. Ore. *Theory of graphs.* American Mathematical Society, Providence, RI, 1962.
54. V. Petrović and C. Thomassen. Kings in $k$-partite tournaments. *Discrete Math.*, 98(3):237–238, 1991.
55. L. Rédei. Ein kombinatorischer Satz. *Acta Litt. Szeged*, 7:39–43, 1934.
56. Z.M. Song. Complementary cycles in bipartite tournaments. *J. Nanjing Inst. Tech.*, 18:32–38, 1988.
57. Z.M. Song. Complementary cycles of all lengths in tournaments. *J. Combin. Theory Ser. B*, 57(1):18–25, 1993.
58. M. Tewes. *In-tournaments and Semicomplete Multipartite Digraphs.* PhD thesis, Lehrstuhl II für Mathematik, RWTH Aachen, 1998.

59. M. Tewes, L. Volkmann, and A. Yeo. Almost all almost regular $c$-partite tournaments with $c \geq 5$ are vertex pancyclic. *Discrete Math.*, 242:201–228, 2002.
60. L. Volkmann. A talk at The Second Krakow Conference of Graph Theory, 1994.
61. L. Volkmann. All regular multipartite tournaments that are cycle complementary. *Discrete Math.*, 281(1-3):255–266, 2004.
62. L. Volkmann. Cycles in multipartite tournaments: results and problems. *Discrete Math.*, 245(1):19–53, 2002.
63. L. Volkmann. Hamiltonian paths containing a given arc, in almost regular bipartite tournaments. *Discrete Math.*, 285:359–364, 2004.
64. L. Volkmann. Longest paths through an arc in strong semicomplete multipartite digraphs. *Discrete Math.*, 258:331–337, 2002.
65. L. Volkmann. Multipartite tournaments: a survey. *Discrete Math.*, 307(24):3097–3129, 2007.
66. L. Volkmann and S. Winzen. Almost regular multipartite tournaments containing a Hamiltonian path through a given arc. *Discrete Math.*, 285:367–378, 2004.
67. L. Volkmann and S. Winzen. On the connectivity of close to regular multipartite tournaments. *Discrete Appl. Math.*, 154(1):1437–1452, 2006.
68. L. Volkmann and A. Yeo. Hamiltonian paths, containing a given path or collection of arcs, in close to regular multipartite tournaments. *Discrete Math.*, 281(1-3):267–276, 2004.
69. J.-Z. Wang. Cycles of all possible lengths in diregular bipartite tournaments. *Ars Combin.*, 32:279–284, 1991.
70. S. Winzen. *Close to Regular Multipartite Tournaments.* PhD thesis, Aachen University, 2004.
71. A. Yeo. A polynomial time algorithm for finding a cycle covering a given set of vertices in a semicomplete multipartite digraph. *J. Algor.*, 33(1):124–139, 1999.
72. A. Yeo. Diregular $c$-partite tournaments are vertex-pancyclic when $c \geq 5$. *J. Graph Theory*, 32:137–152, 1999.
73. A. Yeo. Hamilton cycles, avoiding prescribed arcs, in close to regular tournaments. *J. Graph Theory*, 32:123–136, 1999.
74. A. Yeo. How close to regular must a semicomplete multipartite digraph be to secure Hamiltonicity? *Graphs Combin.*, 15:481–493, 1999.
75. A. Yeo. One-diregular subgraphs in semicomplete multipartite digraphs. *J. Graph Theory*, 24(2):175–185, 1997.
76. A. Yeo. Paths and cycles containing given arcs, in close to regular multipartite tournaments. *J. Combin. Theory Ser. B*, 97(6):949–963, 2007.
77. A. Yeo. *Semicomplete Multipartite Digraphs.* PhD thesis, Department of Mathematics and Computer Science, Odense University, Denmark, 1998.
78. K.M. Zhang. Vertex even-pancyclicity in bipartite tournaments. *Nanjing Daxue Xuebao Shuxue Bannian Kan*, 1(1):85–88, 1984.
79. K.M. Zhang and Z. Song. Complementary cycles containing a pair of fixed vertices in bipartite tournaments. *Appl. J. Chin. Univ.*, 3:401–407, 1988.
80. G. Zhou and K.M. Zhang. Cycles containing a given arc in regular multipartite tournaments. *Acta Math. Appl. Sinica*, 18:681–684, 2002.

# 8. Quasi-Transitive Digraphs and Their Extensions

**Hortensia Galeana-Sánchez and César Hernández-Cruz**

## 8.1 Introduction

### 8.1.1 General Overview

Initially, quasi-transitive digraphs were studied by Ghouila-Houri in [39] because of their relation to comparability graphs.[1] Nonetheless, in their seminal paper [17] of 1995, Bang-Jensen and Huang began the study of this family in its own right. Through the last 20 years, quasi-transitive digraphs have gained a place among the most studied and better understood families of digraphs. Probably the main reason is the characterization theorem found in [17], which has led to solutions of many (usually difficult) problems.

Also, this is a family containing two very well known classes of digraphs: tournaments (and semicomplete digraphs) and transitive digraphs. It is well known that some interesting problems are very easy to solve for both families, *e.g.*, determining hamiltonicity. The appeal of quasi-transitive digraphs comes from the fact that a lot of problems are hard enough to be interesting, but it is still possible to find results similar to those of tournaments or transitive digraphs, yet, it is by no means trivial to do it.

Since a fair number of the classical problems for digraphs have already been studied for the family of quasi-transitive digraphs, it was a natural step to introduce a new class of digraphs generalizing it. Bang-Jensen introduced the family of 3-quasi-transitive digraphs in the context of strong arc-locally semicomplete digraphs [6]. Afterwards, in the context of $k$-kernels of digraphs, Galeana-Sánchez and Hernández-Cruz began in [48] the study of $k$-quasi-transitive digraphs. It came as a surprise that many nice structural properties of quasi-transitive digraphs have a natural generalization to $k$-quasi-transitive digraphs. This made it possible to generalize some of the classical results of

[1] He proved that a graph $G$ admits a quasi-transitive orientation if and only if it admits a transitive orientation if and only if it is a comparability graph.

H. Galeana-Sánchez and C. Hernández-Cruz (✉)
Instituto de Matemáticas, Universidad Nacional Autónoma de México
Área de la Investigación Científica, Circuito Exterior, Ciudad Universitaria,
C.P. 045010, Coyoacán, México, CDMX, México
e-mail: cesar@matem.unam.mx
 
J. Bang-Jensen and G. Gutin (eds.), *Classes of Directed Graphs*, Springer Monographs in Mathematics, https://doi.org/10.1007/978-3-319-71840-8_8

quasi-transitive digraphs to $k$-quasi-transitive digraphs, proving the latter to be an interesting family of digraphs.

Despite this fact, $k$-quasi-transitive digraphs are harder to handle than quasi-transitive digraphs. For larger values of $k$, their structure becomes increasingly complicated; as a matter of fact, the structure of strong 4-quasi-transitive digraphs is not completely understood. In view of this difficulty, Hernández-Cruz studied the classes of 3- and 4-transitive digraphs, [46, 47] obtaining a complete structural characterization of strong 3-transitive and 4-transitive digraphs. In [61], Wang and Wang proved that 3-quasi-transitive digraphs and 3-transitive digraphs are related in the same way as quasi-transitive and transitive digraphs: the underlying graphs of 3-quasi-transitive digraphs can be oriented as 3-transitive digraphs. This motivated the study of $k$-transitive digraphs on their own.

Finally, after reaching the most general case of the $k$-quasi-transitive digraphs and going back through the $k$-transitive digraphs, very recently the class of transitive digraphs has been considered again in the context of digraph homomorphisms. In [28], Feder, Hell and Hernández-Cruz showed that although many classical problems for digraphs are trivially solved in the class of transitive digraphs, there are many natural problems that are $\mathcal{NP}$-complete when restricted to this family. It is to be expected that both transitive and quasi-transitive digraphs will receive renewed attention in the near future.

As is usual with many mathematical concepts, $k$-quasi-transitive digraphs are not the only interesting generalization of quasi-transitive digraphs. On one hand we have $k$-quasi-transitive digraphs, which are obtained by generalizing the definition of a quasi-transitive digraph. As we have already mentioned, no nice structural characterizations of $k$-quasi-transitive digraphs are known for $k \geq 3$. So, on the other hand, instead of generalizing the definition of quasi-transitive digraphs, we can generalize the structure obtained by the characterization theorem. Following this idea, the notion of totally $\Phi$-decomposable digraphs was first introduced by Bang-Jensen and Gutin in [14], precisely as a tool to study quasi-transitive digraphs. Nonetheless, we can trace the basic idea of this family back to [41], where Gutin used a simpler version of the $\Phi$-decomposable digraphs to find a polynomial algorithm to solve the minimum path factor problem for quasi-transitive digraphs. It has turned out that this family is a common generalization of many interesting classes of digraphs, e.g., quasi-transitive digraphs, round decomposable graphs, directed cographs, etc.

### 8.1.2 Chapter Overview

In Subsection 8.1.3 some terminology and notation is introduced that will be used throughout the rest of the chapter. In Section 8.2 a brief overview of transitive digraphs is presented, including some open problems on digraph homomorphisms. Section 8.3 is devoted to presenting structural properties of quasi-transitive digraphs and some of their generalizations, including the canonical decomposition in Subsection 8.3.1, some structural properties of strong $k$-quasi-transitive digraphs in Subsection 8.3.2 and of $k$-transitive digraphs in Subsection 8.3.3, and recognition theorems of totally $\Phi$-decomposable digraphs for some choices of $\Phi$. Section 8.4 deals with paths and cycles; Subsection 8.4.1 reviews the few known results for hamiltonicity and traceability for $k$-transitive and $k$-quasi-transitive digraphs; Hamiltonicity of quasi-transitive and totally $\Phi$-decomposable digraphs is studied in Subsection 8.4.2 and some variants of vertex-cheapest paths and cycles for quasi-transitive digraphs are studied in Subsections 8.4.3, 8.4.4, 8.4.5, and 8.4.6. The linkage problem is covered in Section 8.5; Subsection 8.5.1 is devoted to $k$-linkages, and Subsection 8.5.2 to weak $k$-linkages. The topic of Section 8.6 is kings and kernels; $k$-kings are covered in Subsection 8.6.1 and $k$-kernels in Subsection 8.6.2. Section 8.7 deals with the Path Partition Conjecture, it has two subsections, Subsection 8.7.1 presents the conjecture and some of its known variants, and Subsection 8.7.2 deals with the known results for them. The last section of the chapter, Section 8.8 covers miscellaneous topics; vertex pancyclicity is covered in Subsection 8.8.1, acyclic spanning subdigraphs in Subsection 8.8.2, orientations of digraphs almost preserving the original diameter in Subsection 8.8.3, sparse subdigraphs with prescribed connectivity in Subsection 8.8.4, and arc-disjoint in-and out-branchings in Subsection 8.8.5.

### 8.1.3 Terminology and Notation

In this subsection, for the reader's convenience, we will recall some terminology and notation that will be used throughout this chapter. Only general concepts will be introduced here; more specific ones will be recalled whenever needed.

Throughout this chapter, walks, paths and cycles in a digraph are always meant to be directed. Let $D$ be a digraph. An arc $uv$ of $D$ is **symmetric** if $vu$ is also an arc of $D$, and **asymmetric** otherwise. Notice that a symmetric arc $uv$ together with the arc $vu$ form a 2-cycle of $D$; both this 2-cycle and the arc $uv$ will sometimes be referred to as a **digon**. When $u, v$ are adjacent vertices of $D$, we will write $\overline{uv}$.

If $X$ and $Y$ are disjoint subsets of vertices of $D$, then $X \rightarrow Y$ means that $X$ **dominates** $Y$, that is, every vertex of $X$ dominates every vertex of $Y$. If additionally there is no arc from $Y$ to $X$, then we say that $X$ **completely dominates** $Y$ and denote this by $X \mapsto Y$. We shall use the same notation when $X$ and $Y$ are disjoint subdigraphs rather than subsets of vertices.

Let $k$ be an integer, $k \geq 2$. A digraph $D$ is $k$**-quasi-transitive** if for every pair of vertices $u, v$ of $D$, the existence of a $(u, v)$-path of length $k$ in $D$ implies that $\overline{uv}$. A **quasi-transitive digraph** is a 2-quasi-transitive digraph. A digraph $D$ is $k$**-transitive** if for every pair of vertices $u, v$ of $D$, the existence of a $(u, v)$-path of length $k$ in $D$ implies $u \rightarrow v$. A **transitive digraph** is a 2-transitive digraph. Recall that if $R$ is a digraph on $r$ vertices $v_1, \ldots, v_r$ and $L_1, \ldots, L_r$ is a collection of distinct (but possibly isomorphic) digraphs, then we denote by $D = R[L_1, \ldots, L_r]$ the digraph with vertex set $V(L_1) \cup V(L_2) \cup \ldots \cup V(L_r)$ and arc set $(\bigcup_{i=1}^{r} A(G_i)) \cup \{g_i g_j : g_i \in V(G_i), g_j \in V(G_j), v_i v_j \in A(D)\}$. If $D = R[L_1, \ldots, L_r]$, then $R, L_1, \ldots, L_r$ are induced subdigraphs of $D$ and we say that $D$ is **decomposable** (into $R, L_1, \ldots, L_r$). Let $\Phi$ be a class of digraphs. A digraph $D$ is **$\Phi$-decomposable** if $D$ is a member of $\Phi$ or $D = H[S_1, \ldots, S_h]$ for some $H \in \Phi$ with $h = |V(H)| \geq 2$ and some choice of digraphs $S_1, S_2, \ldots, S_h$ (we call this decomposition a **$\Phi$-decomposition**). A digraph $D$ is called **totally $\Phi$-decomposable** if either $D \in \Phi$ or there is a $\Phi$-decomposition $D = H[S_1, \ldots, S_h]$ such that $h \geq 2$, and each $S_i$ is totally $\Phi$-decomposable. In this case, if $D \notin \Phi$, a $\Phi$-decomposition of $D$, $\Phi$-decompositions $S_i = H_i[S_{i1}, \ldots, S_{ih_i}]$ of all $S_i$ which are not in $\Phi$, $\Phi$-decompositions of those of $S_{ij}$ which are not in $\Phi$, and so on, form a sequence of decompositions which will be called a **total $\Phi$-decomposition** of $D$. If $D \in \Phi$, we assume that the (unique) total $\Phi$-decomposition of $D$ consists of itself.

If $D$ is a digraph on $n$ vertices, and $S_1, \ldots, S_n$ are digraphs with no arcs, then we say that the composition $H = D[S_1, \ldots, S_n]$ is an **extension of** $D$, or we say that $H$ is a **$D$-extension**. When $D$ belongs to some well-known class of digraphs, we will say that $H$ is an **extended** member of the class, e.g., if $D$ is a semicomplete digraph, we will say that $H$ is an extended semicomplete digraph.

A **$k$-path-$q$-cycle subdigraph** (**$k$-path-$q$-cycle factor**), $\mathcal{F}$, of a digraph $D$ is a (spanning) collection of $k$ paths and $q$ cycles, all disjoint. When $k = 0$, $\mathcal{F}$ is a **$q$-cycle subdigraph** (and a **$q$-cycle factor** if it is spanning) and when $q = 0$, $\mathcal{F}$ is a **$k$-path-subdigraph** (and a **$k$-path-factor** if it is spanning). A $k$-path-$q$-cycle subdigraph in which $q$ may be arbitrary (including zero) is called a **$k$-path-cycle subdigraph**.

A longest path in a digraph $D$ is called a **detour** of $D$. The order of a detour of $D$ is called the **detour order** of $D$ and is denoted by $\mathsf{do}(D)$. For a given digraph $D$, let $\mathsf{do}_k(D)$ denote the maximum number of vertices contained in a $k$-path subdigraph of $D$. A $k$-path subdigraph of $D$ which covers $\mathsf{do}_k(D)$ vertices is called a **maximum** $k$-path subdigraph of $D$. Note that $\mathsf{do}_1(D) = \mathsf{do}(D)$.

The **path-covering number** of a digraph $D$ (denoted by $pc(D)$) is the least positive integer $k$ such that $D$ has a $k$-path factor. The **path-cycle-covering number** of a digraph $D$ (denoted by $pcc(D)$) is the least positive integer $k$ such that $D$ has a $k$-path-cycle factor. The path-cycle-covering number of a digraph can easily be found in polynomial time using, in particular, algorithms on flows in networks [10, 14, 41]. The path-covering number is hard to calculate: note that $pc(D) = 1$ if and only if $D$ has a Hamiltonian path. Thus, the path-covering number problem generalizes the Hamiltonian path problem.

Given a fixed digraph $H$, an **$H$-colouring** of a digraph $D$ is a **homomorphism of $D$ to $H$**, i.e., a mapping $f : V(D) \to V(H)$ such that $f(u)f(v)$ is an arc of $H$ whenever $uv$ is an arc of $D$. The **$H$-colouring problem** asks whether an input digraph $D$ admits an $H$-colouring. In the **list $H$-colouring problem** the input $D$ comes equipped with lists $L(u) \subseteq V(H), u \in V(D)$, and the homomorphism $f$ must also satisfy $f(u) \in L(u)$ for all vertices $u$. Finally, the **$H$-retraction problem** is a special case of the list $H$-colouring problem, in which each list is either $L(u) = \{u\}$ or $L(u) = V(H)$. Note that the $H$-colouring problem is a special case of the $H$-retraction problem, in which each $L(u) = V(H)$. The **dichromatic number** of a digraph $D$ is the least integer $\chi(D)$ such that $V(D)$ admits a partition into $\chi(D)$ acyclic sets. Notice that if every arc of $D$ is symmetric, then the dichromatic number of $D$ coincides with the (usual) chromatic number of the underlying graph of $D$.

## 8.2 Transitive Digraphs

A digraph $D$ is defined to be **transitive** if for any three *distinct* vertices $u, v, w$, the existence of the arcs $uv, vw$ implies the existence of the arc $uw$. Note that an acyclic digraph is transitive if and only if its arcs define a transitive relation in the usual sense. However, a digraph with a directed cycle is transitive if and only if its reflexive closure (i.e., adding all loops) defines a transitive relation. This peculiarity means that, for instance, when taking a transitive closure of a digraph we omit any loops that would exist in a transitive closure as a binary relation.

Acyclic transitive digraphs have a particularly nice structure, namely, they are exactly those digraphs whose reflexive closure is a reflexive partial order. It is well known that each transitive digraph $D$ is obtained from an acyclic transitive digraph $J$ by **replication**, whereby each $j \in V(J)$ is replaced by $k_j \geq 0$ vertices forming a complete digraph, so that if $ij$ is an arc in $J$, then all $k_i$ vertices replacing $i$ dominate in $D$ all $k_j$ vertices replacing $j$. Note that all $k_j$ vertices replacing $j$ have exactly the same in- and out-neighbours in $D$ (except that each of them does not dominate itself). Note that the strong components of a transitive digraph $D$ are complete digraphs.

The observations in the preceding paragraph are often stated in terms of contraction[2] of the strong components of a transitive digraph, in order to obtain an acyclic transitive digraph, rather than using the replication operation to obtain an arbitrary transitive digraph from an acyclic one. Of course, both points of view are equivalent, but usually this observation is stated in the following way.

**Proposition 8.2.1** *Let $D$ be a digraph with an acyclic ordering $D_1, \dots, D_p$ of its strong components. The digraph $D$ is transitive if and only if the following holds:*

[2] Contraction is defined in Section 1.4 for directed multigraphs. We can obtain a digraph instead of a directed multigraph by deleting spare parallel arcs after contraction.

1. *Each digraph $D_i$, $i \in [p]$ is complete,*
2. *the digraph $H$ obtained from $D$ by contraction of $D_1, \ldots, D_p$ is a transitive oriented graph, and*
3. *$D = H[D_1, \ldots, D_p]$, where $p = |V(H)|$.*

Notice that Proposition 8.2.1 can be restated as saying that every transitive digraph is totally $\Psi_0$-decomposable, where $\Psi_0$ is the family of all acyclic digraphs and all the complete digraphs. Obviously, for a digraph $D$, the digraph $H$ of Proposition 8.2.1 (which is the same as the digraph $J$ in the above construction by replication), is simply the strong component digraph of $D$. From here, and using the fact that the strong components of a transitive digraph are complete digraphs, one can directly verify that some problems are easy to solve when restricted to transitive digraphs. Recall that the strong component digraph can be constructed in $O(|V| + |A|)$-time. A necessary condition for a digraph $D$ to be Hamiltonian is that $D$ is strong. In the case of transitive digraphs, this condition is also sufficient, since every transitive strong digraph is a complete digraph, and thus Hamiltonian. Hence, hamiltonicity can be verified in linear time for transitive digraphs. Every transitive digraph $D$ has a kernel; to construct one, it suffices to choose one vertex from every terminal component of $D$. Thus, it can be verified in constant time whether a transitive digraph has a kernel, one can be constructed in linear time, and the exact number of different kernels can be calculated in linear time. An acyclic transitive digraph $J$ clearly has dichromatic number equal to one, and it follows from the description of the structure of an arbitrary transitive digraph given by replication that the dichromatic number of an arbitrary transitive digraph $D$ obtained from an acyclic transitive $J$ by vertex substitutions is equal to the maximum value $k_j$ of the size of any replacing set of vertices. Therefore, the dichromatic number of a transitive digraph equals the size of its largest strong component. Again, the dichromatic number of a transitive digraph can be determined in linear time. We could go on, enumerating problems which are $\mathcal{NP}$-complete in the general digraph case and become polynomial time solvable when restricted to transitive digraphs. Nonetheless, it is more revealing to exhibit a very natural problem that remains $\mathcal{NP}$-complete even when restricted to transitive digraphs.

In [29], it is shown that there are bipartite graphs $H$ such that the $H$-retraction problem is $\mathcal{NP}$-complete. Hence, the following result of Feder, Hell and Hernández-Cruz shows that there are digraphs $D$ such that the $D$-homomorphism problem is $\mathcal{NP}$-complete, even when restricted to transitive inputs.

**Theorem 8.2.2** ([28]) *If $H$ is a bipartite graph such that the $H$-retraction problem is $\mathcal{NP}$-complete, then there exists a digraph $H'$ such that the $H'$-homomorphism problem is $\mathcal{NP}$-complete, even when restricted to transitive digraphs.*

Before proving Theorem 8.2.2, we will describe how the digraph $H'$ can be obtained from a bipartite graph $H$. Let $H$ be a bipartite graph with its bipartition given by a set of white vertices and a set of black vertices, with at most $n$ black and at most $n$ white vertices. We form the digraph $H'$ as follows (see Figure 8.1). We first orient all edges of $H$ from the white vertices to the black vertices. Let $P_i$ be a directed path with $n+2$ vertices, in which the first, and the $(i+1)$-st, vertex have been duplicated (replicated once). Let $R_i$ also be a directed path with $n+2$ vertices, in which the the last, and the $(i+1)$-st, vertex have been duplicated. We identify the last vertex of each $P_i$ with the $i$-th white vertex (if any) of $H$ and the first vertex of each $R_i$ with the $i$-th black vertex (if any) of $H$. Then $H'$ is obtained from the resulting digraph by taking the transitive closure. It is easy to see that the added paths ensure that the only homomorphism of $H'$ to itself is the identity. Also consider a directed path $P$ with $n+2$ vertices with only the first vertex duplicated, and a directed path $R$ with $n+2$ vertices and only the last vertex duplicated. Note that $P$ admits a homomorphism to each $P_i$ and $R$ admits a homomorphism to each $R_i$. For future reference, we define the *level* of the $j$-th vertex of $P$ or $P_i$ to be $j$, and the *level* of the $j$-th vertex of $R$ or $R_i$ to be $n+2+j$; in this we assume the duplicated vertices to have the same level. Note that this forces all white vertices to have level $n+2$ and all black vertices to have level $n+3$.

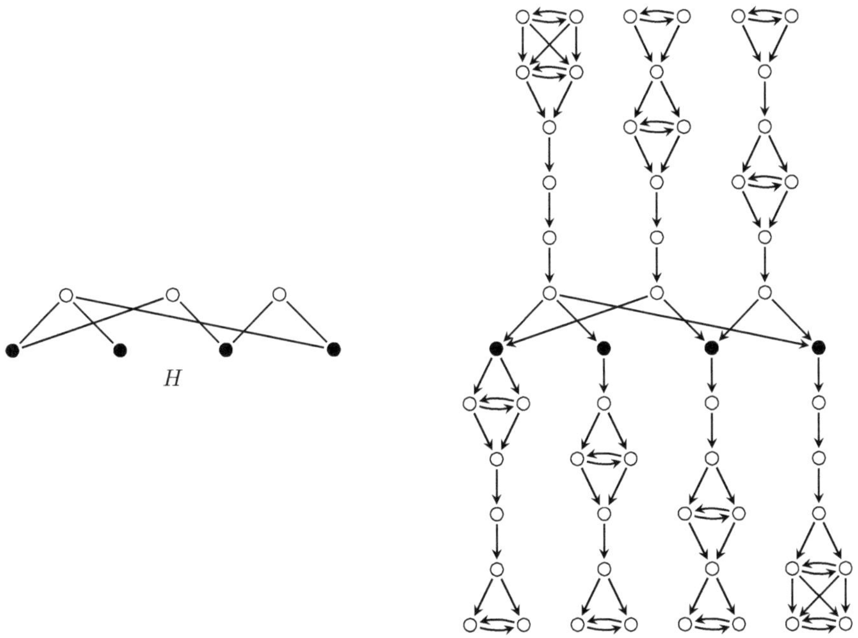


**Figure 8.1** The construction of $H'$ from $H$ used for Theorem 8.2.2. The digraph $H'$ is obtained by taking the transitive closure of the digraph on the right.

**Proof of Theorem** 8.2.2**:** Suppose $G$ is an instance of the $H$-retraction problem, i.e., a bipartite graph containing $H$ as a subgraph with lists $\{x\}$ for each (black and white) vertex $x$ of $H$, and lists $V(H)$ for all other (black and white) vertices of $G$. We construct an instance $G'$ of the $H'$-colouring problem by orienting all edges of $G$ from the white vertices to the black vertices, attaching paths $P_i$ and $R_j$ to the vertices of $H$ as in the construction of $H'$, and then (for the vertices not in $H$) we identify the last vertex of a (separate) copy of $P$ to each white vertex of $G$ not in $H$, and identify the first vertex of a (separate) copy of $R$ to each black vertex of $G$ not in $H$, and finally we take the transitive closure. Now it is easy to see that each homomorphism of $G'$ to $H'$ preserves the level of vertices, and that $G'$ admits an $H'$-colouring if and only if $G$ admits a retraction to $H$.

Moreover, the above construction of $H'$ ensures that it is itself transitive. Thus we have the following fact.

**Corollary 8.2.3** ([28]) *There exists a transitive digraph $H'$ such that the $H'$-homomorphism problem is $\mathcal{NP}$-complete even when restricted to transitive digraphs.*

In view of Corollary 8.2.3, a natural interesting problem is the following.

**Problem 8.2.4** *Characterize the transitive digraphs $H$ such that the $H$-homomorphism problem restricted to transitive inputs is polynomial time solvable.*

Although Problem 8.2.4 may look innocuous, it may be very hard indeed. Recall that Feder and Vardi proved in [29] that in order to classify all constraint satisfaction problems, it is enough to classify all the digraph homomorphism problems. In [28], Feder, Hell and Hernández-Cruz propose the problem of determining whether for any relational structure $H$, a (transitive) digraph $H'$ exists such that the constraint satisfaction problem for $H$ is polynomially equivalent to the $H'$-homomorphism problem for transitive digraphs.

## 8.3 Structural Properties

As mentioned before, the main appeal of quasi-transitive digraphs comes from the fact that their structure is very well understood. Throughout this section, we will consider structural properties of quasi-transitive, $k$-transitive and $k$-quasi-transitive digraphs. Also, some results regarding the recognition of $\Phi$-decomposable digraphs for particular cases of $\Phi$ are included. We begin by presenting the classical results due to Bang-Jensen and Huang from [17].

### 8.3.1 Quasi-Transitive Digraphs

The nice results that have been obtained for quasi-transitive digraphs and all the attention this family and its generalizations have received are principally a consequence of the recursive characterization theorem given by Bang-Jensen and Huang in [17]. The main purpose of this subsection is to reproduce the proof of this theorem, including the lemmas needed, many of which are interesting on their own.

**Proposition 8.3.1** ([17]) *Let $D$ be a quasi-transitive digraph. Suppose that $P = x_0x_1 \dots x_n$ is a shortest $(x_0, x_n)$-path. Then, the subdigraph induced by $V(P)$ is a semicomplete digraph and $x_j \to x_i$ for every $1 \leq i+1 < j \leq k$, unless $n = 3$, in which case the arc between $x_0$ and $x_n$ may be absent.*

**Proof:** The cases $k \in \{2, 3, 4, 5\}$ are easily verified. The proof for the case $k \geq 6$ is by induction on $k$ with the case $k = 5$ as the basis. By induction, each of $D[\{x_0, \dots, x_{k-1}\}]$ and $D[\{x_1, \dots, x_k\}]$ is a semicomplete digraph and $x_j \to x_i$ for any $1 < j - i < k - 2$. Hence, $x_2$ dominates $x_0$ and $x_k$ dominates $x_2$, and the minimality of $P$ implies that $x_k$ dominates $x_0$. □

**Corollary 8.3.2** ([17]) *If a quasi-transitive digraph $D$ has an $(x, y)$-path but $x$ does not dominate $y$, then either $y \to x$, or there exists vertices $u, v \in V(D) - \{x, y\}$ such that $x \to u \to v \to y$ and $y \to u \to v \to x$.*

**Proof:** Consider a minimal $(x, y)$-path and apply Proposition 8.3.1. □

**Lemma 8.3.3** ([17]) *Suppose that $A$ and $B$ are distinct strong components of a quasi-transitive digraph $D$ with at least one arc from $A$ to $B$. Then $A \mapsto B$.*

**Proof:** Suppose $A$ and $B$ are distinct strong components such that there exists an arc from $A$ to $B$. Then, for every choice of $x \in A$ and $y \in B$, there exists a path from $x$ to $y$ in $D$. Since $A$ and $B$ are distinct strong components, none of the alternatives in Corollary 8.3.2 can hold, and hence $x \to y$. □

Proposition 8.3.1 and Lemma 8.3.3 will be generalized in the following sections for $k$-quasi-transitive digraphs. On the other hand, the following lemma does not have any known generalizations for $k$-quasi-transitive digraphs when $k \geq 3$.

**Lemma 8.3.4** ([17]) *Let $D$ be a strong quasi-transitive digraph on at least two vertices. Then the following holds:*

*(a)* $\overline{UG(D)}$ *is disconnected;*
*(b) If $S$ and $S'$ are two subdigraphs of $D$ such that $\overline{UG(S)}$ and $\overline{UG(S')}$ are distinct connected components of $\overline{UG(D)}$, then either $S \mapsto S'$ or $S' \mapsto S$, or both $S \to S'$ and $S' \to S$, in which case $|V(S)| = |V(S')| = 1$.*

**Proof:** The statement *(b)* can be easily verified from the definition of a quasi-transitive digraph and the fact that $S$ and $S'$ are completely adjacent in $D$. We prove *(a)* by induction on $|V(D)|$. Statement *(a)* is trivially true when $|V(D)| \in \{2, 3\}$. Assume that it holds when $|V(D)| < n$, where $n > 3$.

Suppose that there is a vertex $z$ such that $D - z$ is not strong. Then, there is an arc from (to) every terminal (initial) strong component of $D - z$ to (from) $z$. Since $D$ is quasi-transitive, the last fact and Lemma 8.3.3 imply that $X \to Y$ for every initial (terminal) strong component $X$ $(Y)$ of $D - z$. Similar arguments show that each strong component of $D-z$ either dominates some terminal component or is dominated by some initial component of $D-z$ (intermediate strong components satisfy both). These facts imply that $z$ is adjacent to every vertex in $D - z$. Therefore, $\overline{UG(D)}$ contains a component consisting of the vertex $z$, implying that $\overline{UG(D)}$ is disconnected, and *(a)* follows.

Assume that there is a vertex $v$ such that $D - v$ is strong. Since $D$ is strong, it contains an arc $vw$ from $v$ to $D - v$. By induction, $\overline{UG(D - v)}$ is not connected. Let $S$ and $S'$ be connected components of $\overline{UG(D - v)}$ such that $w \in S$ and $S \to S'$ (here we use *(b)* and the fact that $D - v$ is strong). Then $v$ is completely adjacent to $S'$ in $D$ (as $v \to w$). Hence, $\overline{UG(S')}$ is a connected component of $\overline{UG(D)}$ and the proof is complete. □

In the following subsections we will see that, for some values of $k$, there are nice characterizations of strong $k$-transitive and $k$-quasi-transitive digraphs. Also it is even possible to show that the strong components of, for example, a 3-quasi-transitive digraph, are related in a very special way. Nonetheless, it is difficult to obtain a characterization fully describing the structure of those families, mainly because, for sufficiently small induced subdigraphs, the $k$-quasi-transitivity becomes irrelevant. The following theorem gives a complete characterization of quasi-transitive digraphs, which makes members of this family easier to deal with. Notice that, since the characterization is recursive, it provides an excellent structure to apply mathematical induction in this class of digraphs.

**Theorem 8.3.5 (Bang-Jensen, Huang** [17]**)** *Let $D$ be a digraph which is quasi-transitive.*

- *If $D$ is not strong, then there exists a transitive oriented graph $T$ with vertices $\{u_1, u_2, \ldots, u_t\}$ and strong quasi-transitive digraphs $H_1, H_2, \ldots, H_t$ such that $D = T[H_1, H_2, \ldots, H_t]$, where $H_i$ is substituted for $u_i$, $i \in \{1, 2, \ldots, t\}$.*
- *If $D$ is strong, then there exists a strong semicomplete digraph $S$ with vertices $\{v_1, v_2, \ldots, v_s\}$ and quasi-transitive digraphs $Q_1, Q_2, \ldots, Q_s$ such that $Q_i$ is either a vertex or is non-strong and $D = S[Q_1, Q_2, \ldots, Q_s]$, where $Q_i$ is subsituted for $v_i$, $i \in \{1, 2, \ldots, s\}$.*

**Proof:** Suppose that $D$ is not strong and let $H_1, \ldots, H_t$ be the strong components of $D$. According to Lemma 8.3.3, if there is an arc between $H_i$ and $H_j$, then either $H_i \mapsto H_j$ or $H_j \mapsto H_i$. Now, if $H_i \mapsto H_j \mapsto H_k$, then, by quasi-transitivity, $H_i \mapsto H_k$. So, by contracting each $H_i$ to a vertex $h_i$, we get a transitive oriented graph $T$ with vertices $h_1, \ldots, h_t$. This shows that $D = T[H_1, \ldots, H_t]$.

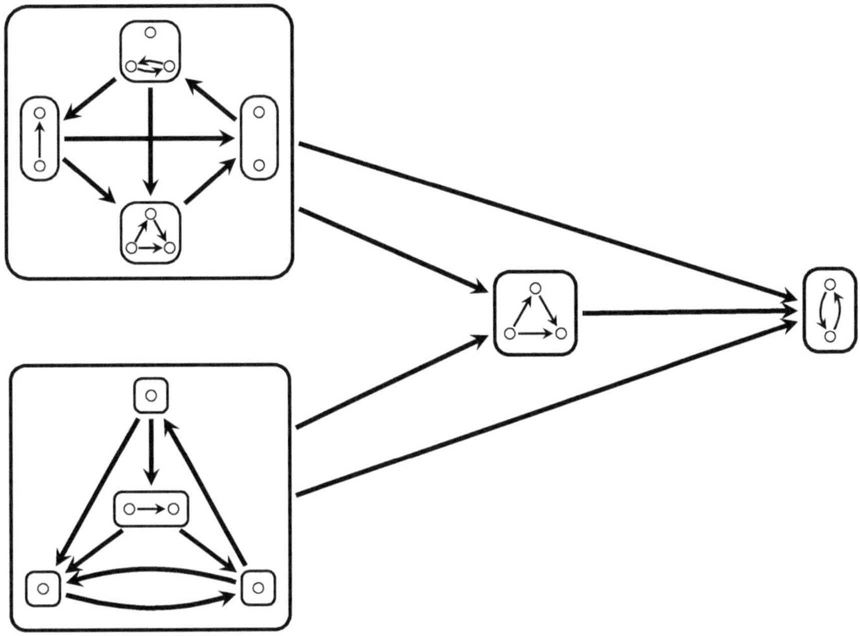

**Figure 8.2** The canonical decomposition of a non-strong quasi-transitive digraph. Big arcs between different boxed sets indicate that there is a complete domination in the direction shown.

Suppose that $D$ is strong. Let $Q_1, \ldots, Q_s$ be the subdigraphs of $D$ such that each $\overline{UG(Q_i)}$ is a connected component of $\overline{UG(D)}$. According to Lemma 8.3.4*(a)*, each $Q_i$ is either non-strong or just a single vertex. By Lemma 8.3.4 *(b)*, we obtain a strong semicomplete digraph $S$ if each $Q_i$ is contracted to a vertex. This shows that $D = S[Q_1, \ldots, Q_s]$. □

The decomposition described by Theorem 8.3.5 is called the **canonical decomposition** of the quasi-transitive digraph $D$. The canonical decomposition of a non-strong quasi-transitive digraph is illustrated in Figure 8.2.

### 8.3.2 $k$-Quasi-Transitive Digraphs

So far, there are no known characterizations of $k$-quasi-transitive digraphs for $k \geq 3$. Even if we restrict ourselves to strong digraphs, only strong 3-quasi-transitive digraphs have a simple complete characterization. Despite this fact, there are some structural results valid for any $k \geq 3$ that have been useful to study $k$-quasi-transitive digraphs.

Despite its simplicity, it could be said that the following result is the cornerstone of the study of $k$-quasi-transitive digraphs; it was proved by Galeana-Sánchez and Hernández-Cruz in [48]. Notice that it can be regarded as a generalization of Corollary 8.3.2.

**Lemma 8.3.6** ([48]) *Let $k$ be an integer with $k \geq 2$. If $D$ is a $k$-quasi-transitive digraph, and for $u, v \in V(D)$ there is a $(u, v)$-path in $D$, then each of the following holds:*

1. *If $d(u, v) = k$, then $d(v, u) = 1$.*
2. *If $d(u, v) = k + 1$, then $d(v, u) \leq k + 1$.*
3. *Assume $d(u, v) = r \geq k + 2$. If $k$ is even or $k$ and $r$ are both odd, then $d(v, u) = 1$; if $k$ is odd and $r$ is even, then $d(v, u) \leq 2$.*

**Proof:** Let $P = x_0, \ldots, x_r$ be a path of length $r = k + j$, $j \geq 0$. Observe that the $k$-quasi-transitivity of $D$ and the fact that $d(u, v) = r$ imply that $x_r \to x_j$. This handles *1.* and *2.*

To prove *3.*, we will proceed by induction on $j$. For $j = 2$, the existence of the $k$-path $x_r x_2 P x_k x_0$ implies $x_r \to x_0$. For $j = 3$, the existence of the $k$-path $x_r x_3 P x_{k+1} x_1$ implies $x_r x_1$. Considering the $k$-path $x_r x_1 P x_k$, we get $x_r \to x_k$. When $k$ is odd, we already have $d(x_r, x_0) \leq 2$. For even $k$, we will prove by induction on $i$ that $x_r \to x_{k-2i}$ for every $0 \leq i \leq \frac{k}{2}$. We already have $x_r \to x_k$, so suppose that $x_r \to x_{k-2i}$ for some $0 < i < \frac{k}{2}$. Now, the existence of the $k$-path $x_r x_{k-2i} P x_k x_0 P x_{k-2(i+1)}$ implies $x_r \to x_{k-2(i+1)}$. In particular, $x_r \to x_0$

So, suppose $j > 3$. By the induction hypothesis, if $k$ is even, or both $k$ and $r$ are odd, we obtain $x_r \to x_2$. Hence, $x_r x_2 P x_k x_0$ is a $k$-path, and thus $x_r \to x_0$. If $k$ is odd and $r$ is even, by the induction hypothesis we have $x_r \to x_1$. So, $x_r x_1 P x_k$ is a $k$-path, the existence of which implies $x_r \to x_k$. Since we already had $x_k \to x_0$, we conclude $d(x_r, x_0) \leq 2$. □

Proposition 8.3.1 was generalized to $k$-quasi-transitive digraphs by Wang and Zhang (when $k$ is even) [62] and by Alva-Samos and Hernández-Cruz (when $k$ is odd) [1]. Its proof is long and technical, and thus will be omitted.

**Proposition 8.3.7** ([1, 62]) *Let $k \geq 3$ be an integer and let $D$ be a $k$-quasi-transitive digraph. Suppose that $P = x_0 x_1 \ldots x_r$ is a shortest $(x_0, x_r)$-path with $r \geq k + 2$ in $D$.*

- *If $k$ is even, then $D[V(P)]$ is a semicomplete digraph and $x_j \to x_i$ for $1 \leq i + 1 < j \leq r$.*
- *If $k$ is odd, then $D[V(P)]$ is either a semicomplete digraph and $x_j \to x_i$ for $1 \leq i + 1 < j \leq r$, or $D[V(P)]$ is a semicomplete bipartite digraph and $x_j \to x_i$ for $1 \leq i + 1 < j \leq r$ and $i \not\equiv j \pmod 2$.* □

In a quasi-transitive digraph $D$, Lemma 8.3.3 tells us that for two different strong components $A$ and $B$, if $A$ reaches $B$, then $A \mapsto B$. Unfortunately, this is not true for $k$-quasi-transitive digraphs when $k \geq 3$. Nonetheless, there are some results resembling this behaviour. The following simple (but very useful) result was originally proved by Hernández-Cruz while studying $k$-transitive digraphs, [47].

**Lemma 8.3.8** ([47]) *Let $k$ be an integer, $k \geq 2$, let $D$ be a $k$-quasi-transitive digraph, and let $C = v_0v_1 \dots v_{r-1}v_0$ be a directed cycle in $D$ with $r \geq k$. For any $v \in V(D) - V(C)$, if $v \to v_i$ and $(V(C), v) = \varnothing$, then $v \to v_{i+(k-1)}$; if $v_i \to v$ and $(v, V(C)) = \varnothing$, then $v_{i-(k-1)} \to v$, where the subscripts are taken modulo $r$.*

**Proof:** It suffices to prove the first statement, the second one is obtained by noting that reversing every arc of a $k$-quasi-transitive digraphs yields a $k$-quasi-transitive digraph.

The path $vv_iCv_{i+(k-1)}$ has length exactly $k$, and thus, $\overline{vv_{i+(k-1)}}$. But $(V(C), v) = \varnothing$, hence $v \to v_{i+(k-1)}$. □

Our previous lemma is complemented by the following result due to Wang and Zhang. Although both results have very simple proofs, they have some very nice consequences on the structure of $k$-quasi-transitive digraphs.

**Lemma 8.3.9** ([62]) *Let $k$ be an integer with $k \geq 2$, and let $D$ be a strong $k$-quasi-transitive digraph. Suppose that $C = v_0v_1 \dots v_{r-1}v_0$ is a cycle of length $r$, with $r \geq k$, in $D$. Then, for any $v \in V(D) - V(C)$, $v$ and $C$ are adjacent.*

**Proof:** Since $D$ is strong, $v$ must reach $C$ and vice versa. Let $P$ be a shortest path from $v$ to $C$, and assume without loss of generality that the endpoint of $P$ is $v_0$. If the length of $P$ is $s$, and $s \leq k$, then $vPv_0Cv_{k-s}$ is a $k$-path, which implies $\overline{vv_{k-s}}$. If $k < s$, then by Lemma 8.3.6, $v_0$ reaches $v$ at distance at most two. If $v_0 \to v$, then we are done. Otherwise, there is a vertex $u$ in $D$ such that $v_0 \to u \to v$. Either $u \in V(C)$, and the desired result is obtained, or $v_{r-(k-2)}Cv_0uv$ is a $k$-path in $D$, implying $v_{r-(k-2)} \to v$. □

As an example of how the previous two lemmas can be used to obtain nice structural results for $k$-quasi-transitive digraph, we present the following proposition, which is their immediate consequence.

**Proposition 8.3.10** ([62]) *Let $k$ be an integer with $k \geq 2$, let $D$ be a strong $k$-quasi-transitive digraph, and let $C = v_0v_1 \dots v_{r-1}v_0$ be a cycle of length $r$ with $r \geq k$ in $D$. Suppose that $r$ and $k-1$ are coprime. For any $v \in V(D) - V(C)$, if $(V(C), v) = \varnothing$, then $v \mapsto V(C)$; if $(v, V(C)) = \varnothing$, then $V(C) \mapsto v$.*

We finish our discussion of general $k$-quasi-transitive digraphs with some results that give us a lot of information on the structure of $k$-quasi-transitive digraphs with diameter at least $k + 2$. Unfortunately, the proofs of these results are long and technical and thus will be omitted.

**Lemma 8.3.11** ([62]) *Let $k$ be an even integer with $k \geq 4$, and let $D$ be a strong $k$-quasi-transitive digraph. Suppose that $P = v_0v_1 \dots v_{k+2}$ is a shortest $(v_0, v_{k+2})$-path in $D$. For any $v \in V(D) - V(P)$, if $(v, V(P)) \neq \varnothing$ and $(V(P), v) \neq \varnothing$, then either $v$ is adjacent to every vertex of $V(P)$, or*

$\{v_{k+2}, v_{k+1}, v_k, v_{k-1}\} \mapsto v \mapsto \{v_0, v_1, v_2, v_3\}$. *In particular, if* $k = 4$, *then* $v$ *is adjacent to every vertex of* $V(P)$. □

**Theorem 8.3.12** ([62]) *Let* $k$ *be an even integer with* $k \geq 4$, *and let* $D$ *be a strong* $k$*-quasi-transitive digraph. Suppose that* $P = v_0 \dots v_{k+2}$ *is a shortest* $(v_0, v_{k+2})$*-path. Then, the subdigraph induced by* $V(D) - V(P)$ *is a semicomplete digraph.* □

Notice that, in particular, it follows from Proposition 8.3.7, Lemma 8.3.11, and Theorem 8.3.12, that a 4-quasi-transitive digraph of diameter at least 6 is a semicomplete digraph. As a more general case, the previous results can be condensed in the following theorem.

**Theorem 8.3.13** *Let* $k$ *be an even integer with* $k \geq 4$, *and let* $D$ *be a strong* $k$*-quasi-transitive digraph. Then,* $V(D)$ *admits a partition* $(V_1, V_2)$ *such that* $V_i$ *induces a semicomplete digraph for* $i \in \{1, 2\}$, *and* $D[V_1]$ *is Hamiltonian.*

When $k$ is odd, Alva-Samos and Hernández-Cruz [1], through a similar development of technical lemmas, obtained the following analogue of Theorem 8.3.13.

**Theorem 8.3.14** *Let* $k$ *be an odd integer,* $k \geq 3$, *and let* $D$ *be a strong* $k$*-quasi-transitive digraph. Then,* $V(D)$ *admits a partition* $(V_1, V_2)$ *such that:*

- *If* $D$ *is bipartite, then* $D[V_i]$ *is a semicomplete bipartite digraph for* $i \in \{1, 2\}$;
- *Else,* $D[V_i]$ *is a semicomplete digraph,* $i \in \{1, 2\}$.

*In either case,* $D[V_1]$ *is Hamiltonian.*

In particular, it is also noted in [1] that a strong 5-quasi-transitive digraph of diameter at least 7 is either a semicomplete bipartite digraph or a semicomplete digraph.

To finish our discussion of the structure of $k$-quasi-transitive digraphs, we present the well understood structure of 3-quasi-transitive digraphs. Although a complete characterization telling us the exact structure of 3-quasi-transitive digraphs does not exist, a lot of information can be put together from the existing characterization of strong 3-quasi-transitive digraphs from [34], and the way the strong components relate to each other described in [64].

Let $F_i$ be the graph on $i + 3$ vertices, consisting of a directed 3-cycle $xyzx$, together with $i$ vertices, $v_1, \dots, v_i$, such that $yv_jz$ is a directed path for each $1 \leq j \leq i$, see Figure 8.3. Define the family $\mathcal{F}$ as $\mathcal{F} = \{F_i \colon i \geq 1\}$. Due to space constraints, we will not give the proof of the following theorem, originally proved by Galeana-Sánchez, Goldfeder, and Urrutia.

**Theorem 8.3.15** ([34]) *Let* $D$ *be a strong* 3*-quasi-transitive digraph. Then* $D$ *is one of the following.*

1. *A semicomplete digraph.*
2. *A semicomplete bipartite digraph.*
3. *An element of the family $\mathcal{F}$ described above.* □

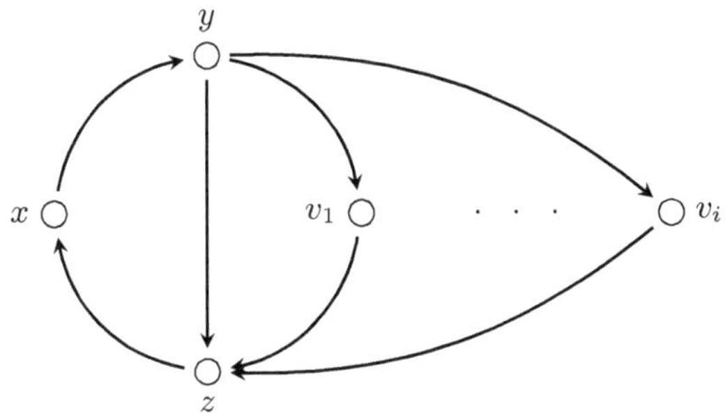


**Figure 8.3** The digraph $F_i$ of the family $\mathcal{F}$.

Theorem 8.3.15 can be complemented with the following result, due to Wang and Wang, found in [64].

**Lemma 8.3.16** ([64]) *Let $D_1$ and $D_2$ be two distinct non-trivial strong components of a 3-quasi-transitive digraph, with at least one arc from $D_1$ to $D_2$. Then, either $D_1 \mapsto D_2$, or $D_1 \cup D_2$ is a semicomplete bipartite digraph.* □

It is not hard to see that, given two strong components $D_1$ and $D_2$ of a 3-quasi-transitive digraph $D$ such that $D_1$ reaches $D_2$, there is an arc from $D_1$ to $D_2$ unless $D_1$ reaches $D_2$ in distance exactly 2, and both $D_1$ and $D_2$ consist of a single vertex. Thus, Lemma 8.3.16 becomes very useful when dealing with non-strong 3-quasi-transitive digraphs.

### 8.3.3 $k$-Transitive Digraphs

It is clear from the definition of both $k$-transitive and $k$-quasi-transitive digraphs that members of these classes having a small order do not really have any organized structure. Nonetheless, as the order increases, a nice structure emerges. As we have seen in the previous subsection, for $k$-quasi-transitive digraphs, the existence of two vertices at distance $k+2$ is sufficient for the rest of the digraph to organize as almost a semicomplete digraph (when $k$ is even). In this section we will see that the tipping point for a $k$-transitive digraph $D$ seems to be the existence of a "long enough" cycle; this will be sufficient for the digraph to be a complete digraph, or an extended cycle.

For $k = 3$, this point is easily reached, and thus, the structure of 3-transitive digraphs is easy to describe. But even for $k = 4$, it becomes hard to obtain a complete description of all 4-transitive digraphs; a classification of 4-transitive strong digraphs is given in this case. We begin with a couple of results that show the importance of cycles in $k$-transitive digraphs.

Results on 3- and 4-transitive digraphs are due to Hernández-Cruz. In this subsection we will present a new, shorter proof of Theorem 8.3.19.

Observe that the proof of Lemma 8.3.8 also yields the following result.

**Proposition 8.3.17** ([47]) *Let $k \geq 2$ be an integer, $D$ a $k$-transitive digraph and $C = v_0 v_1 \dots v_{r-1} v_0$ a directed cycle in $D$ with $r \geq k$. If $v \in V(D) - V(C)$ is such that $v \to v_0$, then $v \to S = \{v_i \big| i \in (k-1)\mathbb{Z}_r\}$.*

Observe that under the same assumptions as in Proposition 8.3.17, if $v_0 \to v$, we can conclude that $S \to v$. This follows from the fact that reversing all the arcs of a $k$-transitive digraph yields a $k$-transitive digraph, and applying Proposition 8.3.17. So, in this subsection we will refer to either result as Proposition 8.3.17.

**Lemma 8.3.18** *Let $D$ be a strong digraph. If the circumference of $D$ is 2, then the underlying graph of $D$ is a tree.*

**Proof:** Assuming that the circumference of $D$ is 2, it is easy to verify that every arc of $D$ is a digon. Thus, between any pair of vertices there is exactly one path, and hence, the underlying graph of $D$ is a tree. □

Recall that $\vec{C}_3$ is the directed cycle on three vertices, and let $C_3^\star$ and $C_3^{\star\star}$ be the directed 3-cycle with exactly one symmetric arc and the directed 3-cycle with exactly two symmetric arcs, respectively. Now we give the characterization of strong 3-transitive digraphs due to Hernández-Cruz, although with a new, simpler proof.

**Theorem 8.3.19** ([46]) *If $D$ is a 3-transitive strong digraph, then $D$ is one of the following:*

1. *A complete biorientation of a complete graph;*
2. *A complete biorientation of a complete bipartite graph; or*
3. *$\vec{C}_3$, $C_3^\star$ or $C_3^{\star\star}$.*

**Proof:** We begin by observing that every strong digraph with fewer than four vertices is either complete, complete bipartite or one of $\vec{C}_3$, $C_3^\star$ or $C_3^{\star\star}$. Thus, we can assume that $D$ has at least four vertices.

**Claim 1.** If the circumference of $D$ is 2, then $D$ is a complete biorientation of a star, and hence, a complete biorientation of a complete bipartite graph.

**Proof of Claim 1.** It follows from Lemma 8.3.18 that $D$ is a complete biorientation of a tree. Since $D$ is 3-transitive, the diameter of $D$ should be strictly less than 3. Hence, the underlying graph of $D$ is a tree of diameter 2, i.e., a star. □

**Claim 2.** If $D$ contains a directed odd cycle, then $D$ is a complete biorientation of a complete graph.

**Proof of Claim 2.** It can be proved inductively that if $D$ contains an odd cycle, then it contains a directed 3-cycle, $C$. Since $D$ has at least four vertices, there exists a vertex $v \in V(D) \setminus V(C)$. Since $D$ is 3-transitive and strong, there must be an arc from $v$ to $C$ and one arc from $C$ to $v$. It follows from Proposition 8.3.17 that $v \rightarrow C$ and $C \rightarrow v$. But now, any two vertices of $C$ together with $v$ induce a 3-cycle (with some symmetric arcs), and the same argument can be used to prove that $v$ is adjacent to any vertex in $D$ through a digon. Since $v$ was chosen arbitrarily outside a 3-cycle, $D$ is a complete biorientation of a complete digraph. □

**Claim 3.** If every directed cycle of $D$ is even, then $D$ is a complete biorientation of a complete bipartite graph.

**Proof of Claim 3.** First, notice that under these assumptions, $D$ is bipartite.

By Claim 1., we may assume that $D$ contains a cycle of length at least 4. Again, it can be proved inductively that $D$ contains a 4-cycle, $C$. One can directly verify that every arc in a 4-cycle of a 3-transitive digraph is a digon. Consider a 2-colouring of $C$ with colours black and white. If there are no more vertices in $D$, then we are done. Otherwise, let $v$ be a vertex of $D$ not in $C$. Since $D$ is 3-transitive and strong, then there is at least one arc from $v$ to $C$ and vice versa. Observe that both arcs join $v$ to only black or only white vertices, otherwise $D$ would not be bipartite. Suppose without loss of generality that $v$ is adjacent to a black vertex in $C$. We will recursively colour all the vertices of $D$ to obtain a bipartition such that every white vertex is adjacent through digons to every black vertex. Proposition 8.3.17 implies that there are digons between $v$ and every black vertex in $C$, so, colour $v$ white. Now, any four vertices of $D$ already coloured, two black and two white, induce a symmetric 4-cycle in $D$. Repeating the argument, it can be shown that every vertex of $D$ not already coloured is either adjacent through digons to every black vertex, and we colour it white, or to every white vertex, and we colour it black. □

Since the cases are exhaustive, the result now follows from Claims 1–3. □

Although more complicated than classifying strong transitive digraphs, strong 3-transitive digraphs are still easy to classify. Nonetheless, as the value of $k$ grows, this task becomes increasingly difficult. In fact, 4 is the largest value of $k$ such that strong $k$-transitive digraphs are characterized. Next, we reproduce the characterization theorem due to Hernández-Cruz found in [47]. The proof, although not very difficult, is lengthy and technical, so we omit it.

**Theorem 8.3.20** ([47]) *Let $D$ be a strong 4-transitive digraph. Then exactly one of the following possibilities holds.*

1. *$D$ is a complete digraph.*
2. *$D$ is a 3-cycle extension.*
3. *$D$ has circumference 3, a 3-cycle extension as a spanning subdigraph with cyclical partition $\{V_0, V_1, V_2\}$, at least one symmetrical arc exists in $D$ and for every symmetrical arc $v_i v_{i+1} \in A(D)$, with $v_j \in V_j$ for $j \in \{i, i+1\}$ (mod 3), $|V_i| = 1$ or $|V_{i+1}| = 1$.*
4. *$D$ has circumference 3, $UG(D)$ is not 2-edge-connected and $\{S_1, S_2, \ldots, S_r\}$ are the vertex sets of the maximal 2-edge connected subgraphs of $UG(D)$, with $S_i = \{u_i\}$ for every $2 \leq i \leq r$ and such that $D[S_1]$ has a 3-cycle extension with cyclical partition $\{V_0, V_1, V_2\}$ as a spanning subdigraph. A vertex $v_0 \in V_0$ (without loss of generality) exists such that $v_0 u_j, u_j v_0 \in A(D)$ for every $2 \leq j \leq n$. Also $|V_0| = 1$ and $D[S_1]$ has the structure described in 1. or 2., depending on the existence of symmetrical arcs.*
5. *A complete biorientation of a 5-cycle.*
6. *$D$ is a complete biorientation of the star $K_{1,r}$, $r \geq 3$.*
7. *$D$ is a complete biorientation of a tree with diameter 3.*
8. *$D$ is a strong digraph of order less than or equal to 4 not included in the previous families.*

For values of $k$ greater than 4, there are no known structural characterizations for strong $k$-transitive digraphs. As we have already mentioned above, this situation may be a consequence of the fact that every digraph on less than $k+1$ vertices, and every digraph without paths of length $k$, are $k$-transitive digraphs, so small $k$-transitive digraphs are difficult to characterize. In spite of this fact, it has been observed that the existence of some structures in a strong $k$-transitive digraph is enough to guarantee that the whole digraph will have a nice structure. Hernández-Cruz and Montellano-Ballesteros proved that $k$-transitive digraphs with cycles of length at least $k$ have a very nice structure. The proofs of the following theorems are several pages long, so they will be omitted; it would be a nice problem to find short proofs for both of them.

**Theorem 8.3.21** ([49]) *Let $k \geq 2$ be an integer, and let $D$ be a strong $k$-transitive digraph. Suppose that $D$ contains a cycle of length $r$ such that the g.c.d. of $r$ and $k-1$ is $d$, and $r \geq k+1$. Then the following hold.*

1. *If $d = 1$, then $D$ is a complete digraph.*
2. *If $d \geq 2$, then $D$ is either a complete digraph, a complete bipartite digraph, or a $d$-cycle extension.* □

**Theorem 8.3.22** ([49]) *Let $k \geq 2$ be an integer, and let $D$ be a strong $k$-transitive digraph of order at least $k+1$. If $D$ contains a cycle of length $k$, then $D$ is a complete digraph.* □

It follows from Theorems 8.3.21 and 8.3.22 that a strong $k$-transitive digraph is not likely to grow disorganizedly. On one hand, we have that every "sufficiently small" digraph is $k$-transitive. On the other hand, if a strong $k$-transitive digraph has a large enough circumference, its structure becomes very well determined. So a natural question arises: what happens if a strong $k$-transitive digraph has circumference less than $k$ but at least $k+1$ vertices? Is there a proportion between order and circumference which allows us to say something about the structure of a strong $k$-transitive digraph? Theorem 8.3.22 seems to be the most simple case of such a result. Following this idea, there is a partial result due to Wang.

**Theorem 8.3.23** ([59]) *Let $D$ be a strong $k$-quasi-transitive digraph with $k \geq 4$, and let $C$ be a cycle of length $k-1$. Then, for every $v \in V(D) \setminus V(C)$, the sets $(v, V(C))$ and $(V(C), v)$ are non-empty.* □

**Proof:** Since reversing every arc of a $k$-transitive digraph yields a $k$-transitive digraph, we only need to show $(v, V(C)) \neq \varnothing$. Let $C = v_0 \dots v_{k-2}v_0$ be a $(k-1)$-cycle. Since $D$ is strong, there exists a path from $v$ to $C$. Let $P = u_0 \dots u_s$ be a shortest path from $v$ to $C$, where $s \geq 1$, $u_0 = v$ and $u_s \in V(C)$. Without loss of generality, assume that $u_s = v_0$. We prove that $u_0$ dominates some vertex of $V(C)$ by induction on the length $s$ of $P$. It clearly holds for $s = 1$. Thus, we assume that $s \geq 2$. Note that $u_1 \dots u_s$ is a path of length $s-1$. By the induction hypothesis, there is a vertex $v_i \in V(C)$ such that $u_1 \rightarrow v_i$. Then $u_0u_1v_iCv_{i-1}$ is a path of length $k$ in $D$, which implies $u_0 \rightarrow v_{i-1}$. □

### 8.3.4 Totally $\Phi$-Decomposable Digraphs

The structure of totally $\Phi$-decomposable digraphs is already determined from its definition and the choice of $\Phi$. Thus, instead of studying their structure, we will show that for some choices of $\Phi$, totally $\Phi$-decomposable digraphs can be recognized in polynomial time.

As we will have already mentioned, Theorem 8.3.5 is the turning point on the study of quasi-transitive digraphs; it will let us construct polynomial algorithms for Hamiltonian paths and cycles in quasi-transitive digraphs, and also solve more general problems in this class of digraphs. This theorem shows that quasi-transitive digraphs are totally $\Phi$-decomposable, where $\Phi$ is the union of extended semicomplete and transitive digraphs. Since both extended semicomplete digraphs and transitive digraphs are special subclasses of much wider classes of digraphs, it is natural to study totally $\Phi$-decomposable digraphs, where $\Phi$ is a much more general class of digraphs than the union of extended semicomplete and transitive digraphs. However, our choice of candidates for the class $\Phi$ should be restricted in such a way that we can still construct polynomial algorithms for some important problems such as the Hamiltonian cycle problem, using properties of digraphs in $\Phi$.

This idea was first used by Bang-Jensen and Gutin [13] to introduce the following classes of digraphs:

**Definition 8.3.24**

- $\Phi_0$ *is the union of all semicomplete multipartite digraphs, all connected extended locally semicomplete digraphs and all acyclic digraphs,*
- $\Phi_1$ *is the union of all semicomplete bipartite digraphs, all connected extended locally semicomplete digraphs and all acyclic digraphs,*
- $\Phi_2$ *is the union of all connected extended locally semicomplete digraphs and all acyclic digraphs, and*
- $\Phi_3$ *is the union of all semicomplete digraphs and all acyclic digraphs.*

Note that we have $\Phi_3 \subset \Phi_2 \subset \Phi_1 \subset \Phi_0$ and that all four classes are closed under taking extensions.

A class $\Phi$ of digraphs is **hereditary** if $D \in \Phi$ implies that every induced subdigraph of $D$ is in $\Phi$. Observe that every $\Phi_i$, $0 \leq i \leq 3$, is a hereditary class. The following results are due to Bang-Jensen and Gutin.

**Lemma 8.3.25** ([13]) *Let $\Phi$ be a hereditary class of digraphs. If a given digraph $D$ is totally $\Phi$-decomposable, then every induced subdigraph $D'$ of $D$ is totally $\Phi$-decomposable. In other words, total $\Phi$-decomposability is a hereditary property.*

**Proof:** By induction on the number of vertices of $D$. The claim is obviously true if $D$ has fewer than 3 vertices.

If $D \in \Phi$, then our claim follows from the fact that $\Phi$ is hereditary. So, we may assume that $D = R[H_1, \ldots, H_r]$, $r \geq 2$, where $R \in \Phi$ and each of $H_1, \ldots, H_r$ is totally $\Phi$-decomposable.

Let $D'$ be an induced subdigraph of $D$. If there is an index $i$ such that $V(D') \subseteq V(H_i)$, then $D'$ is totally $\Phi$-decomposable by induction. Otherwise, $D' = R'[T_1, \ldots, T_{r'}]$, where $r \geq 2$ and $R' \in \Phi$, is the subdigraph of $R$ induced by those vertices $i$ of $R$, whose $H_i$ has a non-empty intersection with $V(D')$ and the $T_j$'s are the corresponding $H_i$'s restricted to the vertices of $D'$. Observe that $R' \in \Phi$, since $\Phi$ is hereditary. Moreover, by induction, each $T_j$ is totally $\Phi$-decomposable, hence so is $D'$. □

The following result gives a polynomial time algorithm for verifying $\Phi_i$-decomposability, $i \in \{0, 1, 2, 3\}$. Its proof can be found in [9].

**Lemma 8.3.26** ([13]) *There exists an $O(mn + n^2)$-algorithm for checking if a digraph $D$ with $n$ vertices and $m$ arcs has a decomposition $D = R[H_1, \ldots, H_r]$, $r \geq 2$, where $H_i$ is an arbitrary digraph and the digraph $R_i$ is either acyclic or semicomplete multipartite or semicomplete bipartite or connected extended locally semicomplete.* □

The previous lemma can now be used to obtain the main result of this section. Again, its proof can be found on [9].

**Theorem 8.3.27** ([13]) *There exists an $O(n^2m + n^3)$-algorithm for checking if a digraph with $n$ vertices and $m$ arcs is totally $\Phi_i$-decomposable, for $i \in \{0, 1, 2, 3\}$.*

## 8.4 Hamiltonian, Longest and Vertex-Cheapest Paths and Cycles

In this section we will study the Hamiltonian path and cycle problems, as well as some problems in weighted digraphs generalizing them. The subsections on quasi-transitive digraphs and totally $\Phi$-decomposable digraphs in this section are strongly based on Sections 6.7 and 6.8 of [9], where this subject has received a full treatment. We begin by considering the few existing results for $k$-transitive and $k$-quasi-transitive digraphs.

### 8.4.1 $k$-Transitive and $k$-Quasi-Transitive Digraphs

Since strong 3-transitive and 4-transitive digraphs are completely characterized, it suffices to make a case by case analysis for these families of digraphs (using Theorems 8.3.19 and 8.3.20) to completely characterize Hamiltonian 3- and 4-transitive digraphs. This analysis can be summarized in the following result. We say that a $k$-cycle extension $D = C_k[S_1, \ldots, S_k]$ is **balanced** if $|S_i| = |S_j|$ for every $i \neq j$, and non-balanced, otherwise.

**Theorem 8.4.1** *If $D$ is a strong 3-transitive digraph, then $D$ is Hamiltonian if and only if it is not a complete bipartite digraph $D = (X, Y)$ with $|X| \neq |Y|$.*

*If $D$ is a strong 4-transitive digraph, then $D$ is Hamiltonian if and only if it is a complete digraph, a balanced 3-cycle extension, a symmetrical 5-cycle, or a semicomplete digraph on at most 4 vertices.* □

It follows from Theorem 8.4.1 that hamiltonicity for 3-transitive and 4-transitive digraphs can be determined in linear time: Hamiltonian members of these families can be easily recognized through their in-degree and out-degree sequences. In view of this fact, the following problem is proposed.

**Problem 8.4.2** *For all values of $k \geq 5$, determine the complexity of the Hamiltonian cycle problem for the class of $k$-transitive digraphs.*

Considering the results for $k \in \{2, 3, 4\}$, it does not seem too adventurous to conjecture that hamiltonicity of a $k$-transitive digraph could be determined in linear time for every integer $k \geq 2$. From Theorems 8.3.21 and 8.3.22, easy to verify sufficient conditions for the existence of a Hamiltonian cycle in a strong $k$-transitive digraph can be derived: A $k$-transitive digraph containing a cycle of length at least $k$ is Hamiltonian unless it is a non-balanced extended cycle.

For 3-quasi-transitive digraphs, Theorem 8.3.15 also provides enough information to completely characterize Hamiltonian members of this family.

**Theorem 8.4.3** *If $D$ is a strong 3-quasi-transitive digraph, then $D$ is Hamiltonian if and only if one of the following hold:*

- *$D$ is semicomplete,*
- *$D$ is semicomplete bipartite with a cycle factor, or*
- *$D$ is the member of the family $\mathcal{F}$ of order 4 (see Figure 8.3).*

**Proof:** Clearly, all the digraphs mentioned in the statement of the theorem are Hamiltonian. Using Theorem 8.3.15 we can rule out the remaining cases for a strong 3-quasi-transitive digraph.

We know that a strong semicomplete bipartite digraph is Hamiltonian if and only if it has a cycle factor (see Theorem 7.4.1), and clearly, every digraph in $\mathcal{F}$ of order greater than 4 is not Hamiltonian. □

It follows from Theorems 8.4.3 and 7.4.1 that hamiltonicity can be verified for 3-quasi-transitive digraphs in time $O(n^{2.5})$. So, the following question comes to mind.

**Problem 8.4.4** *Let $k$ be an integer, $k \geq 4$. Is it true that hamiltonicity can be determined for the class of $k$-quasi-transitive digraphs in polynomial time?*

Regarding Hamiltonian paths, Wang and Zhang gave a sufficient condition for traceability when $k$ is even, [62].

**Theorem 8.4.5** ([62]) *Let $k$ be an even integer with $k \geq 4$ and $D$ be a strong $k$-quasi-transitive digraph. If* $\mathrm{diam}(D) \geq k+2$, *then $D$ has a Hamiltonian path.*

**Proof:** Since $\mathrm{diam}(D) \geq k+2$, there exist $u, v \in V(D)$ such that $d(u,v) = k+2$. Let $P = x_0 \dots x_{k+2}$ be a shortest $(u,v)$-path where $u = x_0$ and $v = x_{k+2}$. By Lemma 8.3.6, $x_{k+2} \rightarrow x_0$. Let $C$ be the cycle $C = x_0 \dots x_{k+2} x_0$ and $H = D[V(D) - V(C)]$. By Proposition 8.3.7 and Theorem 8.3.12, $D[V(C)]$ and $H$ are both semicomplete digraphs. It is well known that there is a Hamiltonian path in every semicomplete digraph. Let $Q = y_0 \dots y_p$ be a Hamiltonian path in $H$. By Lemma 8.3.9, for any $y_i \in V(Q)$, $y_i$ is adjacent to $C$. If there exists an $x_j \in V(C)$ such that $x_j \rightarrow y_0$, then $x_{j+1} C x_j y_0 Q$ is a Hamiltonian path in $D$. Now assume $(V(C), y_0) = \varnothing$. Note that $k-1$ and $k+3$ are coprime.[3] According to Proposition 8.3.10, $y_0 \mapsto V(C)$, and thus, either there is an $x_j \in V(C)$ such that $x_j \rightarrow y_1$ and therefore $y_0 x_{j+1} C x_j y_1 Q$ is a Hamiltonian path in $D$, or $y_1 \mapsto V(C)$. Continuing in this way, we can conclude that either $D$ has a Hamiltonian path, or $V(H) \mapsto V(C)$. But since $D$ is strong, $(V(C), V(H)) \neq \varnothing$. So $D$ has a Hamiltonian path. □

Notice that, since every complete bipartite digraph is $k$-quasi-transitive for any odd integer $k \geq 3$, it is not possible to obtain a result similar to Theorem 8.4.5 for odd values of $k$.

---

[3] Recall that the g.c.d. of two integers is their least positive linear combination. Clearly, 4 is a linear combination of $k-1$ and $k+3$, but since $k$ is even, and $k-1 \not\equiv k+3 \pmod 3$, the least positive linear combination of $k-1$ and $k+3$ is 1.

The following question was proposed by the same authors.

**Problem 8.4.6** ([62]) *Let $k$ be an even integer, $k \geq 4$, and suppose that $diam(D) \geq k+2$. Is there a Hamiltonian cycle in $D$?*

It follows from the remark after Theorem 8.3.12 that Problem 8.4.6 has a positive answer for $k = 4$.

### 8.4.2 Hamiltonian Cycles in quasi-transitive digraphs and Totally $\Phi$-Decomposable Digraphs

Hamiltonicity is one of the most studied topics in both graphs and digraphs. Having a family as nice as quasi-transitive digraphs, it is natural to have a lot of results for this class regarding both Hamiltonian paths and cycles, many of which come from the study of semicomplete digraphs and its corresponding hamiltonicity results. Since Chapter 2 is devoted to tournaments and semicomplete digraphs, we will not elaborate on the results regarding these digraph classes, but we will restate some of them.

As mentioned in the introduction to this chapter, totally $\Phi$-decomposable digraphs generalize the structure of quasi-transitive digraphs. Thus, it is common to find that the techniques used to prove certain results for quasi-transitive digraphs can be adapted to study this more general family of digraphs. In particular, the methods developed in [17] by Bang-Jensen and Huang, and in [41] by Gutin, to characterize Hamiltonian and traceable quasi-transitive digraphs as well as to construct polynomial algorithms for verifying the existence of Hamilton paths and cycles in quasi-transitive digraphs, can be easily generalized to much wider classes of digraphs [11]. Thus, in this subsection, along with quasi-transitive digraphs, we consider totally $\Phi$-decomposable digraphs for various families $\Phi$ of digraphs.

Recall that a digraph $D$ is an **extended semicomplete digraph** if it can be obtained from some semicomplete digraph $S$ by substituting independent sets for the vertices of $S$.

Recall that the decompositions given by Theorem 8.3.5 are called canonical decompositions. The following characterization of Hamiltonian quasi-transitive digraphs is due to Bang-Jensen and Huang [17].

**Theorem 8.4.7** ([17]) *A strong quasi-transitive digraph $D$ with canonical decomposition $D = S[Q_1, Q_2, \ldots, Q_s]$ is Hamiltonian if and only if it has a cycle factor $\mathcal{F}$ such that no cycle of $\mathcal{F}$ is a cycle of some $Q_i$.*

**Proof:** Clearly, a Hamilton cycle in $D$ crosses every $Q_i$. Thus, it suffices to show that if $D$ has a cycle factor $\mathcal{F}$ such that no cycle of $\mathcal{F}$ is a cycle of some $Q_i$, then $D$ is Hamiltonian. Observe that $V(Q_i) \cap \mathcal{F}$ is a path factor $\mathcal{F}_i$ of $Q_i$ for every $i \in [s]$. For every $i \in [s]$, delete the arcs between end-vertices of all paths in $\mathcal{F}_i$ except for the paths themselves, and then perform the operation of path-contraction for all paths in $\mathcal{F}_i$. As a result, one obtains an extended semicomplete digraph $S'$ (since $S$ is semicomplete). Clearly, $S'$ is strong and has a cycle factor. Hence, by Theorem 7.10.1, $S'$ has a Hamilton cycle $C$. After replacing every vertex of $S'$ with the corresponding path from $\mathcal{F}$, we obtain a Hamilton cycle in $D$. □

Similarly to Theorem 8.4.7, one can prove the following characterization of traceable quasi-transitive digraphs. This result is also due to Bang-Jensen and Huang.

**Theorem 8.4.8** ([17]) *A quasi-transitive digraph $D$ with at least two vertices and with canonical decomposition $D = R[G_1, G_2, \ldots, G_r]$ is traceable if and only if it has a 1-path-cycle factor $\mathcal{F}$ such that no cycle or path of $\mathcal{F}$ is completely in some $D[V(G_i)]$.* □

Theorems 8.4.7 and 8.4.8 do not imply polynomial algorithms to verify hamiltonicity and traceability, respectively. The following characterization of Hamiltonian quasi-transitive digraphs is given implicitly in the paper [41] by Gutin:

**Theorem 8.4.9** (Gutin [41]) *Let $D$ be a strong quasi-transitive digraph with canonical decomposition $D = S[Q_1, Q_2, \ldots, Q_s]$. Let $n_1, \ldots, n_s$ be the orders of the digraphs $Q_1, Q_2, \ldots, Q_s$, respectively. Then $D$ is Hamiltonian if and only if the extended semicomplete digraph $S' = S[\overline{K}_{n_1}, \overline{K}_{n_2}, \ldots, \overline{K}_{n_s}]$ has a cycle subdigraph which covers at least $\mathrm{pc}(Q_j)$ vertices of $\overline{K}_{n_j}$ for every $j \in [s]$.*

**Proof:** Suppose that $D$ has a Hamilton cycle $H$. For every $j \in [s]$, $V(Q_j) \cap H$ is a $k_j$-path factor $\mathcal{F}_j$ of $Q_j$. By the definition of the path covering number, we have $k_j \geq \mathrm{pc}(Q_j)$. For every $j \in [s]$, the deletion of the arcs between end-vertices of all paths in $\mathcal{F}_j$ except for the paths themselves, and then path-contraction of all paths in $\mathcal{F}_j$, transforms $H$ into a cycle of $S'$ having at least $\mathrm{pc}(Q_j)$ vertices of $\overline{K}_{n_j}$ for every $j \in [s]$.

Suppose now that $S'$ has a cycle subdigraph $\mathcal{L}$ containing $p_j \geq \mathrm{pc}(Q_j)$ vertices of $\overline{K}_{n_j}$ for every $j \in [s]$. Since $S'$ is a strong extended semicomplete digraph, by Theorem 7.10.2, $S'$ has a cycle $C$ such that $V(C) = V(\mathcal{L})$. Clearly, every $Q_j$ has a $p_j$-path factor $\mathcal{F}_j$. Replacing, for every $j \in [s]$, the $p_j$ vertices of $\overline{K}_{n_j}$ in $C$ with the paths of $\mathcal{F}_j$, we obtain a Hamiltonian cycle in $D$. □

Theorem 8.4.9 can be used to show that the Hamilton cycle problem for quasi-transitive digraphs is polynomial time solvable.

**Theorem 8.4.10 (Gutin** [41]) *There is an $O(n^4)$ algorithm which, given a quasi-transitive digraph $D$, either returns a Hamiltonian cycle in $D$ or verifies that no such cycle exists.* ◇

The approach used in the proofs of Theorems 8.4.9 and 8.4.10 in [41] can be generalized to a much wider class of digraphs, as was observed by Bang-Jensen and Gutin [11]. We follow the main ideas of [11].

Recall the definition of $\Phi_0, \Phi_1, \Phi_2, \Phi_3$ in Definition 8.3.24 and the fact that, for each of these classes, in time $O(n^4)$, one can check if a given digraph $D$ is totally $\Phi_i$-decomposable ($i \in \{0,1,2,3\}$) and (in case it is so) construct a total decomposition of $D$. Moreover, Theorem 8.3.5 implies that every quasi-transitive digraph is totally $\Phi_3$-decomposable.

**Theorem 8.4.11** *Let $\Phi$ be an extension-closed class of digraphs, i.e., $\Phi^{ext} = \Phi$, including the trivial digraph $\overline{K}_1$ on one vertex. Suppose that for every digraph $H \in \Phi$ we have* $\mathrm{pcc}(H) = \mathrm{pc}(H)$. *Let $D$ be a totally $\Phi$-decomposable digraph. Then, given a total $\Phi$-decomposition of $D$, the path covering number of $D$ can be calculated and a minimum path factor found in time $O(n^4)$.*

**Proof:** We prove this theorem by induction on $n$. For $n = 1$ the claim is trivial.

Let $D$ be a totally $\Phi$-decomposable digraph and let $D = R[H_1, \dots, H_r]$ be a $\Phi$-decomposition of $D$ such that $R \in \Phi$, $r = |V(R)|$ and every $H_i$ (of order $n_i$) is totally $\Phi$-decomposable. A $\mathrm{pc}(D)$-path factor of $D$ restricted to every $H_i$ corresponds to a disjoint collection of some $p_i$ paths covering $V(H_i)$. Hence, we have $\mathrm{pc}(H_i) \le p_i \le n_i$. Therefore, arguing similarly to the proof of Theorem 8.4.9, we obtain

$$\mathrm{pc}(D) = \min\{\mathrm{pc}(R[\overline{K}_{p_1}, \dots, \overline{K}_{p_r}]) :\ \mathrm{pc}(H_i) \le p_i \le n_i,\ i \in [r]\}.$$

Since $\Phi$ is extension-closed, and since, for every digraph $Q \in \Phi$, $\mathrm{pc}(Q) = \mathrm{pcc}(Q)$, we obtain

$$\mathrm{pc}(D) = \min\{\mathrm{pcc}(R[\overline{K}_{p_1}, \dots, \overline{K}_{p_r}]) :\ \mathrm{pc}(H_i) \le p_i \le n_i,\ i \in [r]\}. \tag{8.1}$$

Given the lower and upper bounds $\mathrm{pc}(H_i)$ and $n_i$ ($i \in [r]$), the recursive formula (8.1) allows us to find $\mathrm{pc}(D)$ in time $O(n^3)$. To show this, it suffices to demonstrate how to find, in time $O(n^3)$, the minimum in formula 8.1 given all the values of $pc(H_i)$ (and $n_i$). Construct a network $N_R$ containing the digraph $R$ and two additional vertices (source and sink) $s$ and $t$ such that $s$ and $t$ are adjacent to every vertex of $V(R)$ and the arcs between $s$ ($t$, respectively) and $R$ are oriented from $s$ to $R$ (from $R$ to $t$, respectively). Associate with each vertex $v_i$ of $R$ (corresponding to $H_i$ in $D$) the lower and upper bounds $pc(H_i)$ and $n_i$ ($1 \le i \le r$) on the amount of flow that can pass through $v_i$. It is not difficult to see that the minimum value, $m$, of a feasible flow from $s$ to $t$ in $N_R$, is related to the minimum in 8.1, i.e. $pc(D)$, as follows: $pc(D) = \max\{1, m\}$ (for further details, see [41]).

Let $T(n)$ be the time needed to find the path covering number of a totally $\Phi$-decomposable digraph of order $n$. Then, by (8.1),

$$T(n) = O(n^3) + \sum_{i=1}^{r} T(n_i).$$

Furthermore, $T(1) = O(1)$. Hence $T(n) = O(n^4)$. □

As we know, $\mathrm{pc}(D) = \mathrm{pcc}(D)$ for every semicomplete multipartite digraph $D$ (see Theorem 7.5.2), for every extended locally semicomplete digraph $D$ (by Theorem 5.8.1 in [8]) and every acyclic digraph $D$ (which is trivial). Therefore, Theorems 8.4.11 and 8.3.27 imply the following theorem of Bang-Jensen and Gutin:

**Theorem 8.4.12** ([12]) *The path covering number can be calculated in time $O(n^4)$ for digraphs that are totally $\Phi_0$-decomposable.* □

**Corollary 8.4.13** ([12]) *One can verify whether a totally $\Phi_1$-decomposable digraph is Hamiltonian in time $O(n^4)$.*

**Proof:** Let $D = R[H_1, \ldots, H_r]$, $r = |R|$, be a decomposition of a strong digraph $D$ ($r \geq 2$). Then, $D$ is Hamiltonian if and only if the following family $\mathcal{S}$ of digraphs contains a Hamiltonian digraph:

$$\mathcal{S} = \{R[\overline{K}_{p_1}, \ldots, \overline{K}_{p_r}] : \ \mathrm{pc}(H_i) \leq p_i \leq |V(H_i)|, \ i \in [r]\}.$$

Now suppose that $D$ is a totally $\Phi_1$-decomposable digraph. Then, every digraph of the form $R[\overline{K}_{p_1}, \ldots, \overline{K}_{p_r}]$ is in $\Phi_1$. We know (see Theorem 7.4.1 and Theorem 5.8.1 in [8]) that every digraph in $\Phi_1$ is Hamiltonian if and only if it is strong and contains a cycle factor. Thus, all we need is to verify whether there is a digraph in $\mathcal{S}$ containing a cycle factor. It is easily seen that there is a digraph in $\mathcal{S}$ containing a cycle factor if and only if there is a circulation in the network formed from $R$ by adding lower bounds $\mathrm{pc}(H_i)$ and upper bounds $|V(H_i)|$ to the vertex $v_i$ of $R$ for every $i \in [r]$. Since the lower bounds can be found in time $O(n^4)$ (see Theorem 8.4.11) and the existence of a circulation checked in time $O(n^3)$ (a feasible circulation, if one exists, can be found by just one max flow calculation in an $(s,t)$-flow network obtained from our network, see [9, Exercise 4.31]), we obtain the required complexity $O(n^4)$. □

Since every quasi-transitive digraph is totally $\Phi_1$-decomposable this theorem immediately implies Theorem 8.4.10. Note that the minimum path factors in Theorem 8.4.11 can be found in time $O(n^4)$. Also, a Hamiltonian cycle in a Hamiltonian totally $\Phi_1$-decomposable digraph can be constructed in time $O(n^4)$.

### 8.4.3 Vertex-Cheapest Paths and Cycles

For the remainder of this section, we consider problems that generalize the Hamilton path and cycle problems in a significant way. We prove that the problems of finding vertex-cheapest paths and cycles in vertex-weighted quasi-transitive digraphs are polynomial time solvable. The values of the weights can be any reals, positive or negative. Thus, we can conclude that the longest and shortest path and cycle problems for quasi-transitive digraphs are polynomial time solvable. The same result holds for acyclic digraphs as the only non-trivial problem from the above four is the longest path problem and it is well-known that it can be solved in polynomial time (see e.g. [9, Theorem 3.3.5]). Notice that for the quasi-transitive digraphs three of the above four problems are non-trivial (the shortest and longest cycles and longest path) and, in fact, much more difficult than the longest path problem for acyclic digraphs as the reader can see in the rest of this subsection. It appears that the problems are non-trivial even for semicomplete digraphs. Theorems 7.10.4 and 6.17.16 were proved by Bang-Jensen, Gutin and Yeo for extended semicomplete and locally semicomplete digraphs.

The approach described in the previous subsection seems too weak to allow us to construct polynomial time algorithms for vertex-cheapest paths and cycles in quasi-transitive digraphs. A more powerful method that leads to such algorithms was first suggested by Bang-Jensen, Gutin and Yeo [15] and, in the rest of this section, we describe this method.

Recall that the cost of a subset of vertices is the sum of the costs of its vertices and the cost of a subdigraph is the sum of the costs of its vertices. For a digraph $D$ of order $n$ and $i \in [n]$ we define $mp_i(D)$ ($mpc_i(D)$) to be the minimum cost of an $i$-path ($i$-path-cycle) subdigraph of $D$. We set $mp_0(D) = 0$ and $mpc_0(D)$ is zero if $D$ has no negative cycle and otherwise it is the minimum cost of a cycle subdigraph in $D$ which can be found using minimum cost flows. Note that $mp_0(D)$ and $mpc_0(D)$ always exist as we may take single vertices as paths and we always have $mpc_i(D) \leq mp_i(D)$. For any digraph $D$ with at least one cycle we denote by $mc(D)$ the minimum cost of a cycle in $D$.

Let $D = (V, A)$ be a digraph and let $X$ be a non-empty subset of $V$. We say that a cycle $C$ in $D$ is an $X$-**cycle** if $C$ contains all vertices of $X$. In the remaining subsections, we consider the following problems for a digraph $D = (V, A)$ with $n$ vertices and real-valued costs on the vertices:

(P1) Determine $mp_i(D)$ for all $i \in [n]$.
(P2) Find a cheapest cycle in $D$ or determine that $D$ has no cycle.

Clearly, problems (P1) and (P2) are $\mathcal{NP}$-hard as determining the numbers $mp_1(D)$ and $mc(D)$ generalize the Hamiltonian path and cycle problems (assign cost $-1$ to each vertex of $D$). The problem (P2) can be solved in time $O(n^3)$ when all costs are non-negative using an all pairs shortest path calculation. The problems (P1) and (P2) were solved in [14] for the special case when all costs are non-negative. However, the approach of [14] cannot be used or modified to work with negative costs. Bang-Jensen, Gutin and Yeo [15] managed to obtain an approach suitable for arbitrary real costs.

### 8.4.4 Minimum Cost $k$-Path-Cycle Subdigraphs

Although this chapter is intended to be almost self-contained, in order to present the main results of this subsection, we need certain notions and results on network flows. We refer the reader to Section 1.9 of this book for basic terminology, and to chapter 4 of [9] for the proofs of the results we will state. As in the aforementioned chapter of [9], we will allow capacities and costs on the vertices in our networks. This makes it easier to model certain problems for digraphs and it is easy to transform such a network into one where all capacities and costs are on the arcs (see Subsection 4.2.4 of [9] for details). With these remarks in mind, the following lemma of Bang-Jensen, Gutin and Yeo follows directly from Lemma 4.2.4 and Proposition 4.10.7 in [9].

**Lemma 8.4.14** ([15]) *Let $N = (V, A)$ be a network with source $s$ and sink $t$, capacities on arcs and vertices and a real-valued cost $c(v)$ for each vertex $v \in V$. For all integers $i$ such that there exists a feasible $(s,t)$-flow of value $i$ in $N$, let $f_i$ be a minimum cost $(s,t)$-flow in $N$ of value $i$ and let $c(f_i)$ be the cost of $f_i$. Then, for all $i$ where all of $f_{i-1}, f_i, f_{i+1}$ exist, we have*

$$c(f_{i+1}) - c(f_i) \geq c(f_i) - c(f_{i-1}). \tag{8.2}$$

◇

Recall that a cycle subdigraph of a digraph $D$ is a collection of vertex-disjoint cycles of $D$. The following two results are also due to Bang-Jensen, Gutin and Yeo.

**Lemma 8.4.15** ([15]) *Let $D = (V, A)$ be a digraph with real-valued cost function $c$ on the vertices. In time $O(n(m+n \log n))$ we can determine the number $mpc_0(D)$ and find a cycle subdigraph of cost $mpc_0(D)$ if $mpc_0(D) < 0$.*

**Proof:** Let $H(w)$ be the digraph on 4 vertices $w_1, w_2, w_3, w_4$ and the following arcs $w_1w_2, w_2w_1, w_2w_3, w_3w_4, w_4w_3$. Let $D^* = (V^*, A^*)$ be obtained from $D$ as follows: replace every vertex $v$ by the digraph $H(v)$. Furthermore, for every original arc $uv \in A$, $D^*$ contains the arc $u_4v_1$. There are no costs on the vertices and all arcs have cost 0 except the arcs of the form $v_2v_3$ which have cost $c(v)$. Observe that $mpc_0(D)$ is precisely the minimum cost of a spanning cycle subdigraph in $D^*$. Let $V^* = \{x_1, x_2, \ldots, x_{4n}\}$. Construct a bipartite graph $B$ with partite sets $L = \{\ell_1, \ldots, \ell_{4n}\}$ and $R = \{r_1, \ldots, r_{4n}\}$, in which $\ell_i r_j$ is an edge if and only if $x_i x_j \in A^*$. Moreover, the cost of $\ell_i r_j$ is equal to the cost of $x_i x_j$. Observe that a minimum cost perfect matching in $B$ corresponds to a minimum cost cycle subdigraph in $D^*$. We can find a minimum cost perfect matching in $B$ in time $O(n(m + n \log n))$, see the remark after the proof of Theorem 11.1 in [51]. Using the transformation from $B$ to $D^*$, we can compute the minimum cost of a spanning cycle subdigraph $F$ in $D^*$ in time $O(n(m + n \log n))$. If this cost is negative, we can find a minimum cost cycle subdigraph of $D$ within the same time. □

**Lemma 8.4.16** ([15]) *Let $D = (V, A)$ be a vertex-weighted digraph.*

*(a) In total time $O(n^2m + n^3)$ we can determine the numbers $\{mpc_1(D), mpc_2(D), \ldots, mpc_n(D)\}$ and find $j$-path-cycle subdigraphs $F_j$, $j \in \{1, 2, \ldots, n\}$, where $F_j$ has cost $mpc_j(D)$.*
*(b) The costs $mpc_i(D)$ satisfy the following inequality for every $i \in [n-1]$:*

$$mpc_{i+1}(D) - mpc_i(D) \geq mpc_i(D) - mpc_{i-1}(D). \tag{8.3}$$

**Proof:** Form a network $N(D)$ from $D$ by adding a pair $s, t$ of new vertices along with arcs $\{(s, v), (v, t) : v \in V\}$. Let all vertices and all arcs of $D$ have lower bound 0 and capacity 1. Let $c(s) = c(t) = 0$, let each other vertex of $N(D)$ inherit its cost from $D$ and let all arcs have cost 0.

Suppose $F_j$ is a $j$-path-cycle subdigraph of $D$. Using $F_j$ we can obtain a feasible flow $f_j$ of value $j$ in $N(D)$ if we assign $f_j(a) = 1$ to all arcs $a$ in $F_j$ and those arcs $a$ of $N(D)$ that start (terminate) at $s$ ($t$) and terminate (start) at the initial (terminal) vertex of a path in $F_j$, and $f_j(a) = 0$ for all other arcs of $N(D)$. Similarly, we can transform a feasible integer-valued $(s, t)$-flow of value $j$ in $N(D)$ into a $j$-path-cycle subdigraph of $D$ (see Theorem 4.3.1 in [9]).

Notice that $N(D)$ has a feasible integer-valued $(s, t)$-flow of value $k$ for any integer $k \in \{0, 1, \ldots, n\}$. Thus it follows from the observations above that for every $j \in \{0, 1, \ldots, n\}$ the value $mpc_j(D)$ is exactly the minimum cost of a flow of value $j$ in $N(D)$. Now (8.2) implies that the inequality (8.3) is valid.

It remains to prove (a). It follows from Lemma 8.4.15 that we can find a minimum cost flow $f$ of value 0 in time $O(n^3)$. Now we can use the Buildup algorithm from Subsection 4.10.2 in [9] starting from $f$. Using the Buildup algorithm we can find feasible integer-valued flows $f_j$ for all $j \in [n]$, such that $f_j$ is a minimum cost feasible $(s, t)$-flow of value $j$ in $N(D)$, in total time $O(n^2m)$ (the complexity of obtaining $f_{j+1}$ starting from $f_j$ is $O(nm)$). This proves (a). □

### 8.4.5 Cheapest *i*-Path Subdigraphs in Quasi-Transitive Digraphs

Theorem 7.5.4, regarding semicomplete multipartite digraphs, will play an important role in our algorithms. The next theorem due to Bang-Jensen, Gutin and Yeo shows that (P1) is polynomially solvable for quasi-transitive digraphs.

**Theorem 8.4.17** ([15]) *Let $D = (V, A)$ be a vertex-weighted quasi-transitive digraph. Then the following holds:*

*(a) In total time $O(n^2m + n^3)$ we can find for every $i \in [n]$, the value of $mp_i(D)$ and an $i$-path subdigraph $F_i$ of cost $mp_i(D)$.*

*(b) For all $i \in [n-1]$ we have*

$$mp_{i+1}(D) - mp_i(D) \geq mp_i(D) - mp_{i-1}(D). \tag{8.4}$$

**Proof:** We prove (b) by induction on $n$. The statement vacuously holds for $n = 1$ and is easy to verify for $n = 2$ (recall that, by definition, $mp_0(D) = 0$). This proves the basis of induction and we now assume that $n \geq 3$.

By Theorem 8.3.5, $D$ has a decomposition $D = T[Q_1, \dots, Q_t]$, $t = |T| \geq 2$, where $T$ is an acyclic digraph or a semicomplete digraph. Let $D' = T[\overline{K}_{n_1}, \dots, \overline{K}_{n_t}]$ be obtained from $D$ by deleting all arcs inside each $Q_i$, $i \in [t]$. Assign costs to the vertices $v_1^k, \dots, v_{n_k}^k$ of $\overline{K}_{n_k}$, as follows:

$$c'(v_j^k) = mp_j(Q_k) - mp_{j-1}(Q_k).$$

By the induction hypothesis (b) holds for $Q_k$ implying that we have

$$c'(v_j^k) \leq c'(v_{j+1}^k) \text{ for every } j \geq 1. \tag{8.5}$$

Let $F_i'$ be an $i$-path-cycle subdigraph of $D'$. If $T$ is acyclic, then $D'$ is acyclic and, thus, $F_i'$ is an $i$-path subdigraph of $D'$. If $T$ is semicomplete, then $D'$ is extended semicomplete and, thus, by Theorem 7.5.1 and Theorem 7.5.4, we may assume that $F_i'$ is an $i$-path subdigraph of $D'$. Hence, $mp_i(D') = mpc_i(D')$ and it follows from Lemma 8.4.16(b) that (8.4) holds for $D'$. Thus it suffices to prove that $mp_i(D) = mp_i(D')$.

Let $F_i'$ be an $i$-path subdigraph of $D'$ and let $p_k$ denote the number of vertices from $\overline{K}_{n_k}$ which are covered by $F_i'$. Since all vertices of $\overline{K}_{n_k}$ are similar it follows from (8.5) that we may assume (by making the proper replacements if necessary) that $F_i'$ includes $v_1^k, \dots, v_{p_k}^k$. For each $k$, replace the vertices $v_1^k, \dots, v_{p_k}^k$ in $F_i'$ by a $p_k$-path subdigraph of $Q_k$ with cost $mp_{p_k}(Q_k) = \sum_{i=1}^{p_k} c'(v_i^k)$. As a result, we obtain, from $F_i'$, an $i$-path subdigraph $F_i$ of $D$ for which we have $c'(F_i') = \sum_{k=1}^{t} mp_{p_k}(Q_k) = c(F_i)$ and, thus, $c(F_i) = c'(F_i')$. Reversing the process above it is easy to get, from an $i$-path subdigraph of $D$, an $i$-path subdigraph $F_i'$ of $D'$ such that $c(F_i) = c'(F_i')$. This shows that $mp_i(D) = mp_i(D')$ and hence (8.4) holds for $D$ by the remark above.

We prove the complexity by induction on $n$. Let $m'$ be the number of arcs in $D'$ and recall that all these arcs are also in $D$. Clearly when a digraph $H$ has $|V(H)| \leq 2$ we can choose a constant $c_1$ so that we can determine the numbers $mp_i(H)$, $i = 1, 2, \dots, |V(H)|$, in time at most $c_1|V(H)|^2(|A(H)| + |V(H)|)$. Now assume by induction that for each $Q_i$ we can determine the desired numbers inside $Q_i$ in time at most $c_1 n_i^2(m_i + n_i)$. This means that we can find the numbers $mp_i(Q_j)$ for all $j \in [t]$ and $i \in [n_j]$ in total time

$$\sum_{j=1}^{t} c_1 n_j^2(m_j + n_j) \leq c_1 n^2 \sum_{j=1}^{t}(m_j + n_j) = c_1 n^2(m - m' + n).$$

By Lemma 8.4.16(a), Theorems 7.5.1 and 7.5.4, there is a constant $c_2$ such that in total time at most $c_2n^2(m'+n)$ we can find, for every $j \in [n]$, a $j$-path-cycle subdigraph of cost $mp_j(D')$ in $D'$. It follows from the way we construct $F_i$ above from $F'_i$ that if we are given for each $k \in [t]$ and each $1 \le j \le n_k$ a $j$-path subdigraph in $Q_k$ of cost $mp_j(Q_k)$, then we can construct all the path subdigraphs $F_r$, $1 \le r \le n$, in time at most $c_3n^3$ for some constant $c_3$. Hence the total time needed by the algorithm is at most

$$c_1n^2(m-m'+n)+c_2n^2(m'+n)+c_3n^3 =$$
$$c_1n^2(m+n)+(c_2-c_1)n^2m'+(c_2+c_3)n^3,$$

which is at most $c_1n^2(m+n)$ for $c_1$ sufficiently large. □

The next theorem, also due to Bang-Jensen, Gutin and Yeo, is an easy consequence of Theorem 8.4.17 (assign all vertices cost $-1$).

**Theorem 8.4.18** ([15]) *One can find a longest path in any quasi-transitive digraph in time $O(n^2m+n^3)$.* □

Sometimes, one is interested in finding path subdigraphs that include a maximum number of vertices from a given set $X$ or avoid as many vertices of $X$ as possible. We consider a minimum cost extension of this problem in the next result.

**Theorem 8.4.19** ([15]) *Let $D=(V,A)$ be a vertex-weighted quasi-transitive digraph and let $X \subseteq V$ be non-empty. Let $p_j$ be the maximum possible number of vertices from $X$ in a $j$-path subdigraph and let $q_j$ be the maximum possible number of vertices from $X$ not in a $j$-path subdigraph. In total time $O(n^2m+n^3)$ we can find, for all $j \in [n]$, a cheapest $j$-path subdigraph which includes $p_j$ (avoids $q_j$, respectively) vertices of $X$.*

**Proof:** Let $C=\sum_{v\in V}|c(v)|$ and subtract $C+1$ from the cost of every vertex in $X$. Now, for each $j \in [n]$, every cheapest $j$-path subdigraph $F_j$ must cover as many vertices from $X$ as possible, i.e., $p_j$ vertices. Furthermore, since the new cost of $F_j$ is exactly the original one minus $p_j(C+1)$, cheapest $j$-path subdigraphs covering $p_j$ vertices from $X$ are preserved under this transformation. Now the 'including' part of the claim follows from Theorem 8.4.17(a). The 'avoiding' part can be proved similarly, by adding $C+1$ to every vertex of $X$. □

### 8.4.6 Finding a Cheapest Cycle in a Quasi-Transitive Digraph

Bang-Jensen, Gutin and Yeo obtained the following:

**Theorem 8.4.20** ([15]) *For quasi-transitive digraphs with vertex-weights the minimum cost cycle problem can be solved in time $O(n^5 \log n)$.*

**Proof:** Let $D$ be a quasi-transitive digraph. If $D$ is not strong, then we simply look at the strong components, so assume that $D$ is strong. By Theorem 8.3.5, $D = T[Q_1, \ldots, Q_t]$, where $T$ is a strong semicomplete digraph, and each $Q_i$ is either a single vertex or a non-strong quasi-transitive digraph.

Suppose we have found a minimum cost cycle $C_i$ in each $Q_i$ which contains a cycle. Then clearly the minimum cost of a cycle in $D$ is given by $\min(\min_i(c(C_i)), c(C))$, where $C$ is a minimum cost cycle among those intersecting at least two $Q_i$'s. Hence it follows that applying this approach recursively we can find the minimum cost cycle in $D$. Now we show how to compute a minimum cost cycle $C$ as above.

Let $D'$ be defined as in the proof of Theorem 8.4.17 including the vertex-costs. It is easy to show using the same approach as when we converted between $i$-path subdigraphs of $D'$ and $D$ in the proof of Theorem 8.4.17, that the cost of $C$ is precisely $mc(D')$. Now it follows from Theorem 7.10.4 that we can find the cycle $C$ in time $O(n^3m + n^4 \log n)$.

Since we can construct $D'$, including finding the costs for all the vertices in time $O(n^2m + n^3)$ by Theorem 8.4.17, and there are at most $O(n)$ recursive calls, the approach above will lead to a minimum cost cycle of $D$ in time $O(n^4m+n^5 \log n)$. In fact, we can bound the first term as we did in the proof of Theorem 8.4.17 and obtain $O(n^3m + n^5 \log n) = O(n^5 \log n)$ rather than $O(n^4m + n^5 \log n)$. This completes the proof. □

## 8.5 Linkages

It is a well-known fact that it is easy to check (e.g., using flows) whether a directed multigraph $D = (V, A)$ has $k$ (arc)-disjoint paths $P_1, \ldots, P_k$ from a subset $X \subseteq V$ to another subset $Y \subseteq V$, and we can also find such paths efficiently. On many occasions (e.g., in practical applications) we need to be able to specify the initial and terminal vertices of each $P_i$, $1 \leq i \leq k$, that is, we wish to find a so-called **linkage** from $X = \{x_1, \ldots, x_k\}$ to $Y = \{y_1, \ldots, y_k\}$ such that $P_i$ is an $(x_i, y_i)$-path for every $1 \leq i \leq k$. This problem is considerably more difficult and is in fact $\mathcal{NP}$-complete already when $k = 2$.

Recall that, for a digraph $D = (V, A)$ with distinct vertices $x, y$ we denote by $\kappa_D(x, y)$ the largest integer $k$ such that $D$ contains $k$ internally disjoint $(x, y)$-paths. When discussing intersections between paths $P, Q$ we will often use the phrase 'let $u$ be the first (last) vertex on $P$ which is on $Q$'. By this we mean that if, say, $P$ is an $(x, y)$-path, then $u$ is the only vertex of $P[x, u]$ ($P[u, y]$) which is also on $Q$.

Let $x_1, x_2, \ldots, x_k, y_1, y_2, \ldots, y_k$ be distinct vertices of a digraph $D$. A **$k$-linkage** from $(x_1, x_2, \ldots, x_k)$ to $(y_1, y_2, \ldots, y_k)$ in $D$ is a system of vertex-disjoint paths $P_1, P_2, \ldots, P_k$ such that $P_i$ is an $(x_i, y_i)$-path in $D$.[4] A digraph

[4] Sometimes we allow that the paths may share one or both of their end-vertices, i.e., $V(P_i) \cap V(P_j) \subseteq \{x_i, y_i, x_j, y_j\}$ whenever $i \neq j$, where $x_i = y_j$ or $x_i = x_j$ is possible.

$D = (V, A)$ is **$k$-linked** if it contains a $k$-linkage from $(x_1, x_2, \ldots, x_k)$ to $(y_1, y_2, \ldots, y_k)$ for every choice of distinct vertices $x_1, x_2, \ldots, x_k, y_1, y_2, \ldots, y_k$. The $k$-DISJOINT PATHS problem is defined as follows.

> $k$-DISJOINT PATHS
> **Input:** A digraph $D = (V, A)$ and distinct vertices $s_1, \ldots, s_k, t_1, \ldots, t_k$.
> **Question:** Does $D$ contain vertex disjoint paths $P_1, \ldots, P_k$ such that $P_i$ is an $(s_i, t_i)$-path for $i \in [k]$?

Fortune, Hopcroft and Wyllie [30] showed that if we impose no restriction on the input, then the $k$-DISJOINT PATHS problem is $\mathcal{NP}$-complete already for $k = 2$. This motivates the study of subclasses of digraphs for which the problem is polynomial-time solvable.

From the algorithmic point of view, the 2-DISJOINT PATHS problem for semicomplete digraphs has already been solved by Bang-Jensen and Thomassen in Theorem 2.5.6. The proof of this result in [21] is highly non-trivial. The basic approach is divide and conquer and several non-trivial results and steps are needed to make the algorithm work. Now we show that the 2-DISJOINT PATHS problem can be solved in polynomial time for quite large classes of digraphs which can be obtained by starting from semicomplete digraphs and then performing certain substitutions. The algorithm we describe uses the polynomial algorithm from Theorem 2.5.6 for the case of semicomplete digraphs as a subroutine. The results in this section are due to Bang-Jensen [5].

**Theorem 8.5.1** ([5]) *Let $D = F[S_1, S_2, \ldots, S_f]$ where $F$ is a strong digraph on $f \geq 2$ vertices and each $S_i$ is a digraph with $n_i$ vertices and let $x_1, x_2, y_1, y_2$ be distinct vertices of $D$. There exist semicomplete digraphs $T_1, \ldots, T_f$ such that $V(T_i) = V(S_i)$ for all $i \in [f]$, and the digraph $D' = F[T_1, T_2, \ldots, T_f]$ has vertex-disjoint $(x_1, y_1)$-, $(x_2, y_2)$-paths if and only if $D$ has such paths. Furthermore, given $D$ and $x_1, x_2, y_1, y_2$, $D'$ can be constructed in time $O(n^2)$, where $n$ is the number of vertices of $D$.*

**Proof:** If $D$ has the desired paths, then so does any digraph obtained from $D$ by adding arcs. Hence if $D$ has the desired paths, then trivially $D'$ exists and can be constructed in time $O(n^2)$ once we know a pair of disjoint $(x_1, y_1)$-, $(x_2, y_2)$-paths.

If no $S_i$ contains both of $x_1, y_1$ or both of $x_2, y_2$, then it is easy to see that $D$ has the desired paths if and only if it has such paths which do not use an arc inside any $S_j$. Thus in this case we can add arcs arbitrarily inside each $S_i$ to obtain a $D'$ which satisfies the requirement.

Suppose next that some $S_i$ contains all of the vertices $x_1, x_2, y_1, y_2$. If there is an $(x_j, y_j)$-path $P$ in $S_i - \{x_{3-j}, y_{3-j}\}$, $j \in \{1, 2\}$, then it follows from that fact that $F$ is strong that $D$ has the desired paths and we can find such a pair in time $O(n^2)$. Thus, by our initial remark, we may assume that there is no $(x_j, y_j)$-path $P$ in $S_i - \{x_{3-j}, y_{3-j}\}$ for $j \in \{1, 2\}$. Now it is easy to see that $D$ has the desired paths if and only if it has such paths which do not use an arc inside any $S_j$. Thus we can replace $S_i$ by a tournament in which $x_1$ and $x_2$ both have no out-neighbours in $S_i - \{x_1, x_2\}$ and every other $S_k$ by an arbitrary tournament on the same vertex set. Clearly the digraph $D'$ obtained in this way satisfies the requirement.

Suppose now without loss of generality that $x_1, y_1 \in V(S_j)$ for some $j$ but $x_2 \notin V(S_j)$. Suppose first that $y_2 \in V(S_j)$. If there is no $(x_1, y_1)$-path in $S_j - y_2$, then $D$ has the desired paths if and only if it has such paths which do not use an arc inside any $S_i$ and we can construct $D'$ by adding arcs in $S_j$ in such a way that no $(x_1, y_1)$-path avoiding $y_2$ is created (that is, $y_2$ will still separate $x_1$ from $y_1$ in $D'[V(S_j)]$) and arbitrary arcs in every other $S_i$. On the other hand, if $S_j - y_2$ contains an $(x_1, y_1)$-path avoiding $y_2$, then it follows from the fact that $F$ is strong that $D$ has the desired paths and hence $D'$ exists, as remarked above. Hence we may assume that $y_2 \notin V(S_j)$.

If $S_j$ contains an $(x_1, y_1)$-path which does not cover all the vertices of $S_j$, then it follows from the fact that $F$ is strong that $D$ has the desired paths. Thus we may assume that either $S_j$ has no $(x_1, y_1)$-path, or every $(x_1, y_1)$-path in $S_j$ contains all the vertices of $S_j$. In the last case we may assume that $V(S_j)$ separates $x_2$ from $y_2$. Now $D$ has the desired paths if and only if it has such a pair which does not use any arcs from $S_j$. Thus in both cases we can construct $D'$ by replacing $S_j$ by a tournament with no $(x_1, y_1)$-path and every other $S_i$ by an arbitrary tournament on the same vertex set, except in the case when $x_2$ and $y_2$ belong to some $S_i$, $i \neq j$. In this case we replace that $S_i$ by a tournament with no $(x_2, y_2)$-path (by the remark above we may assume that $S_i$ has no $(x_2, y_2)$-path).

It follows from the considerations above that $D'$ can be constructed in time $O(n^2)$. □

Recall that Theorem 8.3.5 gives the canonical decomposition for quasi-transitive digraphs. Hence we can apply Theorem 8.5.1 to these digraphs.

**Theorem 8.5.2** ([5]) *There exists a polynomial-time algorithm for the* 2-DISJOINT PATHS *problem for quasi-transitive digraphs.*

**Proof:** Let $D$ be a quasi-transitive digraph and $x_1, x_2, y_1, y_2$ specified distinct vertices for which we want to determine the existence of vertex-disjoint $(x_1, y_1)$-,$(x_2, y_2)$-paths. First check that $D - \{x_i, y_i\}$ contains an $(x_{3-i}, y_{3-i})$-path for $i \in \{1, 2\}$. If not, then we stop. Now it follows from Theorem 8.3.5 that either $x_1, x_2, y_1, y_2$ are all in the same strong component of $D$, or the paths exist. For example, if $D$ is not strong and $y_1$, say, is not in the same strong component as $x_1$ then, by Theorem 8.3.5, $x_1$ and $y_1$ belong to different sets $W_i, W_j$ in the canonical decomposition $D = Q[W_1, \ldots, W_{|Q|}]$, where $Q$ is a transitive digraph. Hence $x_1 \rightarrow y_1$ and the desired paths clearly exist.

Thus we may assume that $D$ is strong. Let $D = S[W_1, W_2, \ldots, W_{|S|}]$ be the canonical decomposition of $D$. Now apply Theorem 8.5.1 and construct the digraph $D'$ which has the desired paths if and only if $D$ does. As remarked in Theorem 8.5.1, $D'$ can be constructed in polynomial time. By the construction of $D'$ (replacing each $W_i$ by a semicomplete digraph) it follows that $D'$ is a semicomplete digraph and hence we can apply the polynomial algorithm of Theorem 2.5.6 to $D'$ in order to decide the existence of the desired paths in $D$. The algorithm of Theorem 2.5.6 can be used to find vertex-disjoint $(x_1, y_1)$-, $(x_2, y_2)$-paths in $D'$ if they exist and given these paths it is easy to construct the corresponding paths in $D$ (it suffices to take minimal paths). □

By inspecting the proof of Theorem 8.5.1 it is not difficult to see that the following much more general result is true. The main point is that in the proof of Theorem 8.5.1 we either find the desired paths or decide that they exist if and only if there are such paths that use no arcs inside any $S_i$. Hence instead of making each $T_i$ semicomplete, we may just as well make it an independent set, by deleting all arcs inside $S_i$.

**Theorem 8.5.3** ([5]) *Let $\Phi$ be a class of strongly connected digraphs, let $\Phi^{ext}$ denote the class of all extensions of graphs in $\Phi$ and let*

$$\Phi^* = \{F[D_1, \dots, D_{|F|}] : F \in \Phi, \text{ each } D_i \text{ is an arbitrary digraph}\}.$$

*There is a polynomial algorithm for the* 2-DISJOINT PATHS *problem in $\Phi^*$ if and only if there is a polynomial algorithm for the* 2-DISJOINT PATHS *problem for all digraphs in $\Phi^{ext}$.* □

This result shows that studying extensions of digraphs can be quite useful.

One example of such a class $\Phi$, for which Theorem 8.5.3 applies, is the class of strong semicomplete digraphs. This follows from the fact that we can reduce the 2-DISJOINT PATHS problem for extended semicomplete digraphs to the case of semicomplete digraphs in the same way as we did for quasi-transitive digraphs in the proof of Theorem 8.5.2. Hence the 2-DISJOINT PATHS problem is polynomially solvable for all digraphs that can be obtained from strong semicomplete digraphs by substituting arbitrary digraphs for vertices. It is important to note here that $\Phi$ must consist only of strong digraphs, since it is not difficult to reduce the 2-DISJOINT PATHS problem for arbitrary digraphs (which is $\mathcal{NP}$-complete) to the 2-DISJOINT PATHS problem for those digraphs that can be obtained from the digraph $H$ consisting of just an arc $uv$ by substituting arbitrary digraphs for the vertex $v$.

The proof of the following easy lemma is left to the reader. Note that four is the best possible, as can be seen from the complete biorientation of the undirected graph consisting of a 4-cycle $x_1x_2y_1y_2x_1$ and a vertex $z$ joined to each of the four other vertices.

**Lemma 8.5.4** *Let $D$ be a digraph of the form $D = \vec{C}_2[S_1, S_2]$, where $S_i$ is an arbitrary digraph on $n_i$ vertices, $i = 1, 2$. If $D$ is 4-strong, then $D$ is 2-linked.* $\diamond$

Theorem 2.5.1 gives a sufficient condition for a semicomplete digraph to be 2-linked in terms of its strong connectivity. The same condition turns out to be sufficient for quasi-transitive digraphs.

Before proving our final results of this subsection, we will be needing a structural theorem regarding $k$-strong digraphs due to Bang-Jensen.

**Lemma 8.5.5** ([5]) *Let $D = F[S_1, \ldots, S_f]$ where $F$ is a strong digraph on $f \geq 2$ vertices, each $S_i$ is a digraph with $n_i$ vertices, and $F$ has as few vertices as possible among all non-trivial decompositions of $D$ of this kind. Let $D_0 = F[\overline{K}_{n_1}, \ldots, \overline{K}_{n_f}]$ be the digraph obtained from $D$ by deleting every arc which lies inside some $S_i$, and let $S$ be a minimal (with respect to inclusion) separating set of $D_0$. Then $S$ is also a separating set of $D$, unless each of the following holds:*

*(a)* $S = \bigcup_{j \neq i} V(S_j)$ *for some* $1 \leq i \leq f$,
*(b)* $D[S_i]$ *is a strong digraph, and*
*(c)* $D = \vec{C}_2[S, S_i]$.

*In particular, if $F$ has at least three vertices, then $D$ is $k$-strong if and only if $D_0$ is $k$-strong.*

**Theorem 8.5.6** ([5]) *Let $k \geq 4$ be a natural number and let $F$ be a digraph on $f \geq 2$ vertices with the property that every $k$-strongly connected digraph of the form $F[T_1, T_2, \ldots, T_f]$, where each $T_i$, $i \in [f]$, is a semicomplete digraph, is 2-linked. Let $D = F[S_1, S_2, \ldots, S_f]$, where $S_i$ is an arbitrary digraph on $n_i$ vertices for all $i \in [f]$. If $D$ is $k$-strongly connected, then $D$ is 2-linked.*

**Proof:** Let $D = F[S_1, S_2, \ldots, S_f]$, where $S_i$ is an arbitrary digraph on $n_i$ vertices for each $i \in [f]$, be given. By Lemma 8.5.4 we may assume that $D$ cannot be decomposed as $D = \vec{C}_2[R_1, R_2]$, where $R_1$ and $R_2$ are arbitrary digraphs. Construct $D'$ as described in Theorem 8.5.1. Note that by Lemma 8.5.5, $\kappa(D') = \kappa(D)$. Thus $D'$ is $k$-strong and using Theorem 8.5.1 and the assumption of the theorem we conclude that $D$ is 2-linked. $\square$

**Corollary 8.5.7** ([5]) *Every 5-strong quasi-transitive digraph is 2-linked.*

**Proof:** By Theorem 8.3.5, every strong quasi-transitive digraph is of the form $D = F[S_1, S_2, \ldots, S_f]$, $f = |F|$, where $F$ is a strong semicomplete digraph and each $S_i$ is a non-strong quasi-transitive digraph on $n_i$ vertices. By Lemma 8.3.4 and the connectivity assumption, $|F| \geq 3$. Note that for any choice of semicomplete digraphs $T_1, \ldots, T_f$ the digraph $D' = F[T_1, T_2, \ldots, T_f]$ is semicomplete. Hence the claim follows from Theorem 8.5.6 and the fact that, by Theorem 2.5.12, every 5-strong semicomplete digraph is 2-linked. (Since $F$ has at least three vertices, it follows from Lemma 8.5.5 that $\kappa(D') = \kappa(D)$.) $\square$

### 8.5.1 $k$-Linkages

As mentioned at the beginning of this section, since the $k$-DISJOINT PATHS problem is already $\mathcal{NP}$-complete for $k = 2$, the restriction of this problem to particular classes of digraphs has been studied by many authors. It turns out that, for some families, the problem can be solved in polynomial time when $k$ is fixed. For example, consider Theorems 3.4.1, 2.5.7, and 2.5.11.

Recall that a digraph $D$ is decomposable if there exist a digraph $R$ on $r$ vertices, and distinct (but possibly isomorphic) digraphs $L_1, \ldots, L_r$, such that $D = R[L_1, \ldots, L_r]$. In this section we will study the $k$-DISJOINT PATHS problem in decomposable digraphs. As a consequence, we will obtain polynomial algorithms to solve the $k$-DISJOINT PATHS problem in quasi-transitive digraphs and extended semicomplete digraphs. The results of this section are due to Bang-Jensen, Christiansen, and Maddaloni [7].

Let $D = S[M_1, ..., M_s]$ be a decomposable digraph and let $P$ be a path in $D$. We say that $P$ is **$D$-internal** if $P \subseteq M_i$ for some $i$, and we say that $P$ is **$D$-external** otherwise. When $D$ is clear from the context we just call the path **internal** or **external**. Similarly we say that a pair $(s, t) \in V(D) \times V(D)$ is internal if $s, t \in V(M_i)$ for some $i$, and is external otherwise.

Let $\Pi = \{(s_1, t_1), ..., (s_k, t_k)\}$ be a set of $k$ pairs of distinct terminals. A **$\Pi$-linkage** is a collection $L$ of $k$ disjoint paths $P_i$, $i \in [k]$, such that $P_i$ is an $(s_i, t_i)$-path. If a $\Pi$-linkage $L$ exists in the digraph $D$ we say that $L$ is a linkage for $(D, \Pi)$

**Lemma 8.5.8** ([7]) *Let $D = S[M_1, ..., M_s]$ be a decomposable digraph and $\Pi$ a set of pairs of terminals. Then $(D, \Pi)$ has a linkage if and only if it has a linkage whose external paths do not use any arc of $D[M_i]$ for $i \in [s]$.* □

Let $D$ be a digraph with vertex set $v_1, v_2, \ldots, v_n$ and let $K$ be another digraph. By **blowing up $v_i$ into K in D** we mean the operation that substitutes the digraph $K$ for the vertex $v_i$ in $D$, that is, creates the digraph $D' = D[\{v_1\}, \ldots, \{v_{i-1}\}, K, \{v_{i+1}\}, \ldots, \{v_n\}]$. We say that a class of digraphs $\Phi$ is **closed with respect to blow-up** if for any $D \in \Phi$, for every integer $m$ and for every $v \in V(D)$, there exists a digraph $K$ on $m$ vertices such that the blowing up of $v$ into $K$ results in a digraph which is still in $\Phi$.

**Lemma 8.5.9** ([7]) *If the class $\Phi$ is closed with respect to the blowing-up operation, $S \in \Phi$ and $D = S[M_1, ..., M_s]$, then it is possible to replace the arcs inside each $M_i$, $i \in [s]$, with other arcs, so that the resulting digraph is in $\Phi$.* □

We say that a class of digraphs $\Phi$ is **a linkage ejector** if

1. There exists a polynomial algorithm $\mathcal{A}_\Phi$ to find a total $\Phi$-decomposition of every totally $\Phi$-decomposable digraph.
2. There exists a polynomial algorithm $\mathcal{B}_\Phi$ for solving the $k$-DISJOINT PATHS problem on $\Phi$. The running time depends (possibly exponentially) on $k$ but the algorithm is polynomial when $k$ is fixed.

3. The class $\Phi$ is closed with respect to blow-up and there exists a polynomial algorithm $\mathcal{C}_\Phi$ wich given a totally $\Phi$-decomposable digraph $D = S[M_1, ..., M_s]$, constructs a digraph of $\Phi$ by replacing the arcs inside each of the $M_i$'s, as in Lemma 8.5.9.

**Theorem 8.5.10** *Let $\Phi$ be a linkage ejector. For every fixed $k$, there exists a polynomial algorithm to solve the $k$-*DISJOINT PATHS *problem on totally $\Phi$-decomposable digraphs.* □

Recall that, by Theorem 8.3.5, quasi-transitive digraphs are totally $\Phi_3$-decomposable. The following result of Bang-Jensen, Christiansen, and Maddaloni deals with this class of digraphs, which also includes, for example, extended semicomplete digraphs.

**Lemma 8.5.11** ([7]) *The class $\Phi_3$ is a linkage ejector.* □

We thus obtain the following corollary of Theorem 8.5.10

**Theorem 8.5.12** *For every fixed $k$, there exists a polynomial algorithm to solve the $k$-*DISJOINT PATHS *problem on quasi-transitive digraphs and extended semicomplete digraphs.*

### 8.5.2 Weak $k$-Linkages

Note that for this subsection we allow both parallel arcs and loops and (for simplicity) we still use the name digraph rather than directed pseudograph.

Let $D = (V, A)$ be a digraph and let $s_1, \ldots, s_k, t_1, \ldots, t_k$ be a collection of (not necessarily distinct) vertices of $D$. A **weak k-linkage** from $(s_1, \ldots, s_k)$ to $(t_1, \ldots, t_k)$ is a collection of $k$ arc-disjoint paths $P_1, \ldots, P_k$ such that, for each $i \in [k]$, $P_i$ is an $(s_i, t_i)$-path if $s_i \neq t_i$ and a proper cycle containing $s_i$ if $s_i = t_i$.

| |
|---|
| WEAK $k$-LINKAGE<br>**Input:** A digraph $D = (V, A)$ and not necessarily distinct vertices $s_1, \ldots, s_k, t_1, \ldots, t_k$.<br>**Question:** Does $D$ contain a weak $k$-linkage from $(s_1, \ldots, s_k)$ to $(t_1, \ldots, t_k)$? |

It is well-known that the WEAK $k$-LINKAGE problem is $\mathcal{NP}$-complete already when $k = 2$ [30].

Until recently, results regarding the WEAK $k$-LINKAGE problem were limited, both in number and depth. In Section 3.4, the case of acyclic digraphs is discussed, and Section 2.5 presents a brief evolution of this problem, with an obvious emphasis on the class of semicomplete digraphs. In particular, the results of Fradkin and Seymour found in [31] (and included in Section 2.5) mark a turning point in the scope of families for which nice results can be obtained. In this section we present the results obtained by Bang-Jensen and Maddaloni in [19] regarding the WEAK $k$-LINKAGE problem on decomposable digraphs. We begin with a result which is implicitly stated in [31]. See Section 2.5 for the definition and a brief discussion of the concept of cutwidth.

**Theorem 8.5.13** (Fradkin–Seymour [31]) *For every natural number $\theta$ the* WEAK $k$-LINKAGE *problem is polynomial for every fixed $k$, when we consider digraphs with cutwidth at most $\theta$.*

The following easy consequence will be used in our algorithms.

**Theorem 8.5.14** *For every natural number $p$ the* WEAK $k$-LINKAGE *problem is polynomial, for every fixed $k$, when we consider digraphs with at most $p$ directed cycles.*

**Proof:** Let $D$ be a digraph with at most $p$ directed cycles. Then the cutwidth of $D$ is at most $p$: we may delete an arbitrary arc from each of the at most $p$ cycles to get a digraph with cutwidth 0, so $D$ has cutwidth at most $p$. Now the claim follows from Theorem 8.5.13. □

Assume we want to decide the existence of a weak $k$-linkage from the vertices $(s_1, ..., s_k)$ to the vertices $(t_1, ..., t_k)$. We will denote by $\Pi$ the list of pairs[5] $(s_1, t_1), \ldots, (s_k, t_k)$. In the rest of this subsection we will think of $\Pi$ both as a list of $k$ pairs and as a collection of all the terminals $s_1, \ldots, s_k, t_1, \ldots, t_k$.

We say that $D$ has a weak $\Pi$-linkage if it contains a weak $k$-linkage from $(s_1, \ldots, s_k)$ to $(t_1, \ldots, t_k)$. We sometimes also say that $(D, \Pi)$ has a weak linkage.

As in the previous subsection, we will use the term **blow up** of $v_i$ into a digraph $K$ (in $D$) meaning the composition $D[v_1, ..., v_{i-1}, K, v_{i+1}, ..., v_n]$.

Recall that we allow multiple arcs (and loops) in our digraphs, and also that $\mu_D(u, v)$ denotes the number of arcs from a vertex $u$ to a vertex $v$. We will assume throughout the rest of this subsection, unless otherwise stated, that $k$ denotes the number of pairs to be linked. An instance of the problem $(D, \Pi)$ is equivalent to $(D', \Pi)$ where $V(D') = V(D)$ and for every $u, v \in V(D')$ one has $\mu_{D'}(u, v) = \min(\mu_D(u, v), k)$. Therefore from now on **we will only consider digraphs $D$ with**

$$\mu_D(u, v) \leq k \quad \forall u, v \in V(D)$$

while studying the WEAK $k$-LINKAGE problem.

Let $D = (V, A)$ be a digraph and $H$ an induced subdigraph of $D$. We say that $H$ is a **module** if for every $a, b \in V(H)$, $v \in V(D \setminus H)$ we have that $\mu_D(v, a) = \mu_D(v, b)$ and $\mu_D(a, v) = \mu_D(b, v)$. We say that $H$ is a **clean**

[5] Note that the same pair (or the same vertex) may appear more than once in the list and we may have $s_i = t_i$.

**module** with respect to $\Pi$ if it is a module containing no terminals of $\Pi$. The concept of module yields an alternative definition of a decomposable graph. A digraph $D$ is **decomposable** if $D = S[H_1, ..., H_s]$, for some digraph $S$, with $s = |V(S)| \geq 2$ and some choice of disjoint modules $H_1, ..., H_s$. In this case $S$ is called the **quotient digraph** (of $D$) induced by $H_1, ..., H_s$.

The algorithms developed in this subsection rely on the following fundamental fact, the proof of which we will omit: a weak linkage need not use any arc inside clean modules. As mentioned earlier, the results of this section are due to Bang-Jensen and Maddaloni.

**Lemma 8.5.15** ([19]) *Let $D$ be a digraph, $\Pi$ a list of $k$ terminal pairs and $H \subset D$ a clean module with respect to $\Pi$. Let $D'$ be the contraction of $H$ into a single vertex $h$. Then $D$ has a weak $\Pi$-linkage if and only if $D'$ has a weak $\Pi$-linkage.* □

The following result is an immediate consequence of the proof of Lemma 8.5.15 (see [19]).

**Lemma 8.5.16** ([19]) *Let $\Pi$ be a list of terminal pairs and $H \subset D$ be a clean module with respect to $\Pi$. For every weak linkage $P'_1, ..., P'_k$ of $(D, \Pi)$, there exists another weak linkage $P_1, ..., P_k$ such that $P'_i = P_i$ on $D \setminus H$, and for $i = 1, ..., k$, $A(P_i \cap H) = \varnothing$.*

As in the previous subsection, given a decomposable digraph $D = S[H_1, ..., H_s]$ and a path $P$ we say that $P$ is **internal** if $P \subseteq H_j$ for some $H_j$, and we say that $P$ is **external** otherwise.[6]

Similarly, we say that a pair $(s,t)$ is an **internal pair** if $s, t \in H_j$ for some $j$, and we say that $(s,t)$ is an **external pair** otherwise.

If a module $H$ is not clean, i.e. it contains terminals, then some of the arcs in $A(H)$ may be necessary to guarantee a weak linkage. See Figure 8.4. The following lemma shows that, in a precise sense, a weak linkage need not use too many arcs inside a given module. Together with Lemma 8.5.16, this will allow a polynomial brute-force algorithm (Theorem 8.5.19).
For technical reasons that will become clear later, we consider the more general case where a set of arcs $F$ has been deleted from $D$.

**Lemma 8.5.17** ([19]) *Let $D = S[H_1, ..., H_s]$ be a decomposable digraph, let $\Pi'$ be a list of $h$ terminal pairs and let $F$ be a set of arcs in $D$ satisfying $d^-_F(v), d^+_F(v) \leq r$ for all $v \in V(D)$. If $(D \setminus F, \Pi')$ has a weak linkage, then it has a weak linkage $P_1, ..., P_h$ such that we have $|V(\bigcup_{i \in \mathcal{E}} P_i \cap H_j)| \leq 2h(h+r)$, for every $j \in \{1, ..., s\}$, where $\mathcal{E}$ denotes the set of indices $i$ for which $P_i$ is external.*

Note that from the previous proof we have that for every $j \in \{1, \ldots, s\}$ and every $i \in \mathcal{E}$, $|A(P_i \cap H_j)| < 2(h+r)$.

---

[6] Note that an external path may still start and end in the same module $H_j$.

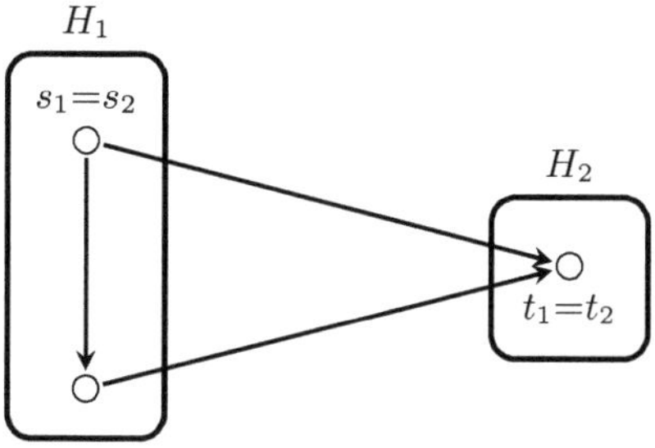


**Figure 8.4** An example with $|\Pi| = 2$, the only weak $\Pi$-linkage uses the arc inside $H_1$.

**Lemma 8.5.18** ([19]) *Let $\mathcal{C}$ be a class of digraphs for which there exists an algorithm $\mathcal{A}$ to decide the* WEAK $k$-LINKAGE *problem, whose running time is bounded by $f(n,k)$. Let $D = (V, A)$ be a digraph, $\Pi$ a list of $k$ pairs of terminals and $F \subseteq V \times V$ such that $D' := (V, A \cup F)$ is a member of $\mathcal{C}$. There exists an algorithm $\mathcal{A}^-$, whose running time is bounded by $f(n, k + |F|)$, to decide whether $D$ has a weak $\Pi$-linkage.*

**Proof:** Suppose $F = \{s'_1t'_1, ...., s'_{k'}t'_{k'}\}$, where $k' = |F|$, and let $\Pi' = (s'_1, t'_1), ..., (s'_{k'}, t'_{k'})$. $D$ has a weak $\Pi$-linkage if and only if $D'$ has a weak $(\Pi \cup \Pi')$-linkage, from which the claim follows. Indeed, if $D$ has a weak $\Pi$ linkage, then this extends to a weak $(\Pi \cup \Pi')$-linkage of $D'$ by simply taking the arcs $s'_it'_i$ as $(s'_i, t'_i)$-paths. If $D'$ has a weak $\Pi \cup \Pi'$-linkage $L$, it is easy to see that $A(L) \setminus F$ contains a weak $\Pi$-linkage of $D$. □

Given a digraph $D$ and a non-negative integer $c$, let $D(c)$ denote the set of digraphs that can be obtained from $D$ by first adding any number of arcs parallel to the already existing ones and then blowing up $b$ vertices, with $0 \le b \le c$, to digraphs of size less than or equal to $c$ each. We say that a class of digraphs $\Phi$ is **bombproof** if there exists a polynomial algorithm $\mathcal{A}_\Phi$ to find a total $\Phi$-decomposition of every totally $\Phi$-decomposable digraph and, for every integer $c$, there exists a polynomial algorithm[7] $\mathcal{B}_\Phi$ to decide the WEAK $k$-LINKAGE problem for the class

$$\Phi(c) := \bigcup_{D \in \Phi} D(c).$$

The following theorem of Bang-Jensen and Maddaloni is the main result in [19].

**Theorem 8.5.19** ([19]) *Let $\Phi$ be a bombproof class of digraphs. There is a polynomial algorithm $\mathcal{M}$ which takes as input a 5 tuple $[D, k, k', \Pi, F]$, where $D$ is a totally $\Phi$-decomposable digraph, $k, k'$ are natural numbers with $k' \le k$, $\Pi$ is a list of $k'$ terminal pairs and $F \subseteq A(D)$ is a set of arcs satisfying*

[7] Note that the running time of $\mathcal{B}_\Phi$ may depend heavily on $c$.

$$d^-_F(v), d^+_F(v) \leq k - k' \textit{for all } v \in V(D).$$
$$|F| \leq (k - k')2k$$

*and decides whether $D \setminus F$ contains a weak $\Pi$-linkage.* □

For the sake of brevity we will omit the proof of Theorem 8.5.19. Nonetheless, we present a description for the proposed polynomial algorithm $\mathcal{M}$.

1. If $\Pi = \varnothing$ output that a solution exists and return.
2. Run $\mathcal{A}_\Phi$ to find a total $\Phi$-decomposition of $D = S[H_1, ..., H_s]$.
3. If this decomposition is trivial, that is $D = S$, then $D \in \Phi \subseteq \Phi(1)$, so run $\mathcal{B}^-_\Phi$ on $(D \setminus F, \Pi)$ to decide the problem and return.
4. Find among $H_1, ..., H_s$ those modules $K_1, ..., K_l$ that contain at least one terminal. Let $D'$ be obtained by contracting all the modules distinct from $K_1, ..., K_l$. Let $F'$ be the set of arcs obtained from $F$ after the contraction.
5. Let $\Pi^e \subseteq \Pi$ ($\Pi^i \subseteq \Pi$) be the list of external (internal) pairs $(s_q, t_q)$ in $\Pi$.
6. For every partition of $\Pi^i = \Pi_1 \cup \Pi_2$ look for external paths linking the pairs in $\Pi^e \cup \Pi_1$ and internal paths linking the pairs in $\Pi_2$. This is done in the following way:

   a) If $\Pi^e \cup \Pi_1 = \varnothing$, then for $i = 1, ..., l$: run $\mathcal{M}$ recursively on input $[K_i, k, k'_i, \Pi \cap K_i, F \cap A(K_i)]$, where $\Pi \cap K_i$ denotes the list of terminal pairs that lie inside $K_i$ and $k'_i$ is the number of those pairs.
   b) If $\Pi^e \cup \Pi_1 \neq \varnothing$, let $k'_i$ be the number of pairs in $\Pi_2 \cap K_i$. We do the following for each possible choice of $l$ vertex sets $W_i \subseteq V(K_i)$, $i = 1, ..., l$, of size $\min\{|V(K_i)|, 2(k' - k'_i)(k - k')\}$ and arc sets[8] $F_i \subseteq A(K_i[W_i]) \setminus F$, $i = 1, ..., l$, with $F_i$ satisfying

   $$d^-_{(F \cap A(K_i)) \cup F_i}(v), d^+_{(F \cap A(K_i)) \cup F_i}(v) \leq k' - k'_i.$$
   $$|F_i| \leq 2(k' - k'_i)(k - k').$$

   - For every module $K_i$ remove all the vertices of $V(K_i) \setminus W_i$ and then all remaining arcs except those in $F_i$.
   - Define $D''$ to be the digraph obtained from $D'$ with this procedure.
   - Run $\mathcal{B}^-_\Phi$ on $(D'' \setminus F', \Pi^e \cup \Pi_1)$.
   - For $i = 1, ..., l$, run $\mathcal{M}$ recursively on input $[K_i, k, k'_i, \Pi_2 \cap K_i, (F \cap A(K_i)) \cup F_i]$.

   If at step 6(a) all the instances examined are linked or at step 6(b), there is a choice of $W_i, F_i$, $i = 1, ..., l$, such that all instances examined are linked, then output that a weak linkage exists and return.
7. If all choices of $\Pi_1, \Pi_2$ have been considered, without verifying the existence of any weak linkage, then output that no weak linkage exists.

[8] $K_i[W_i]$ is the subdigraph of $K_i$ induced by $W_i$.

Taking $k' = k$ and running the previous algorithm on input $[D, k, k, \Pi, \varnothing]$ where $D$ is any totally $\Phi$-decomposable digraph and $\Pi$ is a list of $k$ terminal pairs from $V(D)$, we obtain the main result of this subsection.

**Theorem 8.5.20** *Let $\Phi$ be a bombproof class of digraphs. For every fixed $k$ there exists a polynomial algorithm for the* WEAK $k$-LINKAGE *problem for the totally $\Phi$-decomposable digraphs.*

Based on the recursive structure given by the canonical decomposition for quasi-transitive digraphs (Theorem 8.3.5), Bang-Jensen and Maddaloni proved that there is a polynomial algorithm for the WEAK $k$-LINKAGE problem on quasi-transitive digraphs [19]. Recall that Theorem 8.3.5 can be restated to say that quasi-transitive digraphs are totally $\Phi_3$-decomposable.

**Lemma 8.5.21** *The class $\Phi_3$ is bombproof.*

**Proof:** We can get a polynomial algorithm for the total $\Phi_3$-decomposition from Theorem 8.3.27. Given a positive integer $c$ and a digraph $D \in \Phi_3$, consider a digraph in $D' \in D(c)$: if $D$ is semicomplete, then $D'$ misses no more than $c^3$ arcs to be semicomplete. If $D$ is acyclic, then $D'$ has at most $O(c^{c+1})$ cycles or $O(c \cdot (ck)^c)$ if there are (at most $k$) parallel arcs, because all the cycles must lie in one of the blown up subdigraphs. By Theorem 2.5.5 and Lemma 8.5.18 in the first case and Theorem 8.5.14 in the second case, there is a polynomial algorithm to decide the WEAK $k$-LINKAGE problem in $D(c)$ and hence in $\Phi_3(c)$. Thus we can conclude that $\Phi_3$ is bombproof. □

**Theorem 8.5.22** *For every fixed $k$ there exists a polynomial algorithm for the* WEAK $k$-LINKAGE *problem for quasi-transitive digraphs.*

**Proof:** It follows from Theorem 8.3.5 that quasi-transitive digraphs are totally $\Phi_3$-decomposable. By Lemma 8.5.21 $\Phi_3$ is bombproof, hence we can apply Theorem 8.5.20. □

We can apply Theorem 8.5.20 to another class of digraphs; extended semicomplete digraphs are clearly totally $\Phi_3$-decomposable. Hence, from Theorem 8.5.20, we have the following

**Theorem 8.5.23** *For every fixed $k$ there exists a polynomial algorithm for the* WEAK $k$-LINKAGE *problem for extended semicomplete digraphs.* □

## 8.6 Kings and Kernels

The existence of $k$-kings was one the first problems to be explored for quasi-transitive digraphs. As a matter of fact, the concept of $k$-king was first introduced in [16] for the purpose of studying quasi-transitive digraphs. In families of digraphs closed under the reversal of every arc, like quasi-transitive digraphs, the study of $k$-kings is closely related to the study of $(k+1)$-kernels: a $k$-king in the reversal of $D$ is a $(k+1)$-kernel of $D$.

After spending some years dormant, this subject has received a lot of attention lately. Surprisingly, many of the nice existing results for kings in quasi-transitive digraphs admit natural generalizations to $k$-quasi-transitive digraphs.

### 8.6.1 Kings

A **$k$-king** in a digraph $D$ is a vertex $u$ such that $d(u,v) \leq k$ for every $v \in V(D) - u$ (it is a **$k$-dominating vertex**). A **king** is a 2-king. The study of kings in digraphs began with the mathematical sociologist Landau, who proved that every vertex of maximum out-degree in a tournament is a king, [53] (see Theorem 2.2.12). Nonetheless, the term king was introduced by Maurer in [54], where he used tournaments to model dominance in flocks of chickens. Some of the classical results on $k$-kings in digraphs can be consulted in [9], and Section 2.2 includes the most relevant results for tournaments.

Most of the main results in this section rely on several technical lemmas, so we prefer to omit them for the sake of presentation.

In [16], Theorem 8.3.5 is used extensively by Bang-Jensen and Huang to prove the first results on the existence and number of 3-kings in quasi-transitive digraphs. The main results can be condensed in the following theorem.

**Theorem 8.6.1** ([16]) *Let $D$ be a quasi-transitive digraph. Then we have*

1. *$D$ has a 3-king if and only if it has an out-branching.*
2. *If $D$ has a 3-king, then the following holds:*
    a) *Every vertex in $D$ of maximum out-degree is a 3-king.*
    b) *If $D$ has no vertex of in-degree zero, then $D$ has at least two 3-kings.*
    c) *If the unique initial strong component of $D$ contains at least three vertices, then $D$ has at least three 3-kings.*

**Sketch of Proof.** Clearly, the existence of an out-branching is necessary. To prove the converse, assume that $D$ has an out-branching. This implies that $D$ has a unique initial strong component. Since the strong components digraph of a quasi-transitive digraph is transitive, a vertex is a 3-king of $D$ if and only if it is a 3-king of the unique initial component of $D$. So, we may assume that $D$ is strong.

Let $D = S[Q_1, \dots, Q_s]$ be the decomposition of $D$ given by Theorem 8.3.5. Since $S$ is semicomplete, every vertex of $S$ belongs to a 3-cycle of $S$. Thus, for every $1 \leq i \leq s$, each vertex in every $Q_i$ has distance at most 3 to every other vertex in $Q_i$. Assume without loss of generality that $Q_1$ corresponds to a vertex $s_1$ of maximum out-degree in $S$. Then $s_1$ is a 2-king in $S$, and hence every vertex in $Q_1$ is a 3-king of $D$.

Observe that the vertices of maximum out-degree in $D$ must belong to the $Q_i$'s corresponding to the vertices of maximum out-degree in $D$. If there are no vertices of in-degree zero in $D$, there are at least two vertices of maximum out-degree in $S$. □

From the previous argument it can also be observed that $D$ has a 2-king if and only if $|V(Q_i)| = 1$ for some $Q_i$ corresponding to a 2-king of $S$. Also, similar argumentation leads to other, more specific, results regarding the distribution of 3-kings in a quasi-transitive digraphs. As an example consider the following proposition from [16]; a **non-king** is a vertex which is not a 3-king.

**Proposition 8.6.2** *Let $D$ be a quasi-transitive digraph which contains a 3-king but no vertex of in-degree zero. Every non-king is dominated by at least three 3-kings, unless the initial component of $D$ is a 2-cycle, in which case every non-king is dominated by exactly two 3-kings.* □

After this first wave of results, most of the study of $k$-kings was restricted to multipartite tournaments for several years. It was not until 2012 that $k$-quasi-transitive digraphs were introduced in [48], and the following generalization of the first item of Theorem 8.6.1 was proved between [48] and [37] by Galeana-Sánchez, Hernández-Cruz and Juárez-Camacho.

**Proposition 8.6.3** ([37, 48]) *Let $k \geq 2$ be an integer. If $D$ is a $k$-quasi-transitive digraph, then $D$ has a $(k+1)$-king if and only if it has a unique initial strong component.* □

Proposition 8.6.3 was then the starting point for studying kings in $k$-quasi-transitive digraphs. Further generalizations to Theorem 8.6.1 were obtained, but also some strengthenings. Recall that we know exactly when a quasi-transitive digraph has a 2-king; a similar situation was described by Wang and Meng for $k$-quasi-transitive digraphs.

**Theorem 8.6.4** ([60]) *Let $k \geq 4$ be an integer. If $D$ is a $k$-quasi-transitive digraph, then $D$ has a $k$-king if and only if it has a unique strong component which is not isomorphic to an extended $(k+1)$-cycle $\vec{C}[E_0, \ldots, E_k]$, where each $E_i$ is an independent set on at least two vertices.* □

Now that we know exactly when a $k$-king exists, it is natural to ask for the minimum number of $k$-kings in a $k$-quasi-transitive digraph. The following theorem of Wang and Zhang deals with this question.

**Theorem 8.6.5** ([63]) *Let $k \geq 5$ be an integer, and let $D$ be a strong $k$-quasi-transitive digraph with at least two vertices. If $D$ is not isomorphic to an extended $(k+1)$-cycle, then $D$ has at least two $k$-kings.* □

It should be noted that this is the best possible result in terms of the number of $k$-kings in a $k$-quasi-transitive digraph. Consider the digraph

$H = \vec{C}_{k+1}[\{x_0\}, \{x_1\}, E_2, E_3, \ldots, E_k]$, where $E_i$ is an independent set with at least two vertices. Let $D$ be obtained from $H$ by adding the arc $x_1x_0$. Clearly $D$ is a $k$-quasi-transitive digraph and it is not isomorphic to an extended $(k+1)$-cycle. It is not difficult to check that there are exactly two $k$-kings in $D$, namely, $x_0$ and $x_1$.

Given the previous discussion, it is natural to give further consideration to $(k+1)$-kings in $k$-quasi-transitive digraphs. Unfortunately, unlike the case of quasi-transitive digraphs, it is not true that every vertex of maximum out-degree in a $k$-quasi-transitive digraph is a $(k+1)$-king. As noted in [37], the only vertex of maximum out-degree in the 4-transitive digraph with vertex set $\{v_1, v_2, v_3, v_4\}$, and arc set $\{v_1v_2, v_2v_3, v_2v_4, v_3v_4, v_4v_3\}$, is not a 5-king. Nonetheless, there are some simple conditions that will ensure that every vertex of maximum out-degree in a $k$-quasi-transitive digraphs is a $(k+1)$-king. The following condition was given by Wang and Zhang in [63]; recall that a $k$-king $u$ in a digraph $D$ is *strict* if there exists a vertex $v$ such that $d(u, v) = k$.

**Theorem 8.6.6** ([63]) *Let $k \geq 2$ be an integer and let $D$ be a $k$-quasi-transitive digraph.*

1. *If $D$ is strong, then every vertex of maximum out-degree in $D$ is a $(k+1)$-king.*
2. *If $D$ has a strict $(k+1)$-king, then every vertex of maximum out-degree in $D$ is a $(k+1)$-king.*

□

As can be observed in Theorem 8.6.8, the number of $(k+1)$-kings in $k$-quasi-transitive digraphs can be very large, compared to the number of $k$-kings. The proof uses the following theorem regarding 4-kings in semicomplete bipartite digraphs, which can be obtained from the analogous result due to Koh and Tan, on bipartite tournaments (Theorem 7.12.2), [50].

**Theorem 8.6.7** *Let $D$ be a semicomplete bipartite digraph with a unique initial strong component. If there is no 3-king in $D$, then there are at least eight 4-kings in $D$.*

Let us point out that Theorem 8.6.7 was first stated by Wang in [63], and she cited [50] by Koh and Tan as the source of this result. Nonetheless, as mentioned above, Koh and Tan only proved this result for bipartite tournaments (Theorem 7.12.2). Wang does not give any argument in [63] to extend this result to semicomplete bipartite digraphs, so we will give a short one here.

Let $D$ be a semicomplete bipartite digraph with only one initial strong component $S$, and without 3-kings. By Theorem 7.1.1, every strong component $C$ of $D$ contains a strong spanning subgraph $C'$, which is a bipartite tournament. Replacing every strong component $C$ by $C'$ in $D$ results in a bipartite tournament with a unique strong component and without a 3-king. We can now apply Theorem 7.12.2 to this bipartite tournament to obtain the desired eight 4-kings. When we put back the deleted arcs, we will still have eight 4-kings in $D$, and, since they are in an initial strong component, they are 4-kings of all of $D$.

The following result was proved by Galeana-Sánchez, Hernández-Cruz and Juárez-Camacho, for $k = 2$, and by Wang and Zhang for $k \geq 3$.

**Theorem 8.6.8** ([37, 63]) *Let $k \geq 2$ be an integer, and let $D$ be a $k$-quasi-transitive digraph. If there is no $k$-king in $D$, then the number of $(k+1)$-kings in $D$ is at least $2k+2$.*

**Proof:** It suffices to prove the result for strong digraphs. For $k = 2$, consider the decomposition $D = S[Q_1, \ldots, Q_s]$ given by Theorem 8.3.5. Since $D$ is strong, the semicomplete digraph $S$ is also strong, and thus, by Corollary 2.2.14, it has at least three 2-kings. Since $D$ does not have 2-kings, it follows from the observation made after Theorem 8.6.1 that every $Q_i$ corresponding to a 2-king of $S$ has at least two vertices. Each of these vertices is a 3-king, and thus, $D$ has at least six 3-kings.

The case $k = 3$ can be directly verified using Theorems 8.3.15 and 8.6.7. Finally, for $k \geq 4$, as $D$ has no $k$-king, it must be isomorphic to an extended $(k+1)$-cycle, by Theorem 8.6.4. Every partite set in this cycle extension must have at least two vertices, otherwise there would be a $k$-king in $D$. Since every vertex of $D$ is a $(k+1)$-king, the number of $(k+1)$-kings is at least $2k+2$.

□

### 8.6.2 $(k, \ell)$-Kernels

A **kernel** in a digraph $D$ is an independent set $K$ such that every vertex not in $K$ dominates some vertex in $K$. Kernels in digraphs were introduced by von Neumann and Morgenstern while studying cooperative games [57]. Since then, digraph kernels have been studied in many contexts, including list colouring, game theory and graph perfectness [25], mathematical logic [22], and complexity theory [58].

There are many generalizations of this concept, one that has been widely studied and which relates to the kings from the previous subsection is the following. A subset $K$ of $V(D)$ is **$k$-independent** if the distance between every pair of vertices of $K$ is at least $k$, and it is **$\ell$-absorbing** if for every vertex not in $K$, it reaches a vertex in $K$ at distance at most $\ell$; if $\ell = 1$, we simply say that $K$ is **absorbing**. A **$(k, \ell)$-kernel** in the digraph $D$ is a $k$-independent and $\ell$-absorbing subset of $V(D)$. A **$k$-kernel** is a $(k, k-1)$-kernel, and thus, a 2-kernel is a kernel. The decision problem $k$-KERNEL has an arbitrary digraph $D$ as an input, and asks whether $D$ has a $k$-kernel. When $k = 2$, the corresponding problem will be referred to only as KERNEL.

Chvátal proved that KERNEL is $\mathcal{NP}$-complete [26]. Later Fraenkel proved that this problem remains $\mathcal{NP}$-complete even when restricted to planar digraphs with $\Delta \leq 3$, $\Delta^+, \Delta^- \leq 2$, [32]. Recently, Hell and Hernández-Cruz proved that it is also $\mathcal{NP}$-complete when restricted to digraphs with 3-colourable underlying graph (as opposed to the fact that every bipartite digraph has a kernel) [45]. Given the nice structure of quasi-transitive digraphs, it is not a surprise that members of this family having a kernel admit a simple characterization. One such characterization was given by Hell and Hernández-Cruz in [45].

**Theorem 8.6.9** ([45]) *Let $D$ be a strong quasi-transitive digraph. Then $D$ has a kernel if and only if there is an absorbing vertex in $D$.*

**Proof:** We only prove the non-trivial implication. Let $K$ be a kernel of $D$. Since $K$ is independent, it follows from Lemma 8.3.4 that it must be contained in $V(S)$ for some connected component of $\overline{UG(D)}$. Recalling that $D$ is strongly connected, there must be at least one connected component $S' \neq S$ of $\overline{UG(D)}$ such that $V(S) \rightarrow V(S')$. Since $K \subseteq S$, it must be the case that $V(S') \rightarrow V(S)$. Hence, Lemma 8.3.4 implies that $|V(S)| = 1$, and thus $|K| = 1$. If $K = \{v\}$, then $v$ is an absorbing vertex of $D$. □

In [48] Galeana-Sánchez and Hernández Cruz observe that, in order for a $k$-quasi-transitive digraph $D$ to have a $k$-kernel, it suffices to construct a $k$-kernel for every terminal strong component of $D$. In particular, this applies to kernels and quasi-transitive digraphs, and it allows us to conclude the following observation, which appears implicitly in [45].

**Corollary 8.6.10** *Let $D$ be a quasi-transitive digraph. Then $D$ has a kernel if and only if every terminal strong component contains an absorbing vertex.*

Hence, we obtain a polynomial time algorithm for the problem KERNEL restricted to the class of quasi-transitive digraphs.

**Corollary 8.6.11** *The problem* KERNEL *restricted to the class of quasi-transitive digraphs can be solved in polynomial time. Also, if a kernel exists, it can be constructed in polynomial time.*

**Proof:** Let $D = (V, A)$ be a digraph such that $|V| = n$ and $|A| = m$. The strong components digraph of $D$ can be obtained in time $O(n + m)$ and it can have at most $O(n)$ terminal strong components. For every terminal component $C$, it can be verified in time $O(n + m)$ if an absorbing vertex exists: it suffices to construct the out-degree sequence of $C$. Hence, the kernel problem can be decided in time $O(n^2 + nm)$. If $D$ has a kernel, it can be found in the same time. □

In order to obtain an analogous result for 3-kernels in 3-quasi-transitive digraphs, we need the following result of Hell and Hernández-Cruz.

**Proposition 8.6.12** ([45]) *It can be determined in linear time whether a semicomplete bipartite digraph has a 3-kernel. Also, if a 3-kernel exists, it can be found in linear time.*

This suffices to obtain the desired result. The proof of the following theorem implicitly uses the structure given in Theorem 8.3.15. Recall that the digraph $F_n$ has vertex set $\{x, y, z, v_1 \dots, v_n\}$, and its arcs are such that $xyzx$ is a directed cycle, and $yv_iz$ is a directed path for every $1 \leq i \leq n$ (see Figure 8.3).

**Theorem 8.6.13** ([45]) *The problem* 3-KERNEL *restricted to the class of* 3-*quasi-transitive digraphs can be decided in polynomial time. Also, if a* 3-*kernel exists, it can be constructed in polynomial time.*

**Proof:** Let $D = (V, A)$ be a digraph such that $|V| = n$ and $|A| = m$. The strong components digraph of $D$ can be constructed in time $O(n + m)$ and it can have at most $O(n)$ terminal strong components. For every semicomplete bipartite terminal component, according to Proposition 8.6.12, it can be verified if it has a 3-kernel and, if so, a 3-kernel can be found, both in time $O(n + m)$. For each semicomplete terminal component, a 3-kernel (a 2-king) can be found in time $O(n + m)$. For every terminal component isomorphic to $F_n$, a 3-kernel can be constructed in constant time. Hence, the 3-kernel problem can be decided in time $O(n^2 + nm)$. If $D$ has a 3-kernel, it can be found in the same time. □

In view of Corollary 8.6.11 and Theorem 8.6.13, the following natural question was stated by Hell and Hernández-Cruz in [45].

**Problem 8.6.14** ([45]) *Is* $k$-KERNEL *polynomial time solvable for* $k$-*quasi-transitive digraphs?*

In [48] and [37], the existence of $r$-kernels for $r \geq k+2$ was proved for every $k$-quasi-transitive digraph by Galeana-Sánchez, Hernández-Cruz and Juárez-Camacho. It was also proved in [36] that every quasi-transitive digraph has an $r$-kernel for $r \geq 3$, and in [48] it was proved that every 3-quasi-transitive digraph contains a 4-kernel, so it was natural to conjecture the existence of a $(k + 1)$-kernel for every $k$-quasi-transitive digraph. This conjecture was later proved by Wang and Zhang [63].

**Theorem 8.6.15** ([63]) *Let $D$ be a $k$-quasi-transitive digraph with $k \geq 2$. Then $D$ has a $(k + 1)$-kernel.*

**Proof:** We will only prove the case $k \geq 4$. For the cases $k \in \{2, 3\}$ we refer the reader to [36, 48].

Note that it suffices to choose $(k+1)$-kernels for every terminal component of $D$. To achieve this, consider the digraph $\overleftarrow{D}$ (called the **converse** of $D$) which is obtained from $D$ by reversing every arc. Theorem 8.6.4 guarantees that every initial component of $\overleftarrow{D}$ either contains a $k$-king, or is isomorphic to an extended $(k+1)$-cycle. The $k$-kings in the initial components of $\overleftarrow{D}$ become $(k+1)$-kernels in the terminal components of $D$. Since the reversal of an extended $(k+1)$-cycle is again an extended $(k+1)$-cycle, and it is clear that any partite set of an extended $(k+1)$-cycle is a $(k+1)$-kernel for it, we can choose a $(k+1)$-kernel for every terminal component of $D$. □

A $(k+1)$-cycle is a $k$-quasi-transitive digraph without a $k$-kernel. Thus, Problem 8.6.14 asks whether the first integer $r$ such that the $r$-KERNEL problem is not trivial when restricted to $k$-quasi-transitive digraphs yields a polynomial time solvable $r$-KERNEL problem. Notice that $k$-transitive digraphs always have a $k$-kernel, so, the first interesting kernel problem for $k$-transitive digraphs is $(k-1)$-KERNEL. Regarding this problem, Hernández-Cruz characterized 3-transitive digraphs with a kernel [46].

**Theorem 8.6.16** ([46]) *A 3-transitive digraph has a kernel if and only if none of its terminal components is isomorphic to a 3-cycle.*

**Proof:** Necessity is trivial to verify, we will only prove sufficiency. We will proceed by induction on the number of strong components of $D$. If $D$ is strong, the result can be verified directly by exploring the possibilities in Theorem 8.3.19. Let $D$ be a non-strong 3-transitive digraph, and let $S$ be an initial component of $D$. By induction hypothesis, $D - S$ has a kernel $N$. If $S$ is not a complete bipartite digraph, then either $S$ consists of a single vertex, or it contains a subdigraph isomorphic to $\vec{C}_3$. In the former case, either the only vertex in $S$ is absorbed by $N$, and we are done, or it is not, and we can add it to $N$ to obtain a kernel for $D$. If $S$ contains an isomorphic copy of $\vec{C}_3$, and since at least one vertex from $S$ reaches at least one vertex from some initial component of $D$, say $S'$, then Proposition 8.3.17 implies that $S \mapsto S'$. But $S' \cap N \neq \varnothing$, thus, every vertex of $S$ is absorbed by $N$.

If $S = (X, Y)$ is a complete bipartite digraph, we will consider three cases. If neither $X$ nor $Y$ is absorbed by $N$, then $N \cup X$ is a kernel of $D$. If some vertex of $X$ is absorbed by $N$, it follows from Proposition 8.3.17 that every vertex of $X$ is absorbed by $N$. If $Y$ is also absorbed by $N$, then $N$ is a kernel of $D$. Else, none of the vertices of $Y$ is absorbed by $D$, and thus, $N \cup Y$ is a kernel of $D$. □

Inspired by Theorem 8.6.16, Wang proved the following general result for strong $k$-transitive digraphs.

**Theorem 8.6.17** ([59]) *Let $D$ be a strong $k$-transitive digraph with $k \geq 4$. Then $D$ has a $(k-1)$ kernel if and only if it is not isomorphic to a $k$-cycle.*

This, again observing Theorem 8.6.16, led to the following conjecture.

**Conjecture 8.6.18** ([59]) *Let $k \geq 3$ be an integer. If $D$ is a $k$-transitive digraph, then $D$ has a $(k-1)$-kernel if and only if has no terminal strong component isomorphic to a $k$-cycle.*

In [38], García-Vázquez and Hernández-Cruz provided support to Conjecture 8.6.18 by proving it true for $k = 4$. Additionally, in the same paper the authors characterized 4-transitive digraphs having a kernel. The characterization relies heavily on a characterization of strong 4-transitive digraphs having a kernel, found in the same paper (see Subsection 8.3.3). It is trivial to observe that a $k$-kernel for a $k$-transitive digraph consists of a disjoint union of $k$-kernels for each of its terminal components. Conjecture 8.6.18 can be reformulated as follows: A $k$-transitive digraph $D$ has a $(k-1)$-kernel if and only if each of its terminal components has a $(k-1)$-kernel. To prove the aforementioned characterization of 4-transitive digraphs with a kernel, the authors actually prove that a 4-transitive digraph has a kernel if and only if every terminal component has a kernel. So, the following questions come to mind.

**Problem 8.6.19** *Let $k \geq 4$ be an integer and let $D$ be a $k$-transitive digraph. Is it true that $D$ has a $(k-2)$-kernel if and only if every terminal component of $D$ has a $(k-2)$-kernel?*

*If so, which is the least value of $r$ for $2 \leq r \leq k-3$ such that $D$ has an $r$-kernel if and only if every terminal component of $D$ has an $r$-kernel?*

An affirmative answer to the first question in Problem 8.6.19 would imply that it suffices to solve the $(k-2)$-kernel problem for strong $k$-transitive digraphs to obtain a solution for all $k$-transitive digraphs.

Finally, another particular case of $(k, \ell)$-kernels is that of quasi-kernels. A **quasi-kernel** is simply a $(2, 2)$-kernel. Chvátal and Lovász proved that every digraph has a quasi-kernel. So, a question that has been raised by Gutin, Koh, Tay and Yeo [42] is the following: Which digraphs contain (at least) a pair of disjoint quasi-kernels? Clearly, a digraph which has a pair of disjoint quasi-kernels cannot contain vertices of out-degree zero, since every such vertex is included in every quasi-kernel. Unfortunately, this obvious necessary condition is not sufficient in general for a digraph to have a pair of disjoint quasi-kernels. Examples of digraphs which have neither vertices of out-degree zero nor a pair of disjoint quasi-kernels are given in [42]. Nonetheless, Heard and Huang proved that this condition is indeed sufficient in the class of quasi-transitive digraphs [44]. We need the following result, which is found in [44].

**Proposition 8.6.20** ([44]) *Every semicomplete digraph $D$ with no vertices of out-degree zero contains two vertices $x, y$ such that $\{x\}$ and $\{y\}$ are both quasi-kernels of $D$.*

We begin with strong quasi-transitive digraphs.

**Proposition 8.6.21** ([44]) *Every strong quasi-transitive digraph without vertices of out-degree zero contains a pair of disjoint quasi-kernels.*

**Proof:** Let $D$ be a strong quasi-transitive digraph without vertices of out-degree zero. Let $D = S[H_1, \ldots, H_s]$ be the canonical decomposition of $D$ (Theorem 8.3.5). Since $D$ has no vertices of out-degree zero, $S$ must contain at least two vertices and hence contain no vertices of out-degree zero. By Proposition 8.6.20, there are two vertices $x, y$ such that $\{x\}$ and $\{y\}$ are both quasi-kernels of $S$. Suppose that $H_i$ and $H_j$ are the two digraphs which substitute $x$ and $y$, respectively, in the composition. Let $Q$ and $Q'$ be quasi-kernels of $H_i$ and $H_j$, respectively. Then, it is easy to see that $Q$ and $Q'$ are disjoint quasi-kernels of $D$. □

We now turn to the non-strong case.

**Proposition 8.6.22** ([44]) *Every non-strong quasi-transitive digraph without vertices of out-degree zero contains a pair of disjoint quasi-kernels.*

**Proof:** Let $D$ be a non-strong quasi-transitive digraph without vertices of out-degree zero. Let $D = T[H_1, \ldots, H_t]$ be the canonical decomposition of $D$ (Theorem 8.3.5). Let $\{u_1, \ldots, u_t\}$ be the vertex set of $T$, and, without loss of generality, suppose that $u_1, \ldots, u_r$ are the sinks of $T$. Note that $\{u_1, \ldots, u_r\}$ is a kernel of $T$. Since $D$ does not contain vertices of out-degree zero, neither do any $H_i$, $1 \leq i \leq r$. By Proposition 8.6.21, each $H_i$ contains two disjoint quasi-kernels, say $Q_{i,1}$ and $Q_{i,2}$, $1 \leq i \leq r$. It is not hard to verify that $Q_1 = \bigcup_{i=1}^{r} Q_{i,1}$ and $Q_2 = \bigcup_{i=1}^{r} Q_{i,2}$ are disjoint quasi-kernels of $D$. □

Combining Propositions 8.6.21 and 8.6.22, we have the following:

**Theorem 8.6.23** *Every quasi-transitive digraph without vertices of out-degree zero contains a pair of disjoint quasi-kernels.*

## 8.7 The Path Partition Conjecture

### 8.7.1 The Conjecture

Recall that a longest path in a digraph $D$ is called a **detour** of $D$. The order of a detour of $D$ is called the **detour order** of $D$ and is denoted by $\mathsf{do}(D)$ . The Gallai–Roy–Vitaver Theorem states that the chromatic number of the underlying graph of a digraph $D$ is at most $\mathsf{do}(D)$. In 1982 Laborde, Payan and Xuong posed the following conjecture, which extends this theorem in a natural way.

**Conjecture 8.7.1** ([52]) *Every digraph $D$ contains an independent set $X$ such that* $\mathsf{do}(D - X) < \mathsf{do}(D)$.

Conjecture 8.7.1 has proved to be a very difficult problem, and only a handful of partial results have been obtained. Nonetheless, it has not received as much attention as one of its generalizations. The following conjecture is probably the best known among the related path partition problems, it is known as the **Path Partition Conjecture**.

**Conjecture 8.7.2 (Path Partition Conjecture)** *[52] For every digraph $D$ and every choice of positive integers $\ell_1, \ell_2$ such that $\mathsf{do}(D) = \ell_1 + \ell_2$, there exists a partition of $D$ into two digraphs, $D_1$ and $D_2$, such that $\mathsf{do}(D_i) \leq \ell_i$ for $i \in \{1, 2\}$.*

Clearly, Conjecture 8.7.1 is the particular case of Conjecture 8.7.2 when $\ell_1 = 1$ and $\ell_2 = \mathsf{do}(D) - 1$.

A seemingly stronger version of the conjecture is stated in [24]. Bondy attributes it to Laborde *et al.* [52] although only the undirected version of Conjecture 8.7.2 is explicitly mentioned there.

**Conjecture 8.7.3** ([24]) *For every digraph $D$ and every choice of positive integers $\ell_1, \ell_2$ such that $\mathsf{do}(D) = \ell_1 + \ell_2$, there exists a partition of $D$ into two digraphs, $D_1$ and $D_2$, such that $\mathsf{do}(D_i) = \ell_i$ for $i \in \{1, 2\}$.*

There is another problem also found in [52] which is a stronger version of Conjecture 8.7.1, but somehow this conjecture, sometimes referred to as the **Strong Laborde–Payan–Xuong Conjecture**, has received even less attention than Conjecture 8.7.1.

**Conjecture 8.7.4** ([52]) *Every digraph $D$ contains an independent set $X$ such that that $\mathsf{do}(D - X) < \mathsf{do}(D)$, and has the additional property that every vertex in $X$ is the beginning of some detour of $D$.*

One further extension of Conjecture 8.7.1 has been considered by Galeana-Sánchez and Gómez in [35]. A path $P = x_0x_1 \dots x_n$ is non-augmentable if for every $v \in V(D) - V(P)$, and for every $0 \leq i \leq n-1$, $vx_0 \dots x_n$, $x_0 \dots x_n v$ and $x_0 \dots x_i v x_{i+1} \dots x_n$ are not paths. Clearly, every detour is non-augmentable, so, if true, Conjecture 8.7.1 would be an immediate consequence of the following conjecture, which appears implicitly in the paper of Galeana-Sánchez and Gómez [35] but has never been explicitly stated.

**Conjecture 8.7.5** ([35]) *Every digraph $D$ contains an independent set which intersects every non-augmentable path of $D$.*

### 8.7.2 Known Results

There are some partial results supporting each of the aforementioned conjectures, principally, Conjectures 8.7.1 and 8.7.2; we refer the reader to [2, 20, 33, 35, 38, 64]. Most of the existing results prove some of these conjectures for restricted families of digraphs; in most cases, generalizations of tournaments.

In [20], Conjecture 8.7.2 is considered for the family of quasi-transitive digraphs. There, Bang-Jensen, Nielsen and Yeo prove the following theorem. Recall that $\mathsf{do}_k(D)$ is the maximum number of vertices contained in a $k$-path subdigraph of a digraph $D$.

**Theorem 8.7.6** ([20]) *Let $D$ be a quasi-transitive digraph or a strong extended semicomplete digraph, and let $q$ be any positive integer. Then there exists a partition, $(A, B)$, of $V(D)$ such that the following holds:*

1. $\mathsf{do}(D[A]) \leq q$;
2. $\mathsf{do}_k(D[B]) \leq \mathsf{do}_k(D) - q$ *for all* $k \in \{1, \dots, |V(B)|\}$, *provided* $\mathsf{do}_k - q \geq 0$.

Although Theorem 8.7.6 implies that Conjecture 8.7.1 is also true for quasi-transitive digraphs, it does not give us information on any of the other conjectures mentioned in the previous subsection. In [35], Galeana-Sánchez and Gómez proved Conjecture 8.7.5 to be true for quasi-transitive digraphs. Again, the proof of this result relies heavily on Theorem 8.3.5, which is also used by the following necessary lemma.

**Lemma 8.7.7** ([35]) *Let $H$ be a digraph such that $H = D[H_1, \dots, H_n]$, where $D$ is a transitive acyclic digraph with vertex set $\{v_1, \dots, v_n\}$, and $H_i$ are arbitrary digraphs for $1 \leq i \leq n$. If $\mathcal{I}_i$ is a maximal independent set intersecting every non-augmentable path of $H_i$, $1 \leq i \leq n$, then $\mathcal{I} = \bigcup_{i=1}^{n} \mathcal{I}_i$ is a maximal independent set that intersects every non-augmentable path in $H$.*

**Proof:** Since $D$ is acyclic and transitive, its set of vertices of in-degree zero, $S$, is a maximal independent set intersecting every non-augmentable path of $D$.

Let $P$ be a non-augmentable path in $H$. It is not hard to verify that the contraction[9] $P' = P/\{H_1, \dots, H_n\}$ is a non-augmentable path of $D$, hence, $S$ intersects $P'$. Also, if $P$ uses at least one vertex from $H_i$, then it should be the case that $P \cap H_i$ is a non-augmentable path of $H_i$; otherwise, $P$ could be augmented.

Thus, if we let $\mathcal{I}$ be the union of the $\mathcal{I}_j$'s corresponding to the vertices in $S$, then $\mathcal{I}$ is a maximal independent set intersecting every non-augmentable path of $H$. □

We will only present the general idea of the proof of the following theorem, due to its length and technical arguments.

**Theorem 8.7.8** ([35]) *Let $D$ be a quasi-transitive digraph. There exists a maximal independent set $\mathcal{I}$ of $D$ that intersects every non-augmentable path in $D$. Moreover, if $D$ is strong with decomposition $D = S[Q_1, \dots, Q_s]$, and $\mathcal{I}_i \subseteq V(Q_i)$ is a maximal independent set intersecting every non-augmentable path in $Q_i$, for $1 \leq i \leq s$, then each $\mathcal{I}_i$ is also a maximal independent set intersecting every non-augmentable path in $D$.*

**Idea of Proof.** The proof is by induction on $|V(D)|$. If $|V(D)| = 1$, the result is clearly true.

[9] See Section 1.4.

If $D$ is not strong, the result follows from Lemma 8.7.7 and the induction hypothesis.

If $D$ is strong, then, by Theorem 8.3.5, $D = S[Q_1, \ldots, Q_s]$, with $S$ strong semicomplete and each $Q_i$ a non-strong quasi-transitive digraph or a single vertex. Let $P$ be a non-augmentable path of $D$. Recall that a path in a semicomplete digraph is non-augmentable if and only if it is Hamiltonian. Thus, $P$ must intersect every $Q_i$. If $Q_i$ is a single vertex, then it is a maximal independent set intersecting every non-augmentable path of $D$. Else, by induction hypothesis there is a maximal independent set $S_i$ intersecting every non-augmentable path in $Q_i$. The proof finishes with an analysis of cases to show that $S_i$ intersects every non-augmentable path of $D$. □

Wang and Wang attacked Conjecture 8.7.5 in [64]. Their main result relevant to our interests in this chapter is the following.

**Theorem 8.7.9** ([64]) *If $D$ is a 3-quasi-transitive digraph, then there exists an independent set intersecting every non-augmentable path in $D$.*

**Proof:** If $D$ is strong, then, using the characterization given in Theorem 8.3.15, it is easy to verify that every maximal independent set intersects every non-augmentable path in $D$. Therefore, assume that $D$ is not strong and let $D_0, \ldots, D_k$ be its strong components. Let $D_0, \ldots, D_s$ be the initial strong components and let $F_i$ be a maximal independent set of $D_i$, for $1 \leq i \leq s$. Let $Z = V(D) - \bigcup_{i=0}^{s} V(D_s)$ and define $W$ as

$$W = \{x \in Z\colon \text{ there exists a non-augmentable path in } D \text{ starting at } x\}.$$

Observe that $W$ is either independent or empty. If $|W| \leq 1$, there is nothing to prove. Assume $|W| \geq 2$, and suppose for a contradiction that there is a pair $x, y$ of adjacent vertices in $W$. By the definition of $W$, $x$ and $y$ must belong to the same strong component, say $D_j$. Since $N^-(V(D_j))$ is non-empty, we may choose a vertex $u \in N^-(V(D_j))$. If $D_j$ is non-bipartite, then, by Lemma 8.3.16, $u \mapsto D_j$, and hence $u \mapsto x$, a contradiction. If $D_j$ is bipartite, then $x$ and $y$ must belong to different parts. Hence, by Lemma 8.3.16, $u$ and one of $x$ and $y$ are adjacent, a contradiction.

Let $F = F_0 \cup \cdots \cup F_s \cup W$. It is not difficult to deduce that $F$ is an independent set in $D$. Let $P$ be a non-augmentable path of $D$ with initial vertex $x_0$. If $x_0$ does not belong to any initial component, then $x \in W$. Else, $x_0$ belongs to an initial component $D_0$ of $D$. If $D_0$ is semicomplete, then it is not hard to observe that $P \cap D_0$ is a Hamiltonian path of $D_0$, and thus, $P$ must intersect $F_0$. If $D_0$ is complete bipartite, then $F_0$ is some part of $D_0$, so, $F_0$ intersects $P \cap D_0$. If $D_0$ is an element of the family $\mathcal{F}$ (see Theorem 8.3.15), then it is easy to verify that $F_0$ intersects $P \cap D_0$. □

Theorem 8.7.9 settles Conjecture 8.7.5 (which implies Conjecture 8.7.1) for 3-quasi-transitive digraphs. Arroyo and Galeana-Sánchez proved Conjecture 8.7.2 for strong 3-quasi-transitive digraphs in [2].

**Theorem 8.7.10** ([2]) *Let $D$ be a strong 3-quasi-transitive digraph. Consider two positive integers $\ell_1 \geq \ell_2$ such that $\ell_1 + \ell_2 = \mathsf{do}(D)$. Then there exists a partition $(A, B)$ of $V(D)$ such that $\mathsf{do}(D[A]) \leq \ell_1$ and $\mathsf{do}(D[B]) \leq \ell_2$.*

**Proof:** Since the conjecture is easy to verify for semicomplete and bipartite digraphs, it follows from Theorem 8.3.15 that it only remains to show its validity in the digraphs of the family $\mathcal{F}$.

Let $D$ be a digraph in the family $\mathcal{F}$. Notice that $4 \leq \mathsf{do}(D) \leq 5$, hence, it is easy to verify that, for every choice of $\ell_1, \ell_2$ such that $\ell_1 + \ell_2 = \mathsf{do}(D)$, the partition $(\{y, z\}, V(D) - \{y, z\})$ (see Figure 8.3) has the required property. □

Since every 3-transitive digraph is also 3-quasi-transitive, Theorems 8.7.9 and 8.7.10 also cover the 3-transitive case. Thus, the first interesting case for $k$-transitive digraphs is $k = 4$. For 4-transitive digraphs only Conjecture 8.7.1 has been explored; García-Vázquez and Hernández-Cruz proved it true for 4-transitive digraphs [38]. Again, the proof of the following theorem involves a technical analysis of various cases, and thus, only an idea of the proof method will be given.

**Theorem 8.7.11** ([38]) *For every 4-transitive digraph $D$ there exists an independent set intersecting every longest path of $D$.*

**Idea of Proof.** It is possible to prove that a 4-transitive digraph has a kernel if and only if every terminal strong component has a kernel. Also, using Theorem 8.3.20, it is not hard to characterize the strong 4-transitive digraphs having a kernel.

Let $D$ be a 4-transitive digraph. Using the aforementioned characterization of the strong 4-transitive digraphs having a kernel, it is possible to find a minimal subset $S$ of $V(D)$ such that $D - S$ has a kernel $K$. The set $K$ is precisely the stable set we are looking for. □

To finish this section, we present a table with the values of $k$ for which each of the discussed conjectures is known to be valid in $k$-transitive and $k$-quasi-transitive digraphs, and their corresponding strongly connected versions. In the columns of the table, LPX stands for Laborde–Payan–Xuong (Conjecture 8.7.1), SLPX for Strong LPX (Conjecture 8.7.4), NALPX for Non-Augmentable LPX (Conjecture 8.7.5), PPC for the Path Partition Conjecture (Conjecture 8.7.2), and SPPC for Strong PPC (Conjecture 8.7.3).

| | LPX | SLPX | NALPX | PPC | SPPC |
|---|---|---|---|---|---|
| Strong $k$-transitive | $k \leq 4$ | $k \leq 4$ | $k \leq 3$ | $k \leq 3$ | $k = 2$ |
| $k$-transitive | $k \leq 4$ | $k \leq 3$ | $k \leq 3$ | $k = 2$ | $k = 2$ |
| Strong $k$-quasi-transitive | $k \leq 3$ | | $k \leq 3$ | $k \leq 3$ | |
| $k$-quasi-transitive | $k \leq 3$ | | $k \leq 3$ | $k = 2$ | |

## 8.8 Miscellaneous

### 8.8.1 Vertex Pancyclicity

Pancyclicity is one of the properties that first comes to mind when thinking of tournaments.

Recall from Theorem 2.2.9 that every strong semicomplete digraph is vertex-pancyclic. As a generalization of tournaments, and semicomplete digraphs, it is natural to ask whether a Hamiltonian quasi-transitive digraph is vertex-pancyclic. In [17], Bang-Jensen and Huang use the similarities between extended semicomplete digraphs and quasi-transitive digraphs to derive results on pancyclic and vertex-pancyclic quasi-transitive digraphs. In this section we present a brief summary of these results.

A digraph $D$ is **triangular with partition** $V_0, V_1, V_2$ if the vertex set of $D$ can be partitioned into three disjoint sets $V_0, V_1, V_2$ with $V_0 {\mapsto} V_1 {\mapsto} V_2 {\mapsto} V_0$. Note that this is equivalent to saying that $D = \vec{C}_3[D[V_0], D[V_1], D[V_2]]$.

Gutin [40] characterized pancyclic and vertex-pancyclic extended semicomplete digraphs. Clearly no extended semicomplete digraph of the form $D = \vec{C}_2[\overline{K}_{n_1}, \overline{K}_{n_2}]$ with at least 3 vertices is pancyclic since all cycles are of even length. Hence we must assume that there are at least 3 parts in order to get a pancyclic extended semicomplete digraph. It is also easy to see that the (unique) strong 3-partite extended semicomplete digraph on 4 vertices is not pancyclic (since it has no 4-cycle). These observations together with Theorem 7.10.8 completely characterize pancyclic and vertex-pancyclic extended semicomplete digraphs. It is not difficult to see that Theorem 7.10.8 extends Theorem 1.5.1, since no semicomplete digraph on $n \geq 5$ vertices satisfies any of the exceptions from (a) and (b).

The next two lemmas of Bang-Jensen and Huang [17] concern cycles in triangular digraphs. They are used in the proof of Theorem 8.8.3, which characterizes pancyclic and vertex-pancyclic quasi-transitive digraphs.

**Lemma 8.8.1** ([17]) *Suppose that $D$ is a triangular digraph with a partition $V_0, V_1, V_2$ and suppose that $D$ is Hamiltonian. If $D[V_1]$ contains an arc $xy$ and $D[V_2]$ contains an arc $uv$, then every vertex of $V_0 \cup \{x, y, u, v\}$ is on cycles of lengths $3, 4, \ldots, n$.* □

**Lemma 8.8.2** ([17]) *Suppose that $D$ is a triangular digraph with a partition $V_0, V_1, V_2$ and $D$ has a Hamiltonian cycle $C$. If $D[V_0]$ contains an arc of $C$ and a path $P$ of length 2, then every vertex of $V_1 \cup V_2 \cup V(P)$ is on cycles of lengths $3, 4, \ldots, n$.* □

It is easy to check that a strong quasi-transitive digraph on 4 vertices is pancyclic if and only if it is a semicomplete digraph. For $n \geq 5$ we have the following characterization due to Bang-Jensen and Huang:

**Theorem 8.8.3** ([17]) *Let $D = (V, A)$ be a Hamiltonian quasi-transitive digraph on $n \geq 5$ vertices.*

(a) *$D$ is pancyclic if and only if it is not triangular with a partition $V_0, V_1, V_2$, two of which induce digraphs with no arcs, such that either $|V_0| = |V_1| = |V_2|$, or no $D[V_i]$ $(i = 0, 1, 2)$ contains a path of length 2.*

(b) *$D$ is not vertex-pancyclic if and only if $D$ is not pancyclic or $D$ is triangular with a partition $V_0, V_1, V_2$ such that one of the following occurs:*

  (b1) *$|V_1| = |V_2|$, both $D[V_1]$ and $D[V_2]$ have no arcs, and there exists a vertex $x \in V_0$ such that $x$ is not contained in any path of length 2 in $D[V_0]$ (in which case $x$ is not contained in a cycle of length 5).*

  (b2) *one of $D[V_1]$ and $D[V_2]$ has no arcs and the other contains no path of length 2, and there exists a vertex $x \in V_0$ such that $x$ is not contained in any path of length 1 in $D[V_0]$ (in which case $x$ is not contained in a cycle of length 5);*

□

### 8.8.2 Acyclic Spanning Subgraphs

It is well known that a semicomplete digraph $T$ contains an $(x, y)$-Hamiltonian path if and only if there is a spanning acyclic subgraph $S$ (not necessarily induced) such that $S$ contains an $(x, z)$-path and a $(z, y)$-path for each vertex $z$ of $T$, cf. [56]. This also follows from the fact that semicomplete digraphs are path-mergeable, see [3] and Section 6.2.

It follows from the characterization in Theorem 8.4.7 that a quasi-transitive digraph $D$ may not have a Hamiltonian path even if it is highly connected and has a path $P$ such that $D - P$ has a cycle factor (see [17] for such an example). On the other hand, Bang-Jensen and Huang proved in [17] that if a quasi-transitive digraph has a unique initial and a unique terminal strong component then we can always guarantee the existence of such an acyclic spanning subgraph.

**Theorem 8.8.4** ([17]) *Suppose that $D$ is a quasi-transitive digraph having both in- and out-branchings. Then $D$ has a spanning acyclic subgraph $S$ with a source $x$ and a sink $y$ such that for each vertex $z$ of $D$, $D$ contains an $(x, z)$-path and a $(z, y)$-path.* □

**Corollary 8.8.5** *Every strong quasi-transitive digraph has a spanning acyclic subdigraph $S$ with a source $x$ and a sink $y$ such that, for each vertex $z$ of $D$, $S$ contains an $(x, z)$-path and a $(z, y)$-path.* □

### 8.8.3 Orientations of Digraphs Almost Preserving Diameter

Recall that an **orientation** of a digraph $D$ is a spanning subdigraph of $D$ obtained from $D$ by deleting exactly one arc from every 2-cycle. Chvátal and Thomassen [27] proved that the problem of checking whether a given undirected graph has an orientation of diameter 2 is $\mathcal{NP}$-complete, and the upper bound on the diameter of an orientation of an undirected graph obtained in [27] is far from the best possible for many classes of undirected graphs (recall that undirected graphs may be regarded as digraphs where every arc is symmetric).

We have already seen many problems which have very nice solutions for the class of quasi-transitive digraphs, e.g., hamiltonicity, existence of kernels, $k$-linkages and weak $k$-linkages, which are $\mathcal{NP}$-complete in the general case, are polynomial time solvable for quasi-transitive digraphs. The study of minimum diameter orientations of quasi-transitive digraphs is not an exception; a surprisingly good bound on the minimum diameter of an orientation of a quasi-transitive digraph holds. Before stating the main results of this section, we will recall a result due to Boesch and Tindell which extends Robbins' Theorem.

**Theorem 8.8.6** ([23]) *A strong digraph $D$ has no strong orientation if and only if there is a pair $x, y$ of vertices in $D$ such that the deletion of the arcs $xy, yx$ leaves $D$ disconnected.*

Applying Theorem 8.8.6 it is easy to see that every strong quasi-transitive digraph of order $n \geq 3$ has a strong orientation. For a digraph $D$, let $\mathrm{diam}_{\min}(D)$ denote the minimum diameter of an orientation of $D$. The following result is due to Gutin and Yeo [43].

**Theorem 8.8.7** ([43]) *If $D$ is a strong quasi-transitive digraph, then*

$$\mathrm{diam}_{\min}(D) \leq \max\{3, \mathrm{diam}(D)\}.$$

The upper bound of this theorem is sharp as one can see from the following example. Let $T_k$, $k \geq 3$, be a (transitive) tournament with vertices $x_1, x_2, ..., x_k$ and arcs $x_i x_j$ for every $1 \leq i < j \leq k$. Let $y$ be a vertex not in $T_k$, which dominates all vertices of $T_k$ but $x_k$ and is dominated by all vertices of $T_k$ but $x_1$. The resulting semicomplete digraph $D_{k+1}$ has diameter 2. However, the deletion of any arc of $D_{k+1}$ between $y$ and the set $\{x_2, x_3, ..., x_{k-1}\}$ leaves a digraph with diameter 3. Indeed, if we delete $yx_i$, $2 \leq 2 \leq k-1$, then a shortest $(x_k, x_i)$-path becomes of length 3.

### 8.8.4 Sparse Subdigraphs with Prescribed Connectivity

A spanning $k$-(arc)-strong subdigraph $D'$ of a directed multigraph $D$ is called a **certificate** for the $k$-(arc)-strong connectivity of $D$. A problem of practical interest is the following. Let $D = (V, A)$ be a $k$-(arc)-strong directed multigraph and let $c$ be a cost function on $A$ (possibly $c(a) = 1$ for all $a \in A$). What is the minimum cost of a $k$-(arc)-strong spanning subdigraph $D'$ of $D$? An **optimal certificate** for $k$-(arc)-strong connectivity in $D$ is a spanning $k$-(arc)-strong subdigraph $D'$ of minimum cost. Finding such an optimal certificate is a hard problem already when $k = 1$ and $c \equiv 1$. This follows from the fact that the optimal certificate for the strong connectivity of $D$ has $|V|$ arcs if and only if $D$ has a Hamilton cycle.

When $c \equiv 1$, we have the problem of finding an optimal certificate for strong connectivity. We call this the MINIMUM SPANNING STRONG SUBDIGRAPH problem (MSSS, see [18]).

For the case of quasi-transitive digraphs, we begin with a lower bound. Recall that the path-covering number of a digraph $D$, $pc(D)$, is the least positive integer $k$ such that $D$ has a $k$-path factor. For a strong quasi-transitive digraph $D$ we define $\mathrm{pc}^*(D)$ to be equal to 0 if $D$ is Hamiltonian, and $\mathrm{pc}^*(D) = \mathrm{pc}(D)$ otherwise. The optimal solution to the MSSS problem for quasi-transitive digraphs was given by Bang-Jensen, Huang, and Yeo. The proof can be found in [9].

**Theorem 8.8.8** ([18]) *Every minimum spanning strong subdigraph of a quasi-transitive digraph has precisely $n + \mathrm{pc}^*(D)$ arcs. Furthermore, we can find a minimum spanning strong subdigraph in time $O(|V|^4)$.*

A **directed cactus** is a strongly connected digraph in which each arc is contained in exactly one cycle.

Palbom [55] studied the complexity of various problems related to spanning directed cactii in digraphs. It is not difficult to check whether a given digraph is a cactus, but Palbom proved that deciding whether a digraph contains a spanning cactus is an $\mathcal{NP}$-complete problem [55].

Since the directed spanning cactus problem (the problem of determining whether a digraph contains a spanning cactus) is trivial for locally in-semicomplete digraphs, and easy for path-mergeable digraphs, but already non-trivial for extended semicomplete digraphs (see, Exercises 12.17 and 12.20 in [9]), the following problem comes as a natural next step in this subject.

**Problem 8.8.9** ([9]) *Determine the complexity of the spanning directed cactus problem for quasi-transitive digraphs.*

### 8.8.5 Arc-Disjoint In- and Out-Branchings

We now consider the problem ARC-DISJOINT IN- AND OUT-BRANCHINGS: Given a digraph $D$ and vertices $u, v$ (not necessarily distinct), decide whether $D$ has a pair of arc-disjoint branchings $B_u^+, B_v^-$ such that $B_u^+$ is an out-branching rooted at $u$ and $B_v^-$ is an in-branching rooted at $v$. Recall from Theorem 2.12.19 that Thomassen proved that ARC-DISJOINT IN- AND OUT-BRANCHINGS is $\mathcal{NP}$-complete for general digraphs.

In [4], Bang-Jensen proved that a tournament $T$ has arc-disjoint in- and out-branchings rooted at some vertex $v$ if and only if there is no arc that must be on all out-branchings from $v$ and all in-branchings to $v$, see Corollary 2.12.21. In [17], Bang-Jensen and Huang considered digraphs having a vertex $v$ which is adjacent to every other vertex; they obtained a characterization of digraphs having arc-disjoint in- and out-branchings rooted at $v$. As a consequence, they obtained the following result.

**Theorem 8.8.10** ([17]) *Let $D$ be a strong digraph and $v$ a vertex of $D$ such that $V(D) = \{v\} \cup N^+(v) \cup N^-(v)$. There is a polynomial algorithm to decide if $D$ has arc-disjoint in- and out-branchings $F_v^-, F_v^+$ rooted at $v$.*

The previous result can be combined with the following lemma to obtain a polynomial algorithm to decide if a quasi-transitive digraph $D$ has arc-disjoint in- and out-branchings rooted at a given vertex $v$.

**Lemma 8.8.11** ([17]) *Let $D$ be a quasi-transitive digraph and $v \in V(D)$ a vertex of $D$. Then $D$ contains arc-disjoint branchings $F_v^+, F_v^-$ rooted at $v$ if and only if $D' = D[\{v\} \cup N^-(v) \cup N^+(v)]$ has arc-disjoint branchings $F_v'^+, F_v'^-$ rooted at $v$.* □

**Theorem 8.8.12** ([17]) *Let $D$ be a strong quasi-transitive digraph, and $v$ a vertex of $D$. If $B = \{B_1 \ldots, B_k\}$ ($C = \{C_1, \ldots, C_r\}$) denote the set of terminal (initial) components in $D[N^+(v)]$ ($D[N^-(v)]$), then $D$ contains a pair of arc-disjoint branchings $F_v^+, F_v^-$ such that $F_v^+$ is an out-branching rooted at $v$ and $F_v^-$ is an in-branching rooted at $v$ if and only if there exist two disjoint arc sets $A_B, A_C \subset A(D)$ such that all arcs in $A_B \cup A_C$ go from $N^+(v)$ to $N^-(v)$ and every component in $B_i \in B$ ($C_j \in C$) is incident with an arc from $A_B$ ($A_C$).* □

From here, the following result settling the problem for quasi-transitive digraphs is obtained.

**Corollary 8.8.13** ([17]) *There is a polynomial algorithm to decide if a quasi-transitive digraph $D$ has arc-disjoint in- and out-branchings rooted at a given vertex $v$.* □

As noted in Section 2.12, already for semicomplete digraphs, the problem of finding arc-disjoint in- and out-branchings becomes much harder when $u \neq v$. Even the class of semicomplete digraphs is still lacking a polynomial time algorithm to decide this problem when $u \neq v$.

## References


1. J. Alva-Samos and C. Hernández-Cruz. $k$-quasi-transitive digraphs of large diameter. submitted.
2. A. Arroyo and H. Galeana-Sánchez. The path partition conjecture is true for some generalizations of tournaments. *Discrete Math.*, 313(3):293–300, 2013.
3. J. Bang-Jensen. Digraphs with the path-merging property. *J. Graph Theory*, 20(2):255–265, 1995.

4. J. Bang-Jensen. Edge-disjoint in- and out-branchings in tournaments and related path problems. *J. Combin. Theory Ser. B*, 51(1):1–23, 1991.
5. J. Bang-Jensen. Linkages in locally semicomplete digraphs and quasi-transitive digraphs. *Discrete Math.*, 196(1-3):13–27, 1999.
6. J. Bang-Jensen. The structure of strong arc-locally semicomplete digraphs. *Discrete Math.*, 283(1-3):1–6, 2004.
7. J. Bang-Jensen, T. M. Christiansen, and A. Maddaloni. Disjoint paths in decomposable digraphs. *J. Graph Theory*, 85(2):545–567, 2017
8. J. Bang-Jensen and G. Gutin. *Digraphs: Theory, Algorithms and Applications.* Springer-Verlag, London, 2000.
9. J. Bang-Jensen and G. Gutin. *Digraphs: Theory, Algorithms and Applications.* Springer-Verlag, London, 2nd edition, 2009.
10. J. Bang-Jensen and G. Gutin. Finding maximum vertex weight paths and cycles in $\Phi$-decomposable digraphs, using flows in networks. Technical report 51, Department of Mathematics and Computer Science, Odense University, Denmark, 1993.
11. J. Bang-Jensen and G. Gutin. Generalizations of tournaments: A survey. *J. Graph Theory*, 28:171–202, 1998.
12. J. Bang-Jensen and G. Gutin. On the complexity of hamiltonian path and cycle problems in certain classes of digraphs. *Discrete Appl. Math.*, 95:41–60, 1999.
13. J. Bang-Jensen and G. Gutin. Paths and cycles in extended and decomposable digraphs. *Discrete Math.*, 164(1-3):5–19, 1997.
14. J. Bang-Jensen and G. Gutin. Vertex heaviest paths and cycles in quasi-transitive digraphs. *Discrete Math.*, 163(1-3):217–223, 1997.
15. J. Bang-Jensen, G. Gutin, and A. Yeo. Finding a cheapest cycle in a quasi-transitive digraph with real-valued vertex costs. *Discrete Optim.*, 3:86–94, 2006.
16. J. Bang-Jensen and J. Huang. Kings in quasi-transitive digraphs. *Discrete Math.*, 185(1-3):19–27, 1998.
17. J. Bang-Jensen and J. Huang. Quasi-transitive digraphs. *J. Graph Theory*, 20(2):141–161, 1995.
18. J. Bang-Jensen, J. Huang, and A. Yeo. Strongly connected spanning subgraphs with the minimum number of arcs in quasi-transitive digraphs. *SIAM J. Discrete Math.*, 16:335–343, 2003.
19. J. Bang-Jensen and A. Maddaloni. Arc-disjoint paths in decomposable digraphs. *J. Graph Theory*, 77(2):89–110, 2014.
20. J. Bang-Jensen, M.H. Nielsen, and A. Yeo. Longest path partitions in generalizations of tournaments. *Discrete Math.*, 306(16):1830–1839, 2006.
21. J. Bang-Jensen and C. Thomassen. A polynomial algorithm for the 2-path problem for semicomplete digraphs. *SIAM J. Discrete Math.*, 5:366–376, 1992.
22. M. Bezem, C. Grabmayer, and M. Walicki. Expressive power of digraph solvability. *Ann. Pure Appl. Logic*, 163(3):200–213, 2012.
23. F. Boesch and R. Tindell. Robbins's theorem for mixed multigraphs. *Amer. Math. Mon.*, 87(9):716–719, 1980.
24. J.A. Bondy. Basic graph theory: paths and circuits. In *Handbook of combinatorics, Vol. 1, 2*, pages 3–110. Elsevier, Amsterdam, 1995.
25. E. Boros and V. Gurvich. Perfect graphs, kernels, and cores of cooperative games. *Discrete Math.*, 306(19):2336–2354, 2006.
26. V. Chvátal. On the computational complexity of finding a kernel. Technical report CRM-300, Centre de recherches mathématiques, Université de Montréal, 1973.

27. V. Chvátal and C. Thomassen. Distances in orientations of graphs. *J. Combin. Theory Ser. B*, 24(1):61–75, 1978.
28. T. Feder, P. Hell, and C. Hernández-Cruz. Colourings, Homomorphisms, and Partitions of Transitive Digraphs. *European J. Combin.*, 60:55–65, 2017.
29. T. Feder and M.Y. Vardi. The computational structure of monotone monadic SNP and constraint satisfaction: a study through Datalog and group theory. *SIAM J. Comput.*, 28(1):57–104, 1998.
30. S. Fortune, J.E. Hopcroft, and J. Wyllie. The directed subgraph homeomorphism problem. *Theor. Comput. Sci.*, 10:111–121, 1980.
31. A. Fradkin and P.D. Seymour. Edge-disjoint paths in digraphs with bounded independence number. *J. Combin. Theory Ser. B*, 110:19–46, 2015.
32. A.S. Fraenkel. Planar kernel and Grundy with $d \leq 3$, $d_{out} \leq 2$, and $d_{in} \leq 2$ are $NP$-complete. *Discrete Appl. Math.*, 3(4):257–262, 1981.
33. M. Frick, S. van Aardt, G. Dlamini, J. Dunbar, and O. Oellermann. The directed path partition conjecture. *Discuss. Math. Graph Theory*, 25(3):331–343, 2005.
34. H. Galeana-Sánchez, I.A. Goldfeder, and I. Urrutia. On the structure of strong 3-quasi-transitive digraphs. *Discrete Math.*, 310(19):2495–2498, 2010.
35. H. Galeana-Sánchez and R. Gómez. Independent sets and non-augmentable paths in generalizations of tournaments. *Discrete Math.*, 308(12):2460–2472, 2008.
36. H. Galeana-Sánchez and C. Hernández-Cruz. $k$-kernels in generalizations of transitive digraphs. *Discuss. Math. Graph Theory*, 31(2):293–312, 2011.
37. H. Galeana-Sánchez, C. Hernández-Cruz, and M.A. Juárez-Camacho. $k$-kernels in $k$-transitive and $k$-quasi-transitive digraphs. *Discrete Math.*, 313(22):2582–2591, 2013.
38. P. García-Vázquez and C. Hernández-Cruz. Some results on 4-transitive digraphs. *Discuss. Math. Graph Theory*, 37(1):117–129, 2017.
39. A. Ghouila-Houri. Caractérisation des graphes non orientés dont on peut orienter les arětes de manière à obtenir le graphe d'une relation d'ordre. *C. R. Acad. Sci. Paris*, 254:1370–1371, 1962.
40. G. Gutin. Characterizations of vertex pancyclic and pancyclic ordinary complete multipartite digraphs. *Discrete Math.*, 141(1-3):153–162, 1995.
41. G. Gutin. Polynomial algorithms for finding Hamiltonian paths and cycles in quasi-transitive digraphs. *Australas. J. Combin.*, 10:231–236, 1994.
42. G. Gutin, K.M. Koh, E.G. Tay, and A. Yeo. On the number of quasi-kernels in digraphs. *J. Graph Theory*, 46(1):48–56, 2004.
43. G. Gutin and A. Yeo. Orientations of digraphs almost preserving diameter. *Discrete Appl. Math.*, 121(1-3):129–138, 2002.
44. S. Heard and J. Huang. Disjoint quasi-kernels in digraphs. *J. Graph Theory*, 58(3):251–260, 2008.
45. P. Hell and C. Hernández-Cruz. On the complexity of the 3-kernel problem in some classes of digraphs. *Discuss. Math. Graph Theory*, 34(1):167–185, 2013.
46. C. Hernández-Cruz. 3-transitive digraphs. *Discuss. Math. Graph Theory*, 32(3):205–219, 2012.
47. C. Hernández-Cruz. 4-transitive digraphs I: The structure of strong 4-transitive digraphs. *Discuss. Math. Graph Theory*, 33(2):247–260, 2013.
48. C. Hernández-Cruz and H. Galeana-Sánchez. $k$-kernels in $k$-transitive and $k$-quasi-transitive digraphs. *Discrete Math.*, 312(16):2522–2530, 2012.
49. C. Hernández-Cruz and J.J. Montellano-Ballesteros. Some remarks on the structure of strong $k$-transitive digraphs. *Discuss. Math. Graph Theory*, 34(4):651–671, 2014.

50. K.M. Koh and B.P. Tan. Number of 4-kings in bipartite tournaments with no 3-kings. *Discrete Math.*, 154(1-3):281–287, 1996.
51. B. Korte and J. Vygen. *Combinatorial Optimization.* Springer, Berlin, 2000.
52. J.M. Laborde, C. Payan, and N.H. Xuong. Independent sets and longest directed paths in digraphs. *Teubner-Texte Math.*, 59:173–177, 1983.
53. H.G. Landau. On dominance relations and the structure of animal societies III. The condition for a score structure. *Bull. Math. Biophys.*, 15:143–148, 1953.
54. S.B. Maurer. The king chicken theorems. *Math. Mag.*, 53(2):67–80, 1980.
55. A. Palbom. Complexity of the directed spanning cactus problem. *Discrete Appl. Math.*, 146(1):81–91, 2005.
56. C. Thomassen. Hamiltonian-connected tournaments. *J. Combin. Theory Ser. B*, 28(2):142–163, 1980.
57. J. von Neumann and O. Morgenstern. *Theory of Games and Economic Behavior.* Princeton University Press, Princeton, NJ, 1944.
58. M. Walicki and S. Dyrkolbotn. Finding kernels or solving SAT. *J. Discrete Algorithms*, 10:146–164, 2012.
59. R. Wang. $(k-1)$-kernels in strong $k$-transitive digraphs. *Discuss. Math. Graph Theory*, 35(2):229–235, 2015.
60. R. Wang and W. Meng. $k$-kings in $k$-quasitransitive digraphs. *J. Graph Theory*, 79(1):55–62, 2015.
61. R. Wang and S. Wang. Underlying graphs of 3-quasi-transitive and 3-transitive digraphs. *Discuss. Math. Graph Theory*, 33(2):429–435, 2013.
62. R. Wang and H. Zhang. Hamiltonian paths in $k$-quasi-transitive digraphs. *Discrete Math.*, 339(8):2094–2099, 2016.
63. R. Wang and H. Zhang. $(k+1)$-kernels and the number of $k$-kings in $k$-quasi-transitive digraphs. *Discrete Math.*, 338(1):114–121, 2015.
64. S. Wang and R. Wang. Independent sets and non-augmentable paths in arc-locally in-semicomplete digraphs and quasi-arc-transitive digraphs. *Discrete Math.*, 311(4):282–288, 2011.

# 9. Digraphs of Bounded Width

**Stephan Kreutzer and O-joung Kwon**

## 9.1 Introduction

Structural parameters for undirected graphs such as the **path-width** or **tree-width** of graphs have played a crucial role in developing a structure theory for graphs based on the minor relation and they have also found many algorithmic applications. Starting in the late 1990s, several ideas for generalizing this theory to digraphs have appeared. Broadly, for the purpose of this chapter, we distinguish these approaches into three categories: *tree-width inspired*, *rank-width inspired* and *density based*. The tree-width inspired approaches are based on the idea of generalizing the concept of undirected tree-width (or path-width) to digraphs. The various attempts, which we will discuss below, all have the goal of generalizing some natural property or some natural characterization of tree-width of undirected graphs to digraphs. In the same way as tree-width can be seen as a global connectivity measure for undirected graphs, the various versions of a directed analogue of tree-width measure global connectivity in digraphs. However, on digraphs, connectivity can be measured in many different natural ways. It turns out that equivalent characterizations of tree-width on undirected graphs yield different concepts on digraphs, with different properties, advantages and disadvantages. We will outline the most prominent of these concepts in Section 9.2 below.

The "tree-width inspired" approaches have in common that they define new classes of digraphs using structural parameters for digraphs which can not also be explained by structural parameters of the underlying undirected graphs. In particular, classes $\mathcal{C}$ of digraphs of, e.g., bounded DAG-width, do not automatically have bounded undirected tree-width (in the sense that the class of undirected graphs obtained from $\mathcal{C}$ by ignoring the direction of arc has bounded tree-width).

Another feature that almost all of these approaches have in common is that the class of DAGs has low width in all these definitions. This is a consequence of the fact that these approaches measure strong connectivity in

S. Kreutzer (✉) and O. Kwon
Chair for Logic and Semantics, Technical University Berlin,
Sekr TEL 7-3 Ernst-Reuter Platz 7, 10587 Berlin, Germany
e-mail: stephan.kreutzer@tu-berlin.de

 
J. Bang-Jensen and G. Gutin (eds.), *Classes of Directed Graphs*, Springer
Monographs in Mathematics, https://doi.org/10.1007/978-3-319-71840-8_9

various forms. Unfortunately, this does have problematic algorithmic consequences, as many $\mathcal{NP}$-hard computational problems remain hard on acyclic digraphs, and hence remain hard on classes of bounded width in these measures. Therefore, research in algorithmic applications of digraph width measures has tried to develop width measures for separating the class of DAGs into easy and hard instances. The next two types of digraph width measures achieve this goal.

A different approach to digraph width measures is taken in the definition of directed versions of **rank-width** [82] (a graph measure broadly equivalent to **clique-width** [25]).

Clique-width can naturally be defined on digraphs and it was indeed defined this way right from the beginning. However, algorithms for computing clique-width are not based on clique-width but on rank-width of graphs. Rank decompositions can be computed efficiently [82] and from a rank decomposition a clique-width decomposition can be computed.

In order to translate concepts from undirected rank-width, such as **vertex-minors**, to the directed setting, Kanté developed concepts of rank width for digraphs such as **bi-rank-width** and **$\mathbb{F}_4$-rank-width** [56]. This approach has led to a theory of directed rank-width with connections to other types of digraphs. A feature that distinguishes this approach from the tree-width inspired approaches above is that if a class of digraphs has bounded directed clique or rank-width then the class of underlying undirected graphs also has bounded undirected clique width. As a consequence, any graph property definable in **monadic second order logic** can be decided in linear time on any class of digraphs of bounded bi-rank-width [26]. Another consequence of this fact is that the class of DAGs no longer has bounded width. Those DAGs have low width in the tree-width inspired approaches has led to problems for algorithmic applications of tree width based directed width measures as several interesting computational problems remain hard on DAGs. This problem therefore does not appear in classes of bounded bi-rank-width etc.

Whereas on undirected graphs, classes of graphs of bounded tree-width also have bounded clique-width, in the directed setting these concepts are incomparable. We will present the concepts of directed rank-widths in Section 9.9.

A third, and final, approach to digraph width measures covered in this chapter are concepts based on density arguments. In their quest for a solid mathematical definition of "sparse" classes of graphs, Nešetřil and Ossona de Mendez defined classes of graphs of **bounded expansion** and classes which are **nowhere dense** [74, 75]. These concepts can be generalized to digraphs as well and lead to a surprisingly elegant theory. We will cover the resulting theory in Section 9.6.

*Overview.* The remaining chapter is organized as follows. In Section 9.2 we cover the *tree-width inspired* width measures. In particular, we will briefly introduce *graph searching games* (Section 9.2.1), which provide an intuitive way of defining graph and digraph decompositions, introduce some of the more prominent digraph decompositions (Section 9.2.4 and Theorem 9.2.13) and compare them with respect to generality (Section 9.2.5).

In Section 9.3, we provide a brief overview of the existing structure theory for digraphs based on directed tree-width. In particular we review known obstructions to directed tree width. This also leads to a fixed-parameter algorithm for computing directed tree-decompositions which, together with some algorithmic applications, we present in Section 9.4 and 9.5.

In Sections 9.6 to 9.8 we cover the relatively recent theory of density based width measures: classes of digraphs of bounded expansion (Section 9.7) and nowhere dense classes of digraphs (Section 9.8).

Finally, in the last part of the chapter, Section 9.9, we present the concepts of digraph width measures based on rank-width.

## 9.2 Tree-Width Inspired Width Measures

In this section we will present some of the best known tree-width inspired width measures for digraphs. Many of them can be explained in terms of **graph searching games** and these games provide an intuitive way to understand these measures. We will therefore first give a brief overview of graph searching games, also known as **Cops and Robber games**.

### 9.2.1 Graph Searching Games

Graph searching games have been studied intensively in graph theory and they have found a wide range of applications. See [2, 38, 68] for surveys on the subject. Here we will only review the absolute basics needed for our exposition of digraph width measures.

A graph searching game is played on a graph by two players, often called the **cops** and the **robber** or the **searchers** and the **fugitive**. The general goal for the cops is to catch the robber, whereas the robber tries to evade. The cop player controls a number of cops each of which occupies a single vertex of the graph. The robber also occupies a single vertex. In every round of the game, the cop player can move some of the cops from their current position to new positions on the graph or he can place new cops on the graph. However, he first has to announce his move and lift up all cops he wants to move, releasing their current position. Then the robber can react to this by changing his own position. The rules for the robber movement differ between the various types of graph searching games. Finally, the cops are placed on their new positions. If any cop is placed on the vertex occupied by the robber, then the cops win. Otherwise, if the robber can escape forever, he wins.

More formally, given a graph $G = (V(G), E(G))$, a current position in the game can be described by a pair $(X, v)$, where $X \subseteq V(G)$ are the vertices occupied by the cops and $v \in V(G)$ is the vertex occupied by the robber. A single round of the play can therefore be described as a move from a position $(X, v)$ to a new position $(X', v')$. The game always starts at a position $(\emptyset, v)$, for some vertex $v \in V(G)$.

In most games of interest to us, the cops can move freely, i.e. from the current position $(X, v)$ they can move to any new position $X'$. The robber is more restricted and the various restrictions on the movement of the robber define different variations of the game. To give an example, the game corresponding exactly to tree-width is played on an undirected graph. From a current position $(X, v)$, once the cops announce their move to $X'$, the robber can choose any vertex $v'$ reachable from $v$ in the graph $G - (X \cap X')$, i.e. any position reachable from $v$ by a path not occupied by a cop that remains on the board.

Given a play $(X_i, v_i)_{0 \leq i < l}$, for some $l \in \mathbb{N} \cup \{\omega\}$, we can define the **width** of the play as $\max\{|X_i| : 0 \leq i < l\}$. In this way, any graph searching game defines a graph parameter assigning to every graph or digraph $G$ the minimal number $k$ such that the cops have a winning strategy against the robber on $G$ of width at most $k$. A trivial strategy for the cop player to win on any given graph is to put a cop on every single vertex of the graph. Hence, the width is always well defined and it is the minimal number of cops required for a winning strategy that yields an interesting graph parameter.

Graph searching games can be classified in many different ways. An important distinction is whether the cops can always see the robber, called **visible graph searching**, or whether they need to search the graph without knowing where the robber is. This is referred to as **invisible graph searching**. It is known that in the game variant above where the robber can move along any cop free path, the graph parameter defined by the visible variant is exactly tree-width whereas the invisible variant defines path-width [15, 92].

An important concept in graph searching is **monotonicity**. Monotonicity restricts the winning strategies for the cops. We distinguish two forms of monotonicity: **cop monotonicity** and **robber monotonicity**. In a **cop-monotone** strategy, the cop player is not allowed to place a cop on a vertex that had already been occupied by a cop in the past. That is, once a cop is lifted from a vertex $v \in V(G)$, no cop can later on be placed on $v$. In a **robber-monotone** strategy, the cops have to play in a way such that once, at any particular point in the play, a vertex $v$ is not reachable for the robber, it has to remain unreachable for the rest of the play. More precisely, let $(X_i, v_i)_{0 \leq i < n}$ be a play, for some $n \in \mathbb{N} \cup \{\omega\}$. For all $i$ let $R_i$ be the set of vertices available to the robber starting from $v_i$ in $G - X_i$. The play is robber-monotone, if $R_j \subseteq R_i$ for all $0 \leq i < j$. It is known, that on undirected graphs, in the visible and the invisible graph searching games, the cops have a winning strategy of width $k$ if, and only if, they have a cop- and robber-monotone winning strategy of width $k$. For digraphs, this will often not be the case and monotone and non-monotone versions will define different parameters, see e.g. [1, 69].

There is a natural correspondence between winning strategies of the cops in a graph searching game and graph decompositions. For instance, in the visible graph searching game on undirected graphs described above, a winning strategy for the cop player can be seen as a tree with the initial position in the game as the root and a child for every possible move of the robber and the corresponding move of the cops. This monotone winning strategy tree immediately defines a tree decomposition of the graph of width one less than the width of the winning strategy. Conversely, a tree decomposition of width $k$ of a graph immediately defines a winning strategy for the cop player of width $k+1$. It is this natural correspondence between winning strategies and graph decompositions that makes graph searching games an elegant characterization of width measures for graphs and digraphs.

### 9.2.2 Decompositions of Directed Graphs

In the following sections we will define several width measures of directed graphs. All of them are defined in terms of a **decomposition** of digraphs. The type of decompositions will vary but in general they will all have a common structure. A decomposition of a digraph $D$ consists of a digraph $T$, usually a tree or a DAG, and a labeling function $\beta$ assigning to every vertex of $T$ a subset of vertices of $D$. Furthermore, there is a **guarding function** $\gamma$ that assigns to every arc or to every vertex (or both) a **guard**. Usually, a guard is also a set of vertices. The role of the guard of an arc $e \in A(T)$ is that if $e = (u, v) \in A(T)$ and $X := \bigcup\{\beta(t) : t \text{ is reachable from } v \text{ in } T - e\}$, then $\gamma(e)$ controls connectivity between $X$ and the rest of $D$. Control can mean that there is no path from $X$ to any vertex outside of $X$ in $D - \gamma(e)$, or that there is no strong component in $D - \gamma(e)$ containing a vertex of $X$ and a vertex not in $X$. The various types of decompositions defined in the sequel are obtained by varying the type of guards and the type of the decomposition structure $T$.

**Definition 9.2.1 (Strong and Weak Guarding)** *Let $D$ be a digraph and let $X, Y \subseteq V(D)$ be sets.*

1. *We say that $Y$ **strongly guards** $X$, or is a **strong guard** of $X$, if every directed walk starting and ending in $X$ which contains a vertex of $V(D) \setminus X$ also contains a vertex of $Y$. In other words, $X \setminus Y$ is the union of the vertex sets of some set of strong components of $D - Y$.*
2. *We say that $Y$ **weakly guards** $X$, or is a **weak guard** of $X$, if every arc $e = (u, v) \in A(D)$ with $u \in X \setminus Y$ has $v \in X \cup Y$.*

As an example for the two notions of guarding in the previous definition, consider the set $X := \{3, 4, 5\}$ of vertices in the digraph shown in Figure 9.1 a): The set $\{6, 9\}$ is a weak guard for $X$. The set $\{6\}$ containing only the vertex 6 is already a strong guard, as every path from $X$ to itself that does contain any vertex not in $X$ must go through 6. But $\{6\}$ is not a weak guard for $X$.

The names strong and weak guards come from the intuition that strong guards control strong components, i.e. strong connectivity, whereas weak guards control directed paths and therefore weak reachability. Of course, every weak guard is also strong and therefore weak guarding is the more restrictive concept of guards.

Note that for every set $X$ of vertices in a digraph $G$ there is a uniquely defined minimal weak guard, which consists of every vertex in $G \setminus X$ which is an out-neighbour of a vertex in $X$. But there can be many distinct and even disjoint minimal strong guards.

**Definition 9.2.2 (Abstract Digraph Decomposition)** *Let $D$ be a digraph. An **abstract digraph decomposition** of $D$ is a triple $(T, \beta, \gamma)$, where $T$ is a digraph, $\beta : V(T) \rightarrow 2^{V(D)}$ and $\gamma : A(T) \rightarrow 2^{V(D)}$ such that $\bigcup\{\beta(t) : t \in V(T)\} = V(D)$.*

*For every $t \in V(T)$ we define*

$$T_t := T\big[\{s \in V(T) : s \text{ is reachable from } t \text{ by a directed path in } T\}\big]$$

*as the subgraph of $T$ induced by the vertices reachable from $t$. Furthermore, if $S \subseteq T$ then we define $\beta(S) := \bigcup\{\beta(s) : s \in V(S)\}$.*

*With every decomposition we will define a **width** $w(t)$ for every $t \in V(T)$. The **width** $w(T, \beta, \gamma)$ is then defined as $\max\{w(t) : t \in V(T)\}$.*

*Finally, for every $v \in V(D)$, we define $\beta^{-1}(v) := \{t \in V(T) : v \in \beta(t)\}$.*

*Sometimes, guards are more naturally associated with vertices of $T$ instead of its arcs and hence $\gamma$ is a function from $V(T)$ into $2^{V(D)}$. We call such abstract decompositions **node guarded**.*

Several decompositions below use rooted directed trees as underlying digraph.

**Definition 9.2.3** *A **rooted directed tree**[1] is a digraph obtained from an undirected tree by selecting a vertex $r$ as a root and orienting every arc away from the root vertex $r$.*

### 9.2.3 Tree-Width Based Digraph Width Measures

In this section we describe some of the most prominent tree-width inspired digraph decompositions proposed in the literature. Throughout the section we will illustrate the different decompositions by the following example digraph shown in Figure 9.1 a).

The first generalization of tree-width to digraphs proposed in the literature was directed tree-width [54, 84].

[1] This is also called an out-tree.

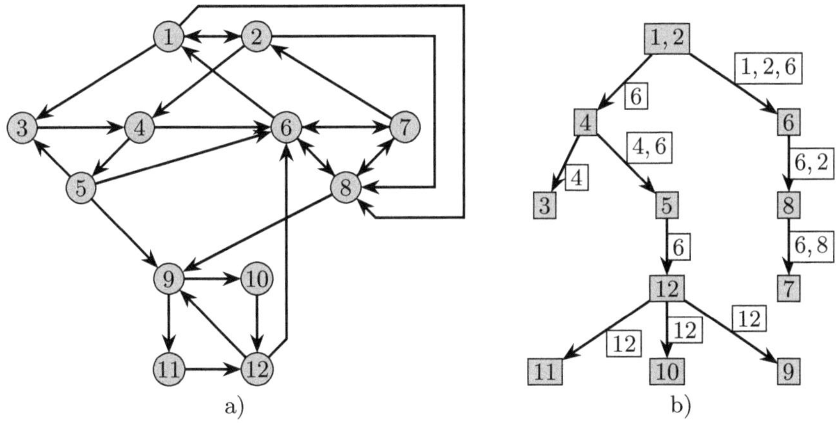


**Figure 9.1** a) An example digraph $D$ and b) a directed tree decomposition of $D$ of width 2.

**Definition 9.2.4 (Directed Tree Decompositions)** *A* ***directed tree decomposition*** *of a digraph $D$ is an abstract digraph decomposition $(T, \beta, \gamma)$ such that $T$ is a rooted directed tree, $\{\beta(t) : t \in V(T)\}$ is a partition of $V(D)$ into non-empty sets and for every $e = (s, t) \in A(T)$, $\gamma(e)$ is a strong guard of $\beta(T_t)$.*

*For every $t \in V(T)$ we define $\Gamma(t) := \beta(t) \cup \bigcup_{e \sim t} \gamma(e)$ and we define the* ***width*** *$w(t)$ as $w(t) := \big|\Gamma(t)\big| - 1$, where $e \sim t$ means that the arc $e$ is incident to $t$.*

See Figure 9.1 b) for an illustration of a directed tree decomposition of the digraph in Figure 9.1. The figure also demonstrates some of the (algorithmically) problematic aspects of directed tree decompositions: The guard 6 on the branch from the root to node labelled by 12 is actually a vertex that is being decomposed in an entirely different subtree of the root. Hence, directed tree decompositions can use vertices in a guard that are contained in strong components which are part of different subtrees. This can cause problems in algorithmic applications. Furthermore, arcs of the digraph $D$ can cross between subtrees in the directed tree decomposition, something that cannot happen in the undirected case. This happens for instance with the arc $(8, 9) \in A(D)$. Finally, on a branch of a directed tree decomposition from its root to a leaf it could happen that a vertex $w$ is contained in a guard of an arc $e = (u, v)$, it then disappears from the next arc of the branch and then reappears later on as a guard on the branch. This phenomenon does not appear in the decomposition on Figure 9.1 but can happen in general.

The second problem, that arcs can cross subtrees – but only in one direction – is an intrinsic feature of directed decompositions. If this were disallowed then we would essentially speak about undirected tree decompositions. The first and the third problem, however, are unavoidable. We will see next the concept of D-decompositions, which are similar to directed tree decompositions but avoid these problems. However, it was shown in [3] that there are classes of digraphs of bounded directed tree-width but unbounded D-width (see Section 9.2.5). The examples separating the two concepts precisely use these properties of guards containing vertices from different strong components as well as guards reappearing along branches, showing that these properties of directed tree decompositions are unavoidable.

In [90], Safari suggests D-width as another structural complexity measure. The definition of D-decompositions is perhaps the most natural translation of undirected tree decompositions to the directed settings in terms of strong connectivity. However, as we will see below, in terms of structural properties, it is directed tree-width that shares most structural properties of undirected tree-width. In the following definition we give a slightly different version of D-decompositions. But the width defined by this concept differs from the original definition at most by a factor of 2. See Figure 9.2 for an illustration.

**Definition 9.2.5 (D-Decompositions)** *A* ***D-decomposition*** *of a digraph $D$ is an abstract digraph decomposition $(T, \beta, \gamma)$ such that $T$ is a rooted directed tree, $\gamma(e) := \beta(u) \cap \beta(v)$ for every $e = (u, v) \in A(T)$ and $\gamma(e)$ is a strong guard of $\beta(T_v)$ and $\beta^{-1}(v)$ induces a non-empty subtree of $T$ for every $v \in V(D)$.*

*For every $t \in V(T)$ we define the* ***width*** *$w(t)$ as $w(t) := |\beta(t)|$.*

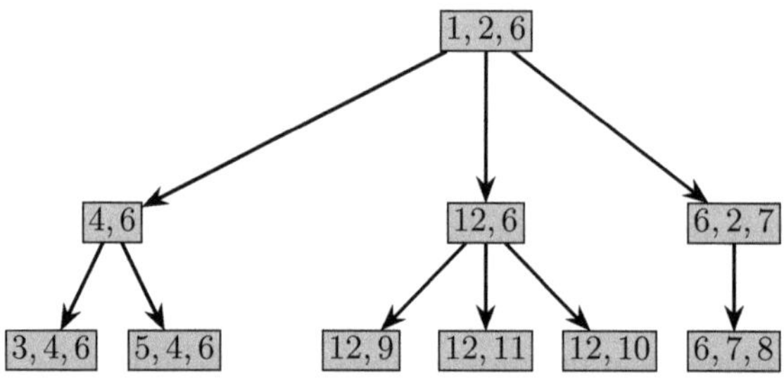


**Figure 9.2** A D-decomposition of width 3 of the digraph in Figure 9.1.

Directed tree-width and D-width are related to each other because they both correspond to the same graph searching game, the game where the robber can only stay within a strong component, but they are related to different type of strategies for the cops. The next following three decompositions are based on a different form of game, where the robber can follow any directed path. DAG-width was defined in [11] and independently in [77], cf. [12]. See Figure 9.3 a) for an illustration.

**Definition 9.2.6 (DAG Decompositions** [12]**)** *Let $D$ be a digraph. A* ***DAG-decomposition*** *of $D$ is an abstract digraph decomposition $(T, \beta, \gamma)$ such that:*

1. *$T$ is a DAG.*
2. *$\gamma(e) = \beta(u) \cap \beta(v)$, for every arc $e = (u, v) \in A(T)$, and $\gamma(e)$ is a weak guard of $\beta(T_v)$.*
3. *$\beta(a) \cap \beta(c) \subseteq \beta(b)$ for every triple $a, b, c \in V(T)$ such that $a, b, c$ appear in this order on some directed path in $T$.*
4. *For every root $t \in V(T)$, $\beta(T_t) = N^+[\beta(T_t)]$.*

*For every $t \in V(T)$ we define the width $w(t) := |\beta(t)|$.*

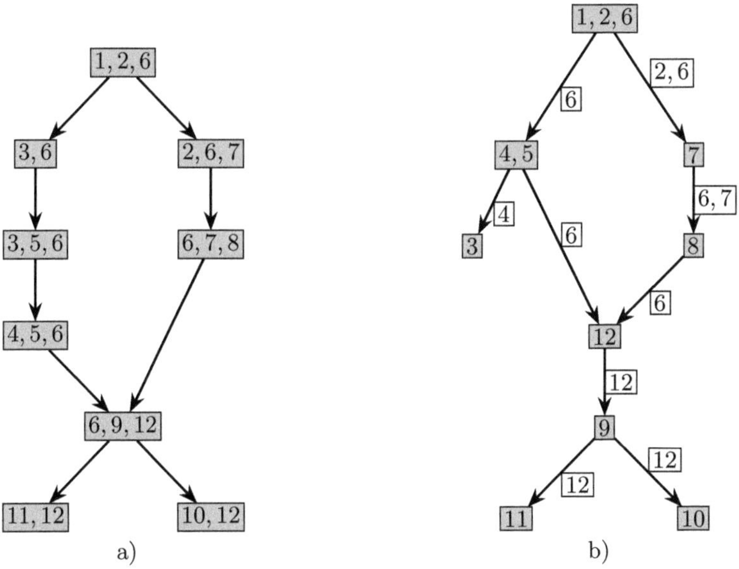


**Figure 9.3** a) A DAG-decomposition and b) a Kelly-decomposition of the digraph in Figure 9.1, both of width 3.

A related width measure is Kelly-width which is based on so-called Kelly-decompositions. It was introduced in [51] to overcome some problems of DAG-decompositions. See Figure 9.3 b) for an illustration.

**Definition 9.2.7 (Kelly Decompositions** [51]**)** *A* ***Kelly-decomposition*** *of a digraph $D$ is a node-guarded abstract decomposition $(T, \beta, \gamma)$ so that*

1. *$T$ is a DAG.*
2. *$\{\beta(t) : t \in V(T)\}$ is a partition of $V(D)$ into non-empty subsets.*
3. *$\gamma(t)$ is a weak guard of $\beta(T_t)$ for every $t \in V(T)$.*

**Figure 9.4** A directed path decomposition of the digraph in Figure 9.1 of width 3.

4. *For all $s \in V(T)$ there is a linear order $<_s$ on its children $t_1, \dots, t_p$ so that for all $1 \leq i \leq p$, $\gamma(t_i) \subseteq \beta(s) \cup \gamma(s) \cup \bigcup_{j <_s i} \beta(V(T_{t_j}))$.*
5. *Similarly, there is a linear order $<_r$ on the roots such that $\gamma(r_i) \subseteq \bigcup_{j <_r i} \beta(V(T_{r_j}))$.*

*The width $w(t)$ of a vertex $t \in V(T)$ is defined as $\beta(t) \cup \gamma(t)$.*

Note that the number of nodes in a Kelly-decomposition is at most the number of vertices of the decomposed digraphs, as the bags form a partition. This is not the case for DAG-decompositions and we will see below that DAG-decompositions of optimal width $k$ can become super-polynomially large, i.e. have number of bags proportional to $n^{k+1}$ (see [3]). See Section 9.4.1 for details.

Finally, we introduce the concept of directed path decompositions, introduced by Robin Thomas in the mid-90s but unpublished. See [8, 9] for published references. See Figure 9.4 for an illustration.

**Definition 9.2.8 (Directed Path Decompositions)** *A **directed path decomposition** of a digraph $D$ is a DAG-decomposition $(T, \beta, \gamma)$ of $D$ such that $T$ is a directed path.*

Every type of decomposition introduced above naturally defines a digraph width measure, summarized in the following definition.

**Definition 9.2.9 (Directed Width Measures)** *Let $D$ be a digraph. The **directed tree-width** $\mathrm{dtw}(D)$ of $D$ is defined as the minimum width of any directed tree decomposition of $D$. Analogously, the **D-width** D-width$(D)$, **DAG-width** dag-width$(D)$, **Kelly-width** Kelly-width$(D)$ and the **directed path-width** $\mathrm{dpw}(D)$ are defined as the minimum width of the corresponding decomposition of $D$.*

*A class $\mathcal{C}$ of digraphs has bounded directed tree-width if there is a constant $c \geq 0$ such that $\mathrm{dtw}(D) \leq c$ for every $D \in \mathcal{C}$. Classes of bounded width for other width measures are defined analogously.*

Digraphs with no directed cycles longer than a fixed constant form an example of a class of digraphs with bounded DAG-width, Kelly-width and directed tree-width. This follows from the following results by Bang-Jensen and Christiansen, respectively, Kintali.

**Theorem 9.2.10** *[7] For every natural number $p$, every $D$ digraph having no directed cycle of length more than $p$ has DAG-width at most $p$ and this is best possible.*

**Theorem 9.2.11** *[63] For every natural number p, every D digraph having no directed cycle of length more than p has directed tree-width and Kelly-width at most $p + 1$.*

We close this section by mentioning two other digraph width measures which do not fall naturally within the framework of abstract decompositions. The first is the **DAG-depth**, defined in [41]. To define it, we need the concept of reachability component. Let $D$ be a digraph. For $v \in V(D)$ we define $\mathrm{Reach}_D(v) := \{u \in V(D) : u \text{ is reachable from } v \text{ by a directed path in } D\}$. A **reachability component** is a subgraph of $D$ induced by an inclusion-wise maximal non-empty set in $\{\mathrm{Reach}_D(v) : v \in V(D)\}$, i.e. an inclusion-wise maximal induced subgraph with only one initial strong component (see Section 1.5 for the definition of an initial component).

**Definition 9.2.12 (DAG-depth)** *Let D be a digraph. The* ***DAG-depth*** $\text{dag-depth}(D)$ *of D is inductively defined as follows: if* $|V(D)| = 0$*, then* $\text{dag-depth}(D) = 0$*. If D has a single reachability component, then we let* $\text{dag-depth}(D) = 1 + \min\{\text{dag-depth}(D{-}v) : v \in V(D)\}$*. Otherwise, if* $D_1, \ldots, D_c$ *are the reachability components of D for some* $C > 1$*, then* $\text{dag-depth}(D) := \max\{\text{dag-depth}(D_i) : 1 \leq i \leq c\}$.

There are various other width measures for digraphs that have been defined in the literature, for instance **oriented tree-width**, **Kenny-width**, **entanglement**, **cycle rank** and others, see e.g. [3, 13, 14, 41, 55].

### 9.2.4 Alternative Characterizations of Digraph Width Measures

In the previous section we have defined several width measures for directed graphs based on variations of digraph decompositions. Many of these measures can also be defined equivalently and the equivalent definitions yield additional insights and intuition about the corresponding width measures.

All width measures defined above can be characterized by graph searching games. We have already covered the basics of graph searching games in Section 9.2.1. For digraphs, two main variants of games have emerged, depending on the ability of the robber to move. Let $(X, v)$ be the current position in a graph searching game on a digraph $D$. Suppose the cops announce to move from $X$ to $X'$. In the **strong reachability game**, the robber can choose any new position $v'$ within the strongly connected component of $D - (X \cap X')$ that contains $v$. In the **weak reachability game**, the robber can choose any position $v'$ that is reachable from $v$ in $D - (X \cap X')$. Combining this distinction with the distinction between a visible and an invisible robber yields a range of graph searching games on directed graphs that can be used to give game based characterizations of the width measures introduced above.

**Theorem 9.2.13** *Let $D$ be a digraph and $k \in \mathbb{N}$.*

1. *If* $\mathrm{dtw}(D) \leq k$, *then $k$ cops have a robber monotone winning strategy in the visible strong cops and robber game on $D$. Conversely, if $k$ cops have a winning strategy in this game on $D$, robber-monotone or not, then* $\mathrm{dtw}(D) \leq 3k+2$. *If $k$ cops have a winning strategy in the visible strong cops and robber game on $D$, then $3k+2$ cops have a robber-monotone winning strategy on $D$ [1, 54].*
2. *$D$ has DAG-width $\leq k$ if, and only if, $k$ cops have a cop-monotone winning strategy on $D$ if, and only if, $k$ cops have a robber-monotone winning strategy on $D$ in the visible weak reachability game [12].*
3. *$D$ has Kelly-width $\leq k$ if, and only if, $k$ cops have robber-monotone winning strategy on $D$ in the invisible inert weak reachability game. Here, in the inert game variant the robber can only move when the cop player announces to place a cop on the current robber position [51].*
4. *$D$ has directed path-width $k$ if, and only if, $k$ cops have a cop-monotone winning strategy on $D$ if, and only if, $k$ cops have robber-monotone winning strategy on $D$ in the invisible weak reachability game [8].*
5. *$D$ has DAG-depth $\leq k$ if, and only if, the cop player has a winning strategy with at most $k$ cops in the visible weak reachability game in which he never moves any cop, i.e. in every round the cop player has to use new cops [41].*

Part (1) of the previous theorem follows from the observation that any directed tree decomposition of a digraph of width $k$ yields a winning strategy for $k+1$ cops. Part (2) – (3), on the other hand, follow from Theorem 9.3.8 below, as a haven of order $k$ yields a winning strategy for the robber against fewer than $k$ cops. See below for details.

We close this section by giving an alternative characterization of Kelly-width in terms of elimination ordering and partial $k$-DAGs.

**Definition 9.2.14 (Directed elimination ordering** [51]**)** *An **elimination order** $\sqsubseteq$ for a digraph $D$ is a linear order on $V(D)$. For a vertex $v$ define $V_{v\sqsubseteq} := \{u \in V : v \sqsubseteq u\}$. The **support** of a vertex $v$ with respect to $\sqsubseteq$ is*

$$\mathrm{supp}_{\sqsubseteq}(v) := \{u \in V_{v\sqsubseteq} : \textit{there is } v' \in \mathrm{Reach}_{G-V_{v\sqsubseteq}}(v) \textit{ with } (v',u) \in E\}\,.$$

*The **width** of an elimination order $\sqsubseteq$ is* $\max_{v\in V} |\mathrm{supp}_{\sqsubseteq}(v)|$.

The name elimination ordering originates in the following equivalent way of defining the width of an elimination ordering based on an explicit elimination process. Let $D$ be a digraph and let $\sqsubseteq$ be a linear order on $V(D)$. Let $(v_0, v_1, \ldots, v_{n-1})$ be the enumeration of $V(D)$ with respect to $\sqsubseteq$. We define $G_0^{\sqsubseteq} := G$ and $G_{i+1}^{\sqsubseteq}$ as the graph obtained from $G_i^{\sqsubseteq}$ by deleting $v_i$ and adding (if necessary) new arcs $(u,v)$ if $(u,v_i),(v_i,v) \in E(G_i^{\sqsubseteq})$ and $u \neq v$. $G_i^{\sqsubseteq}$ is the **directed elimination graph at step $i$ with respect to $\sqsubseteq$.**

Now it is readily verified that the width of the elimination order $\sqsubseteq$ is the maximum over all $i$ of the out-degree of $v_i$ in $G_i^{\sqsubseteq}$.

**Definition 9.2.15 ((Partial) $k$-DAG** [51]) *The class of $k$-DAGs is defined recursively as follows:*

- *A complete digraph with $k$ vertices is a $k$-DAG.*
- *A $k$-DAG with $n+1$ vertices can be constructed from a $k$-DAG $H$ with $n$ vertices by adding a vertex $v$ and arcs satisfying the following:*
  - *there are at most $k$ arcs from $v$ to $H$ and*
  - *if $X$ is the set of endpoints of the arcs added in the previous sub-condition, then there is an arc from $u \in V(H)$ to $v$ if $(u, w) \in E(H)$ for all $w \in X \setminus \{u\}$. Note that if $X = \emptyset$, this condition is true for all $u \in V(H)$.*

*A* ***partial*** *$k$-**DAG** is a subgraph of a $k$-DAG.*

**Theorem 9.2.16** ([51]) *Let $G$ be a digraph. The following are equivalent:*

1. *$G$ has Kelly-width at most $k+1$.*
2. *$G$ has a directed elimination ordering of width $\leq k$.*
3. *$k+1$ cops have a robber-monotone winning strategy to capture an inert invisible robber.*
4. *$G$ is a partial $k$-DAG.*

Further characterizations of classes of digraphs of bounded width have been given in terms of forbidden subgraphs and forbidden minors, e.g. in [65], where Kintali and Zhang characterized partial 1-DAGs in terms of forbidden directed minors. See also [33, 73].

### 9.2.5 Comparing Directed Width Measures

In this section we compare the width measures introduced in the previous section with respect to generality. In particular, we are interested in the question whether classes of digraphs of bounded width with respect to one measure automatically have bounded width in another measure. As we will see, the width measures introduced above form the partial order shown in Figure 9.5.

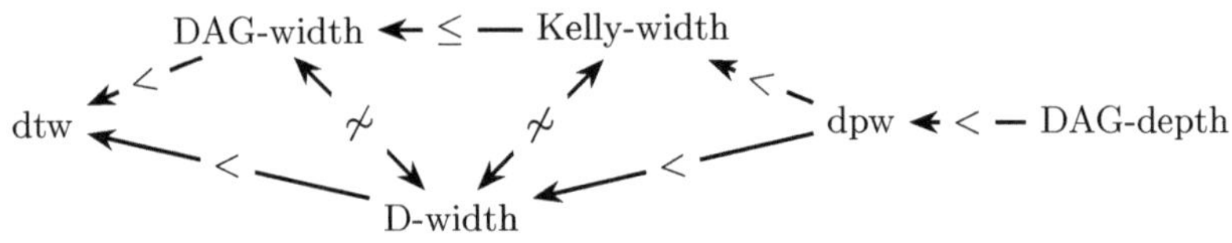


**Figure 9.5** The relation between different measures. An arrow labelled by "$<$" means bounded only in one direction, an arrow labelled by "$\leq$" means bounded at least in one direction. A bidirected arrow labelled "$\not<$" means not bounded in any direction.

**Lemma 9.2.17** ([12])

1. *Every class of digraphs of bounded DAG-width has bounded directed tree-width.*
2. *Conversely, there are classes of digraphs of bounded directed tree-width and unbounded DAG-width.*

Part 1 can easily be seen by considering the game characterization of DAG width and directed tree-width (see Theorem 9.2.13): the set of positions the robber can choose at any particular time in the directed tree-width game is a subset (proper or not) of the set of positions he can choose in the DAG width game. Hence, if $k$ cops can catch the robber in the latter, they can also do so in the former.

Towards Part 2, let $T_t$ be a complete directed binary tree of height $t$, i.e. a tree with all arcs oriented away from the root towards the leaves and every vertex has two or zero successors. Furthermore, every path from the root to a leaf has length $t$. Now add to $T_t$ an arc from every vertex $v \in V(T_t)$ to every ancestor $u \in V(T_t)$ of $v$, i.e. to every $u \neq v \in V(T_t)$ on the unique path from the root $r$ of $T_t$ to $v$. We call this a **tree with back arcs**.

It is not hard to see that two cops can catch the robber on this tree for any value of $t$ in the directed tree-width game: they just start with one cop at the root $r$. Then the robber has to decide into which of the two subtrees he wants to move. The cops can then put the second cop on the root of this subtree, i.e. on the successor $v$ of $r$ which is the root of the subtree containing the robber. If the robber is on this vertex $v$, he can only move further down into the subtree, i.e. into a subtree rooted at a successor $v'$ of $v$. Once the cop on $v$ is in place, the first cop on the root can be lifted and moved to $v'$. The cops continue in this way chasing the robber down. This is possible because once a cop is on $v$, every path that starts at the subtree of $v$ containing the robber and which ends in this subtree but has an inner vertex outside of this tree has to go through $v$. Hence, even with only one cop on $v$ the robber can no longer leave the subtree rooted at $v$.

In the DAG-width game, however, the robber can simply follow a directed path. In this game, to chase the robber down the tree the cops need to occupy the entire path from the root of $T_t$ to the root of the current subtree containing the robber. This results in a strategy using $t$ cops. With a little extra work one can show that there is no other, substantially better strategy. Hence, the DAG width of $T_t$ is proportional to $t$. See [12] for details.

The next result we state is that the DAG-width of a digraph is bounded by a function of its Kelly-width. The question whether DAG-width and Kelly-width of a class of digraphs are mutually bounded is equivalent to the question whether the monotone cop numbers of the DAG-width and Kelly-width game on digraphs are bounded by each other. This is a long open problem in the theory of graph searching games. A partial answer was finally given by Rabinovich [3, 83] who introduced the concept of weak monotonicity in the DAG-width game and proved that every strategy for $k$ cops in the Kelly-width game can be translated into a weakly monotone strategy for $k$ cops in the DAG-width game. Furthermore, any winning strategy for $k$ cops in the weakly monotone game can be translated into a monotone strategy for $k^2$ cops in the DAG width game. This implies the following lemma.

**Lemma 9.2.18** *Every class $\mathcal{C}$ of digraphs with bounded Kelly-width has bounded DAG-width.*

The converse of the lemma is still open and it is related to one of the biggest open problems in graph searching, namely whether the monotonicity costs for Kelly- and DAG-width games are bounded, i.e. if there is a function $f : \mathbb{N} \to \mathbb{N}$ such that for every digraph $D$, if $k$ cops have a winning strategy on $D$ in the Kelly-game then they also have a robber-monotone winning strategy on $D$ using at most $f(k)$ cops (likewise for DAG-width games).

The next result we mention relates directed path-width to Kelly-width. Again it follows immediately from the game characterizations of the width measures that Kelly-width is more general than directed path width. Towards the converse, it can again be shown through the game connection that if $\mathcal{C}$ is a class of bidirected digraphs, i.e. digraphs where for every arc $(u, v)$ also the reverse arc $(v, u)$ is present, the DAG-width, Kelly-width, directed tree-width and D width all coincide with the undirected tree-width of the class $\mathcal{C}'$ of graphs obtained from $\mathcal{C}$ by replacing every directed arc by an undirected arc (removing duplicates). Furthermore, the directed path width of $\mathcal{C}$ equals the path-width of $\mathcal{C}'$ and the DAG-depth equals the tree-depth. As, for instance, the class of trees has unbounded path width but bounded tree-width, the next lemma follows.

**Lemma 9.2.19** *Every class of digraphs of bounded directed path-width has bounded Kelly-width. Conversely, there are classes of digraphs of bounded Kelly-width but unbounded directed path-width.*

Finally, we compare D-width to the other classes. When D-width was introduced, it was conjectured to be equivalent to directed tree-width in the sense that classes of digraphs have bounded D-width if, and only if, they have bounded directed tree-width [90, Page 756]. The observation that bounded

D-width implies bounded directed tree-width is easily seen. One of the main differences between directed tree decompositions and D-decompositions is the concept of guarding. As the example in Figure 9.1 b) and the discussion in the paragraphs following the example show, the guard of an arc $e$ can contain vertices which are contained in bags of an entirely different part of the tree decompositions. Also, along a branch of the directed tree decomposition, vertices can appear in a guard, then disappear from the guards and then reappear later. This leads to strategies for the cop-player which are not cop-monotone. This "external" guarding as well as the non-monotonicity is not possible in D-decompositions. Amiri et al. [3] manage to exploit these differences to exhibit classes of digraphs of bounded directed tree-width but where an unbounded number of cops is needed for the cop-monotone visible strong cops and robber game. This already implies that the D-width is also unbounded. They also exhibited classes of digraphs where a bounded number of cops have cop-monotone winning strategies but where the D-width is still unbounded.

**Lemma 9.2.20** *([3][90])*

1. *Every class of digraphs of bounded D-width has bounded directed tree-width.*
2. *Conversely, there are classes of digraphs of bounded directed tree-width with unbounded D-width.*

In [3], D-width is shown to be incomparable to Kelly and DAG-width.

**Lemma 9.2.21**

1. *There are classes of digraphs of bounded D-width and unbounded Kelly- and DAG-width.*
2. *There are classes of digraphs of bounded DAG-width unbounded D-width.*
3. *Every class of digraphs of bounded DAG-depth has bounded directed path-width but the converse is false.*

Finally, it can again be shown using the game characterization that classes of digraphs of bounded directed path-width have bounded D-width and also bounded Kelly-width. The converse fails in both cases as explained above: the class of trees has bounded tree-width but unbounded path-width in the undirected case and replacing in trees arcs by two directed arcs in opposite directions separates directed path-width from D- and Kelly-width.

To separate directed path-width from DAG-depth note that the class of directed paths has directed path-width 2 but unbounded DAG-depth. On the other hand, one can show that if a digraph $D$ has no path longer than $t$, then this implies that $t+1$ cops can win the invisible cops and robber game on $D$ and hence, by the game characterization of directed path-width in [8] and [50], the directed path-width is also at most $t+1$.

## 9.3 Structure Theory for Directed Graphs Based on Directed Minors

Originally, Robertson and Seymour introduced the tree-width of undirected graphs as part of their monumental graph minor project culminating in the proof of Wagner's conjecture. At the heart of this project is a very powerful structure theorem explaining what can be said about a graph $G$ knowing that it does not contain a fixed graph $H$ as a minor. One simple reason for this could be that the tree width of $G$ is too small. But $G$ may fail to contain $H$ as a minor even if the tree-width is very high. Therefore the major part of the graph minors project deals with graphs of very high tree-width that do not contain a fixed $H$ as a minor. For this, one needs to understand what information can be gained about a graph knowing that its tree width is very high. The most fundamental result in this context is the **excluded grid theorem** in [89] stating that any graph of sufficiently high tree-width contains a large grid as a minor. Once this grid is found one can then analyze how the rest of the graph attaches to this grid which eventually leads to the local structure theorem and furthermore to the full structure theorem mentioned before.

With the introduction of directed tree-width, Reed, Robertson, Seymour and Thomas initiated the programme of generalizing this structure theory from undirected graphs to digraphs. Again, a major challenge is to understand what information can be obtained about a digraph knowing that its directed tree-width is very high, i.e. what can we say about **obstructions** to small directed tree-width. Consequently, the main open conjecture in the initial papers is the directed analogue of the excluded grid theorem, which, however, was only proved more than a decade after directed tree-width was introduced. In this section we present several powerful duality results between directed tree-width and various forms of obstructions. These results are not only interesting from a structural perspective but have found important algorithmic applications. We will comment on these applications in Section 9.4 below.

We begin by establishing a few fundamental properties of directed tree decompositions. Let $(T, \beta, \gamma)$ be a directed tree decomposition of a digraph $D$.

The next lemma follows easily from the definition of directed tree decompositions and establishes a connection between decompositions and strong separators, i.e. sets of vertices separating strongly connected components into smaller components.

**Lemma 9.3.1** *Let $\mathcal{T} := (T, \beta, \gamma)$ be a directed tree decomposition of a digraph $D$.*

1. *For every $e \in E(T)$, $\gamma(e)$ is a strong separator in $D$, i.e. if $S_1, S_2$ are the two components of $T - e$, then every strong component of $D - \gamma(e)$ is either contained in $\beta(S_1)$ or $\beta(S_2)$.*

2. *If $t \in V(T)$ and $T_1, \ldots, T_s$ are the components of $T - t$, then every strong component of $D - \Gamma(t)$ is contained in exactly one $\beta(T_i)$ for some $i$.*

**Definition 9.3.2** *Let $D$ be a digraph and $W \subseteq V(D)$.*

1. *A **balanced $W$-separator** is a set $S \subseteq V(D)$ such that every strong component of $D - S$ contains at most $\frac{|W|}{2}$ vertices of $W$. The **order** of the separator is $|S|$.*
2. *The set $W \subseteq V(D)$ is **$k$-linked** if $D$ does not contain a balanced $W$-separator of order $k$.*

We show first that in a digraph of directed tree-width at most $k-1$ every set has a balanced separator of order $k$, i.e. $D$ does not contain a $k$-linked set.

**Lemma 9.3.3** *Let $D$ be a digraph of directed tree-width at most $k-1$. Then every set $W \subseteq V(D)$ has a balanced $W$-separator of order at most $k$.*

We sketch the proof of the lemma. See [70, 84] for details. Let $(T, \beta, \gamma)$ be a directed tree decomposition of $D$ of order $k$. For every arc $e = (u, v) \in A(T)$ let $C_1, \ldots, C_l$ be the strong components of $D - \gamma(e)$ containing an element of $W$. If none of the $C_i$ contains more than $\frac{1}{2}|W|$ elements of $W$, then $\gamma(e)$ is a balanced $W$-separator and we are done. Otherwise, by Lemma 9.3.1, one of the two components $T_u, T_v$ of $T - e$ contains the (unique) component $C_i$ containing more than half of the elements of $W$. We orient $e$ towards $u$ if $C_i$ is contained in $\beta(T_u)$ and towards $v$ otherwise. This defines an orientation of $T$ and as $T$ is a tree there must be a vertex $t \in V(T)$ such that all incident arcs point towards $t$. It is easily seen that $\Gamma(t)$ is a balanced $W$-separator.

The next theorem establishes an even more precise relation between $k$-linked sets and the directed tree-width.

**Theorem 9.3.4** ([54]) *Every digraph $D$ either has directed tree-width at most $3k+2$ or contains a set $W$ which is $k$-linked and is a witness that $D$ has directed tree-width at least $k$.*

We give the proof of this theorem as it will be the basis of an FPT algorithm[2] for computing, for a given digraph $D$ a directed tree decomposition whose width is an approximation of the directed tree-width of $D$. in Section 9.4.

[2] By an **FPT algorithm** we mean an algorithm with running time $f(k) \cdot n^c$, for some function $f$ and constant $c$, where $n$ is the input size and $k$ is a parameter defined in the definition of the problem the algorithm solves. See Section 1.11 for details.

*Proof.* To prove the theorem we inductively construct a directed tree decomposition $(T, \beta, \gamma)$ of $D$. We maintain the property that for every inner vertex $t \in V(T)$, $|\Gamma(t)| \leq 3k+2$ and for every arc $e \in E(T)$, $|\gamma(e)| \leq 2k+1$.

Either this process will succeed and therefore produce a directed tree decomposition of the required width or it will fail at some point at which we obtain a $k$-linked set.

We initialize the construction by the trivial directed tree decomposition $\mathcal{T} := \big(T, \beta, \gamma\big)$, where $T$ is the tree with one node $r$ and $\beta(r) := V(D)$. Clearly this satisfies the invariant above.

Now suppose $\mathcal{T} = (T, \beta, \gamma)$ has already been constructed. If $\mathcal{T}$ does not contain a leaf $t \in V(T)$ with $|\Gamma(t)| > 3k+2$, then we are done. So let $t \in V(T)$ be such a leaf.

Let $e$ be the arc incident with $t$. By construction, $|\gamma(e)| \leq 2k+1$. If $\gamma(e)$ is $k$-linked, we are done. Otherwise, let $S$ be a balanced $\gamma(e)$-separator of order at most $k$. Let $v \in \beta(t)$ be an arbitrary vertex and let $X := S \cup \{v\}$. By construction, $|X| \leq k+1$, $X \cap \beta(t) \neq \emptyset$ and every strong component $C$ of $D - X$ contains at most $\frac{1}{2}|\gamma(e)| \leq k$ elements of $\gamma(e)$. Let $C_1, \dots, C_s$ be the strong components of $D - (X \cup \gamma(e))$. By the definition of a directed tree decomposition, either $V(C_i) \subseteq \beta(t)$ or $V(C_i) \cap \beta(t) = \emptyset$, for all $1 \leq i \leq s$. Let $D_1, \dots, D_l$ be the components among $\{C_1, \dots, C_s\}$ with $V(C_i) \subseteq \beta(t)$. For each $D_i$, let $D'_i$ be the component of $D - X$, such that $V(D_i) \subseteq V(D'_i)$ and let $W_i = \big(\gamma(e) \cap V(D'_i)\big) \cup X$. Then

$$|W_i| \leq |\gamma(e) \cap V(D'_i)| \cup |X| \leq k + k + 1 = 2k+1$$

and $D_i$ is also a strong component of $D - W_i$.

We extend $\mathcal{T}$ as follows to obtain a new decomposition $\mathcal{T}' := (T', \beta', \gamma')$: add new vertices $t_1, \dots, t_l$ and arcs $e_i := (t, t_i)$ to $T$, for all $1 \leq i \leq l$, and set $\beta'(t) := X \cap \beta(t)$, $\beta'(t_i) := V(D_i)$ and $\gamma'(e_i) := W_i$. For all other nodes $t$ and arcs $e$ we set $\beta'(t) := \beta(t)$ and $\gamma'(e) := \gamma(e)$. It is easily seen that $\mathcal{T}'$ is a directed tree-decomposition of $D$ maintaining the invariant above. In particular, $|\beta'(t)| \leq |X| \leq k+1$ and $|\gamma'(e_i)| \leq 2k+1$. Furthermore, $\gamma'(e_i) \subseteq X \cup \gamma(e)$ and thus $\Gamma'(t) = \beta'(t) \cup \gamma'(e) \cup \bigcup\{\gamma'(e_i) : 1 \leq i \leq s\} \subseteq X \cup \gamma(e)$. It follows that $|\Gamma'(t)| \leq k+1+2k+1 = 3k+2$. Furthermore, as $D_1, \dots, D_l$ are strong components of $D - (X \cup \gamma(e))$, the conditions of directed tree decompositions are still satisfied. □

A consequence of the proof of the previous lemma is that if a digraph $D$ has directed tree-width at most $k$ then it also has a directed tree decomposition of width at most $3k+2$ which has a particularly nice form.

**Definition 9.3.5 (Nice Directed Tree Decomposition)** *Let $D$ be a digraph. A directed tree decomposition $(T, \beta, \gamma)$ of $D$ is* ***nice*** *if*

*a) for all $e = (s, t) \in A(T)$ the set $\beta(T_t)$ is a strong component of $G - \gamma(e)$ and*

*b) if $t \in V(T)$ and $s_1, \dots, s_l$ are the children of $t$ in $T$, then $\bigcup_{1 \leq i \leq l} \beta(s_i) \cap \bigcup_{e \sim t} \gamma(e) = \emptyset$.*

Nice decompositions are easier to work with in algorithmic applications and we will use them in the applications in Section 9.4. One immediate consequence of this definition is the following lemma which is algorithmically useful.

**Lemma 9.3.6** *Let $(T, \beta, \gamma)$ be a directed tree-decomposition of a digraph $D$. For every $t \in V(T)$ there is an ordering $<_t$ on the successors $s_1, \ldots, s_k$ of $t$ in $T$ so that if $s_i <_t s_j$, then $D$ does not contain any arc $e = (u, v) \in E(D)$ with $u \in \beta(T_{s_i})$ and $v \in \beta(T_{s_j})$.*

For now we go back to the study of obstructions for directed tree width. We have already seen that a $k$-linked set is an obstruction to small directed tree-width. The next obstruction we study are known as **havens**. In the sequel, for any set $X$ and $k \geq 0$, we denote the set of all subsets of $X$ of order less than $k$ by $[X]^{<k}$.

**Definition 9.3.7** *Let $D$ be a digraph. A **haven** of $D$ of **order** $k$ is a function $h : [V(D)]^{<k} \to 2^{V(G)}$ assigning to every set $X$ of fewer than $k$ vertices a strong component of $G - X$ such that if $Y \subseteq X \subseteq V(D)$ with $|X| < k$, then $h(X) \subseteq h(Y)$.*

It is easily seen that any $k$-linked set $W$ in a digraph $D$ defines a haven of order $k$: for every set $X \subseteq V(D)$ of order at most $k$ define $h(X)$ as the (unique) strong component of $D-X$ containing more than half of the elements of $W$. It is straightforward to verify that this satisfies the haven axioms. Hence, we obtain the following theorem.

**Theorem 9.3.8** ([54])

1. *If $G$ is a digraph of tree-width at most $k$, then $G$ does not contain a haven of order $k$.*
2. *Conversely, if $G$ does not contain a haven of order $k$, then $G$ has tree-width at most $3k + 2$.*

We now define a sequence of other obstructions for directed tree-width, originally defined in [84].

**Definition 9.3.9** *A **bramble** in a digraph $D$ is a set $\mathcal{B}$ of strongly connected subgraphs of $D$ such that for any pair $B, B' \in \mathcal{B}$, either $V(B) \cap V(B') \neq \emptyset$ or there are arcs $e, e'$ linking $B$ and $B'$ in both directions. A bramble $\mathcal{B}$ is **strong** if $V(B) \cap V(B') \neq \emptyset$ for all $B, B' \in \mathcal{B}$.*

*A **cover**, or **hitting set**, of $\mathcal{B}$ is a set $X \subseteq V(D)$ such that $X \cap V(B) \neq \emptyset$ for all $B \in \mathcal{B}$. The **order** of $\mathcal{B}$ is the minimum size of a cover of $\mathcal{B}$.*

The last type of obstruction we consider are **well-linked sets**.

**Definition 9.3.10** *Let $D$ be a digraph. A set $W \subseteq V(D)$ is* ***well-linked*** *if for any $X, Y \subseteq W$ with $|X| = |Y|$ there are $|X| = |Y|$ pairwise vertex disjoint paths from $X$ to $Y$ in $G - (W \setminus (X \cup Y))$.*

The next lemma, proved in [84], connects the various forms of obstructions we have seen so far. See [70] for details.

**Lemma 9.3.11** *Let $D$ be a digraph and let $k \geq 0$.*

1. *If $D$ contains a $k$-linked set, then it contains a strong bramble of order $k+1$.*
2. *If $D$ contains a bramble $\mathcal{B}$ of order $k$, then $D$ contains a well-linked set of order $k$.*
3. *If $D$ contains a well-linked set of order $4k+1$, then $D$ contains a $k$-linked set.*

*Proof.* To show Part (1), it is not hard to see that a $k$-linked set $W$ in a digraph $D$ defines a bramble of order $k+1$: for every set $X \subseteq V(D)$ of at most $k$ vertices add to the bramble to the unique strong component of $D - X$ containing more than half of the vertices of $W$. It is readily verified that this indeed yields a strong bramble.

For (2) one can show that every minimum size cover of a bramble must be well-linked.

Part (3) is slightly more technical and we refer, e.g. to [70] for details. □

As explained at the beginning of this section, one of the most fundamental theorems in Robertson and Seymour's graph minor project is the excluded grid theorem. In the mid-90s, Reed [85] and Johnson et al. [54] conjectured an analogous theorem for directed graphs, i.e. that any digraph of sufficiently high directed tree-width should contain a large cylindrical grid as a **butterfly minor**.

**Definition 9.3.12 (Butterfly minor)** *Let $D$ be a digraph and let $e = (u, v) \in A(D)$. The digraph $D/e$ obtained from $D$ by* **contracting** *$e$ is defined as the digraph with vertex set $V(D) \setminus \{u, v\} \cup \{x_{u,v}\}$, where $x_{u,v}$ is a fresh vertex. The edges of $D/e$ are the same as the edges of $D$ except for the edges with $u$ or $v$ as endpoint. Any such edge $(w, w')$ or $(w', w)$, where $w \in \{u, v\}$ and $w' \notin \{u, v\}$ is replaced by an edge $(x_{u,v}, w')$ and $(w', x_{u,v})$ resp.*

*A* ***butterfly contraction*** *is the operation of contracting an edge $e = (u, v)$ where $u$ has out-degree 1 or $v$ has in-degree 1. A digraph $H$ is said to be a* ***butterfly minor*** *of a digraph $D$, written $H \preceq^b D$, if it can be obtained from a subgraph of $D$ by a series of butterfly contractions.*

**Definition 9.3.13 (cylindrical grid)** *A* ***cylindrical grid*** *of order $k$, for some $k \geq 1$, is a digraph $G_k$ consisting of $k$ directed cycles $C_1, \dots, C_k$, pair-*

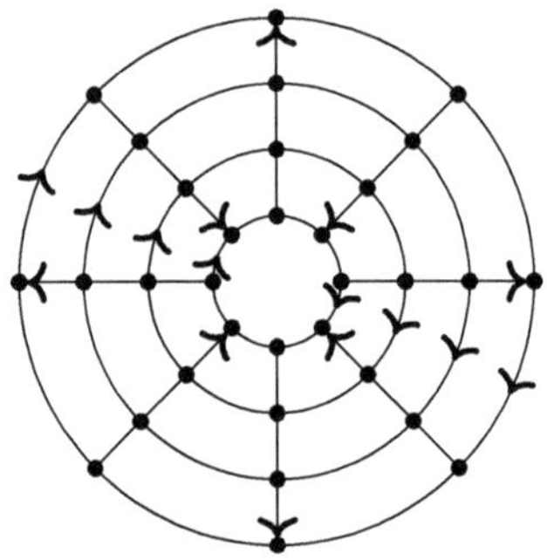

**Figure 9.6** Cylindrical grid $G_4$.

*wise vertex disjoint, together with a set of* $2k$ *pairwise vertex disjoint paths* $P_1, \ldots, P_{2k}$ *such that*

- *each path* $P_i$ *has exactly one vertex in common with each cycle* $C_j$*,*
- *the paths* $P_1, \ldots, P_{2k}$ *appear on each* $C_i$ *in this order*
- *for odd* $i$ *the cycles* $C_1, \ldots, C_k$ *occur on all* $P_i$ *in this order and for even* $i$ *they occur in reverse order* $C_k, \ldots, C_1$*.*

See Figure 9.6 for an illustration of $G_4$. The conjecture by Reed, Johnson, Robertson, Seymour and Thomas was confirmed by Kawarabayashi and Kreutzer in [61].

**Theorem 9.3.14 (The directed grid theorem** [61]**)** *There is a function* $f : \mathbb{N} \to \mathbb{N}$ *such that every digraph of directed tree-width at least* $f(k)$ *contains a cylindrical grid of order* $k$ *as a butterfly minor.*

## 9.4 Complexity of Directed Width Measures and Algorithmic Applications

In this section we describe some of the algorithmic applications of directed width measures. In particular, we will see that some $\mathcal{NP}$-complete graph problems can be solved efficiently on classes of digraphs of bounded width. As these applications usually require the computation of the associated decompositions, we first consider the complexity of computing digraph decompositions in the next section. The main algorithmic applications are presented in Section 9.5 below.

### 9.4.1 Complexity of Directed Width Measures

We first show that for essentially all width measures defined above, the associated decision problem is $\mathcal{NP}$-hard. This follows from the following observation.

**Theorem 9.4.1** *Let $G$ be an undirected graph and let $D$ be the digraph obtained from $G$ by replacing each arc $\{u, v\}$ by two arcs $(u, v)$ and $(v, u)$*[3]*. Then* $\text{tw}(G) + 1 = \text{dtw}(D) + 1 = \text{dag-width}(D) = \text{Kelly-width}(D) = \text{D-width}(D)$, *where* $\text{tw}(G)$ *denotes the tree-width of $G$.*

*Furthermore, the tree-depth of $G$ equals the DAG-depth of $D$ and the path-width of $G$ equals the directed path-width of $D$ minus* 1.

The case for directed tree-width was proved in [54, (2.1)]. The equalities for DAG-width and Kelly-width follow immediately from the corresponding game characterizations. For directed path-width and D-width there are direct translations of the corresponding decompositions and for DAG-depth it follows immediately from the definition of DAG depth and tree-depth.

Deciding the tree-width, the tree-depth and the path-width of a graph $G$ is $\mathcal{NP}$-complete (see e.g. [5]) and hence the decision problems for the directed width measures is $\mathcal{NP}$-hard. For all width measures except DAG-width, the decomposition defining the width are of polynomial size in the size of the input graph and hence the problems are even $\mathcal{NP}$-complete. For DAG-width this is not the case, as we shall see below.

**Corollary 9.4.2** *Deciding the DAG-depth, the directed tree-width, the directed path-width, the D-width and the Kelly-width of a digraph is $\mathcal{NP}$-complete. Deciding the DAG-width of a digraph is $\mathcal{NP}$-hard.*

Right from the definition, the number of bags in a DAG decomposition of a digraph $D$ is not restricted to be polynomial in the size of the decomposed digraph. And in fact, it was shown in [3], that there are classes of digraphs where DAG decompositions of optimal width require super-polynomially many bags, i.e. there is no fixed degree polynomial bounding the number of bags of a DAG-decomposition in the number of vertices of the digraph. In particular, this rules out that optimal DAG-decompositions can be computed by an FPT algorithm parameterized by the DAG-width. To make matters worse, it was also shown in [3], that there is no polynomial size approximation of an optimal DAG decomposition with an additive constant error in the width. Furthermore, the problem of deciding the DAG-width of a digraph turned out to be much harder than deciding any of the other width measures.

**Theorem 9.4.3** ([3]) *The problem, given a digraph $G$ and a number $k \geq 0$, whether the DAG-width of $G$ is at most $k$, is* PSPACE*-complete.*

### 9.4.2 Computing Directed Graph Decompositions

We have seen that deciding directed width measures is computationally hard. However, a range of algorithms have appeared for computing decompositions

---

[3] Thus $D$ is the complete biorientation of $G$.

which approximate the optimal width. Here we present some of these approximation algorithms.

**Directed Tree-Width.** The first algorithm we present below is an FPT approximation algorithm[4] for directed tree-width that follows from [54]. See Section 1.11 for details on parameterized algorithms. The proof of Theorem 9.3.4 showing the duality between havens and directed tree-width can easily be made algorithmic using the notion of weakly balanced separations. Recall the definition of a **nice** directed tree decomposition (Definition 9.3.5).

**Theorem 9.4.4** *There is an algorithm with running time* $\mathcal{O}(3^{2k+2} \cdot k \cdot |A(D)| \cdot |V(D)|)$ *which, on input* $D$ *and* $k \geq 1$*, either computes a nice directed tree-decomposition of* $D$ *of width at most* $5k + 10$ *or a weakly* $k$*-linked set* $W$*.*

*Sketch.* Essentially, the proof of Theorem 9.3.4 already yields an algorithm for computing directed tree decompositions. The only problem is that balanced separators cannot be computed efficiently. However, in the proof balanced $W$-separators can be replaced by weakly balanced $W$-separations. Here, a **weakly balanced** $W$**-separation** is a triple $(X, S, Y)$ of pairwise disjoint sets $X, Y \subseteq W$ of order $0 < |X|, |Y| \leq \frac{3}{4}|W|$ and $S \subseteq V(D)$ such that $W = X \cup (S \cap W) \cup Y$ and there is no directed path from $X$ to $Y$ in $D - S$. The **order** of the separation is $|S|$.

Adapting the algorithm in [34, Corollary 11.22] to the directed setting one can show that there is an algorithm running in time $\mathcal{O}(3^{2k+2}k|A(D)|)$ which, given as input a digraph $D$, a number $k \geq 1$ and a set $W \subseteq V(D)$ of size $2k + 2$, computes a weakly balanced $W$-separation of order at most $k$ if such a separation exists.

Using weakly balanced separations instead of balanced separators in the proof of Theorem 9.3.4 yields an algorithm with the running time as stated in the theorem, at the expense of increasing the width of the constructed directed tree decomposition to $(4k + 1) + (k + 1) = 5k + 2$. □

The previous algorithm yields a fixed-parameter approximation algorithm for directed tree-width. Kintali, Kothari and Kumar designed a polynomial time approximation algorithm of directed tree-width up to $\log n$-factors.

**Theorem 9.4.5** ([64]) *There exists a polynomial time approximation algorithm that, given a digraph* $D$*, computes a directed tree decomposition of* $D$*, whose width is at most* $\mathcal{O}(\log^{\frac{3}{2}} |V(D)| \cdot \mathrm{dtw}(D))$.

**DAG-Width and Kelly-Width.** To date, directed tree-width is the only tree-width inspired width measure which can be computed (approximately)

[4] By an FPT approximation algorithm we mean an algorithm running in time $f(k) \cdot n^c$, for some function $f$ and a constant $c$, which given a number $k$ and a digraph $D$ computes a directed tree decomposition of $D$ of width $O(k)$ or determines that the directed tree-width of $D$ is $> k$.

by FPT algorithms on general digraphs. For DAG-width an XP-algorithm is known for computing an optimal width decomposition.

**Theorem 9.4.6** ([12]) *Given a digraph $D$ of DAG-width at most $k$, a DAG decomposition of $D$ of width at most $k$ can be computed in time $|D|^{\mathcal{O}(k)}$.*

For Kelly-width it is still open whether optimal decompositions can be computed by an XP-algorithm. The reason is that DAG-width is defined by a cops and robber game with a visible robber, i.e. a game of perfect information. Kelly-width, on the other hand, is defined by an invisible robber game and hence by a game with imperfect information, which are computationally harder. Hence, the game characterization does not immediately yield an XP-algorithm. However, there are explicit algorithms known for computing Kelly decompositions.

**Theorem 9.4.7** ([51]) *The Kelly-width of a digraph with $n$ vertices can be determined in time $\mathcal{O}^*(2^n)$ and space $\mathcal{O}^*(2^n)$, or in time $\mathcal{O}^*(4^n)$ and polynomial space, where $\mathcal{O}^*(f(n))$ means that polynomial factors are suppressed.*

Furthermore, the Kelly-width of a digraph can be approximated up to a $\log n$ factor.

**Theorem 9.4.8** ([64]) *There exists a polynomial time approximation algorithm that, given a digraph $D$, computes a Kelly decomposition of $D$, whose width is $\mathcal{O}(\log^{\frac{3}{2}} n \cdot \text{Kelly-width}(D))$.*

Finally, for small values of $k$, efficient and explicit algorithms for deciding the Kelly-width and computing corresponding decompositions were given, e.g. in [73].

**Directed Path-Width.** The situation for directed path-width is similar to the case of Kelly width.

**Theorem 9.4.9** ([66, 97])

1. *There is an algorithm which, given a digraph $D$ and $k \in \mathbb{N}$ as input, computes a directed path decomposition of $D$ of width $k$, if it exists, in time $\mathcal{O}(|D|^{k+1} \cdot |A(D)|)$.*
2. *There is an algorithm computing a directed path-decomposition of a digraph $D$ of optimal width in time $\mathcal{O}^*(1.89^n)$, where $\mathcal{O}^*$ means that polynomial factors are suppressed.*

It is still open whether computing optimal directed path decompositions is fixed-parameter tractable. However, Fomin and Pilipczuk [37] exhibited FPT algorithms for computing optimal width path decompositions on tournaments and semi-complete digraphs.

## 9.5 Applications of Tree-Width Inspired Directed Width Measures

On classes of undirected graphs of bounded tree width $\mathcal{NP}$-hard problems from a very broad spectrum of areas and types of problems have been shown to become efficiently solvable, often even in linear time for any fixed upper bound on the tree-width. In particular, Courcelle [23] proved that every problem definable in **monadic second-order logic ($\mathbf{MSO_2}$)** can be solved in linear time on bounded tree-width classes (of undirected graphs). **Monadic second-order logic($\mathbf{MSO_2}$)** is a logic extending plain first-order logic by quantification over sets of edges and sets of vertices of a graph. It is very powerful logical language in which many graph problems such as 3-Colourability, Hamiltonian paths and -cycles, $k$-disjoint paths, perfect matchings and many more can be expressed very naturally. See [24] for details on monadic second-order logic and its variants $MSO_2$ and $MSO_1$ used below.

For directed graphs, Ganian et al. [43] showed that no such broad $MSO_2$ based algorithm theory is possible for tree width inspired width measures. Essentially, under some technical conditions, they showed that if one wants tractability of all MSO definable problems on classes of bounded width with respect to some width measure that translates undirected tree-width to digraphs (defined as having a graph searching game characterization similar to tree-width), then the only width achieving this undirected tree width. This establishes a general limit of tractability for digraph width measures based on tree-width but allows for algorithmic applications more specific to directed graphs.

Directed width measures, especially directed tree-width, have found various applications in the design of algorithms: in database theory, Bagan et al. [6] proved that **simple regular path queries** can be evaluated in polynomial time on graph databases of bounded directed tree-width (whereas the problem is intractable in general). In the area of Boolean networks, Tamaki [96] conducted experiments on computing attractors in Boolean networks. It turned out that for networks of small directed path-width he was able to handle networks which were significantly larger than what can be handled by standard tools. Another example motivated by practical applications is given in [94], where Sheppard investigates digraphs obtained from DNA sequencing by hybridization. In this method a digraph is constructed where vertices correspond to so-called **k-mers**. An important algorithmic problem in this context is finding Hamiltonian paths. It was shown in [94] that the digraphs occurring in this context usually have very small DAG-width so that polynomial time algorithms for computing Hamiltonian paths on digraphs of small DAG-width (see below) become applicable. In general, the most intensively studied applications of directed width measures are for routing problems in directed graphs. We present some of these applications in the following sections.

### 9.5.1 Disjoint Paths and Linkage Problems in Digraphs of Bounded Width

One of the main applications of directed width measures is to routing problems in digraphs. We will see various examples where directed tree-width is used in algorithms for solving various forms of directed disjoint paths problems. In particular, we will show that problems such as the directed Hamiltonian path problem or the $k$-disjoint paths problem can be solved in polynomial time on classes of digraphs of bounded directed tree-width.

> $k$-DISJOINT PATHS
> **Input:** A digraph $G$ and terminals $s_1, t_1, s_2, t_2, \ldots, s_k, t_k$
> **Question:** Does $D$ have $k$ pairwise internally vertex disjoint paths $P_1, \ldots, P_k$ such that $P_i$ is from $s_i$ to $t_i$ for $i = 1, \ldots, k$?

The $k$-disjoint paths problem on directed and undirected graphs is well-known to be $\mathcal{NP}$-complete. But whereas on undirected graphs, the problem is fixed-parameter tractable, it is $\mathcal{NP}$-complete on directed graphs even for $k = 2$, as shown by Fortune, Hopcroft and Wyllie [39]. See Section 1.6.

**Theorem 9.5.1** ([39]) *The $k$-*DISJOINT PATHS *problem is $\mathcal{NP}$-complete for all $k \geq 2$.*

Furthermore, as shown by Slivkins [95], the $k$-DISJOINT PATHS problem is already $W[1]$-hard on DAGs. But Johnson, Robertson, Seymour and Thomas [54] proved that it can be solved in polynomial time for every fixed value of $k$ on any fixed class $\mathcal{C}$ of digraphs of bounded directed tree-width.

**Definition 9.5.2** *A* ***linkage*** *in a digraph $D$ is a set $\mathcal{L}$ of pairwise internally vertex disjoint directed paths. The* **order** *$|\mathcal{L}|$ is the number of paths in $\mathcal{L}$ and its* ***size*** *is $|V(\mathcal{L})|$, where $V(\mathcal{L}) := |\bigcup_{P \in \mathcal{L}} V(P)|$.*

*Let $\sigma := \{(s_1, t_1), \ldots, (s_k, t_k)\}$ be a set of $k$ pairs of vertices in $D$. A* ***$\sigma$-linkage*** *is a linkage $\mathcal{L} := \{P_1, \ldots, P_k\}$ of order $k$ such that $P_i$ links $s_i$ to $t_i$.*

The first algorithmic result we establish is the following theorem.

**Theorem 9.5.3** ([54])*Let $D$ be a digraph and $(T, \beta, \gamma)$ be a directed tree decomposition of $D$ of width $w$. Let $k, l \geq 1$ and let $\sigma$ be a set of $k$ pairs of vertices in $D$. It can be decided in time $|V(D)|^{\mathcal{O}(k+w)}$ whether $D$ contains a $\sigma$-linkage of size $l$.*

**Problem 9.5.4** *Can the previous theorem be improved to fixed-parameter tractability in the directed tree-width, for any fixed number $k$? I.e. does there exist for every fixed $k$ an algorithm running in time $f(\mathrm{dtw}(G)) \cdot |V(G)|^c$, for some constant $c$ and function $f$, both depending on $k$, that decides whether $G$ has a $\sigma$-linkage for any set $\sigma$ of at most $k$ source/terminal pairs?*

The theorem can also be extended to weighted digraphs, see e.g. [70] for details. Combined with the algorithm for computing directed tree decompositions in Theorem 9.4.4, the theorem immediately implies the following corollary.

**Corollary 9.5.5** *The Hamiltonian cycle, the Hamiltonian path and, for all $k$, the $k$-*DISJOINT PATHS *problem can be solved in polynomial time on any class $\mathcal{C}$ of digraphs of bounded directed tree-width.*

We sketch the proof of Theorem 9.5.3. Recall that for any digraph $D$ and $S \subseteq V(D)$, $D[S] := (S, A(D) \cap S \times S)$ denotes the subdigraph of $D$ induced by $S$. Similarly, if $\mathcal{L}$ is a linkage in $D$ and $S' \subseteq V(D)$, we write $\mathcal{L}[S']$ for the **projection** of $\mathcal{L}$ onto $D[S']$, i.e. the linkage $\{P \cap D[S'] : P \in \mathcal{L}\}$. The algorithm is based on the following observation. Let $D$ be a digraph and let $S \subseteq V(D)$ be a set of vertices. For $k \geq 0$ we say that $S$ is **$k$-protected** if there is a strong guard $Z \subseteq V(D)$ of $S$ of order $|Z| \leq k$. Note that if $(T, \beta, \gamma)$ is a directed tree decomposition of a digraph $D$ of width $k-1$ and $t \in V(D)$, then $\beta(T_t) = \bigcup\{\beta(t') :: t' \text{ is reachable from } t \text{ in } T\}$ is $k$-protected. In particular, if $e = (s,t)$ is an arc in $E(T)$ then we can take $Z := \gamma(e)$ as a witness for $\beta(T_t)$ being $k$-protected. The main observation is now the following.

**Lemma 9.5.6** *Let $D$ be a digraph and let $S \subseteq V(D)$. Let $k, w \geq 0$ and let $\mathcal{L}$ be a linkage of order $k$ in $D[S]$.*

*If $S' \subseteq S$ is $w$-protected, then $\mathcal{L}[S']$ has order at most $k + w$.*

*Proof.* Let $P_1, \cdots, P_k$ be the paths in $\mathcal{L}$ and let $Z \subseteq V(D)$ be such that $|Z| \leq w$ and every directed path in $D$ starting and ending in $S'$ which is not entirely contained in $D[S']$ contains a vertex of $Z$. It follows that if $P_i[S']$ is the union of $j$ directed paths, then $|V(P_i) \cap Z| \geq j - 1$. Hence, $\mathcal{L}[S']$ has order at most $k + w$. □

The previous lemma is the basis for a dynamic programming algorithm for solving the linkage problem in Theorem 9.5.3. Given a digraph $D$ and a directed tree decomposition $(T, \beta, \gamma)$ of $D$ of width $w-1$, the algorithm proceeds as follows. For every $t \in V(T)$ and every tuple $\sigma := \big((u_1, v_1), \ldots, (u_s, v_s)\big)$ of pairs of vertices in $\beta(T_t)$, for some $s \leq k + w$, it computes the set of all $l \leq |V(D)|$ such that $G[\beta(T_t)]$ contains a $\sigma$-linkage of size $l$. As shown in [54], this can be done by dynamic programming. Clearly, once this information is computed for the root of $T$, the linkage problem for $D$ can be answered for every tuple $\sigma = \big((u_1, v_1), \ldots, (u_k, v_k)\big)$ of order $k$. This completes the sketch of the proof of Theorem 9.5.3.

In the terminology of parameterized complexity, see Section 1.11, the previous result shows that the $k$-disjoint paths problem is in XP with parameter $k + w$, where $w$ is the directed tree-width of the input digraph. Unless FPT=$W[1]$, this cannot be improved to fixed-parameter tractability (FPT) in the parameter $k$ for every fixed width $w$, as Slivkins [95] showed that the disjoint paths problem is $W[1]$-hard already on DAGs, which have directed tree-width 0.

We close this section by mentioning an algorithmic meta theorem by Oliveira Oliveira generalizing the previous linkage algorithm.

**Theorem 9.5.7** ([29]) *Let $\Omega$ be a finite commutative semigroup. Let $\varphi$ be an $MSO_2$ sentence and let $k, w \in \mathbb{N}$. There is a computable function $f : \mathbb{N}^3 \to \mathbb{N}$ such that, given a weighted digraph $D = (V, E, \omega : E(D) \to \Omega)$ of directed tree-width $w$, a positive integer $l < |V|$ and an element $\alpha \in \Omega$, one can count in time $f(\varphi, w, k) \cdot |D|^{O(k \cdot (w+1))}$ the number of subgraphs $H$ of $D$ simultaneously satisfying the following four properties:*

1. *$H \models \varphi$.*
2. *$H$ is the union of $k$ directed paths.*
3. *$H$ has $l$ vertices.*
4. *$H$ has weight $\omega(H) = \alpha$.*

*In fact, one can even choose a semigroup of size polynomial in $D$.*

### 9.5.2 Linkages in General Digraphs

The results in the previous section exhibit algorithms for linkage type problems on digraphs of small directed tree-width. However, the machinery of directed tree decompositions and obstructions to low directed tree-width can also be used to obtain results for general digraphs.

Given the $\mathcal{NP}$-hardness of the $k$-Disjoint paths problem already for $k = 2$, it is natural to consider relaxations of the problem in order to obtain polynomial time algorithms. One relaxation that has been studied in the literature is to allow congestions. Let $\sigma := \big((s_1, t_1), \ldots, (s_k, t_k)\big)$ be a $k$-tuple of pairs of vertices in a digraph $D$ and let $c \geq 1$. A set $P_1, \ldots, P_k$ of directed paths in $D$ is a **$\sigma$-linkage with congestion** $c$ if, for all $1 \leq i \leq k$, the path $P_i$ links $s_i$ to $t_i$ and furthermore, every vertex of $D$ is contained in at most $c$ paths. For $c = 2$ we call the linkage **half-integral** and for $c = 4$ it is a **quarter-integral** linkage.

**Problem 9.5.8** *Does there exist, for every fixed integer $k \geq 1$, a polynomial algorithm which, given a digraph $D$ and a tuple $\sigma := \big((s_1, t_1), \ldots, (s_k, t_k)\big)$ as input, decides correctly whether $D$ contains a half-integral $\sigma$-linkage.*

However, partial results are known. In [60], Kawarabayashi, Kobayashi and Kreutzer show the following result for quarter-integral linkages.

**Theorem 9.5.9** ([60]) *For every fixed $k \geq 1$ there is a polynomial time algorithm for deciding the following problem.*

| |
|---|
| QUARTER-INTEGRAL DISJOINT PATHS<br>**Input:** *A digraph $D$ and terminals $s_1, t_1, s_2, t_2, \ldots, s_k, t_k \in V(D)$.*<br>**Find:** *a quarter-integral linkage of $(s_1, t_1), \ldots, (s_k, t_k)$ or conclude that $D$ does not contain disjoint paths $P_1, \ldots, P_k$ such that $P_i$ is from $s_i$ to $t_i$, for $i \in [k]$.* |

The proof of the previous theorem in [60] precedes the proof of the directed excluded grid theorem (Theorem 9.3.14). Using Theorem 9.3.14, the result can be improved to third-integral linkages. The main idea of the proof is to use the duality between directed tree width and cylindrical grids. Roughly, the algorithm works as follows. If the directed tree-width is small then it uses a simple adaptation of the algorithm in Theorem 9.5.3 to solve the problem optimally. Otherwise, Theorem 9.3.14 implies that $D$ contains a large cylindrical grid $C$. If there is a linkage $L_1$ from $s_1, \ldots, s_k$ to $C$ and a linkage $L_2$ from $C$ to $t_1, \ldots, t_k$ then $L_1, C$ and $L_2$ can be used to construct a third-integral linkage linking $s_i$ to $t_i$, for all $1 \leq i \leq k$. Otherwise, by Menger's theorem, there must be a low order separation from, say, $s_1, \ldots, s_k$. The separation does not rule out the existence of a quarter-integral solution but it can sometimes be used to rule out a fully integral solution (which would then be the second outcome of the theorem). If a fully integral solution is not ruled out by this construction, then the problem can be reduced to a smaller instance. In this way, one either gets a third-integral solution or the algorithm certifies that there are no fully disjoint paths linking the sources to the targets.

As mentioned above, it is still an open problem whether the result can be improved to half-integral solutions and, more importantly, whether it can further be improved so that the negative answer also rules out the existence of a half-integral solution.

As a first significant step in this direction, Edwards, Muzi and Wollan proved a polynomial time algorithm for the half-integral linkage problem for highly connected digraphs.

**Theorem 9.5.10** ([31]) *For all integers $k \geq 1$, there exists a value $L(k)$ such that every strongly $L(k)$-connected graph is half-integrally $k$-linked. Moreover, there exists an absolute constant $c$ such that given an instance $(D, (s_1, t_1), \ldots, (s_k, t_k))$ of the half-integral disjoint path problem, where $D$ is $L(k)$-connected, we can find a solution in time $\mathcal{O}(|V(D)|^c)$.*

We close the section by mentioning further applications of these techniques beyond classes of digraphs of small directed tree-width. Fomin and Pilipczuk [37] showed that for tournaments the $k$-arc disjoint paths problem fixed-parameter tractable. Their algorithm uses directed path-width. They first show that on tournaments directed path-width can be decided by an FPT algorithm. They then use a duality of directed path-width and an obstruction called jungles which was proved in [20, 40].

Finally, the concepts of directed tree-width, or more specifically, its dual notion of well-linked sets, have played a decisive role in the study of approximation algorithms for symmetric routing on planar digraphs. See Section 5.2.

### 9.5.3 The Erdős-Pósa Property for Directed Graphs

A classical result by Erdős and Pósa states that there is a function $f : \mathbb{N} \to \mathbb{N}$ such that for every $k$, every graph $G$ contains $k$ pairwise vertex disjoint cycles or a set $T$ of at most $f(k)$ vertices such that $G - T$ is acyclic.

There is a natural generalization of this result to arbitrary graphs: a graph $H$ has the Erdős-Pósa property if there is a function $f : \mathbb{N} \to \mathbb{N}$ such that every graph $G$ either has $k$ disjoint copies of $H$ as a minor or contains a set $T$ of at most $f(k)$ vertices such that $H$ is not a minor of $G - T$. As an application of the undirected excluded grid theorem, Robertson and Seymour [89] proved that a graph $H$ has the Erdős-Pósa-property in this sense if, and only if, $H$ is planar.

The Erdős-Pósa property can also be defined for digraphs. Younger [101] conjectured that there is a function $f : \mathbb{N} \to \mathbb{N}$ such that for every $k$ every digraph either has $k$ disjoint directed cycles or a set of at most $f(k)$ vertices intersecting every directed cycle. The conjecture was proved by Reed, Robertson, Seymour and Thomas in [86]. In fact, the concept of directed tree-width originated in the work on Younger's conjecture.

Again this can be generalised to arbitrary digraphs, based on directed minors (see Section 9.6.1 for the definition of butterfly and topological minors): a digraph $H$ has the **Erdős-Pósa property** for topological (butterfly) minors if there is a function $f : \mathbb{N} \to \mathbb{N}$ such that for all $k \geq 0$, every digraph $D$ either contains $k$ disjoint subgraphs each containing $H$ as a topological (butterfly) minor or there is a set $S \subseteq V(D)$ of at most $f(k)$ vertices such that $D - S$ does not contain $H$ as a topological (butterfly) minor. In [4], Amiri, Kawayabashi, Kreutzer and Wollan used the directed excluded grid theorem (Theorem 9.3.14) to show the following characterization of strongly connected digraphs with the Erdős-Pósa property.

**Theorem 9.5.11** *Let $H$ be a strongly connected digraph.*

1. *$H$ has the Erdős-Pósa property for butterfly minors if, and only if, there is a cylindrical grid $G_c$, for some constant $c = c(H)$, such that $H \preceq^b G_c$.*
2. *$H$ has the Erdős-Pósa property for topological minors if, and only if, there is a cylindrical wall $G_c$, for some constant $c = c(H)$, such that $H \preceq^t G_c$.*

*Furthermore, for every fixed strongly connected digraph $H$ satisfying these conditions and every $k$ there is a polynomial time algorithm which, given a digraph $G$ as input, either computes $k$ disjoint (butterfly or topological) models of $H$ in $G$ or a set $S$ of $\leq h(k)$ vertices such that $G - S$ does not contain a model of $H$.*

The previous theorem settles the case for strongly connected digraphs. It would be interesting to get a similar characterization also for general digraphs. This may be much harder to get as in this case the techniques based on directed tree-width will no longer be as useful as for strongly connected digraphs. An intermediate case could be **vertex cyclic** digraphs which are digraphs in which every strong component is non-trivial, i.e. contains more than a single vertex. In [4], some special cases of vertex-cyclic digraphs are solved, but the general problem remains open.

**Problem 9.5.12**

1. *Characterize the class of vertex-cyclic digraphs which have the Erdős-Pósa property.*
2. *Characterize in general the class of digraphs which have the Erdős-Pósa property.*
3. *What is the complexity of deciding, given a digraph $H$, whether it has the Erdős-Pósa property?*

## 9.6 Density Based Width Measures

In this section we introduce the second type of directed width measures covered in this chapter: width measures based on directed minors and density arguments. For this, we first need to define the notions of directed minors used in this section.

### 9.6.1 Directed Minors

On undirected graphs, one usually distinguishes between two types of minors: **topological minors**, obtained by subdividing edges and deleting edges or vertices, and general **minors**, obtained by a sequence of edge and vertex deletion and arc contraction.

Topological minors have a straight forward generalization to directed graphs.

**Definition 9.6.1** *A* ***subdivision*** *of a digraph $D$ is obtained by replacing some arcs of $D$ by pairwise internally vertex disjoint directed paths respecting the directions of the replaced arcs. For $r \geq 0$, $H$ is an $r$-**subdivision** of $D$ if we can replace some arcs of $H$ by paths of length at most $r+1$ to obtain $D$.*

*For digraphs $D, H$, we say that $H$ is a* ***directed topological minor*** *of $D$, denoted by $H \preceq^t D$, if $D$ contains a subdivision of $H$ as a subgraph. We write $H \preceq^t_r D$ and call $H$ an $r$-**shallow topological minor** of $D$ if $D$ contains a $2r$-subdivision of $H$ as a subgraph.*

The reason we define $r$-shallow topological minors as $2r$-subdivisions is that this corresponds more closely to $r$-shallow directed minors defined below.

For general directed minors, several alternative and not necessarily equivalent definitions have been considered in the literature. The most popular among these are **butterfly minors**, defined in Definition 9.3.12 above.

For undirected graphs, the notion of minors that are obtained by a series of vertex and edge deletions and edge contractions can equivalently be defined in terms of minor models. In the directed setting these two notions are different (every butterfly minor is also a directed minor but not vice versa) [72].

**Definition 9.6.2** *A digraph $H$ has a* ***directed model*** *in a digraph $D$ if there is a function $\delta$ mapping vertices $v \in V(H)$ of $H$ to sub-graphs $\delta(v) \subseteq D$ and arcs $e \in E(H)$ to arcs $\delta(e) \in E(D)$ such that if $v \neq u$ then $\delta(v) \cap \delta(u) = \emptyset$ and if $e = (u, v)$ and $\delta(e) = (u', v')$ then $u' \in \delta(u)$ and $v' \in \delta(v)$.*

*For $v \in V(H)$ let $\mathrm{in}(\delta(v)) := V(\delta(v)) \cap \bigcup_{e=(u,v)\in E(H)} V(\delta(e))$ and $\mathrm{out}(\delta(v)) := V(\delta(v)) \cap \bigcup_{e=(v,w)\in E(H)} V(\delta(e))$.*

*Furthermore, we require that for every $v \in V(H)$*

1. *there is a directed path in $\delta(v)$ from any $u \in \mathrm{in}(\delta(v))$ to every $u' \in \mathrm{out}(\delta(v))$;*
2. *there is at least one source vertex $s_v \in \delta(v)$ that reaches every element of $\mathrm{out}(\delta(v))$;*
3. *there is at least one sink vertex $t_v \in \delta(v)$ that can be reached from every element of $\mathrm{in}(\delta(v))$.*

*We write $H \preceq^d D$ if $H$ has a directed model in $D$ and call $H$ a* ***directed minor*** *of $D$. We call the sets $\delta(v)$ for $v \in V(H)$ the* ***branch-sets*** *of the model.*

Note that the conditions (2) and (3) in the previous definition are implies by Condition (1) for vertices of in- and out-degree $> 0$. They serve the purpose to ensure that sinks and sources in $H$ are represented by a single vertex in $D$ together with paths connecting this vertex to its (in- or out-) neighbours.

**Definition 9.6.3** *For $r \geq 0$, a digraph $H$ is a directed* ***depth-$r$ minor*** *of a digraph $D$, denoted as $H \preceq^d_r D$, if there exists a directed model of $H$ in $D$ in which the length of all the paths in the branch-sets of the model are bounded by $r$.*

We close the section by relating the different concepts of minors to each other. It is not hard to see that for all digraphs $H, D$,

$$H \preceq^t D \quad \Rightarrow \quad H \preceq^b D \quad \Rightarrow \quad H \preceq^d D.$$

The same relation extends to shallow minors:

**Lemma 9.6.4** *For all digraphs $H, D$ and $r \geq 0$: $H \preceq^t_r D$ implies $H \preceq^d_r D$.*

Bipartite digraphs will play a special role on the rest of this section.

**Definition 9.6.5** *A* ***bipartite digraph*** *is a digraph $D = (A\dot{\cup}B, E)$ whose vertex set is partitioned into two sets $A$ and $B$ and $E \subseteq A \times B$.*

For bipartite digraphs, the concepts of butterfly minors and directed models coincide. In the following lemma, an **in-branching** is a digraph obtained from an undirected tree by orienting all edges towards a root node $r$. Analogously, in an **out-branching** all arcs are oriented away from the root, i.e. and out-branching is a rooted directed tree. See Section 1.8.

**Lemma 9.6.6 (see [72])** *If $H$ is a bipartite digraph with $H \preceq^d D$, we can choose the branch-sets of the model of $H$ in $D$ to be in- or out-branchings. In this case $H \preceq^d D \Leftrightarrow H \preceq^b D$.*

### 9.6.2 Width Measures Defined by Shallow Directed Minors and Bounded Edge Densities

Following [71, 72] (see [74, 75] for the undirected case), we define classes of digraphs of bounded expansion, nowhere crownful classes and classes which are nowhere dense. We first need some additional notation.

**Definition 9.6.7** *Let $G$ be a digraph and let $r \geq 0$. The* ***greatest reduced average degree of rank*** *$r$ (short* ***grad****) of $G$, denoted $\nabla_r(G)$ is*

$$\nabla_r(G) := \max\left\{\frac{|E(H)|}{|V(H)|} : H \preceq^d_r G\right\}$$

*and its* ***topological greatest average degree of rank*** *$r$ (short* ***top-grad****) is*

$$\tilde{\nabla}_r(G) := \max\left\{\frac{|E(H)|}{|V(H)|} : H \preceq^t_r G\right\}.$$

A **crown of order** $q$ is a digraph $S_q$ with vertex set $\{v_i : 1 \leq i \leq q\} \cup \{v_{i,j} : 1 \leq i < j \leq q\}$ and arcs $\{(v_{i,j}, v_i), (v_{i,j}, v_j) : 1 \leq i < j \leq q\}$.

**Definition 9.6.8** *Let $\mathcal{C}$ be a class of digraphs.*

1. *$\mathcal{C}$ has* ***bounded expansion*** *if there is a function $f : \mathbb{N} \to \mathbb{N}$ such that $\nabla_r(D) \leq f(r)$ for all $r \geq 0$ and $D \in \mathcal{C}$.*
2. *$\mathcal{C}$ is* ***nowhere crownful*** *if for every $r$, there exists a $q = q(r)$ so that $S_q \not\preceq^d_r D$ for all all $D \in \mathcal{C}$.*
3. *$\mathcal{C}$ is* ***directed nowhere dense*** *if for every $r$, there exists an $n$ and an acyclic tournament $T_n$ so that $T_n \not\preceq^d_r D$ for all $D \in \mathcal{C}$.*
4. *$\mathcal{C}$ is* ***directed somewhere dense*** *if there is an $r \geq 0$ so that the set of depth $r$ minors of $\mathcal{C}$ contains arbitrarily large tournaments.*

It can be shown that a class $\mathcal{C}$ of digraphs is directed somewhere dense if, and only if, it is not directed nowhere dense. Furthermore, the property of being directed nowhere dense is more general than being nowhere crownful and also more general than bounded expansion.

On the other hand, classes of digraphs of bounded expansion and nowhere crownful classes are incomparable. In particular, as shown in [72], nowhere crownful classes and even crown-minor free classes can be very dense.

**Theorem 9.6.9** *For every $\epsilon$, there exists a $q = q(\epsilon)$, such that for every $n$, there exists an $S_q$-minor-free digraph on $2n$ vertices that has arc density at least $\Omega(n^{\frac{1}{2}-\epsilon})$.*

It follows that there are classes of digraphs which are $S_q$-crown-minor free but do not have bounded expansion. Conversely, the class of crowns $S_q$, $q \geq 0$, has bounded expansion.

On the other hand, for the definition of bounded expansion, the precise notion of directed minor we use is not important, as shown by the following theorem proved in [71].

**Theorem 9.6.10** *A class $\mathcal{C}$ of digraphs has bounded expansion if and only if there is a function $f : \mathbb{N} \to \mathbb{N}$ such that for all $r \in \mathbb{N}$ it holds that $\widetilde{\nabla}_r(D) \leq f(r)$ for all $D \in \mathcal{C}$.*

## 9.7 Classes of Directed Bounded Expansion

Classes of digraphs of bounded expansion can be characterized in many different ways. The various characterizations yield a varied set of algorithmic techniques that can be used in the design of algorithms on bounded expansion classes of digraphs. In the following we will present some of the more promising structural properties of bounded expansion classes.

### 9.7.1 Generalised Colouring Numbers

The **colouring number** $\mathrm{col}(G)$ of an undirected graph $G$ is the smallest integer $k$ such that there is a linear order $\sqsubset$ on the vertex set of $D$ for which each vertex $v$ has **back-degree** at most $k-1$, i.e. at most $k-1$ neighbours $u$ with $u \sqsubset v$. It is well-known that for any graph $G$, the chromatic number $\chi(G)$ satisfies $\chi(G) \leq \mathrm{col}(G)$.

Three natural generalization of the colouring number are the series $\mathrm{adm}_r$, $\mathrm{col}_r$ and $\mathrm{wcol}_r$ of **generalised colouring numbers** defining the **admissibility, colouring number** and **weak colouring numbers** introduced by Kierstead and Yang [62] (see Dvořák [30] for the general definition of $\mathrm{adm}_r$) in the context of colouring games and marking games on graphs. Note that the colouring number is equivalent to the **degeneracy** of a graph. As proved by Zhu [102], these invariants can be used to characterize bounded expansion classes of undirected graphs.

The directed versions of the above invariants have been defined in [71] where it was shown that classes of directed bounded expansion can be characterized by bounds on the generalised colouring numbers.

Let $D$ be a digraph. By $\Pi(D)$ we denote the set of all strict linear orders on $V(D)$. For $\sqsubset \in \Pi(G)$, we write $u \sqsubseteq v$ if $u \sqsubset v$ or $u = v$. Let $u, v \in V(D)$, let $\sqsubset \in \Pi(D)$ and let $r \geq 0$.

The vertex $u$ is **weakly $r$-reachable** from $v$ with respect to $\sqsubset$, if there is a directed path $P$ of length $\ell$, $0 \leq \ell \leq r$, connecting $u$ and $v$ (in either direction) such that $u$ is the smallest among the vertices of $P$ (with respect to $\sqsubset$). By $\mathrm{WReach}_r[D, \sqsubset, v]$ we denote the set of vertices that are weakly $r$-reachable from $v$ w.r.t. $\sqsubset$.

The vertex $u$ is **strongly $r$-reachable** from $v$ with respect to $\sqsubset$, if there is a directed path $P$ of length $\ell$, $0 \leq \ell \leq r$, connecting $u$ and $v$ (in either direction) such that $u \sqsubseteq v$ and $v \sqsubset w$ for all internal vertices $w$ of $P$. Let $\mathrm{SReach}_r[D, \sqsubset, v]$ be the set of vertices that are strongly $r$-reachable from $v$ w.r.t. $\sqsubset$. Note that we have $v \in \mathrm{SReach}_r[D, \sqsubset, v] \subseteq \mathrm{WReach}_r[D, \sqsubset, v]$.

We also need a third type of colouring number, the admissibility. For a non-negative integer $r$, the **$r$-admissibility** $\mathrm{adm}_r[D, \sqsubset, v]$ of $v$ w.r.t. a linear order $\sqsubset \in \Pi(D)$ is the maximum size $k$ of a family $\{P_1, \ldots, P_k\}$ of directed paths of length at most $r$ with one end $v$ and the other end at a vertex $w$ with $w \sqsubset v$, and which satisfies $V(P_i) \cap V(P_j) = \{v\}$ for all $1 \leq i < j \leq k$. As for $r > 0$ we can always let the paths end in the first vertex smaller than $v$, we can assume that the internal vertices of the paths are larger than $v$.

**Definition 9.7.1** ([71]) *Let $D$ be a digraph. For a non-negative integer $r$, we define the* ***weak $r$-colouring number*** $\mathrm{wcol}_r(D)$*, the* ***$r$-colouring number*** $\mathrm{col}_r(D)$ *and the* ***$r$-admissibility*** *of $D$ as*

$$\begin{aligned}
\mathrm{wcol}_r(D) &:= \min_{\sqsubset \in \Pi(D)} \max_{v \in V(D)} \big|\mathrm{WReach}_r[D, \sqsubset, v]\big|, \\
\mathrm{col}_r(D) &:= \min_{\sqsubset \in \Pi(D)} \max_{v \in V(D)} \big|\mathrm{SReach}_r[D, \sqsubset, v]\big|. \\
\mathrm{adm}_r(D) &:= \min_{\sqsubset \in \Pi(D)} \max_{v \in V(D)} \mathrm{adm}_r[D, \sqsubset, v].
\end{aligned}$$

The following theorem relates these measures to each other.

**Theorem 9.7.2** ([71]) *Let $D$ be a digraph and let $r \geq 1$. Then* $\mathrm{adm}_r(D) \leq \mathrm{col}_r(D) \leq \mathrm{wcol}_r(D)$*. Furthermore,*

$$\mathrm{col}_r(D) \leq 2 \cdot (\mathrm{adm}_r(D) - 1)^r + 1 \quad \textit{and} \quad \mathrm{wcol}_r(D) \leq 2 \cdot \mathrm{adm}_r(D)^r.$$

The generalised colouring numbers can also be used to characterize bounded expansion classes of digraphs.

**Theorem 9.7.3** ([71]) *For every digraph $D$ and every $r \in \mathbb{N}$ it holds that* $\mathrm{adm}_r(D) < 6r^3 \nabla_r(D)^4$. *Conversely, for every digraph $D$ and every $r \in \mathbb{N}$ it holds that* $\widetilde{\nabla}_r(D) \leq 16(\mathrm{adm}_{2r}(D) + 1)$.

**Corollary 9.7.4** ([71]) *A class $\mathcal{C}$ of digraphs has bounded expansion if, and only if, there is a function $f : \mathbb{N} \to \mathbb{N}$ such that* $\mathrm{wcol}_r(D) \leq f(r)$ *for all $D \in \mathcal{C}$ and all $r \geq 1$.*

A useful property of admissibility is that for every graph $D$ from a bounded expansion class $\mathcal{C}$ an order $\sqsubset$ on $V(D)$ witnessing that the $r$-admissibility is small can be computed efficiently.

**Theorem 9.7.5** ([71]) *Let $\mathcal{C}$ be a class of digraphs of bounded expansion. There is a function $g$ such that for all $r \geq 0$ and all $D \in \mathcal{C}$ we can compute an optimal order for* $\mathrm{adm}_r(D)$ *in time* $g(r) \cdot n^{\mathcal{O}(1)}$, *where* $n := |V(D)|$.

### 9.7.2 Neighbourhood Complexity

We continue the study of structural properties of bounded expansion classes by defining a directed version of neighbourhood complexity, a measure that has very successfully been used in the connection to classes of undirected bounded expansion [87].

**Definition 9.7.6** *Let $D$ be a digraph, let $X \subseteq V(D)$ and let $r \geq 1$. The* ***distance-$r$ out-neighbourhood complexity of $X$ in $D$****, denoted $\nu^+(D, X)$, is defined by*

$$\nu^+(D, X) = \max_{H \subseteq D, X \subseteq V(H)} \left|\{N_r^+(v) \cap X : v \in V(H)\}\right| .$$

Analogously, one can define the **distance-$r$ in-neighbourhood complexity** when using $N_r^-(v)$ and the **distance-$r$ mixed neighbourhood complexity** when using $(N_r^+(v) \cup N_r^-(v))$ in the above definition.

Closure under subgraphs in the above definition is required to characterize sparse graph classes. Classically, this closure is not part of the definition, when it is e.g. used to define classes of bounded VC-dimension [91, 93, 99].

Bounded neighbourhood complexity is not equivalent to directed bounded expansion but at least classes of directed bounded expansion have bounded neighbourhood complexity.

**Theorem 9.7.7** ([71]) *Let $\mathcal{C}$ be a class of digraphs of bounded expansion. Then for all $r \geq 1$ there exists $k \geq 1$ such that for all $D \in \mathcal{C}$ and $X \subseteq V(D)$ we have $\nu_r^+(D, X) \leq |X|^k$. The same statement holds for in-neighbourhood complexity and mixed neighbourhood complexity.*

### 9.7.3 A Splitter Game for Classes of Digraphs of Bounded Expansion

In this section we establish a very useful property of bounded expansion classes of digraphs based on a directed version of a game, known as the **splitter game**, originally introduced as a characterization of nowhere dense classes of undirected graphs in [46].

We first need the following definition. The $r$-**strong-neighbourhood** of $v$, denoted by $\tilde{N}_{G,r}(v)$, or just $\tilde{N}_r(v)$ if $G$ is understood, is defined as the set of vertices $u$ in $G$ such that $G$ contains a closed walk of length at most $2r$ containing $u$ and $v$.

Let $G$ be a digraph and let $\ell, m, r \geq 0$. The $(\ell, m, r)$-**strong directed splitter game** on $G$ is played by two players, **Connector** and **Splitter**, as follows. Let $G_0 := G$. In round $i+1$ of the game, Connector picks a vertex $v_{i+1} \in V(G_i)$. Then Splitter chooses a subset $W_{i+1} \subseteq V(G_i)$ with $|W_{i+1}| \leq m$. Define $G_{i+1}$ as the induced subgraph of $G_i$ with $V(G_{i+1}) = \tilde{N}_{G_i,r}(v_{i+1}) \setminus W_{i+1}$. Splitter wins if $V(G_{i+1}) = \emptyset$. Otherwise the game continues to the next round. If Splitter has not won after $\ell$ rounds, then Connector wins.

A **strategy** for Splitter is a function $f$ associating to every partial play $(v_1, W_1, \dots, v_s, W_s)$ with associated sequence $G_0, \dots, G_s$ and every move $v_{s+1} \in V(G_s)$ by Connector a move $W_{s+1} \subseteq V(G_s)$ with $|W_{s+1}| \leq m$ for Splitter. A strategy $f$ is a **winning strategy** for Splitter if she wins every play in which she follows the strategy $f$. If such a winning strategy exists, we say that Splitter **wins** the $(\ell, m, r)$-directed splitter game on $G$.

The splitter game cannot be used as a characterization of bounded expansion as Splitter wins the $(1, 1, 1)$-strong splitter game on every acyclic digraph, but the class of acyclic digraphs does not have bounded expansion. But on every class of bounded expansion Splitter always has constant length winning strategies. This, together with neighbourhood covers introduced in the following section, can be used to define a bounded depth decomposition of graph from bounded expansion classes.

**Theorem 9.7.8** ([71]) *Let $D$ be a graph, let $r \in \mathbb{N}$ and let $\ell = \mathrm{wcol}_{4r}(G)$. Then splitter wins the $(\ell, 1, r)$-strong splitter game.*

### 9.7.4 Neighbourhood Covers

Neighbourhood covers of small radius and small size play a key role in the design of many data structures for distributed systems. There is also a deep connection between sparse neighbourhood covers of small radius and sparse graph spanners of low stretch. In this section we will see that classes of digraphs of bounded expansion admit sparse strong neighbourhood covers which can be computed by a fixed-parameter algorithm.

Let $r \in \mathbb{N}$. A **strong $r$-neighbourhood cover** $\mathcal{X}$ of a digraph $D$ is a mapping $\mathcal{X} : V(D) \to 2^{V(D)}$ such that $D[\mathcal{X}(v)]$ is strongly connected and $\tilde{N}_r(v) \subseteq \mathcal{X}(v)$. We call each $D[\mathcal{X}(v)]$ a **cluster** of $\mathcal{X}$.

The **radius** of a cluster $C := D[\mathcal{X}(v)]$ is defined as the minimal $r \in \mathbb{N}$ for which there is a vertex $w \in V(C)$ and for every $w \in V(C)$, the cluster $C$ contains a directed path of length at most $r$ from $w$ to $v$ and a directed path of length at most $r$ from $v$ to $w$. The **radius** $\mathrm{rad}(\mathcal{X})$ of a cover $\mathcal{X}$ is the maximum radius of any of its clusters.

The **degree** $d^{\mathcal{X}}(v)$ of $v$ in $\mathcal{X}$ is the number of clusters that contain $v$. The **maximum degree** $\Delta(\mathcal{X})$ of $\mathcal{X}$ is $\Delta(\mathcal{X}) = \max_{v \in V(G)} d^{\mathcal{X}}(v)$.

**Theorem 9.7.9** ([71]) *Let $\mathcal{C}$ be a class of digraphs of bounded expansion. There are functions $f, h : \mathbb{N} \to \mathbb{N}$ such that for all $r \in \mathbb{N}$ and all graphs $D \in \mathcal{C}$, there exists a strong $r$-neighbourhood cover of radius at most $4r$ and maximum degree at most $f(r)$ and this cover can be computed in time $h(r) \cdot n^{\mathcal{O}(1)}$.*

### 9.7.5 Constant-Factor Approximation Algorithms for Strong Dominating Sets

In this section we give an algorithmic application of the bounded expansion classes in proving that strong dominating sets can be approximated up to a constant factor on any class $\mathcal{C}$ of directed bounded expansion.

**Definition 9.7.10 (Strong $r$-Dominating Sets)**

1. *Let $r \geq 1$ and let $D$ be a digraph. A vertex $v \in V(D)$ **strongly-$r$-dominates** a vertex $u \in V(D)$ if there is a closed walk of length at most $2r$ in $D$ containing $u$ and $v$.*
2. *A **strong-$r$-dominating set** is a set $X \subseteq V(D)$ such that every vertex in $D$ is strongly dominated by a vertex in $X$.*
3. *The **strong $r$-domination number** of $D$, denoted $sdom_r(D)$, is the minimum size of a strong $r$-dominating set of $D$.*

Note that if $D$ is a digraph obtained from an undirected graph $G$ by replacing every edge $e$ in $G$ by two arcs with the same endpoints but opposite orientation, then any strong-$r$-dominating set in $D$ is an $r$-dominating set in $G$ and vice versa. This explains the choice of the length $2r$ in Part (1) of the previous definition. It follows that deciding the strong-$r$-domination number of a digraph $D$ is $\mathcal{NP}$-complete.

**Theorem 9.7.11** *Let $\mathcal{C}$ be a class of digraphs of directed bounded expansion. Let $r \geq 1$. There is a polynomial time constant factor approximation algorithm for strong $r$-dominating sets. More precisely, for every value of $r$, there is an algorithm running in time $g(r) \cdot n^{\mathcal{O}(1)}$ for some function $g$ which, on input $D \in \mathcal{C}$ computes a strong-$r$-dominating set $X \subseteq V(G)$ of order at most* $\mathrm{wcol}_{4r}(D)^2 \cdot \mathrm{sdom}_r(D)$.

The proof is based on computing a linear order witnessing that the $4r$-weak colouring number of the input digraph $D$ is bounded. Following this order, a suitable greedy strategy can be shown to produce a strong $r$-dominating set of order $\mathrm{wcol}_{4r}(D)^2 \cdot k$ and an **obstruction** witnessing that there is no strong $r$-dominating set of order $k$. Hence, the approximation factor is $\mathrm{wcol}_{4r}(D)^2$, which is a constant on bounded expansion classes. Here, an **$r$-obstruction set** is a set $X \subseteq V(D)$ such that for any distinct $x, y \in X$, there are *no* two closed directed walks $W_1, W_2 \subseteq V(D)$, each of length at most $2r$, such that $W_1 \cap W_2 \neq \emptyset$ and $x \in W_1$ and $y \in W_2$.

As no two distinct vertices of an obstruction set lie on a closed walk of length at most $2r$, no two vertices from the set can be strongly $r$-dominated by a single vertex. Hence, if $D$ contains an obstruction set of order $k$ then $D$ does not contain a strong $r$-dominating set of order $< |X|$.

A similar strategy was used by Dvořák in [30] to design a constant factor approximation algorithm for dominating sets on classes of undirected graphs of bounded expansion.

## 9.8 Nowhere Crownful Classes of Digraphs

We close our exposition of density and minor based width measures by giving another algorithmic application for dominating sets, this time on nowhere crownful classes. Towards this aim, we introduce the notion of directed uniformly quasi-wide classes and show that this concept yields an equivalent characterization of nowhere crownful classes of digraphs.

**Definition 9.8.1** *Let $D$ be a digraph and $d \in \mathbb{N} \cup \{0\}$. A set $U \subseteq V(D)$ is $d$-**scattered** if there is no $v \in V(D)$ and $u_1 \neq u_2 \in U$ such that $v$ has distance at most $d$ to both $u$ and $u'$.*

Note that any subset of $V(D)$ is 0-scattered since $v$ is the only vertex of distance zero from itself.

**Definition 9.8.2** *A class $\mathcal{C}$ of digraphs is **uniformly quasi-wide** if there are functions $s : \mathbb{N} \to \mathbb{N}$ and $N : \mathbb{N} \times \mathbb{N} \to \mathbb{N}$ such that for every $D \in \mathcal{C}$ and all $d, m \in \mathbb{N}$ and $W \subseteq V(D)$ with $|W| > N(d, m)$ there is a set $S \subseteq V(D)$ with $|S| \leq s(d)$ and $U \subseteq W$ with $|U| = m$ such that $U$ is $d$-scattered in $G - S$. $s, N$ are called the margin of $\mathcal{C}$.*

*If $s$ and $N$ are computable then we call $\mathcal{C}$* **effectively uniformly quasi-wide**.

The next theorem was shown in [72].

**Theorem 9.8.3** *A class $\mathcal{C}$ of digraphs is nowhere crownful if, and only if, it is directed uniformly quasi-wide.*

We demonstrate one algorithmic application of uniformly quasi-wideness by sketching the following theorem. A **directed dominating set** in a digraph $D$ is a set $X \subseteq V(D)$ such that $N_D^+(X) \cup X = V(D)$.

**Theorem 9.8.4** ([72])*Let $\mathcal{C}$ be a class of digraphs which is nowhere crownful. Then the directed dominating set problem is fixed-parameter tractable on $\mathcal{C}$.*

Let $\mathcal{C}$ be nowhere crownful. Given a digraph $D \in \mathcal{C}$ and a number $k$, we compute a directed dominating set $X$ of order $k$, if it exists, as follows. We let $W = V(D)$ be the set of vertices still to be dominated. As $\mathcal{C}$ is uniformly quasi-wide, if $W$ is large enough, we can compute a constant-size set $S$ of vertices and a 1-scattered set $A \subseteq W$ of order $k+1$ in $D - S$. As no vertex not in $S$ can dominate two vertices in $A$, it follows that any set $X$ of vertices dominating every vertex in $W$ needs to contain a vertex in $S$. As $S$ has constant size we can try each choice of a vertex $v \in S$ for the set $X$. For any such choice we recurse with the parameter $k-1$ and the set $W' := W \setminus N^+(v)$ of vertices we still need to dominate. This yields a natural recursion where in each recursion step the parameter is decreased. If at some point the set $W$ is too small to contain a large 1-scattered set, then we can use brute force to compute a set of order $k$ dominating $W$.

Similarly, one can show that on nowhere crownful classes of digraphs, the directed independent dominating set problem, the dominating out-branching problem and the independent set problem as well as their distance-$d$-versions are fixed-parameter tractable.

## 9.9 Rank-Width Inspired Width Measures

In this section, we introduce directed versions of clique-width and rank-width. The motivation of **clique-width** comes from the observation that many algorithmic problems are tractable on classes of graphs that can be recursively decomposable along vertex partitions $(A, B)$ where the number of neighbourhood types between $A$ and $B$ is small. Different from tree-width based width measures, acyclic digraphs have arbitrary large directed clique-width, and clique-width separates the class of acyclic digraphs into easy and hard instances for some algorithmic problems.

When clique-width was first introduced, no FPT approximation algorithm for generating a clique-width expression was known. Oum and Seymour [82] first devised an FPT approximation algorithm for undirected clique-width, using an equivalent width parameter called **rank-width**. While clique-width expressions describe how to generate a graph using certain graph operations, rank-width decompositions generalize decomposition scheme called branch-decompositions [88]. Courcelle and Engelfriet [24] argued that directed clique-width can be approximated using undirected rank-width.

**Bi-rank-width** and **$\mathbb{F}_4$-rank-width** are two natural generalizations of rank-width for directed graphs, introduced by Kanté [56] and Kanté and Rao [59]. They can also be used to approximate directed clique-width. The other motivation of these parameters is on related graph containment relations **vertex-minor** and **pivot-minor**. Because clique-width and rank-width may increase by removing edges or contracting edges, these parameters are not well fit to minor structure theory. Instead, vertex-minor and pivot-minor relations have been studied together with rank-width [80, 81], and provide some structural results, sometimes generalising results on tree-width. Kanté and Rao [59] explained how to generalize these concepts to directed graphs, and generalised known results to directed graphs.

We present FPT approximation algorithms in Subsection 9.9.3. In Subsection 9.9.4, we present algorithmic applications of directed clique-width and bi-rank-width. We discuss structural results on these graph containment relations in Subsection 9.9.5.

### 9.9.1 Directed Clique-Width

Courcelle, Engelfriet and Rozenberg [25] introduced **clique-width** for both undirected graphs and directed graphs. For a digraph $D = (V, A)$ and a function $\mathrm{lab} : V \to \{1, 2, \dots, k\}$, the triple $(V, A, \mathrm{lab})$ is called a **$k$-labeled digraph**. The function lab is called a **labeling** of $D$, and for each $v \in V$, $\mathrm{lab}(v)$ is called its label.

**Definition 9.9.1 (Directed clique-width)** *For a positive integer $k$, the class $\mathrm{dcw}_k$ of $k$-labeled digraphs is recursively defined as follows.*

1. *The digraph on a single vertex $v$ with label $i$ in $\{1, 2, \dots, k\}$ is in $\mathrm{dcw}_k$. We denote by $\bullet_{i,v}$ the operation creating such a vertex.*
2. *Let $D_1 = (V_1, A_1, \mathrm{lab}_1) \in \mathrm{dcw}_k$ and $D_2 = (V_2, A_2, \mathrm{lab}_2) \in \mathrm{dcw}_k$ be two $k$-labeled digraphs on disjoint vertex sets. Let $D_1 \oplus D_2 := (V, A, \mathrm{lab})$ where $V := V_1 \cup V_2$, $A := A_1 \cup A_2$ and*
$$\mathrm{lab}(v) := \begin{cases} \mathrm{lab}_1(v) & \text{if } v \in V_1, \\ \mathrm{lab}_2(v) & \text{if } v \in V_2, \end{cases}$$
*for every $v \in V$. We have $D_1 \oplus D_2 \in \mathrm{dcw}_k$.*
3. *Let $D = (V, A, \mathrm{lab}) \in \mathrm{dcw}_k$ be a $k$-labeled digraph, and $i, j \in \{1, 2, \dots, k\}$ be two distinct integers. Let $\rho_{i \to j}(D) := (V, A, \mathrm{lab}')$ where*
$$\mathrm{lab}'(v) := \begin{cases} \mathrm{lab}(v) & \text{if } \mathrm{lab}(v) \neq i, \\ j & \text{if } \mathrm{lab}(v) = i, \end{cases}$$
*for every $v \in V$. We have $\rho_{i \to j}(D) \in \mathrm{dcw}_k$.*
4. *Let $D = (V, A, \mathrm{lab}) \in \mathrm{dcw}_k$ be a $k$-labeled digraph, and $i, j \in \{1, 2, \dots, k\}$ be two distinct integers. Let $\alpha_{i,j}(D)$ be the digraph obtained from $D$ by adding all arcs $(a, b)$ where $\mathrm{lab}(a) = i$ and $\mathrm{lab}(b) = j$. We have $\alpha_{i,j}(D) \in \mathrm{dcw}_k$.*

*The **directed clique-width** of a digraph* $D = (V, A)$*, denoted by* $\mathrm{dcw}(D)$*, is the minimum integer* $k$ *such that there is a* $k$*-labeling* lab *of* $D$ *where* $(V, A, \mathrm{lab}) \in \mathrm{dcw}_k$*. **Directed clique-width k-expressions** are expressions which recursively construct a graph with the four graph operations in 1-4.*

The difference between directed clique-width and undirected clique-width is on the function $\alpha_{i,j}$; for undirected clique-width, this function adds undirected edges between all pairs $(v, w)$ where $\mathrm{lab}(v) = i$ and $\mathrm{lab}(w) = j$. We can naturally represent a directed clique-width $k$-expression as a tree-structure; an example is depicted in Figure 9.7. We call this tree a **directed clique-width** $k$**-expression tree**.

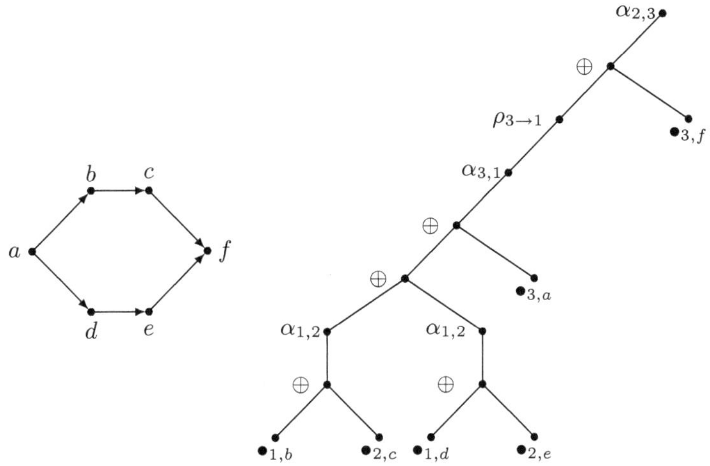


**Figure 9.7** An example of a directed clique-width 3-expression tree, which expresses $\alpha_{2,3}((\rho_{3\to1}(\alpha_{3,1}((\alpha_{1,2}(\bullet_{1,b} \oplus \bullet_{2,c}) \oplus \alpha_{1,2}(\bullet_{1,d} \oplus \bullet_{2,e})) \oplus \bullet_{3,a}))) \oplus \bullet_{3,f})$.

Wanke [100] introduced a similar width parameter NLC-width. In NLC-width expressions, we add edges between two labeled graphs at once after taking disjoint union, while we add edges one by one between two vertex subsets with single labels in clique-width expressions. Gurski, Wanke and Yilmaz [47] generalised this parameter to directed graphs.

**Definition 9.9.2 (Directed NLC-width)** *For a positive integer* $k$*, the class* $\mathrm{dNLC}_k$ *of* $k$*-labeled digraphs is recursively defined as follows.*

1. *The digraph on a single vertex* $v$ *with label* $i$ *in* $\{1, 2, \ldots, k\}$ *is in* $\mathrm{dNLC}_k$*. We denote by* $\bullet_{i,v}$ *the operation creating such a vertex.*
2. *Let* $D_1 = (V_1, A_1, \mathrm{lab}_1) \in \mathrm{dNLC}_k$ *and* $D_2 = (V_2, A_2, \mathrm{lab}_2) \in \mathrm{dNLC}_k$ *be two* $k$*-labeled digraphs on disjoint vertex sets, and* $\overrightarrow{S}, \overleftarrow{S} \subseteq \{1, 2, \ldots, k\} \times$

$\{1, 2, \ldots, k\}$ *be two relations. Let* $D_1 \times_{\overrightarrow{S}, \overleftarrow{S}} D_2 := (V, A, \mathrm{lab})$ *be the labeled graph where* $V := V_1 \cup V_2$, $A := A_1 \cup A_2 \cup \overrightarrow{A} \cup \overleftarrow{A}$ *with*

$$\overrightarrow{A} = \{(v, w) | v \in V_1, w \in V_2, (\mathrm{lab}_1(v), \mathrm{lab}_2(w)) \in \overrightarrow{S}\},$$
$$\overleftarrow{A} = \{(w, v) | v \in V_1, w \in V_2, (\mathrm{lab}_1(v), \mathrm{lab}_2(w)) \in \overleftarrow{S}\},$$

*and*

$$\mathrm{lab}(v) := \begin{cases} \mathrm{lab}_1(v) & \text{if } v \in V_1, \\ \mathrm{lab}_2(v) & \text{if } v \in V_2, \end{cases}$$

*for every* $v \in V$. *We have* $D_1 \times_{\overrightarrow{S}, \overleftarrow{S}} D_2 \in \mathrm{dNLC}_k$.

3. *Let* $D = (V, A, \mathrm{lab}) \in \mathrm{dNLC}_k$ *and* $R : \{1, 2, \ldots, k\} \to \{1, 2, \ldots, k\}$ *be a function. Let* $\circ_R(D) = (V, A, \mathrm{lab}')$ *be the labeled graph where* $\mathrm{lab}'(v) = R(\mathrm{lab}(v))$ *for every* $v \in V$. *We have* $\circ_R(D) \in \mathrm{dNLC}_k$.

*The* ***directed NLC-width*** *of a digraph* $D = (V, A)$, *denoted by* $\mathrm{dnlcw}(D)$, *is the minimum integer* $k$ *such that there is a* $k$*-labeling* lab *of* $D$ *where* $(V, A, \mathrm{lab}) \in \mathrm{dNLC}_k$. ***Directed NLC-width k-expressions*** *are expressions which recursively construct a graph with the three graph operations in 1-3.*

Gurski, Wanke and Yilmaz [47] derived a relationship between directed clique-width and directed NLC-width.

**Theorem 9.9.3** ([47]) *For every digraph* $D$, *the parameters* $\mathrm{dcw}(D)$ *and* $\mathrm{dnlcw}(D)$ *are related as follows:* $\mathrm{dnlcw}(D) \leq \mathrm{dcw}(D) \leq 2\mathrm{dnlcw}(D)$.

One example of digraph classes having bounded directed clique-width is the class of **directed cographs**. This class is a directed variant of the class of undirected cographs. The term cograph stands for complement reducible graph [22], representing the property that the complement of a cograph is again a cograph. Directed cographs are graphs that can be recursively defined as follows:

1. Every single vertex is a directed cograph.
2. If $D_1, \ldots, D_k$ are directed cographs, then the disjoint union of $D_1, \ldots, D_k$ is a directed cograph.
3. If $D_1 = (V_1, A_1), \ldots, D_k = (V_k, A_k)$ are directed cographs, then the digraph obtained from the disjoint union of $D_1, \ldots, D_k$ by adding all arcs $(v, w)$ where $v \in V_i$, $w \in V_j$, and $1 \leq i < j \leq k$, is a directed cograph.
4. If $D_1, \ldots, D_k$ are directed cographs, then the digraph obtained from the disjoint union of $D_1, \ldots, D_k$ by adding all arcs $(v, w)$ where $v \in V_i$, $w \in V_j$, and $i, j \in \{1, \ldots, k\}$, is a directed cograph.

We observe that the complement of a directed cograph is again a directed cograph.

**Theorem 9.9.4** ([47]) *A digraph is a directed cograph if and only if it has directed NLC-width at most* 1.

Theorem 9.9.4 implies that every directed cograph has directed clique-width at most 2. However, as far as we know, no complete characterization of digraphs of directed clique-width at most 2 is known. We refer to Section 11.6 for more information about directed cographs.

Directed clique-width is incomparable with directed tree-width. In particular, acyclic digraphs have unbounded directed clique-width. A discussion about it is presented in the next subsection. The complete biorientations of undirected complete graphs are directed cographs, but have unbounded directed tree-width.

**Lemma 9.9.5**

1. *There are classes of digraphs of bounded directed tree-width and unbounded directed clique-width.*
2. *There are classes of digraphs of bounded directed clique-width and unbounded directed tree-width.*

For fixed $k \geq 2$, it is open whether one can recognize graphs of directed clique-width at most $k$ in polynomial time. This is also an open problem for undirected clique-width with $k \geq 4$, and when $k = 3$, it was solved by Corneil, Habib, Lanlignel, Reed and Rotics [21].

**Problem 9.9.6** *For an integer $k \geq 2$, can we recognize digraphs of directed clique-width at most $k$ in polynomial time?*

### 9.9.2 Bi-Rank-Width and $\mathbb{F}_4$-Rank-Width

Rank-width of undirected graphs is a parameter equivalent to clique-width, in a sense that one is bounded if and only if the other is bounded. The rank of a matrix has a role in counting the number of neighborhood types between two vertex sets. To see this, we consider two disjoint vertex sets $A$ and $B$ in an undirected graph $G = (V, E)$, and an $A \times B$-matrix $M$ where for $a \in A$ and $b \in B$, $M[a, b] = 1$ if $a$ is adjacent to $b$, and $M[a, b] = 0$ otherwise. If the rank of $M$ over the binary field is $k$, then there are at most $2^k$ sets in $\{N_G(v) \cap B | v \in A\}$. Rank-width measures the decomposability along vertex partitions with small rank values of such matrices.

Kanté and Rao [59] introduced two directed versions of rank-width, called **bi-rank-width** and **$\mathbb{F}_4$-rank-width**. Kanté and Rao further generalized these notions to $\mathbb{F}$-edge-colored graphs; that is, graphs whose edges are labeled by elements of a fixed finite field $\mathbb{F}$. Since these generalizations are out of scope of this book, we concentrate on specializations for digraphs. A difference of two notions is that when $(A, B)$ is a vertex partition, bi-rank-width is based on a function summing up ranks of two binary matrices, one for arcs from $A$ to $B$ and the other for arcs from $B$ to $A$, while $\mathbb{F}_4$-rank-width is based on a function measuring all arcs together, using the field $\mathbb{F}_4$.

For a field $\mathbb{F}$ and a matrix $M$, $\mathbb{F}$-rank$(M)$ is the rank of the matrix $M$ over the field $\mathbb{F}$. We denote by $\mathbb{F}_4$ the field on 4 elements $\{0, 1, a, a^2\}$ where $a^3 = 1$ and $a^2 + a + 1 = 0$. We denote by $\mathbb{F}_2$ the binary field.

Let $D = (V, A)$ be a digraph. The **out-neighborhood matrix** $M_D^+$ is the $V \times V$-matrix such that for $v, w \in V$, $M_D^+[v, w] = 1$ if and only if $(v, w) \in A$. The $\mathbb{F}_4$-**adjacency matrix** of $D$ is the $V \times V$-matrix $M_D^4$ where for $v, w \in V$,

$$M_D^4[v, w] := \begin{cases} a & \text{if } (v, w) \in A \text{ and } (w, v) \notin A, \\ a^2 & \text{if } (v, w) \notin A \text{ and } (w, v) \in A, \\ 1 & \text{if } (v, w) \in A \text{ and } (w, v) \in A, \\ 0 & \text{otherwise.} \end{cases}$$

We define functions $\text{bicutrk}_D, \text{cutrk}_D^4 : 2^V \to \mathbb{Z}$ such that for every $S \subseteq V$,

- $\text{bicutrk}_D(S) = \mathbb{F}_2\text{-rank}(M_D^+[S, V \setminus S]) + \mathbb{F}_2\text{-rank}(M_D^+[V \setminus S, S])$,
- $\text{cutrk}_D^4(S) = \mathbb{F}_4\text{-rank}(M_D^4[S, V \setminus S])$.

We define branch-decomposition and $f$-width for symmetric submodular functions $f$. A function $f : X \to Y$ is **symmetric** if for $S \subseteq X$, $f(S) = f(X \setminus S)$. A function $f : X \to Y$ is **submodular** if it satisfies that for $A, B \subseteq X$, $f(A \cap B) + f(A \cup B) \leq f(A) + f(B)$. A tree is **subcubic** if it has at least two vertices and every internal node has degree 3.

**Definition 9.9.7 (Branch-decomposition)** *Let $V$ be a finite set and let $f : 2^V \to \mathbb{Z}$ be a symmetric submodular function. A branch-decomposition of $V$ is a pair $(T, L)$, where $T$ is a subcubic tree and $L$ is a bijection from $V$ to the set of leaves of $T$. For an edge e in $T$, $T - e$ induces a partition $(X_e, Y_e)$ of the leaves of $T$. The **$f$-width** of e is defined as $f(L^{-1}(X_e))$, and the **$f$-width** of a branch-decomposition $(T, L)$ is the maximum $f$-width over all edges of $T$. The **$f$-width** of $V$ is the minimum $f$-width over all branch-decompositions of $V$. If $|V| \leq 1$, then $V$ admits no branch-decomposition and the $f$-width of $V$ is defined to be 0.*

**Definition 9.9.8 (Bi-rank-width and $\mathbb{F}_4$-rank-width)** *Let $D = (V, A)$ be a digraph. The **bi-rank-width** of $D$, denoted by $\text{birw}(D)$, is the $\text{bicutrk}_D$-width of $V$, and the **$\mathbb{F}_4$-rank-width** of $D$, denoted by $\text{rw}^4(D)$, is the $\text{cutrk}_D^4$-width of $V$.*

Note that the functions $\text{bicutrk}_D$ and $\text{cutrk}_D^4$ are submodular. This can be shown using a property of the rank function of a matrix in Proposition 9.9.9. There are several proofs of it; for instance see Truemper [98].

**Proposition 9.9.9** *Let $M$ be an $X \times Y$-matrix over a field $\mathbb{F}$. Then for all $X_1, X_2 \subseteq X$ and $Y_1, Y_2 \subseteq Y$, we have*

$$\begin{aligned} &\mathbb{F}\text{-}\mathit{rank}(M[X_1 \cup X_2, Y_1 \cap Y_2]) + \mathbb{F}\text{-}\mathit{rank}(M[X_1 \cap X_2, Y_1 \cup Y_2]) \\ \leq\ &\mathbb{F}\text{-}\mathit{rank}(M[X_1, Y_1]) + \mathbb{F}\text{-}\mathit{rank}(M[X_2, Y_2]). \end{aligned}$$

Kanté [56] proved that bi-rank-width and $\mathbb{F}_4$-rank-width are equivalent up to a constant factor.

**Lemma 9.9.10** ([56]) *For a digraph* $D = (V, A)$, $\mathrm{rw}^4(D) \leq \mathrm{birw}(D) \leq 4\mathrm{rw}^4(D)$.

Digraphs of bi-rank-width at most 2 are digraphs that are completely decomposable with respect to split decomposition introduced by Cunningham [28]. As a similar concept, Kanté and Rao [58] introduced displit decompositions and showed that digraphs of $\mathbb{F}_4$-rank-width at most 1 are digraphs that are completely decomposable with respect to displit decomposition. Both results provide polynomial-time algorithms for recognizing digraphs of bi-rank-width at most 2 or digraphs of $\mathbb{F}_4$-rank-width at most 1.

Acyclic digraphs and tournaments have unbounded bi-rank-width. The grid-like example in Figure 9.8 is acyclic and its underlying undirected graph has large rank-width; Jelínek [53] proved that the undirected $n \times n$-grid has rank-width exactly $n-1$. A branch-decomposition of a directed graph with small bicutrk-width is also a branch-decomposition of small undirected rank-width and it means that acyclic digraphs have unbounded bi-rank-width. To see that tournaments have unbounded bi-rank-width, we can modify the example in Figure 9.8 into a tournament, such that

- for every two non-adjacent vertices in a column, we add an arc from the higher one to the lower one,
- for every two non-adjacent vertices contained in distinct columns, we add an arc from the right one to the left one.

One can verify that in every its branch-decomposition, there is a vertex partition with high bicutrk value.

**Lemma 9.9.11** *The family of acyclic graphs and the family of tournaments have unbounded bi-rank-width. Thus, these families have unbounded directed clique-width and unbounded* $\mathbb{F}_4$*-rank-width.*

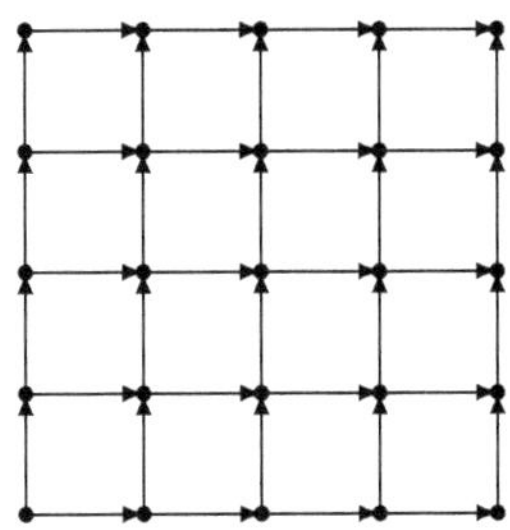

**Figure 9.8** Acyclic graphs that have large bi-rank-width.

### 9.9.3 Computing Rank-Decompositions

We provide FPT approximation algorithms for bi-rank-width and $\mathbb{F}_4$-rank-width. These can be used to obtain an approximated clique-width expression when a graph has small directed clique-width. Oum and Seymour [82] provided a general FPT approximation algorithm on symmetric submodular functions. By adapting the idea of the result of Oum and Seymour, we present FPT approximation algorithms for bi-rank-width and $\mathbb{F}_4$-rank-width.

Let $V$ be a finite set and let $f : 2^V \to \mathbb{Z}$ be a symmetric submodular function. A vertex subset $W \subseteq V$ is called an **$f$-well-linked set** if for every partition $(X, Y)$ of $W$ and every $Z$ with $X \subseteq Z \subseteq V \setminus Y$, we have $f(Z) \geq \min(|X|, |Y|)$. Oum and Seymour showed that $f$-well-linked sets are obstructions for graphs of bounded $f$-width.

**Proposition 9.9.12** *1. There exists an algorithm that, given a digraph $D = (V, A)$ and an integer $k$, runs in time $\mathcal{O}(8^k \cdot poly(|V|))$ either constructs a branch-decomposition of* $\mathrm{cutrk}_D^4$*-width at most $3k+1$, or concludes that* $\mathrm{rw}^4(D) > k$.

*2. There exists an algorithm that, given a digraph $D = (V, A)$ and an integer $k$, runs in time $\mathcal{O}(8^k \cdot poly(|V|))$ either constructs a branch-decomposition of* $\mathrm{bicutrk}_D$*-width at most $12k + 4$, or concludes that* $\mathrm{birw}(D) > k$.

*Proof.* We claim that if there is a $\mathrm{cutrk}_D^4$-well-linked set of size $3k+1$, then $D$ has $\mathbb{F}_4$-rank-width at least $k+1$. Suppose there is a $\mathrm{cutrk}_D^4$-well-linked set $W$ of size $3k+1$ with respect to $\mathrm{cutrk}_D^4$, and $D$ admits a branch-decomposition $(T, L)$ of $\mathrm{cutrk}_D^4$-width at most $k$. We proceed to find a vertex partition $(A_1, A_2)$ induced by some edge in $T$ where $\frac{|W|}{3} < |W \cap A_1| \leq \frac{2|W|}{3}$. We subdivide an edge of $T$, and regard the new vertex as a root node. For each node $t \in V(T)$, let $\mu(t)$ be the number of leaves of $T$ that are descendants of $t$ and mapped to a vertex of $W$ by $L$. We choose a node $t$ that is farthest from the root node such that $\mu(t) > \frac{|W|}{3}$. By the choice of $t$, for each child $t'$ of $t$, $\mu(t') \leq \frac{|W|}{3}$. Therefore, $\frac{|W|}{3} < \mu(t) \leq \frac{2|W|}{3}$. Let $e$ be the edge connecting $t$ and its parent. Clearly, the vertex partition $(A_1, A_2)$ of $D$ induced by $e$ satisfies that for each $i \in \{1, 2\}$, $\frac{|W|}{3} < |A_i \cap W| \leq \frac{2|W|}{3}$. Since $W$ is a $\mathrm{cutrk}_D^4$-well-linked set, we have $\mathrm{cutrk}_D^4(A_1) \geq \max(|W \cap A_1|, |W \cap A_2|) > \frac{|W|}{3} > k$. This contradicts our assumption.

We describe an algorithm that either finds a $\mathrm{cutrk}_D^4$-well-linked set of size $3k+1$ or constructs a branch-decomposition of $\mathrm{cutrk}_D^4$-width at most $3k+1$. In the first case, by the above claim, we conclude that $D$ has $\mathbb{F}_4$-rank-width at least $k+1$.

When we have a mapping $g$ from $V(D)$ to a tree, we say that $g^{-1}(w)$ for $w \in V(D)$ is assigned to the node $w$. Choose a vertex $v$ of $D$ and start with a tree with two nodes where one contains $v$ and the other contains all vertices of $V \setminus \{v\}$. Recursively choose a node $t$ containing more than one vertex, and let $A$ be the vertex set assigned to $t$. If $\mathrm{cutrk}_D^4(A) < 3k+1$, then we choose any vertex $a \in A$, and construct a new tree obtained by adding two nodes $t_1$ and $t_2$ and edges $t_1t$, $t_2t$ to $T$, and assigning $a$ to $t_1$ and all vertices in $A \setminus \{a\}$ to $t_2$. Clearly, we have $\text{cutrk}_D^4(A \setminus \{a\}) \leq 3k+1$.

Now we assume $\text{cutrk}_D^4(A) = 3k+1$. In this case, we find a vertex set $B \subseteq V \setminus A$ such that $|B| = 3k+1$ and $\mathbb{F}_4\text{-rank}(M_D^4[A,B]) = 3k+1$. We can find such a set by enumerating a column basis of the matrix $M_D^4[A, V \setminus A]$.

We check whether $B$ is a $\text{cutrk}_D^4$-well-linked set or not. For this, we take all vertex partitions $(B_1, B_2)$ of $B$, and check for every $Z$ with $B_1 \subseteq Z \subseteq V \setminus B_2$, $\text{cutrk}_D^4(Z) \geq \min(|B_1|, |B_2|)$. We can check this using the submodular function minimization algorithm [52]. If $Z$ is a $\text{cutrk}_D^4$-well-linked set of size $3k+1$, then we output that $D$ has $\mathbb{F}_4$-rank-width at least $k+1$. Otherwise, the procedure outputs a vertex partition $(B_1, B_2)$ of $B$ and a vertex subset $Z$ with $B_1 \subseteq Z \subseteq V \setminus B_2$ where $\text{cutrk}_D^4(Z) < \min(|B_1|, |B_2|)$.

We observe that $A \cap Z$ and $A \setminus Z$ are non-empty. If $A \cap Z = \emptyset$, then $\text{cutrk}_D^4(Z) = \text{cutrk}_D^4(Z \setminus A)$. On the other hand, we have

$$\begin{aligned}\text{cutrk}_D^4(Z \setminus A) &= \mathbb{F}_4\text{-rank}(M_D^4[Z \setminus A, A \cup (V \setminus Z)]) \\ &\geq \mathbb{F}_4\text{-rank}(M_D^4[B_1, A \setminus Z]) \\ &= |B_1| > \text{cutrk}_D^4(Z),\end{aligned}$$

which is a contradiction. Therefore, $A \cap Z \neq \emptyset$ and for a similar reason, $A \setminus Z \neq \emptyset$. We construct a tree obtained by adding two nodes $t_1$ and $t_2$ and adding edges $t_1t, t_2t$ to $T$, and assigning $A \cap Z$ to $t_1$ and $A \setminus Z$ to $t_2$. We observe

$$\begin{aligned}\text{cutrk}_D^4(A) + |B_2| &> \text{cutrk}_D^4(A) + \text{cutrk}_D^4(Z) \\ &\geq \text{cutrk}_D^4(A \cap Z) + \text{cutrk}_D^4(A \cup Z) \\ &= \text{cutrk}_D^4(A \cap Z) + \text{cutrk}_D^4(V \setminus (A \cup Z)) \\ &\geq \text{cutrk}_D^4(A \cap Z) + \mathbb{F}_4\text{-rank}(M_D^4[B_2, A]) \\ &= \text{cutrk}_D^4(A \cap Z) + |B_2|.\end{aligned}$$

This implies that $\text{cutrk}_D^4(A \cap Z) \leq \text{cutrk}_D^4(A) \leq 3k+1$. For a similar reason, we also have $\text{cutrk}_D^4(A \setminus Z) \leq 3k+1$.

Doing this procedure recursively, we obtain either a branch-decomposition of $\text{cutrk}_D^4$-width at most $3k+1$, or conclude that $D$ has $\mathbb{F}_4$-rank-width at least $k+1$. For bi-rank-width, we first run the above algorithm for $\mathbb{F}_4$-rank-width. If it returns that $D$ has $\mathbb{F}_4$-rank-width at least $k+1$, then we can return that it has bi-rank-width at least $k+1$, by Lemma 9.9.10. If the algorithm outputs a branch-decomposition of $\text{cutrk}_D^4$-width at most $3k+1$, then this is also a branch-decomposition of $\text{bicutrk}_D$-width at most $4(3k+1) = 12k+4$, by Lemma 9.9.10. □

Later, Oum [78] investigated an FPT approximation algorithm for undirected rank-width that runs in time $\mathcal{O}(8^k \cdot n^4)$, by replacing the submodular function minimization algorithm with an elementary algorithm that fits to rank-width. Oum [79] raised an open problem whether it can be further reduced to $\mathcal{O}(c^k \cdot n^3)$ for some constant $c$. We ask the same questions for bi-rank-width and $\mathbb{F}_4$-rank-width on digraphs. Note that when allowing $c^{k^2}$ in the parameter part, one can obtain $\mathcal{O}(n^3)$ running time due to Hliněný [48].

**Problem 9.9.13** *Is there a constant-factor FPT approximation algorithm for bi-rank-width or $\mathbb{F}_4$-rank-width that runs in time $\mathcal{O}(c^k n^3)$ for some constant $c$?*

Kanté and Rao [59] observed that as an application of the result of Hliněný and Oum [49], there are also exact FPT algorithms for both parameters. Briefly, Hliněný and Oum developed an exact FPT algorithm for partitioned matroids with respect to matroid branch-width, and then applied to rank-width. This application is also possible for bi-rank-width or $\mathbb{F}_4$-rank-width. Note that the function $g(k)$ in Theorem 9.9.14 is triple exponential.

**Theorem 9.9.14** ([59])

1. *There exists an algorithm that, given a digraph $D = (V, A)$ and an integer $k$, runs in time $g(k)|V|^3$ for some function $g$ and either constructs a branch-decomposition of $\mathrm{cutrk}_D^4$-width at most $k$, or concludes that $\mathrm{rw}^4(D) > k$.*
2. *There exists an algorithm that, given a digraph $D = (V, A)$ and an integer $k$, runs in time $g(k)|V|^3$ for some function $g$ and either constructs a branch-decomposition of $\mathrm{bicutrk}_D$-width at most $k$, or concludes that $\mathrm{birw}(D) > k$.*

We observe that a branch-decomposition of bounded $\mathrm{bicutrk}_D$-width can be efficiently translated to directed clique-width expression. A similar observation for undirected rank-width and clique-width was discussed by Oum and Seymour [82]. We remark that Courcelle and Engelfriet [24, Proposition 6.8] proved that one can approximate directed clique-width using undirected rank-width.

**Lemma 9.9.15** *For a digraph $D = (V, A)$, $\frac{\mathrm{birw}(D)}{2} \leq \mathrm{dcw}(D) \leq 2^{\mathrm{birw}(D)+1} - 1$. Moreover, given a digraph $D$ and its branch-decomposition of $\mathrm{bicutrk}_D$-width $k$, one can construct a directed clique-width $(2^{k+1} - 1)$-expression in time $\mathcal{O}(4^k|V|^3)$.*

*Proof.* We prove that $\mathrm{birw}(D) \leq 2\mathrm{dcw}(D)$. If $|V| = 1$, then $\mathrm{birw}(D) = 0$ and $\mathrm{dcw}(D) = 1$, and the statement holds. We may assume $|V| \geq 2$. Let $k = \mathrm{dcw}(D)$ and let $T$ be a directed clique-width $k$-expression tree of $D$. Note that this tree is a tree with maximum degree 3, and each leaf node is a node introducing a vertex of $D$. We choose an edge $e = uv$ of $T$ where $u$ is a child of $v$. The constructed graph $D_u$ at node $u$ is a $k$-labeled graph, and each vertex set in $D_u$ having the same label has the same out-neighborhood and in-neighborhood to $V \setminus V(D_u)$. This means that $\text{bicutrk}_D(V(D_u)) \leq 2k$, and it shows that $\text{birw}(D) \leq 2k$.

Now, we prove that given a branch-decomposition of $D$ of bi-rank-width $k$, one can construct a directed clique-width $(2^{k+1} - 1)$-expression in time $\mathcal{O}(4^k|V|^3)$. This also proves the inequality $\text{dcw}(D) \leq 2^{\text{birw}(D)+1} - 1$. Let $(T, L)$ be a given branch-decomposition of $D$ of $\text{bicutrk}_D$-width $k$. We choose an edge of $T$ and subdivide this edge with adding a new node $r$, and we consider $T$ as a tree with the root node $r$. For each node $t$ of $T$, let $D_t = (V_t, A_t)$ be the digraph induced by the set of vertices of $D$ that are mapped to a descendant of $t$. Let $\sim_t$ be the equivalent class on $V_t$ such that $v \sim_t w$ if and only if $N_D^+(v) \cap (V \setminus V_t) = N_D^+(w) \cap (V \setminus V_t)$ and $N_D^-(v) \cap (V \setminus V_t) = N_D^-(w) \cap (V \setminus V_t)$. We denote by $V_t/\sim_t$ be the set of equivalent classes. We note that since $D$ has $\text{bicutrk}_D$-width $k$, there are at most $2^k$ equivalent classes in $V_t/\sim_t$ for each node $t$.

We prove by induction on the number of descendants of $T$ that $D_t = (V_t, A_t)$ has a labeling $\text{lab}_t$ satisfying that

1. $(V_t, A_t, \text{lab}_t)$ can be constructed by a directed clique-width $(2^{k+1} - 1)$-expression,
2. $\text{lab}_t$ is a $2^k$-labeling of $D_t$,
3. for $v, w \in V_t$, if $v$ and $w$ are contained in distinct classes of $V_t/\sim_t$, then $\text{lab}_t(v) \neq \text{lab}_t(w)$, and
4. $\{v \in V_t : \text{lab}_t(v) = 1\}$ is exactly the set of vertices in $V_t$ having no in-neighborhood and no out-neighborhood in $V \setminus V_t$.

If $t$ is a leaf node, then it is clear. Assume that $t$ is not a leaf, and let $t_1$ and $t_2$ be the two children of $t$. By induction hypothesis, for each $i \in \{1, 2\}$, $D_{t_i}$ has a labeling $\text{lab}_i$ satisfying the conditions. For $j > 1$, we change each label $j$ of $V_{t_1}$ to $j + (2^k - 1)$, and then take the disjoint union of $D_{t_1}$ and $D_{t_2}$. Then we add arcs between $D_{t_1}$ and $D_{t_2}$ according to the adjacency relation between $D_{t_1}$ and $D_{t_2}$. Note that when we add an arc from $v_1 \in V_{t_1}$ to $v_2 \in V_{t_2}$, we add all arcs from the vertices in the class of $V_{t_1}/\sim_{t_1}$ containing $v_1$ and to the vertices in the class of $V_{t_2}/\sim_{t_2}$ containing $v_2$.

For each $i \in \{1, 2\}$, we relabel $V_{t_i}$ according to the class $V_t/\sim_t$. This is possible as $V_{t_i}/\sim_{t_i}$ is a refinement of $V_t/\sim_t$ on $V_{t_i}$. Then we relabel $V_{t_1}$ according to the labeling of $V_{t_2}$ so that the resulting labeling $\text{lab}_t$ on $V_t$ satisfies that

- for $v, w \in V_t$, if $v$ and $w$ are contained in distinct classes of $V_t/\sim_t$, then $\text{lab}_t(v) \neq \text{lab}_t(w)$, and
- $\{v \in V_t : \text{lab}_t(v) = 1\}$ is exactly the set of vertices in $V_t$ having no in-neighborhood and no out-neighborhood in $V \setminus V_t$.

This concludes the proof. □

### 9.9.4 Algorithmic Applications

We present algorithms based on directed clique-width or bi-rank-width. Courcelle, Makowsky and Rotics [26] showed that every problem expressible in $MSO_1$ logic can be solved in polynomial time on graphs of bounded directed clique-width.

**Theorem 9.9.16** ([26]) *Every problem expressible in $MSO_1$ logic is fixed parameter tractable with respect to the parameter directed clique-width.*

For many problems, we can design a dynamic-programming algorithm with running time much better than one guaranteed by Theorem 9.9.16. For instance, the problem of finding a minimum dominating set can be solved in time $2^{\mathcal{O}(k)}n^{\mathcal{O}(1)}$ when a directed clique-width $k$-expression is given.

**Theorem 9.9.17** *Given a digraph $D = (V, A)$ and its directed clique-width $k$-expression, one can compute a minimum directed dominating set of $D$ in time $2^{\mathcal{O}(k)}n^{\mathcal{O}(1)}$.*

We briefly present how to formulate table indices. Let $\phi$ be the given $k$-expression defining $D$, and let $T$ be the labeled rooted tree induced by $\phi$. For every node $t$ of $T$, let $D_t$ be the subgraph of $D$ defined at node $t$, and for each $i \in \{1, \ldots, k\}$, let $D_t[i]$ be the subgraph of $D_t$ induced on the set of vertices with label $i$.

The property of the constructed graph $D_t$ at some node $t$ is that two vertices in a same label class have same in-neighbors and out-neighbors in $V \setminus V(D_t)$. In the table of dynamic programming, we store the information that label classes that are completely dominated by some vertices of $D_t$, and label classes containing a vertex taken as a dominating set. We can recursively check whether a proper dominating set exists with given these information and a fixed size. This is similar to one developed for undirected case by Kobler and Rotics [67].

For many problems whose solutions can be locally checked, we can similarly design dynamic programming algorithms for problems on digraphs of bounded clique-width, which runs in FPT time. However, it becomes different when a solution requires some global property such as connectivity. For instance, Fomin, Golovach, Lokshtanov and Saurabh [36] proved that the problem of testing whether there is a hamiltonian cycle is $W[1]$-hard parameterized by clique-width, and later the same authors proved that this problem does not admit an algorithm with running time $n^{o(k)}$ under the ETH assumption [35]. On the other hand, it can be solved in time $n^{\mathcal{O}(k^2)}$, similar to the undirected case [32].

**Theorem 9.9.18** *Given a digraph $D = (V, A)$ and its directed clique-width $k$-expression, one can test whether $D$ contains a hamiltonian cycle in time $n^{\mathcal{O}(k^2)}$.*

We briefly explain the idea of Theorem 9.9.18. If $D$ contains a hamiltonian cycle, then its restriction on $D_t$ forms a partition of $D_t$ into vertex-disjoint paths unless $D_t \neq D$. Thus, if we have all possible partitions of $D_t$ into vertex-disjoint paths for each node $t$, then at the last node, we can test whether there is a hamiltonian cycle. One could observe that if there are two partitions into paths where for every pair $(i, j)$ of integers in $\{1, 2, \ldots, k\}$, the number of paths from $D_t[i]$ to $D_t[j]$ is equal, then they have the same role in generating a hamiltonian cycle. Thus, in the indices of tables, we are given some integer for every pair of integers, and we check whether there is a partition into paths meeting this condition. Using this table scheme, we can solve it in time $n^{\mathcal{O}(k^2)}$. Bergounoux, Kanté and Kwon [10] announced that the running time can be further improved to $n^{\mathcal{O}(k)}$.

There are more interesting problems that can be solved in FPT or XP time parameterized by clique-width. For instance, Parity Game can be solved in polynomial time on digraphs of constant directed clique-width [76]. We refer to [41] for more examples.

One issue of using directed clique-width is that if we approximate directed clique-width using Lemma 9.9.15 from obtained rank-decomposition, it is unavoidable single-exponential blow-up on the parameter. Thus, designing an algorithm directly using branch-decompositions of small bi-rank-width is an interesting problem. Ganian, Hliněný and Obdržálek [44] used parsing trees for rank-width to design XP algorithms for several problems such as Graph Coloring, Chromatic Polynomial, and Hamiltonian Path problems. More examples can be found in [42, 45].

### 9.9.5 Vertex-Minors and Pivot-Minors

We introduce **pivot-minor** and **vertex-minor** relations in digraphs. These containment relations are defined using graph operations **pivoting** and **local complementation**, respectively. In undirected graphs, local complementation at a vertex $v$ is an operation to replace the neighborhood of $v$ with its complement. Local complementation was introduced in the study of circle graphs [19], 2-regular Eulerian digraphs and isotropic systems [17, 18] by Bouchet. Pivoting also came up in the study of graphic representations of isotropic systems [17], and it is represented as three successive local complementations at $v$, $w$, $v$ on two adjacent vertices $v$ and $w$. Bouchet [16] observed that the cut-rank function does not change when applying local complementation [17], and based on this property, Oum [80, 81] investigated several structural results related to rank-width. Later, Kanté and Rao [59] extended the notion of local complementation and pivoting to digraphs.

We introduce here the pivoting operation in a digraph. Let $M$ be a $V \times V$-matrix on $\mathbb{F}_4$, and let $x, y$ be distinct elements in $V$ such that $M[x, y] \neq 0$. The matrix $M * (x, y)$ is a $V \times V$-matrix such that $(M * (x, y))[z, z] := 0$ for all $z \in V$, and for all $s, t \in V \setminus \{x, y\}$ with $s \neq t$,

- $(M * (x,y))[s,t] := M[s,t] - \frac{M[s,x]\cdot M[y,t]}{M[y,x]} - \frac{M[s,y]\cdot M[x,t]}{M[x,y]}$,
- $(M * (x,y))[x,t] := \frac{M[y,t]}{M[y,x]}$, $\quad (M * (x,y))[y,t] := -\frac{M[x,t]}{M[x,y]}$,
- $(M * (x,y))[s,x] := -\frac{M[s,y]}{M[x,y]}$, $\quad (M * (x,y))[s,y] := \frac{M[s,x]}{M[y,x]}$,
- $(M * (x,y))[x,y] := -\frac{1}{M[y,x]}$, $\quad (M * (x,y))[y,x] := -\frac{1}{M[x,y]}$,

where all equations are computed over $\mathbb{F}_4$. For an arc $(v,w)$ of a digraph $D$, a digraph obtained by **pivoting** $vw$ is defined as the digraph whose $\mathbb{F}_4$-adjacency matrix is $M_D^4 * (v,w)$, and it is denoted by $D \wedge vw$. A digraph $H$ is a **pivot-minor** of a digraph $D$ if $H$ can be obtained from $D$ by a sequence of pivotings and vertex deletions. We observe that pivot operations do not change the function $\mathrm{cutrk}_D^4$.

**Lemma 9.9.19** ([59]) *Let $D = (V,A)$ be a digraph. Every pivot operation does not change the function $\mathrm{cutrk}_D^4$, and thus, if a digraph $H$ is a pivot-minor of $D$, then $\mathrm{rw}^4(H) \leq \mathrm{rw}^4(D)$.*

*Proof.* Let $(x,y)$ be an arc of $D$, and let $X \subseteq V$ and $Y = V \setminus X$. It is enough to prove that $\mathrm{cutrk}_D^4(X) = \mathrm{cutrk}_{D\wedge xy}^4(X)$. Without loss of generality, we may assume $x \in X$. We divide cases depending on whether $y \in X$ or not. First assume that $y \in X$, and let $X' := X \setminus \{x,y\}$. In this case, we have

$$\begin{aligned}
&\mathbb{F}_4\text{-rank}(M_{D\wedge xy}^4[X,Y]) \\
&= \mathbb{F}_4\text{-rank}\begin{pmatrix} \frac{1}{M_D^4[y,x]}\cdot M_D^4[y,Y] \\ \frac{-1}{M_D^4[x,y]}\cdot M_D^4[x,Y] \\ M_D^4[X',Y] - \frac{M_D^4[X',x]\cdot M_D^4[y,Y]}{M_D^4[y,x]} - \frac{M_D^4[X',y]\cdot M_D^4[x,Y]}{M_D^4[x,y]} \end{pmatrix} \\
&= \mathbb{F}_4\text{-rank}\begin{pmatrix} \frac{1}{M_D^4[y,x]}\cdot M_D^4[y,Y] \\ \frac{-1}{M_D^4[x,y]}\cdot M_D^4[x,Y] \\ M_D^4[X',Y] \end{pmatrix} = \mathbb{F}_4\text{-rank}(M_D^4[X,Y]).
\end{aligned}$$

Now, we assume that $y \notin X$, and let $X' := X \setminus \{x\}$ and $Y' := Y \setminus \{y\}$. Then we have

$$
\begin{aligned}
&\mathbb{F}_4\text{-rank}(M^4_{D\wedge xy}[X,Y]) \\
&= \mathbb{F}_4\text{-rank}\begin{pmatrix} -\frac{1}{M^4_D[y,x]} & \frac{-1}{M^4_D[y,Y']}\cdot M^4_D[y,x] \\ \frac{M^4_D[X',x]}{M^4_D[y,x]} & M^4_D[X',Y'] - \frac{M^4_D[X',x]\cdot M^4_D[y,Y']}{M^4_D[y,x]} - \frac{M^4_D[X',y]\cdot M^4_D[x,Y']}{M^4_D[x,y]} \end{pmatrix} \\
&= \mathbb{F}_4\text{-rank}\begin{pmatrix} -\frac{1}{M^4_D[y,x]} & \frac{-1}{M^4_D[y,Y']}\cdot M^4_D[y,x] \\ 0 & M^4_D[X',Y'] - \frac{M^4_D[X',y]\cdot M^4_D[x,Y']}{M^4_D[x,y]} \end{pmatrix} \\
&= \mathbb{F}_4\text{-rank}\begin{pmatrix} -\frac{1}{M^4_D[y,x]} & 0 \\ 0 & M^4_D[X',Y'] - \frac{M^4_D[X',y]\cdot M^4_D[x,Y']}{M^4_D[x,y]} \end{pmatrix} \\
&= \mathbb{F}_4\text{-rank}\begin{pmatrix} M^4_D[x,y] & 0 \\ M^4_D[X',y] & M^4_D[X',Y'] - \frac{M^4_D[X',y]\cdot M^4_D[x,Y']}{M^4_D[x,y]} \end{pmatrix} \\
&= \mathbb{F}_4\text{-rank}\begin{pmatrix} M^4_D[x,y] & M^4_D[x,Y'] \\ M^4_D[X',y] & M^4_D[X',Y'] \end{pmatrix} = \mathbb{F}_4\text{-rank}(M^4_D[X,Y]).
\end{aligned}
$$

□

Kanté [57] showed that digraphs of bounded $\mathbb{F}_4$-rank-width are well-quasi-ordered under the pivot-minor operation. Note that the class of digraphs of bounded $\mathbb{F}_4$-rank-width is not well-quasi-ordered under the induced subdigraph operation. The set of all directed cycles is such an example.

**Theorem 9.9.20** ([57]) *Every pivot-minor closed class of digraphs of $\mathbb{F}_4$-rank-width at most $k$ is well-quasi-ordered under the pivot-minor relation.*

In undirected case, it is an open problem whether graphs are well-quasi-ordered under the undirected version of pivot-minor relation. If this holds, then it would imply the graph minor theorem which say that graphs are well-quasi-ordered under the minor relation. We ask the same question for directed graphs.

**Problem 9.9.21** *Is the set of digraphs well-quasi-ordered under the pivot-minor relation?*

It is open whether we can check whether a fixed graph $H$ is a pivot-minor of a graph $G$ for undirected graphs. Courcelle and Oum [27] proved that this problem is solvable in polynomial time when underlying graphs have bounded rank-width. Results from [57] imply that the same question for directed graphs is solvable in polynomial time when underlying digraphs have bounded $\mathbb{F}_4$-rank-width. We ask a question for general digraphs, as for undirected graphs.

**Problem 9.9.22** *For every fixed digraph $H$, is there a polynomial time algorithm testing whether a digraph $G$ contains $H$ as a pivot-minor?*

For a vertex $v$ in a digraph $D = (V, A)$, the $\mathbb{F}_4$-**local complementation** at $v$, denote by $D * v$, is the operation to take the digraph with the $\mathbb{F}_4$-adjacency matrix $M'$ where

- for $x, y \in V$ with $x \neq y$, $M'[x,y] = M_D^4[x,y] + M_D^4[x,z]M_D^4[z,y]$,
- for $x \in V$, $M'[x,x] = 0$.

A digraph $H$ is an $\mathbb{F}_4$-**vertex-minor** of $D$ if $H$ can be obtained from $D$ by a sequence of local complementations and vertex deletions. Note that as in the undirected case, it is satisfied that $D \wedge vw = D * v * w * v$ [59].

**Lemma 9.9.23** ([59]) *Let $D = (V, A)$ be a digraph. Every $\mathbb{F}_4$-local complementation does not change the function* $\text{cutrk}_D^4$, *and thus if $H$ is an $\mathbb{F}_4$-vertex-minor of a digraph $D$, then* $\text{rw}^4(H) \leq \text{rw}^4(D)$.

*Proof.* Let $D = (V, A)$ be a digraph and $x$ be a vertex of $D$. Let $X \subseteq V$. We may assume that $x \in X$ as $\text{cutrk}_D^4(X) = \text{cutrk}_D^4(V \setminus X)$. For each $y \in X$, the $\mathbb{F}_4$-local complementation at $x$ results in adding a multiple of the row indexed by $x$ to the row indexed by $y$. Therefore, we have $\text{cutrk}_{D*x}^4(X) = \text{cutrk}_D^4(X)$. □

Kanté and Rao [59] proved that the size of a minimal vertex-minor or pivot-minor obstruction for digraphs of $\mathbb{F}_4$-rank-width at most $k$ is bounded by a function of $k$.

**Theorem 9.9.24** ([59])

1. *For each positive integer $k$, there is a set $\mathcal{C}_k^v$ of directed graphs each having at most $(6^{k+1} - 1)/5$ vertices, such that a digraph has $\mathbb{F}_4$-rank-width at most $k$ if and only if it has no $\mathbb{F}_4$-vertex-minor isomorphic to digraphs in $\mathcal{C}_k^v$.*
2. *For each positive integer $k$, there is a set $\mathcal{C}_k^p$ of directed graphs each having at most $(6^{k+1} - 1)/5$ vertices, such that a digraph has $\mathbb{F}_4$-rank-width at most $k$ if and only if it has no pivot-minor isomorphic to digraphs in $\mathcal{C}_k^p$.*

A similar variant of local complementation can be defined in a way that it preserves the bi-rank-width of a digraph. For a vertex $v$ in a digraph $D = (V, A)$, the $\mathbb{F}_2$-**local complementation** at $v$, denote by $D *_2 v$, is the operation to take the digraph with the out-neighborhood matrix $M'$ where

- for $x, y \in V$ with $x \neq y$, $M'[x,y] = M_D^+[x,y] + M_D^+[x,z]M_D^+[z,y]$,
- for $x \in V$, $M'[x,x] = 0$.

A digraph $H$ is an $\mathbb{F}_2$-**vertex-minor** of $D$ if $H$ can be obtained from $D$ by a sequence of local complementations and vertex deletions.

**Lemma 9.9.25** ([59]) *Let $D = (V, A)$ be a digraph. Every $\mathbb{F}_2$-local complementation does not change the function* $\text{bicutrk}_D$, *and thus if $H$ is an $\mathbb{F}_2$-vertex-minor of a digraph $D$, then* $\text{birw}(H) \leq \text{birw}(D)$.

*Proof.* Let $D = (V, A)$ be a digraph and $x$ be a vertex of $D$. Let $X \subseteq V$. We may assume that $x \in X$. In the matrix $M_D^+[X, V \setminus X]$, for each $y \in X \setminus \{x\}$, the $\mathbb{F}_2$-local complementation at $x$ results in adding a multiple of the row indexed by $x$ to the row indexed by $y$. Therefore, we have $\text{bicutrk}_{D*x}(X) = \text{bicutrk}_D(X)$. □

Kanté and Rao [59] discussed that their generalization of pivot operation for edge-colored graphs does not fit to bi-rank-width. Also, they observed that digraphs of bounded bi-rank-width are not well-quasi-ordered under the $\mathbb{F}_2$-vertex-minor relation. The set of digraphs whose underlying graphs are even cycles such that each vertex has either in-degree 2 or out-degree 2 has bounded bi-rank-width and is not well-quasi-ordered by the $\mathbb{F}_2$-vertex-minor relation. Any $\mathbb{F}_2$-local complementation at a vertex of such cycle does not create any new arc, and thus, it is implied by the observation that such cycles are not well-quasi-ordered under the induced subdigraph relation. Furthermore, we can observe that all of such cycles are $\mathbb{F}_2$-vertex-minor obstructions for digraphs of bi-rank-width at most 1. Thus, we could not expect an upper bound on the size of $\mathbb{F}_2$-vertex-minor obstructions for digraphs of bounded bi-rank-width as in Theorem 9.9.24.

## References


1. I. Adler. Directed tree-width examples. *J. Combin. Theory Ser. B*, 97(5):718–725, 2007.
2. B. Alspach. Searching and sweeping graphs: a brief survey. *Matematiche (Catania)*, 59:5–37, 2006.
3. S. Amiri, L. Kaiser, S. Kreutzer, R. Rabinovich, and S. Siebertz. Graph searching games and width measures for directed graphs. In *STACS 2015: 32nd International Symposium on Theoretical Aspects of Computer Science*, volume 30 of *LIPIcs*, pages 34–47, 2015.
4. S. Amiri, K. Kawarabayashi, S. Kreutzer, and P. Wollan. The Erdős-Posa property for directed graphs. *CoRR*, arXiv:1603.02504, 2016.
5. S. Arnborg, D. Corneil, and A. Proskurowski. Complexity of finding embeddings in a $k$-tree. *SIAM J. Matrix Analysis Applications*, 8(2):277–284, 1987.
6. G. Bagan, A. Bonifati, and B. Groz. A trichotomy for regular simple path queries on graphs. In *PODS 2013: 32nd Symposium on Principles of Database Systems*, pages 261–272, New York, NY, USA, 2013. ACM.
7. J. Bang-Jensen and T.M. Larsen. DAG-width and circumference of digraphs. *J. Graph Theory*, 82:194–206, 2016.
8. J. Barát. Directed path-width and monotonicity in digraph searching. *Graphs Combin.*, 22(2):161–172, 2006.
9. J. Barat, P. Hajnal, Y. Lin, and A. Yang. On the structure of graphs with path-width at most two. *Studia Scientiarum Mathematicarum Hungarica*, 49(2):211–222, 2012.
10. B. Bergougnoux, M.M. Kanté, and O. Kwon. An optimal XP algorithm for hamiltonian cycle on graphs of bounded clique-width. In *WADS 2017: 15th International Symposium on Algorithms and Data Structures*, volume 10389 of *Lect. Notes Comput. Sci.*, pages 121–132, 2017.

11. D. Berwanger, A. Dawar, P. Hunter, and S. Kreutzer. DAG-width and parity games. In *STACS 2006: Symp. on Theoretical Aspects of Computer Science*, volume 3884 of *Lect. Notes Comput. Sci.*, pages 524–536. Springer, 2006.
12. D. Berwanger, A. Dawar, P. Hunter, S. Kreutzer, and J. Obržálek. DAG-width and parity games. *J. Combin. Theory Ser. B*, 102(4):900–923, 2012.
13. D. Berwanger and E. Grädel. Entanglement – a measure for the complexity of directed graphs with applications to logic and games. In *LPAR 2005: Logic for Programming, Artificial Intelligence, and Reasoning*, Lect. Notes Artif. Intel., pages 209–223, 2005.
14. D. Berwanger, E. Grädel, L. Kaiser, and R. Rabinovich. Entanglement and the complexity of directed graphs. *Theor. Comput. Sci.*, 463:2–25, 2012.
15. D. Bienstock and P.D. Seymour. Monotonicity in graph searching. *J. Algor.*, 12(2):239–245, 1991.
16. A. Bouchet. Connectivity of isotropic systems. In *Combinatorial Mathematics: 3rd International Conference (New York, 1985)*, volume 555 of *Ann. New York Acad. Sci.*, pages 81–93, New York, 1989. New York Acad. Sci.
17. A. Bouchet. Graphic presentations of isotropic systems. *J. Combin. Theory Ser. B*, 45(1):58–76, 1988.
18. A. Bouchet. Isotropic systems. *European J. Combin.*, 8(3):231–244, 1987.
19. A. Bouchet. Reducing prime graphs and recognizing circle graphs. *Combinatorica*, 7(3):243–254, 1987.
20. M. Chudnovsky, A. Scott, and P.D. Seymour. Disjoint paths in tournaments. *Adv. Math.*, 270:582–597, 2015.
21. D.G. Corneil, M. Habib, J. Lanlignel, B. Reed, and U. Rotics. Polynomial-time recognition of clique-width $\leq 3$ graphs. *Discrete Appl. Math.*, 160(6):834–865, 2012.
22. D.G. Corneil, H. Lerchs, and L.S. Burlingham. Complement reducible graphs. *Discrete Appl. Math.*, 3(3):163–174, 1981.
23. B. Courcelle. Graph rewriting: An algebraic and logic approach. In J. van Leeuwen, editor, *Handbook of Theoretical Computer Science*, volume 2, pages 194 – 242. Elsevier, 1990.
24. B. Courcelle and J. Engelfriet. *Graph structure and monadic second-order logic*. Cambridge Univ. Press, Cambridge, 2012. A language-theoretic approach, With a foreword by M. Nivat.
25. B. Courcelle, J. Engelfriet, and G. Rozenberg. Handle-rewriting hypergraph grammars. *J. Comput. System Sci.*, 46(2):218–270, 1993.
26. B. Courcelle, J.A. Makowsky, and U. Rotics. Linear time solvable optimization problems on graphs of bounded clique-width. *Theory Comput. Syst.*, 33(2):125–150, 2000.
27. B. Courcelle and S. Oum. Vertex-minors, monadic second-order logic, and a conjecture by Seese. *J. Combin. Theory Ser. B*, 97(1):91–126, 2007.
28. W.H. Cunningham. Decomposition of directed graphs. *SIAM J. Algebraic Discrete Methods*, 3(2):214–228, 1982.
29. M. de Oliveira Oliveira. An algorithmic metatheorem for directed treewidth. *Discrete Appl. Math.*, 204:49–76, 2016.
30. Z. Dvořák. Constant-factor approximation of the domination number in sparse graphs. *European J. Combin.*, 34(5):833–840, 2013.
31. K. Edwards, I. Muzi, and P. Wollan. Half-integral linkages in highly connected directed graphs. *CoRR*, arXiv:1611.01004, 2016.
32. W. Espelage, F. Gurski, and E. Wanke. How to solve NP-hard graph problems on clique-width bounded graphs in polynomial time. In *WG 2001: Graph-theoretic concepts in computer science*, volume 2204 of *Lect. Notes Comput. Sci.*, pages 117–128. Springer, Berlin, 2001.

33. H. Fernau and D. Meister. Digraphs of bounded elimination width. *Discrete Appl. Math.*, 168:78–87, 2014.
34. J. Flum and M. Grohe. *Parameterized Complexity Theory.* Springer, 2006. ISBN 3-54-029952-1.
35. F.V. Fomin, P.A. Golovach, D. Lokshtanov, and S. Saurabh. Almost optimal lower bounds for problems parameterized by clique-width. *SIAM J. Comput.*, 43(5):1541–1563, 2014.
36. F.V. Fomin, P.A. Golovach, D. Lokshtanov, and S. Saurabh. Intractability of clique-width parameterizations. *SIAM J. Comput.*, 39(5):1941–1956, 2010.
37. F.V. Fomin and M. Pilipczuk. Jungles, bundles and fixed-parameter tractability. In *SODA 2013: 24th Annual ACM-SIAM Symposium on Discrete Algorithms*, pages 396–413, 2013.
38. F.V. Fomin and D.M. Thilikos. An annotated bibliography on guaranteed graph searching. *Theor. Comput. Sci.*, 399(3):236–245, 2008.
39. S. Fortune, J.E. Hopcroft, and J. Wyllie. The directed subgraph homeomorphism problem. *Theor. Comput. Sci.*, 10:111–121, 1980.
40. A. Fradkin and P.D. Seymour. Edge-disjoint paths in digraphs with bounded independence number. *J. Combin. Theory Ser. B*, 110:19–46, 2015.
41. R. Ganian, P. Hliněný, J. Kneis, A. Langer, J. Obdržálek, and P. Rossmanith. Digraph width measures in parameterized algorithmics. *Discrete Appl. Math.*, 168:88–107, 2014.
42. R. Ganian, P. Hliněný, J. Kneis, A. Langer, J. Obdržálek, and P. Rossmanith. On digraph width measures in parameterized algorithmics. In *IWPEC 2009: Parameterized and Exact Computation*, volume 5917 of *Lect. Notes Comput. Sci.*, pages 185–197. Springer, Berlin, 2009.
43. R. Ganian, P. Hliněný, J. Kneis, D. Meister, J. Obdržálek, P. Rossmanith, and S. Sikdar. Are there any good digraph width measures? *J. Combin. Theory Ser. B*, 116:250–286, 2016.
44. R. Ganian, P. Hliněný, and J. Obdržálek. A unified approach to polynomial algorithms on graphs of bounded (bi-)rank-width. *European J. Combin.*, 34(3):680–701, 2013.
45. R. Ganian, P. Hliněný, and J. Obdržálek. Better algorithms for satisfiability problems for formulas of bounded rank-width. *Fund. Inform.*, 123(1):59–76, 2013.
46. M. Grohe, S. Kreutzer, and S. Siebertz. Deciding first-order properties of nowhere dense graphs. In *STOC 2014: 46th Annual ACM Symposium on Theory of Computing*, pages 89–98. ACM, 2014.
47. F. Gurski, E. Wanke, and E. Yilmaz. Directed NLC-width. *Theor. Comput. Sci.*, 616:1–17, 2016.
48. P. Hliněný. A parametrized algorithm for matroid branch-width. *SIAM J. Comput.*, 35(2):259–277, loose erratum, 2005.
49. P. Hliněný and S. Oum. Finding branch-decompositions and rank-decompositions. *SIAM J. Comput.*, 38(3):1012–1032, 2008.
50. P. Hunter. Losing the +1: Directed path-width games are monotone. Unpublished note, available from Feb. 2017 at www.cs.ox.ac.uk/paul.hunter/papers/losing.pdf, 2006.
51. P. Hunter and S. Kreutzer. Digraph measures: Kelly decompositions, games, and ordering. *Theor. Comput. Sci.*, 399(3):206–219, 2008.
52. S. Iwata, L. Fleischer, and S. Fujishige. A combinatorial strongly polynomial algorithm for minimizing submodular functions. *J. Assoc. Comput. Mach.*, 48:761–777, 2001.
53. V. Jelínek. The rank-width of the square grid. *Discrete Appl. Math.*, 158(7):841–850, 2010.

54. T. Johnson, N. Robertson, P.D. Seymour, and R. Thomas. Directed Tree-Width. *J. Combin. Theory Ser. B*, 82(1):138–154, 2001.
55. L. Kaiser, S. Kreutzer, R. Rabinovich, and S. Siebertz. Directed width measures and monotonicity of directed graph searching. *CoRR*, arXiv:1408.4745, 2014.
56. M.M. Kanté. The rank-width of directed graphs. arXiv:0709.1433, September 2007.
57. M.M. Kanté. Well-quasi-ordering of matrices under Schur complement and applications to directed graphs. *European J. Combin.*, 33(8):1820–1841, 2012.
58. M.M. Kanté and M. Rao. Directed rank-width and displit decomposition. In C. Paul and M. Habib, editors, *WG 2009: Graph-Theoretic Concepts in Computer Science*, volume 5911 of *Lect. Notes Comput. Sci.*, pages 214–225. Springer Berlin Heidelberg, 2010.
59. M.M. Kanté and M. Rao. The rank-width of edge-coloured graphs. *Theory Comput. Syst.*, 52(4):599–644, 2013.
60. K. Kawarabayashi, Y. Kobayashi, and S. Kreutzer. An excluded half-integral grid theorem for digraphs and the directed disjoint paths problem. In *STOC 2014: The ACM Symposium on Theory of Computing*, pages 70–78, New York, NY, USA, 2014. ACM.
61. K. Kawarabayashi and S. Kreutzer. The directed grid theorem. In *STOC 2015: 47th Annual ACM on Symposium on Theory of Computing*, pages 655–664. ACM, 2015.
62. H.A. Kierstead and D. Yang. Orderings on graphs and game coloring number. *Order*, 20:255–264, 2003.
63. S. Kintali. Directed width parameters and circumference of digraphs. *Theor. Comput. Sci.*, 659:83–87, 2017.
64. S. Kintali, N. Kothari, and A. Kumar. Approximation algorithms for digraph width parameters. *Theor. Comput. Sci.*, 562:365 – 376, 2015.
65. S. Kintali and Q. Zhang. Forbidden directed minors and kelly-width. *Theor. Comput. Sci.*, 662:40 – 47, 2017.
66. K. Kitsunai, Y. Kobayashi, K. Komuro, H. Tamaki, and T. Tano. Computing directed pathwidth in $O(1.89^n)$ time. *Algorithmica*, 75(1):138–157, 2016.
67. D. Kobler and U. Rotics. Edge dominating set and colorings on graphs with fixed clique-width. *Discrete Appl. Math.*, 126(2-3):197–221, 2003.
68. S. Kreutzer. Graph searching games. In K.R. Apt and E. Grädel, editors, *Lectures in Game Theory for Computer Scientists*, chapter 7, pages 213–263. Cambridge University Press, 2011.
69. S. Kreutzer and S. Ordyniak. Digraph decompositions and monotonicity in digraph searching. *Theor. Comput. Sci.*, 412(35):4688–4703, 2011.
70. S. Kreutzer and S. Ordyniak. Width-measures for directed graphs and algorithmic applications. In *Quantitative graph theory*, Discrete Math. Appl. (Boca Raton), pages 181–231. CRC Press, Boca Raton, FL, 2015.
71. S. Kreutzer, R. Rabinovich, S. Siebertz, and G. Weberstädt. Structural properties and constant factor-approximation of strong distance-$r$ dominating sets in sparse directed graphs. In *STACS 2017: 34th International Symposium on Theoretical Aspects of Computer Science*, 2017.
72. S. Kreutzer and S. Tazari. Directed nowhere dense classes of graphs. In *SODA 2012: 23rd Annual ACM-SIAM Symposium on Discrete Algorithms*, pages 1552–1562, 2012.
73. D. Meister, J.A. Telle, and M. Vatshelle. Recognizing digraphs of Kelly-width 2. *Discrete Appl. Math.*, 158(7):741–746, 2010.

74. J. Nešetřil and P.O. de Mendez. Grad and classes with bounded expansion I–III. *European J. Combin.*, 29, 2008. Series of 3 papers appearing in volumes (3) and (4).
75. J. Nešetřil and P.O. de Mendez. On nowhere dense graphs. *European J. Combin.*, 32(4):600–617, 2011.
76. J. Obdržálek. Clique-width and parity games. In *CSL 2007/EACSL 2007: 21st International Conference, and the 16th Annuall Conference on Computer Science Logic*, volume 4646 of *Lect. Notes Comput. Sci.*, pages 54–68, Berlin, Heidelberg, 2007. Springer-Verlag.
77. J. Obdržálek. DAG-width: connectivity measure for directed graphs. In *SODA 2006: Symp. on Discrete Algorithms*, pages 814–821, 2006.
78. S. Oum. Approximating rank-width and clique-width quickly. *ACM Trans. Algorithms*, 5(1):Art. 10, 20, 2009.
79. S. Oum. Rank-width: Algorithmic and structural results. *Discrete Appl. Math.*, 2016.
80. S. Oum. Rank-width and vertex-minors. *J. Combin. Theory Ser. B*, 95(1):79–100, 2005.
81. S. Oum. Rank-width and well-quasi-ordering. *SIAM J. Discrete Math.*, 22(2):666–682, 2008.
82. S. Oum and P.D. Seymour. Approximating clique-width and branch-width. *J. Combin. Theory Ser. B*, 96(4):514–528, 2006.
83. R. Rabinovich. *Graph Complexity Measures and Monotonicity*. PhD thesis, RWTH Aachen University, 2013.
84. B. Reed. Introducing directed tree-width. *Elec. Notes Discrete Math.*, 3:222 – 229, 1999.
85. B. Reed. Tree width and tangles: A new connectivity measure and some applications. In R.A. Bailey, editor, *Surveys in Combinatorics*, volume 241, pages 87–162. Cambridge University Press, 1997.
86. B. Reed, N. Robertson, P.D. Seymour, and R. Thomas. Packing directed circuits. *Combinatorica*, 16(4):535–554, 1996.
87. F. Reidl, F.S. Villaamil, and K. Stavropoulos. Characterising bounded expansion by neighbourhood complexity. arXiv:1603.09532, 2016.
88. N. Robertson and P.D. Seymour. Graph minors. X. Obstructions to tree-decomposition. *J. Combin. Theory Ser. B*, 52(2):153–190, 1991.
89. N. Robertson and P.D. Seymour. Graph minors V. Excluding a planar graph. *J. Combin. Theory Ser. B*, 41(1):92–114, 1986.
90. M.A. Safari. D-width: A more natural measure for directed tree width. In *MFCS 2005: Mathematical Foundations of Computer Science*, number 3618 in Lect. Notes Comput. Sci., pages 745 – 756, 2005.
91. N. Sauer. On the density of families of sets. *J. Combin. Theory Ser. A*, 13(1):145–147, 1972.
92. P.D. Seymour and R. Thomas. Graph searching and a min-max theorem for tree-width. *J. Combin. Theory Ser. B*, 58(1):22–33, 1993.
93. S. Shelah. A combinatorial problem; stability and order for models and theories in infinitary languages. *Pacific J. Math.*, 41(1):247–261, 1972.
94. B. Sheppard. DNA sequencing by hybridization. Master's thesis, Memorial University of Newfoundland, 2014.
95. A. Slivkins. Parameterized tractability of edge-disjoint paths on directed acyclic graphs. *SIAM J. Discrete Math.*, 24(1):146–157, 2010.
96. H. Tamaki. A directed path-decomposition approach to exactly identifying attractors of boolean networks. In *ISCIT 2010: International Symposium on Communications and Information Technologies*, pages 844–849, Oct 2010.

97. H. Tamaki. A polynomial time algorithm for bounded directed pathwidth. In *Graph-Theoretic Concepts in Computer Science*, volume 6986 of *Lecture Notes in Computer Science*, pages 331–342. Springer Berlin Heidelberg, 2011.
98. K. Truemper. *Matroid decomposition.* Academic Press, Inc., Boston, MA, 1992.
99. V.N. Vapnik and A.Y. Chervonenkis. On the uniform convergence of relative frequencies of events to their probabilities. In V. Vovk, H. Papadopoulos, and A. Gammerman, editors, *Measures of Complexity*, pages 11–30. Springer, 2015.
100. E. Wanke. $k$-NLC graphs and polynomial algorithms. *Discrete Appl. Math.*, 54(2):251 - 266, 1994.
101. D.H. Younger. Graphs with interlinked directed circuits,. In *Midwest symposium on circuit theory 2*, pages XVI 2.1–XVI 2.7, 1973.
102. X. Zhu. Colouring graphs with bounded generalized colouring number. *Discrete Math.*, 309(18):5562–5568, 2009.

# 10. Digraphs Products

**Richard H. Hammack**

For our purposes, a **digraph product** is a binary operation $D * D'$ on digraphs, for which $V(D * D')$ is the Cartesian product $V(D) \times V(D')$ of the vertex sets of the factors. There are many ways to define such products. But if we insist on the algebraic property of associativity, and demand that the projections to factors respect adjacency, then we are left with just four products, known as the **standard products**. One of these is the Cartesian product, introduced in Chapter 1. We review it now, and introduce the three other products.

## 10.1 The Four Standard Associative Products

The four standard digraph products are the **Cartesian product** $D \,\square\, D'$, the **direct product** $D \times D'$, the **strong product** $D \boxtimes D'$, and the **lexicographic product** $D \circ D'$. Each has vertex set $V(D) \times V(D')$. Their arcs are

$$\begin{aligned}
A(D \,\square\, D') &= \{(x, x')(y, y') \mid xy \in A(D), x' = y', \text{ **or** } x = y, x'y' \in A(D')\}, \\
A(D \times D') &= \{(x, x')(y, y') \mid xy \in A(D) \text{ **and** } x'y' \in A(D')\}, \\
A(D \boxtimes D') &= A(D \,\square\, D') \cup A(D \times D'), \\
A(D \circ D') &= \{(x, x')(y, y') \mid xy \in A(D), \text{ **or** } x = y \text{ and } x'y' \in A(D')\}.
\end{aligned}$$

In each case $D$ and $D'$ are called **factors** of the product. In drawing products, we often align the factors roughly horizontally and vertically (like $x$- and $y$-coordinate axes) below and to the left of the vertex set $V(D) \times V(D')$, so that $(x, x')$ projects vertically to $x \in V(D)$ and horizontally to $x' \in V(D')$. This is illustrated in Figure 10.1, showing examples of the Cartesian, direct and strong products. The lexicographic product is illustrated in Figure 10.2.

The definitions reveal immediately that the Cartesian, direct and strong products are commutative in the sense that the map $(x, x') \mapsto (x', x)$ yields isomorphisms $D \,\square\, D' \to D' \,\square\, D$, $D \times D' \to D' \times D$, and $D \boxtimes D' \to D' \boxtimes D$. However, Figure 10.2 shows that the lexicographic product is not commutative.

R. H. Hammack (✉)
Department of Mathematics and Applied Mathematics,
Virginia Commonwealth University, P.O. Box 842014, Richmond, VA, USA
e-mail: rhammack@vcu.edu



J. Bang-Jensen and G. Gutin (eds.), *Classes of Directed Graphs*, Springer Monographs in Mathematics, https://doi.org/10.1007/978-3-319-71840-8_10

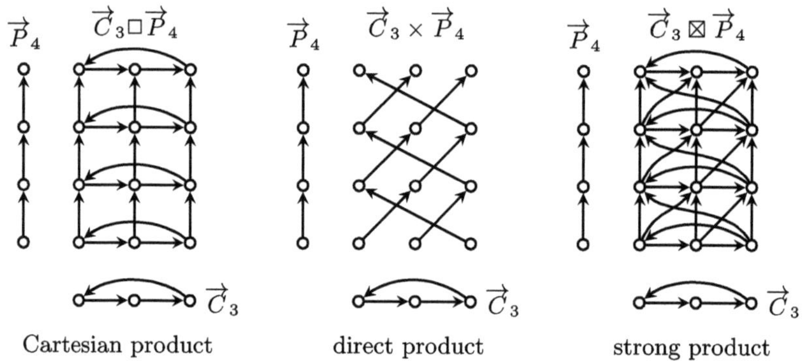


**Figure 10.1** The three standard associative, commutative products

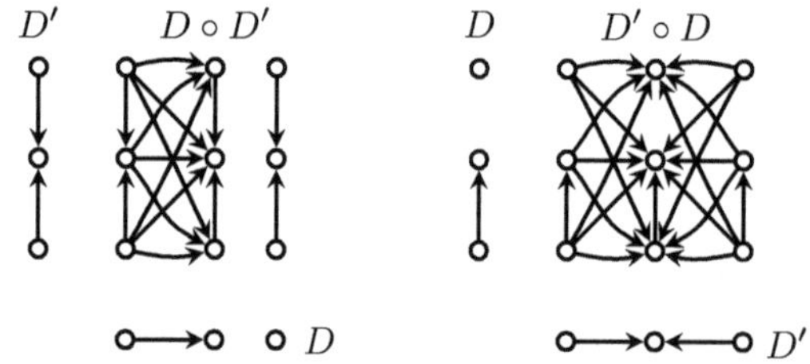


**Figure 10.2** The lexicographic product. Note $D \circ D' \not\cong D' \circ D$.

It is also easy to check that all four standard products are associative in the sense that the identification $(x,(y,z)) = ((x,y),z)$ yields equalities

$$\begin{aligned} D_1 \Box (D_2 \Box D_3) &= (D_1 \Box D_2) \Box D_3, \\ D_1 \times (D_2 \times D_3) &= (D_1 \times D_2) \times D_3, \\ D_1 \boxtimes (D_2 \boxtimes D_3) &= (D_1 \boxtimes D_2) \boxtimes D_3, \\ D_1 \circ (D_2 \circ D_3) &= (D_1 \circ D_2) \circ D_3. \end{aligned}$$

Thus we may unambiguously define products of more than two factors without regard to grouping. The product definitions extend as follows.

The vertex set of the Cartesian product $D_1 \Box \cdots \Box D_n$ is the Cartesian product of sets $V(D_1) \times \cdots \times V(D_n)$. The arcs of the product are all pairs $(x_1,\ldots,x_n)(y_1,\ldots,y_n)$, where $x_i y_i \in A(D_i)$ for some index $i \in [n]$, and $x_j = y_j$ for all $j \neq i$.

The direct product $D_1 \times \cdots \times D_n$ has vertices $V(D_1) \times \cdots \times V(D_n)$ and arcs $(x_1,\ldots,x_n)(y_1,\ldots,y_n)$, where $x_i y_i \in A(D_i)$ for all $i \in [n]$.

The strong product $D_1 \boxtimes \cdots \boxtimes D_n$ has vertices $V(D_1) \times \cdots \times V(D_n)$, and $(x_1,\ldots,x_n)(y_1,\ldots,y_n)$ is an arc provided $x_i = y_i$ or $x_i y_i \in A(D_i)$ for all $i \in [n]$, and $x_i y_i \in A(D_i)$ for at least one $i \in [n]$. Note the containment

$$A(D_1 \Box \cdots \Box D_n) \cup A(D_1 \times \cdots \times D_n) \subseteq A(D_1 \boxtimes \cdots \boxtimes D_n),$$

which is only guaranteed to be an equality when $n = 2$. (As in the definition on page 467.)

Extending the lexicographic product to more than two factors, we see that $D_1 \circ \cdots \circ D_n$ has vertices $V(D_1) \times \cdots \times V(D_n)$ and $(x_1, \ldots, x_n)(y_1, \ldots, y_n)$ is an arc of the product provided that there is an index $i \in [n]$ for which $x_i y_i \in A(D_i)$, while $x_j = y_j$ for any $1 \le j < i$.

We define the $n$th powers with respect to the four products as

$$D^{\square n} = D \,\square\, D \,\square \cdots \square\, D, \qquad D^{\times n} = D \times D \times \cdots \times D,$$
$$D^{\boxtimes n} = D \boxtimes D \boxtimes \cdots \boxtimes D, \qquad D^{\circ n} = D \circ D \circ \cdots \circ D,$$

where there are $n$ factors in each case.

A digraph **homomorphism** $\varphi : D \to D'$ is a map $\varphi : V(D) \to V(D')$ for which $xy \in A(D)$ implies $\varphi(x)\varphi(y) \in A(D')$. We call $\varphi$ a **weak homomorphism** if $xy \in A(D)$ implies $\varphi(x)\varphi(y) \in A(D')$ or $\varphi(x) = \varphi(y)$. A homomorphism is a weak homomorphism, but not conversely. For each $k \in [n]$, define the **projection** $\pi_k : V(D_1) \times \cdots \times V(D_n) \to V(D_k)$ as $\pi_k(x_1, \ldots, x_n) = x_k$. It is straightforward to verify that each projection $\pi_k : D_1 \times \cdots \times D_n \to D_k$ is a homomorphism, and $\pi_k : D_1 \,\square \cdots \square\, D_n \to D_k$ and $\pi_k : D_1 \boxtimes \cdots \boxtimes D_n \to D_k$ are weak homomorphisms. In general, only the first projection $\pi_1 : D_1 \circ \cdots \circ D_n \to D_1$ of a lexicographic product is a weak homomorphism. Although we will not undertake such a demonstration here, it can be shown that $\square$, $\times$, $\boxtimes$ and $\circ$ are the only associative products for which at least one projection is a weak homomorphism (or homomorphism) and each arc of each factor is a projection of an arc in the product. See [18] for details in the class of graphs. (The arguments apply equally well to digraphs.)

For products written as $D \,\square\, H$, we write the projections as $\pi_D$ and $\pi_H$.

We continue with some algebraic properties of the four products. Denote the disjoint union of digraphs $D$ and $H$ as $D + H$. The following distributive laws are immediate:

$$(D+H) \,\square\, K = D \,\square\, K + H \,\square\, K, \qquad (D+H) \times K = D \times K + H \times K,$$
$$(D+H) \boxtimes K = D \boxtimes K + H \boxtimes K, \qquad (D+H) \circ K = D \circ K + H \circ K.$$

The corresponding left-distributive laws also hold, *except* in the case of the lexicographic product, where generally $D \circ (H+K) \neq D \circ H + D \circ K$. Regarding this, the next proposition tells the whole story.

**Proposition 10.1.1** *We have $D \circ (H + K) \cong D \circ H + D \circ K$ if and only if $D$ has no arcs.*

**Proof:** If $D$ has no arcs, then the definition of the lexicographic product shows that both $D \circ (H+K)$ and $D \circ H + D \circ K$ are $|V(D)|$ copies of $H + K$.

Conversely, suppose $D \circ (H+K) \cong D \circ H + D \circ K$, so both digraphs have the same number of arcs. Note that in general

$$|A(D \circ H)| = |A(D)| \cdot |V(H)|^2 + |V(D)| \cdot |A(H)|, \tag{10.1}$$

where the first term counts arcs $(x,x')(y,y')$ with $xy \in A(D)$, and the second term counts such arcs with $x=y$. Using this to count the arcs of $D\circ(H+K)$, and again to count those of $D\circ H + D\circ K$, we see that $|A(D)|=0$. □

The trivial digraph $K_1$ is a unit for $\Box$, $\boxtimes$ and $\circ$, that is, $K_1 \times D = D$, $K_1 \boxtimes D = D$, and $K_1 \circ D = D = D \circ K_1$ (by identifying $(1,x)=x=(x,1)$ for all $x \in V(D)$). However, this does not work for the direct product because $K_1 \times D$ has no arcs, even if $D$ does. But if we admit loops and let $K_1^*$ be a loop on one vertex, then $K_1^*$ is the unique digraph for which $K_1^* \times D = D$. For this reason we often regard the direct product as a product on the class of digraphs with loops allowed, especially when dealing with issues of unique prime factorization, where the existence of a unit is crucial.

As mentioned above, the lexicographic product is the only one of the four standard products that is not commutative. However, if $D = H^{\circ n}$ and $D' = H^{\circ m}$ are lexicographic powers of the same digraph $H$, then we do of course get $D \circ D' = D' \circ D$. Another way that $D$ and $D'$ can commute is if they are both transitive tournaments, in which case we have

$$TT_n \circ TT_m = TT_{mn} = TT_m \circ TT_n. \qquad (10.2)$$

To verify this, order the vertices of $TT_n$ as $v_1, v_2, \ldots, v_n$ with $v_iv_j \in A(TT_n)$ provided $i<j$. Order those of $TT_m$ as $w_1, w_2, \ldots, w_m$ with $w_kw_\ell \in A(TT_m)$ provided $k<\ell$. Order the set $V(TT_n) \times V(TT_m)$ lexicographically, that is, $(v_i,w_k) < (v_j,w_\ell)$ if $i<j$, or $i=j$ and $k<\ell$. The definition of $\circ$ reveals that $TT_n \circ TT_m$ has an arc $(v_i,w_k)(v_j,w_\ell)$ if and only if $(i,k)<(j,\ell)$. Therefore $TT_n \circ TT_m = TT_{mn}$.

Certainly also $\overleftrightarrow{K}_n \circ \overleftrightarrow{K}_m = \overleftrightarrow{K}_{mn} = \overleftrightarrow{K}_m \circ \overleftrightarrow{K}_n$ where $\overleftrightarrow{K}_n$ is the complete biorientation of $K_n$. And if $D_n$ is its complement (i.e., the arcless digraph on $n$ vertices) then $D_n \circ D_m = D_{mn} = D_m \circ D_n$. In fact, these are the only situations in which the lexicographic product commutes, as discovered by Dörfler and Imrich [8].

**Theorem 10.1.2** *Two digraphs commute with respect to the lexicographic product if and only if they are both lexicographic powers of the same digraph, or both transitive tournaments, or both complete symmetric digraphs, or both arcless digraphs.*

We close this section with another property of the lexicographic product. Denote by $\overline{D}$ the complement of the digraph $D$, that is, the digraph on $V(D)$ with $xy \in A(\overline{D})$ if and only if $xy \notin A(D)$. The equation

$$\overline{D\circ H} = \overline{D} \circ \overline{H} \qquad (10.3)$$

is easily confirmed. No other standard product has this property.

The remainder of the chapter is organized as follows. Sections 10.2 and 10.3 treat distance and connectedness for the four products. Sections 10.4,

10.5 and 10.6 deal with kings and kernels, Hamiltonian issues, and invariants. The final five sections consider algebraic questions of cancellation and unique prime factorization. Section 10.7 covers some preliminary material on homomorphisms and quotients that is used in the following section on cancellation. Section 10.9 covers prime factorization for $\Box$ and $\circ$. The cases $\times$ and $\boxtimes$ are treated in Sections 10.10 and 10.11.

## 10.2 Distance

Recall that the **distance** $\text{dist}_D(x, y)$ between two vertices $x, y \in V(D)$ is the length of the shortest directed path from $x$ to $y$, or $\infty$ if no such path exists. This is not a metric in the usual sense, as generally $\text{dist}_D(x, y) \neq \text{dist}_D(y, x)$. Let $\text{dist}'_D(x, y)$ be the length of the shortest $(x, y)$-path in $D$ (not necessarily directed). This *is* a metric.

We begin by recording the distance formulas for each of the four products. These formulas are nearly identical to the corresponding formulas for graphs; here we adapt the proofs of Chapter 5 of Hammack, Imrich and Klavžar [18] to digraphs. Our proofs will use the fact that if $p : D \to H$ is a weak homomorphism, then $\text{dist}_D(x, y) \geq \text{dist}_H\big(p(x), p(y)\big)$, and similarly for $\text{dist}'$. This holds because if $P$ is an $(x, y)$-dipath (or path) in $D$, then the projection of any arc of $P$ is either an arc of $H$ or a single vertex of $H$. The projections that are arcs constitute a $(p(x), p(y))$-diwalk (or walk) in $H$ of length not greater than the length of $P$. (In fact, its length is the length of $P$ minus the number of its arcs that are mapped to single vertices.)

**Proposition 10.2.1** *In a Cartesian product $D = D_1 \Box \cdots \Box D_n$, the distance between vertices $(x_1, \ldots, x_n)$ and $(y_1, \ldots, y_n)$ is*

$$\text{dist}_D\big((x_1, \ldots, x_n), (y_1, \ldots, y_n)\big) = \sum_{1 \leq i \leq n} \text{dist}_{D_i}(x_i, y_i).$$

*For the strong product $D = D_1 \boxtimes \cdots \boxtimes D_n$, the distance is*

$$\text{dist}_D\big((x_1, \ldots, x_n), (y_1, \ldots, y_n)\big) = \max_{1 \leq i \leq n} \{\text{dist}_{D_i}(x_i, y_i)\}.$$

*The same formulas hold when* dist *is replaced with* $\text{dist}'$.

**Proof:** By associativity, it suffices to prove the statements for the case $n = 2$.

First consider the Cartesian product $D = D_1 \Box D_2$. To begin, suppose $\text{dist}_D((x_1, x_2), (y_1, y_2))$ is finite. Take a $((x_1, x_2), (y_1, y_2))$-dipath $P$ of length $\text{dist}_D((x_1, x_2), (y_1, y_2))$. By definition of the Cartesian product, any arc of $P$ is mapped to an arc in $D_1$ or $D_2$ by one of the two projections $\pi_1$ and $\pi_2$, and to a single vertex by the other. It follows that $\pi_1$ maps $P$ to an $(x_1, y_1)$-diwalk in $D_1$ of length (say) $d_1$, and $\pi_2$ maps $P$ to an $(x_2, y_2)$-diwalk in $D_2$ of length $d_2$, with $\text{dist}_D((x_1, x_2), (y_1, y_2)) = d_1 + d_2 \geq \text{dist}_{D_1}(x_1, y_1) + \text{dist}_{D_2}(x_2, y_2)$.

In particular this means the proposition holds if $\text{dist}_{D_1}(x_1, y_1) = \infty$ or $\text{dist}_{D_2}(x_2, y_2) = \infty$. If they are both finite, take a shortest $(x_1, y_1)$-dipath $P_1$ in $D_1$ and a shortest $(x_2, y_2)$-dipath $P_2$ in $D_2$. Then $D_1 \,\Box\, D_2$ has a dipath

$$(P_1 \times \{x_2\}) + (\{y_1\} \times P_2)$$

from $(x_1, x_2)$ to $(y_1, y_2)$, of length $\text{dist}_{D_1}(x_1, y_1) + \text{dist}_{D_2}(x_2, y_2)$. Therefore $\text{dist}_D((x_1, x_2), (y_1, y_2)) \le \text{dist}_{D_1}(x_1, y_1) + \text{dist}_{D_2}(x_2, y_2)$. Equality holds by the previous paragraph.

Now consider the strong product $D_1 \boxtimes D_2$. As each $\pi_i : D_1 \boxtimes D_2 \to D_i$ is a weak homomorphism, it follows that $\text{dist}_D((x_1, x_2), (y_1, y_2)) \ge \text{dist}_{D_i}(x_i, y_i)$ for $i = 1, 2$, so $\text{dist}_D((x_1, x_2), (y_1, y_2)) \ge \max_{1 \le i \le 2}\{\text{dist}_{D_i}(x_i, y_i)\}$.

Thus, if at least one $\text{dist}_{D_i}(x_i, y_i)$ is infinite, then $\text{dist}_D((x_1, x_2), (y_1, y_2)) = \infty$, and the proposition follows. Otherwise, take a shortest $(x_1, y_1)$-dipath $x_1 a_1 a_2 a_3 \ldots a_p y_1$ in $D_1$ and a shortest $(x_2, y_2)$-dipath $x_2 b_1 b_2 b_3 \ldots b_q y_2$ in $D_2$. Say $p \ge q$. We get the following $((x_1, x_2), (y_1, y_2))$-dipath in $D_1 \boxtimes D_2$:

$$(x_1, x_2)(a_1, b_1)(a_2, b_2)(a_3, b_3) \ldots (a_q, b_q)(a_{q+1}, y_2)(a_{q+2}, y_2) \ldots (a_p, y_2)(y_1, y_2).$$

Its length is $\text{dist}_{D_1}(x_1, y_1) = \max\{\text{dist}_{D_i}(x_i, y_i)\} \ge \text{dist}_D((x_1, x_2), (y_1, y_2))$. The reverse inequality was established in the previous paragraph.

The arguments for $\text{dist}'$ are identical, but replacing each occurrence of the word "diwalk" with "walk," and "dipath" with "path." □

The situation for the direct product is quite different. It requires the following useful result concerning directed walks in a direct product.

**Proposition 10.2.2** *A direct product $D = D_1 \times \cdots \times D_n$ has a diwalk of length $k$ from $(x_1, \ldots, x_n)$ to $(y_1, \ldots, y_n)$ if and only if each $D_i$ has a diwalk of length $k$ from $x_i$ to $y_i$.*

**Proof:** Suppose $D$ has a diwalk $W$ from $(x_1, x_2, \ldots, x_n)$ to $(y_1, y_2, \ldots, y_n)$, of length $k$. As each projection $\pi_i : G \to G_i$ is a homomorphism, $W$ projects to an $(x_i, y_i)$-diwalk of length $k$ in each $D_i$.

Conversely, if each factor $D_i$ has a diwalk $x_i x_i^1 x_i^2 x_i^3 \ldots x_i^{k-1} y_i$ of length $k$, then by the definition of the direct product, $D$ has a diwalk

$$(x_1, \ldots, x_n)(x_1^1, \ldots, x_n^1)(x_1^2, \ldots, x_n^2) \ldots (x_1^{k-1}, \ldots, x_n^{k-1})(y_1, \ldots, y_n)$$

of length $k$. □

**Proposition 10.2.3** *In a direct product $D = D_1 \times \cdots \times D_n$, the distance between two vertices $(x_1, \ldots, x_n)$ and $(y_1, \ldots, y_n)$ is*

$$\text{dist}_D\big((x_1, \ldots, x_n), (y_1, \ldots, y_n)\big) = \min \left\{ k \in \mathbb{Z} \;\middle|\; \begin{array}{c} \textit{each } D_i \textit{ has an} \\ (x_i, y_i) - \textit{diwalk} \\ \textit{of length } k \end{array} \right\},$$

*or $\infty$ if no such $k$ exists.*

**Proof:** Let $\text{dist}_D\big((x_1,\ldots,x_n),(y_1,\ldots,y_n)\big) = d$. Let $d'$ equal the smallest $k$ for which each $D_i$ has an $(x_i, y_i)$-diwalk of length $k$, or $\infty$ if no such $k$ exists. We must show $d = d'$.

If $d = \infty$, then $d \geq d'$. But also $d \geq d'$ when $d < \infty$, by Proposition 10.2.2.

On the other hand, if $d' = \infty$, then $d \leq d'$. And again $d \leq d'$ when $d' < \infty$, by Proposition 10.2.2. Thus $d = d'$. □

Distance in the lexicographic product requires a new definition. Given a vertex $x$ of a digraph $D$, let $\xi_D(x)$ be the length of a shortest non-trivial dicycle containing $x$, or $\infty$ if no such dicycle exists. Let $\xi'_D(x)$ be the length of the shortest non-trivial cycle containing $x$. We first state the distance formula for lexicographic products $D_1 \circ D_2$ having just two factors (a consequence of Theorem 4 of Szumny, Włoch and Włoch [54]).

**Proposition 10.2.4** *The distance formula for the lexicographic product is*

$$\text{dist}_{D_1\circ D_2}\big((x_1,x_2),(y_1,y_2)\big) = \begin{cases} \text{dist}_{D_1}(x_1,y_1) & \textit{if } x_1 \neq y_1 \\ \min\big\{\,\xi_{D_1}(x_1),\ \text{dist}_{D_2}(x_2,y_2)\big\} & \textit{if } x_1 = y_1. \end{cases}$$

*The formula also holds with* dist *and* $\xi$ *replaced with* dist$'$ *and* $\xi'$.

**Proof:** Suppose $x_1 \neq y_1$. Then, as the projection $\pi_1$ is a weak homomorphism, we have $\text{dist}_{D_1\circ D_2}\big((x_1,x_2),(y_1,y_2)\big) \geq \text{dist}_{D_1}(x_1,y_1)$. On the other hand, given a shortest $(x_1,y_1)$-dipath $x_1a_1a_2\ldots a_py_1$ in $D_1$, we construct a dipath $(x_1,y_1)(a_1,y_2)(a_2,y_2)(a_3,y_2)\ldots(y_1,y_2)$ in $D_1\circ D_2$ of length $\text{dist}_{D_1}(x_1,y_1)$, so $\text{dist}_{D_1\circ D_2}\big((x_1,x_2),(y_1,y_2)\big) = \text{dist}_{D_1}(x_1,y_1)$.

Now suppose $x_1 = y_1$. Take a shortest $((x_1,x_2),(y_1,y_2))$-dipath $P$ in $D_1 \circ D_2$. Because $\pi_1$ is a weak homomorphism, $\pi_1(P)$ is either a closed diwalk in $D_1$ beginning and ending at $x_1$ that is no longer than $P$, or it is the single vertex $x_1$. In the first case, $\text{dist}((x_1,x_2),(y_1,y_2)) \geq \xi_{D_1}(x_1)$. In the second, $P$ lies in the fiber $\{x_1\}\circ D_2 \cong D_2$, and its length is no less than $\text{dist}_{D_2}(x_2,y_2)$. Thus $\text{dist}_{D_1\circ D_2}\big((x_1,x_2),(y_1,y_2)\big) \geq \min\big\{\,\xi_{D_1}(x_1),\ \text{dist}_{D_2}(x_2,y_2)\big\}$.

Conversely, if $D_1$ has a closed dicycle $x_1a_1a_2\ldots a_px_1$, then $D_1 \circ D_2$ has a dipath $(x_1,y_1)(a_1,y_2)(a_2,y_2)(a_3,y_2)\ldots(x_1,y_2)$ of the same length. And if $D_2$ has an $(x_2,y_2)$-dipath $P$, then $\{x_1\}\circ P$ is a $((x_1,x_2),(y_1,y_2))$-dipath in $D_1\circ D_2$. Thus $\text{dist}_{D_1\circ D_2}\big((x_1,x_2),(y_1,y_2)\big) \leq \min\big\{\,\xi_{D_1}(x_1),\ \text{dist}_{D_2}(x_2,y_2)\big\}$.

The proof is the same for dist$'$. □

**Corollary 10.2.5** *Suppose* $(x_1,x_2,\ldots,x_n)$ *and* $(y_1,y_2,\ldots,y_n)$ *are* distinct *vertices of* $D = D_1\circ D_2\circ\cdots\circ D_n$, *and let* $k \in [n]$ *be the smallest index for which* $x_k \neq y_k$. *Then*

$$\text{dist}_D\big((x_1,x_2,\ldots,x_n),(y_1,y_2,\ldots,y_n)\big) = \min\big\{\xi_{D_1}(x_1),\ \xi_{D_2}(x_2),\ \ldots,\ \xi_{D_{k-1}}(x_{k-1}),\ \text{dist}_{D_k}(x_k,y_k)\big\}.$$

*(For* $k = 1$ *this is* $\text{dist}_D\big((x_1,x_2,\ldots,x_n),(y_1,y_2,\ldots,y_n)\big) = \text{dist}_{D_1}(x_1,y_1)$. *In any case, the distance does not depend on any factor* $D_i$ *with* $k < i \leq n$.*)* *The formula also holds with* dist *and* $\xi$ *replaced with* dist$'$ *and* $\xi'$.

**Proof:** If $n = 2$, this is just a restatement of Proposition 10.2.4. If $n > 2$, then applying Proposition 10.2.4 to $D_1 \circ (D_2 \circ \cdots \circ D_n)$ yields

$$\text{dist}_D\big((x_1, x_2, \ldots, x_n), (y_1, y_2, \ldots, y_n)\big) = \\ \min\left\{\xi_{D_1}(x_1), \text{dist}_{D_2 \circ \cdots \circ D_n}\big((x_2, \ldots x_n), (y_2 \ldots y_n)\big)\right\},$$

and we proceed inductively. □

## 10.3 Connectivity

We now apply the results of the previous section to connectivity of the four products. Our first result characterizes connectivity and strong connectivity of three of our four products. The proofs are straightforward, with appeals to the distance formulas of Section 10.2 as needed. The parenthetical words *(strongly)* and *(strong)* in the proposition can be deleted to obtain parallel results on connectedness. (Recall that a digraph is **connected** if any two of its vertices can be joined by a [not necessarily directed] path.)

**Theorem 10.3.1** *Suppose $D_1, \ldots, D_n$ are digraphs. Then:*

1. *The Cartesian product $D_1 \,\Box\, \cdots \,\Box\, D_n$ is (strongly) connected if and only if each factor $D_i$ is (strongly) connected. More generally, the (strong) components of a product $D_1 \,\Box\, \cdots \,\Box\, D_n$ are the subgraphs $X_1 \,\Box\, \cdots \,\Box\, X_n$ for which each $X_i$ is a (strong) component of $D_i$.*
2. *The strong product $D_1 \boxtimes \cdots \boxtimes D_n$ is (strongly) connected if and only if each factor $D_i$ is (strongly) connected. More generally, the (strong) components of a product $D_1 \boxtimes \cdots \boxtimes D_n$ are the subgraphs $X_1 \boxtimes \cdots \boxtimes X_n$ for which each $X_i$ is a (strong) component of $D_i$.*
3. *The lexicographic product $D_1 \circ \cdots \circ D_n$ of non-trivial digraphs is (strongly) connected if and only if the first factor $D_1$ is (strongly) connected. More generally, the (strong) components of a product $D_1 \circ \cdots \circ D_n$ are the subgraphs $X_1 \circ D_2 \circ \cdots \circ D_n$, where $X_1$ is a non-trivial strong component of $D_1$, as well as*
$$X_1 \circ X_2 \circ \cdots \circ X_k \circ D_{k+1} \circ \cdots \circ D_n,$$
*where $X_i$ is a trivial (strong) component of $D_i$ for $1 \le i < k$, and $X_k$ is a non-trivial strong component of $D_k$ (unless $k = n$, in which case $X_k$ is allowed to be trivial).*

Theorem 10.3.1 is a key to understanding the interconnections between the strong components of products. Recall that the **strong component digraph** of a digraph $D$ is the acyclic digraph $\text{SC}(D)$ whose vertices are the strong components of $D$, with an arc directed from $X$ to $Y$ precisely

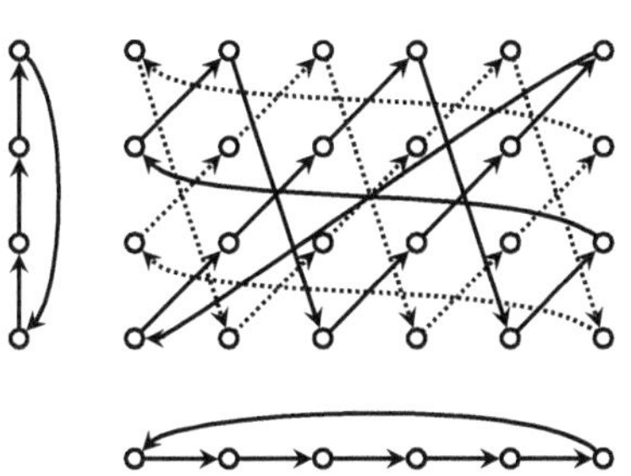


**Figure 10.3** The direct product of strongly connected graphs is not necessarily strongly connected.

when $D$ has an arc from $X$ to $Y$. Thus $\mathrm{SC}(D)$ carries information on the interconnections between the various strong components. The SC operator respects the Cartesian and strong products in the sense that $\mathrm{SC}(D \,\Box\, H) = \mathrm{SC}(D) \,\Box\, \mathrm{SC}(H)$ and $\mathrm{SC}(D \boxtimes H) = \mathrm{SC}(D) \boxtimes \mathrm{SC}(H)$. Indeed, the pairwise projection map $X \mapsto (\pi_D(X), \pi_H(X))$ sending strong components $X$ in the product to pairs of strong components in the factors is an isomorphism in both cases $\Box$ and $\boxtimes$ (as is easily checked).

Also, if every strong component of $D$ is non-trivial, then $\mathrm{SC}(D \circ H) = \mathrm{SC}(D)$. This is so because Theorem 10.3.1 says the strong components of $D \circ H$ have form $X \circ H$, where $X$ is a strong component of $G$. From the definition of $\circ$, the projection $X \circ H \mapsto X$ is an isomorphism $\mathrm{SC}(D \circ H) \to \mathrm{SC}(D)$. (But this breaks down if $D$ has a trivial strong component $X = \{x_0\}$ and $H$ has at least two strong components $Y$ and $Z$, because then the distinct strong components $X \circ Y$ and $X \circ Z$ are both mapped to $X$.)

There is no result analogous to Theorem 10.3.1 for the direct product. Indeed, Figure 10.3 shows a direct product of two strong digraphs that is not even connected: Here $\vec{C}_4 \times \vec{C}_6 = 2\vec{C}_{12}$, where the coefficient 2 means the product is 2 disjoint copies of $\vec{C}_{12}$. In fact, it is easy to verify the formula

$$\vec{C}_m \times \vec{C}_n = \gcd(m,n)\vec{C}_{\mathrm{lcm}(m,n)} \tag{10.4}$$

(which is an instance of Theorem 10.3.2 below).

Despite the fact that a direct product of strongly connected digraphs need not be strongly connected, the converse *is* true: if $D_1 \times \cdots \times D_n$ is strongly connected, then each $D_i$ must be strongly connected. This is a consequence of the fact that the projection maps are homomorphisms. Given two vertices $x_i, y_i$ of $D_i$, take $(x_1, \ldots, x_n), (y_1, \ldots, y_n) \in V(D_1 \times \cdots \times D_n)$. Any diwalk joining these two vertices projects to a diwalk joining $x_i$ to $y_i$.

Additional conditions on the factors that guarantee the product is strongly connected were first spelled out by McAndrew [37]. For a digraph $D$, let $d(D)$ be the greatest common divisor of the lengths of all closed diwalks in $D$.

**Theorem 10.3.2** *If $D_1, D_2, \ldots, D_n$ are strongly connected digraphs, then the number of strong components of the direct product $D_1 \times D_2 \times \cdots \times D_n$ is*

$$\frac{d(D_1) \cdot d(D_2) \cdots d(D_n)}{lcm\big(d(D_1), d(D_2), \ldots, d(D_n)\big)}.$$

*Consequently, $D_1 \times D_2 \times \cdots \times D_n$ is strongly connected if and only if each $D_i$ is strongly connected and the numbers $d(D_1), \ldots, d(D_n)$ are relatively prime.*

Notice how this theorem agrees with Equation (10.4) and Figure 10.3. The proof is constructive and gives a neat description of the strong components.

**Proof:** We need only prove the first statement. Assume each factor $D_i$ is strongly connected, and let $D = D_1 \times \cdots \times D_n$.

For each index $i \in [n]$, put $d_i = d(D_i)$, and fix a vertex $a_i \in V(D_i)$. Define functions $f_i : V(D_i) \to \{0, 1, 2, \ldots, d(D_i) - 1\}$ so that $f_i(v)$ is the length (mod $d_i$) of an $(a_i, v)$-diwalk $W$. To see that this does not depend on $W$, let $W'$ be any other $(a_i, v)$-diwalk. Let $Z$ be a $(v, a_i)$-diwalk. Then the concatenations $W + Z$ and $W' + Z$ are closed $(a_i, a_i)$-diwalks, and $d_i$ divides both of their lengths $|W + Z|$ and $|W' + Z|$. Thus $d_i$ divides the difference $|W| - |W'|$ of their lengths, so $|W|$ and $|W'|$ have the same length, modulo $d_i$. Hence $f$ is well defined.

Regard $f_i(v)$ as a coloring of vertex $v$, so $D_i$ is $d_i$-colored. Now, to each vertex $x = (x_1, \ldots, x_n)$ of $D$, assign the $n$-tuple $f(x) = (f_1(x_1), \ldots, f_n(x_n))$. Regard the distinct $n$-tuples as colors, so $D$ is colored with $d_1 d_2 \cdots d_n$ colors.

Take a vertex $b = (b_1, \ldots, b_n)$ of $D$, and let $X_b$ be the strong component of $D$ that contains $b$. If $x = (x_1, \ldots, x_n)$ is in $X_b$, then $D$ has a $(b, x)$-diwalk of length (say) $k$. By Proposition 10.2.2, each $D_i$ has a $(b_i, x_i)$-diwalk of length $k$. As $b_i$ is colored $f_i(b_i)$, it follows from the definition of $f_i$ that $x_i$ is colored $f_i(x_i) = f_i(b_i) + k \pmod{d_i}$. Thus every vertex $x$ of $X_b$ has a color of form $f(x) = (f_1(b_1) + k, \ldots, f_n(b_n) + k)$ for some non-negative integer $k$. (Where the arithmetic in the $i$th coordinate is done modulo $d_i$.)

Suppose for the moment that the converse is true: If $x \in V(D)$ and $f(x) = (f_1(b_1)+k, \ldots, f_n(b_n)+k)$ for some non-negative $k$, then $x$ belongs to $X_b$. (We will prove this shortly.) Combined with the previous paragraph, this means $V(X_b)$ consists precisely of those vertices colored $(f_1(b_1) + k, \ldots, f_n(b_n) + k)$ for some non-negative integer $k$. There are precisely $\text{lcm}\big(d_1, \ldots, d_n\big)$ such colors. In summary, $D$ has $d_1 d_2 \cdots d_n = d(D_1) \cdot d(D_2) \cdot \cdots \cdot d(D_n)$ color classes, and any strong component of $D$ is the union of $\text{lcm}\big(d(D_1), \ldots, d(D_n)\big)$ of them. Thus $D$ has

$$\frac{d(D_1) \cdot d(D_2) \cdots d(D_n)}{\text{lcm}\big(d(D_1), d(D_2), \ldots, d(D_n)\big)}$$

strong components, and the theorem follows.

It remains to prove the assertion made above, namely that if the vertex $b = (b_1, \ldots, b_n)$ belongs to a strong component $X_b$, then any vertex colored

$(f_1(b_1)+k,\ldots,f_n(b_n)+k)$ belongs to $X_b$. Thus let $x=(x_1,\ldots,x_n)$ be colored $f(x)=(f_1(b_1)+k,\ldots,f_n(b_n)+k)$. That is, each $x_i$ has color $f_i(x_i)=f_i(b_i)+k \pmod{d_i}$. We need to prove that $D$ has both a $(b,x)$-diwalk and an $(x,b)$-diwalk. By Proposition 10.2.2, it suffices to show that there is a positive integer $K$ for which each $D_i$ has a $(b_i,x_i)$-diwalk of length $K$. (And also a $K'$ for which each $D_i$ has a $(x_i,b_i)$-diwalk of length $K'$.) The following claim assures that this is possible.

**Claim.** Suppose vertices $b_i, x_i \in V(D_i)$ have colors $f_i(b_i)$ and $f_i(b_i)+k$, respectively. Then there is an integer $M_i$ such that for all $m_i \geq M_i$ there is a $(b_i,x_i)$-diwalk of length $m_i d_i + k$. Also there is an integer $M_i'$ such that for all $m_i \geq M_i'$ there is an $(x_i,b_i)$-diwalk of length $m_i d_i - k$.

Once the claim is established, we can put $m_i = Ld_1d_2\cdots d_n/d_i$, where $L$ is large enough that each $m_i$ exceeds the maximum of all the $M_i$ and $M_i'$. Then $m_i d_i = Ld_1d_2\cdots d_n$ for each $i\in[n]$, and the claim then gives the required diwalks of lengths $K=Ld_1d_2\cdots d_n+k$ and $K'=Ld_1d_2\cdots d_n-k$.

To prove the claim, let vertices $b_i$ and $x_i$ of $D_i$ have colors $f_i(b_i)$ and $f_i(b_i)+k$, respectively. Because $D_i$ is strongly connected, $D_i$ has a $(b_i,x_i)$-diwalk $W$. Moreover, we may assume $|W|\geq k$, by concatenating with $W$ (if necessary) arbitrarily many closed $(x_i,x_i)$-diwalks. Because $b_i$ has color $f_i(b_i)$ and $x_i$ has color $f_i(b_i)+k$, it follows that $W$ has length $\ell d_i + k$ for some non-negative integer $\ell$.

By definition of $d_i$, there are dicycles $C_1,C_2,\ldots,C_s$ in $D_i$ for which $d_i=\gcd(|C_1|,|C_2|,\ldots,|C_s|)$. Select a vertex $c_i$ of each $C_j$. Let $P_0$ be a $(b_i,c_1)$-diwalk, let $P_s$ be a $(c_s,x_i)$-diwalk, and for each $j\in[s-1]$ let $P_j$ be a $(c_j,c_{j+1})$-diwalk. See Figure 10.4. By the same reasoning used for $W$, the diwalk $W'=P_0+P_1+\cdots+P_s$ has length $\ell' d_i+k$ for some non-negative $\ell'$.

By choice of the $C_i$, there are integers $u_j$ for which $\sum_{j=1}^{s} u_j|C_j| = d_i$. Let $u=\max\{|u_1|,\ldots,|u_s|\}$. Put $w=\sum_{j=1}^{s}\frac{|C_j|}{d_i}$, which is a positive integer because $d_i$ divides each $|C_j|$. We will show that $M_i=\ell'+w+w^2u$ satisfies the requirements of the claim: Let $m_i\geq M_i$. By the division algorithm

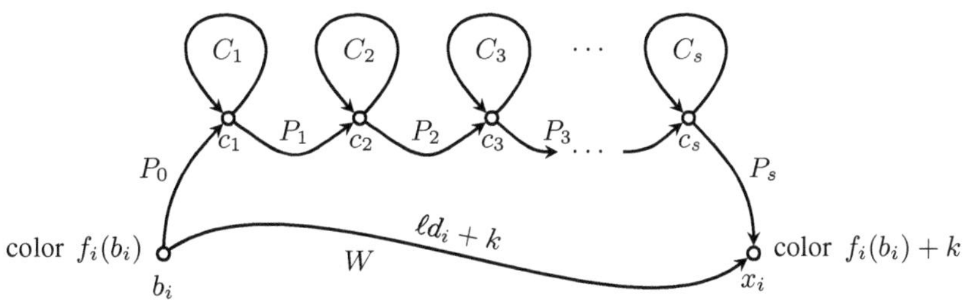


**Figure 10.4** The diwalk $W$ has length $\ell d_i+k$, and the diwalk $W'=P_0+P_1+\cdots+P_s$ has length $\ell' d_i+k$.

$$m_i - \ell' = qw + r \quad \text{with} \quad 0 \le r < w. \tag{10.5}$$

For each $j \in [s]$, put $v_j = q + ru_j$. Note that each $v_j$ is positive because

$$\begin{aligned} v_j &= \left(\frac{qw+r}{w} - \frac{r}{w}\right) + ru_i \\ &> \left(\frac{m_i - \ell'}{w} - 1\right) - wu && \text{(by (10.5) and } u \ge u_i) \\ &\ge \frac{M_i - \ell'}{w} - 1 - wu && \text{(because } m_i \ge M_i) \\ &= 0 && \text{(by definition of } M_i). \end{aligned}$$

Thus we may construct a diwalk

$$W'' = P_0 + v_1C_1 + P_1 + v_2C_2 + P_2 + v_3C_3 + P_3 + \cdots + v_sC_s + P_s,$$

where $v_jC_j$ is $C_j$ concatenated with itself $v_j$ times. The length of $W''$ is

$$\begin{aligned} |W''| &= \sum_{j=0}^{s} |P_j| + \sum_{j=1}^{s} v_j|C_j| \\ &= |W'| + \sum_{j=1}^{s} (q + ru_j)|C_j| && \text{(def. of } W', \text{ and } v_j) \\ &= \ell' d_i + k + q\sum_{j=1}^{s} |C_j| + r\sum_{j=1}^{s} u_j|C_j| && \text{(recall } |W'| = \ell' d_i + k) \\ &= \ell' d_i + k + qd_iw + rd_i && \text{(def. of } w \text{ and choice of } u_j) \\ &= \ell' d_i + k + (qw + r)d_i \\ &= \ell' d_i + k + (m_i - \ell')d_i && \text{(Equation (10.5))} \\ &= m_id_i + k. \end{aligned}$$

Thus for any $m_i \ge M_i$ we have constructed a $(b_i, x_i)$-diwalk $W''$ in $D_i$ of length $m_id_i + k$, and this completes the first part of the claim. By a like construction (reversing the walks in Figure 10.4, which is possible because $D_i$ is strong) there is also a $(x_i, b_i)$-diwalk $W'''$ in $D_i$ of length $m_id_i - k$. This completes the proof of the claim, and also the proof of the Theorem. □

The issue of connectedness of direct products is even more subtle than that of strong connectedness. Despite the contributions [3], [21] and [22], more than 50 years elapsed between McAndrew's result on strong connectedness (Theorem 10.3.2) and the eventual characterization of connectedness by Chen and Chen [5], which we now examine. To begin the discussion, note that because all projections of $D = D_1 \times \cdots \times D_n$ to factors are homomorphisms, if $D$ is connected, then each factor $D_i$ is connected too. The converse is generally false, as demonstrated by $\overrightarrow{P}_2 \times \overrightarrow{P}_2$. Laying out the exact additional conditions on the factors that ensure that the product is connected requires several definitions.

A matrix $A$ is **chainable** if its entries are non-negative, and it has no rows or columns of zeros, and there are no permutation matrices $M$ and $N$ for which $MAN$ has block form

$$MAN = \begin{bmatrix} A_1 & 0 \\ 0 & A_2 \end{bmatrix}.$$

For a positive integer $\ell$, we say $A$ is **$\ell$-chainable** if $A^\ell$ is chainable. A digraph is **$\ell$-chainable** if its adjacency matrix is $\ell$-chainable.

Given a walk $W$ from $x$ to $y$ in a digraph $D$, its **weight** $w(W)$ is the integer $m - n$, where in traversing $W$ from $x$ to $y$, we encounter $m$ arcs in forward orientation and $n$ arcs in reverse orientation. The **weight** $w(D)$ of the digraph $D$ is the greatest common divisor of the weights of all closed walks in $D$, or 0 if all closed walks have weight 0.

Space limitations prevent inclusion of the proof of the following theorem. It can be found in [5].

**Theorem 10.3.3** *Suppose $D_1, \ldots, D_n$ are connected digraphs. Then:*

1. *If no $w(D_i)$ is zero, then $D_1 \times \cdots \times D_n$ is connected if and only if both of the following conditions hold:*
   - $\gcd\big(w(D_1), \ldots, w(D_n)\big) = 1$,
   - *If some $D_i$ has a vertex of in-degree 0 (respectively out-degree 0) then no $D_j$ $(j \neq i)$ has a vertex of out-degree 0 (respectively in-degree 0).*
2. *If some $w(D_i)$ is zero, then $D_1 \times \cdots \times D_n$ is connected if and only if the other $D_j$ $(j \neq i)$ are $\ell$-chainable, where $\ell = \mathrm{diam}(D_i)$.*

We conclude this section with characterizations of unilateral connectedness of the four products. Recall that a digraph is **unilaterally connected** if for any two of its vertices $x, y$ there exists an $(x, y)$-diwalk or a $(y, x)$-diwalk. (Because this relation on vertices is not symmetric, and thus not an equivalence relation, there is no notion of unilateral components.) Note that strongly connected digraphs are unilaterally connected, but not conversely.

**Theorem 10.3.4** *A Cartesian product of digraphs is unilaterally connected if and only if one factor is unilaterally connected and the others are strongly connected. This is also true for the strong product.*

For a proof, see the solution of Exercise 32.4 of Hammack, Imrich and Klavžar [18]. See the solution of Exercise 32.5 for a proof of the next result.

**Theorem 10.3.5** *A lexicographic product of digraphs is unilaterally connected but not strongly connected if and only if each factor is unilaterally connected, and the first factor is not strongly connected.*

Finally, we have a characterization of unilaterally connected direct products due to Harary and Trauth [21].

**Theorem 10.3.6** *A direct product $D_1 \times \cdots \times D_n$ is unilaterally connected if and only if each of the following holds:*

- *At most one factor $D_i$ is unilaterally connected but not strongly connected.*
- *$D_1 \times \cdots \times D_{i-1} \times D_{i+1} \times \cdots \times D_n$ is strongly connected.*
- *$D_1 \times \cdots D_{i-1} \times C \times D_{i+1} \times \cdots \times D_n$ is strongly connected for each strong component $C$ of $D_i$.*

## 10.4 Neighborhoods, Kings and Kernels

The structures of vertex neighborhoods in digraph products are clear from the definitions. For instance, $N^+_{D \,\Box\, D'}(x, y) = \big(N^+_D(x) \times \{y\}\big) \cup \big(\{x\} \times N^+_{D'}(y)\big)$, etc. For future reference we record two particularly useful formulas, namely

$$N^+_{D\times D'}(x, y) = N^+_D(x) \times N^+_{D'}(y), \tag{10.6}$$

$$N^+_{D\boxtimes D'}[(x, y)] = N^+_D[x] \times N^+_{D'}[y]. \tag{10.7}$$

These also hold with the out-neighborhoods $N^+$ replaced by in-neighborhoods $N^-$, and extend to arbitrarily many factors.

Recall that a $k$-**king** in a digraph is a vertex $x$ for which there is an $(x, y)$-dipath of length no greater than $k$ for all vertices $y$ of the digraph. The next proposition follows from the distance properties in Section 10.2.

**Proposition 10.4.1** *Let $D_1$ and $D_2$ be digraphs. Then:*

1. *$(x_1, x_2)$ is a $k$-king in $D_1 \boxtimes D_2$ if and only if each $x_i$ is a $k$-king in $D_i$.*
2. *$(x_1, x_2)$ is a $k$-king in $D_1 \,\Box\, D_2$ if and only if each $x_i$ is a $k_i$-king in $D_i$, where $k_1 + k_2 = k$.*
3. *$(x_1, x_2)$ is a $k$-king in $D_1 \circ D_2$ if and only if $x_1$ is a $k$-king in $D_1$, and $x_2$ is a $k$-king in $D_2$ or $\xi_{D_1}(x_1) \le k$ (where $\xi$ is as defined on page 473).*
4. *If $(x_1, x_2)$ is a $k$-king in $D_1 \times D_2$, then each $x_i$ is a $k$-king in $D_i$.*

This proposition is due to students P. LaBarr, M. Norge and I. Sanders, directed by D. Taylor [40]. Concerning statement 4, no characterization of kings in direct products is known.

Recall that a $(k, l)$-**kernel** of a digraph $D$ is a subset $J \subseteq V(D)$ for which $\text{dist}_D(x, y) \ge k$ for all distinct $x, y \in J$, and to any $x \notin J$ there is a $y \in J$ with $\text{dist}_D(x, y) \le l$. Szumny, Włoch and Włoch [54] explored $(k, l)$-kernels in so-called $D$-joins. Their Theorem 8 implies the following characterization for the lexicographic product. (They also enumerate all $(k, l)$-kernels in $D_1 \circ D_2$.)

**Proposition 10.4.2** *Let $l \ge k \ge 2$. Then $J^* \subseteq V(D_1 \circ D_2)$ is a $(k, l)$-kernel if and only if $D_1$ has a $(k, l)$-kernel $J$ with $J^* = \bigcup_{x\in J}\{x\} \times J_x$, where*

- $J_x$ *is a* $(k,l)$*-kernel of* $D_2$ *if* $\xi_{D_1}(x) > l$ *and* $\mathrm{dist}_{D_2}(y,x) > l$ *for* $y \neq x$*, or*
- $J_x$ *is a single vertex of* $D_2$ *if* $\xi_{D_1}(x) < k$*, or*
- $\mathrm{dist}_{D_2}(x,y) \geq k$ *for all distinct* $x,y \in J_x$ *otherwise.*

The case $k > l$ is open. No characterizations are known for the other products, though Kwaśnik [33] proved the following.

**Proposition 10.4.3** *Let* $D_1$ *and* $D_2$ *be digraphs, and let* $J_i$ *be a* $(k_i,l_i)$*-kernel of* $D_i$ *for each* $i = 1,2$*.*

1. $J_1 \times J_2$ *is a* $\big(\min\{k_1,k_2\},\, l_1+l_2\big)$*-kernel of* $D_1 \,\square\, D_2$ *(for* $k_1,k_2 \geq 2$*).*
2. $J_1 \times J_2$ *is a* $\big(\min\{k_1,k_2\}, \max\{l_1,l_2\}\big)$*-kernel of* $D_1 \boxtimes D_2$*.*

See [59] for corresponding results for generalized products. Finally, we remark that Lakshmi and Vidhyapriya [34] characterize kernels in Cartesian products of tournaments with directed paths and cycles.

## 10.5 Hamiltonian Properties

Hamiltonian properties of digraphs have been studied extensively. The following four theorems are among the results proved in the book [44] by Schaar, Sonntag and Teichert.

**Theorem 10.5.1** *If* $D_1$ *and* $D_2$ *are Hamiltonian digraphs, then* $D_1 \boxtimes D_2$ *and* $D_1 \circ D_2$ *are Hamiltonian. If, in addition,* $D_1$ *is Hamiltonian connected, and* $|D_1| \geq 3$ *and* $|D_2| \geq 4$*, then* $D_1 \,\square\, D_2$ *is Hamiltonian.*

The above additional conditions on the factors of a Cartesian product are necessary, as evidenced by the next theorem of Erdős and Trotter [57].

**Theorem 10.5.2** *The Cartesian product* $\overrightarrow{C}_p \,\square\, \overrightarrow{C}_q$ *is Hamiltonian if and only if there are non-negative integers* $d_1,d_2$ *for which* $d_1+d_2 = \gcd(p,q) \geq 2$ *and* $\gcd(p,d_1) = \gcd(q,d_2) = 1$*.*

Recall that a digraph is **traceable** if it has a Hamiltonian path. It is **homogeneously traceable** if each of its vertices is the initial point of some Hamiltonian path.

**Theorem 10.5.3** *If digraphs* $D_1$ *and* $D_2$ *are homogeneously traceable, then so are* $D_1 \,\square\, D_2$*,* $D_1 \boxtimes D_2$ *and* $D_1 \circ D_2$*.*

**Theorem 10.5.4** *If* $D_1$ *is homogeneously traceable and* $D_2$ *is traceable, then* $D_1 \,\square\, D_2$ *and* $D_1 \boxtimes D_2$ *are traceable. If* $D_1$ *and* $D_2$ *are traceable, then so is* $D_1 \circ D_2$*.*

A digraph is **Hamiltonian decomposable** if it has a family of Hamiltonian dicycles such that every arc of the digraph belongs to exactly one of the dicycles. Ng [39] gives the most complete result among digraph products.

**Theorem 10.5.5** *If $D_1$ and $D_2$ are Hamiltonian decomposable digraphs, and $|V(D_1)|$ is odd, then $D_1 \circ D_2$ is Hamiltonian decomposable.*

At present it is not known if the assumption of odd order can be removed.

**Conjecture 10.5.6** *If $D_1$ and $D_2$ are Hamiltonian decomposable digraphs, then $D_1 \circ D_2$ is Hamiltonian decomposable.*

By Theorem 10.5.2, a Cartesian product of Hamiltonian decomposable digraphs is not necessarily Hamiltonian decomposable. This is also the case for the strong product, as is illustrated by $\overleftrightarrow{K_2} \boxtimes \overleftrightarrow{K_2} = \overleftrightarrow{K_4}$.

**Problem 10.5.7** *Determine conditions under which a Cartesian or strong product of digraphs is Hamiltonian decomposable.*

A solution to this problem may shed light on the longstanding conjecture that a Cartesian product of Hamiltonian decomposable *graphs* is Hamiltonian decomposable. See Section 30.2 of [18] and the references therein.

Despite these difficulties, there *has* been progress on Cartesian products of biorientations of graphs. Stong [52] proved that complete biorientations of odd-dimensional hypercubes decompose into $2m+1$ Hamiltonian cycles, and the same is true for $\overleftrightarrow{C}_{n_1} \square \cdots \square \overleftrightarrow{C}_{n_m} \square \overleftrightarrow{K}_2$ provided $n_i \geq 3$ and $m > 2$.

Hamiltonian results for direct products of digraphs are scarce. Keating [28] proves that if $D_1$ and $D_2 \times \overrightarrow{C}_{|D_1|}$ are Hamiltonian decomposable, then so is $D_1 \times D_2$. Paulraja and Sivasankar [41] establish hamilton decompositions in direct products of biorientations of special classes of graphs.

## 10.6 Invariants

Here we collect various results on invariants of digraph products, beginning with the chromatic number and proceeding to domination and independence.

The chromatic number $\chi(D)$ of a digraph $D$ is the chromatic number of the underlying graph of $D$. For the Cartesian and lexicographic products, the underlying graph of the product is the product of the underlying graph of the factors. Thus for $\square$ and $\circ$, the chromatic number of products of digraphs coincides with that of products of graphs. This has been well-studied. See Chapter 26 of [18] for a survey.

The situation for the direct and strong products is different. For example, $\chi(G \times H) \leq \min\{\chi(G), \chi(H)\}$ is straightforward, whether $G$ and $H$ are graphs or digraphs. The celebrated **Hedetniemi conjecture** asserts that $\chi(G \times H) = \min\{\chi(G), \chi(H)\}$ for all graphs $G$ and $H$. But if $G$ and $H$ are digraphs, then it is quite possible that $\chi(G \times H) < \min\{\chi(G), \chi(H)\}$, as was first noted by Poljak and Röld [43]. More recently, Bessy and Thomassé [2] exhibit a 5-chromatic digraph $D$ for which $\chi(D \times TT_5) = 3$, and Tardif [55] gives digraphs $G_n$ and $H_n$ for which $\chi(G_n) = n$, $\chi(H_n) = 4$ and $\chi(G_n \times H_n) = 3$. Poljak and Röld introduced the functions

$$f(n) = \min\{\chi(G \times H) \mid G \text{ and } H \text{ are } n\text{-chromatic digraphs }\},$$
$$g(n) = \min\{\chi(G \times H) \mid G \text{ and } H \text{ are } n\text{-chromatic graphs}\},$$

and showed that if $g$ is bounded above, then the bound is at most 16. This bound was improved to 3 in [42].

Notice that $f(n) \leq g(n) \leq n$, and Hedetniemi's conjecture is equivalent to the assertion $g(n) = n$. Certainly if $g$ is bounded, then so is $f$. Interestingly, the converse is true. Tardif [56] proved that $f$ and $g$ are either both bounded or both unbounded. Thus Hedetniemi's conjecture is false if $f$ is bounded.

There is an oriented version of the chromatic number, defined on oriented graphs, that is, digraphs with no 2-cycles. A **oriented $k$-coloring** of such a digraph $D$ is a map $c : V(D) \to [k]$ with the property that $c(x) \neq c(y)$ whenever $xy \in A(D)$, and, in addition, the existence of an arc from one color class $X_1$ to another color class $X_2$ implies that there are no arcs from $X_2$ to $X_1$. The smallest such $k$ is called the **oriented chromatic number** of $D$, denoted $\chi_o(D)$. Equivalently, this is the smallest $k$ for which there is a homomorphism from $D$ to an oriented graph of order $k$. The oriented chromatic number $\chi_o(G)$ of a graph $G$ is the maximum oriented chromatic number of all orientations of $G$. For a survey, see Sopena [51]. Tight bounds on this invariant are rare, even for simple classes of graphs. Aravind, Narayanan and Subramanian [1] show $\chi_o(G \,\Box\, P_n) \leq (2n-1)\chi_o(G)$, and $\chi_o(G \,\Box\, C_n) \leq 2n\chi_o(G)$, as well as $8 \leq \chi_o(P_2 \boxtimes P_n) \leq 11$ and $10 \leq \chi_o(P_3 \boxtimes P_n) \leq 67$. There appears to have been no other work with this invariant on products other than some progress on grids [10, 53].

A **dominating set** in a digraph $D$ is a subset $S \subseteq V(D)$ with the property that for any $y \in V(D) - S$ there exists some $x \in S$ for which $xy \in A(D)$. The **domination number** $\gamma(D)$ is the size of a smallest dominating set. Domination in digraphs has not been studied as extensively as in graphs. As computing the domination number of a graph is $\mathcal{NP}$-hard [13], the same is true for digraphs. (Consider the complete biorientation of an arbitrary graph.) Thus we can expect exact formulas only for products of special classes of digraphs. Liu *et al.* [35] and Shaheen [45–47] consider the case of Cartesian products of directed paths and cycles. For example, Shaheen proves

$$\gamma\left(\vec{P}_m \,\Box\, \vec{P}_n\right) = m + \left\lceil\frac{m-1}{3}\right\rceil\left\lceil\frac{n-2}{3}\right\rceil\left\lceil\frac{m}{3}\right\rceil + \left\lceil\frac{m}{3}\right\rceil\left\lceil\frac{n-3}{3}\right\rceil + \left\lceil\frac{m+1}{3}\right\rceil\left\lceil\frac{n-4}{3}\right\rceil,$$

provided $m, n > 3$, and separate formulas are given for $m \leq 3$. Similar results for the strong product of grid graphs are considered in [48].

Concerning independence, note that (as for the chromatic number) questions of independence in Cartesian and lexicographic products of digraphs coincide with the same questions for graphs. So only the direct and strong product of digraphs are not covered by the theory of *graph* products. Despite this, there appears to have been little work done with them. But one interesting application deserves mention. The Gallai–Milgram theorem [12] says that the vertices of any digraph with independence number $n$ can be partitioned into $n$ parts, each of which is the vertex set of a directed path (see also Theorem 1.8.4). Hahn and Jackson [14] conjectured that this theorem is the best possible in the sense that for each positive $n$ there is a digraph with independence number $n$, and such that removing the vertices of any $n-1$ directed paths still leaves a digraph with independence number $n$. Bondy, Buchwalder and Mercier [4] used lexicographic products to construct such digraphs for $n = 2^a 3^b$. (The general conjecture was proved by Fox and Sudakov [11].)

Finally, we briefly visit the notion of the **exponent** $\exp(D)$ of a digraph $D$, which is the least positive integer $k$ for which any two vertices of $D$ are joined by a diwalk of length $k$ (or $\infty$ if no such $k$ exists). We say $D$ is **primitive** if its exponent is finite. Wielandt [58] proved that the exponent of a primitive digraph on $n$ vertices is bounded above by $n^2-n+1$, and established a family $W_n$ of digraphs for which this bound is attained. Kim, Song and Hwang [32] showed that if $D_1$ and $D_2$ have order $n_1$ and $n_2$, respectively, then $\exp(D_1 \square D_2) \leq n_1 n_2 - 1$, and this upper bound can be attained only if $\gcd(n_1, n_2) = 1$. Moreover, if $n_1 = n_2 = n$, then $\exp(D_1 \square D_2) \leq n^2 - n + 1$, and the bound is attained only for $D_1 = \overrightarrow{C}_n$ and $D_2 = W_n$. In [30] they compute the exponents of Cartesian products of cycles, and also show that if $D_1$ is a primitive graph and $D_2$ is a strong digraph, then

$$\exp(D_1 \square D_2) = \exp(D_1) + \mathrm{diam}(D_2).$$

This work continues in [29], which proves $\exp(D_1 \boxtimes D_2) \leq n_1 + n_2 - 2$, with equality for dicycles. Concerning the direct product, the same authors [31] show that for a primitive digraph $D$ there is an integer $m$ for which

$$\begin{aligned} \mathrm{diam}(D) < \mathrm{diam}(D^{\times 2}) < \mathrm{diam}(D^{\times 3}) < \cdots \\ &< \mathrm{diam}(D^{\times m}) \\ &= \mathrm{diam}(D^{\times m+1}) = \cdots = \exp(D). \end{aligned}$$

## 10.7 Quotients and Homomorphisms

Here we set up the notions needed in the subsequent sections on cancellation and prime factorization of digraphs. Some of that material is most naturally phrased within the class of digraphs in which loops are allowed. With this in mind, let $\mathscr{D}$ denote the set of (isomorphism classes of) digraphs without loops, and let $\mathscr{D}_0$ be the set of digraphs in which loops are allowed. Thus $\mathscr{D} \subset \mathscr{D}_0$. We admit as an element of $\mathscr{D}$ the empty digraph $O$ with $V(O) = \emptyset$.

This section's main theme is that a digraph is completely determined, up to isomorphism, by the number of homomorphisms into it. Recall that a **homomorphism** $f : D \to D'$ between digraphs $D, D' \in \mathscr{D}_0$ is a map $f : V(D) \to V(D')$ for which $xy \in A(D)$ implies $f(x)f(y) \in A(D')$. Also $f$ is a **weak homomorphism** if $xy \in A(D)$ implies $f(x)f(y) \in A(D')$ or $f(x) = f(y)$.

The set of all homomorphisms $D \to D'$ is denoted $\mathrm{Hom}(D, D')$, and the set of weak homomorphisms $D \to D'$ is $\mathrm{Hom}_w(D, D')$. A homomorphism is **injective** if it is injective as a map from $V(D)$ to $V(D')$. We denote the set of all injective homomorphisms $D \to D'$ as $\mathrm{Inj}(D, D')$. (Necessarily $\mathrm{Inj}(D, D')$ is also the set of injective weak homomorphisms $D \to D'$.) Let $\hom(D, D') = |\mathrm{Hom}(D, D')|$ be the number of homomorphisms $D \to D'$. Similarly, $\hom_w(D, D') = |\mathrm{Hom}_w(D, D')|$, and $\mathrm{inj}(D, D') = |\mathrm{Inj}(D, D')|$.

We will need several notions of digraph quotients. For a digraph $D$ in $\mathscr{D}$ and a partition $\Omega$ of $V(D)$, the **quotient** $D/\Omega$ **in** $\mathscr{D}$ is the digraph in $\mathscr{D}$ whose vertices are the partition parts $U \in \Omega$, and with an arc from $U$ to $V$ if $U \neq V$ and $D$ has an arc $uv$ with $u \in U$ and $v \in V$. Notice the map $D \to D/\Omega$ sending $u$ to the element $U \in \Omega$ with $u \in U$ is a *weak homomorphism*.

On the other hand, if $D \in \mathscr{D}_0$, then the **quotient** $D/\Omega$ **in** $\mathscr{D}_0$ is as above, but with a loop $UU \in A(D/\Omega)$ whenever $D$ has an arc with both endpoints in $U$. The map $D \to D/\Omega$ sending $u$ to the element $U \in \Omega$ that contains $u$ is a *homomorphism*. See Figure 10.5.

The remaining results in this section (at least in the class $\mathscr{D}_0$) are from Lovász [36]. See also Hell and Nešetřil [23] for a very readable account. The statements concerning weak homomorphisms were developed by Culp in [7].

**Lemma 10.7.1** *For a digraph $D$, let $\mathscr{P}$ be the set of all partitions of $V(D)$.*

1. *If $D, G \in \mathscr{D}_0$, then* $\hom(D, G) = \sum_{\Omega \in \mathscr{P}} \mathrm{inj}(D/\Omega, G)$ *(quotients in $\mathscr{D}_0$).*
2. *If $D, G \in \mathscr{D}$, then* $\hom_w(D, G) = \sum_{\Omega \in \mathscr{P}} \mathrm{inj}(D/\Omega, G)$ *(quotients in $\mathscr{D}$).*

**Proof:** For the first part, put $\Upsilon = \{(\Omega, f) \mid \Omega \in \mathscr{P},\ f \in \mathrm{Inj}(D/\Omega, G)\}$, so $|\Upsilon| = \sum_{\Omega \in \mathscr{P}} \mathrm{inj}(D/\Omega, G)$. It suffices to show a bijection $\theta : \mathrm{Hom}(D, G) \to \Upsilon$. Define $\theta$ to be $\theta(f) = (\Omega, f^*)$, where $\Omega = \{f^{-1}(x) \mid x \in V(G)\} \in \mathscr{P}$, and $f^* : D/\Omega \to G$ is defined as $f^*(U) = f(u)$, for $u \in U$. By construction $\theta$ is an injective map to $\Upsilon$. For surjectivity, take any $(\Omega, f) \in \Upsilon$, and note that $\theta$ sends the composition $D \to D/\Omega \xrightarrow{f} G$ to $(\Omega, f^*)$.

The proof of part 2 is the same, except that $\mathrm{Hom}(D, G)$ and $\hom(D, G)$ are replaced by $\mathrm{Hom}_w(D, G)$ and $\hom_w(D, G)$, and quotients are in $\mathscr{D}$. □

**Proposition 10.7.2** *The isomorphism class of a digraph is determined by the number of homomorphisms into it, in the following senses.*

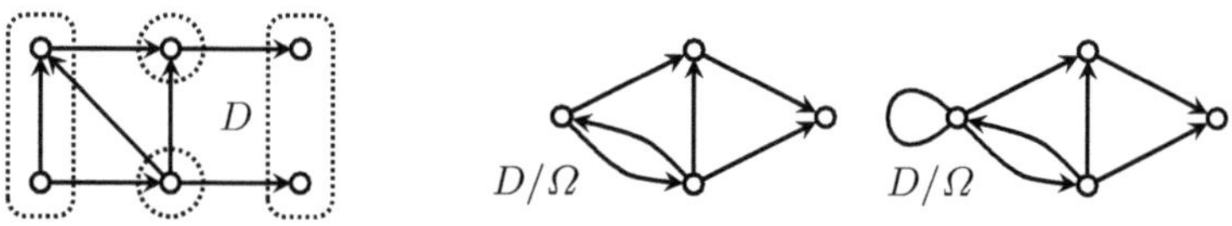


**Figure 10.5** Left: a digraph $D$ and a partition $\Omega$ of $V(D)$. Center: the quotient $D/\Omega$ in $\mathscr{D}$. Right: the quotient $D/\Omega$ in $\mathscr{D}_0$.

1. *If* $G, H \in \mathscr{D}_0$ *and* $\hom(X,G) = \hom(X,H)$ *for all* $X \in \mathscr{D}_0$, *then* $G \cong H$.
2. *If* $G, H \in \mathscr{D}$ *and* $\hom_w(X,G) = \hom_w(X,H)$ *for all* $X \in \mathscr{D}$, *then* $G \cong H$.
3. *If* $G, H \in \mathscr{D}$ *and* $\hom(X,G) = \hom(X,H)$ *for all* $X \in \mathscr{D}$, *then* $G \cong H$.

**Proof:** For the first statement, say $\hom(X,G) = \hom(X,H)$ for all $X \in \mathscr{D}_0$. Our strategy is to show that this implies $\mathrm{inj}(X,G) = \mathrm{inj}(X,H)$ for every $X$. Then the theorem will follow because we get $\mathrm{inj}(H,G) = \mathrm{inj}(H,H) > 0$ and $\mathrm{inj}(G,H) = \mathrm{inj}(G,G) > 0$, so there are injective homomorphisms $G \to H$ and $H \to G$, whence $G \cong H$.

We use induction on $|X|$ to show $\mathrm{inj}(X,G) = \mathrm{inj}(X,H)$. If $|X| = 1$, then

$$\mathrm{inj}(X,G) = \hom(X,G) = \hom(X,B) = \mathrm{inj}(X,H).$$

If $|X| > 1$, Lemma 10.7.1 (1) applied to $\hom(X,G) = \hom(X,H)$ yields

$$\sum_{\Omega \in \mathscr{P}} \mathrm{inj}(X/\Omega, G) = \sum_{\Omega \in \mathscr{P}} \mathrm{inj}(X/\Omega, H).$$

Let $T$ be the trivial partition of $V(X)$ consisting of $|X|$ singleton sets. Then $X/T = X$ and the above equation becomes

$$\mathrm{inj}(X,G) + \sum_{\Omega \in \mathscr{P} - T} \mathrm{inj}(X/\Omega, G) \;=\; \mathrm{inj}(X,H) + \sum_{\Omega \in \mathscr{P} - T} \mathrm{inj}(X/\Omega, H).$$

By the induction hypothesis, $\mathrm{inj}(X/\Omega, G) = \mathrm{inj}(X/\Omega, H)$ for all non-trivial partitions $\Omega$. Consequently $\mathrm{inj}(X,G) = \mathrm{inj}(X,H)$, completing the proof.

The second statement is proved in exactly the same way, but using $\hom_w$ instead of hom, and part 2 of Lemma 10.7.1 instead part 1.

Finally, part 3 follows immediately from part 1, because if $G, H \in \mathscr{D}$ and $X \in \mathscr{D}_0 - \mathscr{D}$, then $X$ has a loop, but neither $G$ nor $H$ has one, so $\hom(X,G) = 0 = \hom(X,H)$. □

Observe that hom and $\hom_w$ factor neatly over the direct and strong products:

**Proposition 10.7.3** *Suppose* $X, D$ *and* $G$ *are digraphs.*

1. *If* $X, D, G \in \mathscr{D}_0$, *then* $\hom(X, D \times G) = \hom(X,D) \cdot \hom(X,G)$.
2. *If* $X, D, G \in \mathscr{D}$, *then* $\hom_w(X, D \boxtimes G) = \hom_w(X,D) \cdot \hom_w(X,G)$.

**Proof:** The map $\mathrm{Hom}(X, D \times G) \to \mathrm{Hom}(X, D) \times \mathrm{Hom}(X, G)$ given by $f \mapsto (\pi_D f, \pi_G f)$ is injective. And it is surjective because any $(f_D, f_G)$ in the codomain is the image of $x \mapsto (f_D(x), f_G(x))$, which is a homomorphism by definition of the direct product. This establishes the first statement, and the second follows analogously. □

As an application, we get a quick result for direct and strong powers.

**Corollary 10.7.4** *If $D, G \in \mathscr{D}_0$, then $D^{\times n} \cong G^{\times n}$ if and only if $D \cong G$. Also, if $D, G \in \mathscr{D}$, then $D^{\boxtimes n} \cong G^{\boxtimes n}$ if and only if $D \cong G$.*

**Proof:** If $D \cong G$, then clearly $D^{\times n} \cong G^{\times n}$. Conversely, if $D^{\times n} \cong G^{\times n}$, then Proposition 10.7.3 gives $\hom(X, D)^n = \hom(X, G)^n$, so $\hom(X, D) = \hom(X, G)$ for any $X \in \mathscr{D}_0$. Thus $D \cong G$, by Proposition 10.7.2. Apply a parallel argument to the strong product. □

## 10.8 Cancellation

Given a product $* \in \{\square, \boxtimes, \times, \circ\}$ the **cancellation problem** seeks the conditions under which $D * G \cong D * H$ implies $G \cong H$ for digraphs $D, G$ and $H$. If this is the case, we say that *cancellation holds*; otherwise it *fails*. Obviously cancellation fails if $D$ is the empty digraph, for then $D * G = O = D * H$ for any $G$ and $H$. We will see that cancellation holds for each of the products $\square$, $\boxtimes$ and $\circ$ provided $D \neq O$. The situation for the direct product is much more subtle; it is reserved for the end of the section.

As in the previous section, $\mathscr{D}$ is the class of digraphs (without loops) and $\mathscr{D}_0$ is the class of digraphs that may have loops. Our first result concerns the strong product. The proof approach is from Culp [7].

**Theorem 10.8.1** *Let $D, G$ and $H$ be digraphs (without loops), with $D \neq O$. If $D \boxtimes G \cong D \boxtimes H$, then $G \cong H$.*

**Proof:** Let $D \boxtimes G \cong D \boxtimes H$. Proposition 10.7.3 says that for any digraph $X$,

$$\hom_w(X, D) \cdot \hom_w(X, G) = \hom_w(X, D) \cdot \hom_w(X, H).$$

If $D \neq O$, then $\hom_w(X, D) > 0$ (constant maps are weak homomorphisms), so $\hom_w(X, G) = \hom_w(X, H)$. Proposition 10.7.2 (2) yields $G \cong H$. □

Theorem 10.8.1 applies only to $\mathscr{D}$. Indeed, cancellation over $\boxtimes$ fails in $\mathscr{D}_0$. Consider the case where $D$ is a single vertex with a loop, and $H = K_1$. Then $D \boxtimes D = D = D \boxtimes H$, but $D \not\cong H$.

Echoing Theorem 10.8.1, we get a partial cancellation result for the direct product [36]. The proof is the same but uses part (1) of Proposition 10.7.2 instead of part (2), plus the fact that any constant map from $X$ to a vertex with a loop is a homomorphism. The result is due to Lovász [36].

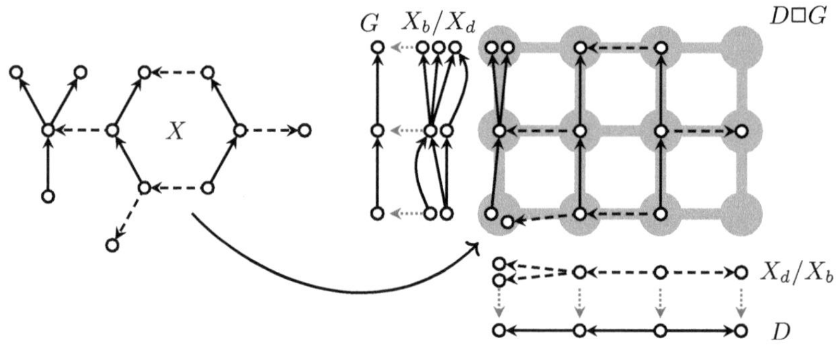


**Figure 10.6** Each homomorphism $X \to D \,\Box\, G$ is encoded as an arc 2-coloring of $X$ with colors dashed and bold, and homomorphisms $X_d/X_b \to D$ and $X_b/X_d \to G$.

**Theorem 10.8.2** *Suppose $D, G, H \in \mathscr{D}_0$, and $D$ has a loop. If $D \times G \cong D \times H$, then $G \cong H$.*

Proposition 10.7.3 has no analogue for the Cartesian product, so to deduce cancellation for it we must count our homomorphisms indirectly. The proof of the next theorem is new. A different approach uses unique prime factorization; see the remarks in Chapter 23 of [18].

**Theorem 10.8.3** *Let $D, G$ and $H$ be digraphs (without loops), with $D \neq O$. If $D \,\Box\, G \cong D \,\Box\, H$, then $G \cong H$.*

**Proof:** The proof has two parts. First we derive a formula for $\hom(X, D \,\Box\, G)$. Then we use it to show $D \,\Box\, G \cong D \,\Box\, H$ implies $\hom(X, G) = \hom(X, H)$ for all $X \in \mathscr{D}$, whence Proposition 10.7.2 yields $G \cong H$.

Our counting formula uses an arc 2-coloring scheme, shown in Figure 10.6. Given a 2-coloring of $A(X)$ by the colors *dashed* and *bold*, let $X_d$ be the spanning subdigraph of $X$ whose arcs are the dashed arcs, and let $X_b$ be the spanning subdigraph whose arcs are bold. Let $X_b/X_d$ be the contraction in $\mathscr{D}_0$ of $X_b$ in which each connected component of $X_d$ is collapsed to a vertex. Specifically, $V(X_b/X_d)$ is the set of connected components of $X_d$, and

$$A(X_b/X_d) = \{UV \mid X \text{ has a } \textbf{bold} \text{ arc from } U \text{ to } V\}.$$

Define $X_d/X_b$ analogously, as the contraction of $X_d$ by the connected components of $X_b$. Note that $X_b/X_d$ (resp. $X_d/X_b$) has a loop at $U$ if and only if the subdigraph of $X$ induced on $U$ has a bold (resp. dashed) arc.

Let $\mathscr{C}$ be the set of all arc 2-colorings of $X$ by colors dashed and bold. We claim that there is a disjoint union

$$\mathrm{Hom}(X, D \,\Box\, G) = \bigcup_{\mathscr{C}} \mathrm{Hom}(X_d/X_b, D) \times \mathrm{Hom}(X_b/X_d, G).$$

Indeed, any $f \in \mathrm{Hom}(X, D \,\square\, G)$ corresponds to a 2-coloring in $\mathscr{C}$ and a pair $(\pi_D f, \pi_G f) \in \mathrm{Hom}(X_d/X_b, D) \times \mathrm{Hom}(X_b/X_d, G)$, as follows. For any $xy \in A(X)$, either $\pi_D f(x) = \pi_D f(y)$ and $\pi_G f(x)\pi_G f(y) \in A(G)$, or $\pi_D f(x)\pi_D f(y) \in A(D)$ and $\pi_G f(x) = \pi_G f(y)$. Color $xy$ bold in the first case and dashed in the second. One verifies that $\pi_D f$ is a well-defined homomorphism $X_d/X_b \to D$, and similarly for $\pi_G f : X_b/X_d \to D$, and it is easy to check that $f \mapsto (\pi_D f, \pi_G f)$ is injective. For surjectivity, note that for any arc 2-coloring of $X$ and pair $(f_D, f_G) \in \mathrm{Hom}(X_d/X_b, D) \times \mathrm{Hom}(X_b/X_d, G)$, there is an associated $f \in \mathrm{Hom}(X, D \,\square\, G)$ defined as $f(x) = (f_D(U), f_G(V))$, where $x \in U, V$.

It follows that we can count the homomorphisms from $X$ to $D \,\square\, G$ as

$$\hom(X, D \,\square\, G) = \sum_{\mathscr{C}} \hom(X_d/X_b, D) \cdot \hom(X_b/X_d, G). \tag{10.8}$$

This completes the first part of the proof.

For the second step, suppose $D \,\square\, G \cong D \,\square\, H$ and $X$ is arbitrary. We will show $\hom(X, G) = \hom(X, H)$ by induction on $|X|$. If $|X| = 1$, then $\hom(X, G) = |G| = |H| = \hom(X, H)$. Otherwise, by Equation (10.8),

$$\sum_{\mathscr{C}} \hom(X_d/X_b, D)\cdot\hom(X_b/X_d, G) = \sum_{\mathscr{C}} \hom(X_d/X_b, D)\cdot\hom(X_b/X_d, H).$$

By induction, $\hom(X_b/X_d, G) = \hom(X_b/X_d, H)$ for all colorings with at least one dashed edge. Thus, for the coloring where all edges are bold, we get

$$\hom(X_d/X_b, D) \cdot \hom(X_b/X_d, G) = \hom(X_d/X_b, D) \cdot \hom(X_b/X_d, H).$$

But then $X_d/X_b$ has no arcs, so $\hom(X_d/X_b, D) > 0$. Also, $X_b/X_d = X$, so we get $\hom(X, G) = \hom(X, H)$. Finally, Proposition 10.7.2 says $G \cong H$. □

Next we aim our homomorphism-counting program at the lexicographic product and bag a particularly strong cancellation law. We use a coloring scheme like that in Figure 10.6. For a homomorphism $X \to D \circ G$, arcs mapping to fibers over vertices of $D$ are colored bold, and all other arcs are colored dashed. Equation (10.8) adapts as

$$\hom(X, D \circ G) = \sum_{\mathscr{C}} \hom(X_d/X_b, D) \cdot \hom(X_b, G). \tag{10.9}$$

Verification is left as an exercise. Using this, we can prove right- and left-cancellation for the lexicographic product.

**Lemma 10.8.4** *Suppose $D, G$ and $H$ are digraphs (without loops) and $D \neq O$. If $G \circ D \cong H \circ D$, then $G \cong H$. If $D \circ G \cong D \circ H$, then $G \cong H$.*

**Proof:** Say $G \circ D \cong H \circ D$. We will get $G \cong H$ by showing $\hom(X, G) = \hom(X, H)$ for any $X$. If $|X| = 1$, then $\hom(X, G) = |G| = |H| = \hom(X, H)$.

Let $|X| > 1$ and assume $\hom(X', G) = \hom(X', H)$ whenever $|X'| < |X|$. As $\hom(X, G \circ D) = \hom(X, H \circ D)$, Equation 10.9 gives

$$\sum_{\mathscr{C}} \hom(X_d/X_b, G) \cdot \hom(X_b, D) = \sum_{\mathscr{C}} \hom(X_d/X_b, H) \cdot \hom(X_b, D).$$

Now, $\hom(X_d/X_b, G) = \hom(X_d/X_b, H)$ unless all arcs of $X$ are dashed, in which case $X_d/X_b = X$ and $X_b$ is the arcless digraph on $V(X)$. From this, the above equation reduces to $\hom(X, G) \cdot |D|^{|G|} = \hom(X, H) \cdot |D|^{|G|}$, and then $G \cong H$ by Proposition 10.7.2. For the second statement, Equation 10.9 gives

$$\sum_{\mathscr{C}} \hom(X_d/X_b, D) \cdot \hom(X_b, G) = \sum_{\mathscr{C}} \hom(X_d/X_b, D) \cdot \hom(X_b, H),$$

and we reason as in the first case. □

We now discuss a notion that leads to a much stronger cancellation law. A subdigraph $X$ of $D$ is said to be **externally related** if for each $b \in V(D) - V(X)$ the following holds: if there is an arc from $b$ to a vertex of $X$, then there are arcs from $b$ to every vertex in $X$; and if there is an arc from a vertex of $X$ to $b$, then there are arcs from every vertex of $X$ to $b$. (In the context of graphs, see Section 10.2 of [18], and the references therein.)

Given a vertex $a = (x_1, x_2) \in V(G \circ D)$, let $D^a$ denote the subdigraph of $G \circ D$ induced on the vertices $\{(x_1, x) \mid x \in V(D)\}$. We call $D^a$ the **$D$-layer** through $a$. The definition of the lexicographic product implies $D^a \cong D$, and that each $D^a$ is externally related in $G \circ D$. Note that each $D^a$ is also an induced subdigraph of $G \circ D$. All of these ideas are used in the proof of the next theorem, which was first proved by Dörfler and Imrich [8].

**Theorem 10.8.5** *Let $D, G, H$ and $K$ be non-empty digraphs (without loops). If $G \circ D \cong H \circ K$ and $|D| = |K|$, then $G \cong H$ and $D \cong K$.*

**Proof:** We prove this under the assumption that either $D$ is disconnected, or that both $D$ and its complement $\overline{D}$ are connected. Once proved, this implies the general result, because if $D$ is connected and $\overline{D}$ is disconnected, then we can use Equation 10.3 to get $\overline{G} \circ \overline{D} \cong \overline{H} \circ \overline{K}$. Then $\overline{G} \cong \overline{H}$ and $\overline{D} \cong \overline{K}$, and the theorem follows.

Take an isomorphism $\varphi : G \circ D \to H \circ K$.

We first claim that for any $D$-layer $D^a$, the image $\pi_H\varphi(D^a)$ is either an arcless subdigraph of $H$ (i.e., one or more vertices of $H$), or it is a single arc of $H$. Indeed, suppose it has an arc. Then $\varphi(D^a)$ has an arc $cd$ with $\pi_H(c) \neq \pi_H(d)$. We will show that if $\varphi(D^a)$ has a vertex $x$ with $\pi_H(x) \notin \{\pi_H(c), \pi_H(d)\}$, then all arcs $x''y, yx''$ are present in $\varphi(D^a)$, for any vertex $y \in \varphi(D^a) \cap (K^c \cup K^d)$ and $x'' \in K^x$. This will contradict our assumption about $D$, because it implies that $\varphi(D^a)$ (hence also $D$) is connected, but its complement is disconnected, for in $\overline{\varphi(D^a)}$ it is impossible to find a path from $x$ to $c$ or $d$. Thus let $x$ be as stated above. Select vertices $c', d', x'$ in $H \circ K - \varphi(D^a)$ with $\pi_H(c') = \pi_H(c)$, $\pi_H(d') = \pi_H(d)$ and $\pi_H(x') = \pi_H(x)$. (Possible because $|\varphi(D^a)| = |K|$, and the existence of the arc $cd$ means that no $K$-layer is contained in $\varphi(D^a)$.) The definition of $\circ$ implies $cd', c'd \in A(H \circ K)$. In turn, $c'x, xd' \in A(H \circ K)$ because $\varphi(D^a)$ is externally related. By definition of $\circ$ we get $cx', x'd \in A(H \circ K)$, and then also $x'c, dx' \in A(H \circ K)$ because $\varphi(D^a)$ is externally related. From this, the definition of $\circ$ implies that for any vertex $x''$ of $K^x$ and $y$ of $K^c \cup K^d$ we have $yx''$, $x''y \in A(H \circ K)$. The claim is proved. Now we break into cases.

**Case 1.** Suppose $D$ is disconnected. Then $\pi_H\varphi(D^a)$ is never an arc because then every vertex of $\varphi(D^a)$ in the fiber over the tail of the arc would be adjacent to every vertex in the fiber over the tip, making $\varphi(D^a)$ connected. It follows that $\varphi$ maps components of $D$-layers into components of $K$-layers. Further, $\varphi$ maps each component of a $D$-layer *onto* a component of a $K$-layer: Suppose to the contrary that $C$ is a component of $D^a$ and $\varphi(C)$ is a proper subgraph of a component of $K^{\varphi(a)}$. Take a vertex $x$ of $K^{\varphi(a)} - \varphi(C)$ that is adjacent to or from $\varphi(C)$. Then $\varphi^{-1}(x)$ is adjacent to or from $C \subsetneq D^a$, which is externally related, so $\varphi^{-1}(x)$ is adjacent to or from every vertex of $D^a$. Consequently $x$ is adjacent to or from every vertex of $\varphi(D^a)$. But then any vertex of $y$ of $\varphi(D^a)$ must be contained in $K^{\varphi(a)}$, for otherwise it is adjacent to or from $x \in V(K^{\varphi(a)})$, and hence also to or from $\varphi(C)$, which is impossible. Thus $K^{\varphi(a)}$ contains $\varphi(D^a)$ as well as $x$, contradicting $|K^{\varphi(a)}| = |\varphi(D^a)|$.

Thus each component of a $D$-layer is isomorphic to a component of a $K$-layer, and conversely, as $\varphi$ is bijective. As there are $|G|$ $D$-layers (all isomorphic to $D$), and just as many $K$-layers (isomorphic to $K$), we conclude $D \cong K$. Thus $G \circ D \cong H \circ D$, and Lemma 10.8.4 implies $G \cong H$.

**Case 2.** Suppose $D$ is connected and its complement is connected. If $\varphi$ maps a $D$-layer to a $K$-layer, then $D \cong K$ and Lemma 10.8.4 implies $G \cong H$. Otherwise $\pi_H\varphi(D^a)$ is an arc for every layer $D^a$. Thus we can define a map $f : G \to H$ by declaring $f(x)$ to be the tail of the arc $\pi_H\varphi(\pi_G^{-1}(x))$. We will finish the proof by showing $f$ is an isomorphism. (For then $G \cong H$, and $D \cong K$, by Lemma 10.8.4.) We will show that $f$ is injective; once this is done the isomorphism properties are simple consequences of the definitions. Suppose to the contrary that $f$ is not injective, which means that for some $a \neq b$ we have $\pi_H\varphi(D^a) = wy$ and $\pi_H\varphi(D^b) = wz$. Say the vertex set of $\varphi(D^a)$ is $A_w \cup Ay$ with $\pi_H(A_w) = w$ and $\pi_H(A_y) = y$. Likewise the vertex set of $\varphi(D^b)$ is $B_w \cup B_z$ with $\pi_H(B_w) = w$ and $\pi_H(B_z) = z$. Then there are arcs from each vertex of $\varphi(B_w)$ to each vertex of $\varphi(B_z)$, and then by definition of $\circ$ there are arcs from each vertex of $\varphi(A_w)$ to each vertex of $\varphi(B_z)$. For the same reasons there are arcs from $\varphi(A_w)$ to $\varphi(A_y)$, and thus from $\varphi(B_w)$ to $\varphi(A_y)$. From this we conclude that in $G \circ D$ there are arcs from every vertex of $G^a$ to every vertex of $G^b$, and arcs from every vertex of $G^b$ to every vertex of $G^a$. Hence there are arcs from $\varphi(B_z)$ to $\varphi(A_w)$, so there is an arc from $\varphi(B_z)$ to $\varphi(B_w)$. Thus $\pi_H\varphi(G^b)$ contains two arcs $wz$ and $zw$, contradicting the fact that this projection is a single arc. □

We get a quick corollary concerning lexicographic powers.

**Corollary 10.8.6** *If $G, H \in \mathscr{D}$, then $G^{\circ n} \cong H^{\circ n}$ if and only if $G \cong H$.*

Having cancellation laws for the strong, Cartesian and lexicographic products, we devote the remainder of this section to the direct product. The next result due to Lovász [36] is useful in this context.

**Proposition 10.8.7** *Let $D, C, G$ and $H$ be digraphs in $\mathscr{D}_0$. If $C \times G \cong C \times H$ and there is a homomorphism $D \to C$, then $D \times G \cong D \times H$.*

**Proof:** As $C \times G \cong C \times H$, Proposition 10.7.3 says $\hom(X, C) \cdot \hom(X, G) = \hom(X, C) \cdot \hom(X, H)$ for any $X$. The homomorphism $D \to C$ guarantees $\hom(X, D) = 0$ whenever $\hom(X, C) = 0$. Thus $\hom(X, D) \cdot \hom(X, G) = \hom(X, D) \cdot \hom(X, H)$, so $\hom(X, D \times G) = \hom(X, D \times H)$ by Proposition 10.7.3, and then Proposition 10.7.2 gives $D \times G \cong D \times H$. □

Now observe that cancellation can fail over the direct product. Figure 10.7 shows digraphs $D, G, H \in \mathscr{D}_0$ for which $D \times G \cong 3\overrightarrow{C_3} \cong D \times H$, but $G \ncong H$. Cancellation can also fail in the class of loopless digraphs. For example, note that for graphs we have $K_2 \times 2C_3 = 2C_6 = K_2 \times C_6$, so $\overleftrightarrow{K_2} \times 2\overleftrightarrow{C_3} \cong \overleftrightarrow{K_2} \times \overleftrightarrow{C_6}$.

A digraph $D$ is called a **zero divisor** if there are digraphs $G \ncong H$ for which $D \times H \cong D \times G$. For example, Figure 10.7 shows that $D = \overrightarrow{C_3}$ is a zero divisor, and the equation above shows $\overleftrightarrow{K_2}$ is a zero divisor. The following characterization of zero divisors is due to Lovász [36].

**Theorem 10.8.8** *A digraph $D$ is a zero divisor if and only if there exists a homomorphism $D \to \overrightarrow{C}_{p_1} + \overrightarrow{C}_{p_2} + \cdots + \overrightarrow{C}_{p_k}$ into a disjoint union of directed cycles of distinct prime lengths $p_1, p_2, \ldots, p_k$.*

**Proof:** We will prove only one (the easier) direction. See [36] for the other.

Suppose there is a homomorphism $D \to C = \overrightarrow{C}_{p_1} + \overrightarrow{C}_{p_2} + \cdots + \overrightarrow{C}_{p_k}$, where the $p_i$ are distinct primes. Our plan is to produce non-isomorphic digraphs $G$ and $H$ for which $C \times G = C \times H$, for then Proposition 10.8.7 will insure $D \times G \cong D \times H$, showing $D$ is a zero divisor.

Put $n = p_1 p_2 \cdots p_k$. Let $\mathscr{G}$ be the set of positive divisors of $n$ that are products of an even number of the $p_i$'s, whereas $\mathscr{H}$ is the set of divisors that are products of an odd number of the $p_i$'s. Let $G$ and $H$ be the disjoint unions

$$G = \sum_{d \in \mathscr{G}} d\,\overrightarrow{C}_{\frac{n}{d}} \qquad \text{and} \qquad H = \sum_{d \in \mathscr{H}} d\,\overrightarrow{C}_{\frac{n}{d}}.$$

Clearly $G \ncong H$. As the direct product distributes over disjoint unions, $C \times G = C \times H$ will follow provided $\overrightarrow{C}_{p_i} \times G = \overrightarrow{C}_{p_i} \times H$ for each $p_i$. We establish this with the aid of Equation (10.4), as follows:

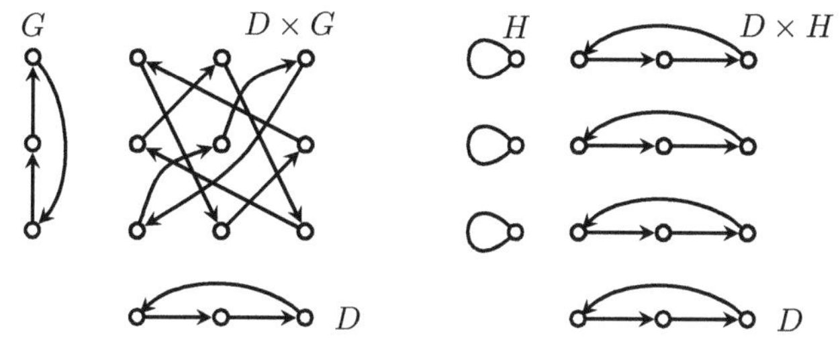


**Figure 10.7** Failure of cancellation over the direct product.

$$\begin{aligned}
\vec{C}_{p_i}\times G=\sum_{d\in\mathscr{G}}\vec{C}_{p_i}\times d\vec{C}_{\frac{n}{d}} &= \sum_{\substack{d\in\mathscr{G}\\ p_i|d}}\vec{C}_{p_i}\times d\vec{C}_{\frac{n}{d}} + \sum_{\substack{d\in\mathscr{G}\\ p_i\nmid d}}\vec{C}_{p_i}\times d\vec{C}_{\frac{n}{d}}\\
&= \sum_{\substack{d\in\mathscr{G}\\ p_i|d}} d\vec{C}_{\frac{p_i n}{d}} + \sum_{\substack{d\in\mathscr{G}\\ p_i\nmid d}} p_i d\vec{C}_{\frac{p_i n}{p_i d}}\\
&= \sum_{\substack{d\in\mathscr{H}\\ p_i\nmid d}} dp_i\vec{C}_{\frac{p_i n}{d}} + \sum_{\substack{d\in\mathscr{H}\\ p_i\mid d}} d\vec{C}_{\frac{p_i n}{d}}\\
&= \sum_{\substack{d\in\mathscr{H}\\ p_i\nmid d}}\vec{C}_{p_i}\times d\vec{C}_{\frac{n}{d}} + \sum_{\substack{d\in\mathscr{H}\\ p_i\mid d}}\vec{C}_{p_i}\times d\vec{C}_{\frac{n}{d}}\\
&= \sum_{d\in\mathscr{H}}\vec{C}_{p_i}\times d\vec{C}_{\frac{n}{d}} = \vec{C}_{p_i}\times H.
\end{aligned}$$

From this, $C\times G=C\times H$, and hence $D\times G=D\times H$, as noted above. □

For example, $\vec{C}_n$ is a zero divisor when $n>1$, as there is a homomorphism $\vec{C}_n\to\vec{C}_p$ for any prime divisor $p$ of $n$. Also, each $\vec{P}_n$ is a zero divisor, as there are homomorphisms $\vec{P}_n\to\vec{C}_p$.

Paraphrasing Theorem 10.8.8, if there are no homomorphisms from $D$ into a union of directed cycles, then $D\times G\cong D\times H$ necessarily implies $G\cong H$. But if there *is* such a homomorphism then $D$ is a zero divisor and there exist non-isomorphic digraphs $G$ and $H$ for which $D\times G\cong D\times H$, as constructed in the proof of Theorem 10.8.8.

Given a digraph $G$ and a zero divisor $D$, a natural problem is to determine all digraphs $H$ for which $G\times D\cong H\times D$. If there is only one such $H$, then necessarily $H\cong G$, and cancellation holds. Thus it is meaningful to ask if there are conditions on $G$ and $D$ that force cancellation to hold, even if $D$ is a zero divisor. For example, if $G=K_1^*$, then $G\times D\cong H\times D$ implies $G\cong H$, regardless of whether $D$ is a zero divisor. What other graphs have this property? We now turn our attention to this type of question, adopting the approach of [15, 19, 20].

For a digraph $G$, let $S_{V(G)}$ denote the symmetric group on $V(G)$, that is, the set of bijections from $V(G)$ to itself. For $\sigma \in S_{V(G)}$, define the **permuted digraph** $G^\sigma$ to be $V(G^\sigma) = V(G)$ and $A(G^\sigma) = \{x\sigma(y) \mid xy \in A(G)\}$. Thus $xy \in A(G)$ if and only if $x\sigma(y) \in A(G^\sigma)$, and $xy \in A(G^\sigma)$ if and only if $x\sigma^{-1}(y) \in E(G)$. Figure 10.8 shows several examples. The upper part shows a digraph $G$ and two of its permuted digraphs. In the lower part, the cyclic permutation (0245) of the vertices of $\overrightarrow{C_6}$ yields a permuted digraph $\overrightarrow{C_6}^{(0245)} = 2\overrightarrow{C_3}$. The permuted digraph $\overrightarrow{C_6}^{(01)}$ is also shown. For another example, note that $G^{\text{id}} = G$ for any digraph $G$. It may be possible that $G^\sigma \cong G$ for some non-identity permutation $\sigma$. For instance, $\overrightarrow{C_6}^{(024)} \cong \overrightarrow{C_6}$.

The significance of permuted digraphs is given by the next proposition. asserting that $D \times G \cong D \times H$ implies that $H$ is a permuted digraph of $G$.

**Proposition 10.8.9** *Let $G, H$ and $D$ be digraphs, where $D$ has at least one arc. If $D \times G \cong D \times H$, then $H \cong G^\sigma$ for some permutation $\sigma \in S_{V(G)}$. As a partial converse, $D \times G \cong D \times G^\sigma$ for all $\sigma \in S_{V(A)}$, provided there is a homomorphism $D \to \overrightarrow{P}_2$.*

**Proof:** Suppose $D \times G \cong D \times H$, and $D$ has at least one arc. Then there is a homomorphism $\overrightarrow{P}_2 \to D$, and Proposition 10.8.7 yields an isomorphism $\varphi : \overrightarrow{P_2} \times G \to \overrightarrow{P_2} \times H$. We may assume $\varphi$ has the form $(\varepsilon, x) \mapsto (\varepsilon, \varphi_\varepsilon(x))$, where $\varepsilon \in \{0,1\} = V(\overrightarrow{P_2})$, and each $\varphi_\epsilon$ is a bijection $V(G) \to V(H)$. (That a $\varphi$ of such form exists is a consequence of Theorem 3 of [36]. However, it is also easily verified in the present setting, when the common factor is $\overrightarrow{P}_2$.) Hence $\varphi_0^{-1}\varphi_1 : V(G) \to V(G)$ is a permutation of $V(G)$. We now show that the map $\varphi_0 : G^{\varphi_0^{-1}\varphi_1} \to H$ is an isomorphism. Simply observe that

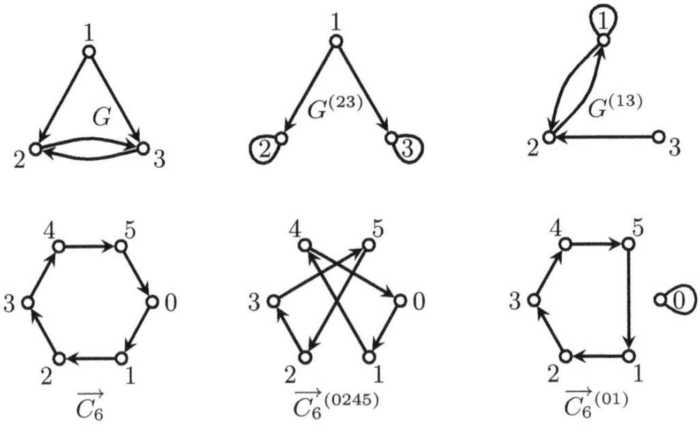


**Figure 10.8** Upper: A digraph $G$ and permuted digraphs $G^\sigma$ for transpositions $\sigma = (23)$ and $\sigma = (13)$. Lower: two permutations of a directed cycle.

$$\begin{aligned} xy \in A(G^{\varphi_0^{-1}\varphi_1}) &\iff x\,(\varphi_0^{-1}\varphi_1)^{-1}(y) \in A(G) \\ &\iff x\,\varphi_1^{-1}\varphi_0(y) \in A(G) \\ &\iff (0,x)\,(1,\varphi_1^{-1}\varphi_0(y)) \in A(\overrightarrow{P_2} \times G) \\ &\iff (0,\varphi_0(x))\,(1,\varphi_1\varphi_1^{-1}\varphi_0(y)) \in A(\overrightarrow{P_2} \times H) \quad (\text{apply}\varphi) \\ &\iff (0,\varphi_0(x))\,(1,\varphi_0(y)) \in A(\overrightarrow{P_2} \times H) \\ &\iff \varphi_0(x)\,\varphi_0(y) \in A(H). \end{aligned}$$

Conversely, let $\sigma \in S_{V(G)}$. Note that the map $\varphi$ defined as $\varphi(0,x) = (0,x)$ and $\varphi(1,x) = (1,\sigma(x))$ is an isomorphism $\overrightarrow{P}_2 \times G \to \overrightarrow{P}_2 \times G^\sigma$ because $(0,x)(1,y) \in A(\overrightarrow{P}_2 \times G)$ if and only if $(0,x)(1,\sigma(y)) \in A(\overrightarrow{P}_2 \times G^\sigma)$. If there is a homomorphism $D \to \overrightarrow{P}_2$, Proposition 10.8.7 gives $D \times G \cong D \times G^\sigma$. □

In general, the full converse of Proposition 10.8.9 is (as we shall see) false. If there is no homomorphism $D \to \overrightarrow{P}_2$, then not every $\sigma$ will yield a digraph $H = G^\sigma$ for which $D \times G \cong D \times H$. In addition, it is possible that $\sigma \neq \tau$ but $G^\sigma \cong G^\tau$. Towards clarifying these issues, we next introduce a group action on $S_{V(G)}$ whose orbits correspond to isomorphism classes of permuted digraphs.

The **factorial** of a digraph $G$ is a digraph $G!$, defined as $V(G!) = S_{V(G)}$, and $\alpha\beta \in A(G!)$ provided that $xy \in A(G) \iff \alpha(x)\beta(y) \in A(G)$ for all pairs $x, y \in V(G)$. To avoid confusion with composition, we will denote arcs $\alpha\beta$ of $G!$ as $[\alpha,\beta]$. Note that $A(G!)$ has a group structure as a subgroup of $S_{V(G)} \times S_{V(G)}$, that is, we can multiply arcs as $[\alpha,\beta][\gamma,\delta] = [\alpha\gamma,\beta\delta]$.

Observe that the definition implies that there is a loop $[\alpha,\alpha]$ at $\alpha \in V(G!)$ if and only if $\alpha$ is an automorphism of $G$. In particular, any $G!$ has a loop at the identity id.

Our first example explains the origins of our term "factorial." Let $K_n^*$ be the complete symmetric digraph with a loop at each vertex, and note that

$$K_n^*! \cong K_{n!}^* \cong K_n^* \times K_{n-1}^* \times K_{n-2}^* \times \cdots \times K_3^* \times K_2^* \times K_1^*.$$

For less obvious computations, it is helpful to keep in mind the following interpretation of $A(G!)$. Any arc $[\alpha,\beta] \in A(G!)$ is a permutation of the arcs of $G$, where $[\alpha,\beta](xy) = \alpha(x)\beta(y)$. This permutation preserves in-incidences and out-incidences in the following sense: Given two arcs $xy, xz$ of $G$ that have a common tail, $[\alpha,\beta]$ carries them to the two arcs $\alpha(x)\beta(y)$, $\alpha(x)\beta(z)$ of $G$ with a common tail. Given two arcs $xy, zy$ with a common tip, $[\alpha,\beta]$ carries them to the two arcs $\alpha(x)\beta(y)$, $\alpha(z)\beta(y)$ of $G$ with a common tip.

Bear in mind, however, that even if the head of $xy$ meets the tail of $yz$, then the arcs $[\alpha,\beta](xy)$ and $[\alpha,\beta](yz)$ need not meet; they can be quite far apart in $G$. To illustrate these ideas, Figure 10.9 shows the effect of a typical $[\alpha,\beta]$ on the arcs incident with a typical vertex $z$ of $G$.

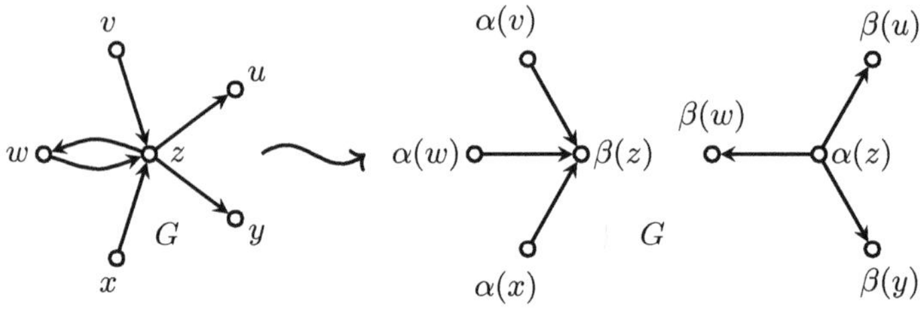


**Figure 10.9** Action of an arc $[\alpha, \beta]$ of $G!$ on the neighborhood of a vertex $z \in V(G)$.

Let's use these ideas to compute the factorial of the transitive tournament $TT_n$, which has distinct out- and in-degrees $0, 1, \ldots, n-1$. The above discussion implies if $[\alpha, \beta] \in A(TT_n!)$, the out-degree of any $x \in V(TT_n)$ equals the out-degree of $\alpha(x)$. Hence $\alpha = \text{id}$. The same argument involving in-degrees gives $\beta = \text{id}$. Therefore $TT_n!$ has $n!$ vertices but only one arc $[\text{id}, \text{id}]$. Figure 10.10 shows $T_3!$, plus two other examples.

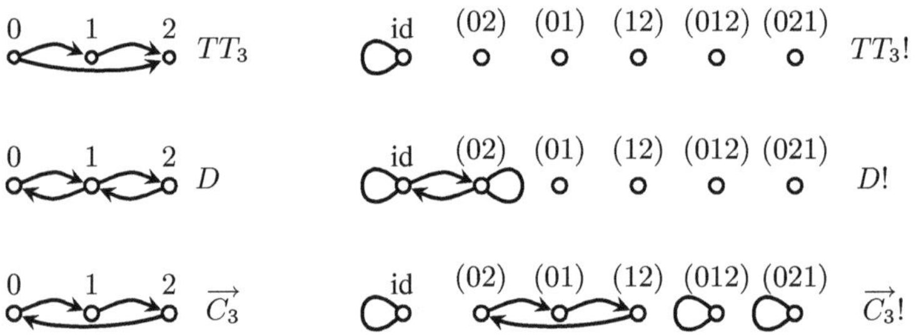


**Figure 10.10** Some digraphs (left) and their factorials (right).

The group $A(G!)$ acts on $S_{V(G)}$ as $[\alpha, \beta] \cdot \sigma = \alpha\sigma\beta^{-1}$, and this determines the situation in which $G^\sigma = G^\tau$.

**Proposition 10.8.10** *If $\sigma, \tau$ are permutations of the vertices of a digraph $G$, then $G^\sigma = G^\tau$ if and only if $\sigma$ and $\tau$ are in the same $A(G!)$-orbit.*

**Proof:** If there is an isomorphism $\varphi : G^\sigma \to G^\tau$, then for any $x, y \in V(G)$,

$$xy \in A(G) \iff x\sigma(y) \in A(G^\sigma) \iff \varphi(x)\varphi\sigma(y) \in A(G^\tau)$$
$$\iff \varphi(x)\tau^{-1}\varphi\sigma(y) \in A(G).$$

This means $[\varphi, \tau^{-1}\varphi\sigma] \in A(G!)$. Then $[\varphi, \tau^{-1}\varphi\sigma] \cdot \sigma = \tau$, so $\sigma$ and $\tau$ are indeed in the same orbit.

Conversely, suppose $\sigma$ and $\tau$ are in the same orbit. Take $[\alpha, \beta] \in A(G!)$ with $\tau = [\alpha, \beta] \cdot \sigma = \alpha\sigma\beta^{-1}$. Then $\alpha : G^\sigma \to G^{\alpha\sigma\beta^{-1}} = G^\tau$ is an isomorphism:

$$\begin{aligned} xy \in A(G^{\sigma}) \iff x\sigma^{-1}(y) \in A(G) &\iff \alpha(x)\beta\sigma^{-1}(y) \in A(G) \\ &\iff \alpha(x)\alpha\sigma\beta^{-1}\beta\sigma^{-1}(y) \in A(G^{\alpha\sigma\beta^{-1}}) \\ &\iff \alpha(x)\alpha(y) \in A(G^{\alpha\sigma\beta^{-1}}). \qquad \square \end{aligned}$$

Given an arc $[\alpha, \beta] \in A(G!)$, we have $[\alpha, \beta] \cdot \beta = \alpha$. The previous proposition then assures $G^{\alpha} \cong G^{\beta}$, and therefore yields the following corollary.

**Corollary 10.8.11** *If two permutations $\sigma, \tau$ are in the same component of $G!$, then $G^{\sigma} \cong G^{\tau}$.*

For a given digraph $G$, the next theorem and corollary characterize the complete set of digraphs $H$ for which $D \times G \cong D \times H$, provided $D$ is a zero divisor that admits a homomorphism into a directed path. Space limitations prohibit inclusion of a proof of the theorem, as well as inclusion of the characterization for general zero divisors $D$. For a full treatment, see Hammack [15].

**Theorem 10.8.12** *Suppose $G$ and $H$ are digraphs, and $D$ is a zero divisor that admits a homomorphism $D \to \overrightarrow{P}_n$. Assume $n \geq 2$ is the smallest such integer. Then $D \times G \cong D \times H$ if and only if $H \cong G^{\sigma}$, where $\sigma$ is a vertex of a diwalk of length $n-2$ in $G!$.*

Given a digraph $G$ and a zero divisor $D$ that admits a homomorphism $D \to \overrightarrow{P}_n$, Theorem 10.8.12 describes a complete collection of digraphs $H$ for which $D \times G \cong D \times H$. Of course it is possible that some (possibly all) of these $H$ are isomorphic. We next describe a means of constructing the exact set of isomorphism classes of such $H$. Combining the previous theorem with Proposition 10.8.10 yields the following.

**Corollary 10.8.13** *Suppose $G$ and $D$ are digraphs, and $D$ is a zero divisor that admits a homomorphism $D \to \overrightarrow{P}_n$. Assume $n \geq 2$ is the smallest such integer. Then the set of distinct (up to isomorphism) digraphs $H$ for which $D \times G \cong D \times H$ can be obtained as follows: Let $\Upsilon_{n-2}$ denote the set of vertices of $G!$ that lie on a directed walk of length $n-2$. Select a maximal set of elements $\sigma_1, \sigma_2, \ldots, \sigma_k \in \Upsilon_{n-1}$ that are in distinct orbits of the $A(G!)$-action on $S_{V(G)}$. Then the digraphs $H$ for which $D \times G \cong D \times H$ are precisely $H \cong G^{\sigma_1}, G^{\sigma_2}, \ldots, G^{\sigma_k}$.*

*Cancellation holds ($D \times G \cong D \times H$ implies $G \cong H$) if and only if $k = 1$.*

According to Theorem 10.8.12, if $D$ admits a homomorphism into $\overrightarrow{P}_2$, then $D \times G \cong D \times H$ if and only if $H \cong G^{\sigma}$, where $\sigma$ is a vertex of $G!$ on a diwalk of length 0. In this case there are no restrictions whatsoever on $\sigma$; it can be any permutation of $V(G)$. Consequently, there can be potentially $|V(G)|!$ different $H \cong G^{\sigma}$.

We close with an application of these results that illustrates an extreme failure of cancellation involving the transitive tournament $TT_n$. We remarked earlier that $TT_n!$ has $n!$ vertices and a single arc [id, id]. Therefore each $A(TT_n!)$-orbit of $S_{V(TT_n)}$ consists of a single permutation. Also $\Upsilon_0 = S_{V(TT_n)}$. Thus, if $D$ is a zero divisor that admits a homomorphism to $\overrightarrow{P}_2$, then there are exactly $n!$ distinct digraphs $TT_n^\sigma$ for which $D \times TT_n \cong D \times TT_n^\sigma$. By Proposition 10.8.9, this is the maximum number possible.

But notice that if we merely replace $D$ with a zero divisor that admits a homomorphism to $\overrightarrow{P}_n$, with $n > 2$, then $\Upsilon_{n-2} = \{\text{id}\}$ and cancellation holds!

## 10.9 Prime Factorization

We mentioned in Section 10.1 that the trivial digraph $K_1$ is a unit for $\Box$, $\boxtimes$ and $\circ$ in the sense that $K_1 \Box D = D$, $K_1 \boxtimes D = D$ and $K_1 \circ D = D$ for any digraph $D$. If $* \in \{\Box, \boxtimes, \circ\}$, we say a digraph $D$ is **prime** over $*$ if $D$ is non-trivial, and for any factoring $D = D_1 * D_2$, one factor $D_i$ is isomorphic to $D$ and the other is $K_1$.

Certainly any non-trivial digraph $D$ has a factoring $D = D_1 * D_2 * \cdots * D_n$, where each $D_i$ is prime (possibly $n = 1$). We call any such factoring a **prime factoring** over $*$. (Note that $n \leq \log_2 |V(D)|$ because a product $D_i * D_j$ always has at least twice as many vertices as either of its factors.)

It is natural to ask whether any prime factoring of a given digraph $D$ is unique up to order and isomorphism of the factors. In general this is false. For $\Box$ and $\boxtimes$, the standard counterexamples arise from the equation

$$(1 + x + x^2)(1 + x^3) = (1 + x^2 + x^4)(1 + x), \tag{10.10}$$

giving two distinct prime factorings of the polynomial $1+x+x^2+x^3+x^4+x^5$ in the semiring $\mathbb{Z}^+[x]$. Let $\overleftrightarrow{Q}_n$ be the complete biorientation of the $n$-cube $Q_n$. For typographical efficiency, let us denote $\overleftrightarrow{Q}_n$ simply as $Q_n$. Then $Q_n = \overleftrightarrow{K}_2^{\Box n}$ (the $n$th Cartesian power of $\overleftrightarrow{K}_2$), and $Q_0 = K_1$. Substituting $Q_1$ for $x$ in Equation (10.10) yields two factorings

$$(Q_0 + Q_1 + Q_2) \Box (Q_0 + Q_3) = (Q_0 + Q_2 + Q_4) \Box (Q_0 + Q_1),$$

of the digraph $Q_0 + Q_1 + Q_2 + Q_3 + Q_4 + Q_5$. It is routine to check that the above factors are prime.

The same idea applies to the strong product. Denote the complete biorientation $\overleftrightarrow{K}_n$ of $K_n$ simply as $K_n$ (a convention we will adhere to for the rest of this section). Note that $K_m \boxtimes K_n = K_{mn}$. Then, as above,

$$(K_1 + K_2 + K_4) \boxtimes (K_1 + K_8) = (K_1 + K_4 + K_{16}) \boxtimes (K_1 + K_2)$$

are two distinct prime factorings of $K_1 + K_2 + K_4 + K_8 + K_{16} + K_{32}$.

Despite these failures of unique prime factorization, *connected* digraphs do factor uniquely over the Cartesian and strong products. For the Cartesian product, this was first proved by Feigenbaum [9], who also gives a polynomial algorithm for finding the prime factors. (More recently, Crespelle and Thierry [6] give a linear algorithm.) Our approach adapts that of Imrich, Klavžar and Rall [25]. Their proof is for graphs; we adapt it here to digraphs.

Convexity is the central ingredient of the proof. A subdigraph $H$ of $D$ is **convex** if any shortest path (not necessarily directed) in $D$ that joins two vertices of $H$ is itself a path in $H$. (There are other notions of convexity. For example, it could be phrased in terms of directed paths; however the one given here is best suited for our present purposes.) The next lemma makes use of $\mathrm{dist}'_D(x, y)$, the length of a shortest $(x, y)$-path in $D$. (See Proposition 10.2.1.)

**Lemma 10.9.1** *A subdigraph $H$ of $D = D_1 \,\Box\, \cdots \,\Box\, D_k$ is convex if and only if $H = H_1 \,\Box\, \cdots \,\Box\, H_k$, where each $H_i$ is a convex subdigraph of $D_i$.*

**Proof:** Suppose $H = H_1 \,\Box\, \cdots \,\Box\, H_k$, with each $H_i$ a convex subdigraph of $D_i$. We claim that any shortest path $P$ joining two vertices $a = (a_1, \dots, a_k)$ and $b = (b_1, \dots, b_k)$ in $H$ lies entirely in $H$. By Proposition 10.2.1, the length of $P$ is the sum of the lengths of the shortest $(a_i, b_i)$-paths $P_i$ in $D_i$ for $i \in [k]$. Because each arc of $P$ projects to an arc in only one factor (and to single vertices in all the others) it follows that each projection $\pi_i(P)$ is a shortest $(a_i, b_i)$ path in $D_i$, and therefore lies entirely in $H_i$, by convexity. Thus $P$ lies entirely in $H = H_1 \,\Box\, \cdots \,\Box\, H_k$, so $H$ is convex.

Conversely, suppose $H$ is convex in $D$. Note $H \subseteq \pi_1(H) \,\Box\, \cdots \,\Box\, \pi_k(H)$. We complete the proof by showing that the inclusion is equality, and each $\pi_i(H)$ is convex in $D_i$.

To see that $\pi_i(H)$ is convex in $D_i$, take vertices $a_i, b_i \in \pi_i(H)$. Let $x_i$ be on a shortest $(a_i, b_i)$-path in $D_i$. We must show $x_i \in \pi_i(H)$. Choose vertices $a = (a_1, \dots, a_k)$ and $b = (b_1, \dots, b_k)$ of $H$ with $\pi_i(a) = a_i$ and $\pi_i(b) = b_i$. Define $x = (x_1, \dots, x_k)$ as follows. For each index $j \neq i$, let $x_j$ be on a shortest $(a_j, b_j)$-path in $D_j$. Thus $\mathrm{dist}'_{D_s}(a_s, b_s) = \mathrm{dist}'_{D_s}(a_s, x_s) + \mathrm{dist}'_{D_s}(x_s, b_s)$ for each $s \in [k]$, and Proposition 10.2.1 gives $\mathrm{dist}'_D(a, b) = \mathrm{dist}'_D(a, x) + \mathrm{dist}'_D(x, b)$. It follows that $x$ is on a shortest $(a, b)$-path in $D$, so $x \in H$ by convexity of $H$. Hence $x_i = \pi_i(x) \in \pi_i(H)$.

Finally, we prove $H \subseteq \pi_1(H) \,\Box\, \cdots \,\Box\, \pi_k(H)$ is equality. Since both sides are connected, it sufficies to show that any vertex $v$ of $\pi_1(H) \,\Box\, \cdots \,\Box\, \pi_k(H)$ at distance 1 from a vertex $x \in V(H)$ is also in $H$. Let $v = (v_1, \dots, v_i, \dots, v_k)$ and $x = (v_1, \dots, v_{i-1}, x_i, v_{i+1}, \dots, v_k)$ be such vertices. As $v$ is in the product of the projections of $H$, there is a $u = (x_1, \dots, x_{i-1}, v_i, x_{i+1}, \dots x_k) \in V(H)$. Proposition 10.2.1 says $\mathrm{dist}'(x, v) + \mathrm{dist}'(v, u) = \mathrm{dist}'(x, u)$, meaning $v$ is on a shortest path joining $x, u \in V(H)$, so $v \in V(H)$ by convexity of $H$. □

Given a vertex $a = (a_1, \dots, a_k)$ of $D_1 \,\Box\, \cdots \,\Box\, D_k$, and an $i \in [k]$, we define $D_i^a$ to be the subgraph of the product induced on the vertices

$(a_1, \ldots, a_{i-1}, x, a_{i+1}, \ldots, a_k)$, where $x \in V(D_i)$. That is,

$$D_i^a = a_1 \,\Box\, \cdots \,\Box\, a_{i-1} \,\Box\, D_i \,\Box\, a_{i+1} \,\Box\, \cdots \,\Box\, a_k.$$

Thus $D_i^a \cong D_i$, and it is a convex subdigraph of the product, by Lemma 10.9.1. We call $D_i^a$ the **$D_i$-layer through** $a$. We are ready for our main results on prime factorization of digraphs over the Cartesian product.

**Theorem 10.9.2** *Connected digraphs factor uniquely into primes over the Cartesian product, up to order and isomorphism of the factors. Specifically, if a digraph $D$ factors into primes as*

$$D = D_1 \,\Box\, \cdots \,\Box\, D_k \quad \textit{and} \quad D = G_1 \,\Box\, \cdots \,\Box\, G_\ell,$$

*then $k = \ell$, and $D_i \cong G_{\sigma(i)}$ for some permutation $\sigma$ of $[k]$.*

**Proof:** As remarked earlier, $D$ has a prime factorization $D = D_1 \,\Box\, \cdots \,\Box\, D_k$. Now suppose $D$ has two prime factorings $D = D_1 \,\Box\, \cdots \,\Box\, D_k$ and $D = G_1 \,\Box\, \cdots \,\Box\, G_\ell$. We may assume $k \geq \ell$. Take an isomorphism

$$\varphi : D_1 \,\Box\, \cdots \,\Box\, D_k \to G_1 \,\Box\, \cdots \,\Box\, G_\ell.$$

Fix $a = (a_1, \ldots, a_k)$, and say $\varphi(a) = b = (b_1, \ldots, b_\ell)$. It suffices to show $k = \ell$, and there is a permutation $\sigma$ of $[k]$ for which $\varphi(D_i^a) = G_{\sigma(i)}^b$ for $1 \leq i \leq k$. (Recall $D_i^a \cong D_i$ and $G_{\sigma(i)}^b \cong G_{\sigma(a)}$.)

To this end, fix $i \in [k]$. As mentioned above, any $D_i^a$ is convex in $D_1 \,\Box\, \cdots \,\Box\, D_\ell$, so $\varphi(D_i^a)$ is convex in $G_1 \,\Box\, \cdots \,\Box\, G_\ell$. Using Lemma 10.9.1,

$$(b_1, \ldots, b_\ell) \in \varphi(D_i^a) = H_1 \,\Box\, \cdots \,\Box\, H_\ell,$$

where each $H_j$ is a convex subgraph of $G_j$. But $D_i \cong D_i^a \cong \varphi(D_i^a)$ is prime, so $H_i = \{b_i\}$ for all but one index, call it $\sigma(i)$. This means $\varphi(D_i^a) \subseteq G_{\sigma(i)}^b$. But then $D_i^a \subseteq \varphi^{-1}(G_{\sigma(i)}^b)$. Now, $G_{\sigma(i)}^b$ is prime, and convex in $G_1 \,\Box\, \cdots \,\Box\, G_\ell$, so also $\varphi^{-1}(G_{\sigma(i)}^b)$ is prime, and convex in $D_1 \,\Box\, \cdots \,\Box\, D_k$. Lemma 10.9.1 gives

$$D_i^a \subseteq \varphi^{-1}(G_{\sigma(i)}^b) = H_1' \,\Box\, \cdots \,\Box\, H_k', \tag{10.11}$$

where each $H_j'$ is a subdigraph of $D_i$, containing $a_j$. Primeness assures all but one $H_j'$ is trivial, and necessarily it is $H_i$ that is nontrivial. Therefore (10.11) implies $D_i^a \subseteq \varphi^{-1}(G_{\sigma(i)}^b) \subseteq D_i^a$, whence $\varphi(D_i^a) = G_{\sigma(i)}^b$.

We claim that the map $\sigma : [k] \to [\ell]$ is injective. If $\sigma(i) = \sigma(j)$, then

$$\varphi(D_i^a) = G_{\sigma(i)}^b = \varphi(D_j^a).$$

Because $G_{\sigma(i)}^b$ is nontrivial (it is prime), it follows that $D_i^a$ and $D_j^a$ have a nontrivial intersection. This means $i = j$, so $\sigma$ is injective. Thus $k \leq \ell$. We have assumed $k \geq \ell$, so $k = \ell$, so $\sigma$ is a permutation. □

Theorem 10.9.2 implies our next result, which describes the structure of isomorphisms between digraphs. The proof uses the notation from the proof of Theorem 10.9.2.

**Theorem 10.9.3** *Let $D$ and $G$ be isomorphic connected digraphs with prime factorizations $D = D_1 \Box \cdots \Box D_k$ and $G = G_1 \Box \cdots \Box G_k$. Then for any isomorphism $\varphi : D \to G$, there is a permutation $\sigma$ of $[k]$ and isomorphisms $\varphi_i : D_{\sigma(i)} \to G_i$ for which*

$$\varphi(x_1, x_2, \ldots, x_k) = \big(\varphi_1(x_{\sigma(1)}), \varphi_2(x_{\sigma(2)}), \ldots, \varphi_k(x_{\pi(k)})\big). \tag{10.12}$$

**Proof:** By Theorem 10.9.2, there is a permutation $\sigma$ of $[k]$ for which $\varphi$ restricts to an isomorphism $D_i^a \to G_{\sigma(i)}^b$ for each index $i$. Replacing $\sigma$ with $\sigma^{-1}$, we can say that, for each $i$, $\varphi$ restricts to an isomorphism $D_{\sigma(i)}^a \to H_i^b$.

To finish the proof, we show that $\pi_i\varphi(x_1, \ldots, x_k)$ depends only on $x_{\sigma(i)}$. Then we can put $\varphi_i(x_{\sigma(i)}) = \pi_i\varphi(x_1, \ldots, x_k)$, which yields Equation (10.12), and it is immediate that the $\varphi_i$ are isomorphisms.

For any $x_{\sigma(i)} \in V(D_{\sigma(i)})$, define the "hyperplane" subdigraph

$$B[x_{\sigma(i)}] := D_1 \Box D_2 \Box \cdots \Box x_{\sigma(i)} \Box \cdots \Box D_k \subseteq D_1 \Box \cdots \Box D_k,$$

whose $\sigma(i)th$ factor is the single vertex $x_{\sigma(i)}$. This subdigraph is convex, so Lemma 10.9.1 says $\varphi(B[x_{\sigma(i)}]) = U_1 \Box \cdots \Box U_k$, with each $U_j$ convex in $G_j$.

Now, $B[x_{\sigma(i)}] \cap D_{\sigma(i)}^a = \{(a_1, a_2, \ldots, x_{\sigma(i)}, \ldots, a_k)\}$. Thus $\varphi(B[x_{\pi(i)}]) = U_1 \Box \cdots \Box U_k$ meets $\varphi(D_{\sigma(i)}^a) = G_i^b = b_1 \Box \cdots \Box G_i \Box \cdots \Box b_k$ at the single vertex $\varphi(a_1, a_2, \ldots, x_{\sigma(i)}, \ldots, a_k)$. This means all vertices in $\varphi(B[x_{\sigma(i)}])$ have the same $i$th coordinate $\pi_i\varphi(a_1, a_2, \ldots, x_{\sigma(i)}, \ldots, a_k)$, so

$$\pi_i\big(\varphi(B[x_{\sigma(i)}])\big) = \pi_i\varphi(a_1, a_2, \ldots, x_{\sigma(i)}, \ldots, a_k).$$

Now, any $(x_1, \ldots, x_{\sigma(i)}, \ldots, x_k) \in V(G)$ belongs to $B[x_{\sigma(i)}]$. Consequently $\pi_i\varphi(x_1, \ldots, x_{\sigma(i)}, \ldots, x_k) = \pi_i\varphi(a_1, \ldots, x_{\sigma(i)}, \ldots, a_k)$, which depends only on $x_{\sigma(i)}$. □

In Theorem 10.9.3, we may relabel each vertex $x$ of $G_i$ with its preimage under the isomorphism $D_i \to G_{\sigma(i)}$ to make this isomorphism an identity map. We record this observation as a useful corollary.

**Corollary 10.9.4** *For an isomorphism $\varphi : D_1 \Box \cdots \Box D_k \to G_1 \Box \cdots \Box G_k$ where each $D_i$ and $G_i$ is prime, the vertices of the $G_i$ can be relabeled so that $\varphi(x_1, x_2, \ldots, x_k) = (x_{\sigma(1)}, x_{\sigma(2)}, \ldots, x_{\sigma(k)})$ for some permutation $\sigma$ of $[k]$.*

We turn now to the lexicographic product. It is not commutative, so we should not expect a prime factorization to be unique up to order of the factors. Indeed this is not so, but there is a fascinating relationship between different prime factorings. Explaining it requires the idea of the **join** $D \oplus G$ of two digraphs with disjoint vertex sets, which is the digraph obtained from $D + G$ by adding arcs from each vertex of $D$ to every vertex of $G$, and from each vertex of $G$ to every vertex of $D$.

Recall that the right-distributive law holds for $\circ$, but there is no general left-distributive law. However, if $K_n$ is the biorientation of the complete graph on $n$ vertices, and $D_n$ (the arcless digraph) is its complement, we do have

$$D_n \circ (G + H) = D_n \circ G + D_n \circ H,$$
$$K_n \circ (G \oplus H) = K_n \circ G \oplus K_n \circ H.$$

The first equation follows from Proposition 10.1.1. The second follows from the first, with the observation that $G \oplus H = \overline{\overline{G} + \overline{H}}$ and $\overline{D \circ D'} = \overline{D} \circ \overline{D'}$ (Equation (10.3)), where the bar denotes the complement.

We see now that unique prime factorization over the lexicographic product can fail in at least two ways: If $q$ is prime and if $D_q \circ G + D_m$ is prime, then

$$(D_q \circ G + D_m) \circ D_q = D_q \circ (G \circ D_q + D_m)$$

are two different prime factorizations of the same graph. We say they are related by a *transposition of a totally disconnected graph.* Analogously, if $K_q \circ G \oplus K_m$ is prime, then

$$(K_q \circ G \oplus K_m) \circ K_q = K_q \circ (G \circ K_q \oplus K_m)$$

are two different prime factorizations of the same graph, and we say they are related by a *transposition of a complete graph.* Also, we call the transition from $TT_m \circ TT_n$ to $TT_n \circ TT_m$ a *transposition of transitive tournaments.* (Recall that transitive tournaments commute, by Equation (10.2).)

Our final theorem of the section is due to Dörfler and Imrich [8].

**Theorem 10.9.5** *Any prime factorization of a digraph over the lexicographic product can be transformed into any other prime factorization by transpositions of totally disconnected graphs, transpositions of complete graphs, and transpositions of transitive tournaments.*

## 10.10 Cartesian Skeletons

The previous section developed prime factorization results for the Cartesian and Lexicographic products. In order to get analogous results for the direct and strong products, we first need to define what is called the *Cartesian skeleton* of a digraph. This is an operator $S$ that transforms a digraph $D$ into a symmetric digraph $S(D)$, and, under suitable conditions, obeys $S(D \times G) = S(D) \,\square\, S(G)$. In the subsequent section we will use it to transform questions about factorizations over $\times$ to the more manageable product $\square$ (which was treated in the previous section).

Our exposition is a generalization to digraphs of Hammack and Imrich [16], which developed $S$ in the setting of graphs. We also draw inspiration from Hellmuth and Marc [24], who devised a similar skeleton operator for which $S(D \boxtimes G) = S(D) \,\square\, S(G)$. The present development is from [17].

We need several definitions. An **antiwalk** in a digraph is a walk in which the orientations of the arcs alternate as the walk is traversed. An **out-antiwalk** is an antiwalk for which the first and last arcs are directed away from the end-vertices of the walk. An **in-antiwalk** is one for which the first and last arcs are directed towards the end-vertices. See Figure 10.11. Notice that in- and out-antiwalks necessarily have even length.

For a digraph $D$, let $D^+$ be the symmetric digraph on $V(D)$ for which $xy, yx \in A(D^+)$ whenever $D$ has an out-antiwalk of length 2 from $x$ to $y$, that is, if $N_D^+(x) \cap N_D^+(y) \neq \emptyset$. See Figure 10.12 (left), where a dotted line between $x$ and $y$ represents two arcs $xy$ and $yx$ in $D^+$. It is immediate from the definitions that $(D \times G)^+ = D^+ \times G^+$. Note that $D^+$ has a loop at each vertex of positive out-degree. We define $D^-$ similarly, where $xy, yx \in A(D^-)$ provided $N_D^-(x) \cap N_D^- \neq \emptyset$. Again, $(D \times G)^- = D^- \times G^-$. Because they are symmetric digraphs, $D^+$ and $D^-$ can be regarded as *graphs.* (In a different context [49, 50], $D^+$ is also called the **competition graph** of $D$.)

Observe that $D^+$ is connected if and only if any two vertices of $D$ are joined by an out-antiwalk in $D$, and $D^-$ is connected if and only if any two vertices of $D$ are joined by an in-antiwalk in $D$.

We now explain how to construct **Cartesian skeletons** $S^+(D)$ and $S^-(D)$ of a digraph $D$ by removing strategic edges from $D^+$ and $D^-$. Given a factoring $D = H \times K$, we say an arc $(h, k)(h', k')$ of $D^+$ is **diagonal** relative to the factoring if it is a loop, or $h \neq h'$ and $k \neq k'$; otherwise it is **Cartesian**. For example, in Figure 10.12, arcs $xz$ and $zy$ of $D^+$ are Cartesian, and arcs $xy$ and $yy$ of $D^+$ are diagonal. We note two intrinsic criteria that tell us if a non-loop arc of $D^+$ is diagonal relative to some factoring of $D$.

1. In Figure 10.12, arc $xy$ of $D^+$ is not Cartesian, and there is a $z \in V(D)$ with
$N_D^+(x) \cap N_D^+(y) \subset N_D^+(x) \cap N_D^+(z)$ and $N_D^+(x) \cap N_D^+(y) \subset N_D^+(y) \cap N_D^+(z)$.
2. In Figure 10.12, arc $x'y'$ of $D^+$ is not Cartesian, and there is a $z' \in V(G)$ with $N_D^+(x') \subset N_D^+(z') \subset N_D^+(y')$.

We will get $S^+(D)$ by removing from $D^+$ all loops, and arcs that meet one of these criteria. Now, these criteria are somewhat dependent on one another. Note $N_D^+(x) \subset N_D^+(z) \subset N_D^+(y)$ implies $N_D^+(y) \cap N_D^+(x) \subset N_D^+(y) \cap N_D^+(z)$. Also, $N_D^+(y) \subset N_D^+(z) \subset N_D^+(x)$ implies $N_D^+(x) \cap N_D^+(y) \subset N_D^+(x) \cap N_D^+(z)$. This allows us to pack the above criteria into the following definition.

**Definition 10.10.1** *An arc $xy$ of $D^+$ is* **dispensable in** $D^+$ *if it is a loop, or if there is some $z \in V(D)$ for which both of the following statements hold:*

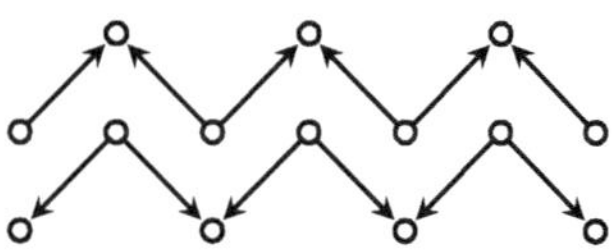

**Figure 10.11** An out-antiwalk (top) and an in-antiwalk (bottom).

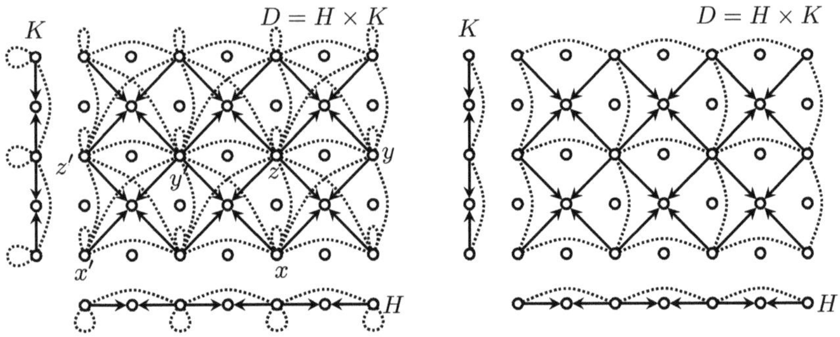


**Figure 10.12** Left: Digraphs $H$, $K$, $H \times K$ (bold), and $H^+$, $K^+$, $(H \times K)^+$ (dotted). Right: Digraphs $H$, $K$, $H \times K$ (bold), and $S^+(H)$, $S^+(K)$, $S^+(H \times K)$ (dotted). Note that $(H \times K)^+ = H^+ \times K^+$, and $S^+(H \times K) = S^+(H) \,\Box\, S^+(K)$.

*1.* $N_D^+(x) \cap N_D^+(y) \subset N_D^+(x) \cap N_D^+(z)$ *or* $N_D^+(x) \subset N_D^+(z) \subset N_D^+(y)$,
*2.* $N_D^+(y) \cap N_D^+(x) \subset N_D^+(y) \cap N_D^+(z)$ *or* $N_D^+(y) \subset N_D^+(z) \subset N_D^+(x)$.

*Similarly, an arc $xy$ of $D^-$ is* **dispensable in** $D^-$ *if the above conditions hold with $N_D^-$ used in the place of $N_D^+$.*

Note that the above statements (1) and (2) are symmetric in $x$ and $y$. The next remark follows from the paragraph preceding the definition. It will be used often.

*Remark* 10.10.2 An arc $xy$ of $D^+$ is dispensable in $D^+$ if and only if there is a $z \in V(D)$ with $N_D^+(x) \subset N_D^+(z) \subset N_D^+(y)$, or $N_D^+(y) \subset N_D^+(z) \subset N_D^+(x)$, or $N_D^+(x) \cap N_D^+(y) \subset N_D^+(x) \cap N_D^+(z)$ and $N_D^+(y) \cap N_D^+(x) \subset N_D^+(y) \cap N_D^+(z)$. The same remark holds for dispensability in $D^-$ (replacing $N^+$ with $N^-$).

Now we come to the main definition of this section.

**Definition 10.10.3** *The* **Cartesian out-skeleton** $S^+(D)$ *of a digraph $D$ is the spanning subgraph of $D^+$ obtained by deleting all arcs that are dispensable in $D^+$. The* **Cartesian in-skeleton** $S^-(D)$ *of $D$ is the spanning subgraph of $D^-$ obtained by deleting all arcs that are dispensable in $D^-$. The* **Cartesian skeleton** $S(D)$ *of $D$ is the graph with vertices $V(D)$ and arcs $A(S(D)) = A(S^+(D)) \cup A(S^-(D))$.*

*Note that each of $D^+$, $D^-$, $S^+(D)$, $S^-(D)$ and $S(D)$ is a symmetric digraph. We thus tend to refer to them as* graphs, *and call their arcs* edges.

As an example, the right side of Figure 10.12 is the same as the left, except that all dispensable edges of $H^+$, $K^+$, and $(H \times K)^+$ are deleted. Thus the remaining dashed edges are $S^+(H), S^+(K)$, and $S^+(H \times K)$. Note that although $S^+(D)$ was defined without regard to the factoring $D = H \times K$, we nonetheless have $S^+(H \times K) = S^+(H) \,\Box\, S^+(K)$. In fact, we will shortly prove that this equation holds for each of $S^+, S^-$ and $S$, under mild restrictions.

These restrictions involve certain equivalence relations on the vertex set of a digraph. Define an equivalence relation $R^+$ on $V(D)$ by declaring $xR^+y$ whenever $N_D^+(x) = N_D^+(y)$. A digraph is called **$\mathbf{R^+}$-thin** if $N_D^+(x) = N_D^+(y)$ implies $x = y$ for all $x, y \in V(D)$, that is, if each $R^+$-class contains exactly one vertex. Similarly, we define $R^-$ and **$\mathbf{R^-}$-thinness** as above, but replacing $N_D^+$ with $N_D^-$. Finally, we say $D$ is **R-thin** if it is both $R^+$ thin and $R^-$-thin. We will need the following.

**Lemma 10.10.4** *Let $H$ and $K$ be digraphs for which all vertices have positive in- and out-degrees. Then $H$ and $K$ are $R^+$-thin (respectively $R^-$-thin) if and only if $H \times K$ is $R^+$-thin (respectively $R^-$-thin). Consequently $H$ and $K$ are $R$-thin if and only if $H \times K$ is $R$-thin.*

**Proof:** Immediate from $N_{H\times K}^+(x, y) = N_H^+(x) \times N_K^+(y)$ (Equation 10.6) and its companion $N_{H\times K}^-(x, y) = N_H^-(x) \times N_K^-(y)$, combined with the fact that no neighborhoods are empty. □

The next lemma and proposition show $S^+(H \times K) = S^+(H) \,\Box\, S^+(K)$ and $S^-(H \times K) = S^-(H) \,\Box\, S^-(K)$ for $R^+$- and $R^-$-thin digraphs. The proofs frequently use the fact that for $D = H \times K$,

$$N_D^+(h,k) \cap N_D^+(h',k') = \big(N_H^+(h) \cap N_H^+(h')\big) \times \big(N_K^+(k) \cap N_K(k')\big),$$

which follows from $N_D^+(h, k) = N_H^+(h) \times N_K^+(k)$ and simple set theory.

**Lemma 10.10.5** *Suppose $D$ is a digraph with a factorization $D = H \times K$. If $D$ is $R^+$-thin, then every arc of $S^+(D)$ is Cartesian with respect to the factorization. Similarly, if $D$ is $R^-$-thin, then every arc of $S^-(D)$ is Cartesian with respect to the factorization.*

**Proof:** We prove only the first statement. The proof of the second is identical, but replaces $N^+$ with $N^-$, and the notion of $R^+$-thinness with $R^-$-thinness.

Let $(h, k)(h', k')$ be a non-Cartesian edge of $D^+$. We need only show that it is dispensable. It is certainly dispensable if it is a loop. Otherwise $h \neq h'$ and $k \neq k'$ Observe:

$$\begin{aligned} N_D^+(h,k) \cap N_D^+(h',k') &= \big(N_H^+(h) \cap N_H^+(h')\big) \times \big(N_K^+(k) \cap N_K(k')\big) \\ &\subseteq \qquad\qquad N_H^+(h) \times \big(N_H^+(k) \cap N_H^+(k')\big) \\ &= N_D^+(h,k) \cap N_D^+(h,k'), \end{aligned}$$

$$\begin{aligned} N_D^+(h',k') \cap N_D^+(h,k) &= \big(N_H^+(h') \cap N_H^+(h)\big) \times \big(N_K^+(k') \cap N_K^+(k)\big) \\ &\subseteq \big(N_H^+(h') \cap N_H^+(h)\big) \times N_K^+(k') \\ &= N_D^+(h',k') \cap N_D^+(h,k'). \end{aligned}$$

If both of these inclusions are proper, then $(h, k)(h', k')$ is dispensable. If one inclusion is equality, then $N_H^+(h) \cap N_H^+(h') = N_H^+(h)$ in the first case or

$N_K^+(k') \cap N_K^+(k) = N_K^+(k')$ in the second. From this, $N_H^+(h) \subseteq N_H^+(h')$ or $N_K^+(k') \subseteq N_K^+(k)$. By $R^+$-thinness,

$$N_H^+(h) \subset N_H^+(h') \quad \text{or} \quad N_K^+(k') \subset N_K^+(k). \tag{10.13}$$

Repeating this argument but interchanging $h$ with $h'$, and $k$ with $k'$,

$$N_H^+(h') \subset N_H^+(h) \quad \text{or} \quad N_K^+(k) \subset N_K^+(k'). \tag{10.14}$$

Inclusions (10.13) and (10.14) show $N_H^+(h) \subset N_H^+(h')$ and $N_K^+(k) \subset N_K^+(k')$, or $N_K^+(k') \subset N_K^+(k)$ and $N_H^+(h') \subset N_H^+(h)$. The first case gives

$$N_H^+(h) \times N_K^+(k) \subset N_H^+(h) \times N_K^+(k') \subset N_H^+(h') \times N_K^+(k'),$$

that is, $N_D^+(h,k) \subset N_D^+(h,k') \subset N_D^+(h',k')$, so $(h,k)(h',k')$ is dispensable. The second case yields $N_D^+(h',k') \subset N_D^+(h,k') \subset N_D^+(h,k)$, with the same conclusion. □

**Proposition 10.10.6** *If $H, K$ are $R^+$-thin digraphs with no vertices of zero out-degree, then $S^+(H \times K) = S^+(H) \,\square\, S^+(K)$. If $H, K$ are $R^-$-thin, with no vertices of zero in-degree, then $S^-(H \times K) = S^-(H) \,\square\, S^-(K)$.*

**Proof:** Again, we prove only the first statement; the proof of the second is entirely analogous.

First we show $S^+(H \times K) \subseteq S^+(H) \,\square\, S^+(K)$. By Lemma 10.10.5, all arcs of $S^+(H \times K)$ are Cartesian, so we need only show $(h,k)(h',k) \in S^+(H \times K)$ implies $hh' \in S^+(H)$. (The same argument will work for arcs $(h,k)(h,k')$.) Thus suppose $hh' \notin S^+(H)$. Then $hh'$ is dispensable in $H^+$, so there is a $z'$ in $V(H)$ for which both of the following conditions hold:

$$\begin{array}{lll} N_H^+(h) \cap N_H^+(h') \subset N_H^+(h) \cap N_H^+(z') & \text{or} & N_H^+(h) \subset N_H^+(z') \subset N_H^+(h') \\ N_H^+(h') \cap N_H^+(h) \subset N_H^+(h') \cap N_H^+(z') & \text{or} & N_H^+(h') \subset N_H^+(z') \subset N_H^+(h). \end{array}$$

Because there are no vertices of zero out-degree, $N_K^+(k) \neq \emptyset$. Thus we can multiply each neighborhood $N_H^+(u)$ above by $N_K^+(k)$ on the right and still preserve the proper inclusions. Then the fact $N_H^+(u) \times N_K^+(k) = N_{H \times K}^+(u,k)$ yields the dispensability conditions (1) and (2), where $x = (h,k)$, $y = (h',k)$ and $z = (z',k)$. Thus $(h,k)(h',k) \notin S^+(H \times K)$.

Now we show $S^+(H) \,\square\, S^+(K) \subseteq S^+(H \times K)$. Take an arc in $S(H) \,\square\, S(K)$, say $(h,k)(h',k)$ with $hh' \in S^+(H)$. We must show that $(h,k)(h',k)$ is not dispensable in $(H \times K)^+$. Suppose it was. Then there would be a vertex $z = (z', z'')$ in $H \times K$ such that the dispensability conditions (1) and (2) hold for $x = (h,k), y = (h',k)$, and $z = (z', z'')$. The various cases are considered below. Each leads to a contradiction.

Suppose $N_D^+(x) \subset N_D^+(z) \subset N_D^+(y)$. This means

$$N_H^+(h) \times N_K^+(k) \subset N_H^+(z') \times N_K^+(z'') \subset N_H^+(h') \times N_K^+(k),$$

so $N_K^+(z'') = N_K^+(k)$. Then the fact that $N_K^+(k) \neq \emptyset$ permits cancellation of the common factor $N_K^+(k)$, so $N_H^+(h) \subset N_H^+(z') \subset N_H^+(h')$, and $hh'$ is dispensable. We reach the same contradiction if $N_D^+(y) \subset N_D^+(z) \subset N_D^+(x)$.

Finally, suppose there is a $z = (z', z'')$ for which both $N_D^+(x) \cap N_D^+(y) \subset N_D^+(x) \cap N_D^+(z)$ and $N_D^+(y) \cap N_D^+(x) \subset N_D^+(y) \cap N_D^+(z)$. Rewrite this as

$$\begin{aligned} N_D^+(h,k) \cap N_D^+(h',k) &\subset N_D^+(h,k) \cap N_D^+(z',z'') \\ N_D^+(h',k) \cap N_D^+(h,k) &\subset N_D^+(h',k) \cap N_D^+(z',z''), \end{aligned}$$

which is the same as

$$\begin{aligned} \big(N_H^+(h) \cap N_H^+(h')\big) \times N_K^+(k) &\subset \big(N_H^+(h) \cap N_H^+(z')\big) \times \big(N_K(k) \cap N_K(z'')\big) \\ \big(N_H^+(h') \cap N_H^+(h)\big) \times N_K^+(k) &\subset \big(N_H^+(h') \cap N_H^+(z')\big) \times \big(N_K^+(k) \cap N_K^+(z'')\big). \end{aligned}$$

Thus $N_K^+(k) \subseteq N_K^+(k) \cap N_K^+(z'')$, so $N_K^+(k) = N_K^+(k) \cap N_K^+(z'')$, whence

$$\begin{aligned} N_H^+(h) \cap N_H^+(h') &\subset N_H^+(h) \cap N_H^+(z') \\ N_H^+(h') \cap N_H^+(h) &\subset N_H^+(h') \cap N_H^+(z'). \end{aligned}$$

Thus $hh'$ is dispensable, a contradiction. □

The next corollary follows from Proposition 10.10.6, Definition 10.10.3, as well as the definition of the Cartesian product. (Recall that a digraph is $R$-thin if it is both $R^+$-thin and $R^-$-thin.)

**Corollary 10.10.7** *Suppose $K$ and $H$ are $R$-thin digraphs, no vertices of which have zero in- or out-degree. Then $S(K \times H) = S(K) \,\square\, S(H)$.*

Because the various skeletons are defined entirely in terms of adjacency structure, we have the following immediate consequence of Definition 10.10.3.

**Proposition 10.10.8** *Any isomorphism $\varphi : D \to D'$ between digraphs, as a map $V(D) \to V(D')$, is also an isomorphism $\varphi : S(D) \to S(D')$.*

We next consider connectivity of $S(G)$. The following lemma is needed.

**Lemma 10.10.9** *Suppose a digraph $D$ has no vertex of zero out-degree, and $x, y \in V(D)$. If $N_D^+(x) \subset N_D^+(y)$, then $D^+$ has an $(x,y)$-path consisting of edges that are non-dispensable in $D^+$. Similarly, if no vertex of $D$ has zero in-degree and and $N_D^-(x) \subset N_D^-(y)$, then $D^-$ has an $(x,y)$-path consisting of edges that are non-dispensable in $D^-$.*

**Proof:** We prove the first statement; the second follows analogously.

Consider the following maximal chain of neighborhoods between $N_D^+(x)$ and $N_D^+(y)$, ordered by proper inclusion. (It is possible that $y_1 = y$.)

$$N_D^+(x) \subset N_D^+(y_1) \subset N_D^+(y_2) \subset N_D^+(y_3) \subset \cdots \subset N_D^+(y_k) \subset N_D^+(y).$$

We claim that $xy_1$ is non-dispensable in $D^+$. Certainly $N_D^+(x) \subset N_D^+(y_1)$ implies $xy_1$ is an edge of $D^+$, because $N_D^+(x) \neq \emptyset$. Also, there is no $z$ for which $N_D^+(x) \cap N_D^+(y_1) \subset N_D^+(x) \cap N_D^+(z)$; otherwise the condition $N_D^+(x) \subset N_D^+(y_1)$ would yield $N_D^+(x) \subset N_D^+(x) \cap N_D^+(z)$, which is impossible. As the chain is maximal, there is no $z$ for which $N_D^+(x) \subset N_D^+(z) \subset N_D^+(y_1)$. Further, $N_D^+(y_1) \subset N_D^+(z) \subset N_D^+(x)$ is impossible, so $xy_1$ is non-dispensable in $D^+$.

The same argument shows that each $y_i y_{i+1}$ is a non-dispensable edge of $D^+$, as is $y_k y$. Thus we have the required path $xy_1y_2 \ldots y_k y$. □

Let us define a digraph to be **anti-connected** if any two of its vertices are joined by an antiwalk of even length. It should be clear that a direct product of digraphs is anti-connected if and only if all of its factors are anti-connected.

**Proposition 10.10.10** *If $D$ is anti-connected, then $S(D)$ is connected.*

**Proof:** Take $x_1, x_2 \in V(S(D)) = V(D)$. Suppose first that they are joined by an (even) out-antiwalk $W$ in $D$. As $E(S(D)) = E(S^+(D)) \cup E(S^-(D))$, and because $D^+$ has an $(x_1, x_2)$-path $P$ on alternate vertices of $W$, it suffices to show that for any dispensable edge $xy$ of $P$, there is an $(x, y)$-path in $D^+$ consisting of non-dispensable edges. In fact, we will prove this for any edge $xy$ of $D^+$. Given such an edge $xy$, define the integer

$$k_{xy} = \max\{\, |N_D^+(u) \cap N_D^+(v)| - |N_D^+(x) \cap N_D^+(y)| \;\mid u, v \in V(D), u \neq v\}.$$

Notice $k_{xy} \geq 0$. (Put $u = x$ and $v = y$.) If $k_{xy} = 0$, then the definition of $k_{xy}$ implies that there is no $z$ for which $N_D^+(x) \cap N_D^+(y) \subset N_D^+(x) \cap N_D^+(z)$ or $N_D^+(y) \cap N_D^+(x) \subset N_D^+(y) \cap N_D^+(z)$. Then $N_D^+(x) \subset N_D^+(z) \subset N_D^+(y)$ is also impossible, as it implies $N_D^+(y) \cap N_D^+(x) \subset N_D^+(y) \cap N_D^+(z)$. Therefore $xy$ is not dispensable if $k_{xy} = 0$.

Take $N > 0$, and assume that whenever $D^+$ has an edge $xy$ with $k_{xy} < N$, there is a $(x, y)$-path in $D^+$ composed of non-dispensable edges. Now suppose $xy$ is dispensable and $k_{xy} = N$. If $N_D^+(x) \subset N_D^+(y)$ or $N_D^+(y) \subset N_D^+(x)$, then we are done, by Lemma 10.10.9, so assume $N_D^+(x) \not\subset N_D^+(y)$ and $N_D^+(y) \not\subset N_D^+(x)$. As $xy$ is dispensable, there is a vertex $z$ with

$$N_D^+(x) \cap N_D^+(y) \subset N_D^+(x) \cap N_D^+(z) \text{ and } N_D^+(y) \cap N_D^+(x) \subset N_D^+(y) \cap N_D^+(z).$$

This implies $N_D^+(x) \cap N_D^+(z) \neq \emptyset \neq N_D^+(y) \cap N_D^+(z)$, so $xz, yz \in E(D^+)$. But it also means

$$|N_D^+(u) \cap N_D^+(v)| - |N_D^+(x) \cap N_D^+(z)| \;<\; |N_D^+(u) \cap N_D^+(v)| - |N_D^+(x) \cap N_D^+(y)|$$

for all $u, v$, so $k_{xz} < k_{xy}$. Similarly, $k_{zy} < k_{xy}$. The induction hypothesis guarantees $(x, z)$- and $(z, y)$-paths of non-dispensable edges in $D^+$, so we have an $(x, y)$-path of non-dispensable edges in $D^+$.

To finish the proof, we must treat the case where $x_1$ and $x_2$ are joined by an in-antiwalk. Just repeat the above argument with $N^-$ and $D^-$. □

## 10.11 Prime Factorings of Direct and Strong Products

Now we turn to prime factorings over the direct product. Recall that the one-vertex digraph $K_1^*$ with a loop is the unit for the direct product, that is, $K_1^* \times D = D$ for every digraph $D$, and $K_1^*$ is the unique digraph with this property. Thus we say a digraph $D$ is **prime over the direct product** if it has more than one vertex, and in any factoring $D = G \times H$, one factor is $K_1^*$ and the other is isomorphic to $D$. For this reason, the entire discussion of prime factorization over the direct product takes place in the class $\mathscr{D}_0$ of digraphs that may have loops.

This section adopts the approach of Hammack and Imrich [17]. The next lemma uses the Cartesian skeleton and unique prime factorization over $\Box$ to deliver a key ingredient to the proof of our unique prime factorization theorem for the direct product (Theorem 10.11.2).

**Lemma 10.11.1** *Suppose $\varphi : D_1 \times \cdots \times D_k \to G_1 \times \cdots \times G_\ell$ is an isomorphism, where all the factors are anti-connected and $R$-thin, and that we have $\varphi(x_1, \ldots, x_k) = \big(\varphi_1(x_1, \ldots, x_k), \varphi_2(x_1, \ldots, x_k), \ldots, \varphi_\ell(x_1, \ldots, x_k)\big)$. If a factor $D_i$ is prime, then exactly one of the functions $\varphi_1, \varphi_2, \ldots, \varphi_\ell$ depends on $x_i$.*

**Proof:** By commutativity and associativity, it suffices to prove the lemma for the case $k = \ell = 2$, and with $D_1$ prime. Thus take an isomorphism $\varphi : D_1 \times D_2 \to G_1 \times G_2$, where $\varphi(x_1, x_2) = \big(\varphi_1(x_1, x_2), \varphi_2(x_1, x_2)\big)$. We will prove the lemma by showing that if it is not the case that exactly one of $\varphi_1$ and $\varphi_2$ depends on $x_1$, then $D_1$ is not prime.

Certainly if neither $\varphi_1$ nor $\varphi_2$ depends on $x_1$, then the fact that $\varphi$ is bijective means that $|V(D_1)| = 1$, so $D_1$ is not prime. Thus assume that both $\varphi_1$ and $\varphi_2$ depend on $x_1$. This means each of $D_1, G_1$, and $G_2$ has more than one vertex. If $D_2$ had only one vertex, then $D_1 \cong G_1 \times G_2$, and $D_1$ would not be prime. Thus each factor $D_1, D_2, G_1$, and $G_2$ has more than one vertex. Taking skeletons, and applying Proposition 10.10.8, we see that $\varphi$ is also an isomorphism $\varphi : S(D_1 \Box D_2) \to S(G_1 \Box G_2)$. Because all factors are $R$-thin (and anti-connectedness implies that all vertices have positive in- and out-degrees), Corollary 10.10.7 applies, and we have an isomorphism

$$\varphi : S(D_1) \Box S(D_2) \to S(G_1) \Box S(G_2). \tag{10.15}$$

Note that $\varphi$ is simultaneously an isomorphism $\varphi : D_1 \times D_2 \to G_1 \times G_2$ and an isomorphism $\varphi : S(D_1) \Box S(D_2) \to S(G_1) \Box S(G_2)$. Because each of $D_1, D_2, G_1$, and $G_2$ is anti-connected, each factor $S(D_1), S(D_2), S(G_1)$, and $S(G_2)$ is connected, by Proposition 10.10.10. Consider prime factorizations

$$\begin{aligned} S(D_1) &= H_1 \Box H_2 \Box \cdots \Box H_k, & S(G_1) &= L_1 \Box L_2 \Box \cdots \Box L_\ell, \\ S(D_2) &= K_1 \Box K_2 \Box \cdots \Box K_m, & S(G_2) &= M_1 \Box M_2 \Box \cdots \Box M_n, \end{aligned}$$

where each factor is prime over $\Box$. Our isomorphism (10.15) becomes

$$\varphi : (H_1 \square \cdots \square H_k) \square (K_1 \square \cdots \square K_m) \to \\ (L_1 \square \cdots \square L_\ell) \square (M_1 \square \cdots \square M_n). \qquad (10.16)$$

Corollary 10.9.4 applies here. In fact, in using it, we may order the factors $H_i$ and $K_i$ and relabel the vertices of the $L_i$ and $M_i$ so that, for some $0 < s < k$ and $0 \le t \le m$, the isomorphism (10.16) has form

$$\varphi :(H_1 \square \cdots \square H_k) \square (K_1 \square \cdots \square K_m) \to \\ \big(H_1 \square \cdots \square H_s \square K_1 \square \cdots \square K_t\big) \square \big(H_{s+1} \square \cdots \square H_k \square K_{t+1} \square \cdots \square K_m\big)$$

and where

$$\varphi((h_1, \dots, h_k),(k_1, \dots, k_m)) = \\ ((h_1, \dots, h_s, k_1, \dots, k_t), (h_{s+1}, \dots, h_k, k_{t+1}, \dots, k_m)).$$

Our assumption that both $\varphi_1$ and $\varphi_2$ depend on $x_1 \in V(G_1)$ forces $0 < s < k$.

We have now labeled the vertices of $D_1$ with $V(H_1 \square \cdots \square H_k)$, and those of $D_2$ with $V(K_1 \square \cdots \square K_m)$. We have labeled vertices of $G_1$ with $V(H_1 \square \cdots \square H_s \square K_1 \square \cdots \square K_t)$, and we have labeled the vertices of $G_2$ with $V(H_{s+1} \square \cdots \square H_k \square \; K_{t+1} \square \cdots \square K_m)$. To tame the notation, we denote a vertex $(h_1, \dots, h_s, h_{s+1}, \dots, h_k) \in V(D_1)$ as $(x, y)$, where $x = (h_1, \dots, h_s)$ and $y = (h_{s+1}, \dots, h_k)$. Similarly, any $(k_1, \dots, k_t, k_{t+1}, \dots, k_m) \in V(D_2)$ is denoted $(u, v)$, where $u = (k_1, \dots, k_t)$ and $v = (k_{t+1}, \dots, k_m)$. With this convention we regard vertices of $G_1$ and $G_2$ as $(x, u)$ and $(y, v)$, respectively, and we have

$$\varphi((x, y), (u, v)) = ((x, u), (y, v)).$$

Remember that this is the same isomorphism $\varphi : D_1 \times D_2 \to G_1 \times G_2$ that we began the proof with; all we have done is relabel the vertices of the factors to put $\varphi$ into a more convenient form.

Now we display a nontrivial factorization $D_1 = S \times S'$. Define digraphs $S$ and $S'$ as follows:

$$V(S) = \{x \mid \big((x, y), (u, v)\big) \in V(D_1 \times D_2)\},$$
$$A(S) = \{xx' \mid \big((x, y), (u, v)\big)\big((x', y'), (u', v')\big) \in A(D_1 \times D_2)\},$$

$$V(S') = \{y \mid \big((x, y), (u, v)\big) \in V(D_1 \times D_2)\},$$
$$A(S') = \{yy' \mid \big((x, y), (u, v)\big)\big((x', y'), (u', v')\big) \in A(D_1 \times D_2)\}.$$

We claim $D_1 = S \times S'$, that is, $(x, y)(x', y') \in A(D_1)$ if and only if $(x, y)(x', y') \in A(S \times S')$. Certainly if $(x, y)(x', y') \in A(D_1)$, there is an arc

$$\big((x, y), (u, v)\big)\big((x', y'), (u', v')\big) \in A(D_1 \times D_2).$$

The definitions of $S$ and $S'$ then imply $(x,y)(x',y') \in A(S \times S')$.

Conversely, suppose $(x,y)(x',y') \in A(S \times S')$. Then $xx' \in A(S)$ and $yy' \in A(S')$. By definition of $S$ and $S'$, this means $D_1 \times D_2$ has arcs

$$\big((x,y''),(u,v)\big)\big((x',y'''),(u',v')\big) \text{ and } \big((x'',y),(u'',v'')\big)\big((x''',y'),(u''',v''')\big).$$

Applying the isomorphism $\varphi$, we see that $G_1 \times G_2$ has arcs

$$\big((x,u),(y'',v)\big)\big((x',u'),(y''',v')\big) \text{ and } \big((x'',u''),(y,v'')\big)\big((x''',u'''),(y',v''')\big).$$

Then $(x,u)(x',u') \in A(G_1)$ and $(y,v'')(y',v''') \in A(G_2)$. Thus $G_1 \times G_2$ has an arc $\big((x,u),(y,v'')\big)\big((x',u'),(y',v''')\big)$. Applying $\varphi^{-1}$ to this, we get

$$\big((x,y),(u,v'')\big)\big((x',y'),(u',v''')\big) \in A(D_1 \times D_2),$$

hence $(x,y)(x',y') \in A(D_1)$. Thus $D_1 = S \times S'$, and the lemma is proved. □

We now can easily prove that anti-connected $R$-thin digraphs factor uniquely into primes over the direct product, up to order and isomorphism of the factors.

**Theorem 10.11.2** *Take any isomorphism $\varphi : D_1 \times \cdots \times D_k \to G_1 \times \cdots \times G_\ell$, where all factors $D_i$ and $G_i$ are anti-connected, $R$-thin, and prime. Then $k = \ell$, and there is a permutation $\sigma$ of $[k]$ and isomorphisms $\varphi_i : D_{\sigma(i)} \to G_i$ for which $\varphi(x_1, x_2, \ldots, x_k) = \big(\varphi_1(x_{\sigma(1)}), \varphi_2(x_{\sigma(2)}), \ldots, \varphi_k(x_{\sigma(k)})\big)$.*

**Proof:** Assume the hypothesis. Note that Lemma 10.11.1 implies that for each $i \in [k]$, exactly one $\varphi_j$ depends on $x_i$. But no $\varphi_j$ is constant, because $\varphi$ is surjective and each $G_i$ has more than one vertex (it is prime). Thus $k \geq \ell$. The same argument applied to $\varphi^{-1}$ gives $\ell \geq k$, therefore $k = \ell$.

Thus each $\varphi_j$ depends on only one $x_i$, call it $x_{\sigma(j)}$. The result follows. □

To see that prime factorization may fail if the hypotheses of this theorem are not met, let $D$ be a closed antiwalk on six vertices, which is not anti-connected. Indeed, we have the non-unique prime factorization

$$D \cong \overrightarrow{P}_2 \times K_3 \cong \overrightarrow{P}_2 \times H,$$

where $H$ is the symmetric path $\overleftrightarrow{P_3}$ of length two with loops at each end.

A careful examination of its proof shows that Theorem 10.11.2 still holds if $R$-thinness is replaced by $R^+$-thinness (respectively, $R^-$-thinness) and the assumption of anti-connectivity is replaced with the condition that any two vertices are joined by an out-antiwalk (respectively, an in-antiwalk), in which case we say the graph is **out-anti-connected** (respectively in-anti-connected). Imrich and Klöckl [26] present a polynomial algorithm that computes the prime factorization of any out-anti-connected $R^+$-thin digraph.

In [27] they weaken (but do not entirely eliminate) the $R^+$-thinness condition.

We can remove the condition of $R$-thinness in Theorem 10.11.2 if we strengthen the connectivity condition. The fundamental work of McKenzie [38] on relational structures yields the following corollary.

**Theorem 10.11.3** *Suppose each pair of vertices of a digraph is joined by* **both** *an in-antiwalk* **and** *an out-antiwalk. Then it has a unique prime factorization over the direct product, up to isomorphism and order of the factors.*

It is not known whether the hypotheses of this theorem can be relaxed to anti-connectivity, nor is there currently an algorithm that finds the prime factors. Any progress would be a welcome contribution.

**Problem 10.11.4** *Find an efficient algorithm that computes the prime factors of a digraph meeting the conditions of Theorem 10.11.3.*

Note that Hellmuth and Marc [24] develop such an algorithm for connected strong products.

Theorem 10.11.3 yields a parallel theorem for the strong product. For a digraph $D$, let $\mathscr{L}(D)$ be the digraph obtained from $D$ by adding a loop to each vertex. If $D_1, \ldots, D_k$ are digraphs without loops, then

$$\mathscr{L}(D_1 \boxtimes \cdots \boxtimes D_k) = \mathscr{L}(D_1) \times \cdots \times \mathscr{L}(D_k), \qquad (10.17)$$

which follows immediately from the definitions. Notice that if $D$ is connected, then $\mathscr{L}(D)$ is automatically anti-connected. In fact, any two of its vertices can be joined by an in-antiwalk and an out-antiwalk, so Theorem 10.11.3 applies to it. And clearly if $D$ and $G$ are digraphs without loops, then $D \cong G$ if and only if $\mathscr{L}(D) \cong \mathscr{L}(G)$.

**Theorem 10.11.5** *Every connected digraph (without loops) has a unique prime factorization over* $\boxtimes$*, up to isomorphism and order of the factors.*

**Proof:** Let $D$ be a connected digraph without loops. Then, as noted above, Theorem 10.11.3 applies to $\mathscr{L}(D)$, so it has a unique prime factorization over the direct product. Because $\mathscr{L}(D)$ has a loop at each vertex, each of its prime factors also have loops at all of their vertices. Thus each prime factor has the form $\mathscr{L}(D_i)$ for some $D_i$ (without loops). Write the prime factorization as

$$\mathscr{L}(D) = \mathscr{L}(D_1) \times \mathscr{L}(D_2) \times \cdots \times \mathscr{L}(D_n), \qquad (10.18)$$

where the $\mathscr{L}(D_i)$ (and hence also each $D_i$) are uniquely determined by $D$.

Now consider any prime factorization

$$D = G_1 \boxtimes G_2 \boxtimes \cdots \boxtimes G_k \qquad (10.19)$$

over the strong product. From this, Equation (10.17) yields

$$\mathscr{L}(D) = \mathscr{L}(G_1) \times \mathscr{L}(G_2) \times \cdots \times \mathscr{L}(G_k). \tag{10.20}$$

Observe that each $\mathscr{L}(G_i)$ is prime over $\times$. Indeed, any factoring of it must have the form $\mathscr{L}(G_i) = \mathscr{L}(H) \times \mathscr{L}(H')$ for digraphs $H, H'$ (without loops), and Equation (10.17) gives $\mathscr{L}(G_i) = \mathscr{L}(H \boxtimes H')$. Hence $G_i \cong H \boxtimes H'$ and primeness of $G_i$ implies one of $H$ or $H'$ is $K_1$, and therefore one of the factors $\mathscr{L}(H)$ or $\mathscr{L}(H')$ is $\mathscr{L}(K_1)$. Thus $\mathscr{L}(G_i)$ is prime.

Comparing prime factorizations (10.18) and (10.20), and applying Theorem 10.11.3, we get $n = k$, and we may assume the ordering is such that $\mathscr{L}(D_i) \cong \mathscr{L}(G_i)$ for each $1 \leq i \leq n$. Consequently, $D_i \cong G_i$ for each $i \in [k]$. But, as was noted above, the $G_i$ are uniquely determined by $D$, so the factorization (10.19) is unique. □

A different approach is taken by Hellmuth and Marc [24], who design and apply a skeleton operator $\mathbb{S}$ satisfying $\mathbb{S}(D \boxtimes D') = \mathbb{S}(D) \,\square\, \mathbb{S}(D')$.

## References


1. N. R. Aravind, N. Narayanan, and C.R. Subramanian. Oriented colouring of some graph products. *Discuss. Math. Graph Theory*, 31(4):675–686, 2011.
2. S. Bessy and S. Thomassé. The categorical product of two 5-chromatic digraphs can be 3-chromatic. *Discrete Math.*, 305(1-3):344–346, 2005.
3. V.D. Blondel, A. Gajardo, M. Heymans, P. Senellart, and P. Van Dooren. A measure of similarity between graph vertices: applications to synonym extraction and web searching. *SIAM Rev.*, 46(4):647–666, 2004.
4. J.A. Bondy, X. Buchwalder, and F. Mercier. Lexicographic products and a conjecture of Hahn and Jackson. *SIAM J. Discrete Math.*, 23(2):882–887, 2009.
5. S. Chen and X. Chen. Weak connectedness of tensor product of digraphs. *Discrete Appl. Math.*, 185:52–58, 2015.
6. C. Crespelle and E. Thierry. Computing the directed Cartesian-product decomposition of a directed graph from its undirected decomposition in linear time. *Discrete Math.*, 338(12):2393–2407, 2015.
7. L.J. Culp and R. Hammack. An isomorphism theorem for digraphs. *Australas. J. Combin.*, 48:205–211, 2010.
8. W. Dörfler and W. Imrich. Das lexikographische Produkt gerichteter Graphen. *Monatsh. Math.*, 76:21–30, 1972.
9. J. Feigenbaum. Directed Cartesian-product graphs have unique factorizations that can be computed in polynomial time. *Discrete Appl. Math.*, 15(1):105–110, 1986.
10. G. Fertin, A. Raspaud, and A. Roychowdhury. On the oriented chromatic number of grids. *Inform. Process. Lett.*, 85(5):261–266, 2003.
11. J. Fox and B. Sudakov. Paths and stability number in digraphs. *Elec. J. Combin.*, 16(1):Note 23, 4, 2009.
12. T. Gallai and A.N. Milgram. Verallgemeinerung eines graphentheoretischen Satzes von Rédei. *Acta Sci. Math. Szeged*, 21:181–186, 1960.
13. M.R. Garey and D.S. Johnson. *Computers and intractability*. W. H. Freeman, San Francisco, 1979.
14. G. Hahn and B. Jackson. A note concerning paths and independence number in digraphs. *Discrete Math.*, 82(3):327–329, 1990.

15. R. Hammack. Cancellation of digraphs over the direct product. *European J. Combin.*, 34(5):846–858, 2013.
16. R. Hammack and W. Imrich. On Cartesian skeletons of graphs. *Ars Math. Contemp.*, 2(2):191–205, 2009.
17. R. Hammack and W. Imrich. Unique prime factorization of digraphs over the direct product. preprint, 2017.
18. R. Hammack, W. Imrich, and S. Klavžar. *Handbook of Product Graphs.* Discrete Mathematics and its Applications (Boca Raton). CRC Press, Boca Raton, FL, second edition, 2011. With a foreword by Peter Winkler.
19. R. Hammack and H. Smith. Zero divisors among digraphs. *Graphs Combin.*, 30(1):171–181, 2014.
20. R. Hammack and K.E. Toman. Cancellation of direct products of digraphs. *Discuss. Math. Graph Theory*, 30(4):575–590, 2010.
21. F. Harary and C.A. Trauth, Jr. Connectedness of products of two directed graphs. *SIAM J. Appl. Math.*, 14:250–254, 1966.
22. D.J. Hartfiel and C.J. Maxson. The chainable matrix, a special combinatorial matrix. *Discrete Math.*, 12:245–256, 1975.
23. P. Hell and J. Nešetřil. *Graphs and Homomorphisms.* Oxford Univ. Press, Oxford, 2004.
24. M. Hellmuth and T. Marc. On the Cartesian skeleton and the factorization of the strong product of digraphs. *Theoret. Comput. Sci.*, 565:16–29, 2015.
25. W. Imrich, S. Klavžar, and D. F. Rall. *Topics in Graph Theory: Graphs and their Cartesian Product.* A K Peters, Ltd., Wellesley, MA, 2008.
26. W. Imrich and W. Klöckl. Factoring directed graphs with respect to the cardinal product in polynomial time. *Discuss. Math. Graph Theory*, 27(3):593–601, 2007.
27. W. Imrich and W. Klöckl. Factoring directed graphs with respect to the cardinal product in polynomial time II. *Discuss. Math. Graph Theory*, 30(3):461–474, 2010.
28. K. Keating. Multiple-ply Hamiltonian graphs and digraphs. In *Cycles in graphs (Burnaby, B.C., 1982)*, volume 115 of *North-Holland Math. Stud.*, pages 81–87. North-Holland, Amsterdam, 1985.
29. B.M. Kim, B.C. Song, and W. Hwang. Exponents and diameters of strong products of digraphs. *Elec. J. Linear Algebra*, 22:1106–1111, 2011.
30. B.M. Kim, B.C. Song, and W. Hwang. The exponent of Cartesian product of cycles. *Appl. Math. Lett.*, 22(4):561–564, 2009.
31. B.M. Kim, B.C. Song, and W. Hwang. The exponent of a digraph and the diameter of its multiple direct product. *Linear Algebra Appl.*, 437(10):2601–2612, 2012.
32. B.M. Kim, B.C. Song, and W. Hwang. Wielandt type theorem for Cartesian product of digraphs. *Linear Algebra Appl.*, 429(4):841–848, 2008.
33. M. Kwaśnik. *$(k,l)$-kernels in graphs and their products.* PhD thesis, Wrocław, 1980.
34. L. Lakshmi and Vidhyapriya. Kernels in Cartesian products of digraphs. *Australas. J. Combin.*, 66(2):229–239, 2016.
35. J. Liu, X. Zhang, and J. Meng. On domination number of Cartesian product of directed paths. *J. Comb. Optim.*, 22(4):651–662, 2011.
36. L. Lovász. On the cancellation law among finite relational structures. *Period. Math. Hungar.*, 1(2):145–156, 1971.
37. M.H. McAndrew. On the product of directed graphs. *Proc. Amer. Math. Soc.*, 14:600–606, 1963.
38. R. McKenzie. Cardinal multiplication of structures with a reflexive relation. *Fund. Math.*, 70(1):59–101, 1971.

39. L.L. Ng. Hamiltonian decomposition of lexicographic products of digraphs. *J. Combin. Theory Ser. B*, 73(2):119–129, 1998.
40. M. Norge, P. Labarr, B. Sanders, and D. Taylor. Kings in products of digraphs. preprint, 2017.
41. P. Paulraja and S. Sivasankar. Directed Hamilton cycle decompositions of the tensor products of symmetric digraphs. *Graphs Combin.*, 25(4):571–581, 2009.
42. S. Poljak. Coloring digraphs by iterated antichains. *Comment. Math. Univ. Carolin.*, 32(2):209–212, 1991.
43. S. Poljak and V. Rödl. On the arc-chromatic number of a digraph. *J. Combin. Theory Ser. B*, 31(2):190–198, 1981.
44. G. Schaar, M. Sonntag, and H.-M. Teichert. *Hamiltonian properties of products of graphs and digraphs*, volume 108 of *Teubner-Texte zur Mathematik [Teubner Texts in Mathematics]*. BSB B. G. Teubner Verlagsgesellschaft, Leipzig, 1988. With German, French and Russian summaries.
45. R. Shaheen. Domination number of toroidal grid digraphs. *Util. Math.*, 78:175–184, 2009.
46. R. Shaheen. On the domination number of Cartesian products of two directed paths. *Intern. J. Contemp. Math. Sci.*, 7(33-36):1785–1790, 2012.
47. R. Shaheen. Total domination number of products of two directed cycles. *Util. Math.*, 92:235–250, 2013.
48. R. Shaheen. Total domination number of strong products of directed cycles. *Util. Math.*, 99:89–100, 2016.
49. M. Sonntag and H.-M. Teichert. Competition hypergraphs of products of digraphs. *Graphs Combin.*, 25(4):611–624, 2009.
50. M. Sonntag and H.-M. Teichert. Products of digraphs and their competition graphs. *Discuss. Math. Graph Theory*, 36(1):43–58, 2016.
51. E. Sopena. Oriented graph coloring. *Discrete Math.*, 229(1-3):359–369, 2001. Combinatorics, graph theory, algorithms and applications.
52. R. Stong. Hamilton decompositions of directed cubes and products. *Discrete Math.*, 306(18):2186–2204, 2006.
53. A. Szepietowski and M. Targan. A note on the oriented chromatic number of grids. *Inform. Process. Lett.*, 92(2):65–70, 2004.
54. W. Szumny, A. Włoch, and I. Włoch. On $(k,l)$-kernels in $D$-join of digraphs. *Discuss. Math. Graph Theory*, 27(3):457–470, 2007.
55. C. Tardif. Chromatic numbers of products of tournaments: fractional aspects of Hedetniemi's conjecture. In *Graphs, morphisms and statistical physics*, volume 63 of *DIMACS Ser. Discrete Math. Theoret. Comput. Sci.*, pages 171–175. Amer. Math. Soc., Providence, RI, 2004.
56. C. Tardif and D. Wehlau. Chromatic numbers of products of graphs: the directed and undirected versions of the Poljak-Rödl function. *J. Graph Theory*, 51(1):33–36, 2006.
57. W.T. Trotter, Jr. and P. Erdős. When the Cartesian product of directed cycles is Hamiltonian. *J. Graph Theory*, 2(2):137–142, 1978.
58. H. Wielandt. Unzerlegbare, nicht negative Matrizen. *Math. Z.*, 52:642–648, 1950.
59. A. Włoch and I. Włoch. On $(k,l)$-kernels in generalized products. *Discrete Math.*, 164(1-3):295–301, 1997. The Second Krakow Conference on Graph Theory (Zgorzelisko, 1994).

# 11. Miscellaneous Digraph Classes

**Yubao Guo and Michel Surmacs**

## 11.1 Introduction

Obviously, there are countless digraph classes, so that any attempt to give a complete overview is doomed to failure. One has to restrict oneself to a selection. Some will be presented in their own chapter or section, some will only be mentioned for some specific results throughout the book and some won't be mentioned at all. As tournaments (`tou`) are arguably the best studied class of digraphs with a rich library of strong results (see Chapter 2), their prominent place in any selection is a given. Unsurprisingly, several authors have tried to generalize the class in different directions in order to obtain larger classes of digraphs while retaining enough structure that most central results on tournaments still hold. Those classes include semicomplete digraphs (`scd`) (see Chapter 2), multipartite tournaments (`mut`) (see Chapter 7) and local tournaments (`lct`) (see Chapter 6). Results on hypertournaments (`hyt`), an extension of tournaments to directed hypergraphs that is not featured in this book, have been obtained by Q. Guo, Y. Guo, Gutin, Kayibi, Khan, Koh, H. Li, R. Li, S. Li, Lu, Ning, Petrović, Pirzada, Ree, Surmacs, Thomassen, Wang, Yang, Yao, Yeo, K.M. Zhang, X. Zhang and Zhou (see, e.g., [77–79, 100–102, 104, 109–111, 126, 145, 154, 166, 173]).

Several of those tournament generalizations have since been generalized themselves, resulting in an array of tournament-related digraph classes. Locally semicomplete digraphs (`lsd`), round digraphs (`rod`), in/out-round digraphs (`ird`), locally in/out-tournaments (`lit`), locally in/out-semicomplete digraphs (`lis`) and path-mergeable digraphs (`pmd`) are considered in Chapter 6. Chapter 7 is dedicated to semicomplete multipartite digraphs (`smd`). Results on transitive digraphs (`trd`), $k$-transitive digraphs (`ktd`), quasi-transitive digraphs (`qtd`) and $k$-quasi-transitive digraphs (`kqt`) can be found in Chapter 8.

In Section 11.8 of this chapter, we will consider another generalization of both semicomplete and semicomplete bipartite digraphs: Arc-locally

Y. Guo and M. Surmacs (✉)
Lehrstuhl C für Mathematik, RWTH Aachen University,
Pontdriesch 12-14, 52062, Aachen, Germany
e-mail: surmacs@mathc.rwth-aachen.de



J. Bang-Jensen and G. Gutin (eds.), *Classes of Directed Graphs*, Springer Monographs in Mathematics, https://doi.org/10.1007/978-3-319-71840-8_11

semicomplete digraphs (`als`). They are themselves generalized by $\mathcal{H}_1$-free digraphs (`h1f`) and $\mathcal{H}_2$-free digraphs (`h2f`) in Section 11.9. The related classes of $\mathcal{H}_3$-free digraphs (`h3f`) and $\mathcal{H}_4$-free digraphs (`h4f`) are also briefly considered.

Of course, there are also digraph classes (fairly) unrelated to tournaments such as acyclic digraphs (`acd`), investigated in Chapter 3. Kernel-perfect digraphs (`kpd`) are mentioned in several results throughout the book, for example in Section 11.7, which is mainly dedicated to perfect digraphs (`ped`), game-perfect digraphs (`gpd`) and weakly game-perfect digraphs (`wgp`).

Many digraph classes appear naturally in applications to other fields such as mathematical logic or computer science. One such class is that of circulant digraphs (`cid`), which have been considered by such authors as Alspach, Burkard, Çela, Parsons, Van Doorn, Woeginger and Yang (see, e.g., [2, 151, 167]). They include regular round digraphs (`rrd`) and are themselves included in the class of Cayley digraphs (`cad`), whose properties have been investigated, for example, by Curran, Gallian, Hamidoune, Parhami, Rankin, Witte, Xiao and Xu (see, e.g., [44, 81, 132, 160–162, 164]).

Two classes which also have applications in the construction of interconnection networks (see [27] for a survey by Bermond, Homobono and Peyrat) are de Bruijn digraphs (`dbd`) and Kautz digraphs (`kad`), which we will consider in Sections 11.4 and 11.5, respectively. Both classes can be defined using the line digraph operator, which will be investigated more closely in Section 11.2 on line digraphs (`lnd`) and Section 11.3 on iterated line digraphs (`ild`).

The closely related minimal series-parallel digraphs (`msp`), series-parallel digraphs (`spd`) and series-parallel partial order digraphs (`spo`), appear in flow diagrams and dependency charts and have an application to the problem of scheduling under constraints. We will consider them briefly in Section 11.6 on directed cographs (`dco`), a generalization of series-parallel partial order digraphs.

Figure 11.1 gives a first overview of how the previously mentioned classes relate to one another. For more structure, we also include the subclasses of loopless line digraphs (`lld`) and loopless iterated line digraphs (`lil`). Class $x$ is included in class $y$ if the depicted digraph contains an $(x, y)$-path. Obviously, neither the list of considered digraph classes nor the relations depicted are necessarily exhaustive.

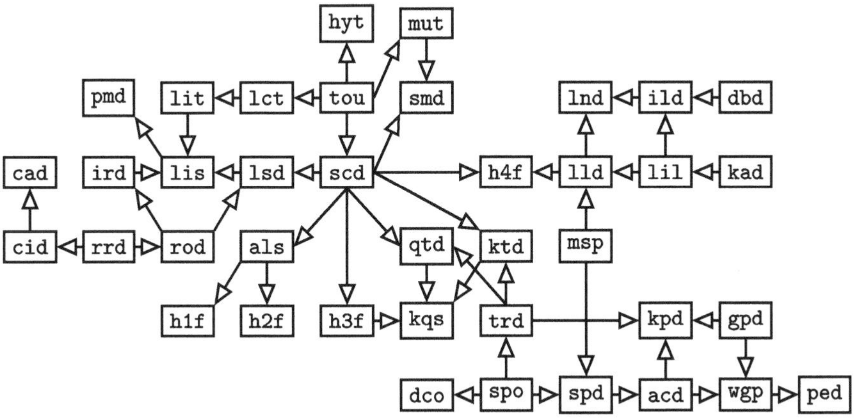


**Figure 11.1** Digraph depicting relations between digraph classes.

Note that we omitted certain digraph classes, such as Euler digraphs (see Chapter 4), planar digraphs (see Chapter 5), digraphs with bounded width (see Chapter 9), digraph products (see Chapter 10) and underlying graphs of digraphs (see Chapter 12), mostly because they intersect many others but are not contained in / do not contain other classes. Intersection digraphs on the other hand include all digraphs, as Beineke and Zamfirescu [23] and Sen, Das, Roy and West [139] proved, which makes their inclusion in the figure redundant. For further results on intersection digraphs and their subclass of interval digraphs, however, we also refer to work by Brown, Busch, Dasgupta, Feder, Francis, Hell, Huang, Lundgren, Müller, Rafiey, Sanyal and Talukdar (see, e.g., [35, 45–47, 62, 120, 138, 140, 141, 159, 168]).

## 11.2 Line Digraphs

Krausz [105] defined the line graph $L(G)$ of a graph $G = (V, E)$ to be the graph with vertex set $E$ and an edge between $e, f \in E$, if and only if $e$ and $f$ are incident in $G$. Since then, differing generalizations of the concept for directed pseudographs have been introduced. The most common definition for the line digraph $L(D)$ of a directed pseudograph $D = (V, A)$ – and the only one we will consider here – is due to Harary and Norman [82]. Corresponding to the undirected version, the vertex set of $L(D)$ is the arc set $A$ of $D$. Due to the orientation of arcs, there is the additional choice of when and how to connect two vertices $a, b \in A$ of $L(D)$, which distinguishes the competing concepts of line digraphs. Here, $(a, b)$ is an arc of $L(D)$ if and only if the head of $a$ coincides with the tail of $b$. In other words, $ab$ is a directed walk of length 2 in $D$. Note that the line digraph $L(D)$ does not contain multiple arcs, but contains a loop, if and only if $D$ contains a loop. Therefore, technically, the line digraph of a directed pseudograph containing a loop is not a digraph, but again a directed pseudograph.

A directed pseudograph $D$ is called a line digraph if $D = L(D')$ for some directed pseudograph $D'$.

The first easy observation Harary and Norman [82] then made is the following.

**Theorem 11.2.1** ([82]) *Let $D$ be a directed pseudograph. Then,*

$$|V(L(D))| = |A(D)| \quad \textit{and} \quad |A(L(D))| = \sum_{v \in V(D)} d_D^-(v) d_D^+(v).$$

Another nice property that directly follows from the definition is the invariance of the minimum and maximum semi-degree under the line digraph operator.

**Proposition 11.2.2** *Let $D = (V, A)$ be a directed pseudograph. Then,*

$$d_{L(D)}^+(xy) = d_D^+(y) \quad \textit{and} \quad d_{L(D)}^-(xy) = d_D^-(x) \quad \textit{for all } xy \in A.$$

*Particularly,*

$$\delta^0(L(D)) = \delta^0(D) \quad \textit{and} \quad \Delta^0(L(D)) = \Delta^0(D).$$

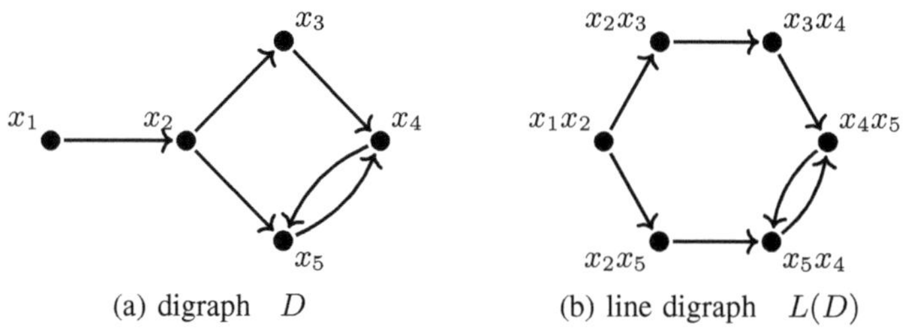


**Figure 11.2** A digraph and its line digraph.

In the following theorem we collect a number of characterizations of line digraphs. Characterization (ii) is among the first results on line digraphs and due to Harary and Norman [82]. Later, Heuchenne [90] found the local criterion (iii) and Richards [137], in (iv) and (v), considered adjacency matrices to determine line digraphs, for which we recall the following definition. For a matrix $M = [m_{ik}] \in \{0,1\}^{n \times n}$, a row $i$ is **orthogonal** to a row $j$ if $\sum_{k=1}^{n} m_{ik} m_{jk} = 0$. One can give a similar definition of orthogonal columns. Conditions (ii) and (iii) have each been rediscovered by several authors, as Hemminger and Beineke [88] found in their survey on line graphs and line digraphs. The proof presented here is also adapted from that survey.

**Theorem 11.2.3** *Let $D = (V, A)$ be a directed pseudograph with vertex set $V = \{1, 2, \ldots, n\}$ and with no multiple arcs and let $M = [m_{ij}]$ be its adjacency matrix (i.e., the $n \times n$-matrix such that $m_{ij} = 1$, if $ij \in A(D)$, and $m_{ij} = 0$, otherwise). Then the following assertions are equivalent:*

*(i) $D$ is a line digraph;*
*(ii) there exist two partitions $\{A_i\}_{i \in I}$ and $\{B_i\}_{i \in I}$ of $V(D)$ such that*

$$A(D) = \bigcup_{i \in I} A_i \times B_i;$$

*(iii) if $vw, uw$ and $ux$ are arcs of $D$, then so is $vx$;*
*(iv) any two rows of $M$ are either identical or orthogonal;*
*(v) any two columns of $M$ are either identical or orthogonal.*

**Proof:** We prove the following implications and equivalences: (i) $\Leftrightarrow$ (ii), (ii) $\Rightarrow$ (iii), (iii) $\Rightarrow$ (iv), (iv) $\Leftrightarrow$ (v), (iv) $\Rightarrow$ (ii).

(i) $\Rightarrow$ (ii). Let $D = L(H)$. For each $v_i \in V(H)$, let $A_i$ and $B_i$ be the sets of in-coming and out-going arcs at $v_i$, respectively. Then the arc set of the subdigraph of $D$ induced by $A_i \cup B_i$ equals $A_i \times B_i$. If $ab \in A(D)$, then there is an $i$ such that $a = v_j v_i$ and $b = v_i v_k$. Hence, $ab \in A_i \times B_i$. The result follows.

(ii) $\Rightarrow$ (i). Let $Q$ be the directed pseudograph with ordered pairs $(A_i, B_i)$ as vertices, and with $|A_j \cap B_i|$ arcs from $(A_i, B_i)$ to $(A_j, B_j)$ for each $i$ and $j$ (including $i = j$). Let $\sigma_{ij}$ be a bijection from $A_j \cap B_i$ to this set of arcs (from $(A_i, B_i)$ to $(A_j, B_j)$) of $Q$. Then the function $\sigma$ defined on $V(D)$ by taking $\sigma$ to be $\sigma_{ij}$ on $A_j \cap B_i$ is a well-defined function of $V(D)$ into $V(L(Q))$, since $\{A_j \cap B_i\}_{i,j \in I}$ is a partition of $V(D)$. Moreover, $\sigma$ is a bijection since every $\sigma_{ij}$ is a bijection. Furthermore, it is not difficult to see that $\sigma$ is an isomorphism from $D$ to $L(Q)$.

(ii) $\Rightarrow$ (iii). If $vw, uw$ and $ux$ are arcs of $D$, then there exist $i, j$ such that $\{u, v\} \subseteq A_i$ and $\{w, x\} \subseteq B_j$. Hence, $(v, x) \in A_i \times B_j$ and $vx \in D$.

(iii) $\Rightarrow$ (iv). Assume that (iv) does not hold. This means that some rows, say $i$ and $j$, are neither identical nor orthogonal. Then there exist $k, h$ such that $m_{ik} = m_{jk} = 1$ and $m_{ih} = 1, m_{jh} = 0$ (or vice versa). Hence, $ik, jk, ih$ are in $A(D)$ but $jh$ is not. This contradicts (iii).

(iv) $\Leftrightarrow$ (v). Both (iv) and (v) are equivalent to the statement:

$$\text{for all } i, j, h, k, \text{ if } m_{ih} = m_{ik} = m_{jk} = 1, \text{ then } m_{jh} = 1.$$

(iv) $\Rightarrow$ (ii). For each $i$ and $j$ with $m_{ij} = 1$, let $A_{ij} = \{h : m_{hj} = 1\}$ and $B_{ij} = \{k : m_{ik} = 1\}$. Then, by (iv), $A_{ij}$ is the set of vertices in $D$ whose row vectors in $M$ are identical to the $i$th row vector, whereas $B_{ij}$ is the set of vertices in $D$ whose column vectors in $M$ are identical to the $j$th column vector (we use the previously proved fact that (iv) and (v) are equivalent).

Thus, $A_{ij} \times B_{ij} \subseteq A(D)$, and moreover $A(D) = \bigcup\{A_{ij} \times B_{ij} : m_{ij} = 1\}$. By the orthogonality condition, $A_{ij}$ and $A_{hk}$ are either equal or disjoint, as are $B_{ij}$ and $B_{hk}$. For a zero row vector $i$ in $M$, let $A_{ij}$ be the set of vertices whose row vector in $M$ is the zero vector, and let $B_{ij} = \emptyset$. Doing the same with the zero column vectors of $M$ completes the partition as in (ii). □

The characterizations (ii)–(v) all imply polynomial algorithms to verify whether a given directed pseudograph is a line digraph. For an example of an effective polynomial algorithm using (ii) to recognize acyclic line digraphs, see [16, Page 42]. Criterion (iii) can also be reformulated to obtain a characterization of line digraphs in terms of forbidden induced subdigraphs.

**Corollary 11.2.4** ([16]) *A directed pseudograph $D$ is a line digraph if and only if $D$ does not contain, as an induced subdigraph, any directed pseudograph that can be obtained from one of the directed pseudographs in Figure 11.3 (dashed arcs are missing) by adding zero or more arcs (other than the dashed ones).*

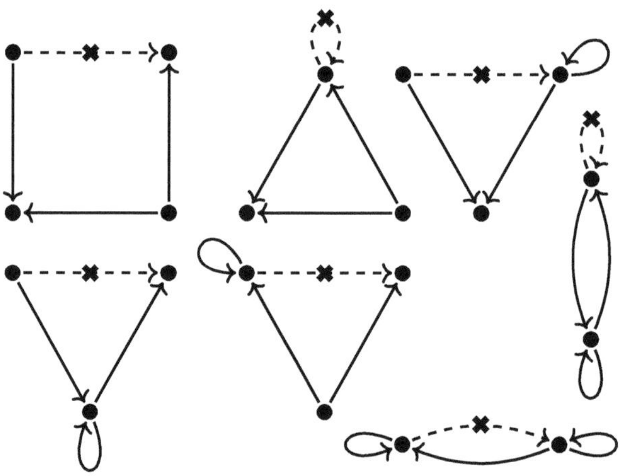

**Figure 11.3** Forbidden directed pseudographs of line digraphs.

Observe that the digraph of order 4 in Figure 11.3 corresponds to the case of distinct vertices in Part (iii) of Theorem 11.2.3, and the two directed pseudographs of order 2 correspond to the cases $x = u \neq v = w$ and $u = w \neq v = x$, respectively.

From Corollary 11.2.4 a simpler characterization of the line digraphs of digraphs (i.e. without loops and multiple arcs) is easily obtained by omitting those forbidden induced subdigraphs that imply said loops or parallel arcs.

More details and further characterizations of special classes of line digraphs can be found in the surveys by Hemminger and Beineke [88] and Prisner [131].

As, for every line digraph $Q$, under '(ii) $\Rightarrow$ (i)' in Theorem 11.2.3 a directed pseudograph $D$ such that $Q = L(D)$ is constructed, it is natural to ask whether $D$ is unique with said property. Harary and Norman [82] answered the question in the negative, but recognized that two directed pseudographs with the same line digraph cannot differ too much, as the following theorem shows.

**Theorem 11.2.5** ([82]) *Let $D$ and $D'$ be directed pseudographs such that $L(D) = L(D')$. Then the directed pseudographs obtained from $D$ and $D'$, by deleting all vertices with in-degree 0 and all vertices with out-degree 0, are isomorphic.*

Prisner [130] found that under certain circumstances, even the consideration of the underlying graph may be enough to determine quasi-uniqueness, a generalization of results due to Villar [152].

**Theorem 11.2.6** ([130]) *Let $D$ and $D'$ be directed pseudographs without parallel arcs and both of minimum semi-degree at least 2. Then $UG(L(D)) \cong UG(L(D'))$ implies that $D$ is isomorphic to $D'$ or its converse.*

Harary and Norman [82] also gave a partial answer to the related question of which directed pseudographs are isomorphic to their line digraph.

**Theorem 11.2.7** ([82]) *Let $D$ be a unilateral (i.e., any two vertices are connected by a directed path in at least one direction) directed pseudograph without multiple arcs. Then $D$ is isomorphic to $L(D)$ if and only if each of its vertices has in-degree 1 or each of its vertices has out-degree 1.*

Aigner [1] then gave a generalization of this result.

**Theorem 11.2.8** ([1]) *Let $D$ be a directed pseudograph without isolated vertices. Then $D$ is isomorphic to $L(D)$, if and only if $D \cong D_1 \cup \ldots \cup D_k$, where the $D_i$s are mutually vertex-disjoint and either $D_i$ consists of a directed cycle and a number (possibly zero) of out-trees, each rooted at a vertex of this cycle or $D_i$ is the converse of such a digraph.*

Harary and Norman [82] provided corresponding examples which show that these characterizations do not hold for general directed pseudographs. Therefore, finding a general characterization is still an open problem.

### 11.2.1 Connectivity

In most applications, connectivity plays a vital role. Thus, Aigner's [1] result that strong connectivity is preserved under the line digraph operator is particularly useful.

**Theorem 11.2.9** ([1]) *Let $D$ be a directed pseudograph without isolated vertices. Then $D$ is strongly connected if and only if $L(D)$is strongly connected. Furthermore, $L(D)$ being unilateral implies $D$ is unilateral.*

Several authors then noted the following (see, e.g., [171]).

**Proposition 11.2.10** *Let $D$ be a directed pseudograph without parallel arcs. Then,*

$$\kappa(L(D)) = \lambda(D).$$

Therefore, by the well-known fact that $\kappa(D) \leq \lambda(D) \leq \delta^0(D)$ for any directed pseudograph $D$ without parallel arcs and Proposition 11.2.2, we have

$$\kappa(D) \leq \lambda(D) = \kappa(L(D)) \leq \lambda(L(D)) \leq \delta^0(L(D)) = \delta^0(D).$$

In other words, application of the line digraph operator can only increase the connectivity, which is one of the reasons it has been used in the construction of interconnection networks (see also the following section on iterated line digraphs). In this context, more refined connectivity concepts, as a measure of reliability, such as super connectivity, introduced by Bauer, Boesch, Suffel and Tindell [21], have been considered. A separator (cut) of a directed pseudograph is called **trivial** if its removal yields a strong component of order 1. In other words, all in-neighbours or all out-neighbours (or the corresponding arcs, respectively) of a vertex are contained in the separator (cut). A directed pseudograph $D$ has **super (vertex-)connectivity** $k$ if $\kappa(D) = k$ and every minimum separator is trivial. Analogously, a $D$ has **super arc-connectivity** $k$ if $\lambda(D) = k$ and every minimum cut is trivial. Obviously, super connectivity implies maximum fault tolerance, in some sense.

Although not every cut of $D$ is a separator of $L(D)$, we still get the following natural-feeling result, due to Cheng, Du, Min, Ngo, Ruan, Sun and Wu [38], which was rediscovered by Zhang, Liu and Meng [171] with a more precise proof.

**Proposition 11.2.11** ([38]) *Let $D$ be a strongly connected directed pseudograph without parallel arcs. Then, $D$ has super arc-connectivity $k$ if and only if $L(D)$ has super connectivity $k$.*

Furthermore, Cheng, *et al.* [38] claimed that super arc-connectivity is preserved by the line digraph operator. Their proof is incorrect and the claim false, as, for example, the complete digraph on 3 vertices is super arc-connected, but its line digraph is not. However, Zhang, *et al.* [171] obtained a weaker version of the claim as a corollary of the following theorem.

**Theorem 11.2.12** ([171]) *Let $D$ be a strongly connected directed pseudograph without parallel arcs with $\delta^0(D) \geq 3$. Then, if $L(D)$ has super connectivity $k$, it also has super arc-connectivity $k$.*

Now, we simply combine Proposition 11.2.11 and Theorem 11.2.12.

**Corollary 11.2.13** ([171]) *Let $D$ be a strongly connected directed pseudograph without parallel arcs with $\delta^0(D) \geq 3$. If $D$ has super arc-connectivity $k$, then $L(D)$ has super arc-connectivity $k$.*

Lü and Xu [115] and Zhang and Zhu [172] published results on even more refined connectivity measures for line digraphs.

### 11.2.2 Diameter

In the previous subsection we have seen that strong connectivity is preserved under the line digraph operator. As a consequence, it is natural to ask whether the distances between vertices increase drastically, since the number of vertices of the line digraph may possibly be up to almost the square of the order of the corresponding digraph. In spite of this fact, Aigner [1] was able to prove that the diameter increases by at most 1 under the line digraph operator.

**Theorem 11.2.14** ([1]) *Let $D$ be a strongly connected directed pseudograph. Then,*

$$\operatorname{diam}(L(D)) = \operatorname{diam}(D) + 1,$$

*unless $D \cong L(D)$ (i.e., $D$ is a directed cycle).*

As we already know that the maximum semi-degree is also invariant, iterated application of the line digraph operator to the right digraphs is predestined to obtain digraphs of high order and comparatively small degree and diameter (cf. Sections 11.4 and 11.5 on de Bruijn and Kautz digraphs, respectively).

### 11.2.3 Kernels, Solutions and Generalizations

Another popular distance related concept are kernels of digraphs. Introduced in the context of game theory by von Neumann and Morgenstern [153], they have since found a wide array of applications in other fields.

A set $N$ of vertices of a digraph $D$ is called a **kernel** of $D$ if $N$ is independent in $D$ and for every vertex $u \in V(D) \setminus N$, there is a vertex $v \in N$ such that $uv \in A(D)$. A **solution** of $D$ is a kernel of the converse of $D$.

Since the introduction of the concept, several generalizations of kernels have been considered, many of which can be described as $(k,l)$-kernels. A set $N$ of vertices of a digraph $D$ is called a **$(k,l)$-kernel** of $D$ if there is no oriented path of length shorter than $k$ between any two distinct vertices of $N$ in $D$ and for every vertex $u \in V(D) \setminus N$, there is a directed path of length at most $l$ from $u$ to a vertex in $N$ in $D$. Now, obviously, a $(2,1)$-kernel is a common kernel. Furthermore, a $(k,k-1)$-kernel is also called a **$k$-kernel**, a $(2,2)$-kernel is called a **quasi-kernel** and a $(k,2(k-1))$-kernel is called a **$k$-quasi-kernel**.

A $(k,l)$-semikernel is defined slightly differently. A set $N$ of vertices of a digraph $D$ is called a **$(k,l)$-semikernel** of $D$ if there is no oriented path of length shorter than $k$ between any two distinct vertices of $N$ in $D$ and for every vertex $u \in V(D) \setminus N$, if there is a directed path of length at most $l$ from a vertex in $N$ to $u$ in $D$, then there is such a path from $u$ to a vertex in $N$. A $(k,k-1)$-kernel is also called a **$k$-semikernel** and a 2-semikernel is also called a **semikernel**.

For all these generalized concepts of kernels, again, a corresponding version of a solution can be defined by considering the converse digraph.

Harminc [83] considered the correlation between solutions of a digraph and its line digraph and found the following.

**Theorem 11.2.15** ([83]) *The cardinality of the system of all solutions of a digraph is equal to the cardinality of the system of all solutions of its line digraph.*

More precisely, for a digraph $D = (V, A)$, he proved that $f : \mathcal{K} \to \mathcal{K}'$, $S \mapsto \{xy \in A \mid x \in S, y \in V\}$, where $\mathcal{K}$ and $\mathcal{K}'$ are the systems of all solutions of $D$ and its line digraph, respectively, is an injective function. Conversely, $g : \mathcal{K}' \to \mathcal{K}$, $H \mapsto X(H) \cup Y(H)$, where $X(H)$ is the set of all tails of arcs in $H$ and $Y(H)$ consists of those vertices of $D$ with out-degree 0 that are not adjacent to any vertices in $X(H)$, is also shown to be injective. Thus, we can easily obtain the kernels of $L(D)$ from the kernels of $D$ and vice versa.

**Proof:** **$f$ is well-defined**: Let $R$ be a solution of $D = (V, A)$. Suppose that $ab \in A(L(D))$ for $a, b \in f(r)$. Then, by the definition of $f$, the tails of both $a$ and $b$ are contained in $R$ and they are connected by the arc $a \in A$, a contradiction to the choice of $R$. Now, let $b \in A \setminus f(R)$. By the definition of $f$, the tail of $b$ is not in $R$ and is therefore dominated by some vertex of $R$ in $D$ via some arc $a \in f(R)$. Hence, $b$ is dominated by $a$ and, all in all, $f(R)$ is a solution of $L(D)$.

**$f$ is injective**: Let $R$ and $S$ be distinct solutions of $D$. Without loss of generality, we may assume that there is a vertex $y \in R \backslash S$. Since $S$ is a solution of $D$, there is a vertex $x \in S$ such that $xy \in A$ and therefore, $xy \in f(S)$. The independence of $R$ implies that $xy \notin f(R)$. Hence, $f(R) \neq f(S)$.

**$g$ is well-defined**: Let $R$ be a solution of $D = (V, A)$. Suppose that there are vertices $x, y \in g(R)$ such that $xy \in A$. If $x \in Y(R)$ or $y \in Y(R)$, the definition of $Y(R)$ immediately implies a contradiction. Thus, we may assume that $x, y \in X(R)$. Consequently, $x$ is the tail of some arc $a \in R$ and $y$ is the tail of some arc $b \in R$. The independence of $R$ implies $xy \notin R$, as $xy$ and $b$ are connected in $L(D)$. Hence, there exists a $c \in R$ that dominates $xy$ in $L(D)$ and, by definition of the line digraph, also dominates $a \in R$, a contradiction to the choice of $R$. Now, let $y \in V \setminus g(R)$. If $y$ is the head of some arc $b \in A$, then, by the definition of $g(R)$, $b \notin R$. Therefore, $b$ is dominated by some $a = xy \in R$ and hence, $y$ is dominated by $x \in X(R) \subseteq g(R)$. If $y$ has out degree 0, by the definition of $g(R)$, $y$ is dominated by some $x \in X(R) \subseteq g(R)$. All in all, $g(R)$ is a solution of $D$.

$g$ **is injective**: Let $R$ and $S$ be distinct solutions of $L(D)$. Without loss of generality, we may assume that there is an arc $b = yz \in R \setminus S$. Therefore, $y \in X(R) \subseteq g(R)$. Since $b \notin S$, there is some arc $a = xy \in S$ that dominates $b$. As $x \in X(S) \subseteq g(S)$, the independence of $g(S)$ implies $y \notin g(S)$. Hence, $g(R) \neq g(S)$. □

As an obvious corollary, we have the following.

**Corollary 11.2.16** ([83]) *A digraph has a solution if and only if its line digraph has a solution.*

The easily seen fact that the converse of $L(D)$ is the line digraph of the converse of the digraph $D$ immediately implies the corresponding results on kernels.

**Corollary 11.2.17** ([83]) *The cardinality of the system of all kernels of a digraph is equal to the cardinality of the system of all kernels of its line digraph.*

**Corollary 11.2.18** ([83]) *A digraph has a kernel if and only if its line digraph has a kernel.*

Since then, utilizing Harminc's functions, several authors have found similar results for the various generalizations of kernels. Galeana-Sánchez, Ramírez and Rincón-Mejía [71] compared the number of semikernels and quasi-kernels of digraphs $D$ with $\delta^-(D) \geq 1$ with the respective numbers of their line digraphs. Galeana-Sánchez and Li [70] proved that Corollary 11.2.18 also holds for semikernels, if $\delta^-(D) \geq 1$, which is a necessary condition, and studied the relationship between the number of $(k,l)$-kernels of certain digraphs and their line digraphs.

Galeana-Sánchez and Gómez [69] provided, amongst other results, a weaker version of 11.2.17 for $(k,l)$-semikernels of certain digraphs, with the use of state splittings.

**Theorem 11.2.19** ([69]) *Let $k \geq 2$, $l \geq 2$ and let $D$ be a digraph with $g(D) \geq k$ and $\delta^-(D) \geq 1$. Then, the cardinality of the system of all $(k,l)$-semikernels of $D$ is less than or equal to the cardinality of the system of all $(k,l)$-semikernels of its line digraphs.*

Shan, Kang and Lu [142], found a generalization of Corollary 11.2.18 for $k$-semikernels of certain digraphs.

**Theorem 11.2.20** ([142]) *Let $D$ be a digraph with $g(D) \geq k \geq 2$ and $\delta^-(D) \geq 1$. Then, $D$ has a $k$-semikernel if and only if its line digraph has a $k$-semikernel.*

Lu, Shan and Zhao [116] proved that Harminc's functions are also well-defined and injective on the respective sets of $(k,l)$-kernels of certain digraphs and thereby obtained the following generalizations of Corollaries 11.2.17 and 11.2.18.

**Theorem 11.2.21** ([116]) *Let $k > l \geq 2$ and let $D$ be a digraph with $g(D) \geq k$ and $\delta^-(D) \geq 1$. Then, the cardinality of the system of all $(k,l)$-kernels of $D$ is equal to the cardinality of the system of all $(k,l)$-kernels of its line digraphs.*

**Theorem 11.2.22** ([116]) *Let $k > l \geq 2$ and let $D$ be a digraph with $g(D) \geq k$ and $\delta^-(D) \geq 1$. Then, $D$ has a $(k,l)$-kernel if and only if its line digraph has a $(k,l)$-kernel.*

Some additional results on kernels and related concepts in generalized line digraphs have been found by Balbuena and Guevara [12] and Guevara, Balbuena and Galeana-Sánchez [76].

### 11.2.4 Branchings

Recall that an in-branching (also called a rooted spanning tree or an arborescence in the literature) is an oriented spanning tree with exactly one vertex (the root) of out-degree 0. For a vertex $x$ of a directed pseudograph $D$, let $\mathrm{IB}_x(D)$ be the number of in-branchings of $D$ rooted at $x$.

Knuth [103] proved the following correlation (in a different form) between in-branchings of a directed pseudograph and those of its line digraph algebraically, using Tutte's Matrix Tree Theorem [147]. Orlin [125] gave a combinatorial proof of the theorem in its present form.

**Theorem 11.2.23** ([103]) *Let $D = (V, A)$ be a directed pseudograph without isolated vertices. Then,*

$$\mathrm{IB}_{xy}(L(D)) = \begin{cases} \mathrm{IB}_y(D) \cdot F, & \textit{if } d^+(y) = 0 \textit{ or } d^-(y) = 1 \\ d^+(y)^{-1}\,\mathrm{IB}_x(D) \cdot F, & \textit{otherwise,} \end{cases}$$

*where $F = \prod_{v \in V} d^+(v)^{d^-(v)-1}$.*

Among other results, Levine [108] found a generating function identity for digraphs with minimum in-degree 1, which implies the following formula for the total number of in-branchings of a line digraph.

**Corollary 11.2.24** ([108]) *Let $D = (V, A)$ be a directed pseudograph with $\delta^-(D) \geq 1$. Then, the number of in-branchings of $L(D)$ is*

$$b \cdot \prod_{v \in V} d^+(v)^{d^-(v)-1},$$

*where $b$ is the number of in-branchings of $D$.*

Bidkhori and Kishore [30] found another proof of the result by constructing an explicit bijection. Furthermore, it can be extended to iterated line digraphs (see Corollary 11.3.9). The following identity due to Orlin [125] implies that Corollary 11.2.24 is also a corollary of Theorem 11.2.23 and also holds for directed pseudographs without isolated vertices.

**Proposition 11.2.25** ([125]) *Let $D = (V, A)$ be a directed pseudograph without isolated vertices. Then, for each $y \in V$,*

$$d^+(y)\, \mathrm{IB}_y(D) \;=\; \sum_{x \in V} a_{xy}\, \mathrm{IB}_x(D),$$

*where $a_{xy}$ is the number of arcs from $x$ to $y$ in $D$.*

Branchings are not only interesting from a theoretical point of view, but, particularly in line digraphs, as a model of interconnection networks, for their practical use in broadcasting algorithms, that is to say, sending a message from one vertex to all others in an efficient manner. In this context, the sheer number of branchings in a digraph is less important than the number of arc-disjoint or independent branchings (for fault-tolerance) and their depth, which is to say the length of a longest directed path between the root and a leaf (for efficiency). Two out-branchings of a directed pseudograph with root $r$ are called **independent** if, for any vertex $x$, the unique paths from $r$ to $x$ are internally disjoint. Hasunuma and Nagamochi [85] studied both the number of independent out-branchings and their depths in line digraphs. Applying the following theorem, they were able to prove the well-known Independent Spanning Tree Conjecture (disproved in general by Huck [91]) for line digraphs.

**Theorem 11.2.26** ([85]) *Let $D$ be a directed pseudograph without parallel arcs and let $r$ be a vertex of $L(D)$. Suppose that for any vertex $v \neq r$ of $L(D)$, there are $k$ internally disjoint paths from $r$ to $v$ in $L(D)$. Then there are $k$ independent out-branchings rooted at $r$ of $L(D)$.*

**Corollary 11.2.27 (Independent Spanning Tree Conjecture** [85]) *Let $D$ be a directed pseudograph without parallel arcs. If $L(D)$ is $k$-strong, then there are $k$ independent out-branchings rooted at any vertex of $L(D)$.*

For considerations of the depth of independent out-branchings, see Theorems 11.3.10 and 11.3.11 in the section on iterated line digraphs.

Du, Lyuu and Hsu [51, 55, 56] introduced the related concept of spreads, prescribing a number of vertex-disjoint paths of certain maximum length between sets of vertices, to combine fault-tolerance and transmission delay considerations in interconnection networks and gave results on (iterated) line digraphs as an example of such networks.

Bermond, Munos and Marchetti-Spaccamela [28] proposed broadcasting algorithms for the (iterated) line digraph of a regular digraph $D$ based on a broadcasting protocol for $D$.

### 11.2.5 Cycles and Trails

Aigner [1] was the first to notice the natural relation between Euler trails in a digraph and Hamiltonian cycles in its line digraph.

**Theorem 11.2.28** ([1]) *Let $D$ be a directed pseudograph without isolated vertices. Then, $L(D)$ is Hamiltonian if and only if $D$ is Eulerian.*

The well-known characterization of Eulerian directed pseudographs and the definition of line digraphs lead to the following characterization of Eulerian line digraphs.

**Theorem 11.2.29** *Let $D$ be a strongly connected directed pseudograph. Then $L(D)$ is Eulerian if and only if $d_D^-(u) = d_D^+(v)$ for each arc $uv$ of $D$.*

For line graphs of strongly connected regular directed pseudographs, Aardenne-Ehrenfest and de Bruijn [150] determined the number of Euler trails contained, a result that can also be derived from Corollary 11.2.24.

**Theorem 11.2.30** ([150]) *Let $D$ be a strongly connected $d$-regular directed pseudograph of order $n$. Then, the number of Euler trails of $L(D)$ is*

$$d^{-1}(d!)^{n(d-1)} \cdot t,$$

*where $t$ is the number of Euler trails of $D$.*

Hasunuma and Otani [86] noted the following lower bound on the number of arc-disjoint Hamiltonian cycles in a regular line digraph.

**Theorem 11.2.31** ([86]) *Let $D$ be a strongly connected $d$-regular directed pseudograph without parallel arcs. Then there are $\lfloor d/2 \rfloor$ arc-disjoint Hamiltonian cycles in $L(D)$.*

As a generalization of pancyclicity (i.e. containing a cycle of every possible length), Imori, Matsumoto and Yamada [96] introduced the similar property of pancircularity. A directed pseudograph $D = (V, A)$ is called **pancircular** if it contains closed trails of length $\ell$ for all $3 \leq \ell \leq |A|$. As a first obvious result, they noted the following consequence of the fact that a cycle in $L(D)$ corresponds to a trail in $D$.

**Proposition 11.2.32** ([96]) *A directed pseudograph is pancircular if and only if its line digraph is pancyclic.*

For regular directed pseudographs, they gave a stronger result.

**Theorem 11.2.33** ([96]) *If a regular directed pseudograph is pancircular, then its line digraph is pancircular.*

Note that pancyclicity is not a sufficient condition in Theorem 11.2.33. Furthermore, it can be iterated (see Corollary 11.3.15).

### 11.2.6 $\mathcal{NP}$-Complete Problems for Line-Digraphs

The following results on $\mathcal{NP}$-completeness were published by Gavril [72]. He proved that several graph problems that are known to be $\mathcal{NP}$-complete on general (di)graphs (see, e.g., [98]), are still $\mathcal{NP}$-complete when restricted to line digraphs. The considered problems are the following.

SIMPLE MAX CUT **Parameter:** $k$
**Input:** An undirected graph $G = (V, E)$ and a positive integer $k$.
**Question:** Does there exist a set of vertices $S \subseteq V$ such that there are at least $k$ edges between $S$ and $V \setminus S$ in $G$?

INDEPENDENT SET **Parameter:** $k$
**Input:** A digraph $D = (V, A)$ and a positive integer $k$.
**Question:** Is there a set of vertices $S \subseteq V$ of size $k$ such that no vertex in $S$ dominates any other vertex in $S$?

VERTEX COVER **Parameter:** $k$
**Input:** A digraph $D = (V, A)$ and a positive integer $k$.
**Question:** Is there a set of vertices $S \subseteq V$ of size at most $k$ such that every vertex not in $S$ either dominates or is dominated by a vertex in $S$?

FEEDBACK VERTEX SET **Parameter:** $k$
**Input:** A digraph $D = (V, A)$ and a positive integer $k$.
**Question:** Is there a set of vertices $S \subseteq V$ of size at most $k$ such that $D - S$ is acyclic?

FEEDBACK ARC SET **Parameter:** $k$
**Input:** A digraph $D = (V, A)$ and a positive integer $k$.
**Question:** Is there a set of arcs $F \subseteq A$ of size at most $k$ such that $D - F$ is acyclic?

The reductions used below are partially based on private communication between Gavril and Knuth.

**Lemma 11.2.34** ([72]) SIMPLE MAX CUT *is reducible to* INDEPENDENT SET *for line digraphs.*

**Proof:** Given an undirected graph $G = (V, E)$ and a positive integer $k$, we consider the complete biorientation $D = \overleftrightarrow{G}$ of $G$ obtained by replacing each edge $\{x, y\}$ of $G$ with the pair $xy, yx$ of arcs. Now, for a cut $(S, V \setminus S)$ of size at least $k$ of $G$, the arc set $\{(x, y) \mid x \in S, y \in V \setminus S\}$ is an independent vertex set of size at least $k$ in $L(D)$. Conversely, for an independent vertex set $F$

of order $k$ of $L(D)$, let $S = \{x \in V \mid (x, y) \in F\}$. Since $F$ is independent in $L(D)$, $y \in V \setminus S$ for all $(x, y) \in F$ and thus, $(S, V \setminus S)$ is a cut of size at least $k$ of $G$. □

**Lemma 11.2.35** ([72]) Independent Set *for line digraphs is reducible to* Vertex Cover *for line digraphs.*

**Proof:** A set of vertices is independent if and only if its complement is a vertex cover. □

**Lemma 11.2.36** ([72]) Feedback Vertex Set *(FVS) for line digraphs is reducible to* Feedback Arc Set *(FAS) for line digraphs.*

**Proof:** Let $D = (V, A)$ be a line digraph. By Theorem 11.2.3, there exist two partitions $\{A_i\}_{i \in I}$ and $\{B_i\}_{i \in I}$ of $V$ such that $A = \cup_{i \in I} A_i \times B_i$. We define the digraph $D' = (V \times \{0, 1\}, A')$ through the partitions $\{A'_i\}_{i \in I'}$ and $\{B'_i\}_{i \in I'}$ of $V'$, where $I' = I \cup V$, $A'_i = \{(x, 1) \mid x \in A_i\}$, $B_i = \{(y, 0) \mid y \in B_i\}$ for $i \in I$ and $A'_i = \{(i, 0)\}$, $B'_i = \{(i, 1)\}$ for $i \in V$, and the arc set $A' = \cup_{i \in I'} A'_i \times B'_i$. Obviously, $D'$ is also a line digraph. Furthermore, a feedback vertex set $S$ of $D$ implies that $\{((x, 0), (x, 1)) \mid x \in S\}$ is a feedback arc set of $D'$. Conversely, a feedback arc set $S$ of $D'$ implies that $\{y \in V \mid ((x, i), (y, j)) \in S\}$ is a feedback vertex set of $D$. □

**Lemma 11.2.37** ([72]) Feedback Arc Set *(FAS) is reducible to* Feedback Vertex Set *(FVS) for line digraphs.*

**Proof:** It is easy to see that an arc set of a digraph is a feedback arc set if and only if it is a feedback vertex set of its line digraph. □

Summarizing the discussion above, we have shown the following.

**Theorem 11.2.38** ([72]) Independent Set, Vertex Cover, Feedback Vertex Set, Feedback Arc Set *are $\mathcal{NP}$-complete for line digraphs.*

Syslo [146] showed that the Travelling Salesman Problem (TSP) – the problem of finding a minimum weight Hamiltonian cycle in a weighted digraph – notorious for being $\mathcal{NP}$-complete in the general case, is solvable in polynomial time in terms of the size of the digraph, for line digraphs with constant arc weights.

By Theorem 11.2.3, we know that line digraphs can be recognized in polynomial time. In contrast, the problem of recognizing underlying graphs of line digraphs is $\mathcal{NP}$-complete, as Chvátal and Ebenegger [41] proved. Prisner [130] qualified the result by giving a polynomial-time algorithm to recognize underlying graphs of line digraphs with minimum semi-degree at least 2.

Poljak and Rödl [129] found that the problem of determining the chromatic number of a line digraph is $\mathcal{NP}$-complete.

### 11.2.7 Independence Number

Since the determination of the independence number of line digraphs is $\mathcal{NP}$-complete by Theorem 11.2.38, Lichiardopol [112] searched for and found an upper bound for the independence number of regular line digraphs.

**Proposition 11.2.39** ([112]) *Let $D$ be a $d$-regular directed pseudograph without parallel arcs, $d \geq 2$. Then,*

$$\alpha(L(D)) \leq \frac{|V(L(D))|}{2}.$$

He then went on to prove that the ratio can be obtained asymptotically for any regular line digraph, by iterated application of the line digraph operator (see Theorem 11.3.16).

### 11.2.8 Chromatic Number

As we have seen in Subsection 11.2.6, the exact determination of the chromatic number of a line digraph is $\mathcal{NP}$-complete. However, Harner and Entringer [84] gave bounds on the chromatic number of the line digraph of a digraph $D$ in terms of the chromatic number $\chi(D)$ of $D$.

**Theorem 11.2.40** ([84]) *Let $D$ be a digraph. Then,*

$$\min\{t \mid \chi(D) \leq 2^t\} \leq \chi(L(D)) \leq \min\{t \mid \chi(D) \leq \binom{t}{\lfloor \frac{t}{2} \rfloor}\},$$

*where the lower bound is sharp.*

Iterated application of the line digraph operator eventually leads to a 3-colourable digraph (see Corollary 11.3.18). For more on the chromatic number of certain line digraphs, see the work of Ochem, Pinlou and Sopena [123, 124, 127, 128].

## 11.3 Iterated Line Digraphs

Since the 1980s, interconnection networks have attracted more and more attention. In their design, for varying technical reasons, it is interesting to find digraphs with certain attributes such as bounded maximum degree, small diameter and good connectivity. Early on, iterated line digraphs were recognized as a potential source to obtain digraphs of large order but fixed degree and diameter that also allow for easy routing, as Fiol, Yebra and Alegre [63] proved.

Iterated line digraphs are, as their name suggests, defined recursively. For some directed pseudograph $D$, the **first-order line digraph** $L^1(D)$ of

$D$ is the line digraph of $D$. For an integer $k \geq 1$, the $(k+1)$th-order line digraph $L^{k+1}(D)$ of $D$ is defined as the line digraph of $L^k(D)$. A directed pseudograph is called a **$k$th-order line digraph** if it is the $k$th-order line digraph of some directed pseudograph and it is called an iterated line digraph if it is a $k$th-order line digraph for some integer $k \geq 1$.

It is not difficult to prove by induction that $L^k(D)$ is isomorphic to the digraph $Q$ whose vertex set consists of directed walks of $D$ of length $k$ and a vertex $v_0v_1 \dots v_k$ (which is a directed walk in $D$) dominates the vertex $v_1v_2 \dots v_kv_{k+1}$ for every $v_{k+1} \in V(D)$ such that $v_kv_{k+1} \in A(D)$. This fact allows for a new perspective that can be useful in proofs and is, for example, the basis for Fiol, Yebra and Alegre's [63] routing algorithm.

While Theorem 11.2.3 provides several concise characterizations of (first-order) line digraphs, the problem is more complicated for higher order iterated line digraphs. Hemminger [87] generalized condition (iii) from Theorem 11.2.3, which he called the **(first) Heuchenne condition**, in the following way. For a positive integer $k$, a directed pseudograph $D$ satisfies the **$k$th Heuchenne condition** if, for any vertices $x, y, u, v \in V(D)$ such that there is a directed walk of length $k$ from $x$ to $u$, from $y$ to $u$ and from $y$ to $v$, there is also a directed walk of length $k$ from $x$ to $v$. He then proposed that a directed pseudograph without multiple arcs is a $k$th-order line digraph if and only if it satisfies the first $k$ Heuchenne conditions. He did not prove his statement, as, at first glance, it seemed to be obvious. Like several other such results on line digraphs, it turned out to be false. While it is true that it is a necessary condition, it is not sufficient, as Beineke and Zamfirescu [23] proved by constructing counterexamples.

They then set out to find further conditions to add to the $k$th Heuchenne condition to obtain a characterization of iterated line digraphs. With this approach, they were able to characterize the line digraphs that also are second-order line digraphs. Sadly, even for $k = 2$, the necessary conditions are much more complicated than for first-order line digraphs, which is why we will not consider them here in detail and why it seems unlikely that a characterization of $k$th-order line digraphs for $k > 2$ can be derived in a similar manner. This assumption is furthermore backed by an attempt by Beineke and Zamfirescu [23] to find a characterization of second-order line digraphs via forbidden subgraphs comparable to Corollary 11.2.4, which, again, needed rather complicated additional conditions that could not be stated in the form of forbidden subgraphs. Still, the problem of characterizing higher order iterated line digraphs, probably by different means, remains open.

To be able to give any sort of general characterization of higher order iterated line digraphs, in the following theorem, Beineke and Zamfirescu [23] considered only a restricted set of directed pseudographs. Their proof of the result given below is a nice example of the natural idea of using induction on the order of the iterated line digraph.

**Theorem 11.3.1** ([23]) *Let $D$ be a directed pseudograph without multiple arcs and vertices of in-degree or out-degree 0. Then $D$ is a $k$th-order line digraph if and only if, for $i = 1, \ldots, k$, the following conditions are satisfied.*

*(1) For any pair of vertices $x, y \in V(D)$, there is at most one directed walk of length $i$ from $x$ to $y$.*
*(2) $D$ satisfies the $i$th Heuchenne condition.*

**Proof:** We establish the sufficiency of these conditions using induction on $k$. The result holds for $k = 1$, and we assume it holds for $k = p$. Assume that $D$ satisfies the hypotheses for $k = p + 1$. Then, by induction hypothesis, $D$ is a line digraph. Let $Q$ be a directed pseudograph such that $D = L(Q)$. Since the removal of isolated vertices in $Q$ does not affect $L(Q)$, we may assume that $Q$ contains no isolated vertices. Suppose that there is a vertex $x$ of in-degree 0 in $Q$. Since $x$ is not isolated, there is an arc $a \in A(Q)$ with $x$ as its tail. Now $a$, as a vertex of $D = L(Q)$, has in-degree 0, a contradiction. Analogously, $Q$ does not contain vertices of out-degree 0.

Suppose now that, for some $i \leq p$ and a pair of vertices $x, y \in V(Q)$, there are two distinct directed walks $P_1$ and $P_2$ of length $i$ from $x$ to $y$. As there is at least one arc $a \in A(Q)$ whose head coincides with $x$ and one arc $b \in A(Q)$ whose tail coincides with $y$, $P_1$ and $P_2$ can be extended to distinct directed walks $P_1'$ and $P_2'$, respectively, of length $i + 2$, by appending both $a$ and $b$. Consequently, $P_1'$ and $P_2'$ imply distinct directed walks of length $i+1 \leq p+1$ from $a$ to $b$ in $D = L(Q)$, a contradiction to the choice of $D$.

Finally, let $x, y, u, v \in V(Q)$ be vertices such that there are directed walks of length $i$ from $x$ to $u$, from $y$ to $u$ and from $y$ to $v$. Again, we find arcs $a, b, c, d \in A(Q)$ such that the head $a$ coincides with $x$, the head of $b$ coincides with $y$, the tail of $c$ coincides with $u$ and the tail of $d$ coincides with $v$. By appending these arcs to the appropriate directed walks of length $i$ in $Q$, we find directed walks of length $i + 1 \leq p + 1$ from $a$ to $c$, from $b$ to $c$ and from $b$ to $d$ in $D = L(Q)$. Since $D$ satisfies the $(p + 1)$th Heuchenne condition, we also obtain a directed walk of length $i + 1$ from $a$ to $d$ in $D = L(Q)$. By the definition of the line digraph operator, $a$ and $d$, said walk implies a directed walk of length $i$ from $x$ to $v$ in $Q$ and hence, $Q$ satisfies the $i$th Heuchenne condition.

All in all, by induction hypothesis, $Q$ is a $p$th-order line digraph and thus, $D = L(Q)$ is a $(p + 1)$th-order line digraph.

The proof of necessity can be derived from Beineke and Zamfirescu's proof [23] of their general characterization of second-order line digraphs. □

Using what they called coreflexive vertex sets, whose definition is tightly linked to the (iterated) Heuchenne condition, Liu and West [113] gave similar characterizations of (iterated) line digraphs, viewed from a new perspective.

Harary and Norman [82] considered the characteristics of high order iterated line digraphs.

**Theorem 11.3.2** ([82]) *Let $D$ be a directed pseudograph.*

*(i)* $L^k(D) = \emptyset$ *for sufficiently large $k$ if and only if $D$ contains no directed cycles.*
*(ii) The order of* $L^k(D)$ *becomes arbitrarily large for sufficiently large $k$, if and only if $D$ contains two directed cycles that are connected by a directed path.*
*(iii) If $D$ contains two cycles, which are not connected by a directed path, then* $L^k(D)$ *is disconnected for sufficiently large $k$.*

As a corollary, Hemminger and Beineke [88] noted the following.

**Corollary 11.3.3** ([88]) *If $D$ is a directed pseudograph such that* $D \cong L^k(D)$ *for some integer $k$, then $D$ is a directed cycle and particularly* $D \cong L(D)$.

### 11.3.1 Connectivity

As mentioned in the previous section, refined connectivity concepts, as a measure of reliability, are of particular importance for interconnection networks and have therefore been studied for line digraphs, in particular. The well-known fact that $\kappa(D) \leq \lambda(D) \leq \delta^0(D)$ for any digraph $D$, for example, motivated the following definition. A strongly connected digraph $D$ is **maximally connected** if $\kappa(D) = \lambda(D) = \delta^0(D)$. We have already seen in the last section that the line digraph operator does not decrease connectivity and therefore, line digraphs of maximally connected digraphs are again maximally connected. Fàbrega and Fiol [60] proved a stronger result for iterated line digraphs, using the following graph invariant. For a given digraph $D$, let $l(D)$ be the largest integer such that, for any two (not necessarily distinct) vertices $x, y \in V(D)$, (a) if $d(x, y) < l(D)$, the shortest path from $x$ to $y$ is unique and there is no such path of length $d(x, y) + 1$; (b) if $d(x, y) = l(D)$, there is only one shortest path from $x$ to $y$. As a corollary of a more general result, they found that the $k$th-order line graph of any digraph with minimum semi-degree at least 2 is maximally connected, for $k$ sufficiently large.

**Theorem 11.3.4** ([60]) *Let $D$ be a digraph with* $\delta^0(D) > 1$. *Then,*

*(a)* $\lambda(L^k(D)) = \delta^0(D)$ *if* $k \geq \mathrm{diam}(D) - 2l(D)$;
*(b)* $\kappa(L^k(D)) = \delta^0(D)$ *if* $k \geq \mathrm{diam}(D) - 2l(D) + 1$.

Fàbrega and Fiol [60] also proved a similar result on super connectivity.

**Theorem 11.3.5** ([60]) *Let $D$ be a digraph with* $\delta^0(D) \geq 3$. *Then,*

*(a)* $L^k(D)$ *is super arc-connected if* $k \geq \mathrm{diam}(D) - 2l(D) + 1$;
*(b)* $L^k(D)$ *is super connected if* $k \geq \mathrm{diam}(D) - 2l(D) + 2$.

As a corollary of Theorem 11.2.12 and Corollary 11.2.13, Zhang, Liu and Meng [171] obtained a related result.

**Corollary 11.3.6** ([171]) *Let $D$ be a strongly connected directed pseudograph without parallel arcs with $\delta^0(D) \geq 3$. If $D$ is super arc-connected, then $L^k(D)$ is super connected and super arc-connected for any positive integer $k$.*

### 11.3.2 Diameter

As previously indicated, Theorem 11.2.14 directly implies an iterated version of the result.

**Corollary 11.3.7** *Let $D$ be a strongly connected directed pseudograph that is not a cycle. Then, for any positive integer $k$,*

$$\mathrm{diam}(L^k(D)) = \mathrm{diam}(D) + k.$$

### 11.3.3 Branchings

For the number $\mathrm{IB}(D)$ of in-branchings of a regular directed pseudograph $D$, Zhang, Zhang and Huang [170] gave the following formula.

**Theorem 11.3.8** ([170]) *Let $D$ be a $d$-regular digraph of order $n$. Then*

$$\mathrm{IB}(L^k(D)) = d^{(d^k-1)n} \cdot \mathrm{IB}(D).$$

Since line digraphs of $d$-regular directed pseudographs of order $n$ are $d$-regular directed pseudographs of order $d^k n$, Theorem 11.3.8 is also an easy corollary of Corollary 11.2.24. Levine [108] was able to extend Corollary 11.2.24 to iterated line digraphs.

**Corollary 11.3.9** ([108]) *Let $D = (V, A)$ be a directed pseudograph with $\delta^-(D) \geq 1$. Then,*

$$\mathrm{IB}(L^k(D)) = \mathrm{IB}(D) \cdot \prod_{v \in V} d^+(v)^{p(k,v)-1},$$

*where $p(k, v)$ is the number of directed walks of length $k$ that end in $v$.*

Xu, Zhang, Ning and Li [163] extended Levine's results to directed pseudographs without isolated vertices.

As in the previous section, Hasunuma and Nagamochi [85] studied independent out-branchings of iterated line digraphs. Their proof of Corollary 11.2.27 can be applied iteratively to obtain a corresponding result on iterated line digraphs. But they were able to prove more.

**Theorem 11.3.10** ([85]) *Let $D$ be an $l$-strong directed pseudograph without parallel arcs such that $l < \delta^0(D)$. Let $c$ be an upper bound on the depths of $l$ arc-disjoint out-branchings rooted at any vertex of $D$. Then there are $l$ independent out-branchings rooted at any vertex of depths at most $k+\log_2 k+c+1$ of $L^k(D)$ such that any vertex except for the root is contained in at most one tree as an internal vertex.*

**Theorem 11.3.11** ([85]) *Let $D$ be an $l$-strong directed pseudograph without parallel arcs such that $l = \delta^0(D) \geq 3$. Let $c$ be an upper bound on the depths of $l$ arc-disjoint out-branchings rooted at any vertex of $D$. Then there are $l$ independent out-branchings rooted at any vertex of depths at most $k+\log_{\sqrt{3}} k+c+1$ of $L^k(D)$.*

### 11.3.4 $(h,p)$-Domination Number

Another concept used in fault-tolerance analysis of interconnection networks is $(h,p)$-domination. Let $D=(V,A)$ be a directed pseudograph and $S \subset V$. Then $S$ is called an **$(h,p)$-domination set** if $D[S]$ is $h$-strong and $|(\{x\} \cup N^-(x)) \cap S| \geq p$ and $|(\{x\} \cup N^+(x)) \cap S| \geq p$ for every vertex $x \in V$. The **$(h,p)$-domination number** $\gamma_{h,p}(D)$ of $D$ is the minimum cardinality of an $(h,p)$-domination set of $D$. $(h,p)$-domination has been studied for iterated line digraphs by Hasunuma and Otani [86]. Particularly interesting are their results on regular iterated line digraphs, which generalized several results for popular interconnection networks.

**Theorem 11.3.12** ([86]) *Let $D$ be a strong $d$-regular directed pseudograph without parallel arcs and $1 \leq p < d$. Then,*

$$\gamma_{h,p}(L^k(D)) = pd^{k-1}|V(D)|$$

*for all $k \geq 2$ and $0 \leq h \leq \min\{p, \lfloor d/2 \rfloor\}$.*

### 11.3.5 Cycles and Trails

Using Theorem 11.3.8, Zhang, Zhang and Huang [170] were able to calculate the number of Euler trails of regular iterated line digraphs.

**Theorem 11.3.13** ([170]) *Let $D$ be a strongly connected $d$-regular digraph of order $n$. Then, the number of Euler trails of $L^k(D)$ is*

$$\frac{(d!)^{nd^k}}{nd^{k+n}} \cdot b,$$

*where $b$ is the number of in-branchings of $D$.*

By iteratively applying Theorem 11.2.30, we obtain the following corollary of it.

**Corollary 11.3.14** *Let $D$ be a strongly connected $d$-regular directed pseudograph of order $n$. Then, the number of Euler trails of $L^k(D)$ is*

$$d^{-k}(d!)^{(d^k-1)n} \cdot t,$$

*where $t$ is the number of Euler trails of $D$.*

Analogous iteration of Proposition 11.2.32 and Theorem 11.2.33 produces the following corollary.

**Corollary 11.3.15** ([96]) *If a regular directed pseudograph $D$ is pancircular, then $L^k(D)$ is pancyclic and pancircular for any positive integer $k$.*

### 11.3.6 Independence Number

In addition to the upper bound in Proposition 11.2.39, Lichiardopol [112] also gave a (far more complicated) lower bound for the independence number of regular iterated line digraphs, which implies that approximately half of the vertices of a regular $k$th-order line digraph are contained in an independent set for $k$ large enough.

**Theorem 11.3.16** ([112]) *Let $D$ be a $d$-regular directed pseudograph without parallel arcs, $d \geq 2$. Then,*

$$\lim_{k\to\infty} \frac{\alpha(L^k(D))}{|V(L^k(D))|} = \frac{1}{2}.$$

### 11.3.7 Chromatic Number

Duffus, Lefmann and Rödl [58] noted that the second-order line digraph of a 4-colourable digraph is 3-colourable, a result that can be generalized as follows.

**Proposition 11.3.17** *Let $D$ be a digraph with $\chi(D) \geq 4$. Then,*

$$\chi(L^2(D)) < \chi(D).$$

**Proof:** Let $c : V(D) \to \{1, \ldots, \chi(D)\}$ be a proper colouring of $D$. We then define a colouring $c' : V(L^2(D)) \to \{1, \ldots, \chi(D) - 1\}$ of $L^2(D)$. For $((u, v)(v, w)) \in V(L^2(D))$, let $c'(((u, v)(v, w))) = c(v)$, if $c(v) \neq \chi(D)$, and $c'(((u, v)(v, w))) = i$ for an arbitrary $i \in \{1, \ldots, \chi(D)\} \setminus \{c(u), c(v), c(w)\}$, otherwise. Suppose two adjacent vertices $((u, v)(v, w))$ and $((v, w)(w, x))$

of $L^2(D)$ receive the same colour. Since $v$ and $w$ are adjacent in $D$, we have $c(v) \neq c(w)$. Therefore, without loss of generality, we may assume that $c(v) \neq \chi(D) = c(w)$. Consequently, $c'(((u,v)(v,w))) = c(v)$ and $c'(((v,w)(w,x))) \in \{1,\ldots,\chi(D)\} \setminus \{c(v),c(w),c(x)\}$, a contradiction. Hence, $c'$ is a proper colouring of $L^2(D)$. □

Proposition 11.3.17 implies that iterated lined digraphs of any digraph eventually become 3-colourable, a fact recognized by Prisner [131].

**Corollary 11.3.18** ([131]) *Let $D$ be a digraph. Then $\chi(L^k(D)) \leq 3$, for $k$ sufficiently large.*

For a digraph with large chromatic number, by Theorem 11.2.40, the chromatic number of its iterated line digraphs decrease much faster than suggested by Proposition 11.3.17.

## 11.4 de Bruijn Digraphs

As previously mentioned, the line digraph operator has been found very useful in the design of interconnection networks because of its specific properties, which are particularly suitable for the following problem: Given positive integers $n$ and $d$, construct a digraph $D$ of order $n$ and maximum out-degree at most $d$ such that the diameter $\mathrm{diam}(D)$ is as small as possible, while the vertex-strong connectivity $\kappa(D)$ is as large as possible. In general, such 2-objective optimization problems do not necessarily have admissible solutions. In this case, however, solutions which (almost) maximize/minimize both objective functions exist and can be constructed via the line digraph operator. They are presented in this and the following section.

For positive integers $d$ and $t$, the **de Bruijn digraph** [48] $D_B(d,t)$ can be defined as the directed pseudograph whose vertices are all words of length $t$ from an alphabet of $d$ letters. There is an arc from a vertex $x$ to a vertex $y$ if and only if the last $t-1$ letters of $x$ coincide with the first $t-1$ letters of $y$ (see Figure 11.4). This definition bears a striking similarity to the alternative definition of iterated line digraphs we gave in the previous section. In fact, if $K_d^\circ$ is the complete digraph on $d$ vertices with a loop at each vertex, then

$$D_B(d,t) \cong L^{t-1}(K_d^\circ).$$

Therefore, all results on iterated line digraphs can be applied to de Bruijn digraphs and many of them have been proven for exactly that purpose. The following proposition is a collection of obvious consequences.

**Proposition 11.4.1** *Let $d$ and $t$ be positive integers. Then the de Bruijn digraph $D_B(d,t)$:*

*(a) has $d^t$ vertices;*

*(b) has in- and out-degree $d$ for every vertex (counting loops);*
*(c) has diameter $t$;*
*(d) has no parallel arcs;*
*(e) has a loop at exactly those vertices represented by repetitions of a single letter;*
*(f) has $\kappa(D_B(d,t)) = \lambda(D_B(d,t)) = d-1$.*

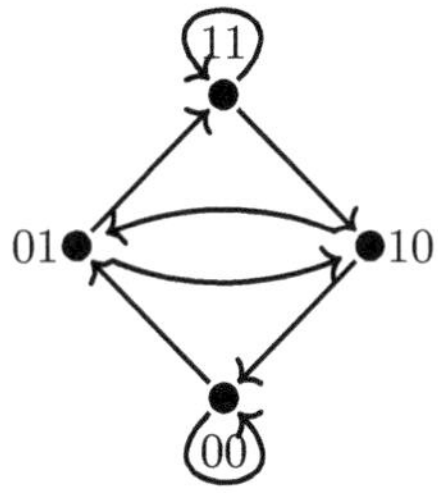


(a) $D_B(2,1) = K_2^\circ$ (b) $D_B(2,2) = L(K_2^\circ)$

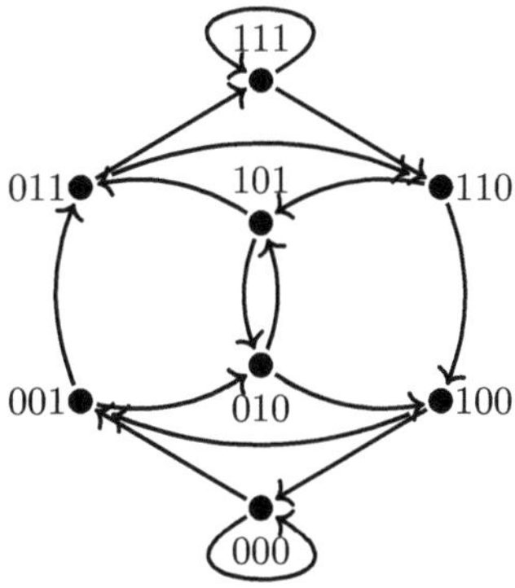


(c) $D_B(2,3) = L^2(K_2^\circ)$

**Figure 11.4** Construction of de Bruijn digraphs via the line digraph operator.

### 11.4.1 Connectivity

As, in Proposition 11.4.1, we have seen that $\kappa(D_B(d,t)) = \lambda(D_B(d,t)) = d-1$, de Bruijn digraphs are almost maximally connected. The connectivity is obviously best possible for $d$-regular digraphs containing a loop.

Furthermore, Soneoka [143] proved that de Bruijn digraphs are super arc-connected by relating the order, degrees and diameter of a de Bruijn digraph.

**Theorem 11.4.2** ([143]) *$D_B(d,t)$ is super arc-connected for all integers $d \geq 2$ and $t \geq 1$.*

Zhang, Liu and Meng [171] obtained the same from a result on iterated line digraphs (see Corollary 11.3.6). In the same manner, they were able to prove the super connectivity of de Bruijn digraphs.

**Corollary 11.4.3** ([171]) *$D_B(d,t)$ is super connected for all integers $d \geq 2$ and $t \geq 1$.*

Lü and Xu [115] and Cheng, Du, Min, Ngo, Ruan, Sun and Wu [38] obtained the result as a corollary of their own results on iterated line digraphs.

### 11.4.2 Diameter

By Proposition 11.4.1, we know that $\text{diam}(D_B(d,t)) = t$. The well-known Moore-bound states for any strongly connected digraph on $n$ vertices with maximum out-degree $d$ and diameter $t$ that

$$n \leq 1 + d + d^2 + \ldots + d^t,$$

where Bridges and Toueg [34] proved that equality is not attained unless $t = 1$ or $d = 1$. The corresponding values for de Bruijn digraphs given in Proposition 11.4.1 now imply the following.

**Proposition 11.4.4** *For all positive integers $d$ and $t$, the de Bruijn digraph $D_B(d,t)$ achieves the minimum value $t$ of diameter for directed pseudographs of order $d^t$ and maximum out-degree at most $d$.*

Furthermore, Imase, Soneoka and Okada [94] noted that the diameter of de Bruijn digraphs is fairly robust against deletion of vertices and/or arcs. They proved that, in $D_B(d,t)$, the diameter increases by at most one if fewer than $d-1$ vertices or arcs are deleted. To prove this result we will use the following lemma.

**Lemma 11.4.5** ([16]) *Let $d$ and $t$ be positive integers and let $x$ and $y$ be distinct vertices of $D_B(d,t)$ such that $x \rightarrow y$. Then, there are $d-2$ internally disjoint $(x,y)$-paths different from $xy$, each of length at most $t+1$.*

**Proof:** Let $x = (x_1, x_2, \ldots, x_t)$ and $y = (x_2, \ldots, x_t, y_t)$. Consider the walk $W_k$ given by $W_k = (x_1, x_2, \ldots, x_t), (x_2, \ldots, x_t, k), (x_3, \ldots, x_t, k, x_2), \ldots, (k, x_2, \ldots, x_t), (x_2, \ldots, x_t, y_t)$, where $k \neq x_1, y_t$. For each $k$, every internal vertex of $W_k$ has coordinates forming the same multiset $M_k = \{x_2, \ldots, x_t, k\}$. Since for different $k$, the multisets $M_k$ are different, the walks $W_k$ are internally disjoint. Each of these walks is of length $t+1$. Therefore, $D_B(d,t)$ contains $d-2$ internally disjoint $(x,y)$-paths $P_k$ with $A(P_k) \subseteq A(W_k)$. Since $k \neq x_1, y_t$, we may form the paths $P_k$ such that none of them coincides with $xy$. □

The result, due to Imase, Soneoka and Okada [94], now states the following.

**Theorem 11.4.6** *[94] For all positive integers $d$ and $t$, from any vertex to any other in $D_B(d,t)$, there are at least $d-1$ internally-disjoint paths, one of which has length at most $t$, and $d-2$ have length at most $t+1$.*

**Proof:** By induction on $t \geq 1$. Clearly, the claim holds for $t = 1$ since $D_B(d,1)$ contains, as spanning subdigraph, $\overleftrightarrow{K}_d$. For $t \geq 2$, we know that

$$D_B(d,t) = L(D_B(d,t-1)). \tag{11.1}$$

Let $x, y$ be a pair of distinct vertices in $D_B(d,t)$ and let $e_x, e_y$ be the arcs of $D_B(d,t-1)$ corresponding to vertices $x, y$ due to (11.1). Let $u$ be the head of $e_x$ and let $v$ be the tail of $e_y$.

If $u \neq v$, by the induction hypothesis, $D_B(d,t-1)$ has $d-1$ internally disjoint $(u,v)$-paths, one of length at most $t-1$ and the others of length at most $t$. The arcs of these paths together with arcs $e_x$ and $e_y$ correspond to $d-1$ internally disjoint $(x,y)$-paths in $D_B(d,t)$, one of length at most $t$ and the others of length at most $t+1$.

If $u = v$, we have $x \rightarrow y$ in $D_B(d,t-1)$. It suffices to apply Lemma 11.4.5 to see that there are $d-1$ internally disjoint $(x,y)$-paths in $D_B(d,t)$, one of length one and the others of length at most $t+1$. □

### 11.4.3 Branchings

Zhang and Lin [169] calculated the total number of in-branchings of de Bruijn digraphs.

**Theorem 11.4.7** *[169] For all positive integers $d$ and $t$, the number of in-branchings of $D_B(d,t)$ is*

$$d^{d^t-1}.$$

Bermond and Fraigniaud [26] and Ge and Hakimi [73] both found $d-1$ independent out-branchings rooted at any vertex of $D_B(d,t)$, while the latter group gave the better estimation of their depths.

**Theorem 11.4.8** *[73] For all positive integers $d$ and $t$, in $D_B(d,t)$, there are $d-1$ independent out-branchings rooted at any vertex of depths at most $\lceil 3t/2 \rceil$.*

As a corollary of Theorem 11.3.10, Hasunuma and Nagamochi [85] obtained the following result.

**Corollary 11.4.9** *[85] For all positive integers $d$ and $t \geq 2$, in $D_B(d,t)$, there are $d-1$ independent out-branchings rooted at any vertex of depths at most $t + \log_2(t-1) + 1$ such that any vertex except for the root is contained in at most one tree as an internal vertex.*

### 11.4.4 $(h, p)$-Domination Number

As an application of Theorem 11.3.12, Hasunuma and Otani [86] calculated the $(h, p)$-domination number for certain de Bruijn digraphs.

**Theorem 11.4.10** *[86] Let $d$ and $p$ be integers such that $d \geq 2$ and $1 \leq p < d$. Then,*

$$\gamma_{h,p}(D_B(d,t)) \;=\; pd^{t-1}$$

*for all $t \geq 3$ and $0 \leq h \leq \min\{p, \lfloor d/2 \rfloor\}$.*

### 11.4.5 Cycles and Trails

Imori, Matsumoto and Yamada [96] obtained the pancyclicity and pancircularity of de Bruijn digraphs as a corollary of their work on iterated line digraphs (see Corollary 11.3.15).

**Corollary 11.4.11** *[96] For all positive integers $d$ and $t$, $D_B(d,t)$ is pancyclic and pancircular.*

Due to Zhang and Lin [169] and, via different method, Zhang, Zhang and Huang [170], we know the exact number of Euler trails contained in de Bruijn digraphs.

**Theorem 11.4.12** *[169] For all positive integers $d$ and $t$, the number of Euler trails of $D_B(d,t)$ is*

$$(d!)^{d^t} d^{-t-1}.$$

Generalizations of de Bruijn digraphs such as **generalized de Bruijn digraphs**, introduced independently by Imase and Itoh [93] and Reddy, Pradhan and Kuhl [134], and **consecutive-$d$ digraphs** suggested by Du, Hsu and Hwang [50] share many of their desirable properties (see, e.g., [36, 49, 53, 54, 95]).

## 11.5 Kautz Digraphs

For positive integers $d$ and $t$, the **Kautz digraph** [99] $D_K(d,t)$ can be obtained from the de Bruijn digraph $D_B(d+1,t)$ by deleting all vertices representing words containing two consecutive identical letters (see Figure 11.5). In particular, Kautz digraphs do not contain loops and are therefore actual digraphs. Fiol, Yebra and Alegre [63] noted that Kautz digraphs, just as de Bruijn digraphs, can be described as iterated line digraphs,

$$D_K(d,t) \;\cong\; L^{t-1}(\overleftrightarrow{K}_{d+1}),$$

where $\overleftrightarrow{K}_{d+1}$ is the complete digraph on $d+1$ vertices. And just as with de Bruijn digraphs, this fact is a widely-used tool in proofs on Kautz digraphs. For example, the following proposition is easily deduced.

**Proposition 11.5.1** *Let $d$ and $t$ be positive integers. Then the Kautz digraph $D_K(d,t)$:*

*(a) has $d^t + d^{t-1}$ vertices;*
*(b) has in- and out-degree $d$ for every vertex;*
*(c) has diameter $t$.*

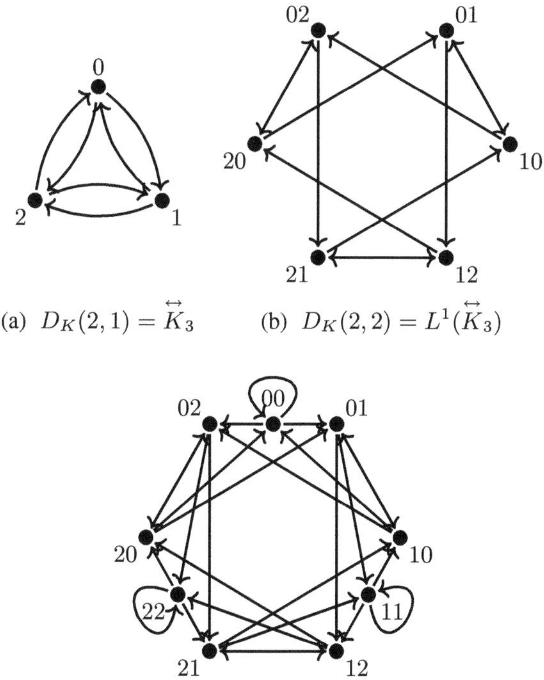


(a) $D_K(2,1) = \overleftrightarrow{K}_3$ (b) $D_K(2,2) = L^1(\overleftrightarrow{K}_3)$

(c) $D_B(3,2) = L^1(K_3^\circ)$

**Figure 11.5** Construction of a Kautz digraph via the line digraph operator or from a de Bruijn digraph.

### 11.5.1 Connectivity

Reddy, Kuhl, Hosseini and Lee [133], as well as Fàbrega, Fiol and Yebra [61] and Imase, Soneoka and Okada [95] independently noted that Kautz digraphs are maximally connected, which is implied by corresponding results on iterated line digraphs.

**Theorem 11.5.2** ([133]) *$D_K(d,t)$ is maximally connected, i.e.* $\kappa(D_K(d,t)) = d$.

In a sense, this result suggests that Kautz digraphs are better than de Bruijn digraphs.

Fàbrega and Fiol [60] obtained the super connectivity and super arc-connectivity of Kautz digraphs as a corollary of their more general results on iterated line digraphs (see Theorem 11.3.5).

**Corollary 11.5.3** ([60]) *$D_K(d,t)$ is super connected and super arc-connected for all integers $d \geq 3$ and $t \geq 2$.*

Soneoka [143] independently proved the super arc-connectivity of Kautz digraphs by relating the order, degrees and diameter of a Kautz digraph. Furthermore, Zhang, Liu and Meng [171] and Lü and Xu [115] realized that super connectivity and super arc-connectivity of Kautz digraphs follows from their respective results on iterated line digraphs.

### 11.5.2 Diameter

By the same reasoning as for Proposition 11.4.4, Reddy, Kuhl, Hosseini and Lee [133] noted that the diameter of Kautz digraphs is minimum for digraphs of their order and degree, making them a solution of the optimization problem mentioned at the beginning of the previous section.

**Proposition 11.5.4** *For all positive integers $d$ and $t$, the Kautz digraph $D_K(d,t)$ achieves the minimum value $t$ of diameter for directed pseudographs of order $d^t + d^{t-1}$ and maximum out-degree at most $d$.*

Du, Hsu and Lyuu [52] improved the results on diameter vulnerabilities due to Reddy, Kuhl, Hosseini and Lee [133] and Imase, Soneoka and Okada [94].

**Theorem 11.5.5** ([52]) *For all positive integers $d$ and $t$, from any vertex to any other in $D_K(d,t)$, there are at least $d$ internally-disjoint paths, one of which has length at most $t$, $d-2$ have length at most $t+1$ and one has length at most $t+2$.*

Furthermore, they determined that, in the worst case, the diameter of $D_K(d,t)$ increases by 1, if fewer than $d-1$ vertices are deleted, and by 2, if $d-1$ vertices are deleted, thereby proving their result to be best possible.

### 11.5.3 Branchings

As an application of Theorem 11.3.13, Zhang, Zhang and Huang [170] gave the number of in-branchings of a Kautz digraph.

**Corollary 11.5.6** ([170]) *For all positive integers $d$ and $t$, the number of in-branchings of $D_K(d,t)$ is*

$$d^{(d+1)d^{t-1}-d-1}(d+1)^d.$$

Just as for de Bruijn digraphs, Ge and Hakimi [73] found the maximum possible number, $d$, of independent out-branchings rooted at any vertex of $D_K(d,t)$.

**Theorem 11.5.7** ([73]) *For all positive integers $d$ and $t$, in $D_K(d,t)$, there are $d$ independent out-branchings rooted at any vertex of depths at most $\lceil 3t/2 \rceil + 1$.*

As a corollary of Theorem 11.3.11, Hasunuma and Nagamochi [85] obtained the following result.

**Corollary 11.5.8** ([85]) *For all positive integers $d$ and $t \geq 2$, in $D_K(d,t)$, there are $d$ independent out-branchings rooted at any vertex of depths at most $t + \log_b t + 1$, where $b = (1+\sqrt{5})/2$, if $d = 2$, and $b = \sqrt{3}$, if $d \geq 3$.*

### 11.5.4 $(h,p)$-Domination Number

Hasunuma and Otani [86] used Theorem 11.3.12 to give the $(h,p)$-domination number for certain Kautz digraphs.

**Corollary 11.5.9** ([86]) *Let $d$ and $p$ be integers such that $d \geq 2$ and $1 \leq p < d$. Then,*

$$\gamma_{h,p}(D_K(d,t)) \;=\; p(d^{t-1} + d^{t-2})$$

*for all $t \geq 3$ and $0 \leq h \leq \min\{p, \lfloor d/2 \rfloor\}$.*

### 11.5.5 Cycles and Trails

As a consequence of their work on iterated line digraphs, Imori, Matsumoto and Yamada [96] obtained the pancyclicity and pancircularity of Kautz digraphs.

**Corollary 11.5.10** ([96]) *For all positive integers $d$ and $t$, $D_K(d,t)$ is pancyclic and pancircular.*

Zhang, Zhang and Huang [170] calculated the number of Euler trails of Kautz digraphs.

**Theorem 11.5.11** ([170]) *For all positive integers $d$ and $t$, the number of Euler trails of $D_K(d,t)$ is*

$$(d!)^{(d+1)d^{t-1}}d^{-d-t}(d+1)^{d-1}.$$

Generalizations of Kautz digraphs such as **Imase–Itoh digraphs**, introduced by Imase and Itoh [92], and **consecutive-$d$ digraphs** suggested by Du, Hsu and Hwang [50] share many of their desirable properties (see, e.g., [36, 49, 53, 54]).

## 11.6 Directed Cographs

A **series-parallel partial order** is a partially ordered set $(X, \leq)$ that can be constructed from a single element using the **series composition** and the **parallel composition** operation. For two disjoint series-parallel partial orders $(X_1, \leq)$ and $(X_2, \leq)$, distinct elements $x, y \in X_1 \cup X_2$ of the series composition have the same order they have in $X_1$ or $X_2$, respectively, if they are both from the same set, and $x \leq y$, if $x \in X_1$ and $y \in X_2$. Elements $x, y \in X_1 \cup X_2$ of the parallel composition are comparable if and only if they are both in $X_1$ or both in $X_2$, and then retain their corresponding order.

A **series-parallel partial order digraph** is a digraph whose vertex set is a series-parallel partial order $(V, \leq)$ and $x \rightarrow y$ if and only if $x \neq y$ and $x \leq y$. More commonly, series-parallel partial orders are represented by **(vertex) series-parallel digraphs**, which can be defined as exactly those digraphs whose transitive closure is a series-parallel partial order digraph, i.e. $x \leq y$, if and only if there is an $(x, y)$-path in the corresponding series-parallel digraph. For some applications it might be desirable to use a particularly sparse representation. A **minimal series-parallel digraph** is a series-parallel digraph for which the removal of any arc alters its transitive closure. Valdes, Tarjan and Lawler [149] defined minimal series-parallel digraphs recursively: The trivial digraph is minimal series-parallel. For two vertex-disjoint minimal series-parallel digraphs $D_1 = (V_1, A_1)$ and $D_2 = (V_2, A_2)$, $P = (V_1 \cup V_1, A_1 \cup A_2)$ is a minimal series-parallel digraph. Furthermore, if $O_1 \subseteq V_1$ is the set of vertices of out-degree zero in $D_1$ and $I_2 \subseteq V_2$ is the set of vertices of in-degree zero in $D_2$, then $S = (V_1 \cup V_1, A_1 \cup A_2 \cup (O_1 \times I_2))$ is a minimal series-parallel digraph. Based on this definition, they defined series-parallel digraphs as exactly those digraphs whose transitive closure equals the transitive closure of a minimal series-parallel digraph.

Among other results, Valdes, Tarjan and Lawler [149] gave a forbidden subdigraph characterization of series-parallel digraphs using the following definition. A digraph is called $N$-free if it does not contain an induced subdigraph on four vertices $\{u, v, w, x\}$ with the arc set $\{vw, uw, ux\}$.

**Theorem 11.6.1** ([148, 149]) *An acyclic digraph is series-parallel, if and only if its transitive closure is $N$-free.*

Note that the $N$-free property is fairly reminiscent of the Heuchenne condition in the characterization of line digraphs (cf. Theorem 11.2.3 (iii)). In particular, Theorem 11.6.1 implies that transitive acyclic line digraphs are series-parallel partial order digraphs.

Valdes, Tarjan and Lawler [149] also found a connection in the opposite direction.

**Theorem 11.6.2** ([149]) *Every minimal series-parallel digraph is a line digraph.*

In fact, they were able to characterize those directed pseudographs whose line digraphs are minimal series-parallel digraphs, which they used in a linear-time recognition algorithm for series-parallel digraphs.

For further results and applications, see, e.g., the work of Monma and Sidney [119], Lawler [107], Baffi and Petreschi [11], Bertolazzi, Cohen, Di Battista, Tamassia and Tollis [29], Rendl [135], Steiner [144] and Möhring [117].

A **cograph**, short for **complement-reducible graph**, is an undirected graph that, like series-parallel partial order digraphs, can be defined recursively: The trivial graph is a cograph. The complement of a cograph is a cograph. And finally, if $G_1 = (V_1, E_1)$ and $G_2 = (V_2, E_2)$ are vertex-disjoint cographs, so is their disjoint union $(V_1 \cup V_2, E_1 \cup E_2)$. There are several further equivalent characterizations of cographs. Particularly, Jung [97] showed that cographs are comparability graphs of series-parallel partial orders $(X, \leq)$, i.e. the graph that contains an edge $xy$ between distinct vertices $x, y \in X$ if and only if $x \leq y$ or $y \leq x$. In other words, if we consider a graph to be a symmetric digraph (i.e. each edge is represented by a directed 2-cycle), then cographs can be defined as the family of digraphs that contains the trivial digraph and is closed under the operations of **disjoint union** and **series**, where, for $h$ disjoint digraphs $D_1, \ldots, D_h$, the disjoint union of $D_1, \ldots, D_h$ is the digraph on the vertex set $\bigcup_{1 \leq i \leq h} V(D_i)$ and the arc set $\bigcup_{1 \leq i \leq h} A(D_i)$, while the series composition of $D_1, \ldots, D_h$ is obtained from the disjoint union by adding all possible arcs between vertices of distinct $D_i$.

Like series-parallel digraphs, cographs can be recognized in linear-time. Corresponding algorithms have been found, e.g., by Corneil, Perl and Stewart [42] and Bretscher, Corneil, Habib and Paul [33].

Finally, we arrive at the eponym of this section. **Directed cographs** generalize both series parallel partial order digraphs and cographs, which is obvious by their recursive definition: The trivial digraph is a directed cograph. Both the disjoint union and the series composition of disjoint directed cographs are directed cographs. Additionally, the **order composition** of $h$ disjoint digraphs $D_1, \ldots, D_h$, which is obtained from the disjoint union by adding the arcs from vertices in $D_i$ to vertices in $D_j$ if and only if $1 \leq i < j \leq h$.

Consistent with the definition of symmetric digraphs, we call an arc $xy \in A(D)$ **symmetric** if $yx \in A(D)$. Otherwise, we call it **asymmetric**.

The **symmetric part** sym($D$) of a digraph $D$ is the spanning subdigraph containing exactly the symmetric arcs of $D$. The **asymmetric part** asym($D$) is defined analogously.

Then, a result due to Bechet, de Groote and Retoré [22] implies that the asymmetric part of a directed cograph is a series-parallel partial order digraph and the symmetric part is a cograph. Furthermore, Crespelle and Paul [43] noted that a forbidden subdigraph characterization can be derived from a result due to Ehrenfeucht and Rozenberg [59].

**Theorem 11.6.3** ([43]) *A digraph is a directed cograph if and only if it does not contain any of the (connected) digraphs depicted in Figure 11.6 as an induced subdigraph.*

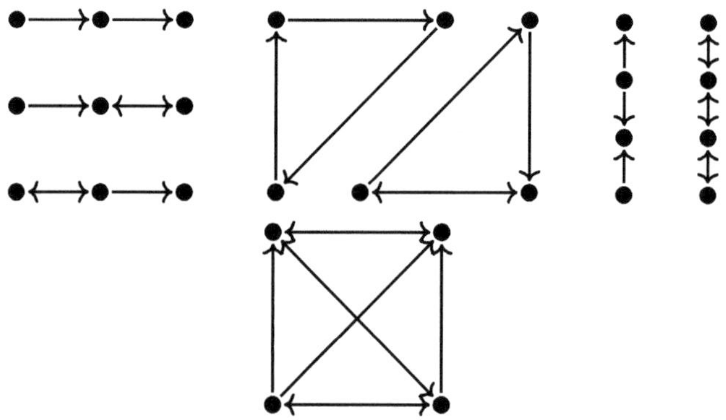

**Figure 11.6** Forbidden subdigraphs for directed cographs.

Consequently, the class of directed cographs is hereditary (that is, an induced subdigraph of a directed cograph is a directed cograph) and closed under complementation. Furthermore, by results due to Möhring and Radermacher [118], directed cographs have a unique representation as a modular decomposition tree, also called a **cotree**. The leaves of the cotree are labelled with the vertices of the directed cograph, while the inner nodes are labelled with the respective operation (disjoint union, series, order) connecting its children (see Figure 11.7).

Using the cotree representation, Crespelle and Paul [43] obtained an optimal algorithm for the Dynamic Recognition and Representation Problem for directed cographs. The input of the problem is a directed cograph with its cotree representation and a series of modifications of the following form: adding/deleting a vertex and its incident arcs or adding/deleting an arc or two symmetric arcs, where all modifications must be valid, i.e., a vertex/arc to be deleted must exist, one to be added must not. If the resulting digraph is again a directed cograph, the algorithm provides its representation, if not, it provides a certificate of that fact, i.e. a forbidden subdigraph.

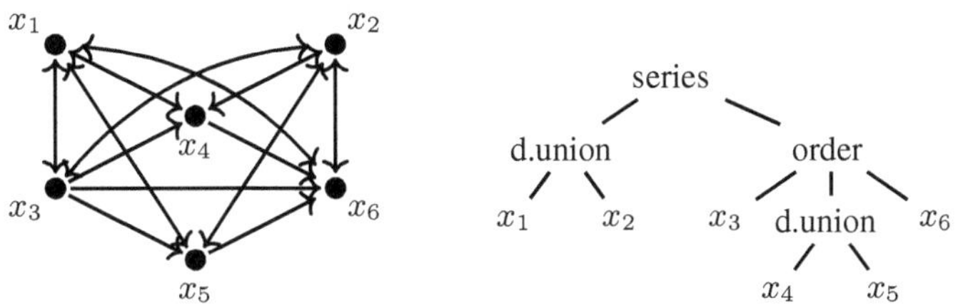


(a) directed cograph (b) corresponding cotree

**Figure 11.7** Cotree representation of a directed cograph.

**Theorem 11.6.4** ([43]) *The* DYNAMIC RECOGNITION AND REPRESENTATION PROBLEM *for directed cographs is solvable in $O(d)$ worst-case time per update, where $d$ is the number of arcs involved in the updating operation. Moreover, if needed, a certificate that the modified digraph is not a directed cograph is provided within the same time complexity.*

For another problem that is solvable in polynomial time, we turn to Bang-Jensen and Maddaloni [19], who considered the WEAK $k$-LINKAGE PROBLEM for directed cographs.

> WEAK $k$-LINKAGE
> **Input:** A digraph $D = (V, A)$ and not necessarily distinct vertices $s_1, \ldots, s_k, t_1, \ldots, t_k$.
> **Question:** Does $D$ contain a weak-$k$-linkage from $(s_1, \ldots, s_k)$ to $(t_1, \ldots, t_k)$?

In fact, they proved that the WEAK $k$-LINKAGE problem is solvable in polynomial time for fixed $k$ for totally $\Phi$-decomposable digraphs, for certain digraph classes $\Phi$.

A digraph $D$ is **totally $\Phi$-decomposable** if either $D \in \Phi$ or $D = P[T_1, \ldots, T_h]$ is composed of a digraph $P \in \Phi$ and pairwise vertex-disjoint totally $\Phi$-decomposable digraphs $T_1, \ldots, T_h$. The recursive definition of totally $\Phi$-decomposable digraphs is valuable in the construction of polynomial algorithms. Of course, the choice of the underlying digraph class $\Phi$ is important. It should be chosen large enough as to assure a rich class of totally $\Phi$-decomposable digraphs, while restricted enough to still allow for polynomial algorithms for important problems in $\Phi$ itself.

One promising class, $\Phi_1$, was introduced by Bang-Jensen and Gutin [17], who, among other results, proved that totally $\Phi_1$-decomposable digraphs are recognizable in polynomial time, another desirable property.

$\Phi_1$ is the union of all semicomplete bipartite digraphs, all connected extended locally semicomplete digraphs and all acyclic digraphs.

The following result is a special case of a broader one due to Bang-Jensen and Maddaloni [19].

**Theorem 11.6.5** ([19]) *For every fixed k there exists a polynomial algorithm for the* WEAK $k$-LINKAGE PROBLEM *for the totally* $\Phi_1$*-decomposable digraphs.*

All that remains is to realize that directed cographs are in fact totally $\Phi_1$-decomposable digraphs, which is fairly obvious by their recursive definition. The trivial digraph (the initial directed cograph), arcless digraphs (realizing disjoint unions) and transitive tournaments (realizing the order composition) are all acyclic, while complete digraphs (realizing the series composition) are particularly connected locally semicomplete digraphs. Thus, we obtain the following corollary.

**Corollary 11.6.6** ([19]) *For every fixed k there exists a polynomial algorithm for the* WEAK $k$-LINKAGE PROBLEM *for directed cographs.*

For more results on totally $\Phi$-decomposable digraphs, see Chapter 8 and for an application of directed cographs in mathematical logic, we refer to the work of Retoré [136].

## 11.7 Perfect Digraphs

First, recall that an undirected graph is called perfect if the chromatic number of every induced subgraph equals its clique number. This property is particularly interesting for its impact on complexity results, as several well-known $\mathcal{NP}$-complete problems, such as the determination of the chromatic number, the clique number or the independence number of a graph, are solvable in polynomial time for perfect graphs (cf. Grötschel, Lovász and Schrijver [75]). Furthermore, the results are actually applicable in practice, since several common graph classes, such as bipartite graphs, chordal graphs, triangulated graphs, interval graphs and comparability graphs, are perfect.

The long-standing Strong Perfect Graph Conjecture, due to Berge [24], after inspiring generations to an array of related research, was finally proven after more than four decades by Chudnovsky, Robertson, Seymour and Thomas [40] and is now known as the Strong Perfect Graph Theorem. It states that a graph is perfect if and only if it contains neither odd holes nor odd antiholes as induced subgraphs, where an odd hole is an induced cycle of odd length at least 5 and an odd antihole is the complement of such a graph. Combined with the corresponding result of Chudnovsky, Cornuéjols, Liu, Seymour and Vušković [39] for graphs without odd holes and odd antiholes, the Strong Perfect Graph Theorem furthermore implies that perfect graphs can be recognized in polynomial time.

Motivated by this breakthrough for undirected perfect graphs, Andres and Hochstättler [9] introduced the class of perfect digraphs and, among other results, gave a Strong Perfect Digraph Theorem. The following additional notation is needed. Particularly, in the context of this book, it has to be pointed out that, in their definition of perfect digraphs, instead of the chromatic number for digraphs, as introduced in the first chapter, Andres and Hochstättler used the dichromatic number, as introduced by Neumann-Lara [122]: A $k$-dicolouring of a digraph is vertex-colouring with $k$ colours such that no directed cycle is monochromatic. The **dichromatic number** $\vec{\chi}(D)$ of a digraph $D$ is the smallest positive integer $k$ such that $D$ admits a $k$-dicolouring. The **clique number** $\omega(D)$ of a digraph $D$ is the order of a largest complete subdigraph of $D$. Now, a digraph is called **perfect** if, for any induced subdigraph, the dichromatic number equals its clique number. Note that a graph is perfect if and only if its complete biorientation (where every edge is replaced by a pair of opposing arcs) is perfect. Therefore, the given concept is a natural extension of perfectness to digraphs.

Recall that the **symmetric part** $\mathrm{sym}(D)$ of a digraph $D$ is the spanning subdigraph containing exactly the symmetric arcs of $D$ (see Figure 11.8(b)). The **asymmetric part** $\mathrm{asym}(D)$ is defined analogously (see 11.8(c)). Now we can state the Strong Perfect Digraph Theorem due to Andres and Hochstättler [9] and give their proof.

**Theorem 11.7.1 (Strong Perfect Digraph Theorem** [9]**)** *A digraph $D$ is perfect if and only if* $\mathrm{sym}(D)$ *(identified with the corresponding undirected graph) is perfect and $D$ does not contain any directed cycle of length at least 3 as an induced subdigraph.*

**Proof:** Assume that $\mathrm{sym}(D)$ is not perfect. Then there is an induced subgraph $G = (V, E)$ of $\mathrm{sym}(D)$ (identified with the corresponding undirected graph) with $\omega(G) < \chi(G)$. Since $\omega(D\langle V\rangle) = \omega(\mathrm{sym}(D\langle V\rangle))$ and $\mathrm{sym}(D\langle V\rangle)$ is the complete biorientation of $G$, we conclude that

$$\omega(D\langle V\rangle) = \omega(\mathrm{sym}(D\langle V\rangle)) = \omega(G) < \chi(G) = \chi(\mathrm{sym}(D\langle V\rangle)) \leq \chi(D\langle V\rangle).$$

Therefore, $D$ is not perfect.

If $D$ contains a directed cycle $C$ of length at least 3 as an induced subdigraph, then $D$ is obviously not perfect, since $\omega(D) = 1 < 2 = \chi(C)$.

Now, assume that $\mathrm{sym}(D)$ is perfect, but $D$ is not. It suffices to show that $D$ contains a directed cycle of length at least 3 as an induced subdigraph. Let $D' = (V', A')$ be an induced subdigraph of $D$ such that $\omega(D') < \chi(D')$. As $\mathrm{sym}(D)$ is perfect, there is a $k$-dicolouring of $\mathrm{sym}(D') = \mathrm{sym}(D)\langle V'\rangle$ with $k = \omega(\mathrm{sym}(D')) = \omega(D')$ colours. By choice of $D'$, this cannot be a $k$-dicolouring of $D'$. Hence, there is a (not necessarily induced) monochromatic directed cycle $C$ of minimal length in $\mathrm{asym}(D')$, which automatically implies its length to be at least 3. $C$ cannot have a symmetric chord, since its head and tail would receive the same colour, in contradiction to the definition of a $k$-dicolouring of $\mathrm{sym}(D')$. By minimality, $C$ cannot have an asymmetric arc. Therefore, $C$ is an induced directed cycle of length at least 3 in $D'$, and thus in $D$. □

Applying the Strong Perfect Graph Theorem, Theorem 11.7.1 can be restated without undirected perfectness, using the following terminology. A **filled odd hole** is a digraph $D$ such that $\mathrm{sym}(D)$ is the complete biorientation of an odd hole. A **filled odd antihole** is defined analogously.

**Corollary 11.7.2** ([9]) *A digraph $D$ is perfect if and only if it does not contain filled odd holes, filled odd antiholes, or any directed cycle of length at least* 3 *as induced subdigraphs.*

Furthermore, Theorem 11.7.1 implies that, for perfect digraphs, the symmetric part determines the validity of a $k$-dicolouring.

**Corollary 11.7.3** ([9]) *If $D$ is a perfect digraph, then every $k$-dicolouring of* $\mathrm{sym}(D)$ *is a $k$-dicolouring of $D$.*

Since the maximum order of an induced acyclic subdigraph of a digraph $D$ also depends solely on $\mathrm{sym}(D)$, as another corollary of Theorem 11.7.1 combined with the respective results on perfect graphs due to Grötschel, Lovász and Schrijver [75], Andres and Hochstättler [9] obtained the following complexity results.

**Corollary 11.7.4** ([9]) *For a perfect digraph $D$, the problems of determining the chromatic number, the clique number and the maximum order of an induced acyclic subdigraph are solvable in polynomial time.*

As a natural follow-up question, Andres and Hochstättler [9] asked whether there are more interesting instances of such problems.

**Problem 11.7.5** ([9]) *Are there any other problems that are $\mathcal{NP}$-complete for general digraphs but solvable in polynomial time for perfect digraphs?*

While the results we have considered so far all indicate that the properties of perfect digraphs are as favourable as those of their undirected counterparts, Andres and Hochstättler [9] had to concede that perfect digraphs lack one central virtue: In contrast to the results of Chudnovsky, Cornuéjols, Liu, Seymour and Vušković [39] on perfect graphs, perfect digraphs cannot be recognized in polynomial time (unless $\mathcal{P} = \mathcal{NP}$).

**Theorem 11.7.6** ([9]) *Deciding whether a digraph is perfect is a* co-$\mathcal{NP}$-*complete problem.*

Their proof is mainly based on a result of Bang-Jensen, Havet and Trotignon [18] stating the co-$\mathcal{NP}$-completeness of determining whether a given digraph does not contain any directed cycle of length at least 3 as an induced subdigraph (cf. Theorem 11.7.1).

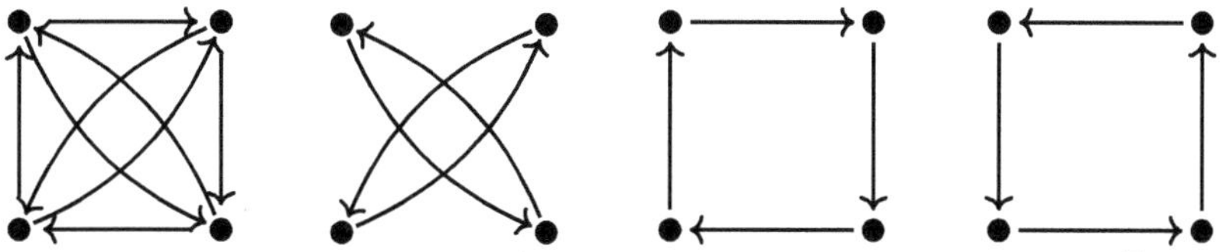


**Figure 11.8** A perfect digraph $D$ with imperfect complement $\bar{D}$.

The loss of another nice property of perfectness in the translation from graphs to digraphs is implied by the following results on kernels due to Andres and Hochstättler [9].

**Theorem 11.7.7** ([9]) *It is $\mathcal{NP}$-complete to decide whether a perfect digraph has a kernel.*

On the other hand, a result due to Boros and Gurvich [31] can be rephrased as follows.

**Corollary 11.7.8** ([9]) *The complement of a perfect digraph is kernel-perfect, i.e. every induced subdigraph has a kernel.*

Therefore, Theorem 11.7.7 and Corollary 11.7.8 imply that complements of perfect digraphs are not necessarily perfect (see, e.g., Figures 11.8(a) and 11.8(d), respectively), in contrast to the result that, for undirected graphs, is well-known as the Weak Perfect Graph Theorem due to Lovász [114].

Still, Andres and Hochstättler [9] were able to prove a similar result.

**Theorem 11.7.9** ([9]) *A digraph $D$ is perfect if and only if its complement $\bar{D}$ is a biorientation of a perfect graph $G$ such that no vertex set of a cycle in* asym($\bar{D}$) *induces a clique in $G$.*

Note that, if we identify an undirected graph $G$ with its complete biorientation $D$, then $\bar{D}$ is the complete biorientation of $\bar{G}$. Furthermore, asym($\bar{D}$) is empty and, in particular, does not contain any cycle. Therefore, Theorem 11.7.9 is a generalization of the Weak Perfect Graph Theorem.

Before we close this section with a variation of perfectness in digraphs, we give another problem posed by Andres and Hochstättler [9].

**Problem 11.7.10** ([9]) *Are there any other problems that are $\mathcal{NP}$-complete* (co-$\mathcal{NP}$*-complete, respectively) for general digraphs that remain so for perfect digraphs?*

For several decades, game-variants of certain graph invariants have become increasingly popular. Colouring games and corresponding game chromatic numbers are certainly among the most prominent. For an undirected graph, we define a maker-breaker game, where both players (starting with maker) take turns assigning a colour to a previously uncoloured vertex from a given finite set of colours such that no two adjacent vertices receive the same colour. The game stops, if either the whole graph is coloured properly, in which case maker wins, or none of the remaining uncoloured vertices can be coloured properly, in which case breaker wins. The **game chromatic number** $\chi_g(G)$ of a graph $G$ is the smallest number of colours for maker to have a winning strategy for the colouring game on $G$, which is well-defined, as $|V(G)|$ colours are obviously sufficient.

Andres [6] extended the concept to digraphs in the following way. The colouring game is now played on a digraph and on each turn a player must choose a vertex to assign a colour to, distinct from the colours of all its in-neighbours. The **game chromatic number** $\chi_g(D)$ of a digraph $D$ is then defined as in the undirected case. Since, for a graph $G$ and its complete biorientation $\overleftrightarrow{G}$, we have

$$\chi_g(G) = \chi_g(\overleftrightarrow{G}),$$

this definition is natural and well-defined.

Yang and Zhu [165] gave another variant of the game chromatic number for digraphs. In their colouring game, on each turn a player must colour a vertex without creating a monochromatic directed cycle. As their colouring rule is weaker than Andres', they called the smallest number of colours for maker to have a winning strategy for their colouring game on a digraph $D$ the **weak game chromatic number**. For the obvious similarity to the dichromatic number, we prefer the name **game dichromatic number**, which we will denote by $\vec{\chi}_g(D)$. Obviously, the game chromatic number of a graph also equals the game dichromatic number of its complete biorientation.

As with many problems, the directed versions have so far received less attention than the undirected game chromatic number, but in addition to the introductory paper [6], there are some results due to Andres [3, 4, 8], Yang and Zhu [165] and Chan, Shiu, Sun and Zhu [37]. Note that the oriented game chromatic number introduced by Nešetril and Sopena [121], while also based on a digraph colouring game, differs greatly from the game dichromatic number considered here. Particularly, it is only defined for orientations of graphs.

Finally, we can give the definitions that motivated our brief excursion. A digraph $D$ is called **game-perfect** if, for any induced subdigraph, the game chromatic number equals its clique number. Analogously, $D$ is **weakly game-perfect** if, for any induced subdigraph, the game dichromatic number equals its clique number. Note that since

$$\omega(D) \le \vec{\chi}(D) \le \vec{\chi}_g(D) \le \chi_g(D)$$

for every digraph $D$, game-perfect digraphs are also weakly game-perfect and weakly game-perfect digraphs are perfect.

The following result due to Andres [8], in combination with Theorem 11.7.7, implies that game-perfect digraphs are a proper subclass of perfect digraphs.

**Theorem 11.7.11** ([8]) *Game-perfect digraphs are kernel-perfect.*

As a natural consequence of their fairly recent introduction by Andres [5], except for Theorem 11.7.11, there are mostly only basic results on game-perfect digraphs so far and a lot of open questions, the most interesting one arguably being the following.

**Problem 11.7.12** ([9]) *Give a characterization of game-perfect digraphs by a set of forbidden induced subdigraphs (analogue to Theorem 11.7.1 and Corollary 11.7.2, respectively).*

For weakly game-perfect digraphs, this problem has been solved by Andres [8].

**Theorem 11.7.13** ([8]) *A digraph $D$ is weakly game-perfect if and only if* $\mathrm{sym}(D)$ *(identified with the corresponding undirected graph) is game-perfect and $D$ does not contain any directed cycle of length at least* 3 *as an induced subdigraph.*

Since game-perfect graphs have previously been characterized by a set of forbidden induced graphs [7], we obtain the following characterization of weakly game-perfect digraphs.

**Corollary 11.7.14** ([8]) *A digraph $D$ is weakly game-perfect if and only if $D$ does not contain a directed cycle of length at least* 3 *or a $C_4$, $P_4$, a triangle star, a $\Xi$-graph, two double fans, two split* 3*-stars, or one of each (see Figure 11.9, where an edge corresponds to a directed* 2*-cycle) as an induced subgraph.*

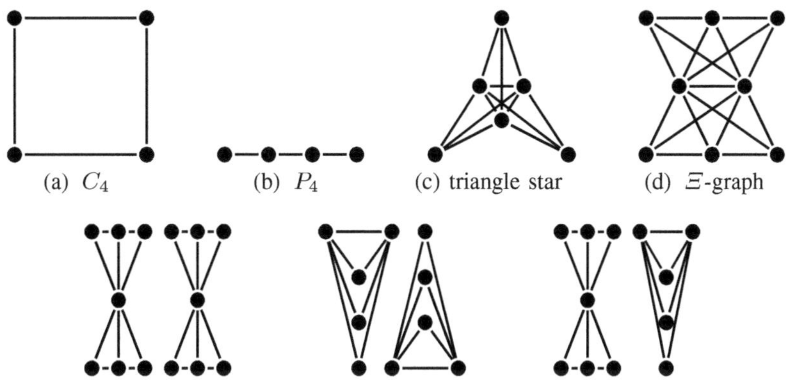

(a) $C_4$ (b) $P_4$ (c) triangle star (d) $\Xi$-graph

(e) two double fans (f) two split 3-stars (g) mixed graph

**Figure 11.9** Forbidden subgraphs for weak game-perfectness.

The research on perfect and (weakly) game perfect digraphs is very much a young and active field and further results are to be expected.

## 11.8 Arc-Locally Semicomplete Digraphs

Arc-locally semicomplete digraphs were initially introduced as arc-local tournament digraphs by Bang-Jensen [13] as a natural analogue of locally semicomplete digraphs (cf. Chapter 6) and a generalization of both semicomplete and semicomplete bipartite digraphs. A digraph is called **arc-locally semicomplete** if, for any pair of adjacent vertices $x$ and $y$, every in-neighbour (out-neighbour, respectively) of $x$ is adjacent to every distinct in-neighbour (out-neighbour, respectively) of $y$.

Although arc-locally semicomplete digraphs can be quite sparse, directed cycles being among the simplest examples, the first result on the class, given by Bang-Jensen [13], suggests that arc-locally semicomplete digraphs, in general, are fairly dense in some sense.

**Lemma 11.8.1** ([13]) *Let $D$ be a connected arc-locally semicomplete digraph and let $D'$ be any non-trivial strong subdigraph of $D$. Every vertex $x \in V(D) \setminus V(D')$ is adjacent to $D'$.*

**Proof:** Suppose some vertex $x \in V(D) \setminus V(D')$ is not adjacent to $D'$. Let $x = x_1x_2 \dots x_n$, $n \geq 3$, $x_n \in V(D')$, be a shortest path between $x$ and $D'$ in $UG(D)$. Let $u \in V(D')$ ($w \in V(D')$) be some vertex which dominates (is dominated by) $x_n$ in $D'$. Now, depending on the orientation of the edge $x_{n-2}x_{n-1}$ in $D$, we conclude that $x_{n-2}$ is adjacent to $u$ or $w$, contradicting the minimality of the path above. □

A common method of proof relating to arc-locally semicomplete digraphs is to show that the considered arc-locally semicomplete digraph either has some desired property, or it is semicomplete or semicomplete bipartite, respectively. The following lemma, which Bang-Jensen [13] derived from Lemma 11.8.1, is particularly useful.

**Lemma 11.8.2** ([13]) *Let $D$ be a connected, non-strong arc-locally semicomplete digraph. If every vertex is on some cycle, then $D$ is semicomplete or semicomplete bipartite.*

Combining Lemma 11.8.2 with the following one, Bang-Jensen [13] provided a characterization of Hamiltonian arc-locally semicomplete digraphs.

**Lemma 11.8.3** ([13]) *Every strong arc-locally semicomplete digraph $D$ having two disjoint cycles covering $V(D)$ is Hamiltonian.*

**Theorem 11.8.4** ([13]) *A strong arc-locally semicomplete digraph $D$ has a Hamiltonian cycle if and only if it has a directed cycle factor.*

**Proof:** One direction is clear, so suppose $D$ is a strong arc-local tournament which has a directed cycle factor. We prove by induction on its order $n$ that $D$ is Hamiltonian. The cases $n = 3, 4, 5$ are trivial, so we proceed to the induction step assuming $n \geq 6$. Let $C_1, \ldots, C_k$, $k \geq 1$, be a directed cycle factor of $D$ chosen such that $k$ is minimum. We claim that $k = 1$, in which case we are done. So suppose that $k \geq 2$. By Lemma 11.8.3 we must have $k \geq 3$. Now it follows, from the induction hypothesis and the minimality of $k$, that no proper subset of $C_1, \ldots, C_k$ can induce a strong digraph. Thus if we delete the vertices of $C_i$ for any $1 \leq i \leq k$, the remaining digraph $D - C_i$ is a non-strong arc-locally semicomplete digraph in which each vertex lies on a cycle and hence, by Lemma 11.8.2, it is semicomplete or semicomplete bipartite. From this and the fact that no proper subset of $C_1, \ldots, C_k$ can induce a strong digraph, we conclude that $k = 3$, and that there is no arc from $C_{i+1}$ to $C_i$ for $i = 1, 2, 3$, indices modulo 3. Now it is easy to see that $D$ has a Hamiltonian cycle, contradicting the choice of $C_1, \ldots, C_k$. □

Moreover, Bang-Jensen [13] proved that the problem of deciding the existence of (and finding) a Hamiltonian cycle can be solved in polynomial time. Corresponding complexity results, based on the following theorem, also hold for Hamiltonian paths.

**Theorem 11.8.5** ([13]) *A connected arc-locally semicomplete digraph $D$ has a Hamiltonian path if and only if it has a path $P$ (where we allow $V(P) = \emptyset$ or $V(P) = V(D)$) such that $D - V(P)$ has a directed cycle factor.*

For a characterization of strong arc-locally semicomplete digraphs, we need the following additional definitions. Let $E_1, \ldots, E_k$ be $k$ disjoint sets of independent vertices, then $\vec{C}_k[E_1, \ldots, E_k]$ is the digraph obtained by substituting the vertex $x_i$ for the vertex set $E_i$ in a $k$-cycle $\vec{C}_k = x_1 \ldots x_k x_1$. In other words, $V(\vec{C}_k[E_1, \ldots, E_k]) = E_1 \cup \ldots \cup E_k$ and

$$xy \in A(\vec{C}_k[E_1, \ldots, E_k]) \quad \Leftrightarrow \quad x \in E_i \text{ and } y \in E_{i+1} (\text{modulo } k)$$

for some $i \in \{1, \ldots k\}$ (see Figure 11.10(a)). We call $\vec{C}_k[E_1, \ldots, E_k]$ an **extended cycle**. Furthermore, for an integer $n \geq 4$, let $F_n$ be the digraph on the vertex set $\{u, v, x_1, \ldots, x_{n-2}\}$ and the arc set $\{uv, vu\} \cup \{x_i u, v x_i \mid 1 \leq i \leq n-2\}$ (see Figure 11.10(b)).

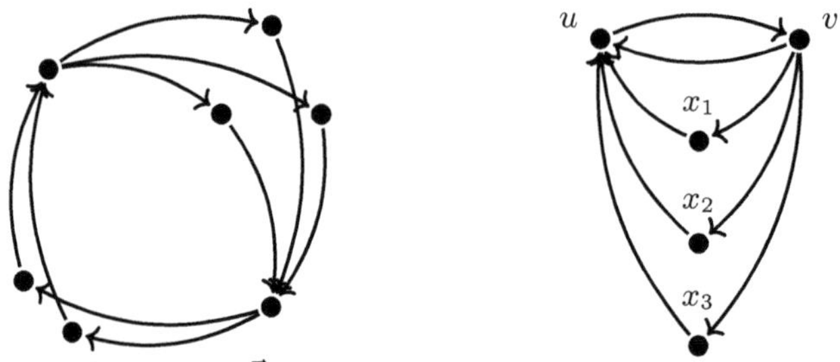


(a) Extended $\vec{C}_4$. (b) $F_5$

**Figure 11.10** Example of an extended cycle and an $F_n$ digraph.

Galeana-Sánchez and Goldfeder [65, 74] and, independently, Wang and Wang [158] completed a previously deficient characterization of strong arc-locally semicomplete digraphs due to Bang-Jensen [15].

**Theorem 11.8.6** ([65, 74, 158]) *Let $D$ be a strong arc-locally semicomplete digraph, then $D$ is either semicomplete, semicomplete bipartite, an extended cycle or isomorphic to $F_n$ for some $n \geq 4$.*

For the following complete characterization of arc-locally semicomplete digraphs due to Galeana-Sánchez and Goldfeder [66] we note that the concept of extended cycles is easily transferable to paths $P_k = x_1 \dots x_k$ and transitive tournaments $TT_k$ on $k$ vertices $x_1, \dots, x_k$ such that $x_i \to x_j$ if and only if $1 \leq i < j \leq k$. Furthermore, we may want to substitute a vertex $x_i$ by a digraph $D_i$ instead of a set of independent vertices $E_i$. $\vec{C}_k[D_1, \dots, D_k]$, for example, is obtained from the digraph $\vec{C}_k[V(D_1) \cup \dots \cup V(D_k)]$ by adding all arcs of $A(D_1) \cup \dots \cup A(D_k)$.

**Theorem 11.8.7** ([66]) *Let $D$ be a connected digraph. Then $D$ is arc-locally semicomplete if and only if it is one of the following:*

*(1) a subdigraph of an extended $P_2$,*
*(2) $P_3[E_1, D', E_1]$, where $D'$ is a semicomplete digraph,*
*(3) $TT_3[E_1, E_n, E_1]$, for some positive integer $n$,*
*(4) $F_n$ for some $n \geq 4$,*
*(5) an extended path or an extended cycle,*
*(6) a semicomplete digraph, or*
*(7) semicomplete bipartite digraph.*

Using Theorem 11.8.7, Arroyo and Galeana-Sánchez [10] verified the Directed Path Partition Conjecture for arc-locally semicomplete digraphs.

**Theorem 11.8.8** ([10]) *Let $D$ be an arc-locally semicomplete digraph. If $D$ contains no path with more than $\lambda$ vertices, then, for every pair $a, b$ of positive integers with $\lambda = a + b$, there exists a partition $(A, B)$ of $V(D)$ such that no path in $D\langle A\rangle$ ($D\langle B\rangle$, respectively) has more than $a$ ($b$, respectively) vertices.*

One particular instance of the Directed Path Partition Conjecture due to Laborde, Payan and Xuong [106] states that every digraph contains a maximal independent set that intersects every longest path. Galeana-Sánchez and Gómez [68] proved a stronger result for arc-locally semicomplete digraphs concerned with a generalization of longest paths. A path $P = x_1 \dots x_k$ is non-augmentable if $P$ cannot be expanded to a path $Q = x_1 \dots x_i y_1 \dots y_\ell x_{i+1} \dots x_k$, $0 \le i \le k$.

**Theorem 11.8.9** ([68]) *Let $D$ be an arc-locally semicomplete digraph. If $\delta^+(D) > 0$, then every maximal independent set intersects every non-augmentable path in $D$.*

Furthermore, Galeana-Sánchez and Gómez [68] constructed an infinite family of arc-locally semicomplete digraphs containing a maximal independent set that does not intersect at least one non-augmentable path. Thus, the degree condition is necessary. For the general case, they found that there is at least one maximal independent set that intersects a particular subset of non-augmentable paths, a result that was extended by Wang and Wang [157].

**Theorem 11.8.10** ([157]) *Let $D$ be an arc-locally semicomplete digraph. Then there exists a maximal independent set intersecting every non-augmentable path in $D$.*

Bang-Jensen and Manoussakis [20] considered a somewhat complementary problem. Instead of a set of vertices intersecting a prescribed set of paths, they were interested in a cycle intersecting a prescribed set of vertices. Combining their result, which is for semicomplete bipartite digraphs, and Theorem 11.8.7 one obtains the following.

**Theorem 11.8.11** ([20]) *Every $k$-strong arc-locally semicomplete digraph has a cycle through any set of $k$ given vertices.*

In [80], Häggkvist and Manoussakis gave examples of $(k-1)$-connected bipartite tournaments with no cycle through some set of $k$ vertices, proving Theorem 11.8.11 best possible.

In contrast to locally semicomplete digraphs (cf. Proposition 6.2.4), arc-locally semicomplete digraphs are not necessarily path-mergeable. However, Bang-Jensen [14] gave a sufficient condition for an arc-locally semicomplete digraph to be path-mergeable. A digraph is **2-path-mergeable**, if, for every pair of vertices $x$ and $y$ and every pair of internally disjoint $(x,y)$-paths $P$ and $P'$ of length at most 2, there is an $(x,y)$-path $P^*$ such that $V(P^*) = V(P) \cup V(P')$.

**Proposition 11.8.12** ([14]) *Every 2-path-mergeable arc-locally semicomplete digraph is path-mergeable.*

A conjecture proposed by Berge and Duchet [25] stating that a graph is perfect, if and only if any normal biorientation is kernel-perfect (where the "only if" part has since been proven by Boros and Gurvich [31] and the "if" part is implied by the Strong Perfect Graph Theorem [32]), inspired Galeana-Sánchez [64] to investigate kernel-perfectness of arc-locally semicomplete digraphs and the relation to perfectness of their underlying graphs.

Adapted from the definition of perfect graphs, a digraph $D$ is called **kernel-perfect** if every induced subdigraph $D'$ contains a kernel, i.e. an independent vertex set $N \subseteq V(D')$ such that, for every $u \in V(D') \setminus N$, there is a $v \in N$ such that $uv \in A(D')$. A digraph is **critical kernel-imperfect** if it is not kernel-perfect, but every induced subdigraph is.

Using the following additional notation, Galeana-Sánchez [64] characterized kernel-perfect arc-locally semicomplete digraphs. A **pseudodiagonal** of a cycle $C$ is an arc whose initial and terminal vertices belong to $V(C)$, but itself is not contained in $A(C)$. A digraph is called **odd-chorded** if every cycle of odd length has at least one pseudodiagonal. Furthermore, we call a digraph **normal** if every semicomplete subdigraph contains a vertex that is a kernel.

**Theorem 11.8.13** ([64]) *Let $D$ be an arc-locally semicomplete digraph. Then, $D$ is a kernel-perfect digraph if and only if $D$ is a normal odd-chorded digraph.*

Note that the proof of Theorem 11.8.13 is based on an incomplete characterization of arc-locally semicomplete digraphs, but the missing case is easily added and thus, the result and those based on it still hold. As a corollary, Galeana-Sánchez [64] verified a conjecture due to Meyniel [57], although disproven for general digraphs, for arc-locally semicomplete digraphs.

**Corollary 11.8.14** ([64]) *Let $D$ be an arc-locally semicomplete digraph. If each odd cycle has at least two pseudodiagonals, then $D$ is a kernel-perfect digraph.*

Furthermore, Galeana-Sánchez [64] found that critical kernel-imperfect arc-locally semicomplete digraphs have a very specific structure.

**Theorem 11.8.15** ([64]) *Let $D$ be an arc-locally semicomplete digraph. Then, $D$ is critical kernel-imperfect if and only if $D \cong C_{2n+1}$, $n \geq 1$ or $D \cong C_n[1, \pm 2, \pm 3, \dots, \pm\lfloor n/2 \rfloor]$, $n \geq 4$, where $C_n[j_1, \dots, j_k]$ is the digraph on the vertex set $\{0, \dots, n-1\}$ and the arc set $\{u, v \mid u - v = j_s \mod n$ for $s = 1, \dots, k\}$.*

The following result of Galeana-Sánchez [64], as a corollary, implies a strong relation between kernel-perfectness of arc-locally semicomplete digraphs and perfectness of their underlying graphs. In fact, kernel-perfectness even implies **strong perfectness**, that is to say, every induced subgraph $G'$ contains an independent vertex set which meets every maximal clique of $G'$.

**Theorem 11.8.16** ([64]) *Let $D$ be an arc-locally semicomplete digraph. If $N$ is a kernel of $D$, then $N$ is an independent set of $UG(D)$ which meets every maximal clique of $UG(D)$.*

**Corollary 11.8.17** ([64]) *Let $D$ be an arc-locally semicomplete digraph. If $D$ is a kernel-perfect digraph, then $UG(D)$ is strongly perfect.*

Galeana-Sánchez [64] was able to extend the result to the following characterization.

**Theorem 11.8.18** ([64]) *Let $D$ be an arc-locally semicomplete digraph.*

*(i) $D$ is a kernel-perfect digraph if and only if $UG(D)$ is a strongly perfect graph.*
*(ii) $D$ is a critical kernel-imperfect digraph if and only if $UG(D)$ is a critically imperfect graph.*

Building on this result, for underlying graphs of normal arc-locally semicomplete digraphs, Galeana-Sánchez [64] proved a variation of the Strong Perfect Graph Theorem (cf. Section 11.7).

**Theorem 11.8.19** ([64]) *Let $D$ be a normal arc-locally semicomplete digraph. Then $UG(D)$ is a strongly perfect graph if and only if it contains no induced subgraph to $C_{2n+1}$, for $n \geq 2$.*

Unlike underlying graphs of line digraphs (cf. Section 11.2.6), those of arc-locally semicomplete digraphs can be recognized in polynomial time, as Bang-Jensen [13] showed by reducing the problem to 2-SAT.

## 11.9 $\mathcal{H}_i$-Free Digraphs

Just as locally semicomplete and quasi-transitive digraphs can be characterized by forbidden induced subdigraphs, so can arc-locally semicomplete digraphs, as Bang-Jensen [15] noted. Let $\mathcal{H}$ denote the digraphs on 4 vertices whose underlying graphs contain two non-adjacent vertices $x$ and $y$ that are connected by a path $P = xuvy$ of length 3. We then distinguish four subsets based on the possible orientations of the path $P$. Let $\mathcal{H}_1$ be those digraphs where $P$ is oriented $x \rightarrow u \leftarrow v \leftarrow y$. $\mathcal{H}_2$ are the digraphs where $P$ is oriented $x \leftarrow u \rightarrow v \rightarrow y$. The subset $\mathcal{H}_3$ contains exactly those digraphs where $P$ is oriented $x \rightarrow u \rightarrow v \rightarrow y$. And finally, let $\mathcal{H}_4$ be the digraphs where $P$ is oriented $x \rightarrow u \leftarrow v \rightarrow y$.

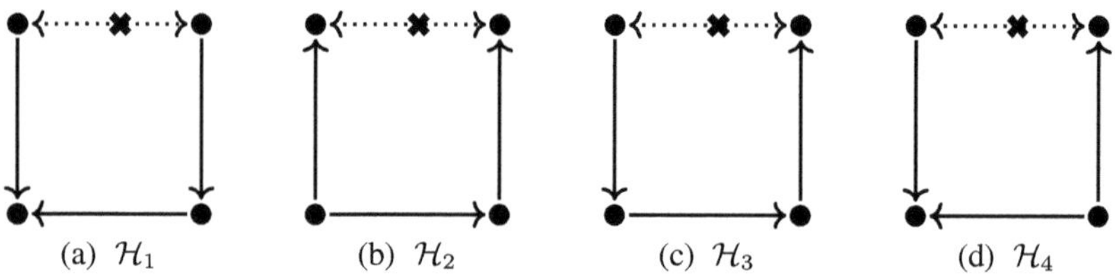

Figure 11.11 Substructures defining $\mathcal{H}_i$ digraphs. The dotted arc with a cross indicates that the two vertices are not adjacent.

Now arc-locally semicomplete digraphs are exactly the $\{\mathcal{H}_1, \mathcal{H}_2\}$-free digraphs, i.e. those which do not contain any induced subdigraph from $\mathcal{H}_1$ or $\mathcal{H}_2$. $\mathcal{H}_1$-free ($\mathcal{H}_2$-free, respectively) digraphs were dubbed **arc-locally in-semicomplete** (**arc-locally out-semicomplete**, respectively) digraphs by Wang and Wang [158]. $\mathcal{H}_3$-free digraphs are also known as **3-quasi-transitive** (see, e.g., [89] or Chapter 8) or **quasi-arc-transitive** (see, e.g., [158]) digraphs and $\mathcal{H}_4$-free digraphs are sometimes called **quasi-antiarc-transitive** (see, e.g., [158]).

In the introductory paper [15], Bang-Jensen conjectured that Theorem 11.8.4 also holds for $\mathcal{H}_i$-free digraphs, $i = 1, \ldots, 4$, a conjecture that was the main motivator for further work on $\mathcal{H}_i$-free digraphs and that has since been verified, as we will see in the remainder of this section.

Since an $\mathcal{H}_2$-free digraph is the converse of an $\mathcal{H}_1$-free digraph, we will limit our considerations to $\mathcal{H}_1$-free digraphs and only remark that analogous results obviously hold for $\mathcal{H}_2$-free digraphs. Wang and Wang [158] extended several structural results on arc-locally semicomplete digraphs to $\mathcal{H}_1$-free digraphs aiming at a generalization of Theorem 11.8.6. For their characterization of strong $\mathcal{H}_1$-free digraphs, we need to define another class of digraphs.

A **T-digraph** is a strong digraph $D = (V, A)$ whose vertex set has a partition (called a **T-partition**) $(V_1, V_2, V_3, V_4)$ such that

(i) $|V_2| = 1$ and one of $V_3$ or $V_4$ is permitted to be empty,
(ii) $D_4 := D\langle V_4\rangle$ is semicomplete,
(iii) $A_{\min} := A(D_4) \cup V_1 \times V_2 \cup V_2 \times V_3 \cup (V_3 \cup V_4) \times V_1 \cup V_4 \times V_3 \subseteq A$,
(iv) $A \subset A_{\min} \cup V_4 \times V_2 \cup V_2 \times (V_1 \cup V_4)$, and
(v) every vertex of $V_2$ is adjacent to every vertex of $V_1 \cup V_4$.

Note that $F_n$, $n \geq 4$ (cf. Theorem 11.8.6), is a T-digraph with T-partition $(\{u\}, \{v\}, \{x_1, \ldots, x_{n-2}\}, \emptyset)$ and the converse of a T-digraph. Now we may give the characterization.

**Theorem 11.9.1** ([158]) *Let $D$ be a strong $\mathcal{H}_1$-free digraph. Then $D$ is either semicomplete, semicomplete bipartite, an extended cycle or a T-digraph.*

So, by Theorem 11.8.6, except for T-digraphs, strong $\mathcal{H}_1$-free digraphs are arc-locally semicomplete digraphs, which implies the following corollary.

**Corollary 11.9.2** ([158]) *Let $D$ be a 2-strong $\mathcal{H}_1$-free digraph. Then $D$ is an arc-locally semicomplete digraph.*

Furthermore, Wang and Wang [158] used Theorem 11.9.1 to verify Bang-Jensen's [15] conjecture stating that Theorem 11.8.4 also holds for $\mathcal{H}_1$-free digraphs.

**Theorem 11.9.3** ([158]) *A strong $\mathcal{H}_1$-free digraph has a Hamiltonian cycle if and only if it has a directed cycle factor.*

Theorem 11.9.1 also implies that the Directed Path Partition Conjecture is true for strong $\mathcal{H}_1$-free digraphs, as Arroyo and Galeana-Sánchez [10] noted.

**Theorem 11.9.4** ([10]) *Let $D$ be a strong $\mathcal{H}_1$-free digraph. If $D$ contains no path with more than $\lambda$ vertices, then, for every pair $a, b$ of positive integers with $\lambda = a + b$, there exists a partition $(A, B)$ of $V(D)$ such that no path in $D\langle A\rangle$ ($D\langle B\rangle$, respectively) has more than $a$ ($b$, respectively) vertices.*

Finally, Wang and Wang [158] proved Theorem 11.8.10 actually not only for arc-locally semicomplete digraphs, but for the larger class of $\mathcal{H}_1$-free digraphs.

**Theorem 11.9.5** ([157]) *Let $D$ be an $\mathcal{H}_1$-free digraph. Then there exists a maximal independent set intersecting every non-augmentable path in $D$.*

For their results (and others) on $\mathcal{H}_3$-free digraphs, also known as 3-quasi-transitive digraphs, we refer to Chapter 8 and therefore turn directly to $\mathcal{H}_4$-free digraphs, whose structure seems much more elaborate than those of $\mathcal{H}_1$-, $\mathcal{H}_2$- and $\mathcal{H}_3$-free digraphs.

Galeana-Sánchez and Goldfeder [67] and, independently, Wang [155] proved Bang-Jensen's [15] conjecture for $\mathcal{H}_4$-free digraphs.

**Theorem 11.9.6** ([67, 155]) *A strong $\mathcal{H}_4$-free digraph has a Hamiltonian cycle if and only if it has a directed cycle factor.*

For strong $\mathcal{H}_i$-free digraphs, $i = 1, 2, 3$, the corresponding theorems were derived from structural results implying a close relation to semicomplete and semicomplete bipartite digraphs. The lack of such results for $\mathcal{H}_4$-free digraphs, particularly of a characterization similar to Theorem 11.9.1, made Galeana-Sánchez and Goldfeder [67] prove Theorem 11.9.6 directly via algebraic methods. Wang [155] on the other hand proved the necessary structure combinatorially.

As a consequence of Theorem 11.9.6, Galeana-Sánchez and Goldfeder [67] obtained that Theorem 11.8.5 also holds for $\mathcal{H}_4$-free digraphs.

**Theorem 11.9.7** ([67]) *A connected $\mathcal{H}_4$-free digraph $D$ has a Hamiltonian path if and only if it has a path $P$ (where we allow $V(P) = \emptyset$ or $V(P) = V(D)$) such that $D - V(P)$ has a directed cycle factor.*

While the Directed Path Partition Conjecture remains open for $\mathcal{H}_4$-free digraphs, Galeana-Sánchez and Gómez [68] proved that, in $\mathcal{H}_4$-free digraphs, not only does a maximal independent set of vertices intersecting every non-augmentable path exist (cf. Theorem 11.8.10), but every maximal independent set has this property.

**Theorem 11.9.8** ([68]) *Let $D$ be an $\mathcal{H}_4$-free digraph. Then every maximal independent set intersects every non-augmentable path in $D$.*

As Galeana-Sánchez and Gómez [68] noted, the Heuchenne condition (cf. Theorem 11.2.3 (iii)) implies $x \to y$ for every oriented path $x \to u \leftarrow v \to y$. Thus, $\mathcal{H}_4$-free digraphs are a generalization of line digraphs (without loops).

**Corollary 11.9.9** ([68]) *Let $D$ be a line digraph. Then every maximal independent set intersects every non-augmentable path in $D$.*

Wang [156] considered a cycle analogue of Theorem 11.9.8. To account for the missing structure of $\mathcal{H}_4$-free digraphs, Wang restricted his studies to a subclass mirroring line digraphs. An $\mathcal{H}_4$-free digraph is called an **$\mathcal{H}_4^*$-free digraph** if every oriented path $x \to u \leftarrow v \to y$ implies $y \to x$. Wang then proceeded to show that these digraphs, under certain conditions, again, are closely related to semicomplete and semicomplete bipartite digraphs.

**Theorem 11.9.10** ([156]) *Let $D$ be a strong $\mathcal{H}_4^*$-free digraph. If $D$ has a directed cycle factor $C_1, \dots, C_t$, $t \geq 2$, then $D$ is either semicomplete, semicomplete bipartite or isomorphic to $D^*$ (see Figure 11.12).*

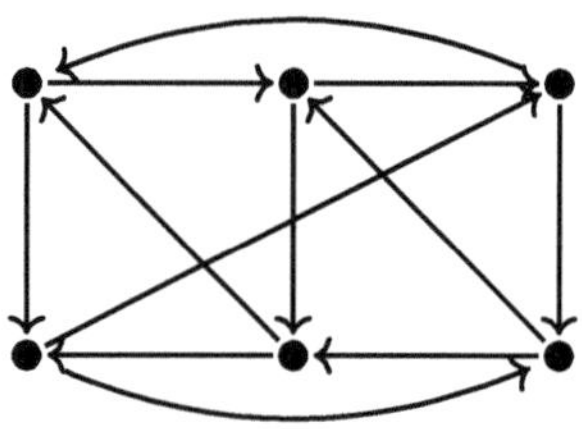

**Figure 11.12** Special $\mathcal{H}_4^*$-free digraph $D^*$.

Finally, Wang's [156] variation of Theorem 11.9.8 reads as follows.

**Theorem 11.9.11** ([156]) *Let $D$ be a strong $\mathcal{H}_4^*$-free digraph. Then there exists a maximal independent set intersecting every longest cycle in $D$.*

## References


1. M. Aigner. On the linegraph of a directed graph. *Mathematische Zeitschrift*, 102(1):56–61, 1967.
2. B. Alspach and T.D. Parsons. Isomorphism of circulant graphs and digraphs. *Discrete Mathematics*, 25(2):97–108, 1979.
3. S.D. Andres. Asymmetric directed graph coloring games. *Discrete Math.*, 309(18):5799–5802, 2009.
4. S.D. Andres. Directed defective asymmetric graph coloring games. *Discrete Appl. Math.*, 158(4):251–260, 2010.
5. S.D. Andres. Game-perfect digraphs. *Math. Methods Oper. Res.*, 76(3):321–341, 2012.
6. S.D. Andres. Lightness of digraphs in surfaces and directed game chromatic number. *Discrete Math.*, 309(11):3564–3579, 2009.
7. S.D. Andres. On characterizing game-perfect graphs by forbidden induced subgraphs. *Contributions to Discrete Mathematics*, 7(1), 2012.
8. S.D. Andres. On kernels in game-perfect digraphs and a characterization of weakly game-perfect digraphs. Technical report feU-dmo043.17, Fakultät für Mathematik und Informatik, Fernuniversität Hagen, Germany, 2017.
9. S.D. Andres and W. Hochstättler. Perfect digraphs. *J. Graph Theory*, 79(1):21–29, 2015.
10. A. Arroyo and H. Galeana-Sánchez. The path partition conjecture is true for some generalizations of tournaments. *Discrete Math.*, 313(3):293–300, 2013.
11. L. Baffi and R. Petreschi. Parallel maximal matching on minimal vertex series parallel digraphs. In *Algorithms, concurrency and knowledge (Pathumthani, 1995)*, pages 34–47. Springer-Verlag, 1995.
12. C. Balbuena and M. Guevara. Kernels and partial line digraphs. *Appl. Math. Lett.*, 23(10):1218–1220, 2010.
13. J. Bang-Jensen. Arc-local tournament digraphs: a generalization of tournaments and bipartite tournaments. Technical report 2, Department of Mathematics and Computer Science, Odense University, Denmark, 1993.
14. J. Bang-Jensen. Digraphs with the path-merging property. *J. Graph Theory*, 20(2):255–265, 1995.
15. J. Bang-Jensen. The structure of strong arc-locally semicomplete digraphs. *Discrete Math.*, 283(1-3):1–6, 2004.
16. J. Bang-Jensen and G. Gutin. *Digraphs: Theory, Algorithms and Applications*. Springer-Verlag, London, 2nd edition, 2009.
17. J. Bang-Jensen and G. Gutin. Paths and cycles in extended and decomposable digraphs. *Discrete Math.*, 164(1-3):5–19, 1997.
18. J. Bang-Jensen, F. Havet, and N. Trotignon. Finding an induced subdivision of a digraph. *Theor. Comput. Sci.*, 443:10–24, 2012.
19. J. Bang-Jensen and A. Maddaloni. Arc-disjoint paths in decomposable digraphs. *J. Graph Theory*, 77(2):89–110, 2014.
20. J. Bang-Jensen and Y. Manoussakis. Cycles through $k$ vertices in bipartite tournaments. *Combinatorica*, 14(2):243–246, 1994.
21. D. Bauer, F. Boesch, C. Suffel, and R. Tindell. Connectivity extremal problems and the design of reliable probabilistic networks. *Theory and Application of Graphs*, pages 89–98, 1981.
22. D. Bechet, P. De Groote, and C. Retoré. A complete axiomatisation for the inclusion of series-parallel partial orders. In *International Conference on Rewriting Techniques and Applications*, pages 230–240. Springer, 1997.
23. L.W. Beineke and C.M. Zamfirescu. Connection digraphs and second-order line digraphs. *Discrete Math.*, 39:237–254, 1982.

24. C. Berge. Färbung von Graphen, deren sämtliche bzw. deren ungerade Kreise starr sind. *Wiss. Z. Martin-Luther-Univ. Halle-Wittenberg Math.-Natur. Reihe*, 10(114):88, 1961.
25. C. Berge and P. Duchet. Recent problems and results about kernels in directed graphs. *Discrete Math.*, 86(1-3):27–31, 1990.
26. J.-C. Bermond and P. Fraigniaud. Broadcasting and gossiping in de Bruijn networks. *SIAM J. Comput.*, 23(1):212–225, 1994.
27. J.-C. Bermond, N. Homobono, and C. Peyrat. Large fault-tolerant interconnection networks. *Graphs Combin.*, 5(1):107–123, 1989.
28. J.-C. Bermond, X. Munos, and A. Marchetti-Spaccamela. A Broadcasting Protocol in Line Digraphs. *J. Parallel Distributed Comput.*, 61(8):1013–1032, 2001.
29. P. Bertolazzi, R.F. Cohen, G. Di Battista, R. Tamassia, and I.G. Tollis. How to draw a series-parallel digraph. *Int. J. Comput. Geom. Appl.*, 4(4):385–402, 1994.
30. H. Bidkhori and S. Kishore. A bijective proof of a theorem of Knuth. *Combin. Prob. Comput.*, 20(1):11–25, 2011.
31. E. Boros and V. Gurvich. Perfect graphs are kernel-solvable. *Discrete Math.*, 159:35–55, 1996.
32. E. Boros and V. Gurvich. Perfect graphs, kernels, and cores of cooperative games. *Discrete Math.*, 306(19):2336–2354, 2006.
33. A. Bretscher, D.G. Corneil, M. Habib, and C. Paul. A simple linear time lexbfs cograph recognition algorithm. *SIAM Journal on Discrete Mathematics*, 22(4):1277–1296, 2008.
34. W.G. Bridges and S. Toueg. On the impossibility of directed Moore graphs. *J. Combin. Theory Ser. B*, 29:339–341, 1980.
35. D.E. Brown, A.H. Busch, and J.R. Lundgren. Interval tournaments. *J. Graph Theory*, 56:72–81, 2007.
36. F. Cao, D.-Z. Du, D.F. Hsu, L. Hwang, and W. Wu. Super line-connectivity of consecutive-d digraphs. *Discrete Math.*, 183(1):27–38, 1998.
37. W.H. Chan, W.C. Shiu, P.K. Sun, and X. Zhu. The strong game colouring number of directed graphs. *Discrete Math.*, 313(10):1070–1077, 2013.
38. X. Cheng, X. Du, M. Min, H.Q. Ngo, L. Ruan, J. Sun, and W. Wu. Super link-connectivity of iterated line digraphs. *Theor. Comput. Sci.*, 304(1):461–469, 2003.
39. M. Chudnovsky, G. Cornuéjols, X. Liu, P.D. Seymour, and K. Vušković. Recognizing Berge graphs. *Combinatorica*, 25(2):143–186, 2005.
40. M. Chudnovsky, N. Robertson, P.D. Seymour, and R. Thomas. The strong perfect graph theorem. *Annals Math.*, pages 51–229, 2006.
41. V. Chvátal and C. Ebenegger. A note on line digraphs and the directed max-cut problem. *Discrete Appl. Math.*, 29(2):165–170, 1990.
42. D.G. Corneil, Y. Perl, and L.K. Stewart. A linear recognition algorithm for cographs. *SIAM J. Comput.*, 14(4):926–934, 1985.
43. C. Crespelle and C. Paul. Fully dynamic recognition algorithm and certificate for directed cographs. *Discrete Appl. Math.*, 154(12):1722–1741, 2006.
44. S.J. Curran and J.A. Gallian. Hamiltonian cycles and paths in Cayley graphs and digraphs – a survey. *Discrete Mathematics*, 156(1-3):1–18, 1996.
45. S. Das, M. Francis, P. Hell, and J. Huang. Recognition and characterization of chronological interval digraphs. *the electronic journal of combinatorics*, 20(3):P5, 2013.
46. S. Das and M. Sen. An interval digraph in relation to its associated bipartite graph. *Discrete Math.*, 122(1-3):113–136, 1993.

47. S. Dasgupta. *On characterizations and structure of interval digraphs and unit probe interval graphs.* PhD thesis, University of Colorado, 2012.
48. N.G. de Bruijn. A combinatorial problem. *Ned. Akad. Wet. Proc.*, 49:758–764, 1946.
49. D.-Z. Du, F. Cao, and D.F. Hsu. De Bruijn digraphs, and Kautz digraphs, and their generalizations. In D.-Z. Du and D.F. Hsu, editors, *Combinatorial network theory*, pages 65–105. Kluwer, Dordrecht, 1996.
50. D.-Z. Du, D.F. Hsu, and F.K. Hwang. Hamiltonian property of $d$-consecutive digraphs. *Math. Comput. Modeling*, 17:61–63, 1993.
51. D.-Z. Du, D.F. Hsu, and Y.-D. Lyuu. Corrigendum to 'Line digraph iterations and connectivity analysis of de Bruijn and Kautz Graphs'. *IEEE Trans. Computers*, 45(7):863, 1996.
52. D.-Z. Du, D.F. Hsu, and Y.-D. Lyuu. On the diameter vulnerability of Kautz digraphs. *Discrete Math.*, 151(1):81–85, 1996.
53. D.-Z. Du, D.F. Hsu, H.Q. Ngo, and G.W. Peck. On connectivity of consecutive-$d$ digraphs. *Discrete Math.*, 257(2):371–384, 2002.
54. D.-Z. Du, D.F. Hsu, and G.W. Peck. Connectivity of consecutive-$d$ digraphs. *Discrete Appl. Math.*, 37:169–177, 1992.
55. D.-Z. Du, Y.-D. Lyuu, and D.F. Hsu. Line digraph iterations and connectivity analysis of de Bruijn and Kautz graphs. *IEEE Trans. Computers*, 42(5):612–616, 1993.
56. D.Z. Du, Y.-D. Lyuu, and D.F. Hsu. Line digraph iterations and the spread concept—with application to graph theory, fault tolerance, and routing. In *WG 1991: Graph-theoretic concepts in computer science*, volume 570 of *Lect. Notes Comput. Sci.*, pages 169–179. Springer, 1992.
57. P. Duchet and H. Meyniel. A note on kernel-critical graphs. *Discrete Math.*, 33(1):103–105, 1981.
58. D. Duffus, H. Lefmann, and V. Rödl. Shift graphs and lower bounds on Ramsey numbers $r_k(l;r)$. *Discrete Math.*, 137(1):177–187, 1995.
59. A. Ehrenfeucht and G. Rozenberg. Primitivity is hereditary for 2-structures. *Theor. Comput. Sci.*, 70(3):343–358, 1990.
60. J. Fàbrega and M.A. Fiol. Maximally connected digraphs. *J. Graph Theory*, 13(6):657–668, 1989.
61. J. Fabrega, M.A. Fiol, J.L.A. Yebra, and I. Alegre. Connectivity and reliable routing algorithms in line digraphs. In *3rd Int. Symp. Applied Informatics*, pages 45–50, 1985.
62. T. Feder, P. Hell, J. Huang, and A. Rafiey. Adjusted interval digraphs. *Elec. Notes Discrete Math.*, 32:83–91, 2009.
63. M.A. Fiol, J.L.A. Yebra, and I. Alegre. Line digraph iteration and the $(d,k)$ digraph problem. *IEEE Trans. Comput.*, C-33:400–403, 1984.
64. H. Galeana-Sánchez. Kernels and perfectness in arc-local tournament digraphs. *Discrete Math.*, 306(19-20):2473–2480, 2006.
65. H. Galeana-Sánchez and I.A. Goldfeder. A classification of arc-locally semicomplete digraphs. *Elec. Notes Discrete Math.*, 34:59–61, 2009.
66. H. Galeana-Sánchez and I.A. Goldfeder. A classification of all arc-locally semicomplete digraphs. *Discrete Math.*, 312(11):1883–1891, 2012.
67. H. Galeana-Sánchez and I.A. Goldfeder. Hamiltonian cycles in a generalization of bipartite tournaments with a cycle factor. *Discrete Math.*, 315:135–143, 2014.
68. H. Galeana-Sánchez and R. Gómez. Independent sets and non-augmentable paths in generalizations of tournaments. *Discrete Math.*, 308(12):2460–2472, 2008.

69. H. Galeana-Sánchez and R. Gómez. ($k$,$l$)-kernels,($k$,$l$)-semikernels, $k$-Grundy functions and duality for state splittings. *Discuss. Math. Graph Theory*, 27(2):359–371, 2007.
70. H. Galeana-Sánchez and X. Li. Semikernels and $(k,l)$-kernels in digraphs. *SIAM J. Discrete Math.*, 11(2):340–346, 1998.
71. H. Galeana-Sánchez, L.P. Ramírez, and H.A. Rincón-Mejía. Semikernels, quasi kernels, and Grundy functions in the line digraph. *SIAM J. Discrete Math.*, 4(1):80–83, 1991.
72. F. Gavril. Some NP-complete problems on graphs. In *11th Conf. on Information Sciences and Systems*, pages 91–95, 1977.
73. Z. Ge and S.L. Hakimi. Disjoint rooted spanning trees with small depths in deBruijn and Kautz graphs. *SIAM J. Comput.*, 26(1):79–92, 1997.
74. I.A. Goldfeder. Una clasificación de las digráficas localmente semicompletas en flechas. *Undergraduate Thesis, Facultad de Ciencias, Universidad Nacional Autónoma de México*, 2008.
75. M. Grötschel, L. Lovász, and A. Schrijver. Polynomial algorithms for perfect graphs. *North-Holland Mathematics Studies*, 88:325–356, 1984.
76. M. Guevara, C. Balbuena, and H. Galeana-Sánchez. Relation between number of kernels (and generalizations) of a digraph and its partial line digraphs. *Elec. Notes Discrete Math.*, 54:265–269, 2016.
77. Y. Guo and M. Surmacs. Pancyclic arcs in Hamiltonian cycles of hypertournaments. *J. Korean Math. Soc*, 51(6):1141–1154, 2014.
78. Y. Guo and M. Surmacs. Pancyclic out-arcs of a vertex in a hypertournament. *Australas. J. Combinatorics*, 61(3):227–250, 2015.
79. G. Gutin and A. Yeo. Hamiltonian paths and cycles in hypertournaments. *J. Graph Theory*, 25(4):277–286, 1997.
80. R. Häggkvist and Y. Manoussakis. Cycles and paths in bipartite tournaments with spanning configurations. *Combinatorica*, 9(1):33–38, 1989.
81. Y.O. Hamidoune. On the connectivity of Cayley digraphs. *European Journal of Combinatorics*, 5(4):309–312, 1984.
82. F. Harary and R.Z. Norman. Some properties of line digraphs. *Rend. Circ. Mat. Palermo*, 9(2):161–168, 1960.
83. M. Harminc. Solutions and kernels of a directed graph. *Mathematica Slovaca*, 32(3):263–267, 1982.
84. C.C. Harner and R.C. Entringer. Arc colorings of digraphs. *J. Combin. Theory Ser. B*, 13(3):219–225, 1972.
85. T. Hasunuma and H. Nagamochi. Independent spanning trees with small depths in iterated line digraphs. *Discrete Appl. Math.*, 110(2):189–211, 2001.
86. T. Hasunuma and M. Otani. On the $(h,k)$-domination numbers of iterated line digraphs. *Discrete Appl. Math.*, 160(12):1859–1863, 2012.
87. R.L. Hemminger. Line digraphs. In *Graph Theory and Applications*, pages 149–163. Springer, 1972.
88. R.L. Hemminger and L.W. Beineke. Line graphs and line digraphs. In L.W. Beineke and R.J. Wilson, editors, *Selected Topics in Graph Theory*, pages 271–305. Academic Press, London, 1978.
89. C. Hernández-Cruz and H. Galeana-Sánchez. $k$-kernels in $k$-transitive and $k$-quasi-transitive digraphs. *Discrete Math.*, 312(16):2522–2530, 2012.
90. C. Heuchenne. Sur une certaine correspondance entre graphs. *Bull. Soc. R. Sci. Liége*, 33:743–753, 1964.
91. A. Huck. Disproof of a conjecture about independent branchings in $k$-connected directed graphs. *J. Graph Theory*, 20(2):235–239, 1995.
92. M. Imase and M. Itoh. Design for directed graphs with minimum diameter. *IEEE Trans. Comput.*, C-32:782–784, 1983.

93. M. Imase and M. Itoh. Design to minimize a diameter on building block networks. *IEEE Trans. Comput.*, C-30:439–443, 1981.
94. M. Imase, I. Soneoka, and K. Okada. A fault tolerant processor interconnection network. *Syst. Comput. Japan*, 17:21–30, 1986.
95. M. Imase, I. Soneoka, and K. Okada. Connectivity of regular directed graphs with small diameter. *IEEE Trans. Comput.*, C-34:267–273, 1985.
96. M. Imori, M. Matsumoto, and H. Yamada. The line digraph of a regular and pancircular digraph is also regular and pancircular. *Graphs Combin.*, 4:235–239, 1988.
97. H.A. Jung. On a class of posets and the corresponding comparability graphs. *Journal of Combinatorial Theory, Series B*, 24(2):125–133, 1978.
98. R.M. Karp. Reducibility among combinatorial problems. In *Complexity of computer computations (Proc. Symp., IBM Thomas J. Watson Res. Center, Yorktown Heights, N.Y., 1972)*, pages 85–103. Plenum, 1972.
99. W.H. Kautz. Bounds on directed $(d,k)$ graphs. Theory of cellular logic networks and machines. *AFCRL-68-0668 Final report*, pages 20–28, 1968.
100. K.K. Kayibi, M.A. Khan, and S. Pirzada. Scores, inequalities and regular hypertournaments. *J. Math. Ineq. Appl.*, 15(2):343–351, 2012.
101. K.K. Kayibi, M.A. Khan, and S. Pirzada. Uniform sampling of $k$-hypertournaments. *Linear and Multilinear Algebra*, 61(1):123–138, 2013.
102. K.K. Kayibi, M.A. Khan, S. Pirzada, and G. Zhou. On scores, losing scores and total scores in hypertournaments. *Elec. J. Graph Theory and Appl.*, 3(1):8–21, 2015.
103. D.E. Knuth. Oriented subtrees of an arc digraph. *J. Combin. Theory*, 3(4):309–314, 1967.
104. Y. Koh and S. Ree. On $k$-hypertournament matrices. *Linear algebra and its applications*, 373:183–195, 2003.
105. J. Krausz. Démonstration nouvelle d'une théorème de Whitney sur les réseaux. *Mat. Fiz. Lapok*, 50(1):75–85, 1943.
106. J.M. Laborde, C. Payan, and N.H. Xuong. Independent sets and longest directed paths in digraphs. *Graphs and other combinatorial topics (Prague, 1982)*, 59:173–177, 1983.
107. E.L. Lawler. Sequencing jobs to minimize total weighted completion time subject to precedence constraints. *Ann. Discrete Math.*, 2:75–90, 1978. Algorithmic aspects of combinatorics (Conf., Vancouver Island, B.C., 1976).
108. L. Levine. Sandpile groups and spanning trees of directed line graphs. *J. Combin. Theory Ser. A*, 118(2):350–364, 2011.
109. H. Li, S. Li, Y. Guo, and M. Surmacs. On the vertex-pancyclicity of hypertournaments. *Discrete Appl. Math.*, 161(16):2749–2752, 2013.
110. H. Li, W. Ning, Y. Guo, and M. Lu. On pancyclic arcs in hypertournaments. *Discrete Appl. Math.*, 215:164–170, 2016.
111. R. Li, X. Zhang, S. Li, Q. Guo, and Y. Guo. The $h$-force set of a hypertournament. *Discrete Appl. Math.*, 169:168–175, 2014.
112. N. Lichiardopol. Independence number of iterated line digraphs. *Discrete Math.*, 293(1):185–193, 2005.
113. X. Liu and D.B. West. Line digraphs and coreflexive vertex sets. *Discrete Math.*, 188(1-3):269–277, 1998.
114. L. Lovász. A characterization of perfect graphs. *J. Combin. Theory Ser. B*, 13(2):95–98, 1972.
115. M. Lü and J.-M. Xu. Super connectivity of line graphs and digraphs. *Acta Mathematicae Applicatae Sinica*, 22(1):43–48, 2006.
116. Q. Lu, E. Shan, and M. Zhao. $(k,l)$-kernels in line digraphs. *J. Shanghai University (English Edition)*, 10(6):484–486, 2006.

117. R.H. Möhring. Computationally tractable classes of ordered sets. In *Algorithms and order*, pages 105–193. Springer, 1989.
118. R.H. Möhring and F.J. Radermacher. Substitution decomposition for discrete structures and connections with combinatorial optimization. *North-Holland mathematics studies*, 95:257–355, 1984.
119. C.L. Monma and J.B. Sidney. A general algorithm for optimal job sequencing with series-parallel constraints. *Math. Oper. Res.*, 4:215–224, 1977.
120. H. Müller. Recognizing interval digraphs and interval bigraphs in polynomial time. *Discrete Appl. Math.*, 78:189–205, 1997.
121. J. Nešetril and E. Sopena. On the oriented game chromatic number. *Elec. J. Combin.*, 8(2), 2001.
122. V. Neumann-Lara. The dichromatic number of a digraph. *J. Combin. Theory Ser. B.*, 33:265–270, 1982.
123. P. Ochem and A. Pinlou. Oriented vertex and arc colorings of partial 2-trees. *Elec. Notes Discrete Math.*, 29:195–199, 2007.
124. P. Ochem, A. Pinlou, and E. Sopena. On the oriented chromatic index of oriented graphs. *J. Graph Theory*, 57(4):313–332, 2008.
125. J.B. Orlin. Line-digraphs, arborescences, and theorems of Tutte and Knuth. *J. Combin. Theory Ser. B*, 25(2):187–198, 1978.
126. V. Petrovic and C. Thomassen. Edge-disjoint Hamiltonian cycles in hypertournaments. *J. Graph Theory*, 51(1):49–52, 2006.
127. A. Pinlou. On oriented arc-coloring of subcubic graphs. *Elec. J. Combinatorics*, 13(3):R69, 2006.
128. A. Pinlou and E. Sopena. Oriented vertex and arc colorings of outerplanar graphs. *Inform. Process. Lett.*, 100(3):97–104, 2006.
129. S. Poljak and V. Rödl. On the arc-chromatic number of a digraph. *J. Combin. Theory Ser. B*, 31(2):190–198, 1981.
130. E. Prisner. Bicliques in graphs. II. Recognizing $k$-path graphs and underlying graphs of line digraphs. In *WG 1996: Graph-theoretic concepts in computer science*, volume 1335 of *Lect. Notes Comput. Sci.*, pages 273–287. Springer, 1997.
131. E. Prisner. Line graphs and generalizations—a survey. *Congr. Numer.*, 116:193–229, 1996. Surveys in graph theory (San Francisco, 1995).
132. R.A. Rankin. A campanological problem in group theory. *Math. Proc. Cambridge Philosophical Soc.*, 44:17–25, 1948.
133. S.M. Reddy, J.G. Kuhl, S.H. Hosseini, and H. Lee. On digraphs with minimum diameter and maximum connectivity. In *20th Annual Allerton Conference*, pages 1018–1026, 1982.
134. S. M. Reddy, D.K. Pradhan, and J. G. Kuhl. Directed graphs with minimal diameter and maximal connectivity. Tech. Rep., School of Engineering, Oakland University, 1980.
135. F. Rendl. Quadratic assignment problems on series-parallel digraphs. *Z. Oper. Res. Ser. A-B*, 30(3):A161–A173, 1986.
136. C. Retoré. Pomset logic as a calculus of directed cographs. Technical Report n3714, Unité de recherche INRIA Rennes, France, 1999.
137. P.I. Richards. Precedence constraints and arrow diagrams. *SIAM Rev.*, 9:548–553, 1967.
138. B.K. Sanyal and M.K. Sen. New characterization of digraphs represented by intervals. *J. Graph Theory*, 22:297–303, 1996.
139. M. Sen, S. Das, A.B. Roy, and D.B. West. Interval digraphs: an analogue of interval graphs. *J. Graph Theory*, 13(2):189–202, 1989.
140. M. Sen, P. Talukdar, and S. Das. Chronological orderings of interval digraphs. *Discrete Math.*, 306(14):1601–1609, 2006.

141. M.K. Sen, B.K. Sanyal, and D.B. West. Representing digraphs using intervals or circular arcs. *Discrete Math.*, 147(1-3):235–245, 1995.
142. E. Shan, L. Kang, and Q. Lu. $k$-semikernels, $k$-quasikernels, $k$-kernels in digraphs and their line digraphs. *Utilitas Math.*, 72:267–277, 2007.
143. T. Soneoka. Super edge-connectivity of dense digraphs and graphs. *Discrete Appl. Math.*, 37:511–523, 1992.
144. G. Steiner. A compact labeling scheme for series-parallel graphs. *Discrete Appl. Math.*, 11(3):281–297, 1985.
145. M. Surmacs. Regular hypertournaments and arc-pancyclicity. *J. Graph Theory*, 84:176–190, 2017.
146. M.M. Syslo. A new solvable case of the traveling salesman problem. *Math. Program.*, 4(1):347–348, 1973.
147. W.T. Tutte. The dissection of equilateral triangles into equilateral triangles. *Proc. Cambridge Philos. Soc.*, 44:463–482, 1948.
148. J. Valdes. Parsing flowcharts and series-parallel graphs. Technical Report STAN-CS-78-682, Computer Science Department, Stanford University, Stanford, Ca., 1978.
149. J. Valdes, R.E. Tarjan, and E.L. Lawler. The recognition of series parallel digraphs. *SIAM J. Comput.*, 11(2):298–313, 1982.
150. T. van Aardenne-Ehrenfest and N.G. de Bruijn. Circuits and trees in oriented linear graphs. *Simon Stevin*, 28:203–217, 1951.
151. E.A. van Doorn. Connectivity of circulant digraphs. *J. Graph Theory*, 10(1):9–14, 1986.
152. J.L. Villar. The underlying graph of a line digraph. *Discrete Appl. Math.*, 37:525–538, 1992.
153. J. von Neumann and O. Morgenstern. *Theory of Games and Economic Behavior.* Princeton University Press, Princeton, NJ, 1944.
154. C. Wang and G. Zhou. Note on the degree sequences of $k$-hypertournaments. *Discrete Math.*, 308(11):2292–2296, 2008.
155. R. Wang. A conjecture on 3-anti-quasi-transitive digraphs. *Discrete Math.*, 322:48–52, 2014.
156. R. Wang. Cycles in 3-anti-circulant digraphs. *Australasian Journal of Combinatorics*, 60(2):158–168, 2014.
157. S. Wang and R. Wang. Independent sets and non-augmentable paths in arc-locally in-semicomplete digraphs and quasi-arc-transitive digraphs. *Discrete Math.*, 311(4):282–288, 2011.
158. S. Wang and R. Wang. The structure of strong arc-locally in-semicomplete digraphs. *Discrete Math.*, 309(23):6555–6562, 2009.
159. D.B. West. Short proofs for interval digraphs. *Discrete Math.*, 178(1-3):287–292, 1998.
160. D.S. Witte. On Hamiltonian circuits in Cayley diagrams. *Discrete Mathematics*, 38(1):99–108, 1982.
161. D. Witte and J.A. Gallian. A survey: Hamiltonian cycles in Cayley graphs. *Discrete Mathematics*, 51(3):293–304, 1984.
162. W. Xiao and B. Parhami. Some mathematical properties of Cayley digraphs with applications to interconnection network design. *Intern. J. Comput. Math.*, 82(5):521–528, 2005.
163. C. Xu, S. Zhang, B. Ning, and B. Li. A note on the number of spanning trees of line digraphs. *Discrete Math.*, 338(5):688–694, 2015.
164. M.-Y. Xu. Automorphism groups and isomorphisms of Cayley digraphs. *Discrete Mathematics*, 182(1-3):309–319, 1998.
165. D. Yang and X. Zhu. Game colouring directed graphs. *Elec. J. Combinatorics*, 17(R11):1, 2010.

166. J. Yang. Vertex-pancyclicity of hypertournaments. *J. Graph Theory*, 63(4):338–348, 2010.
167. Q.F. Yang, R.E. Burkard, E. Cela, and G.J. Woeginger. Hamiltonian cycles in circulant digraphs with two stripes. *Discrete Math.*, 176:233–254, 1997.
168. C.M.D. Zamfirescu. Transformations of digraphs viewed as intersection digraphs. In *Convexity and Discrete Geometry Including Graph Theory*, pages 27–35. Springer, 2016.
169. F. Zhang and G. Lin. On the de Bruijn-Good graphs. *Acta Math. Sinica*, 30(2):195–205, 1987.
170. H. Zhang, F. Zhang, and Q. Huang. On the number of spanning trees and Eulerian tours in iterated line digraphs. *Discrete Appl. Math.*, 73(1):59–67, 1997.
171. Z. Zhang, F. Liu, and J. Meng. Super-connected $n$-th iterated line digraphs. *Oper. Res. Transactions*, 9(2005):35–39, 2005.
172. Z. Zhang and Y. Zhu. Restricted connectivity of line digraphs. *J. Mathematical Study*, 43:107–113, 2010.
173. G. Zhou, T. Yao, and K. Zhang. On score sequences of $k$-hypertournaments. *European J. Combin.*, 21(8):993–1000, 2000.

# 12. Lexicographic Orientation Algorithms

**Jing Huang**

## 12.1 Introduction

Graph orientation, which provides a link between graphs and digraphs, is an actively studied area in the theory of graphs and digraphs. One of the fundamental problems asks whether a given graph admits an orientation that satisfies a prescribed property and to find such an orientation if it exists. A celebrated theorem of Robbins [34] which answers a question of this type states that a graph has a strong orientation if and only if it is 2-edge-connected (i.e., has no bridge). It is easy to check whether a graph is 2-edge-connected and to obtain, using the depth-first search algorithm, a strong orientation of a 2-edge-connected graph, cf. [35].

Which graphs have orientations in which the longest directed path has at most $k$ vertices? Answering this question, Gallai, Roy and Vitaver [13, 37, 47] proved that a graph has such an orientation if and only if it is $k$-colourable. The theorem nicely links orientations and colourings of graphs but it provides little help in finding such orientations. This is due to the fact that the $k$-colouring problem is NP-complete for each $k \geq 3$, cf. [15].

Given a graph $G$, an **orientation** of $G$ is a digraph $D$ obtained from $G$ by replacing every edge $uv$ of $G$ with an arc (i.e., a directed edge that is either $u \to v$ or $v \to u$). Since graphs considered in this chapter are all simple (i.e., having no loops or multiple edges), the digraphs resulting from orientations are **oriented graphs**. Let $\Pi$ be a property of oriented graphs. We say that a graph $G$ is **$\Pi$-orientable** if it admits an orientation that has the property $\Pi$. For a fixed property $\Pi$ the $\Pi$-ORIENTATION PROBLEM is as follows.

| |
|---|
| $\Pi$-ORIENTATION PROBLEM<br>**Input:** A graph $G$.<br>**Find:** A $\Pi$-orientation of $G$ or certify that $G$ is not $\Pi$-orientable. |

J. Huang (✉)
Mathematics and Statistics,
University of Victoria, PO BOX 1700 STN CSC,
Victoria, B.C. V8W 2Y2, Canada
e-mail: huangj@uvic.ca

 
J. Bang-Jensen and G. Gutin (eds.), *Classes of Directed Graphs*, Springer Monographs in Mathematics, https://doi.org/10.1007/978-3-319-71840-8_12

For instance, an oriented graph $D$ is **transitive** if for any three vertices $u, v, w$, $u \to v$ and $v \to w$ imply $u \to w$ in $D$. Thus a graph is **transitively orientable** if it admits an orientation that is a transitive oriented graph. The TRANSITIVE ORIENTATION PROBLEM asks whether a graph is transitively orientable and to find a transitive orientation of the graph if it exists.

Transitively orientable graphs are also known as **comparability graphs**, cf. [18]. Naturally connected to partially ordered sets, comparability graphs are perfect (in Berge's sense) and have been extensively studied, cf. [14, 16–19, 33]. A classical result of Gallai [14] characterizes comparability graphs by forbidden subgraphs (cf. [30] for the English translation). Gallai's characterization however does not immediately imply a polynomial time algorithm for recognizing comparability graphs or finding transitive orientations. But he proved that a graph is a comparability graph if and only if its knotting graph (cf. [14]) is bipartite, and he also gave a procedure for constructing knotting graphs which runs in polynomial time. It follows that comparability graphs can be recognized in polynomial time. Polynomial time algorithms for finding transitive orientations of comparability graphs have been given by Ghouila-Houri [16], Habib, McConnel Paul and Viennot [21], McConnell and Spinrad [32], and Pnueli, Lempel and Even [33].

In [22] Hell and Huang devised a very simple algorithm for determining whether a graph $G$ is a comparability graph and, if it is, finding a transitive orientation of it. The algorithm first constructs the auxiliary graph $G^+$ of the input graph $G$. The auxiliary graph $G^+$ is used to test whether $G$ is a comparability graph and to find, whenever possible, a transitive orientation of $G$. To test whether $G$ is a comparability graph, the algorithm proceeds to find a 2-colouring of $G^+$ using a lexicographic scheme. If the 2-colouring scheme fails, $G$ is not a comparability graph. Otherwise a 2-colouring of $G^+$ is obtained and the algorithm transforms the 2-colouring of $G^+$ into a transitive orientation of $G$. The 2-colourability of $G^+$ alone is sufficient for $G$ to be a comparability graph. Using the lexicographic scheme to find a 2-colouring of $G^+$ is to guarantee that the orientation of $G$ transformed from the 2-colouring is transitive. The time complexity of this algorithm is $O(m\Delta)$ where $m$ and $\Delta$ are the number of edges and the maximum degree of the input graph.

The technique described above for recognizing comparability graphs and obtaining transitive orientations is called the **lexicographic orientation method**. The lexicographic orientation method has also been applied for recognizing several other classes of graphs and finding desired orientations, cf. [22]. An oriented graph $D$ is called a **local tournament** (respectively, **locally transitive local tournament**) if for every vertex $v$, the in-neighbourhood and the out-neighbourhood of $v$ each induces a tournament (respectively, transitive tournament) in $D$, cf. [26]. Local tournaments and locally transitive local tournaments naturally generalize tournaments and transitive tournaments, respectively, cf. [1]. Despite the fact that the class of local tournaments properly contains the class of locally transitive local tournaments, it is proved by Hell and Huang [22] that they share the same class of underlying graphs, that is, a graph is local tournament orientable if and only if it is local transitive tournament orientable (see Corollary 12.2.7).

A graph $G$ is called a **circular arc graph** if it is the intersection graph of a family of circular arcs $I_v, v \in V(G)$, on a circle (i.e., two vertices $u, v$ are adjacent in $G$ if and only if $I_u, I_v$ intersect). The family $I_v, v \in V(G)$, is called a **circular arc representation** of $G$. Circular arc graphs have also been extensively studied by McConnell [31], Spinrad [39], Trotter and Moore [42], and Tucker [43–46].

Circular arc graphs generalize **interval graphs** which are the intersection graphs of intervals on the real line. A circular arc graph (respectively, an interval graph) is called **proper** if the family of circular arcs (respectively, intervals) can be chosen so that none of them is contained in another. Proper circular arc graphs and proper interval graphs are closely related to local tournaments. In fact, as proved by Skrien [38], a connected graph is local tournament orientable if and only if it is a proper circular arc graph (see Corollary 12.2.7). It is proved in [22, 26] that a graph is acyclic local tournament orientable if and only if it is a proper interval graph (see Corollary 12.2.11). Locally transitive local tournament (respectively, acyclic local tournament) orientations are useful in constructing proper circular arc (respectively, proper interval) representations of their underlying graphs, cf. [9]. Thus the lexicographic orientation method simultaneously solves the recognition and the representation problems for proper circular arc graphs and for proper interval graphs.

Let $G$ be a bipartite graph with bipartition $(X, Y)$. Then $G$ is called an **interval containment bigraph** if there is a family of intervals $I_v, v \in X \cup Y$ such that for all $x \in X$ and $y \in Y$, $xy$ is an edge of $G$ if and only if $I_x \supset I_y$. The family of intervals will be referred to as an **interval containment representation** of $G$. Various characterizations of interval containment bigraphs have been obtained by Feder, Hell and Huang [10], Hell and Huang [23], Huang [25], and Spinrad [39], and Trotter and Moore [42]. Interval containment bigraphs are closely related to circular arc graphs. In fact, the complements of interval containment bigraphs are precisely the circular arc graphs of clique covering number two. The lexicographic orientation method can also be used for recognizing interval containment bigraphs and constructing interval containment representations whenever possible.

The lexicographic orientation method has also been applied by Bang-Jensen, Huang and Zhu in [4] to solve some orientation completion problems. A **partially oriented graph** is a mixed graph which may contain both edges and arcs. We use $Q = (V, E \cup A)$ to denote a partially oriented graph where $E$ consists of edges and $A$ consists of arcs. An **orientation completion** of $Q$ is an oriented graph obtained from $Q$ by replacing every edge in $E$ with an arc. For a fixed property $\Pi$ of oriented graphs, the $\Pi$-ORIENTATION COMPLETION PROBLEM is as follows.

| |
|---|
| *Π*-ORIENTATION COMPLETION PROBLEM<br>**Input:** A partially oriented graph $Q = (V, E \cup A)$.<br>**Find:** An orientation of the edges in $E$ to a set of arcs $A'$ so that $Q = (V, A \cup A')$ has property $\Pi$ or certify that no such orientation is possible. |

Clearly, the *Π*-ORIENTATION COMPLETION PROBLEM generalizes the *Π*-ORIENTATION PROBLEM. Robbins' theorem as stated at the beginning of this chapter provides a polynomial time solution to the STRONG ORIENTATION PROBLEM. A result of Boesch and Tindell [5] implies that a partially oriented graph can be completed to a strong oriented graph if and only if it has no bridge and no directed cut. Either a bridge or a directed cut in a partially oriented graph (if any exists) can be detected in polynomial time. Hence the STRONG ORIENTATION COMPLETION PROBLEM is also polynomial time solvable. The orientation completion problem for local tournaments is polynomial time solvable (see Theorem 12.3.4). By slightly modifying the lexicographic orientation method for the orientation problem for acyclic local tournaments, Bang-Jensen, Huang and Zhu [4] proved that the corresponding orientation completion problem is polynomial time solvable (see Theorem 12.3.5). In contrast they [4] showed that the orientation completion problem for locally transitive local tournaments is NP-complete (see Theorem 12.3.14).

Orientation completion problems generalize certain representation extension problems. For example, the REPRESENTATION EXTENSION PROBLEM for proper interval graphs asks whether it is possible to obtain a proper interval representation of a graph $G$ that includes a proper interval representation of an induced subgraph of $G$. This problem has been studied by Klavik, Kratochvil, Otachi, Rutter, Saitoh, Saumell and Vystocil in [28]. As mentioned above, a proper interval representation of a proper interval graph corresponds to an acyclic local tournament orientation of the graph. Thus the representation extension problem for proper interval graphs is just the orientation completion problem for acyclic local tournaments where a partial orientation corresponds to an interval representation of an induced subgraph. The representation extension problem for proper interval graphs was shown to be polynomial time solvable, cf. [28]. The lexicographic orientation method can be applied to show that the orientation completion problem for acyclic local tournaments is polynomial time solvable.

The key notion used in the lexicographic method is the concept of lexicographic order. Suppose $(s_1, s_2, \ldots, s_k), (t_1, t_2, \ldots, t_k)$ are two ordered $k$-tuples over the set $\{1, 2, \ldots, n\}$. We say that $(s_1, s_2, \ldots, s_k)$ is **lexicographically smaller than** $(t_1, t_2, \ldots, t_k)$, provided $s_1 < t_1$ or there exists an $f$ with $1 < f \leq k$ such that $s_f < t_f$ and $s_i = t_i$ for all $i < f$. If $S$ and $T$ are two sets of $k$ elements, we say that $S$ is lexicographically smaller than $T$ provided $(s_1, s_2, \ldots, s_k)$ is lexicographically smaller than $(t_1, t_2, \ldots, t_k)$, where $s_1, s_2, \ldots, s_k$ and $t_1, t_2, \ldots, t_k$ are the elements of $S$ and $T$ listed in increasing order. Suppose $S'$ is a subset of $S$ and $T'$ is lexicographically smaller than $S'$. Then it is easy to see that $T = (S - S') \cup T'$ is lexicographically smaller than $S$. Note that lexicographic orders are linear, and hence any subset of a lexicographically ordered set has a smallest element.

## 12.2 Algorithms for $\Pi$-Orientations

We begin by formalizing the generic idea of the lexicographic orientation algorithm for deciding whether a graph is $\Pi$-orientable and finding (if one exists) such a $\Pi$-orientation of $G$. Let $G$ be the input graph. Define the **auxiliary graph** $G^+$ of $G$ as follows: The vertex set of $G^+$ consists of all ordered pairs $(u, v)$ such that $uv$ is an edge of $G$. Note that each each edge $uv$ of $G$ gives rise to two vertices $(u, v), (v, u)$ and these two vertices are always adjacent in $G^+$. Depending on the property $\Pi$, $G^+$ may contain additional edges, which will be defined for each problemin question.

**Algorithm 1 Generic lexicographic orientation**

*Input:* A graph $G$ with vertices $1, 2, \ldots, n$.
*Output:* A $\Pi$-orientation of $G$ if one exists.

Construct the auxiliary graph $G^+$.
While there exist uncoloured vertices do
  Colour by $A$ the lexicographically smallest uncoloured vertex $(u, v)$
  Use breadth first search to 2-colour (if possible) the connected component of $G^+$ which contains $(u, v)$.
  If some component could not be 2-coloured then report that $G$ is not $\Pi$-orientable.
For every edge $uv \in E$ orient it as $u \rightarrow v$ if $(u, v)$ obtained colour $A$ and otherwise orient it as $v \rightarrow u$.

The purpose of Algorithm 1 is two-fold. First, it determines whether the input graph $G$ is $\Pi$-orientable by verifying the 2-colourability of the auxiliary graph $G^+$. Second, it constructs a $\Pi$-orientation of $G$ in the case when $G^+$ is 2-colourable. The correctness of Algorithm 1 is validated by the two statements described in the following proposition.

**Proposition 12.2.1** *Algorithm 1 is correct if and only if the following two statements hold:*

- *If $G$ is $\Pi$-orientable, then $G^+$ is bipartite.*
- *If $G^+$ is bipartite, then the orientation of $G$ obtained by Algorithm 1 has the property $\Pi$.* □

As a simple example suppose that $\Pi$ is the property of being acyclic and that $G^+$ is the auxiliary graph of $G$ as defined above, which contains no other edges except those between $(u,v)$ and $(v,u)$ for edges $uv$ of $G$. Since every graph is acyclically orientable and $G^+$ is bipartite for every graph $G$, the first statement holds vacuously. According to Step 2, vertex $(u,v)$ of $G^+$ is coloured by $A$ if and only if $u < v$. It follows that the orientation of $G$ obtained by the algorithm is acyclic and hence the second statement holds.

We will show that the above generic lexicographic orientation algorithm can be modified to solve the $\Pi$-orientation problem when $\Pi$ is the property of being a transitive digraph, respectively being a locally transitive local tournament, respectively being an acyclic local tournament. The only modifications involved are on the definition of the auxiliary graph $G^+$. We will also show that it can be applied to recognize interval containment bigraphs and obtain the desired orientations of their complements.

### 12.2.1 Comparability Graphs

For the input graph $G$, we modify the definition of the auxiliary graph $G^+$ as follows: The vertex set of $G^+$ is the same as above (i.e., consisting of ordered pairs $(u,v),(v,u)$ for edges $uv$ of $G$). In $G^+$, every vertex $(u,v)$ is adjacent to $(v,u)$, to any $(w,u)$ such that $v$ and $w$ are not adjacent in $G$, and to any $(v,w)$ such that $u$ and $w$ are not adjacent in $G$. Figure 12.1 shows an example of a graph $G$ and its auxiliary graph $G^+$.

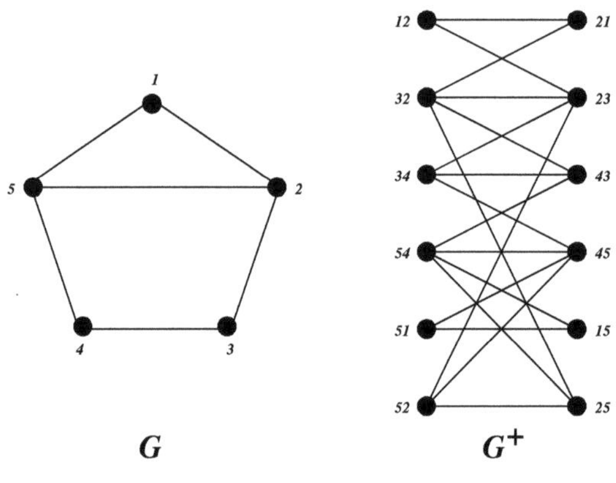


**Figure 12.1** A graph $G$ and its auxiliary graph $G^+$.

Suppose that $G^+$ is bipartite. Colour $G^+$ with two colours $A, B$ and orient each edge $uv$ of $G$ as $u{\rightarrow}v$ whenever $(u,v)$ is coloured $A$. Then for any edges $uv, vw$ with $uw$ being a non-edge of $G$, $(u,v)$ and $(v,w)$ are adjacent and $(w,v)$ and $(v,u)$ are adjacent in $G^+$. Thus $(u,v)$ and $(v,w)$ are coloured by opposite colours and $(w,v)$ and $(v,u)$ are coloured by opposite colours in any 2-colouring of $G^+$. Consequently, we have either $u \rightarrow v$ and $w \rightarrow v$ or $v \rightarrow u$ and $v \rightarrow w$. Therefore we obtain an orientation of $G$ which satisfies the property that $u \rightarrow v$ and $v \rightarrow w$ imply that there is an arc between $u$ and $w$. An oriented graph which has this property is called **quasi-transitive**, cf. [3] and Chapter 8. On the other hand, any quasi-transitive orientation of $G$ corresponds to a colour class of a 2-colouring of $G^+$.

Every transitive oriented graph is quasi-transitive and thus every transitively orientable graph is also quasi-transitively orientable. It was first observed by Ghouila-Houri [16] that every quasi-transitively orientable graph is also transitively orientable. Hence comparability graphs are exactly the quasi-transitively orientable graphs. In particular, if $G^+$ is not bipartite then $G$ is not a comparability graph and hence not transitively orientable. The result of Ghouila-Houri will follow as a byproduct from the lexicographic orientation algorithm, as stated below.

**Theorem 12.2.2** ([22]) *Suppose that $G$ is a comparability graph and that $D$ is an orientation of $G$ obtained by the lexicographic orientation algorithm. Then $D$ is a transitive orientation of $G$.*

**Proof:** Since $G$ is a comparability graph, $G^+$ is bipartite. For each vertex $(u, v)$ of $G^+$, let $C(u, v)$ be the set of all vertices whose distance from $(u, v)$ in $G^+$ is even. It follows from the definition of $G^+$ that if $(x, y), (x', y') \in C(u, v)$ then there exist

$$(x_0, y_0), (x_1, y_0), (x_1, y_1), (x_2, y_1), \ldots, (x_k, y_k) \in C(u, v)$$

such that $(x_0, y_0) = (x, y)$ and $(x_k, y_k) = (x', y')$ and for each $i = 0, 1, \ldots, k-1$, $x_i x_{i+1} \notin E(G)$ and $y_i y_{i+1} \notin E(G)$. The following claim, known as "**The Triangle Lemma**, can be found in the book [18] by Golumbic.

**Claim.** Let $uvwu$ be a 3-cycle in $G$. Suppose that $C(u, v) \neq C(w, v)$ and $C(u, v) \neq C(u, w)$. Then for any $(u', v') \in C(u, v)$, we must have $(w, v') \in C(w, v)$ and $(u', w) \in C(u, w)$.

**Proof of Claim.** Since $(u', v') \in C(u, v)$, there exist

$$(u_0, v_0), (u_1, v_0), (u_1, v_1), (u_2, v_1), \ldots, (u_\ell, v_\ell) \in C(u, v)$$

such that $(u_0, v_0) = (u, v)$ and $(u_\ell, v_\ell) = (u', v')$ and for each $= 0, 1, \ldots, \ell-1$, $u_i u_{i+1} \notin E(G)$ and $v_i v_{i+1} \notin E(G)$. We prove by induction on $\ell$ that $(w, v_\ell) \in C(w, v)$ and $(u_\ell, w) \in C(u, w)$. Assume that $(w, v_{\ell-1}) \in C(w, v)$ and $(u_{\ell-1}, w) \in C(u, w)$. Since $C(u, v) \neq C(w, v) = C(w, v_{\ell-1})$, $wu_\ell \in E(G)$. Since $u_{\ell-1}u_\ell \notin E(G)$, $(u_\ell, w) \in C(u_{\ell-1}, w) = C(u, w)$. Similarly, since $C(u_\ell, v_\ell) \neq C(u_\ell, w)$, $wv \in E(G)$ and since $v_{\ell-1}v_\ell \notin E(G)$, $(w, v_\ell) \in C(w, v_{\ell-1}) = C(w, v)$. □

Suppose to the contrary that $D$ is not transitive. Then there is a triangle $uvwu$ such that $u \rightarrow v$, $v \rightarrow w$ and $w \rightarrow u$ in $D$. Assume that $\{u, v, w\}$ is the lexicographically smallest amongst all such triangles. Without loss of generality assume that $u > v$ and therefore $(u, v)$ was not the first vertex coloured $A$ in its component of $G^+$. It follows that there exists $(u', v') \in$

$C(u,v)$ such that $\{u',v'\}$ is lexicographically smaller than $\{u,v\}$. Since $u \to v$, $v \to w$ and $w \to u$, $C(u,v) \neq C((w,v)$ and $C(u,v) \neq C(u,w)$. Hence by the claim above, $(w,v') \in C(w,v)$ and $(u',w) \in C(u,w)$. Since $u \to v$, $v \to w$ and $w \to u$ in $D$, we must also have $u' \to v'$, $v' \to w$ and $w \to u'$ in $D$. But $\{u',v',w\}$ is lexicographically smaller than $\{u,v,w\}$, which contradicts the choice of $\{u,v,w\}$. □

For $k \geq 1$, a $(2k+1)$**-asteroid** in a graph is a sequence of $2k+1$ vertices

$$u_0, u_1, \ldots, u_{2k}$$

together with $2k+1$ paths

$$P_0, P_2, \ldots, P_{2k}$$

where $P_i$ is a $(u_i, u_{i+1})$-path such that $u_i$ has no neighbours in $P_{i+k}$ (subscripts are modulo $2k+1$) for each $i = 0, 1, \ldots, 2k$. A 3-asteroid is also known as an **asteroidal triple**, which is an important concept for characterizing interval graphs, cf. [29]. It is easy to verify that an odd cycle in $G^+$ corresponds to a $(2k+1)$-asteroid for some $k$ in $\overline{G}$.

**Corollary 12.2.3** *The following statements are equivalent for a graph $G$.*

1. *$G$ is a comparability graph;*
2. *$G$ is transitively orientable;*
3. *$G$ is quasi-transitively orientable;*
4. *$G^+$ is bipartite;*
5. *$\overline{G}$ contains no asteroid.* □

### 12.2.2 Proper Circular Arc Graphs

A **round ordering** of a digraph $D$ is a cyclic ordering $\mathcal{O} = v_1, v_2, \ldots, v_n, v_1$ of the vertices of $D$ such that for each vertex $v_i$ we have $N^+(v_i) = \{v_{i+1}, \ldots, v_{d^+(v_i)+i}\}$ and $N^-(v_i) = \{v_{i-d^-(v_i)}, \ldots, v_{i-1}\}$ where indices are modulo $n$. A digraph which has a round ordering is called **round**. Round digraphs were characterized by Huang in [27]. It is easy to see that if an oriented graph has a round ordering then it is locally transitive. The following theorem, due to Bang-Jensen, asserts that the converse is also true when $D$ is connected.

**Theorem 12.2.4** ([1]) *A connected oriented graph $D$ has a round ordering $\mathcal{O} = v_1, v_2, \ldots, v_n, v_1$ of its vertices if and only if $D$ is a locally transitive local tournament. Furthermore, there is a polynomial algorithm for deciding whether a given oriented graph is round and finding a round ordering if one exists.* □

Suppose that $G$ is a proper circular arc graph and that $I_v, v \in V(G)$, is a proper circular arc representation of $G$. We may assume without loss of generality that if two circular arcs $I_u, I_v$ intersect then either $I_u$ contains the counterclockwise endpoint of $I_v$ or $I_v$ contains the counterclockwise endpoint of $I_u$ (but not both). Orient $G$ in such a way that each edge $uv$ of $G$ is oriented as $u \rightarrow v$ if $I_u$ contains the counterclockwise endpoint of $I_v$. It is easy to see that this is a locally transitive local tournament orientation of $G$. A round ordering of the orientation of $G$ corresponds to the clockwise ordering of clockwise endpoints of the circular arcs in the proper circular arc representation of $G$. Conversely, suppose that $D$ is a connected locally transitive local tournament. Then $D$ has a round ordering by Theorem 12.2.4 and a family of inclusion-free circular arcs $I_v, v \in V(D)$, can be obtained such that $u \rightarrow v$ in $D$ if and only if $I_u$ contains the counterclockwise endpoint of $I_v$, cf. [22, 26]. Thus the underlying graph of $D$ is a proper circular arc graph.

**Theorem 12.2.5** ([22, 26]) *A connected graph is a proper circular arc graph if and only if it is orientable as a locally transitive local tournament.* □

Every locally transitive local tournament is a local tournament. Skrien [38] proved that a connected graph is a proper circular arc graph if and only if it is local tournament orientable. Clearly, a graph (not necessarily connected) is local tournament (respectively, locally transitive local tournament) orientable if and only if so is every connected component of the graph. Therefore a graph $G$ is orientable as a locally transitive local tournament if and only if it is orientable as a local tournament. With this in mind we define the edge set of the auxiliary graph $G^+$ of $G$ as follows: each vertex $(u, v)$ is adjacent to $(v, u)$, to any vertex $(u, w)$ such that $v$ and $w$ are not adjacent in $G$, and to any vertex $(w, v)$ such that $u$ and $w$ are not adjacent in $G$. As in the previous subsection, we see that any local tournament orientation of $G$ gives rise to a 2-colouring of $G^+$ and in case when $G^+$ is 2-colourable the vertices of one colour in any 2-colouring of $G^+$ induce a local tournament orientation of $G$. Not every 2-colouring of $G^+$ induces a locally transitive local tournament orientation of $G$. However, the 2-colouring of $G^+$ produced by the lexicographic orientation algorithm gives a locally transitive local tournament orientation of $G$.

**Theorem 12.2.6** ([22]) *Suppose that $G$ is a proper circular arc graph and that $D$ is an orientation of $G$ obtained by the lexicographic orientation algorithm. Then $D$ is a local transitive tournament orientation of $G$.*

**Proof:** Since $G$ is a proper circular arc graph, $G^+$ is bipartite and hence $D$ is a local tournament. Suppose to the contrary that $D$ is not a locally transitive local tournament. Then there exists a set $\{u, v, w, z\}$ of vertices of $D$ such that $u, v, w$ induce a directed 3-cycle $u \rightarrow v \rightarrow w \rightarrow u$, which either dominates $z$ or is dominated by $z$. Assume that $\{u, v, w, z\}$ is the lexicographically smallest set with this property. Assume further that $z$ dominates $\{u, v, w\}$. (The situation is symmetric when $z$ is dominated by $\{u, v, w\}$.) Without loss of generality assume that $u > v$ and therefore $(u, v)$ was not the first vertex coloured $A$ in its component of $G^+$.

Let $C(u, v)$ (respectively, $C(v, u)$) be the set of all vertices in $G^+$ whose distance from $(u, v)$ in $G^+$ is even (respectively, odd), and let $(u', v') \in C(u, v)$ be the first vertex coloured $A$ in the component of $(u, v)$. Then $\{u', v'\}$ is lexicographically smaller than $\{u, v\}$ and hence $\{u', v', w, z\}$ is lexicographically smaller than $\{u, v, w, z\}$. We show that the subdigraph of $D$ induced by $\{u', v', w, z\}$ also contains a directed 3-cycle which either dominates the fourth vertex or is dominated by the fourth vertex. This contradicts the choice of $\{u, v, w, z\}$ and therefore $D$ is a locally transitive local tournament.

Since $(u', v') \in C(u, v)$, there exist

$$(u_0, v_0), (u_1, v_1), \ldots, (u_\ell, v_\ell)$$

such that

- $(u_0, v_0) = (u, v)$;
- $(u_i, v_i) \in C(u, v)$ when $i$ is even and $(u_i, v_i) \in C(v, u)$ when $i$ is odd;
- $(u_\ell, v_\ell) = (u', v')$ when $\ell$ is even and $(u_\ell, v_\ell) = (v', u')$ when $\ell$ is odd;
- for each $i = 0, 1, \ldots, \ell - 1$, either $u_i = u_{i+1}$ and $v_i v_{i+1} \notin E(G)$ or $v_i = v_{i+1}$ and $u_i u_{i+1} \notin E(G)$.

Let $U_i = \{u_0, u_1, \ldots, u_i\}$ and $V_i = \{v_0, v_1, \ldots, v_i\}$. Note that not all elements in $U_i$ (respectively, $V_i$) are distinct. We use $||U_i||$ (respectively $||V_i||$) to denote the number of distinct elements in $U_i$ (respectively, $V_i$). Observe that $i$ and $||U_i|| + ||V_i||$ have the same parity for each $i$. We claim that in $D$ the following property holds:

- when $||U_i||$ is odd, $\{w, z\} \to u_i \to v$;
- when $||U_i||$ is even, $v \to u_i \to \{w, z\}$;
- when $||V_i||$ is odd, $\{u, z\} \to v_i \to w$;
- when $||V_i||$ is even, $w \to v_i \to \{u, z\}$.

When $i = 0$, we have $||U_0|| = ||V_0|| = 1$ and the property holds. Assume that $i \geq 1$ and the property holds for $i - 1$. We consider only the case when $u_{i-1} = u_i$ and $v_{i-1} v_i \notin E(G)$. (The other case, $v_{i-1} = v_i$ and $u_{i-1} u_i \notin E(G)$, is symmetric.)

Suppose that $i$ is odd. Then $v_i \to u_i = u_{i-1} \to v_{i-1}$. Since $i$ is odd, $||U_i||$ and $||V_i||$ have different parity. Suppose first that $||U_i||$ is odd. Then $||V_{i-1}||$ is also odd. By the inductive hypothesis, $\{w, z\} \to u_{i-1} = u_i \to v$ and $\{u, z\} \to v_{i-1} \to w$. Hence $v_i, w, z$ are in-neighbours of $u_i$. Since $D$ is a local tournament, $v_i$ is adjacent to both $w$ and $z$. Since $z \to v_{i-1} \to w$, we must have $w \to v_i \to z$. Hence $u, v_i$ are both out-neighbours of $w$ and must be adjacent. Since $u \to v_{i-1}$ and $v_{i-1} v_i \notin E(G)$, we have $v_i \to u$. Therefore $w \to v_i \to \{u, z\}$. Suppose that $||U_i||$ is even. Then $||V_{i-1}||$ is also even. By the inductive hypothesis, $v \to u_{i-1} = u_i \to \{w, z\}$ and $w \to v_{i-1} \to \{u, z\}$. Since $v, v_i$ are both in-neighbours of $u_i$, $v, v_i$ are adjacent. Either $v \to v_i$ or $v_i \to v$ in $D$. Assume that $v \to v_i$. (The case when $v_i \to v$ is again symmetric.).

Then $w, v_i$ are out-neighbours of $v$ and hence are adjacent. Since $w \to v_{i-1}$ and $v_{i-1}v_i \notin E(G)$, $v_i \to w$; thus both $v_i$ and $z$ are in-neighbours of $w$. Since $v_{i-1} \to z$ and $v_{i-1}v_i \notin E(G)$, $z \to v_i$. Hence $u, v_i$ are both out-neighbours of $z$ and must be adjacent. Since $v_{i-1} \to u$ and $v_{i-1}v_i \notin E(G)$, we have $u \to v_i$. Therefore $\{u, z\} \to v_i \to w$.

Suppose that $i$ is even. Then $v_{i-1} \to u_{i-1} = u_i \to v_i$. Since $i$ is even, $||U_i||$ and $||V_i||$ have the same parity. Suppose first that $||U_i||$ is odd. Then $||V_{i-1}||$ is even. By the induction hypothesis, $\{w, z\} \to u_{i-1} = u_i \to v$ and $w \to v_{i-1} \to \{u, z\}$. Since $v, v_i$ are both out-neighbours of $u_i$, $v, v_i$ are adjacent. Either $v \to v_i$ or $v_i \to v$ in $D$. Assume that $v \to v_i$. (The case when $v_i \to v$ is again symmetric.) Then $w, v_i$ are out-neighbours of $v$ and hence are adjacent. Since $w \to v_{i-1}$ and $v_{i-1}v_i \notin E(G)$, $v_i \to w$; thus both $v_i$ and $z$ are in-neighbours of $w$. Since $v_{i-1} \to z$ and $v_{i-1}v_i \notin E(G)$, we have $z \to v_i$. Hence $u, v_i$ are both out-neighbours of $z$ and must be adjacent. Since $v_{i-1} \to u$ and $v_{i-1}v_i \notin E(G)$, we have $u \to v_i$. Therefore $\{u, z\} \to v_i \to w$. Suppose now that $||U_i||$ is even. Then $||V_{i-1}||$ is odd. By the inductive hypothesis, $v \to u_{i-1} = u_i \to \{w, z\}$ and $\{u, z\} \to v_{i-1} \to w$. Thus $v_i, w, z$ are out-neighbours of $u_i$. So $v_i$ is adjacent to both $w$ and $z$. Since $z \to v_{i-1} \to w$ and $v_{i-1}v_i \notin E(G)$, $w \to v_i \to z$. Now $u$ and $v_i$ are both out-neighbours of $w$ and must be adjacent. Since $u \to v_{i-1}$ and $v_{i-1}v_i \notin E(G)$, we must have $v_i \to u$. Therefore $w \to v_i \to \{u, z\}$.

If $\ell$ is even, then $(u_\ell, v_\ell) = (u', v')$, and $||U_\ell||$ and $||V_\ell||$ have the same parity. When $||U_\ell||$ and $||V_\ell||$ are both odd, $\{u', v', w\}$ induces a directed cycle and is dominated by $z$; when $||U_\ell||$ and $||V_\ell||$ are both even, $\{w, v', z\}$ induces a directed cycle and is dominated by $u'$. If $\ell$ is odd, then $(u_\ell, v_\ell) = (v', u')$, and $||U_\ell||$ and $||V_\ell||$ have different parity. When $||U_\ell||$ is odd and $||V_\ell||$ is even, $\{w, v', z\}$ induces a directed cycle and dominates $u'$; when $||U_\ell||$ is even and $||V_\ell||$ is odd, $\{u', v', z\}$ induces a directed cycle and dominates $w$. □

Combining Theorems 12.2.5 and 12.2.6 and the remarks made between the two theorems we have the following:

**Corollary 12.2.7** *The following statements are equivalent for a connected graph $G$.*

1. *$G$ is a proper circular arc graph;*
2. *$G$ is local tournament orientable;*
3. *$G$ is locally transitive local tournament orientable;*
4. *$G^+$ is bipartite.* □

Through a careful analysis of the structure of proper circular arc graphs, a full description of all local tournament orientations of a proper circular arc graph was obtained in [24]. Let $G$ be a graph and $uv, u'v'$ be two edges of $G$. We say that $uv, u'v'$ are **implicated** if $(u, v)$ and $(u', v')$ are in the same connected component of $G^+$. The implication relation is an equivalence relation on the set of edges of $G$ and each equivalence class is called an **implication class** of $G$. Call an edge $uv$ in $G$ **balanced** if $N[u] = N[v]$

and **unbalanced** otherwise. It follows from the definition that an edge is balanced if and only if it forms an implication class by itself. In general, two edges of $G$ are implicated with each other if and only if the orientation of one uniquely determines the orientation of the other in any local tournament orientation of $G$.

**Theorem 12.2.8** ([24]) *Let $G$ be a connected proper circular arc graph. Suppose that $C_1, C_2, \ldots, C_k$ are the connected components of $\overline{G}$. Then all unbalanced edges of $G$ within a fixed $C_i$ form an implication class and all unbalanced edges between two fixed $C_i$ and $C_j$ $(i \neq j)$ form an implication class. Moreover, if $\overline{G}$ is not bipartite, then $k = 1$ and all unbalanced edges of $G$ form an implication class.* □

### 12.2.3 Proper Interval Graphs

Proper interval graphs are proper circular arc graphs and hence are locally transitive local tournament orientable. In fact they admit locally transitive local tournament orientations that contain no directed cycles (or equivalently, acyclic local tournament orientations). Indeed, suppose that $G$ is a proper interval graph and that $I_v, v \in V(G)$, is a proper interval representation of $G$. Orient $G$ in such a way that $u \rightarrow v$ if and only if $I_u$ contains the left endpoint of $I_v$. This is an acyclic local tournament orientation of $G$. On the other hand, an acyclic local tournament orientation of $G$ can be efficiently transformed into a proper interval representation of $G$, cf. [26] and [22]. So acyclic local tournament orientations of proper interval graphs are in a sense an orientation formulation of their proper interval representations.

When the input graph $G$ is a proper interval graph (and hence a proper circular arc graph), the lexicographic orientation algorithm using the same auxiliary graph $G^+$ as defined in Subsection 12.2.2 will produce a locally transitive local tournament orientation $D$ of $G$ according to Theorem 12.2.6. But this $D$ may not be acyclic. To make sure that $D$ is also acyclic, we use a **perfect elimination ordering** of $G$ (that is, a vertex ordering $1, 2, \ldots, n$ such that for each $i$ the set of neighbours $j$ of $i$ with $j > i$ induce a complete subgraph of $G$). It is well-known that $G$, which is a chordal graph, must have such an ordering, which can be obtained in time $O(m + n)$ using the algorithm called **Lexicographic Breadth First Search (LBFS)** devised by Rose, Tarjan and Lueker in [36]. We summarize the lexicographic orientation algorithm for finding an acyclic local tournament orientation of a proper intervalgraph.

The proof of correctness of the algorithm makes use of a full description of implication classes of a proper interval graph obtained in [24]. A vertex in a graph $G$ is called **universal** if it is adjacent to every other vertex in $G$.

**Algorithm 2 Lexicographic acyclic local-tournament-orientation**

*Input:* A graph $G$.
*Output:* An acyclic local tournament orientation of $G$.

Find a perfect elimination ordering $1, 2, \ldots, n$ of $G$.
 If $G$ does not have a perfect elimination ordering then report that $G$ is not a proper interval graph.
Construct the auxiliary graph $G^+$.
While there exist uncoloured vertices do
 Colour by $A$ the lexicographically smallest uncoloured vertex $(u, v)$
 Use breadth first search to 2-colour (if possible) the connected component of $G^+$ which contains $(u, v)$.
 If some component could not be 2-coloured then report that $G$ is not a proper interval graph.
Orient each edge $uv$ of $G$ as $u \rightarrow v$ if $(u, v)$ obtained colour $A$ and otherwise orient it as $v \rightarrow u$.
Check whether the resulting oriented graph contains a directed cycle.
If it has a directed cycle then report that $G$ is not a proper interval graph.

**Theorem 12.2.9** ([24]) *Let $G$ be a connected proper interval graph. Then one of the following statements holds:*

- *if $G$ has no universal vertex, then all unbalanced edges of $G$ form an implication class;*
- *if $G$ has universal vertices, then all unbalanced edges incident with universal vertices form an implication class and all other unbalanced edges form an implication class.* □

**Theorem 12.2.10** ([22]) *Suppose that $G$ is a proper interval graph. Then the orientation of $G$ obtained by Corollary 12.2.3 is an acyclic local tournament.*

**Proof:** Assume without loss of generality that $G$ is connected. Suppose first that $G$ has no universal vertex. Then by Theorem 12.2.9, the vertices of $G$ can be partitioned into complete subgraphs $V_1, V_2, \ldots, V_p$ and $G^+$ has the following components: For each pair of vertices $u, v$ in the same $V_i$, there is a separate component consisting of adjacent vertices $(u, v), (v, u)$. In addition, there is one component containing all remaining vertices $(u, v)$ (i.e., $u \in V_i$ and $v \in V_j$ with $i \neq j$). Moreover, in this last component, one colour class contains all vertices $(u, v)$ with $u \in V_i, v \in V_j$ and $i < j$. In this case, the lexicographic orientation algorithm orients each $V_i$ as a transitive tournament and the remaining edges $uv$ as $u \rightarrow v$ either for all $u \in V_i, v \in V_j, i < j$ or for all $u \in V_i, v \in V_j, i > j$. It is clear that the orientation does not contain a directed cycle and hence is an acyclic local tournament.

Suppose now that $G$ has universal vertices and that $1, 2, \ldots, n$ is a perfect elimination ordering of $G$. Then again by Theorem 12.2.9 the vertices of $G$

can be partitioned into complete subgraphs $V_1, V_2, \ldots, V_p$ where $V_m$ with $1 < m < p$ consists of all universal vertices that are in $V_m$ and $V_1 \cup V_p$ consists of all simplicial vertices. The components of $G^+$ are as follows: For each $u, v$ in the same $V_i$, there is a separate component consisting of adjacent vertices $(u, v), (v, u)$. There is again one component consisting of all vertices $(u, v)$ with $u \in V_i$, $v \in V_j$, $i \neq j$, $i \neq m$, and $j \neq m$. One colour class in this component consists of all $(u, v)$ with $u \in V_i$, $v \in V_j$, $i < j$. Finally, there is, for each vertex $w \in V_m$, a component consisting of all vertices $(v, w), (w, v)$ for all $v \in V_i$ with $i \neq m$. One colour class of this component consists of $(u, w), (w, v)$ for all $u \in V_i$ and $v \in V_j$ with $1 \leq i < m$ and $m < j \leq p$. The simplicial vertex 1 is in $V_1$ or $V_p$. The lexicographic orientation algorithm orients each $V_i$ as a transitive tournament and the remaining edges $uv$ as $u \to v$ either for all $u \in V_i, v \in V_j, i < j$ or for all $u \in V_i, v \in V_j, i > j$. The orientation is an acyclic local tournament. □

**Corollary 12.2.11** *The following statements are equivalent for a graph G.*

1. *G is a proper interval graph;*
2. *G is acyclic local tournament orientable.* □

### 12.2.4 Interval Containment Bigraphs

Let $G$ be a bipartite graph with bipartition $(X, Y)$. Recall that $G$ is an interval containment bigraph if there is a family of intervals $I_v, v \in X \cup Y$, such that for all $x \in X$ and $y \in Y$, $xy$ is an edge of $G$ if and only if $I_x \supset I_y$. The family of intervals will be refered to as an interval containment representation of $G$. See Figure 12.2 for an example of an interval containment bigraph and its interval containment representation.

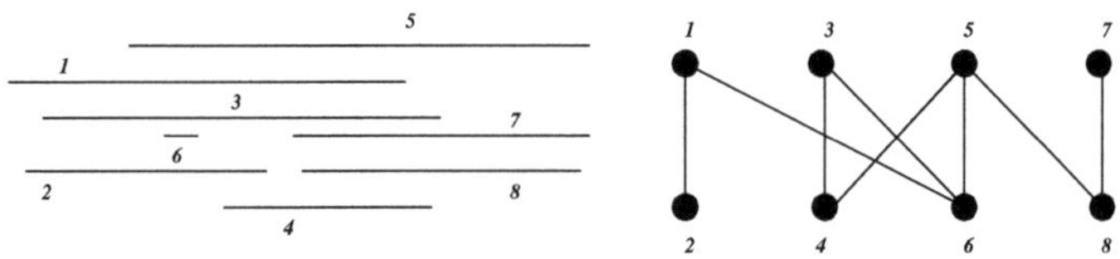


**Figure 12.2** An interval containment bigraph and an interval containment representation.

Suppose that $G$ is an interval containment bigraph and that the collection of intervals $I_v = [\ell_v, r_v], v \in X \cup Y$, form an interval containment representation of $G$. Assume without loss of generality that the ends of the intervals are all distinct. We orient $\overline{G}$ as follows: each edge $uv$ of $\overline{G}$ is oriented as $u \to v$ if $\ell_u < \ell_v$. Clearly, the orientation is acyclic. We claim that it does not contain the digraph in Figure 12.3 as an induced subdigraph.

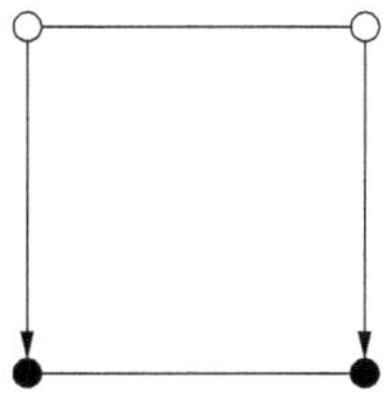

**Figure 12.3** White vertices are in $X$ and black vertices are in $Y$ or the other way around. The orientation between white vertices or between black vertices is not specified and may be in either direction.

Indeed, suppose that $u \to u'$ and $v \to v'$ are oriented edges where the four vertices $u, v, v', u'$ induce a 4-cycle $vv'u'u$ in $\overline{G}$. By the way of orientation we must have $\ell_u < \ell_{u'}$ and $\ell_v < \ell_{v'}$. If $u, v \in X$ and $u', v' \in Y$, then $r_u < r'_u$ and $r_v < r_{v'}$ as $uu', vv' \notin E(G)$. Since $uv', vv' \in E(G)$, we have

$$\ell_u < \ell_{v'} < r_{v'} < r_u \text{ and } \ell_v < \ell_{u'} < r_{u'} < r_v.$$

Hence we have $\ell_v < \ell_{v'} < r_{v'} < r_u < r_{u'} < r_v$ and so $I_v \supset I_{v'}$, a contradiction to the assumption that $vv' \notin E(G)$. If $u, v \in Y$ and $u', v' \in X$, then

$$\ell_{u'} < \ell_v < r_v < \ell_{u'} \text{ and } \ell_{v'} < \ell_u < r_u < r_{v'}.$$

Thus we have $\ell_{v'} < \ell_u < \ell_{u'} < \ell_v < \ell_{v'}$, a contradiction.

Acyclic orientations of the complements of bipartite graphs which do not contain an induced subdigraph in Figure 12.3 may again be viewed as an orientation formulation of interval containment representations of interval containment bigraphs. Thus the recognition and representation problems for interval containment bigraphs become the following:

**Problem 12.2.12** *Given a bipartite graph $G$, does $\overline{G}$ have an acyclic orientation which does not contain one of the digraphs in Figure 12.3 as an induced subdigraph?*

Define the auxiliary graph $G^+$ of $G$ with bipartition $(X, Y)$ as follows: The vertices of $G^+$ are ordered pairs $(v, v'), (v', v)$ with $v \in X, v' \in Y$ and $vv' \notin E(G)$. In $G^+$, each $(v, v')$ is adjacent to $(v', v)$ and for each induced 4-cycle $vv'u'u$ in $\overline{G}$, $(v, v')$ is adjacent to $(u, u')$ and $(v', v)$ is adjacent to $(u', u)$. The above observation simply asserts that if $G$ is an interval containment bigraph then $G^+$ is bipartite.

Suppose that the auxiliary graph $G^+$ of $G$ is bipartite. Colour the vertices of $G^+$ with colours $A, B$ and orient an edge $vv'$ of $\overline{G}$ as $v' \to v$ if $(v, v')$ is coloured $A$ and as $v \to v'$ if $(v', v)$ is coloured $A$. This is a partial orientation of $\overline{G}$; all edges between $X$ and $Y$ are oriented but none of edges in $X$ or in $Y$ is oriented. The definition of $G^+$ implies that any completion of this partial orientation to an orientation of $\overline{G}$ will not contain the digraph in Figure 12.3

as an induced subdigraph. However, there may be no acyclic completion. In order for the partial orientation of $\overline{G}$ to have an acyclic completion, particular 2-colourings of $G^+$ are needed.

We will fix a bipartition $(X, Y)$ of $G$ and use letters without primes for vertices in $X$ and letters with primes for vertices in $Y$.

**Algorithm 3 Lexicographic restricted acyclic orientation**

*Input:* A bipartite graph $G$ with bipartition $(X, Y)$ and vertices $1, 2, \ldots, n$ where vertices of $X$ preceede the vertices of $Y$.
*Output:* An acyclic orientation of $\overline{G}$ that does not contain one of the digraphs in Figure 12.3 as an induced subdigraph.

Construct the auxiliary graph $G^+$ with respect to $(X, Y)$.
While there exist uncoloured vertices do
    Colour by $A$ the lexicographically smallest uncoloured vertex $(\alpha, \beta)$
    Use breadth first search to 2-colour (if possible) the component of $G^+$ which contains $(\alpha, \beta)$.
    If some component could not be 2-coloured then report that $G$ is not an interval containment bigraph.
Orient the edge $vv'$ of $\overline{G}$ as $v' \rightarrow v$ if $(v, v')$ is coloured $A$ and as $v \rightarrow v'$ otherwise.
Complete the partial orientation obtained in Step 3 to an orientation of $\overline{G}$ as follows: orient each edge $uv$ as $u \rightarrow v$ if $N^-(u) \cap Y \subseteq N^-(v) \cap Y$ and orient each edge $u'v'$ as $u' \rightarrow v'$ if $N^+(u') \cap X \supseteq N^+(v') \cap X$.

The correctness of the algorithm above is ensured by the following reformulation of a theorem of Hell and Huang [22].

**Theorem 12.2.13** *Suppose that $G$ is an interval containment bigraph and that $D$ is an orientation of $\overline{G}$ obtained by Theorem 12.2.4. Then $D$ is acyclic and does not contain the digraph in Figure 12.3 as an induced subdigraph.*

**Proof:** We first prove that for any $u, v \in X$, the following properties hold:

- either $N^-(u) \cap Y \subseteq N^-(v) \cap Y$ or $N^-(u) \cap Y \supseteq N^-(v) \cap Y$;
- either $N^+(u) \cap Y \subseteq N^+(v) \cap Y$ or $N^+(u) \cap Y \supseteq N^+(v) \cap Y$.

We prove it by contradiction. So suppose that one of the properties does not hold for some $u, v \in X$. Let $u, v$ be such vertices with the minimum $u + v$. Assume by symmetry that the first property does not hold for $u, v$, that is, there are vertices $u', v' \in Y$ such that

- $u' \rightarrow u$ and $v' \rightarrow v$,
- $vu'$ is not an edge of $\overline{G}$ or $v \rightarrow u'$, and
- $uv'$ is not an edge of $\overline{G}$ or $u \rightarrow v'$.

Observe that at least one of $vu', uv'$ must be an edge of $\overline{G}$; otherwise $(u, u')$ and $(v, v')$ are adjacent vertices of $G^+$ of the same colour $A$, a contradiction. Assume without loss of generality that $uv'$ is an edge of $\overline{G}$. Since $u \rightarrow v'$, the vertex $(u,v')$ was coloured $B$. Hence there exists a vertex $(w,w')$ of colour $A$ such that $wuv'w'$ is an induced 4-cycle of $\overline{G}$. Since $u{\rightarrow}v'$, $w'{\rightarrow}w$. Now we have $w{\rightarrow}w$, $u'{\rightarrow}u$ and $uw'$ is not an edge of $\overline{G}$. This implies that $wu'$ is an edge of $\overline{G}$. If $w{\rightarrow}u'$, then the four vertices $w,u,w',u'$ can be used in the place of $u,v,u',v'$. On the other hand, if $u'{\rightarrow}w$, then $w,v,u',v'$ can be used in the place of $u,v,u',v'$. Therefore we may assume without loss of generality that for the four vertices $u,v,u',v'$, $vu'$ is not an edge of $\overline{G}$. We show that there exist $z,z'$ with $z<u$ such that $z{\rightarrow}u'$ and $v{\rightarrow}z'$. This implies that $y,z$ are two vertices for which one of the above two properties does not hold. This contradicts the choice of $u,v$ because $u+v>z+v$.

Since $u{\rightarrow}v'$, $(u,v')$ was coloured $B$, which implies that $(u,v')$ is not the lexicographically smallest vertex of its component. Let $(z,z')$ be the lexicographically smallest vertex in the component of $(u,v')$. Then there are vertices $(u_i,v_i')$, $i=1,2,\ldots,k$, with $(u_1,v_1')=(u,v')$, $(u_k,v_k')=(z,z')$ and each $u_iv_i'v_{i+1}'u_{i+1}$ is an induced 4-cycle in $\overline{G}$. Note that $u_i{\rightarrow}v_i'$ when $i$ is odd and $v_i'{\rightarrow}u_i$ when $i$ is even. In particular, $k$ must be even. We prove by induction on $k$ that $z=u_k<u_1=u$, $z=u_k{\rightarrow}u'$ and $v{\rightarrow}v_k'=z'$. Note that to show $u_k<u_1=u$ it suffices to prove $u_k\neq u_1=u$. When $k=2$, clearly $u_2\neq u_1$. As $v_2'{\rightarrow}u_2$, $v_1'=v'{\rightarrow}v$ and $u_2v_1'$ is not an edge of $\overline{G}$, $vz'$ is an edge of $\overline{G}$. Since $u'{\rightarrow}u_1$ and neither $u_1v_2'$ nor $vu'$ is an edge of $\overline{G}$, $v{\rightarrow}v_2'=z$. Similarly, as $v_2'{\rightarrow}u_2$, $u'{\rightarrow}u_1$ and $u_1v_2'$ is not an edge of $\overline{G}$, $u_2u'$ is an edge of $\overline{G}$. Since $v_1'{\rightarrow}v$ and neither $u_2v_1'$ nor $vu'$ is an edge of $\overline{G}$, $u_2{\rightarrow}u'$.

Assume that $k>2$ and that, by the induction hypothesis, $u_{k-2}{\rightarrow}u'$ and $v{\rightarrow}v_{k-2}'$. If we can show that $v_{k-1}'{\rightarrow}v$ and $u'{\rightarrow}u_{k-1}$, then we can argue exactly as in the case of $k=2$, to conclude that both $v{\rightarrow}v_k'$ and $u_k{\rightarrow}u'$ and $u_k\neq u$. Thus we can again let $z=u_k, z'=v_k'$ to complete the proof. Since both $u_{k-2}{\rightarrow}u'$ and $u_{k-1}{\rightarrow}v_{k-1}'$ and $u_{k-2}v_{k-1}'$ is not an edge of $\overline{G}$, $u_{k-1}u'$ is an edge of $\overline{G}$. We must have $u'{\rightarrow}u_{k-1}$ as $v{\rightarrow}v_{k-2}'$ and $u_{k-1}vv_{k-2}'u'$ is an induced 4-cycle in $\overline{G}$. Similarly, since $v{\rightarrow}v_{k-2}'$, $u_{k-1}{\rightarrow}v_{k-1}'$ and $u_{k-1}v_{k-2}'$ is not an edge of $\overline{G}$, $vv_{k-1}'$ is an edge of $\overline{G}$. We must have $v_{k-1}'{\rightarrow}v$ as $u_{k-2}{\rightarrow}u'$ and $u_{k-2}u'v_{k-1}'v$ is an induced 4-cycle in $\overline{G}$.

This justifies that the execution of Step 4 of Theorem 12.2.4 is possible. It is easy to verify now that the orientation of $\overline{G}$ obtained by Theorem 12.2.4 is acyclic and does not contain the digraph in Figure 12.3 as an induced subdigraph. □

**Corollary 12.2.14** *The following statements are equivalent for a bipartite graph $G$.*

1. *$G$ is an interval containment bigraph;*
2. *$\overline{G}$ is a circular arc graph of clique covering number two;*
3. *$\overline{G}$ has an acyclic orientation that does not contain as an induced subdigraph the digraph in Figure 12.3;*
4. *$G^+$ is bipartite.* □

## 12.3 Orientation Completion Problems

It is easy to see that a partially oriented graph can be completed to an acyclic oriented graph if and only if it does not contain a directed cycle. Algorithm 1 can be adapted to obtain an acyclic orientation completion of the input partially oriented graph that contains no directed cycle.

We have seen in Section 12.2 that the orientation problem is polynomial time solvable for each of the five classes: quasi-transitive oriented graphs, transitive oriented graphs, local tournaments, locally transitive local tournaments, and acyclic local tournaments. The situation changes for the orientation completion problem. We will show that the orientation completion problem is NP-complete for locally transitive local tournaments, while it remains polynomial time solvable for the other classes.

### 12.3.1 Quasi-transitive and Transitive Orientation Completions

Let $Q = (V, E \cup A)$ be a partially oriented graph. We use $G = UG(Q)$ to denote the underlying graph of $Q$ and $G^+$ to denote the auxiliary graph of $G$ as defined in Subsection 12.2.1. That is, the vertex set of $G^+$ consists of all ordered pairs $(u, v), (v, u)$ for edges $uv \in E(G)$ and in $G^+$ each vertex $(u, v)$ is adjacent to $(v, u)$, to any vertex $(v, w)$ such that $u$ and $w$ are not adjacent in $G$, and to any vertex $(w, u)$ such that $v$ and $w$ are not adjacent in $G$. Thus the arc set $A$ of $Q$ corresponds to a subset $S$ of the vertex set of $G^+$. An orientation completion of $Q$ to a quasi-transitive oriented graph corresponds to a colour class of a 2-colouring of $G^+$ that contains $S$. It follows that $Q$ can be completed to a quasi-transitive oriented graph if and only if the following properties hold:

- $G^+$ is bipartite, and
- no two vertices of $S$ are at an odd distance in $G^+$.

If $G^+$ has these two properties, then it can be 2-coloured such that all vertices of $S$ are of the same colour and the colour class that contains $S$ gives rise to a quasi-transitive orientation completion of $Q$. Finding such a 2-colouring of $G^+$ (if it exists) can be done in linear time. Therefore we have the following:

**Theorem 12.3.1** ([4]) *The orientation completion problem is polynomial time solvable for the class of quasi-transitive oriented graphs.* □

A partially oriented graph that can be completed to a transitive oriented graph cannot contain directed cycles. So the additional assumption of being acyclic is necessary for a partially oriented graph to admit a completion to a transitive oriented graph. But this additional assumption is not sufficient as there are acyclic partially oriented graphs which can be completed to quasi-transitive oriented graphs but not to transitive oriented graphs. Nevertheless, we show that deciding whether a partially oriented graph can be completed to a transitive oriented graph can be done in polynomial time.

A partially oriented graph $Q = (V, E \cup A)$ is called **consentaneous** if the following properties hold: Let $G^+$ be the auxiliary graph of $UG(Q)$ and $S$ correspond to the arc set $A$.

- $G^+$ is bipartite,
- no two vertices of $S$ are at an odd distance in $G^+$, and
- for any two vertices at an even distance in $G^+$, either both are in $S$ or neither.

**Theorem 12.3.2** *Let $Q = (V, E \cup A)$ be a partially oriented graph. Suppose that $UG(Q)$ is a comparability graph and $Q$ is consentaneous. Then $Q$ can be completed to a transitive oriented graph if and only if $Q$ does not contain a directed cycle.*

**Proof:** Let $\sigma$ be a vertex ordering of $UG(Q)$ such that all arcs in $A$ are forward (i.e., $(u,v) \in A$ implies $\sigma^{-1}(u) < \sigma^{-1}(v)$). Obtain an orientation completion of $Q$ using the lexicographic orientation algorithm in Subsection 12.2.1 with respect to $\sigma$. By Theorem 12.2.2 the orientation completion of $Q$ is a transitive oriented graph. □

**Corollary 12.3.3** *The orientation completion problem for the class of transitive oriented graphs is solvable in polynomial time.*

**Proof:** Suppose that a partially oriented graph $Q = (V, A \cup E)$ is given. Let $G = UG(Q)$. If $G^+$ is not bipartite, then the answer is 'no'. Assume that $G^+$ is bipartite. Obtain the minimal consentaneous partial oriented graph $Q' = (V, A' \cup E')$ from $Q$ by orienting (if needed) some edges in $E$. If $Q'$ contains a directed cycle, then the answer is again 'no' by Theorem 12.3.2. Otherwise, $Q'$ contains no directed cycle and we can complete $Q'$ to a transitive oriented graph according to Theorem 12.3.2. This transitive oriented graph is also an orientation completion of $Q$. □

### 12.3.2 Local and Acyclic Local Tournament Orientation Completions

The orientation completion problem for local tournaments can be solved in a similar way as above for the quasi-transitive orientation completion problem.

**Theorem 12.3.4** ([4]) *The orientation completion problem is polynomial time solvable for the class of local tournaments.* □

We consider next the orientation completion problem for the class of acyclic local tournaments. For a partially oriented graph $Q = (V, E \cup A)$, we use $G^+$ to denote the auxiliary graph of $UG(Q)$ as defined in Subsection 12.2.2 and use $S$ to denote the set of vertices of $G^+$ corresponding to the arc set $A$. Again, we call $Q$ **consentaneous** if the following conditions hold:

- $G^+$ is bipartite,
- no two vertices of $S$ are at an odd distance in $G^+$, and
- for any two vertices at an even distance in $G^+$, either both are in $S$ or neither.

**Theorem 12.3.5** ([4]) *Let $Q = (V, E \cup A)$ be a partially oriented graph. Suppose that $UG(Q)$ is a proper interval graph and $Q$ is consentaneous. Then $Q$ can be completed to an acyclic local tournament if and only if $Q$ does not contain a directed cycle.*

**Proof:** If $Q$ contains a directed cycle then it cannot be completed to an acyclic oriented graph and hence not to an acyclic local tournament. For the other direction, we first show that $Q$ admits a perfect elimination ordering $v_1, v_2, \ldots, v_n$ such that all arcs are forward, that is, if $(v_i, v_j)$ is an arc then $i < j$. To obtain such an ordering we apply a modified LBFS beginning with a vertex of out-degree 0, with preferences (in the case of ties) given to vertices having no out-neighbours among unlabeled vertices.

Let $v_1, v_2, \ldots, v_n$ be an ordering obtained by the modified LBFS. According to Rose, Tarjan and Lueker [36], it is a perfect elimination ordering. Suppose that the ordering contains a backward arc. Let $(v_i, v_j) \in A$ be a backward arc having the largest subscript $i$. Since $(v_i, v_j)$ is backward, we have $i > j$. The choice of $v_n$ implies $n > i$. Since $i > j$, at the time of labeling $v_i$ the vertex $v_j$ is an unlabeled out-neighbour of $v_i$. The LBFS rule ensures that $v_i$ is a vertex having the lexicographically largest neighbourhood among the vertices $v_n, \ldots, v_{i+1}$. If the neighbourhood of $v_i$ (among the labeled vertices) is lexicographically larger than the neighbourhood of $v_j$, some vertex $v_\ell$ with $\ell > i$ is adjacent to $v_i$ but not to $v_j$ in $Q$. The assumption that $Q$ is consentaneous implies $(v_\ell, v_i)$ is an arc which is backward with respect to the ordering. This contradicts the choice of $(v_i, v_j)$. Hence $v_i$ and $v_j$ must have the same neighbourhood among the labeled vertices. But then the rule prefers $v_j$ to $v_i$ for the next labeled vertex, unless $v_j$ has an out-neighbour $v_k$ among unlabeled vertices. A similar proof above (when applied to $v_j, v_k$) implies $v_j$ and $v_k$ must have the same neighbourhood among the labeled vertices. Continuing in this way, we obtain a directed cycle, which contradicts the assumption. Hence $v_1, v_2, \ldots, v_n$ is a perfect elimination ordering of $Q$ that contains no backward arcs.

Now we apply the lexicographic orientation algorithm using the perfect elimination ordering to obtain an orientation $D$ of $UG(Q)$. By Theorem 12.2.10 $D$ is an acyclic local tournament. Since the perfect elimination ordering has no backward arc from $A$, the arc set of $D$ contains $A$. Hence $D$ is an orientation completion of $Q$. □

**Corollary 12.3.6** *The orientation completion problem for the class of acyclic local tournaments is solvable in polynomial time.*

**Proof:** Suppose that a partially oriented graph $Q = (V, A \cup E)$ is given. Let $G = UG(Q)$. If $G^+$ is not bipartite, then the answer is 'no'. Assume that $G^+$ is bipartite. Obtain the minimal consentaneous partial oriented graph $Q' = (V, A' \cup E')$ from $Q$ by orienting (if needed) some edges in $E$. If $Q'$ contains a directed cycle, then the answer is again 'no' by Theorem 12.3.5. Otherwise, $Q'$ contains no directed cycle and we can complete $Q'$ to an acyclic local tournament orientation according to Theorem 12.3.5. This acyclic local tournament is also an orientation completion of $Q$. □

**Corollary 12.3.7** ([28]) *The problem of extending partial proper interval representations of proper interval graphs is solvable in polynomial time.*

**Proof:** We show how to reduce the problem of extending partial proper interval representations of proper interval graphs to the orientation completion problem for the class of acyclic local tournaments which is polynomial time solvable according to Corollary 12.3.6. Suppose that $G$ is a proper interval graph and $H$ is an induced subgraph of $G$. Given a proper interval representation $I_v, v \in V(H)$, of $H$ (i.e., a partial proper interval representation of $G$), we obtain an orientation of $H$ in such a way that $(u, v)$ is an arc if and only if $I_u$ contains the left endpoint of $I_v$. The oriented edges together with the remaining edges in $G$ yield a partial orientation of $G$. This partial orientation of $G$ can be completed to an acyclic local tournament if and only if the partial representation of $H$ can be extended to a proper interval representation of $G$. □

### 12.3.3 Locally Transitive Local Tournament Orientation Completions

A cyclic ordering $\mathcal{O} = v_1, v_2, \ldots, v_n, v_1$ of the vertices of a partially oriented graph $Q = (V, E \cup A)$ is called **excellent** if $Q$ has no pair of arcs $v_i \rightarrow v_j$ and $v_s \rightarrow v_t$ (with a possibility that $i = t$ or $s = j$) such that the vertices occur as $v_i, v_t, v_s, v_j$ in the cyclic ordering, cf. [4]. Since a round ordering of an oriented graph is excellent, by Theorem 12.2.4, every connected locally transitive local tournament has an excellent cyclic ordering, cf. [24]. Thus, a necessary condition for completing $Q$ to a locally transitive local tournament is that it has an excellent ordering. It turns out, as we will show, that the problem of determining whether a partially oriented graph has an excellent ordering is polynomially equivalent to the orientation completion problem for locally transitive local tournaments and both problems are NP-complete (Theorem 12.3.14). The presentation below follows the paper [4] by Bang-Jensen, Huang and Zhu.

Let $\mathcal{O} = v_1, v_2, \ldots, v_n, v_1$ be a cyclic ordering of the vertices of a partially oriented graph $P = (V, E \cup A)$. An arc $(v_i, v_j) \in A$ **dominates** an arc $(v_s, v_t) \in A$ with respect to $\mathcal{O}$ if the vertices of the two arcs appear in the order $v_i, v_s, v_t, v_j$ in $\mathcal{O}$, where we can have $i = s$ or $j = t$. An arc $(v_i, v_j) \in A$

dominates an edge $v_pv_q$ if both of the vertices $v_p, v_q$ occur in the interval $[v_i, v_j]$ from $v_i$ to $v_j$ according to $\mathcal{O}$. An arc is **maximal** with respect to $\mathcal{O}$ if it is not dominated by any other arc.

**Lemma 12.3.8** ([4]) *Suppose $P = (V, E \cup A)$ is a partially oriented graph for which the digraph $D = (V, A)$ induced by its arcs has an excellent cyclic ordering $\mathcal{O} = v_1, \ldots, v_n, v_1$ of its vertices. Then $P$ can be completed to an oriented graph $D'$ for which the same cyclic ordering $\mathcal{O}$ is excellent.*

**Proof:** Let $P = (V, E \cup A)$ be a partially oriented graph and let $\mathcal{O} = v_1, v_2, \ldots, v_n, v_1$ be an excellent cyclic ordering of $D$. Let $a_1 = (v_{i_1}, v_{j_1}), a_2 = (v_{i_2}, v_{j_2}), \ldots, a_k = (v_{i_k}, v_{j_k})$ be the maximal arcs of $D$ with respect to $\mathcal{O}$. By the assumption of the lemma, for each arc $a_r$ every arc $(v_p, v_q)$ for which both vertices $v_p, v_q$ occur after in the interval $[v_i, v_j]$ satisfy that the vertices occur in the order $v_{i_r}, v_p, v_q, v_{j_r}$. For each $r \in [k]$ in increasing order and all indices $p, q$ with $v_{i_r}, v_p, v_q, v_{j_r}$ occurring in that order such that $v_pv_q$ is an edge of $P$, we orient this edge as the arc $(v_p, v_q)$. Let $D^* = (V, A \cup A^*)$ be the oriented graph consisting of the original arcs and those edges which we have oriented so far. By construction of $D^*$, $\mathcal{O}$ is an excellent ordering of $D^*$. Hence if no edge of $E$ is still unoriented we are done. It suffices to show that we may orient one of the remaining edges, since then the claim follows by induction on the number of unoriented edges. Let $v_pv_q$ be an edge which was not oriented and orient this as $(v_p, v_q)$. We claim that $\mathcal{O}$ is an excellent ordering of $D^* \cup \{(v_p, v_q)\}$. If not then there is an arc $(v_a, v_b)$ of $D^*$ such that the vertices occur in the order $v_p, v_b, v_a, v_q$ but then the edge $v_pv_q$ is dominated by the arc $(v_a, v_b)$ and hence by one of the arcs $a_1, \ldots, a_k$, contradicting that it was not oriented above. □

**Lemma 12.3.9** ([4]) *An oriented graph $D$ has an excellent cyclic ordering $\mathcal{O}$ if and only if it can be extended to a round local tournament $D^*$ by adding new arcs. In particular, every excellent ordering of $D$ is a round ordering of $D^*$ and conversely.*

**Proof:** Suppose first that $D$ can be extended to a round local tournament $D^*$. According to Theorem 12.2.4 there is a round ordering $\mathcal{O} = v_1, v_2, \ldots, v_n, v_1$ of $V(D^*) = V(D)$. We claim that this ordering is also excellent. If not, then there are arcs $(v_i, v_j)$ and $(v_s, v_t)$ so that the vertices occur in the order $v_i, v_t, v_s, v_j$ according to $\mathcal{O}$. Since $\mathcal{O}$ is a round ordering, we have that $(v_i, v_t)$ and $(v_t, v_j)$ are arcs of $D^*$ but then the neighbours of $v_t$ do not occur correctly according to $\mathcal{O}$, contradiction. So $\mathcal{O}$ is an excellent ordering of $D^*$ and hence also of the subdigraph $D$. To prove the only if part let $\mathcal{O} = v_1, v_2, \ldots, v_n, v_1$ be an excellent cyclic ordering of the oriented graph $D$. It suffices to observe that for every maximal arc $(v_i, v_j)$ with respect to $\mathcal{O}$ and any pair of non-adjacent vertices $v_a, v_b$ in the interval $[v_i, v_j]$ with $v_a$ before $v_b$ we may add the arc $(v_a, v_b)$ and still have an excellent ordering of the resulting oriented graph.

Now the claim follows by induction on the number of such non-adjacent pairs. □

For a given oriented graph $D$ we denote by $D^c$ the partially oriented complete graph obtained from $D$ by adding an edge between each pair of non-adjacent vertices.

**Lemma 12.3.10** ([4]) *If $D$ is a round oriented graph, then $D^c$ can be completed to a locally transitive tournament.*

**Proof:** We prove the statement by induction on the number of vertices in $D$ which are not adjacent to all other vertices. By Theorem 12.2.4, the base case where there is no such vertex is true. So assume that all round oriented graphs on $n$ vertices with at most $k$ vertices as above can be completed to a locally transitive tournament and let $D$ be a round digraph with $k+1$ vertices, each of which has a non-neighbour. Let $\mathcal{O} = v_1, v_2, \ldots, v_n, v_1$ be a round ordering of $D$. W.l.o.g. the vertex $v_1$ has a non-neighbour, so we have that $v_{d^+(v_1)+2} \neq v_{n-d^-(v_1)}$. We claim that there is no arc $(v_p, v_q)$ with $1 \leq q < p < n - d^-(v_1)$. Suppose such an arc does exist. Then we have $p > d^+(v_1)+1$ by the choice of $\mathcal{O}$ and we have $q > 1$ since $v_p$ is not adjacent to $v_1$. But this contradicts the fact that the vertex $v_p$ sees its out-neighbourhood as an interval just after itself according to $\mathcal{O}$ because $v_1$ is not-adjacent to $v_p$. Thus if we add all the arcs $(v_1, v_{d^+(v_1)+2}), \ldots, (v_1, v_{n-d^-(v_1)-1})$ to $D$ the order $\mathcal{O}$ is an excellent ordering of the resulting digraph $D'$. By Lemmas 12.3.8 and 12.3.9 this implies that $D'$ can be extended to a round local tournament $D''$ by adding new arcs. Now the claim follows by induction since $D'$ has fewer vertices with non-neighbours than $D$ does. □

Combining Lemmas 12.3.8, 12.3.9, and 12.3.10 we have the following:

**Lemma 12.3.11** ([4]) *An oriented graph $D$ has an excellent ordering if and only if the partially oriented graph $D^c$ has a completion to a tournament $T$ which is locally transitive. Furthermore, given an excellent ordering of $D$ we can construct $T$ in polynomial time and conversely, given $T$, we can obtain an excellent ordering of $D$ in polynomial time.* □

The following is easy to check.

**Proposition 12.3.12** *Each of the two labellings $X, \bar{X}$ of the same partially oriented complete graph in Figure 12.4 have exactly two completions to a locally transitive tournament. For $X$ these are obtained by orienting the two edges $ab, \alpha\beta$ as either $(b,a), (\beta,\alpha)$ or $(a,b), (\alpha,\beta)$. For $\bar{X}$ they are obtained by orienting the two edges $uv, \alpha\beta$ as either $(v,u), (\alpha,\beta)$ or $(u,v), (\beta,\alpha)$.* □

**Lemma 12.3.13** ([4]) *Consider the partially oriented 6-wheel $W$ in Figure 12.5. Let $D$ be an orientation completion of $W$. Then $D$ does not have an excellent ordering if and only if the three edges $c_{11}c_{12}, c_{21}c_{22}, c_{31}c_{32}$ are oriented as $(c_{11}, c_{12}), (c_{21}, c_{22}), (c_{31}, c_{32})$.*

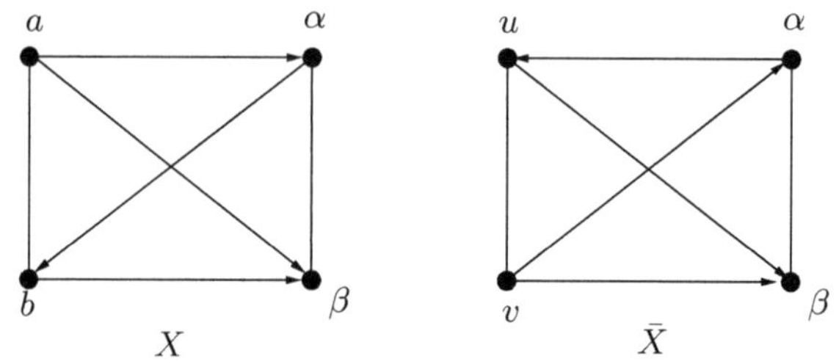


**Figure 12.4** Two different labellings of the same partially oriented complete graph on 4 vertices. For later convenience we name these $X, \bar{X}$.

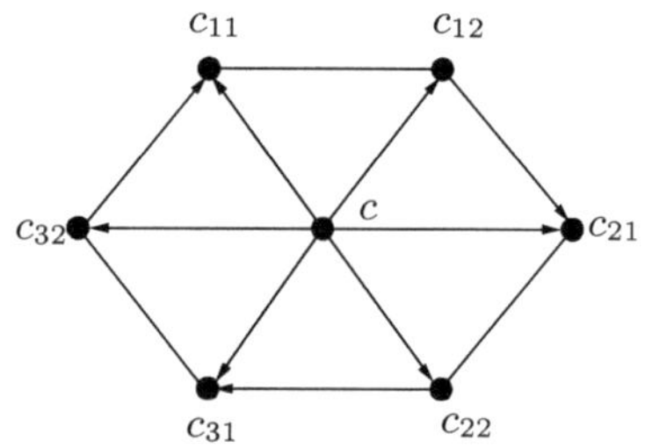


**Figure 12.5** A partially oriented wheel $W$.

**Proof:** If the three edges $c_{11}c_{12}, c_{21}c_{22}, c_{31}c_{32}$ are oriented as $(c_{11}, c_{12})$, $(c_{21}, c_{22})$, $(c_{31}, c_{32})$ then the vertex $c$ has a directed 6-cycle in its out-neighbourhood and hence $D^c$ has no completion to a locally transitive tournament. By Lemma 12.3.9, $D$ has no excellent ordering. On the other hand, if $D$ contains at least one of the arcs $(c_{12}, c_{11}), (c_{22}, c_{21}), (c_{32}, c_{31})$, then $D$ is acyclic. Clearly $D^c$ can be completed to a transitive tournament and hence by Lemma 12.3.11, $D$ has an excellent ordering. □

**Theorem 12.3.14** ([4]) *The following polynomially equivalent problems are NP-complete.*

- *Deciding whether an oriented graph has an excellent ordering.*
- *Deciding whether a given partially oriented complete graph can be completed to a locally transitive tournament.*

**Proof:** We describe polynomial reductions from 3-SAT to these problems.

Let $\mathcal{F}$ be an instance of 3-SAT with variables $x_1, x_2, \ldots, x_n$ and clauses $C_1, C_2, \ldots, C_m$, where each clause is of the form $(\ell_1 \vee \ell_2 \vee \ell_3)$ and each $\ell_i$ is either one of the variables $x_j$ or the negation $\bar{x}_j$ of such a variable.

Let $p_i$ ($q_i$) be the number of times variable $x_i$ ($\bar{x}_i$) occurs as a literal in $\mathcal{F}$. The enumeration of the clauses $C_1, \ldots, C_m$ induces an ordering on the occurrences of the same literal in the formula. Guided by this ordering we now construct a partially oriented graph $H' = H'(\mathcal{F})$ as follows:

Let $X, \bar{X}$ be as in Figure 12.4. For each variable $x_i$ we form the partially oriented graph $X_i$ from $p_i$ copies of $X$ and $q_i$ copies of $\bar{X}$ (these $p_i+q_i$ graphs are vertex disjoint) by identifying all the $\alpha$ vertices and all the $\beta$ vertices and denote these identified vertices by $\alpha(x_i), \beta(x_i)$, respectively. Denote the $p_i$ copies of $a, b$ by $a_{i,1}, \ldots, a_{i,p_i}, b_{i,1}, \ldots, b_{i,p_i}$ and the $q_i$ copies of $u, v$ by $u_{i,1}, \ldots, u_{i,q_i}, v_{i,1}, \ldots, v_{i,q_i}$.

Take $m$ disjoint copies $W_1, W_2, \ldots, W_m$ of the partially oriented 6-wheel from Figure 12.5 where we use $c_i, c^i_{11}, c^i_{12}, c^i_{21}, c^i_{22}, c^i_{31}, c^i_{32}$ to denote the vertices of $W_i$. Make the following association between the literals of $\mathcal{F}$ and the $W_i$'s: If $C_i = (\ell_{i,1} \vee \ell_{i,2} \vee \ell_{i,3})$ we associate the vertices $c^i_{j1}, c^i_{j2}$ with the literal $\ell_{i,j}$ of $C_i$, $j \in [3]$.

Now we make the following vertex identifications. For each clause $C_i = (\ell_{i,1} \vee \ell_{i,2} \vee \ell_{i,3})$ we identify the vertices $c^i_{11}, c^i_{12}, c^i_{21}, c^i_{22}, c^i_{31}, c^i_{32}$ with vertices from the union of the graphs $X_1, \ldots, X_n$ as follows: If $\ell_{i,j} = x_r$ and this is the $h$'th occurrence of variable $x_r$ according to the induced ordering of that literal, then identify $c^i_{j1}$ with $a_{r,h}$ and $c^i_{j2}$ with $b_{r,h}$. If $\ell_{i,j} = \bar{x}_r$ and this is the $t$'th occurrence of $\bar{x}_r$ according to the induced ordering of that literal, then identify $c^i_{j1}$ with $u_{r,t}$ and $c^i_{j2}$ with $v_{r,t}$. Note that even after these identifications each of the subdigraphs $W_1, \ldots, W_m$ are still vertex disjoint.

Clearly we can construct $H'$ in polynomial time from $\mathcal{F}$. Denote by $H$ the oriented graph obtained from $H'$ by deleting all (unoriented) edges. It is easy to check that the in- and out-neighbourhoods of each vertex in $H$ is acyclic.

By Lemma 12.3.11 it suffices to show that $H$ has an excellent ordering if and only if $\mathcal{F}$ is satisfiable.

First suppose that $H$ has an excellent ordering. By Lemma 12.3.11 this means that the partially oriented complete graph $H^c$ has a completion $T$ as a locally transitive tournament. We claim that the following is a satisfying truth assignment: If the edge $\alpha(x_i)\beta(x_i)$ is oriented in $T$ as $(\alpha(x_i), \beta(x_i))$ then let $x_i$ be false and if it is oriented as $(\beta(x_i), \alpha(x_i))$ then let $x_i$ be true. First observe that, by Proposition 12.3.12, this implies that for each $i \in [n]$ the variable $x_i$ is false if and only if each of the edges $a_{i,j}b_{i,j}$, $j \in [p_i]$, are oriented as $(a_{i,j}, b_{i,j})$ and each of the edges $u_{i,r}v_{i,r}$, $r \in [q_i]$, are oriented as $(v_{i,r}, u_{i,r})$.

We now use this to show that each of the clauses of $\mathcal{F}$ are satisfied by our truth assignment. As $T$ is locally transitive, for each of the induced subdigraphs $T[W_j]$, $j \in [m]$, the out-neighbourhood of $c_j$ is acyclic which implies that at least one of three arcs of $H$ which correspond to the literals of $\mathcal{F}$ is oriented as $(c_{j2}, c_{j1})$. If this arc corresponds to the literal $x_s$ then, by the identification rule above, this is an arc of the form $(b_{s,t}, a_{s,t})$, so the variable $x_s$ is true and $C_j$ is satisfied. If the arc corresponds to the literal $\bar{x}_s$ then the identification rule implies that this is an arc of the form $(v_{s,t}, u_{s,t})$, implying that $\bar{x}_s$ is true so again $C_j$ is satisfied. Thus we have shown that $\mathcal{F}$

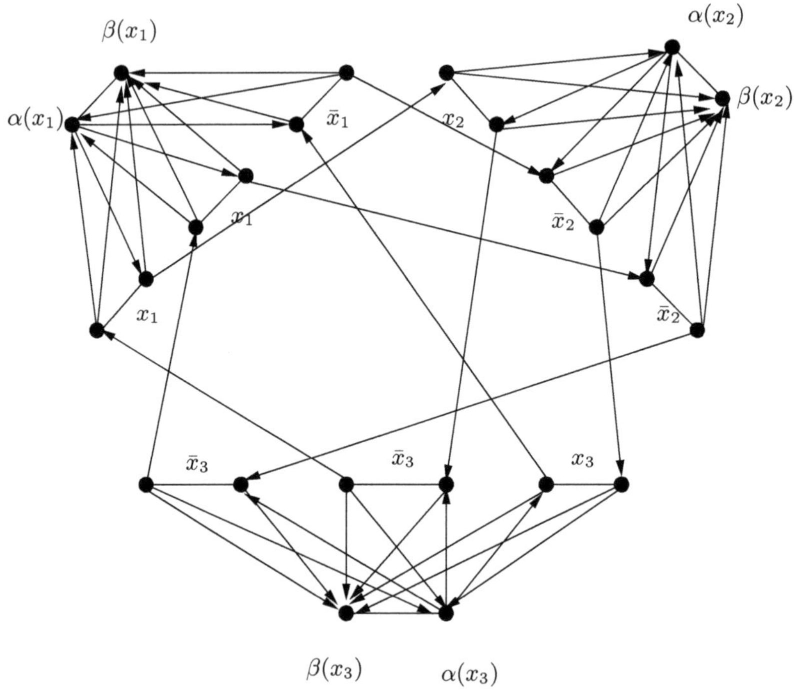


**Figure 12.6** Part of the digraph $H'(\mathcal{F})$ when $\mathcal{F} = (x_1 \vee x_2 \vee \bar{x}_3) \wedge (\bar{x}_1 \vee \bar{x}_2 \vee x_3) \wedge (x_1 \vee \bar{x}_2 \vee \bar{x}_3)$. For better readability the vertices $c_1, c_2, c_3$ are not shown.

is satisfiable if $H^c$ has a locally transitive completion ($H$ has an excellent ordering).

Now suppose that $t : \{x_1, \ldots, x_n\} \to \{true, false\}$ is a satisfying truth assignment for $\mathcal{F}$. We shall use this truth assignment to construct an excellent ordering of the partially oriented graph $H'$. Recall that this is also an excellent ordering of the directed part $H$ of $H'$.

We first orient the edges $\alpha(x_1)\beta(x_1), \ldots \alpha(x_n)\beta(x_n)$ as follows: Orient $\alpha(x_i)\beta(x_i)$ as $(\beta(x_i), \alpha(x_i))$ if $x_i = true$ and as $(\alpha(x_i), \beta(x_i))$ otherwise. Denote by $\hat{H}$ the resulting partially oriented graph. It follows from Proposition 12.3.12, the way we made identifications between vertices of the $W_j$'s and variable vertices and the fact that $t$ is a satisfying truth assignment that we can now orient all the remaining edges of $\hat{H}$ (recall that those correspond to the literals) uniquely so that the resulting full orientation $\overrightarrow{H}$ of $H'$ satisfies that the in- and out-neighbourhood of each vertex is still acyclic.

We now construct an excellent ordering for $\overrightarrow{H}$. Denote by $A(x_i)$ ($B(x_i)$), $i \in [n]$ the set of out-neighbours (in-neighbours) of $\alpha(x_i)$ in $\overrightarrow{H}$. Note that if $t(x_i) = false$, then $A(x_i) = \{b_{i,1}, \ldots, b_{i,p_i}, u_{i,1}, \ldots, u_{i,q_i}, \beta(x_i)\}$, $B(x_i) = \{a_{i,1}, \ldots, a_{i,p_i}, v_{i,1}, \ldots, v_{i,q_i}\}$ and there is no oriented arc from $A(x_i)$ to $B(x_i)$.

Similarly, if $t(x_i) = true$, then $A(x_i) = \{b_{i,1}, \ldots, b_{i,p_i}, u_{i,1}, \ldots, u_{i,q_i}\}$, $B(x_i) = \{a_{i,1}, \ldots, a_{i,p_i}, v_{i,1}, \ldots, v_{i,q_i}, \beta(x_i)\}$ and there is no oriented arc from $B(x_i)$ to $A(x_i)$.

Furthermore, observe that $\beta(x_i)$ has no out-neighbour when $t(x_i) = false$ and precisely one out-neighbour, namely $\alpha(x_i)$ when $t(x_i) = true$. Let $1 \leq i_1 < i_2 < \ldots < i_k \leq n$ and $1 < j_1 < j_2 < \ldots < j_g \leq n$ denote the indices of the true, respectively the false variables. Consider the following cyclic ordering $\mathcal{O}$ of $V(\overrightarrow{H})$:

$$\alpha(x_{i_1}), \alpha(x_{i_2}), \ldots, \alpha(x_{i_k}), c_1, c_2, \ldots, c_m, A(x_{i_1}), \ldots, A(x_{i_k}), B(x_{j_1}), \ldots, B(x_{j_g}), \alpha(x_{j_1}), \ldots, \alpha(x_{j_g}), A(x_{j_1}), \ldots, A(x_{j_g}), B(x_{i_1}), \ldots, B(x_{i_k}), \alpha(x_{i_1}),$$

where the ordering inside each $A(x_i), B(x_i)$ is as according to the way we listed those sets above.

We shall prove that the ordering $\mathcal{O}$ is excellent. Suppose for contradiction that there is a pair of arcs $(v_i, v_j)$ and $(v_s, v_t)$ with the vertices occurring in the order $v_i, v_t, v_s, v_j$ according to $\mathcal{O}$.

- We cannot have $v_i = \alpha(x_{i_f})$ for some $f \in [k]$ because there is no backward arc in the interval of $\mathcal{O}$ from $\alpha(x_{i_f})$ to (the end of) $A(x_f)$ ($\alpha(x_{i_f})$ is only adjacent to vertices in $A(x_{i_f})$). Similarly, we cannot have $v_i$ in the interval $[\alpha(x_{j_1}), \alpha(x_{j_g})]$.
- We cannot have $v_i = c_p$ for some $p \in [m]$ because the only arcs incident to $c_p$ are from $c_p$ to the six vertices which correspond to its three literals and we ordered the $A$ and $B$ sets and $\alpha(x_{j_1}), \ldots, \alpha(x_{j_g})$ in such a way that any arc between them goes forward in the ordering. In particular, there are no backwards arcs with respect to the ordering in the interval $A(x_{i_1}), \ldots, A(x_{i_k}), B(x_{j_1}), \ldots, B(x_{j_g}), \alpha(x_{j_1}), \ldots, \alpha(x_{j_g}), A(x_{j_1}), \ldots, A(x_{j_g}), B(x_{i_1}), \ldots, B(x_{i_k})$.
- We cannot have $v_i$ in the interval $A(x_{i_1}), \ldots, A(x_{i_k})$ since all out-neighbours of those vertices are in the interval $B(x_{i_1}), \ldots, B(x_{i_k})$ and then the remark above implies the claim. Similarly, we cannot have $v_i$ in the interval $A(x_{j_1}), \ldots, A(x_{j_g})$.
- We cannot have $v_i$ in the interval $B(x_{j_1}), \ldots, B(x_{j_g})$ because there are no backward arcs in the interval $B(x_{j_1}), \ldots, B(x_{j_g}), \alpha(x_{j_1}), \ldots, \alpha(x_{j_g}), A(x_{j_1}), \ldots, A(x_{j_g})$ and this contains all out-neighbours of such a $v_i$.
- Finally we cannot have $v_i$ in the interval $B(x_{i_1}), \ldots, B(x_{i_k})$ because all arcs out of a vertex in this interval remain inside the interval $B(x_{i_1}), \ldots, B(x_{i_k}), \alpha(x_{i_1}), \alpha(x_{i_2}), \ldots, \alpha(x_{i_k})$ and there is no backward arc here.

Thus we have shown that $\mathcal{O}$ is excellent and hence, by Lemma 12.3.11, the partially oriented complete graph $H^c$ has a completion to a locally transitive tournament. □

## 12.4 Orientation Sandwich Completion Problems

For a fixed property $\Pi$ of partially oriented graphs, the $\Pi$-SANDWICH PROBLEM is defined as follows:

> $\Pi$-SANDWICH PROBLEM
> **Input:** A pair of partially oriented graphs $Q_1 = (V, E_1 \cup A_1)$ and $Q_2 = (V, E_2 \cup A_2)$.
> **Question:** Is there a partially oriented graph $Q = (V, E \cup A)$ with $E_1 \subseteq E \subseteq E_2$ and $A_1 \subseteq A \subseteq A_2$ which satisfies $\Pi$?

Sandwich problems for partially oriented graphs simultaneously generalize graph sandwich problems and digraph sandwich problems, which have been studied by Golumbic, Kaplan and Shamir in [20]. **Graph sandwich problems** restrict $Q_1, Q_2$ and $Q$ in the above definition to be graphs, while **digraph sandwich problems** restrict them to be digraphs.

Graph sandwich problems are polynomial time solvable for several graph properties, including being bipartite graphs, threshold graphs, split graphs, cographs and Eulerian graphs, and are NP-complete for properties such as being chordal graphs, interval graphs, circle graphs, circular arc graphs, proper circular arc graphs, comparability graphs, co-comparability graphs, and permutation graphs, cf. [20]. Little is known about digraph sandwich problems but for Eulerian digraphs it is proved to be polynomial time solvable by Ford and Fulkerson in [11].

A partially oriented graph $Q = (V, E \cup A)$ is called **mixed Eulerian** if both $(V, E)$ and $(V, A)$ are Eulerian, that is, in $(V, E)$ every vertex has an even degree and in $(V, A)$ every vertex has its in-degree equal to its out-degree. Although both sandwich problems for Eulerian graphs and digraphs are polynomial time solvable, the sandwich problem for mixed Eulerian partially oriented graphs remains open.

**Problem 12.4.1** *Determine the complexity of the sandwich problem for mixed Eulerian partially oriented graphs.*

For a fixed property $\Pi$ of oriented graphs, we define the $\Pi$-ORIENTATION SANDWICH COMPLETION PROBLEM as follows:

> $\Pi$-ORIENTATION SANDWICH COMPLETION PROBLEM
> **Input:** A pair of partially oriented graphs $Q_1 = (V, E_1 \cup A_1)$ and $Q_2 = (V, E_2 \cup A_2)$.
> **Question:** Is there a partially oriented graph $Q = (V, E \cup A)$ with $E_1 \subseteq E \subseteq E_2$ and $A_1 \subseteq A \subseteq A_2$ which can be completed to an oriented graph that satisfies $\Pi$?

Orientation sandwich completion problems generalize orientation completion problems and hence orientation problems. Orientation sandwich completion problems and sandwich problems for partially oriented graphs are closely related. Let $\Pi$ be a property of oriented graphs. A partially oriented graph is said to have property $\Pi^*$ if it can be completed to an oriented graph that has the property $\Pi$. Then the $\Pi$-orientation sandwich completion problem is just the $\Pi^*$-sandwich problem. For instance, suppose that $\Pi$ is the property of being an Eulerian oriented graph, then a partially oriented graph has property $\Pi^*$ if and only if it is mixed Eulerian and thus the $\Pi$-orientation sandwich completion problem is just Problem 12.4.1. As mentioned above, the $\Pi$-orientation completion problem is polynomial time solvable but the $\Pi$-orientation sandwich completion problem is open. Special cases of the $\Pi$-orientation sandwich completion problem have been studied by de Gevigney, Klein, Nguyen and Szigeti [8].

A property $\Pi$ of oriented graphs is called **sup-preservable** if $Q_1 = (V, A_1)$ has the property $\Pi$ and $A_1 \subseteq A_2$ imply that $Q_2 = (V, A_2)$ also has the property $\Pi$. As an example, being $k$-arc-strong is a sup-preservable property for each $k \geq 1$. Let $\Pi$ be a sup-preservable property of oriented graphs. Then the $\Pi$-orientation sandwich completion problem can be reduced to the $\Pi$-orientation completion problem. Indeed, suppose that $Q_1 = (V, E_1 \cup A_1)$ and $Q_2 = (V, E_2 \cup A_2)$ form an instance of the $\Pi$-orientation sandwich completion problem. In order to have a partially oriented graph $Q = (V, E \cup A)$ satisfying $E_1 \subseteq E \subseteq E_2$ and $A_1 \subseteq A \subseteq A_2$, we must have $E_1 \subseteq E_2$ and $A_1 \subseteq A_2$. For any such $Q$, $Q$ can be completed to an oriented graph that has the property $\Pi$ if and only if $Q_2$ can. Hence the $\Pi$-orientation sandwich completion problem reduces to the $\Pi$-orientation completion problem. In particular, the $k$-arc-strong orientation sandwich completion problem reduces to the $k$-arc-strong orientation completion problem for each $k \geq 1$. Each $k$-arc-strong-orientation completion problem can be formulated as a feasible submodular flow problem which is polynomial time solvable (cf. [4]). Consequently, we have the following:

**Theorem 12.4.2** *For each $k \geq 1$, the $k$-arc-strong orientation sandwich completion problem is polynomial time solvable.* □

In contrast, the $k$-strong orientation sandwich completion problem is NP-complete for each $k \geq 3$ as this is shown to be the case for the $k$-strong orientation problem by de Gevigney [7]. Thomassen [41] proved that a graph $G$ has a 2-strong orientation if and only if $G$ is 4-edge-connected and $G - v$ is 2-edge-connected for every vertex $v$. This implies that the 2-strong orientation problem is polynomial time solvable.

**Theorem 12.4.3** ([7, 41]) *The $k$-strong orientation problem is polynomial time solvable when $k \leq 2$ and NP-complete when $k \geq 3$.* □

Thus to complete a dichotomy of $k$-strong orientation completion problems and $k$-strong orientation sandwich completion problems the only case left open is $k = 2$.

**Problem 12.4.4** *Determine the complexity of the 2-strong orientation sandwich completion problem and of the 2-strong orientation completion problem.*

A **directed cycle factor** in a digraph is a spanning subdigraph that is a vertex-disjoint union of directed cycles. The orientation completion problem for the property of having a directed cycle factor is shown to be NP-complete in [4].

**Theorem 12.4.5** ([4]) *It is NP-complete to decide whether a partially oriented graph $Q$ has a completion $D$ with a directed cycle factor.*

**Proof:** It was shown by Bang-Jensen and Casselgren [2] that it is NP-complete to decide whether a bipartite digraph $B$ has a directed cycle-factor consisting of cycles $C_1, C_2, \ldots, C_k$ so that no $C_i$ has length 2. Let $B$ be given and form the partially oriented graph $Q$ from $B$ by replacing the two arcs of each directed 2-cycle by an edge. It is easy to see that $Q$ has a completion with a directed cycle factor if and only if $B$ has a cycle factor with no directed 2-cycle, implying the theorem. □

The complexity of the orientation sandwich completion problem for having directed cycle factors is open.

**Problem 12.4.6** *Determine the complexity of the orientation sandwich completion problem for having directed cycle factors.*

Let $\pi = \{(s_1, t_1), \ldots, (s_k, t_k)\}$ be a set of $k$ pairs of distinct vertices in a (di)graph $H$. A **$\pi$-linkage** in $H$ is a collection of $k$ disjoint paths $R_1, \ldots, R_k$ such that $R_i$ starts in $s_i$ and ends in $t_i$. For a given class $\mathcal{C}$ of digraphs, the **$\mathcal{C}$-$\pi$-linkage completion problem** is defined as follows: given a partially oriented graph $Q = (V, E \cup A)$ and a set $\pi$ of $k$ terminal pairs in $V$, is it possible to complete the orientation of $Q$ so that the resulting oriented graph is in $\mathcal{C}$ and has a $\pi$-linkage?

For general digraphs the $\pi$-linkage problem, and hence also the completion version, is NP-complete already when $k = 2$ and even if the digraph is highly connected [12, 40]. Chudnovsky, Scott and Seymour [6] proved that the $\pi$-linkage problem is polynomial for semicomplete digraphs (that is, digraphs whose underlying graph is complete). This implies that the **tournament-$\pi$-linkage completion** problem is polynomial because such a completion is possible if and only if the digraph that we obtain from the partially oriented graph $Q$ by replacing each undirected edge by a directed 2-cycle is semicomplete and has a $\pi$-linkage (no two paths in a linkage intersect).

**Problem 12.4.7** *What is the complexity of the local-tournament-$\pi$-linkage completion problem when $k \geq 2$ is fixed?*

An oriented graph is called an **in-tournament** if the in-neighbourhood of every vertex induces a tournament. The orientation completion problem for in-tournaments is polynomial time solvable, cf. [4]. The orientation sandwich completion problem for in-tournaments is open.

**Problem 12.4.8** *Determine the complexity of the orientation sandwich completion problem for in-tournaments.*

The orientation problem for the class of acyclic in-tournaments is polynomial time solvable. This follows from the fact that chordal graphs are exactly the graphs which admit acyclic in-tournament orientations. However, the orientation completion problem as well as the orientation sandwich completion problem for acyclic in-tournaments remain open.

**Problem 12.4.9** *Determine the complexity of the orientation sandwich completion problem for acyclic in-tournaments.*

## References


1. J. Bang-Jensen. Locally semicomplete digraphs: a generalization of tournaments. *J. Graph Theory*, 14(3):371–390, 1990.
2. J. Bang-Jensen and C.J. Casselgren. Restricted cycles factors and arc-decompositions of digraphs. *Discrete Appl. Math.*, 193:80–93, 2015.
3. J. Bang-Jensen and J. Huang. Quasi-transitive digraphs. *J. Graph Theory*, 20(2):141–161, 1995.
4. J. Bang-Jensen, J. Huang, and X Zhu. Completing orientations of partially oriented graphs. *J. Graph Theory*, 87(3) 285–304, 2018.
5. F. Boesch and R. Tindell. Robbins's theorem for mixed multigraphs. *Amer. Math. Mon.*, 87(9):716–719, 1980.
6. M. Chudnovsky, A. Scott, and P.D. Seymour. Disjoint paths in tournaments. *Adv. Math.*, 270:582–597, 2015.
7. O.D. de Gevigney. On Frank's conjecture on $k$-connected orientations. Preprint arXiv:1212.4086v1, December 2012.
8. O.D. de Gevigney, S. Klein, V.H. Nguyen, and Z. Szigeti. Sandwich problems on orientations. *J. Brazilian Comput. Soc.*, 18:85–93, 2012.
9. X. Deng, P. Hell, and J. Huang. Linear-time representation algorithms for proper circular-arc graphs and proper interval graphs. *SIAM J. Comput.*, 25(2):390–403, 1996.
10. T. Feder, P. Hell, and J. Huang. List homomorphisms and circular arc graphs. *Combinatorica*, 19:487–505, 1999.
11. L.R. Ford, Jr. and D.R. Fulkerson. *Flows in networks.* Princeton University Press, Princeton, NJ, 1962.
12. S. Fortune, J.E. Hopcroft, and J. Wyllie. The directed subgraph homeomorphism problem. *Theor. Comput. Sci.*, 10:111–121, 1980.
13. T. Gallai. On directed paths and circuits. In *Theory of Graphs (Proc. Colloq., Tihany, 1966)*, pages 115–118. Academic Press, 1968.
14. T. Gallai. Transitiv orientbare Graphen. *Acta Math. Acad. Sci. Hungar*, 18:25–66, 1967.
15. M.R. Garey and D.S. Johnson. *Computers and intractability.* W. H. Freeman, San Francisco, 1979.

16. A. Ghouila-Houri. Caractérisation des graphes non orientés dont on peut orienter les arêtes de manière à obtenir le graphe d'une relation d'ordre. *C. R. Acad. Sci. Paris*, 254:1370–1371, 1962.
17. P.C. Gilmore and A.J. Hoffman. A characterization of comparability graphs and interval graphs. *Canad. J. Math.*, 16:539–548, 1964.
18. M.C. Golumbic. *Algorithmic graph theory and perfect graphs*. Academic Press, New York, 1980. With a foreword by Claude Berge.
19. M.C. Golumbic. The complexity of comparability graph recognition and coloring. *Computing*, 18(3):199–208, 1977.
20. M.C. Golumbic, H. Kaplan, and R. Shamir. Graph sandwich problems. *J. Algor.*, 19:449–473, 1995.
21. M. Habib, R.M. McConnell, C. Paul, and L. Viennot. Lex-BFS and partition refinement, with applications to transitive orientation, interval graphs recognition and consecutive ones testing. *Theor. Comput. Sci.*, 234:59–84, 2000.
22. P. Hell and J. Huang. Lexicographic orientation and representation algorithms for comparability graphs, proper circular arc graphs, and proper interval graphs. *J. Graph Theory*, 20(3):361–374, 1995.
23. P. Hell and J. Huang. Two remarks on circular arc graphs. *Graphs Combin.*, 13:65–72, 1997.
24. J. Huang. On the structure of local tournaments. *J. Combin. Theory, Ser. B*, 63(2):200–221, 1995.
25. J. Huang. Representation characterizations of chordal bipartite graphs. *J. Combin. Theory Ser. B*, 96:673–683, 2006.
26. J. Huang. *Tournament-like oriented graphs*. PhD thesis, School of Computing Science, Simon Fraser University, Canada, 1992.
27. J. Huang. Which digraphs are round? *Australas. J. Combin.*, 19:203–208, 1999.
28. P. Klavik, J. Kratochvil, Y. Otachi, I. Rutter, T. Saitoh, M. Saumell, and T. Vyskocil. Extending partial representations of proper and unit interval graphs. In *SWAT 2014: 14th Scandinavian Symposium and Workshops on Algorithm Theory*, volume 8503 of *Lect. Notes Comput. Sci.*, pages 253–264. Springer, 2014.
29. C.G. Lekkerkerker and J.Ch. Boland. Representation of a finite graph by a set of intervals on the real line. *Fund. Math.*, 51:45–64, 1962.
30. F. Maffray and M. Preissmann. A translation of Gallai's paper: 'Transitiv orientierbare graphen'. In *Perfect Graphs, J.L.R. Alfonsín and B.A. Bruce ed.*, pages 25–66. Wiley, 2001.
31. R.M. McConnell. Linear time recognition of circular-arc graphs. *Algorithmica*, 37:93–147, 2003.
32. R.M. McConnell and J.P. Spinrad. Modular decomposition and transitive orientation. *Discrete Math.*, 201:189–241, 1999.
33. A. Pnueli, A. Lempel, and A. Even. Transitive orientation of graphs and identification of permutation graphs. *Canad. J. Math.*, 23:160–175, 1971.
34. H.E. Robbins. A theorem on graphs with an application to a problem on traffic control. *Amer. Math. Mon.*, 46:281–283, 1939.
35. F.S. Roberts. *Graph Theory and its Applications to Problems of Society*. NSF-CMBS Monograph No. 29. SIAM, Philadelphia, PA, 1978.
36. D.J. Rose, R.E. Tarjan, and G.S. Lueker. Algorithmic aspects of vertex elimination on graphs. *SIAM J. Comput.*, 5:266–283, 1976.
37. B. Roy. Nombre chromatique et plus longs chemins d'un graphe. *Rev. Fr. Inf. Rech. Opér.*, 1(5):129–132, 1967.
38. D.J. Skrien. A relationship between triangulated graphs, comparability graphs, proper interval graphs, proper circular-arc graphs, and nested interval graphs. *J. Graph Theory*, 6(3):309–316, 1982.

39. J. Spinrad. Circular-arc graphs with clique cover number two. *J. Combin. Theory Ser. B*, 44:300–306, 1988.
40. C. Thomassen. Highly connected non-2-linked digraphs. *Combinatorica*, 11(4):393–395, 1991.
41. C. Thomassen. Strongly 2-connected orientations of graphs. *J. Combin. Theory Ser. B*, 110:67–78, 2015.
42. W.T. Trotter and J.I. Moore. Characterization problems for graphs, partially ordered sets, lattices, and families of sets. *Discrete Math.*, 16:361–381, 1976.
43. A.C. Tucker. An efficient test for circular arc graphs. *SIAM J. Comput.*, 9:1–24, 1980.
44. A.C. Tucker. Characterizing circular-arc graphs. *Bull. Amer. Math. Soc.*, 76:1257–1260, 1970.
45. A.C. Tucker. Colouring a family of circular arcs. *SIAM J. Appl. Math.*, 29:439–502, 1975.
46. A.C. Tucker. Matrix characterization of circular-arc graphs. *Pacific J. Math.*, 39:535–545, 1971.
47. L.M. Vitaver. Determination of minimal coloring of vertices of a graph by means of Boolean powers of the incidence matrix. *Dokl. Akad. Nauk SSSR*, 147:758–759, 1962.

# Symbol Index

To shorten and unify the notation, in this index we use the following conventions:
B denotes a bipartite (di)graph.
$C, Ci$ denote cycles (directed, undirected, edge-coloured, oriented).
$D, Di$ denote digraphs, directed multigraphs and directed pseudographs.
$G, Gi$ denote undirected graphs and undirected multigraphs.
$H$ denotes a hypergraph.
$M$ denotes a mixed graph or a matroid.
$P, Pi$ denote path (directed, undirected, edge-coloured, oriented).
$S$ denotes a matrix or a multiset.
$X, Xi$ denote abstract sets or sets of vertices.
$Y, Yi$ denote sets of arcs.






J. Bang-Jensen and G. Gutin (eds.), *Classes of Directed Graphs*, Springer Monographs in Mathematics, https://doi.org/10.1007/978-3-319-71840-8

**G**



**U**

# Author Index



 
J. Bang-Jensen and G. Gutin (eds.), *Classes of Directed Graphs*, Springer Monographs in Mathematics, https://doi.org/10.1007/978-3-319-71840-8

**Z**

# Subject Index



 
J. Bang-Jensen and G. Gutin (eds.), *Classes of Directed Graphs*, Springer Monographs in Mathematics, https://doi.org/10.1007/978-3-319-71840-8

**E**

MIX
Papier aus verantwortungsvollen Quellen
Paper from responsible sources
FSC® C105338

If you have any concerns about our products,
you can contact us on
**ProductSafety@springernature.com**

In case Publisher is established outside the EU,
the EU authorized representative is:
**Springer Nature Customer Service Center GmbH**
**Europaplatz 3, 69115 Heidelberg, Germany**

Printed by Libri Plureos GmbH
in Hamburg, Germany